József Lőrinczi, Fumio Hiroshima and Volker Betz

Feynman–Kac-Type Theorems and Gibbs Measures on Path Space

De Gruyter Studies in Mathematics

Volume 34/1

József Lőrinczi, Fumio Hiroshima and Volker Betz

Feynman–Kac-Type Theorems and Gibbs Measures on Path Space

Volume 1:
Feynman–Kac-Type Formulae and Gibbs Measures

3rd, updated edition

DE
G
DE GRUYTER

Mathematics Subject Classification 2010
Primary: 47D08, 81Q10, 35J10; Secondary: 47-01, 60-01

Authors
Prof. Dr. József Lőrinczi
Alfréd Rényi Institute of Mathematics
Reáltanoda utca 13-15
1053 Budapest
Hungary
lorinczi@renyi.hu

Prof. Dr. Fumio Hiroshima
Kyushu University
Faculty of Mathematics
744 Motooka
Nishiku
819-0395 Fukuoka
Japan
hiroshima@math.kyushu-u.ac.jp

Prof. Dr. Volker Betz
Technische Universität Darmstadt
Fachbereich Mathematik
Schloßgartenstr. 7
64289 Darmstadt
Germany
betz@mathematik.tu-darmstadt.de

ISSN 0179-0986
ISBN 978-3-11-223145-6
ISBN 978-3-11-223146-3 (PDF)
ISBN 978-3-11-223147-0 (E-PUB)
DOI https://doi.org/10.1515/9783112231463

Library of Congress Control Number: 2026942787

Bibliographic information published by the Deutsche Nationalbibliothek
The Deutsche Nationalbibliothek lists this publication in the Deutsche Nationalbibliografie; detailed bibliographic data are available on the Internet at https://dnb.dnb.de.

De Gruyter and Walter de Gruyter GmbH are part of De Gruyter Brill.
www.degruyterbrill.com

Questions about General Product Safety Regulation: productsafety@degruyterbrill.com

Printing: CPI books GmbH, Leck

Contents

Preface to the third edition

We are grateful to many colleagues who carefully read earlier versions of the manuscript and pointed out numerous misprints and errors, which in the present edition have been amended. In particular, we would like to thank Jinqiao Duan, Fritz Gesztesy, Haruto Inoue, Niels Jacob, DongYun Lee, Michael Pang, Michał Ryznar, Soichiro Sakamoto, Itaru Sasaki, Yuki Tsujimoto, Yuma Takabayashi, Kota Ujino, Vanja Wagner, and Yuki Yamamoto.

The first named author is thankful to Giacomo Ascione, Anup Biswas, Atsuhide Ishida, Kamil Kaleta, Itaru Sasaki, Xiaochuan Yang, for recent collaborations which have added new developments in the field, and contributed to the present edition. He also gratefully thanks IHES Bures-sur-Yvette, Scuola Normale Superiore, Naples, Università Sapienza, Rome, Tokyo University of Science, and Great Bay University, Dongguan, for research and lecturing opportunities, providing an environment of sustained interest in this book, and interactions which inspired further work on these subjects and shaped the present edition. Special thanks are due to Isaac Newton Institute, Cambridge, for giving an opportunity to host the 4-month thematic programme "Fractional Differential Equations," having strong and motivating connections with these topics and promoting further research. Also, the yearly series of the workshop "Fractional Calculus, Probability, and Non-Local Operators," which has been a joint project with Gianni Pagnini and Enrico Scalas began at BCAM Bilbao, and which has recently come to its 10th round, has provided an active platform of interactions across the fields which cluster around topics that also include and had an impact on the subject matter of this book.

The second named author was supported in part by JSPS KAKENHI Grant Numbers 20K20886, 20H01808, and 25H00595. The second named author also gratefully acknowledges the kind hospitality of Technische Universität Darmstadt, Hanoi University of Science and Technology, Thammasat University, Paderborn University, the Hausdorff School at the University of Bonn, and the National University of Singapore. In particular, at the Hausdorff School at the University of Bonn, the second named author benefited from very insightful discussions on quantum field theory and path integrals, which were highly encouraging. Part of this work was carried out during the workshop "Quantum Field Theory and Microlocal Analysis" in Roscoff, as well as during the above-mentioned visits. The second named author expresses his sincere thanks to his collaborators, Zied Ammari, Marco Falconi, Benjamin Hinrichs, Oliver Matte, and Tomoyuki Shirai. The second named author is also grateful to Ms. Tsubura Imabayashi for her administrative assistance. Finally, he would like to thank his family, Yuki, Kotaro, and Yui for their warm and constant support.

We both thank Kristin Berber-Nerlinger and Melanie Götz of Walter de Gruyter/Brill for their professionalism and support through the process of the present edition.

June 2026

https://doi.org/10.1515/9783112231463-203

Preface to the second edition

Since the publication of the first edition of this book in 2011 we have received much feedback from our colleagues and students, which kept encouraging us to consider a second edition. In this new edition, we set ourselves the task to amplify the text both by an account of the latest developments in functional integration applied to quantum theory, and to make the perusal of the book more comfortable to our readers by having most of the background material closer at hand. This meant to accept that the length of this book will be more manageable by splitting the material in two parts, each of a volume of its own.

This first volume offers then a detailed discussion of Feynman–Kac type path integral representations and their uses in a study of progressively more elaborate models ranging from classical to relativistic and more generally non-local Schrödinger operators with electrostatic and/or magnetic potentials, to further including spin. The second volume titled “Applications in Rigorous Quantum Field Theory” makes then the passage from applications in quantum mechanics to applications in quantum fields.

The original Feynman–Kac formula, and its many subsequent variants, is a mathematical relation connecting an evolution equation, an operator, and a random process, offering a probabilistic representation of solutions obtained as space-time transformations of an initial function under a semigroup, by running a suitable random process and averaging over its paths. This theory is then a fascinating meeting point of functional analysis, stochastic processes, and stochastic analysis, and it is in continuing development.

Those who are new to these topics and wish to learn about Feynman–Kac techniques may appreciate that in this edition Chapters 2 and 3 are now devoted to a discussion including detailed proofs of essential background material of concepts and facts of measure theory, probability, the theory of continuous-time stochastic processes and martingales, jump processes, and stochastic analysis. The first section of the main Chapter 4 similarly reviews fundamentals of linear operator theory in general, and Schrödinger operators in particular. The bulk of the chapter then continues with a systematic presentation of Feynman–Kac formulae for operators primarily of interest in mathematical quantum theory. Chapter 5 is about the “right hand side” of the Feynman–Kac formula including now a discussion of the related random processes with continuous paths or paths with jump discontinuities, and their distributions obtained as Gibbs measures. Experts will notice that new material includes Feynman–Kac representations of semigroups generated by relativistic and other non-local Schrödinger operators, a use of sojourn times, explicit solutions, eigenfunctions corresponding to embedded eigenvalues, and new results involving jump processes. The bibliography has also been updated, and we recommend the reading of this book with reference to the notes and comments in Chapter 6.

We thank our collaborators Anup Biswas, Takashi Ichinose, Kamil Kaleta, Mateusz Kwaśnicki, Jacek Małecki, Itaru Sasaki, Xiaochuan Yang, with whom we had the oppor-

https://doi.org/10.1515/9783112231463-204

tunity to further develop functional integration methods in the recent years, and who have greatly enriched our experience. We also thank Gonzalo A. Bley, Thomas Norman Dam, Batu Güneysu, Christian Jäh, Nikolai Leonenko, Oliver Matte, Jacob Schach Møller, René Schilling, Stefan Steinerberger, Renming Song, Yuma Takabayashi, Bruno Toaldo, Zoran Vondraček for reflections, suggestions or just interest in our book.

The first named author gratefully thanks IHES, Bures-sur-Yvette, for visiting fellowships through several years, where part of this book has been written, and Professor David Ruelle for sustained interest and discussions. The second named author thanks the kind hospitality of Aarhus University and the International Network Program of the Danish Agency for Science, Technology and Innovation, where part of this work has been done. His work was also financially supported by JSPS KAKENHI Grant Number JP16H03942 and JSPS KAKENHI Grant Number JP16K17612.

Finally, we also express our appreciation to Sabina Dabrowski of Walter de Gruyter, and Ina Talandienė and her team at VTeX UAB, Lithuania, for their professionalism and support.

August 2019

József Lőrinczi
Fumio Hiroshima
Volker Betz

Preface to the first edition

For a long time disciplines such as astronomy, celestial mechanics, the theory of heat, and the theory of electromagnetism counted among the most important sources in the development of mathematics, continually providing new problems and challenges. The main mathematical facts resulting from these fields typically crystallized in linear ordinary or partial differential equations.

In the early 20th century Heisenberg's discovery of a rule showing that the measurement of the position and momentum of a quantum particle give different answers, depending on the order of their measurement, led to the concept of commutation relations, and it made matrix theory find an unexpected application in the natural sciences. This resulted in an explosion of interest in linear structures developed by what we call today linear algebra and linear analysis, among them functional analysis, operator theory, C^*-algebras, and distribution theory. Similarly, quantum mechanics inspired much of group theory, lattice theory, and quantum logic.

Following this initial boom Wigner's beautiful phrase talking of the "unreasonable effectiveness of mathematics" marks a point of reflection on the fact that it is by no means obvious why this kind of abstract and sophisticated mathematics can be expected at all to make faithful descriptions and reliable predictions of natural phenomena. If indeed this can be appreciated as a miracle, as he said, the degree of flexibility that mathematics allows in its uses is perhaps at least a remarkable fact. Feynman has pointed to the different cultures of using mathematics when he remarked that "if all of mathematics disappeared, physics would be set back by exactly one week." While mathematics cannot compete with physics in discovering new phenomena and offering explanations of them, physics continues to depend on the terminology, arsenal, and discipline of mathematics. Mathematical physics, which is part of mathematics and therefore operates by its rules and standards, has set the goal to understand the models of physics in a rigorous way. Mathematical physicists thus find themselves at the borderline, listening to physics and speaking mathematics, at best able to use these functionalities interchangeably.

If one of the imports of early quantum mechanics has been the realization that the basic laws on the atomic scale can be formulated by linear superposition rules, another was a probabilistic interpretation of the wave function. This connection with chance, amplified and strongly advocated by the Copenhagen school, was no less revolutionary than the commutation relations. Richard Feynman has discovered a second connection with probability when he offered a representation of the state of a particle in terms of averages over all of its possible histories from one point in space-time to another. While this was an instance seriously questioning his dictum quoted above and his use of mathematics was in this case very problematic, the potential lying in the description advanced in his work turned out to be far reaching. The mathematician Mark Kac was the first to show that this method could be made sense of and was eminently viable.

Our project to write the present monograph has grown out of the thematic program *At the Interface of PDE, Self-Adjoint Operators and Stochastics: Models with Exclusion* or-

https://doi.org/10.1515/9783112231463-205

ganized by the first named author at the Wolfgang Pauli Institute, Vienna, in 2006. The initial concept was a smaller-scale but up-to-date account of Feynman–Kac-type formulae and their uses in quantum field theory, which we wanted to dedicate to the person all three of us have much to thank to, both scientifically and personally, Herbert Spohn. At that time we secretly meant this as a present for his 60th birthday, and we are glad that we are now able to make this hopefully more mature tribute on the occasion of his 65th birthday!

Apart from the inspiring and pleasant environment at Zentrum Mathematik, Munich University of Technology, where our collaboration has started, and Wolfgang Pauli Institut, University of Vienna, hosting the thematic program, we are indebted to a number of other individuals and institutions. We are thankful for the joint work and friendship of our closest collaborators in this direction, Asao Arai, Massimiliano Gubinelli, Masao Hirokawa, and Robert Adol'fovich Minlos. We also thank Cristian Gérard, Takashi Ichinose, Kamil Kaleta, Jacek Małecki, Annalisa Panati, and Akito Suzuki for ongoing related joint work. It is furthermore a pleasure to thank Norbert Mauser, Director of the Wolfgang Pauli Institute, for supporting the program, and Sergio Albeverio, Rodrigo Bañuelos, Krzysztof Bogdan, Carlo Boldrighini, Thierry Coulhon, Laure Coutin, Cécile DeWitt-Morette, Aernout van Enter, Bill Faris, Gero Friesecke, Jürg Fröhlich, Mohammud Foondun, Hans-Otto Georgii, Alexander Grigor'yan, Martin Hairer, Takeru Hidaka, Wataru Ichinose, Keiichi Itô, Niels Jacob, Hiroshi Kawabi, Tadeusz Kulczycki, Kazuhiro Kuwae, Terry Lyons, Tadahiro Miyao, Hirofumi Osada, Habib Ouerdiane, Sylvie Roelly, Itaru Sasaki, Tomoyuki Shirai, Toshimitsu Takaesu, Setsuo Taniguchi, Josef Teichmann, Daniel Ueltschi, and Jakob Yngvason for many useful discussions over the years. We thank especially Erwin Schrödinger Institut, Vienna, Institut des Hautes Études Scientifiques, Bures-sur-Yvette, and Università La Sapienza, Rome, for repeatedly providing excellent environments to further the project and where parts of the manuscript have been written. Also, we thank the hospitality of Kyushu University, Loughborough University, and Warwick University, for accommodating mutual visits and extended stays.

We are particularly pleased to thank the professionalism, assistance, and endless patience of Robert Plato, Simon Albroscheit, and Friederike Dittberner of Walter de Gruyter. Also, we are grateful to Max Lein for technical support in the initial phase of this project.

February 2011

József Lőrinczi
Fumio Hiroshima
Volker Betz

1 Heuristics and history

1.1 Feynman path integrals and Feynman–Kac formulae

In Feynman's work concepts such as paths and actions, just discarded by the new-born quantum mechanics, witnessed a comeback in the description of the time evolution of quantum particles, proposing an alternative to Heisenberg's matrix mechanics and Schrödinger's wave mechanics. It is worthwhile to briefly review how this turned out to be a viable project and led to a mathematically meaningful approach to quantum mechanics and quantum field theory.

A basic tool in the hands of the quantum physicist is the *Schrödinger equation*

$$\begin{aligned} i\hbar\frac{\partial}{\partial t}\psi(t,x) &= -\frac{\hbar^2}{2m}\Delta\psi(t,x) + V(x)\psi(t,x), \\ \psi(0,x) &= \phi(x), \quad (t,x) \in \mathbb{R}^+ \times \mathbb{R}^3 \end{aligned} \tag{1.1.1}$$

with the *Hamilton operator*

$$H = -\frac{\hbar^2}{2m}\Delta + V \tag{1.1.2}$$

and a complex-valued *wave function* ψ describing a quantum particle of mass m at position x in space at time t. Here $\hbar$ is Planck's constant, V is a potential in whose force field the particle is moving, and ϕ describes the initial state of the quantum particle. The Schrödinger equation can be solved only under exceptional choices of V. However, formally the solution of the equation can be written as

$$\psi(t,x) = (e^{-(it/\hbar)H}\phi)(x)$$

for any V. The difficulty is that for a general choice of V we do not know how to calculate the right-hand side of the equality.

The action of the complex exponential operator on ϕ can be evaluated by using an *integral kernel* G allowing us to write the wave function of the particle at position y at time t based on its knowledge in x at $s < t$, i. e.,

$$\psi(t,y) = \int_{\mathbb{R}^3} G(x,y;s,t)\psi(s,x)dx.$$

This representation can then in principle be iterated from one interval $[s,t]$ to another, and thus our knowledge of the wave function "propagates" from the initial time to any time point in the future. This observation suggests that at least a qualitative analysis of the solution might be possible for a wide choice of V by using this integral kernel, and the key to this approach is the information content of the kernel.

https://doi.org/10.1515/9783112231463-001

Feynman proposed a method of constructing this integral kernel in the following way. Consider first a classical *Hamiltonian* function

$$\mathscr{H}(p,q) = \frac{1}{2m}p^2 + V(q), \quad (p,q) \in \mathbb{R}^3 \times \mathbb{R}^3, \tag{1.1.3}$$

describing a particle at position q moving with momentum p under the same potential V as above. The corresponding *Lagrangian* is obtained through the Legendre transform

$$\mathscr{L}(q,\dot{q}) = \dot{q}p - \mathscr{H}(p,q) = \frac{m}{2}\dot{q}^2 - V(q),$$

where

$$\dot{q} = \frac{\partial \mathscr{H}}{\partial p} = \frac{p}{m}.$$

Classical mechanics says that the particle is equivalently described by the *action functional*

$$\mathscr{A}(q,\dot{q};0,t) = \int_0^t \mathscr{L}(q(s),\dot{q}(s))ds,$$

where $q(s)$ is the path of the particle as a function of the time s. In particular, the equation of motion can be derived by the *principle of least action* by using variational calculus, and it is found that the particle follows a route from a fixed initial to a fixed endpoint which minimizes the action.

In the late 1920s Dirac worked on a Lagrangian formulation of quantum mechanics and arrived at the conclusion that the ideas of the classical Lagrangian theory can be taken over, though not its equations. Specifically, he arrived at a rule he called "transformation theory" allowing to connect the positions at s and t of the quantum particle through the phase factors $e^{(i/\hbar)\mathscr{A}(q,\dot{q};s,t)}$. By dividing up the interval $[s,t]$ into small pieces, Dirac noticed [95–97] that due to the fact that $\hbar$ is of the order of 10^{-34} in standard units, this is so small that the integrands in the exponent are rapidly oscillating functions around zero. Therefore, as a result of the many cancellations, the only significant contribution into the integral is given by those q whose large variations over the subintervals involve small differences in the Lagrangian.

Feynman, while a Princeton research assistant during 1940–1941, turned Dirac's somewhat vague observations into a more definite formulation. He argued that the propagators $G(x,y;s,t)$, which were Dirac's "contact transformations," can be obtained as limits over products of propagators at intermediary time points. This means that with a partition of $[s,t]$ to subintervals $[t_j,t_{j+1}]$ with $t_j = s + jb, j = 0,\dots,n-1, b = (t-s)/n$, the equality

$$G(x,y;s,t) = \lim_{n\to\infty} \prod_{j=0}^{n-1} \int_{\mathbb{R}^3} G(x_j, x_{j+1}; t_j, t_{j+1}) \prod_{j=1}^{n-1} dx_j$$

holds, where

$$G(x_j, x_{j+1}; t_j, t_{j+1}) = \sqrt{\frac{m}{2b\pi i\hbar}} \exp\left(\frac{ib}{\hbar}\mathscr{L}\left(x_{j+1}, \frac{x_{j+1}-x_j}{b}\right)\right).$$

From here it is possible to conclude that

$$(e^{-(it/\hbar)H})(x,y) = G(x,y;0,t) = \text{const} \int_{\Omega_{xy}} \exp\left(\frac{i}{\hbar}\mathscr{A}(q,\dot{q};0,t)\right) \prod_{0\le s\le t} dq(s), \tag{1.1.4}$$

giving a formal expression of the integral kernel of the operator $e^{-(it/\hbar)H}$ for all t. Here $dq(s)$ is the Lebesgue measure, and Ω_{xy} is the set of continuous paths $q(s)$ such that $q(0) = x$ and $q(t) = y$. The right-hand side of the expression is *Feynman's integral*. These results have been obtained first in his 1942 thesis, which remained unpublished; the paper [123] published in 1948, however, reproduces the crucial statements (see also [126]). Feynman was able to apply his approach also to quantum field theory [124, 125].

For all its intuitive appeal and potential, Feynman's method did not win immediate acclaim, not even in the ranks of theoretical physicists. Mathematically viewed, the object Feynman proposed was by no means easy to define. Mathematicians hitherto have not found a satisfactory meaning of Feynman integrals covering a sufficiently wide class of choices of V. For instance, there is no countably additive measure on Ω_{xy} allocating equal weight to every path, and $\mathscr{A}(q,\dot{q};0,t)$ fails to be defined on much of Ω_{xy} since it involves $\dot{q}(s)$. Various ways of trying to cope with these difficulties have been explored. Several authors considered purely imaginary potentials V and masses m leading to a complex measure to define Feynman's integral, but these attempts had to face the problem of infinite total variation [62, 77, 281]. Another suggestion was to define the Feynman integral in terms of Fresnel distributions [93, 4, 94, 63, 67, 366]. Other attempts of constructing solutions of Schrödinger's equation from Feynman's integral were made through piecewise classical paths or broken lines or polygonal paths [133, 187, 192, 181, 368, 200].

In Feynman's days integration theory on the space of continuous functions had already been in existence due to Wiener's work on Brownian motion, initiated in 1923. It was Kac in 1949 who first realized that this was a suitable framework, however, not for the path integral (1.1.4) directly; see also [202, 203]. In contrast to Feynman's expression, the *Feynman–Kac formula* offers an integral representation on path space for the semigroup $e^{-(1/\hbar)tH}$ instead of $e^{-(i/\hbar)tH}$, obtained upon replacing t by $-it$. Why this should actually improve the situation can be seen by the following argument. Analytic continuation $s \mapsto -is$, $ds \mapsto -ids$, and the replacement $\dot{q}(s)^2 \mapsto -\dot{q}(s)^2$ in (1.1.4) lead to the kernel

$$(e^{-tH})(x,y) = \text{const} \int_{\Omega_{xy}} e^{-(1/\hbar)\int_0^t V(q(s))ds} e^{-(m/2\hbar)\int_0^t \dot{q}(s)^2 ds} \prod_{0\le s\le t} dq(s). \tag{1.1.5}$$

Since defining the latter factor as a measure remains to be a problem as

$$\int_{\Omega_{xy}} \prod_{0\le s\le t} dq(s) = \infty \tag{1.1.6}$$

for every t, one might boldly hope that its support $\mathcal{C}_{xy}$ is a set on which the exponential weights typically vanish. If $\dot{q}$ were almost surely plus or minus infinite on this set, then the situation

$$\exp\left(-\frac{1}{\hbar}\int_0^t \frac{m}{2}\dot{q}(s)^2 ds\right)\Bigg|_{\mathcal{C}_{xy}} = 0 \tag{1.1.7}$$

would occur; $\mathcal{C}_{xy}$ would thus be required to consist of continuous but nowhere differentiable paths. Once this can be allowed, zero in (1.1.7) can possibly cancel infinity in (1.1.6), and it would in principle be possible that

$$\exp\left(-\frac{1}{\hbar}\int_0^t \frac{m}{2}\dot{q}(s)^2 ds\right) \prod_{0\le s\le t} dq(s) = 0 \cdot \infty$$

is a well-defined measure. As we will explain below, the miracle happens: The paths of a random process $(B_t)_{t\ge 0}$ called *Brownian motion* just have the required properties and the Feynman–Kac formula

$$(e^{-tH})(x,y) = \int_{C(\mathbb{R}^+,\mathbb{R}^3)} e^{-\int_0^t V(B_s(\omega))ds} d\mathcal{W}^{x,y}_{[0,t]}(\omega), \quad t \ge 0,\ x,y \in \mathbb{R}^3, \tag{1.1.8}$$

rigorously holds provided some technical, but highly satisfying, conditions hold on V. In particular, there is a probability measure supported on the space $\mathcal{C}_{xy}$ identified with the space $C(\mathbb{R}^+,\mathbb{R}^3)$ of continuous functions $\mathbb{R}^+ \to \mathbb{R}^3$, and it can be identified as *Wiener measure* $\mathcal{W}$ conditional on paths leaving from x at time 0 and ending in y at time t. In other words, the kernel of the semigroup at the left-hand side (the object corresponding to $G(x,y;0,t)$ above) can be studied by running a Brownian motion weighted by the potential V.

Kac has actually proven [201] that the *heat equation with dissipation*

$$\frac{\partial f}{\partial t} = \frac{1}{2}\Delta f - Vf, \quad f(x,0) = \phi(x), \tag{1.1.9}$$

is solved by the function

$$f(x,t) = \int_{C(\mathbb{R}^+,\mathbb{R}^3)} e^{-\int_0^t V(B_s(\omega))\,ds} \phi(B_t(\omega))\, d\mathcal{W}^x(\omega). \tag{1.1.10}$$

The above heat equation with dissipation is actually the same as the Schrödinger equation (1.1.1) when analytic continuation $t \mapsto -it$ is made, a fact that has been noticed by Ehrenfest as early as in 1927. In particular, on setting $V = 0$ and formally replacing ϕ with a δ-distribution, the fundamental solution $\Pi(x,t)$ of the standard heat equation is obtained. The relation of this so-called *heat kernel*

$$\Pi(x,t) = \frac{1}{(2\pi t)^{3/2}} \exp\left(-\frac{|x|^2}{2t}\right), \quad t > 0, \tag{1.1.11}$$

is once again close to random processes as it gives the transition probabilities of Brownian motion. The Feynman–Kac formula thus provides a link between an operator semigroup, the heat equation, and Brownian motion. This observation has opened up a whole new conceptual and technical framework pointing to and exploring the rich interplay between self-adjoint operators, probability, and partial differential equations.

1.2 Plan and scope of the book

The Feynman–Kac formula for the operator $H = -\frac{1}{2}\Delta + V$ derived from the heat equation with dissipation can be written in any dimension d as

$$(e^{-tH} f)(x) = \int_{C(\mathbb{R}^+,\mathbb{R}^d)} e^{-\int_0^t V(B_s(\omega))\,ds} f(B_t(\omega))\, d\mathcal{W}^x(\omega),$$

where $(B_t)_{t\geq 0}$ is accordingly an $\mathbb{R}^d$-valued Brownian motion. Due to the connections outlined above, the operator H is called a *Schrödinger operator* with potential V. If the potential is such that H admits an eigenfunction ψ at eigenvalue E, then $e^{-tH}\psi = e^{-Et}\psi$ and thus this eigenfunction can be represented in terms of an average over the paths of Brownian motion given by the right-hand side. This makes it possible to obtain information on the spectral properties of H by probabilistic means. In particular, the Feynman–Kac formula can be used to study the *ground states* (i. e., eigenfunctions at the bottom of the spectrum) of Schrödinger operators to a great detail.

It is natural to ask whether it is possible to derive representations similar to the Feynman–Kac formula for any other operator. In a companion volume we will show that also other operators, used in quantum field theory, allow obtaining Feynman–Kac-type formulae with appropriately modified random processes. On the one hand, this relation will allow us to ask and answer questions relevant to quantum field theory in a rigorous way. On the other hand, we will face situations in which new concepts and methods need to be developed, which thus generates new mathematics.

We split the material of this monograph in two volumes of comparable size, each focussing on one aspect explained above. In the present Volume 1, we discuss Feynman–Kac-type representations for classical and nonlocal (relativistic and other) Schrödinger operators, their related random processes, and explore the wide-ranging applications of this relationship. In Volume 2 we carry out a similar programme for specific operators in quantum field theory.

The present volume consists of six chapters, the last being a section of notes and commentary of the literature. In Chapter 1 we have outlined above the origins of the idea of path integrals applied in quantum mechanics and quantum field theory, due to the physicist Richard Ph. Feynman, and formulated in a mathematically rigorous way by the mathematician Mark Kac.

In Chapter 2 we make preparatory steps to a rigorous development of these ideas, first by offering background material from probability, the theory of random processes, and stochastic analysis. We start from general measure theory, martingales and Markov processes, and a review of Brownian motion. We conclude by a section devoted to basic concepts of Itô integration theory.

In Chapter 3 we give a similar survey of Lévy processes and discuss related concepts such as infinite divisibility, the Lévy–Khintchine representation formula, and Markov and martingale properties of jump processes. We continue with sample path properties, Poisson random measures, the canonical Lévy–Itô decomposition, and the Itô formula for semimartingales. Finally, we discuss exponentials of Lévy processes, capacitary measures, subordinate Brownian motion, and related Bernstein functions.

In Chapter 4 we start by a review of background material of the theory of self-adjoint operators and their semigroups, with a special focus on Schrödinger operators. We present established material in the light of our present goals and framework. The remainder of this main chapter deals then with a systematic presentation of variants of increasing complexity of the Feynman–Kac formula. We cover cases starting from an external scalar potential, and continue by including vector potentials describing interactions with magnetic fields, spin, and relativistic effects. Mathematically this means a range of operators consisting of the sum of the negative Laplacian and a multiplication operator of varying regularity, further including gradient operator terms and pseudo-differential operators such as the square root of the Laplacian and related operators. From the perspective of random processes involved, we proceed from Brownian motion to diffusions ($P(\phi)_1$-processes and Itô diffusions) and subordinate Brownian motion (jump Lévy processes such as relativistic stable and other cases). Apart from first establishing the variants of the Feynman–Kac formula for these choices of operators and processes, we also address their applications in the study of analytic and spectral properties of these operators (properties of integral kernels, positivity improving properties, L^p-smoothing properties, intrinsic ultracontractivity, existence, uniqueness and multiplicity of eigenfunctions, decay of eigenfunctions, number of eigenfunctions with negative eigenvalues, diamagnetic inequalities, spectral comparison inequalities, etc.). From the perspective of quantum mechanics, these results translate into the bound state properties of nonrelativistic or relativistic quantum particles under scalar or magnetic potentials, with or without spin. Finally, we present a class of nonlocal Schrödinger operators given by Bernstein functions of the Laplacian, possibly including magnetic fields and generalized spin, and study their Feynman–Kac formulae and related properties. We apply the stochastic representation of eigenfunctions to study their local behavior and asymptotic properties at infinity, present the explicit solution of an eigenvalue problem, and also discuss the properties of bound states corresponding to an embedded eigenvalue at zero.

In Chapter 5 we focus on the "right-hand side" of the Feynman–Kac formula and its variants. In all of the cases that we consider, the kernel of the operator semigroup can be expressed in terms of an expectation over the paths of a random process with respect to a measure weighted by exponential densities dependent on functionals additive in the time intervals. This structure suggests viewing this as a new probability measure, which can be interpreted as a Gibbs measure on the path space for the potential(s) appearing directly in or derived from the generator of the process. Gibbs measures are well understood in discrete stochastic models and originate from statistical mechanics, where they have been used to model thermodynamic equilibrium states. In our context this interpretation has another relevance, and our main focus here is to make use of the so obtained Gibbs measures in studying analytic and spectral properties of the operators at the left-hand side of the Feynman–Kac formula. In this chapter we define and prove the existence and properties of Gibbs measures on the path space. First, we present this for the case of external potentials and define the framework of $P(\phi)_1$-processes. Then we establish existence and further fundamental results (uniqueness, almost sure path behavior, mixing properties, etc.) for the case when the densities of the Gibbs measure also contain a pair-interaction potential.

In the concluding Chapter 6 we offer a commented guide to the selected bibliography, which we have extended and updated from the previous edition. Undoubtedly, our list of references could be further increased, and we trust that the interested reader will be able to find their way through further literature by relying on this list.

2 Brownian motion

2.1 Concepts and facts of general measure theory and probability

2.1.1 Elements of general measure theory

For a more self-contained presentation of the material, we present some fundamental facts often used in measure theory. Let $\Omega \neq \emptyset$ be a set. We will denote by 2^{Ω} the *power-set* of Ω, i. e., the set of all of its subsets. A collection of subsets $\mathscr{F} \subset 2^{\Omega}$ is called a *σ-field* with respect to Ω if it satisfies

(1) $\Omega \in \mathscr{F}$;
(2) $A \in \mathscr{F}$ implies $A^c \in \mathscr{F}$;
(3) $A_n \in \mathscr{F}, n \in \mathbb{N}$, implies $\bigcup_{n=1}^{\infty} A_n \in \mathscr{F}$.

A pair $(\Omega, \mathscr{F})$ consisting of a nonempty set Ω and σ-field $\mathscr{F}$ is called a *measurable space.* When Ω is equipped with a topology, its *Borel σ-field* is defined by the minimal σ-field including all open sets of Ω and is denoted by $\mathcal{B}(\Omega)$. The pair $(\Omega, \mathcal{B}(\Omega))$ is called a *Borel measurable space.* A measurable space $(\Omega, \mathscr{F})$ is *separable* if $\mathscr{F}$ is generated by a countable collection of subsets in Ω. In particular, if Ω is a locally compact space with a countable basis, then $(\Omega, \mathcal{B}(\Omega))$ is separable. We will assume throughout that Borel measurable spaces are separable.

Let $(\Omega, \mathscr{F})$ be a measurable space. A set function $\mu : \mathscr{F} \to [0, \infty)$ is called a *measure* if

$$\mu\left(\bigcup_{n=1}^{\infty} A_n\right) = \sum_{n=1}^{\infty} \mu(A_n)$$

holds for all sets $A_n \in \mathscr{F}, n \in \mathbb{N}$, such that $A_i \cap A_j = \emptyset$ whenever $i \neq j$. A set function $\mu : \mathscr{F} \to [0, \infty)$ is called a *finitely additive measure* if it satisfies

$$\mu\left(\bigcup_{n=1}^{m} A_n\right) = \sum_{n=1}^{m} \mu(A_n)$$

for any finitely many pairwise disjoint sets $A_n \in \mathscr{F}, n = 1, \dots, m, m \in \mathbb{N}$. A measurable space equipped with a measure μ is a *measure space*, denoted by $(\Omega, \mathscr{F}, \mu)$. If $\mu(\Omega) < \infty$, then μ is called a *finite measure.* If there exists a sequence $(A_n)_{n\in\mathbb{N}} \subset \mathscr{F}$ such that $\Omega = \bigcup_{n=1}^{\infty} A_n$ and $\mu(A_n) < \infty$ for all $n \in \mathbb{N}$, then μ is a *σ-finite measure.* Let μ be a measure on a Borel measurable space $(\Omega, \mathcal{B}(\Omega))$. The measure μ is called *locally finite* if for every point $\omega \in \Omega$ there exists an open neighborhood $U \ni \omega$ such that $\mu(U) < \infty$.

If $\mathscr{A} \subset 2^{\Omega}$ satisfies that

(1) $\Omega \in \mathscr{A}$;
(2) $A \in \mathscr{A}$ implies $A^c \in \mathscr{A}$;
(3) $A_1, \dots, A_n \in \mathscr{A}$ implies $\bigcup_{m=1}^{n} A_m \in \mathscr{A}$,

https://doi.org/10.1515/9783112231463-002

then $\mathscr{A}$ is called a *set-algebra*, or *field*. Whenever $\mathscr{M} \subset 2^\Omega$ satisfies that

(1) for every increasing sequence $(A_n)_{n\in\mathbb{N}} \subset \mathscr{M}$ it follows that $\bigcup_{n=1}^\infty A_n \in \mathscr{M}$;
(2) for every decreasing sequence $(B_n)_{n\in\mathbb{N}} \subset \mathscr{M}$ it follows that $\bigcap_{n=1}^\infty B_n \in \mathscr{M}$,

then $\mathscr{M}$ is called a *monotone family*.

Denote by $\sigma(\mathscr{A})$ the σ-field generated by $\mathscr{A} \subset 2^\Omega$, and by $m(\mathscr{A})$ the monotone family generated by $\mathscr{A}$.

Proposition 2.1. *If $\mathscr{A}$ is a set-algebra, then $m(\mathscr{A}) = \sigma(\mathscr{A})$.*

Proof. Since $\sigma(\mathscr{A})$ is a monotone family, $\sigma(\mathscr{A}) \supset m(\mathscr{A})$ follows. To prove that $\sigma(\mathscr{A}) \subset m(\mathscr{A})$, it is sufficient to show that $m(\mathscr{A})$ is a σ-field. We show that $A, B \in m(\mathscr{A})$ implies that $A \cup B, A \setminus B, B \setminus A \in m(\mathscr{A})$. Define

$$D_A = \left\{B \in 2^\Omega \mid A \cup B, A \setminus B, B \setminus A \in m(\mathscr{A})\right\}.$$

It is easily seen that D_A is a monotone family and $m(\mathscr{A}) \subset D_A$ for all $A \in \mathscr{A}$. This implies that $B \in D_A$ for every $B \in m(\mathscr{A})$ and $A \in \mathscr{A}$. Note that $B \in D_A$ if and only if $A \in D_B$. Hence $B \in D_A$ for every $A \in m(\mathscr{A})$ and $B \in \mathscr{A}$. We see that $\mathscr{A} \subset D_A$ for every $A \in m(\mathscr{A})$, and $m(\mathscr{A}) \subset m(D_A) = D_A$ for all $A \in m(\mathscr{A})$. This shows that $A, B \in m(\mathscr{A})$ implies that $A \cup B, A \setminus B, B \setminus A \in m(\mathscr{A})$. Hence we furthermore have that $A \in m(\mathscr{A})$ implies $A^c \in m(\mathscr{A})$, and $A_j \in m(\mathscr{A}), j \in \mathbb{N}$, implies that $\bigcup_{j=1}^n A_j \in m(\mathscr{A})$ for every $n \in \mathbb{N}$. Since $\bigcup_{j=1}^n A_j \uparrow \bigcup_{j=1}^\infty A_j$ as $n \to \infty$, we have $\bigcup_{j=1}^\infty A_j \in m(\mathscr{A})$. Thus $m(\mathscr{A})$ is a σ-field. □

A family $\mathscr{A} \subset 2^\Omega$ is called a *π-system* if

$$A_1, \dots, A_n \in \mathscr{A} \quad \text{implies} \quad \bigcap_{m=1}^n A_m \in \mathscr{A}.$$

Furthermore, $\mathscr{L} \subset 2^\Omega$ is called a *λ-system* if

(1) $\phi \in \mathscr{L}$;
(2) $A \in \mathscr{L}$ implies $A^c \in \mathscr{L}$;
(3) $A_n \in \mathscr{L}, n \in \mathbb{N}$, and $A_n \cap A_m = \emptyset$ for $n \neq m$ implies $\bigcup_{n=1}^\infty A_n \in \mathscr{L}$.

Let $\lambda(\mathscr{I})$ denote the λ-system generated by $\mathscr{I} \subset 2^\Omega$.

Proposition 2.2 (π–λ theorem). *Let $\mathscr{I}$ be a π-system. Then the following hold:*

(1) $\sigma(\mathscr{I}) = \lambda(\mathscr{I})$.
(2) *If $\mathscr{L}$ is a λ-system and $\mathscr{L} \supset \mathscr{I}$, then $\mathscr{L} \supset \sigma(\mathscr{I})$.*

Proof. Since $\sigma(\mathscr{I})$ is a λ-system, it follows that $\sigma(\mathscr{I}) \supset \lambda(\mathscr{I})$. In order to show that $\sigma(\mathscr{I}) \subset \lambda(\mathscr{I})$, it suffices to see that $\lambda(\mathscr{I})$ is a σ-field. Let $D \subset 2^\Omega$ be a λ-system. It is straightforward to check that D is a π-system if and only if it is a σ-field. Thus it is sufficient to show that $\lambda(\mathscr{I})$ is a π-system, i. e., that $A, B \in \lambda(\mathscr{I})$ implies $A \cap B \in \lambda(\mathscr{I})$. Define the subset $D_B = \{A \in \lambda(\mathscr{I}) \mid A \cap B \in \lambda(\mathscr{I})\}$. We see that $\lambda(\mathscr{I})$ is a π-system if and only if $\lambda(\mathscr{I}) \subset D_B$ for every $B \in \lambda(\mathscr{I})$. We note that D_E is a λ-system for every $E \in \lambda(\mathscr{I})$. Since $\mathscr{I}$ is a π-system, we have $\mathscr{I} \subset D_E$ for all $E \in \mathscr{I}$. This implies $\lambda(\mathscr{I}) \subset \lambda(D_E) = D_E$ for every $E \in \mathscr{I}$. From this it follows that $B \cap E \in \lambda(\mathscr{I})$ for every $B \in \lambda(\mathscr{I})$ and $E \in \mathscr{I}$. Hence $\mathscr{I} \subset D_B$ and $\lambda(\mathscr{I}) \subset D_B$ for every $B \in \lambda(\mathscr{I})$, which proves (1). Next take a λ-system $\mathscr{L}$ and $\mathscr{L} \supset \mathscr{I}$. Then we have $\mathscr{L} = \lambda(\mathscr{L}) \supset \lambda(\mathscr{I}) = \sigma(\mathscr{I})$, and (2) follows. □

Part (1) of Proposition 2.2 is known as the *π–λ theorem.*

The following is a useful extension theorem for finitely additive measures.

Proposition 2.3 (Hopf's extension theorem). *Let $\mathscr{A} \subset 2^\Omega$ be a set-algebra, and a set function $\mu : \mathscr{A} \to [0,\infty)$ be a finitely additive measure. Then there exists a measure $\bar{\mu}$ on $\sigma(\mathscr{A})$ such that*

$$\bar{\mu}(A) = \mu(A), \quad A \in \mathscr{A},$$

if and only if $A_n \in \mathscr{A}$, $n = 1, 2, \dots$, are pairwise disjoint, and $\bigcup_{n=1}^\infty A_n \in \mathscr{A}$ implies that

$$\mu\Big(\bigcup_{n=1}^{\infty} A_n\Big) = \sum_{n=1}^{\infty} \mu(A_n).$$

In addition, if μ is σ-finite, then the extension $\bar{\mu}$ is unique.

We note that the condition $A_n \in \mathscr{A}$, $n = 1, 2, \dots$, are pairwise disjoint, and $\bigcup_{n=1}^\infty A_n \in \mathscr{A}$ implies that $\mu(\bigcup_{n=1}^\infty A_n) = \sum_{n=1}^\infty \mu(A_n)$ above is equivalent to having $\lim_{n\to\infty} \mu(A_n) = 0$ for every decreasing sequence $(A_n)_{n\in\mathbb{N}} \subset \mathscr{A}$ such that $\bigcap_{n=1}^\infty A_n = \emptyset$.

A family of sets $\mathscr{A} \subset 2^\Omega$ is called a *semiring* if

(1) $\emptyset \in \mathscr{A}$;
(2) for every $A, B \in \mathscr{A}$ the set $B \setminus A$ is a finite union of pairwise disjoint sets in $\mathscr{A}$;
(3) $A, B \in \mathscr{A}$ implies $A \cap B \in \mathscr{A}$.

Let $\mathscr{A}$ be a semiring and $A \in \sigma(\mathscr{A})$. Lemma 2.4 below shows that any set from $\sigma(\mathscr{A})$ can be well approximated by sets obtained from finite operations involving sets in $\mathscr{A}$. Denote by

$$A \triangle B = (A \setminus B) \cup (B \setminus A)$$

the *symmetric difference* of A and B.

Lemma 2.4 (Approximation of measures). *Let Ω be a given set, $\mathscr{A} \subset 2^\Omega$ be a semiring, and μ a σ-finite measure on $\sigma(\mathscr{A})$. Then the following hold:*

(1) *For every $A \in \sigma(\mathscr{A})$ with $\mu(A) < \infty$ and $\varepsilon > 0$, there exist $k \in \mathbb{N}$ and pairwise disjoint sets $A_1, \dots, A_k \in \mathscr{A}$ such that $\mu(A \triangle \bigcup_{n=1}^k A_n) < \varepsilon$.*
(2) *For every $A \in \sigma(\mathscr{A})$ and $\varepsilon > 0$, there exist pairwise disjoint sets $\{A_n\}_{n\in\mathbb{N}} \subset \mathscr{A}$ such that $A \subset \bigcup_{n=1}^\infty A_n$ and $\mu(\bigcup_{n=1}^\infty A_n \setminus A) < \varepsilon$.*

Proof. (1) Let $\varepsilon > 0$. By the definition of the measure μ, there exists a covering $B_1, B_2, \ldots \in \mathscr{A}$ such that $A \subset \bigcup_{n=1}^{\infty} B_n$ and $\mu(A) \geq \sum_{n=1}^{\infty} \mu(B_n) - \varepsilon/2$. Since $\mu(A) < \infty$, there exists m such that $\sum_{n=m+1}^{\infty} \mu(B_n) < \varepsilon/2$. In general, for any sets C, D, and E, we have $C \triangle D \subset (C \triangle (D \cup E)) \cup E$. Choosing $C = A$, $D = \bigcup_{n=1}^{m} B_n$ and $E = \bigcup_{n=m+1}^{\infty} B_n$, we have

$$\mu\left(A \triangle \bigcup_{n=1}^{m} B_n\right) \leq \mu\left(A \triangle \bigcup_{n=1}^{\infty} B_n\right) + \mu\left(\bigcup_{n=m+1}^{\infty} B_n\right) \leq \mu\left(\bigcup_{n=1}^{\infty} B_n\right) - \mu(A) + \frac{\varepsilon}{2} \leq \varepsilon.$$

We see that $\bigcup_{n=1}^{m} B_n = B_1 \cup \bigcup_{n=2}^{m} \bigcap_{j=1}^{n-1} (B_n \setminus B_j)$. Note that $C_n = \bigcap_{j=1}^{n-1} (B_n \setminus B_j)$, $n = 2, \ldots, m$, are disjoint sets. Since $\mathscr{A}$ is a semiring, the set $B_n \setminus B_j$ can be obtained as a finite union of pairwise disjoint sets of $\mathscr{A}$, and there exist a finite number $k \in \mathbb{N}$ and pairwise disjoint sets $A_1, \ldots, A_k \in \mathscr{A}$ such that $\bigcup_{n=1}^{m} B_n = \bigcup_{n=1}^{k} A_n$. Then $\mu(A \triangle \bigcup_{n=1}^{k} A_n) < \varepsilon$ follows.

(2) Let $A \in \sigma(\mathscr{A})$ and $E_n \uparrow \Omega$ be such that $E_n \in \sigma(\mathscr{A})$ and $\mu(E_n) < \infty$, for all $n \in \mathbb{N}$. For each n, we can choose a covering $(B_{n,m})_{m \in \mathbb{N}}$ of $A \cap E_n$ with $\mu(A \cap E_n) \geq \sum_{m=1}^{\infty} \mu(B_{n,m}) - \varepsilon/2^n$. Moreover, it is also possible to show that there exist $A_n \in \mathscr{A}$, $n \in \mathbb{N}$, such that $A_1, A_2, \ldots$ are pairwise disjoint and $\bigcup_{n,m=1}^{\infty} B_{n,m} = \bigcup_{n=1}^{\infty} A_n$. Thus

$$\begin{aligned}\mu\left(\bigcup_{n=1}^{\infty} A_n \setminus A\right) &= \mu\left(\bigcup_{n=1}^{\infty} \bigcup_{m=1}^{\infty} B_{n,m} \setminus A\right) \leq \mu\left(\bigcup_{n=1}^{\infty} \bigcup_{m=1}^{\infty} (B_{n,m} \setminus (A \cap E_n))\right) \\ &\leq \sum_{n=1}^{\infty} \left(\left(\sum_{m=1}^{\infty} \mu(B_{n,m})\right) - \mu(A \cap E_n)\right) \leq \varepsilon.\end{aligned}$$

Hence (2) follows. □

With measurable spaces $(\Omega, \mathscr{F})$ and $(S, \mathscr{S})$, we write $f \in \mathscr{F}/\mathscr{S}$ to indicate that the function $f : \Omega \to S$ is *measurable* with respect to $\mathscr{F}$ and $\mathscr{S}$, i. e.,

$$f^{-1}(E) \in \mathscr{F}, \quad E \in \mathscr{S}.$$

Whenever $f \in \mathscr{F}/\mathcal{B}(\mathbb{R}^d)$, f is called a *Borel measurable function.*

Two measurable spaces $(\Omega, \mathscr{F})$ and $(\Omega', \mathscr{F}')$ are said to be *Borel isomorphic* if there exists a bijection $f : \Omega \to \Omega'$ such that $f \in \mathscr{F}/\mathscr{F}'$ and $f^{-1} \in \mathscr{F}'/\mathscr{F}$. A measurable space $(\Omega, \mathscr{F})$ is called a *standard measurable space* if it is Borel isomorphic to one of the measurable spaces

$$(\{1, \ldots, n\}, \mathcal{B}(\{1, \ldots, n\})), \quad (\mathbb{N}, \mathcal{B}(\mathbb{N})), \quad (\mathbb{M}, \mathcal{B}(\mathbb{M})),$$

where $\mathbb{M} = \{0,1\}^{\mathbb{N}} = \{\omega = \{\omega_1, \omega_2, \ldots\} \mid \omega_j = 0, 1\}$. The sets $\{1, \ldots, n\}$ and $\mathbb{N}$ are understood to be endowed with the discrete topology, and $\mathbb{M}$ with the product topology.

We define the integral of a Borel measurable function f on $(\Omega, \mathscr{F}, \mu)$, which will be denoted by any of the symbols

$$\int_\Omega f d\mu = \int_\Omega f(x) d\mu(x) = \int_\Omega f(x) d\mu.$$

Next we review the convergence properties of measures on metric spaces. Let S be a metric space. A σ-finite Borel measure μ on $(S, \mathcal{B}(S))$ is called a *Radon measure* if it is locally finite and inner regular, i. e.,

$$\mu(A) = \sup \left\{ \mu(K) \mid K \subset A \text{ and } K \text{ is compact} \right\}$$

for all $A \in \mathcal{B}(S)$. Note that since μ is locally finite, $\mu(K) < \infty$ for every compact set K.

Definition 2.5 (Weak and vague convergence). Let (S, d) be a metric space.

(1) A sequence of finite measures $(\mu_n)_{n\in\mathbb{N}}$ on $(S, \mathcal{B}(S))$ is *weakly convergent* to a finite measure μ whenever

$$\lim_{n\to\infty} \int_S f d\mu_n = \int_S f d\mu, \tag{2.1.1}$$

for every bounded continuous function f on S.

(2) A sequence of Radon measures $(\mu_n)_{n\in\mathbb{N}}$ on $(S, \mathcal{B}(S))$ is *vaguely convergent* to a Radon measure μ whenever

$$\lim_{n\to\infty} \int_S f d\mu_n = \int_S f d\mu, \tag{2.1.2}$$

for every continuous function f with compact support on S.

In general, a sequence of finite measures $(\mu_n)_{\in\mathbb{N}}$ may not converge to a finite measure in the vague topology. A sufficient condition is the following. Suppose that $\sup_{n\in\mathbb{N}} \mu_n(S) \le m$ and μ_n vaguely converges to μ as $n \to \infty$. Since μ is inner regular, for every $\varepsilon > 0$ there exists a compact set K_ε such that $\mu(S \setminus K_\varepsilon) < \varepsilon$. We have $\mu(K_\varepsilon) = \lim_{n\to\infty} \mu_n(K_\varepsilon) \le m$ and $\mu(S) \le m + \varepsilon$, for every $\varepsilon > 0$. In particular, μ is finite and $\mu(S) \le m$. We note that even though $\mu_n(S) = m$ for all n, in general $\mu(S) \neq m$.

Let $\mathscr{P}_\mathrm{f}(S)$ be the set of finite measures on $(S, \mathcal{B}(S))$, and $\mathscr{P}_\mathrm{R}(S)$ be the set of Radon measures on $(S, \mathcal{B}(S))$. Weak convergence induces the weak topology τ_w on $\mathscr{P}_\mathrm{f}(S)$, which is the weakest topology such that for every bounded continuous function f, the map $\mathscr{P}_\mathrm{f}(S) \ni Q \mapsto \int_S f dQ \in \mathbb{R}$ is continuous. It is known that if S is separable, then $(\mathscr{P}_\mathrm{f}(S), \tau_\mathrm{w})$ is metrizable. Similarly, vague convergence induces the vague topology τ_v on $\mathscr{P}_\mathrm{R}(S)$, which is the weakest topology such that for every continuous function f with compact support, the map $\mathscr{P}_\mathrm{R}(S) \ni Q \mapsto \int_S f dQ \in \mathbb{R}$ is continuous. It is also known that if S is locally compact, then $(\mathscr{P}_\mathrm{R}(S), \tau_\mathrm{v})$ is a Hausdorff space.

The following is an example of a result on vague convergence, known as *Helly's second theorem* or *Helly's selection theorem*. Consider a sequence of finite measures $(\mu_n)_{n\in\mathbb{N}}$ on the measurable space $(\mathbb{R}^d, \mathcal{B}(\mathbb{R}^d))$. If $\sup_{n\in\mathbb{N}} \mu_n(\mathbb{R}^d) < \infty$, then $(\mu_n)_{n\in\mathbb{N}}$ is called *uniformly bounded.*

Proposition 2.6 (Helly's second theorem). *Let $(\mu_n)_{n\in\mathbb{N}}$ be uniformly bounded on the measurable space $(\mathbb{R}^d, \mathcal{B}(\mathbb{R}^d))$. Then there exist a subsequence $(\mu_{n_k})_{n_k\in\mathbb{N}}$ and a measure μ on $(\mathbb{R}^d, \mathcal{B}(\mathbb{R}^d))$ such that*

$$\lim_{n_k\to\infty} \int_{\mathbb{R}^d} f(x)\mu_{n_k}(dx) = \int_{\mathbb{R}^d} f(x)\mu(dx),$$

for every continuous function f with compact support.

Proof. We give an outline of the proof. Let

$$I = \Big\{(a_1, b_1) \times \cdots \times (a_d, b_d) \mid a_j, b_j \in \mathbb{Q},\ j = 1, \dots, d\Big\}$$

be the Cartesian product of d intervals whose endpoints are rational numbers. Let $M = \mathbb{Q}^d$, which is clearly a countable set and its elements can be enumerated as $M = \{m_1, m_2, \dots\}$. Define the function F_n on $\mathbb{R}^d$ by

$$F_n(x) = F_n(x_1, \dots, x_d) = \int_{\mathbb{R}^d} 1_{(-\infty,x_1]\times\cdots\times(-\infty,x_d]}\mu_n(dx),$$

called the *distribution function* of μ_n. Consider the restriction of F_n on M. Since $\{F_n(m_1)\}_{n\in\mathbb{N}}$ is a family of bounded sets of $\mathbb{R}$, there exists a subsequence $F_{n_1}(m_1)$ convergent to the real number $F(m_1)$. From $F_{n_1}(m_2)$ we can select a subsequence $F_{n_2}(m_2)$ which converges to $F(m_2)$, and so on. Repeating this procedure, by a diagonal argument we see that $F_{n_n}(m_k)$ converges to the real number $F(m_k)$ as $n \to \infty$, for every $m_k \in M$. For $x \in \mathbb{R}^d$, define

$$\bar{F}(x) = \sup_{\substack{\xi_1\le x_1,\dots,\xi_d\le x_d \\ (\xi_1,\dots,\xi_d)\in M}} F(\xi).$$

Thus it is seen that there exists a measure μ such that

$$\bar{F}(x) = \int_{\mathbb{R}^d} 1_{(-\infty,x_1]\times\cdots\times(-\infty,x_d]}\mu(dx)$$

and $\mu_{n_n}(A) \to \mu(A)$ for every $A \in I$. This is proven in a similar way to Corollary 2.46 below. Thus it follows that

$$\int_{\mathbb{R}^d} 1_A(x)\mu_{n_n}(dx) \to \int_{\mathbb{R}^d} 1_A(x)\mu(dx), \quad A \in I,$$

as $n \to \infty$. The proof can be completed by a limiting argument. □

Recall that a complete and separable metric space is called a *Polish space.*

Example 2.7 (Compact case). Let (M, d) be a compact metric space, and $C(M)$ denote the set of continuous functions on M. Define the metric

$$\rho(f, g) = \max\left\{|f(m) - g(m) \mid m \in M\right\}$$

on $C(M)$. Then $(C(M), \rho)$ is a Polish space.

Example 2.8 (Noncompact case). Let $C(\mathbb{R})$ (resp. $C([0, \infty))$) be the set of continuous functions on $\mathbb{R}$ (resp. $[0, \infty)$). We introduce the *locally uniform topology* given by the metric

$$d(f, g) = \sum_{k=0}^{\infty} \frac{1}{2^k}\left(\left(\sup_{0 \le |x| \le k} |f(x) - g(x)|\right) \wedge 1\right). \tag{2.1.3}$$

Then $C(\mathbb{R})$ and $C([0, \infty))$ are Polish spaces under this metric.

We discuss some properties of measures on Polish spaces, which will be useful in the study of path measures. In general, it is known that a σ-finite measure ν on a metric space S endowed with a Borel σ-field $\mathcal{B}(S)$ is a *regular measure*, i. e., for arbitrary $A \in \mathcal{B}(S)$ and $\varepsilon > 0$ there exist a closed set K and an open set U such that $K \subset A \subset U$ and $\nu(U \setminus K) < \varepsilon$.

Lemma 2.9. *Let (S, d) be a Polish space with metric d, and μ a measure on $(S, \mathcal{B}(S))$. Take $A \in \mathcal{B}(S)$ such that $\mu(A) < \infty$. Then for every $\varepsilon > 0$ there exists a compact set $K \subset A$ such that $\mu(A \setminus K) < \varepsilon$.*

Proof. Let $\mu_A(\cdot) = \mu(\cdot \cap A)$. The measure μ_A is finite and thus σ-finite. Hence there exists a closed set $K \subset A$ such that

$$\mu_A(A \setminus K) = \mu(A \setminus K) < \varepsilon.$$

Restricting the metric d on K, we see that (K, d) is also a Polish space. Let $\{a_n\}_{n\in\mathbb{N}} \subset K$ be a countable dense subset of K. Hence for every $k \ge 1$ we have $\bigcup_{n=1}^{\infty} \bar{B}_{1/k}(a_n) = K$, where $\bar{B}_r(a)$ denotes the closed ball centered in a with radius r. It can be then checked that $\lim_{N\to\infty} \mu(K \setminus \bigcup_{n=1}^{N} \bar{B}_{1/k}(a_n)) = 0$. There exists N_k such that

$$\mu\left(K \setminus \bigcup_{n=1}^{N_k} \bar{B}_{1/k}(a_n)\right) < \frac{\varepsilon}{2^k}.$$

Let $F = \bigcap_{k=1}^{\infty} \bigcup_{n=1}^{N_k} \bar{B}_{1/k}(a_n) \subset K$, which is closed and precompact. Since S is complete, so is F. This yields that F is compact and

$$\mu(K \setminus F) \leq \sum_{k=1}^{\infty} \mu\Big(K \setminus \bigcup_{n=1}^{N_k} \bar{B}_{1/k}(a_n)\Big) \leq \varepsilon.$$

It is then straightforward that $\mu(A \setminus F) \leq \mu(A \setminus K) + \mu(K \setminus F) \leq 2\varepsilon$. □

We complete this section by recalling Fourier transform, which is an important tool in the analysis of measures. The *Schwartz space* $\mathscr{S}(\mathbb{R}^d)$ is the set of infinitely differentiable complex-valued functions $f(x)$ on $\mathbb{R}^d$ for which

$$\sup_{x \in \mathbb{R}^d} |x^\alpha D^\beta f(x)| < \infty$$

for all $\alpha = (\alpha_1, \ldots, \alpha_d)$ and $\beta = (\beta_1, \ldots, \beta_d)$. Here α_i and β_j are nonnegative integers, and

$$x^\alpha = x_1^{\alpha_1} \cdots x_d^{\alpha_d} \quad \text{and} \quad D^\beta = \frac{\partial^{|\beta|}}{\partial x_1^{\beta_1} \cdots \partial x_d^{\beta_d}} \quad \text{with} \quad |\beta| = \sum_{j=1}^{d} \beta_j.$$

Schwartz space is a dense subspace of $L^2(\mathbb{R}^d)$. The Fourier transform $F : \mathscr{S}(\mathbb{R}^d) \to \mathscr{S}(\mathbb{R}^d)$ is a bijective linear map defined by

$$Ff(k) = \frac{1}{(2\pi)^{d/2}} \int_{\mathbb{R}^d} f(x) e^{-ik \cdot x} dx.$$

For convenience, we use the notation $\hat{f}(k) = Ff(k)$. The inverse Fourier transform is given by

$$\check{f}(x) = F^{-1} f(x) = \frac{1}{(2\pi)^{d/2}} \int_{\mathbb{R}^d} f(k) e^{+ik \cdot x} dk.$$

We have

$$FF^{-1} = F^{-1}F = \mathrm{id}_{\mathscr{S}(\mathbb{R}^d)}$$

and $(Ff, Fg) = (f, g)$, where (f, g) denotes the scalar product on $L^2(\mathbb{R}^d)$. Since $\mathscr{S}(\mathbb{R}^d)$ is dense in $L^2(\mathbb{R}^d)$, Fourier transform F can be extended to a unitary operator $\bar{F}$ on $L^2(\mathbb{R}^d)$. We will use the same notation F instead of $\bar{F}$ in what follows.

2.1.2 Probability measures and limit theorems

A *probability space* is a measure space $(\Omega, \mathscr{F}, P)$ such that $P(\Omega) = 1$, where Ω is called a *sample space*, $\mathscr{F}$ *algebra of events*, and P a *probability measure* on $(\Omega, \mathscr{F})$. An element

$A \in \mathscr{F}$ is often called an *event*. The probability $P(\{\omega \in \Omega \mid \text{conditions}\})$ of an event defined in terms of some conditions will be simply denoted as

$$P(\text{conditions}).$$

An event is said to be *almost sure* with respect to P if it has probability 1 under P. If a property holds for every ω in an almost sure event, then we say that it holds *almost surely* and we write *P-a. s.* or simply *a.s.* The σ-field $\mathscr{F}$ is called *countably determined* if there exists a countable subset $\mathscr{G} \subset \mathscr{F}$ such that if any two probability measures agree on $\mathscr{G}$, then they coincide.

Two events $A, B \in \mathscr{F}$ are called *independent* if $P(A \cap B) = P(A)P(B)$. Two sub-σ-fields $\mathscr{F}_1, \mathscr{F}_2 \subset \mathscr{F}$ are *independent* if all $A \in \mathscr{F}_1$ and $B \in \mathscr{F}_2$ are pairwise independent. The concept of an event occurring infinitely often is formalized by the following result.

Theorem 2.10 (Borel–Cantelli lemma). *Let $(\Omega, \mathscr{F}, P)$ be a probability space, $A_1, A_2, \ldots \in \mathscr{F}$ be events, and write $\bar{A} = \bigcap_{m=1}^{\infty} \bigcup_{n=m}^{\infty} A_n$. The following hold:*
(1) *If $\sum_{n=1}^{\infty} P(A_n) < \infty$, then $P(\bar{A}) = 0$.*
(2) *If $\sum_{n=1}^{\infty} P(A_n) = \infty$ and $\{A_n\}_{n \in \mathbb{N}}$ are independent events, then $P(\bar{A}) = 1$.*

Proof. Note that $B_N = \bigcap_{m=1}^{N} \bigcup_{n=m}^{\infty} A_n$, $N = 1, 2, \ldots$, are decreasing sets as N increases and $\lim_{N \to \infty} B_N = \bar{A}$. Thus

$$P(\bar{A}) = \lim_{N \to \infty} P(B_N) = \lim_{N \to \infty} P\Big(\bigcup_{n=N}^{\infty} A_n\Big) \leq \lim_{N \to \infty} \sum_{n=N}^{\infty} P(A_n) = 0.$$

Hence (1) follows. On the other hand, $C_N = \bigcup_{m=1}^{N} \bigcap_{n=m}^{\infty} A_n^c$, $N = 1, 2, \ldots$, are increasing sets as N increases and $\lim_{N \to \infty} C_N = \bar{A}^c$. Thus we see that

$$P(\bar{A}^c) = \lim_{N \to \infty} P(C_N) = \lim_{m \to \infty} P\Big(\bigcap_{n=m}^{\infty} A_n^c\Big).$$

By the independence of A_n, we furthermore have

$$P\Big(\bigcap_{n=m}^{\infty} A_n^c\Big) = \lim_{k \to \infty} \prod_{n=m}^{k} (1 - P(A_n)) = \lim_{k \to \infty} e^{\sum_{n=m}^{k} \log(1 - P(A_n))} \leq \lim_{k \to \infty} e^{-\sum_{n=m}^{k} P(A_n)} = 0.$$

Thus (2) follows. □

The Borel–Cantelli lemma provides a zero–one law for independent events. Next we discuss a similar type of law for independent σ-fields, using the following concept.

Definition 2.11 (Tail σ-field). Let I be a countable infinite index set and $\{\mathscr{F}_n\}_{n \in I}$ a family of σ-fields. The σ-field

$$\mathcal{T}(\{\mathcal{F}_n\}_{n\in I}) = \bigcap_{\substack{J\subset I\\ \#J<\infty}} \sigma\Big(\bigcup_{j\in I\setminus J} \mathcal{F}_j\Big) \tag{2.1.4}$$

is called the *tail σ-field* of $\{\mathcal{F}_n\}_{n\in I}$. If $\{A_n\}_{n\in I}$ is a family of events, then

$$\mathcal{T}(\{A_n\}_{n\in I}) = \mathcal{T}(\{\sigma(A_n)\}_{n\in I}) \tag{2.1.5}$$

is called the *tail σ-field* of $\{A_n\}_{n\in I}$.

It can be seen that if $J_1 \subset J_2 \subset \cdots$ and $\bigcup_{n=1}^{\infty} J_n = I$, then

$$\mathcal{T}(\{\mathcal{F}_n\}_{n\in I}) = \bigcap_{n=1}^{\infty} \sigma\Big(\bigcup_{m\in I\setminus J_n} \mathcal{F}_m\Big). \tag{2.1.6}$$

In particular, for $I = \mathbb{N}$,

$$\mathcal{T}(\{\mathcal{F}_n\}_{n\in\mathbb{N}}) = \bigcap_{n=1}^{\infty} \sigma\Big(\bigcup_{m=n}^{\infty} \mathcal{F}_m\Big) \tag{2.1.7}$$

is a tail σ-field.

Example 2.12. Let $A_1, A_2, \ldots,$ be events. Then the events $\overline{A} = \bigcap_{n=1}^{\infty}\bigcup_{m=n}^{\infty} A_m$ and $\underline{A} = \bigcup_{n=1}^{\infty}\bigcap_{m=n}^{\infty} A_m$ are in $\mathcal{T}(\{A_n\}_{n\in\mathbb{N}})$. Indeed, $B_n = \bigcap_{m=n}^{\infty} A_m \in \sigma(\{A_m\}_{m\geq N})$, for every $n \geq N$. Thus $\underline{A} \in \sigma(\{A_m\}_{m\geq N})$ for every N, and hence $\underline{A} \in \mathcal{T}(\{A_n\}_{n\in\mathbb{N}})$. The case of $\overline{A}$ is similar.

Theorem 2.13 (Kolmogorov's zero–one law). *Let $(\Omega, \mathcal{F}, P)$ be a probability space, I a countably infinite index set, and $\{\mathcal{F}_n\}_{n\in I}$ an independent family of sub-σ-fields of $\mathcal{F}$. Then the tail σ-field $\mathcal{T}(\{\mathcal{F}_n\}_{n\in I})$ is trivial, i. e., either $P(A) = 0$, or $P(A) = 1$ for all $A \in \mathcal{T}(\{\mathcal{F}_n\}_{n\in I})$.*

Proof. It is sufficient to consider the case $I = \mathbb{N}$. Let

$$\mathcal{B}_n = \Big\{\bigcap_{k=1}^{n} A_k \mid A_1 \in \mathcal{F}_1, \ldots, A_n \in \mathcal{F}_n\Big\}.$$

Let $\mathcal{B} = \bigcup_{n=1}^{\infty} \mathcal{B}_n$, which is a semiring. It is immediate that $\sigma(\mathcal{B}) = \sigma(\bigcup_{n=1}^{\infty} \mathcal{F}_n)$. Let $A \in \mathcal{T}(\{\mathcal{F}_n\}_{n\in\mathbb{N}})$ and $\varepsilon > 0$. By Lemma 2.4, there exist pairwise disjoint sets $F_1, \ldots, F_N \in \mathcal{B}$ such that

$$P(A \triangle (F_1 \cup \cdots \cup F_N)) < \varepsilon. \tag{2.1.8}$$

There exists $n \in \mathbb{N}$ such that $F_1, \ldots, F_N \in \mathcal{B}_n$ and hence $F_1 \cup \cdots \cup F_N \in \sigma(\mathcal{F}_1 \cup \cdots \cup \mathcal{F}_n)$. Obviously, $A \in \sigma(\bigcup_{m=n+1}^{\infty} \mathcal{F}_m)$, hence A is independent of $F_1 \cup \cdots \cup F_n$. Thus

$$\begin{aligned}\varepsilon &> P(A \setminus (F_1 \cup \cdots \cup F_n)) = P(A \cap (F_1 \cup \cdots \cup F_n)^c)\\ &= P(A)(1 - P(F_1 \cup \cdots \cup F_n)) \geq P(A)(1 - P(A) - \varepsilon).\end{aligned}$$

Letting $\varepsilon \downarrow 0$ yields $P(A)(1 - P(A)) = 0$, and the theorem follows. □

There are useful characterizations of the weak convergence of probability measures on metric spaces. Let (S, d) be a metric space, and $C_{\mathrm{b}}(S)$, $C_{\mathrm{u}}(S)$ be the set of bounded continuous functions and the set of uniformly continuous functions on S, respectively.

Lemma 2.14. *Let S be a metric space and $P, P_1, P_2, \dots$ be a sequence of probability measures on $(S, \mathcal{B}(S))$. Then the following are equivalent:*
(1) *$(P_n)_{n\in\mathbb{N}}$ is weakly convergent to P as $n \to \infty$.*
(2) *$\lim_{n\to\infty} \mathbb{E}_{P_n}[f] = \mathbb{E}_P[f]$, for all $f \in C_{\mathrm{b}}(S) \cap C_{\mathrm{u}}(S)$.*
(3) *For every closed set $F \subset S$, $\limsup_{n\to\infty} P_n(F) \le P(F)$.*
(4) *For every open set $U \subset S$, $\liminf_{n\to\infty} P_n(U) \ge P(U)$.*

Proof. (1) $\Longrightarrow$ (2) is trivial. We show (2) $\Longrightarrow$ (3). Let $f \in C_{\mathrm{b}}(\mathbb{R})$ be defined by

$$f(t) = \begin{cases} 1, & t \le 0, \\ 1 - t, & 0 < t \le 1, \\ 0, & t > 1. \end{cases}$$

For a closed set $F \subset S$ and $k > 0$, define $\phi_k(x) = f(kd(x, F))$, where $d(x, F)$ denotes the distance between x and F. We see that $\lim_{k\to\infty} \phi_k(x) = 1_F(x)$. Hence $\phi_k \in C_{\mathrm{b}}(S) \cap C_{\mathrm{u}}(S)$ and $\phi_k \ge 1_F$. It follows then that $\mathbb{E}_{P_n}[\phi_k] \ge P_n(F)$. Letting $n \to \infty$ yields $\mathbb{E}_P[\phi_k] \ge \limsup_{n\to\infty} P_n(F)$ by (2). Letting $k \to \infty$ again yields (3). It is clear that (3) and (4) are equivalent. We show (3) $\Longrightarrow$ (1). Let $f \in C_{\mathrm{b}}(S)$ and suppose that $0 \le f(x) < 1$ without loss of generality. Fix $k \in \mathbb{N}$. Define sets $F_i = \{x \in S \mid f(x) \ge i/k\}$ for $0 \le i \le k$; F_i is closed and a decreasing sequence. It is direct to check that

$$\sum_{i=1}^{k} \frac{i-1}{k} P_n(F_{i-1} \setminus F_i) \le \mathbb{E}_{P_n}[f] \le \sum_{i=1}^{k} \frac{i}{k} P_n(F_{i-1} \setminus F_i).$$

From this we deduce that

$$\frac{1}{k} \sum_{i=1}^{k} P_n(F_i) \le \mathbb{E}_{P_n}[f] \le \frac{1}{k} + \frac{1}{k} \sum_{i=1}^{k} P_n(F_i). \tag{2.1.9}$$

Letting $n \to \infty$ on both sides, we see that by (3),

$$\limsup_{n\to\infty} \mathbb{E}_{P_n}[f] \le \frac{1}{k} + \frac{1}{k} \sum_{i=1}^{k} P(F_i). \tag{2.1.10}$$

Since (2.1.9) is also valid for P_n replaced by P, we have

$$\frac{1}{k} + \frac{1}{k} \sum_{i=1}^{k} P(F_i) \le \frac{1}{k} + \mathbb{E}_P[f]. \tag{2.1.11}$$

By (2.1.10)–(2.1.11), we then have

$$\limsup_{n\to\infty} \mathbb{E}_{P_n}[f] \le \frac{1}{k} + \mathbb{E}_P[f].$$

Taking $k \to \infty$, we have

$$\limsup_{n\to\infty} \mathbb{E}_{P_n}[f] \le \mathbb{E}_P[f].$$

Replacing f by $1 - f$, we obtain (1). □

Let (M, d) be a compact metric space. Define

$$\mathscr{P}(M) = \Big\{P \mid P \text{ is a probability measure on } (M, \mathcal{B}(M))\Big\}.$$

Let $\{f_n\}_{n\in\mathbb{N}}$ be a dense subset of $C(M)_+$, i. e., the set of positive functions in $C(M)$. We define $\tau : \mathscr{P}(M) \times \mathscr{P}(M) \to [0, \infty)$ by

$$\tau(P, Q) = \sum_{n=1}^{\infty} \frac{1}{2^n}\Big(|\mathbb{E}_P[f_n] - \mathbb{E}_Q[f_n]| \wedge 1\Big).$$

This gives a metric on $\mathscr{P}(M)$.

Lemma 2.15. *Let $(\mathscr{P}(M), \tau)$ be as above. Then $(\mathscr{P}(M), \tau)$ is a compact metric space.*

Proof. Let $f \in C(M)$ and $\{f_n\}_{n\in\mathbb{N}}$ be a dense subset of $C(M)_+$. Let $P_n \in \mathscr{P}(M)$, write $\mathbb{E}_{P_n}[f] = l_n(f)$, and consider $a_{n,m} = l_n(f_m)$. Fix $m \ge 1$ and let

$$0 \le a_{n,m} \le \|f_m\|,$$

where $\|f\| = \rho(f, 0)$. Then $(a_{n,m})_{n\in\mathbb{N}}$ is a bounded sequence for each m. By a diagonalization argument, we can find a subsequence $n_1 < n_2 < \cdots$ such that $a_{n_k,m} \to a_m$ as $n_k \to \infty$ for each m. We have

$$\begin{aligned} |l_{n_i}(f) - l_{n_j}(f)| &\le |l_{n_i}(f) - l_{n_i}(f_m)| + |l_{n_i}(f_m) - l_{n_j}(f_m)| + |l_{n_j}(f_m) - l_{n_j}(f)| \\ &\le 2\|f - f_m\| + |a_{n_i,m} - a_{n_j,m}|. \end{aligned} \tag{2.1.12}$$

For $\varepsilon > 0$, we can take m such that $\|f - f_m\| \le \varepsilon$. By (2.1.12), it follows that $(l_{n_i}(f))_{i=1}^{\infty}$ is a Cauchy sequence, and thus there exists $\lim_{n_i\to\infty} l_{n_i}(f) = l(f)$. Hence $l(f)$ satisfies $l(1) = 1$ and $|l(f)| \le \|f\|$. We conclude that $l : C(M)_+ \to \mathbb{R}$ is a positive linear functional. By the *Riesz–Markov representation theorem*, there exists a Baire measure P on M such that

$$l(f) = \int_M f(m)dP.$$

Furthermore, $P \in \mathscr{P}(M)$ and $P_{n_i} \xrightarrow{d} P$ follow. □

Definition 2.16 (Tightness). Let (S, d) be a metric space and P a probability measure on $(S, \mathcal{B}(S))$.

(1) A family $\mathscr{P}$ of probability measures on $(S, \mathcal{B}(S))$ is called *relatively compact* whenever every sequence of elements of $\mathscr{P}$ contains a weakly convergent subsequence.
(2) A family $\mathscr{P}$ of probability measures on $(S, \mathcal{B}(S))$ is called *tight* whenever for every $\varepsilon > 0$ there exists a compact set $K_\varepsilon \subset S$ such that $P(K_\varepsilon) > 1 - \varepsilon$, for every $P \in \mathscr{P}$.

Theorem 2.17 (Prokhorov's theorem). *Let (S, d) be a Polish space. A family of probability measures $\mathscr{P}$ on $(S, \mathcal{B}(S))$ is relatively compact if and only if it is tight.*

Proof. Suppose that $\mathscr{P}$ is tight. Since the metric space (S, d) is separable, there exists a metric $\tilde{d}$ on S such that d and $\tilde{d}$ give the same topology, and $(S, \tilde{d})$ is totally bounded. Thus we may assume without restriction of generality that (S, d) is totally bounded. We denote the completion of (S, d) by $(\hat{S}, \hat{d})$, i. e., we have a map $i : S \to \hat{S}$ such that

(1) $\hat{d}(i(x), i(y)) = d(x, y)$;
(2) $i(S)$ is dense in $\hat{S}$;
(3) $(\hat{S}, \hat{d})$ is complete.

Note that $(\hat{S}, \hat{d})$ is totally bounded and complete, which is equivalent to compactness. Define the image measure on $\hat{S}$ by $\hat{P} = P \circ i^{-1}$, $P \in \mathscr{P}$, and denote the set of image measures induced by i on $\hat{S}$ by $\hat{\mathscr{P}}$. We denote then the set of probability measures on $\hat{S}$ by $\mathscr{Q}$ and have $\hat{\mathscr{P}} \subset \mathscr{Q}$. By Lemma 2.15, $(\mathscr{Q}, \tau)$ is a compact metric space and hence $\hat{\mathscr{P}}$ is precompact, thus there exist $(P_n)_{n\in\mathbb{N}} \subset \mathscr{P}$ and $Q \in \mathscr{Q}$ such that $\hat{P}_n \xrightarrow{\text{d}} Q$ as $n \to \infty$. Since $\mathscr{P}$ is tight, for every $N \geq 1$ there exists a compact set $K_N \subset S$ such that $P_n(K_N) \geq 1 - 1/N$ holds for all $n \geq 1$. Note that $i(K_N) \subset \hat{S}$ is also compact since i is a continuous map. We can see by Lemma 2.14 that

$$Q(i(K_N)) \geq \limsup_{n\to\infty} \hat{P}_n(i(K_N)) = \limsup_{n\to\infty} P_n(K_N) \geq 1 - \frac{1}{N}, \tag{2.1.13}$$

where we used $\hat{P}_n(i(K_N)) = P_n(K_N)$, which follows from i being injective. Write $S_0 = \bigcup_{N=1}^{\infty} K_N$. It follows from (2.1.13) that $Q(i(S_0)) \geq 1$, which implies that $Q(i(S_0)) = 1$. Define the probability measure P on S by $P(A) = Q(i(A \cap S_0))$. For a bounded and uniformly continuous function f on S, we have

$$\int_S f(\omega) dP = \int_{S_0} f(\omega) dP = \int_{\hat{S}} \hat{f}(\hat{\omega}) dQ. \tag{2.1.14}$$

Here $\hat{f}$ is a continuous function on $\hat{S}$ defined in the following way. Let $\hat{\omega} \in \hat{S}$. There exists a sequence $(\omega_n)_{n\in\mathbb{N}} \subset S$ such that $i(\omega_n) \to \hat{\omega}$ as $n \to \infty$. Since f is uniformly continuous, $(f(\omega_n))_{n\in\mathbb{N}}$ is a Cauchy sequence. Thus there exists the limit $a = \lim_{n\to\infty} f(\omega_n) \in \mathbb{C}$. We define $\hat{f}(\hat{\omega}) = a$. In particular, $\hat{f}(i(\omega)) = f(\omega)$ for all $\omega \in S$. It is also seen that $\hat{f}$ is continuous and $\hat{f}$ satisfying (2.1.14) is unique. By (2.1.14), we have

$$\int_S f(\omega)dP_n = \int_{\hat{S}} \hat{f}(\hat{\omega})d\hat{P}_n \to \int_S f(\omega)dP = \int_{\hat{S}} \hat{f}(\hat{\omega})dQ, \quad \text{as } n \to \infty.$$

Hence $\mathscr{P}$ is relatively compact.

Next suppose that $\mathscr{P}$ is relatively compact. Let $\{\omega_n\}_{n\in\mathbb{N}} \subset S$ be a countable dense set. Define $A_{n,N} = \bigcup_{i=1}^N \bar{B}_{1/n}(\omega_i)$, where $\bar{B}_{1/n}(\omega)$ denotes the closed ball of radius $1/n$ centered at ω. Then $A_{n,N} \uparrow S$ as $N \to \infty$ for all n. Define

$$\delta = \sup_{n\in\mathbb{N}} \inf_{N\in\mathbb{N}} \sup_{P\in\mathscr{P}} P(A_{n,N}^c).$$

Thus there exists a natural number n such that for every N there is $P_N \in \mathscr{P}$, satisfying $P_N(A_{n,N}^c) \geq \delta/2$. Since S is precompact, $(P_N)_{N\in\mathbb{N}}$ contains a weakly convergent subsequence $(P_{N_k})_{k\in\mathbb{N}}$, whose weak limit we denote by P. We have by Lemma 2.14,

$$P(A_{n,N}^c) \geq \liminf_{k\to\infty} P_{N_k}(A_{n,N}^c) \geq \liminf_{k\to\infty} P_{N_k}(A_{n,N_k}^c) \geq \frac{\delta}{2}.$$

Since $A_{n,N}^c \downarrow \emptyset$ as $N \to \infty$, $\lim_{N\to\infty} P(A_{n,N}^c) = 0$ follows. Hence $\delta = 0$. Fix $\varepsilon > 0$. Then $\delta = 0$ implies that for every n we can choose a natural number N'_n such that $P(A_{n,N'_n}^c) < \varepsilon/2^n$, for all $P \in \mathscr{P}$. Note that $A = \bigcap_{n=1}^\infty A_{n,N'_n} \subset S$ is totally bounded, so the closure $\bar{A}$ of A is compact. For every $P \in \mathscr{P}$, we have

$$P((\bar{A})^c) \leq P(A^c) \leq \sum_{n=1}^\infty P(A_{n,N'_n}^c) \leq \varepsilon.$$

Hence $P(\bar{A}) = P(S) - P((\bar{A})^c) \geq 1 - \varepsilon$ and $\mathscr{P}$ is tight. □

We introduce on $C(\mathbb{R})$ or $C([0,\infty))$ the locally uniform topology given by the metric (2.1.3). We will be mainly interested in the cases when S is one of these two function spaces. Since both are Polish spaces under the locally uniform topology generated by (2.1.3), any family of probability measures on $(S, \mathcal{B}(S))$ has a weakly convergent sequence if it is tight.

Take $S = C(\mathbb{R})$. Recall that for $T, \delta > 0$ the *modulus of continuity* of $\omega \in S$ is defined by

$$m^T(\omega,\delta) = \max_{\substack{|s-t|\leq\delta \\ -T\leq s,t\leq T}} |\omega(s) - \omega(t)|. \tag{2.1.15}$$

For $S = C([0,\infty))$, the modulus of continuity is defined by replacing $-T$ by 0.

Lemma 2.18. *A set $A \subset C(\mathbb{R})$ has a compact closure in the locally uniform topology if and only if the following two conditions hold:*

$$\sup_{\omega\in A} |\omega(0)| < \infty, \tag{2.1.16}$$

$$\lim_{\delta\downarrow 0} \sup_{\omega\in A} m^T(\omega,\delta) = 0. \tag{2.1.17}$$

Proof. First, we show that
(1) $\omega \mapsto m^T(\omega, \delta)$ is continuous in the locally uniform topology;
(2) $\delta \mapsto m^T(\omega, \delta)$ is nondecreasing;
(3) $\lim_{\delta \downarrow 0} m^T(\omega, \delta) = 0$.

Suppose that on the interval $[-T, T]$ the sequence $(\omega_n)_{n \in \mathbb{N}}$ is uniformly convergent to ω as $n \to \infty$. Let $\sup_{s \in [-T,T]} |\omega_n(s) - \omega(s)| \leq \varepsilon$. We see that

$$m^T(\omega, \delta) \leq \max_{\substack{|s-t| \leq \delta \\ -T \leq s,t \leq T}} (|\omega(s) - \omega_n(s)| + |\omega_n(s) - \omega_n(t)| + |\omega_n(t) - \omega(t)|) \leq 2\varepsilon + m^T(\omega_n, \delta).$$

Similarly, $m^T(\omega_n, \delta) \leq 2\varepsilon + m^T(\omega, \delta)$. $\lim_{n \to \infty} m^T(\omega_n, \delta) = m^T(\omega, \delta)$ follows. Part (2) is trivial, and (3) follows from the fact that $t \mapsto \omega(t)$ is uniformly continuous in $[-T, T]$.

Assume that the closure $\bar{A}$ of A is compact. Since $\bar{A}$ is contained in the union of the open sets $G_n = \{\omega \in C(\mathbb{R}) \mid |\omega(0)| < n\}$, $n = 1, 2, \ldots$, A is contained in a specific G_n. Thus (2.1.16) follows. For $\varepsilon > 0$, let $K_\delta = \{\omega \in \bar{A} \mid m^T(\omega, \delta) \geq \varepsilon\}$. By the continuity of $m^T(\omega, \delta)$ with respect to ω, each K_δ is closed and $K_\delta \subset \bar{A}$. Thus K_δ is compact. Since $\lim_{\delta \downarrow 0} m^T(\omega, \delta) = 0$, we have $\bigcap_{\delta > 0} K_\delta = \emptyset$. This implies $K_{\delta_\varepsilon} = \emptyset$ for some $\delta_\varepsilon > 0$, and thus (2.1.17) follows.

Next assume (2.1.16)–(2.1.17). We will prove that $\bar{A}$ is compact. It suffices to show that every sequence $(\omega_n)_{n \in \mathbb{N}} \subset A$ has a convergent subsequence. Fix $T > 0$ and note that for some $\delta_1 > 0$ we have $m^T(\omega, \delta_1) \leq 1$, for all $\omega \in A$ by (2.1.17). For a fixed integer $m \geq 1$ and $t \in [-T, T] \setminus \{0\}$ with $(m-1)\delta_1 < t \leq m\delta_1 \wedge T$, we have

$$|\omega(t)| \leq |\omega(0)| + \sum_{k=1}^{m-1} |\omega(k\delta_1) - \omega((k-1)\delta_1)| + |\omega(t) - \omega((m-1)\delta_1)| \leq |\omega(0)| + m.$$

If $r \in \mathbb{Q} \cap [-T, T]$, $(\omega_n(r))_{n \in \mathbb{N}}$ is bounded for every r. Let $\{r_0, r_1, \ldots\}$ be the enumeration of $\mathbb{Q} \cap [-T, T]$. Choose $(\omega_n^{(0)})_{n \in \mathbb{N}}$ such that $\omega_n^{(0)}(r_0)$ converges to a limit denoted by $\omega(r_0)$. From $(\omega_n^{(0)})_{n \in \mathbb{N}}$, choose a further subsequence $(\omega_n^{(1)})_{n \in \mathbb{N}}$ such that $\omega_n^{(1)}(r_1)$ converges to a limit denoted by $\omega(r_1)$. Continuing this process, we obtain a diagonal sequence $(\omega_n^{(n)})_{n \in \mathbb{N}}$. Hence $\omega_n^{(n)}(r) \to \omega(r)$ for all $r \in \mathbb{Q} \cap [-T, T]$. Put $\omega_n^{(n)} = \tilde{\omega}_n$. By (2.1.17), for every $\varepsilon > 0$ there exists $\delta(\varepsilon) > 0$ such that $|\tilde{\omega}_n(s) - \tilde{\omega}_n(t)| \leq \varepsilon$, whenever $-T \leq s, t \leq T$, and $|s - t| \leq \delta(\varepsilon)$. The same inequality holds for ω with $|\omega(s) - \omega(t)| \leq \varepsilon$, whenever $s, t \in [-T, T] \cap \mathbb{Q}$ and $|s - t| \leq \delta(\varepsilon)$. Then ω can be extended to a continuous function on $[-T, T]$, denoted by $\bar{\omega}$. We have $\bar{\omega} \in C([-T, T])$. Furthermore, it follows that $|\bar{\omega}(s) - \bar{\omega}(t)| \leq \varepsilon$ whenever $-T \leq s, t \leq T$, and $|s - t| \leq \delta(\varepsilon)$. For n sufficiently large, whenever $t \in [-T, T]$ there is some $r_k \in \mathbb{Q}$ such that $k \leq n$ and $|t - r_k| \leq \delta(\varepsilon)$. For sufficiently large $M \geq n$, we have $|\tilde{\omega}_m(r_j) - \bar{\omega}(r_j)| \leq \varepsilon$, for all $j = 0, 1, 2, \ldots, n$ and $m \geq M$. We then conclude that

$$|\tilde{\omega}_m(t) - \bar{\omega}(t)| \leq |\tilde{\omega}_m(t) - \tilde{\omega}_m(r_k)| + |\tilde{\omega}_m(r_k) - \bar{\omega}(r_k)| + |\bar{\omega}(r_k) - \bar{\omega}(t)| < 3\varepsilon,$$

for all $m \geq M$ and $t \in [-T, T]$. Thus $\tilde{\omega}_n$ converges uniformly on bounded intervals to the function $\bar{\omega} \in C([-T, T])$. □

Theorem 2.19 (Tightness of probability measures). *A sequence $(P_n)_{n\in\mathbb{N}}$ of probability measures on $(C(\mathbb{R}), \mathcal{B}(C(\mathbb{R})))$ is tight if and only if*

$$\lim_{\lambda\uparrow\infty} \sup_{n\in\mathbb{N}} P_n(|\omega(0)| > \lambda) = 0, \tag{2.1.18}$$

$$\lim_{\delta\downarrow 0} \sup_{n\in\mathbb{N}} P_n(m^T(\omega, \delta) > \varepsilon) = 0, \quad T > 0,\ \varepsilon > 0. \tag{2.1.19}$$

Proof. Suppose that $(P_n)_{n\in\mathbb{N}}$ is tight. Given $\varepsilon > 0$, there exists a compact set K with $P_n(K) \geq 1 - \varepsilon$ for every n. According to Lemma 2.18, for sufficiently large $\lambda > 0$ we have $|\omega(0)| \leq \lambda$ for all $\omega \in K$. Thus

$$\begin{aligned} P_n(|\omega(0)| \leq \lambda) &= P_n(|\omega(0)| \leq \lambda, \omega \in K) + P_n(|\omega(0)| \leq \lambda, \omega \in K^c) \\ &\leq P_n(|\omega(0)| \leq \lambda, \omega \in K) + \varepsilon, \end{aligned}$$

and we see that $\lim_{\lambda\uparrow\infty} \sup_{n\in\mathbb{N}} P_n(|\omega(0)| > \lambda) \leq \varepsilon$. Since ε is arbitrary, (2.1.18) follows. According to Lemma 2.18 again, if T and ε are given, then there exists δ_0 such that $m^T(\omega, \delta) \leq \varepsilon$ for $0 < \delta < \delta_0$ and $\omega \in K$. This gives (2.1.19).

Next suppose (2.1.18)–(2.1.19) hold. Given a positive integer T and $\varepsilon > 0$, choose $\lambda > 0$ such that $\sup_{n\in\mathbb{N}} P_n(|\omega(0)| > \lambda) \leq \varepsilon/2^{T+1}$. Also, choose $\delta_k > 0$, $k = 1, 2, \ldots$, such that

$$\sup_{n\in\mathbb{N}} P_n\left(m^T(\omega, \delta_k) > \frac{1}{k}\right) \leq \frac{\varepsilon}{2^{T+k+1}}.$$

Define

$$A_T = \left\{\omega \in C(\mathbb{R}) \,\middle|\, |\omega(0)| \leq \lambda, m^T(\omega, \delta_k) \leq \frac{1}{k}, k = 1, 2, \ldots\right\}$$

and $A = \bigcap_{T=1}^{\infty} A_T$. By Lemma 2.18, the set A is compact. We have

$$\begin{aligned} P_n(A_T) &= 1 - P_n(A_T^c) \geq 1 - P_n(|\omega(0)| > \lambda) - \sum_{k=1}^{\infty} P_n(m^T(\omega, \delta_k) > 1/k) \\ &\geq 1 - \frac{\varepsilon}{2^{T+1}} - \sum_{k=1}^{\infty} \frac{\varepsilon}{2^{T+k+1}} = 1 - \frac{\varepsilon}{2^T} \end{aligned}$$

and

$$P_n(A) = P_n\left(\bigcap_{T=1}^{\infty} A_T\right) = 1 - P_n\left(\bigcup_{T=1}^{\infty} A_T^c\right) \geq 1 - \sum_{T=1}^{\infty} P_n(A_T^c) \geq 1 - \varepsilon$$

follows for all n, hence $(P_n)_{n\in\mathbb{N}}$ is tight. □

In the remaining part of this subsection, we discuss a key extension theorem due to Kolmogorov. Let Ω be a Polish space, J an index set, and $\Lambda \subset J$, $|\Lambda| < \infty$, with $|\cdot|$ denoting the cardinality of a finite set. Consider the family of product probability spaces $(\Omega^\Lambda, \mathcal{B}(\Omega^\Lambda), \mu_\Lambda)$, where $\Omega^\Lambda = \times^{|\Lambda|}\Omega$. An element $\omega \in \Omega^\Lambda$ is regarded as a map $\Lambda \mapsto \Omega$. Take now $\Lambda_1 \subset \Lambda_2 \subset J$ and let

$$\pi_{\Lambda_1\Lambda_2} : \Omega^{\Lambda_2} \to \Omega^{\Lambda_1} \tag{2.1.20}$$

denote the projection defined by $\pi_{\Lambda_1\Lambda_2}(\omega) = \omega\lceil_{\Lambda_1}$. The equality

$$\mu_{\Lambda_1}(E) = \mu_{\Lambda_2}(\pi^{-1}_{\Lambda_1\Lambda_2}(E)) \tag{2.1.21}$$

for $E \in \mathcal{B}(\Omega^{\Lambda_1})$ and $\Lambda_1 \subset \Lambda_2$ with $|\Lambda_1|, |\Lambda_2| < \infty$, is called *Kolmogorov's consistency relation.*

Theorem 2.20 (Kolmogorov's extension theorem). *Let $(\Omega^\Lambda, \mathcal{B}(\Omega^\Lambda), \mu_\Lambda)$, $\Lambda \subset J$, with $|\Lambda| < \infty$, be a family of probability spaces with underlying Polish space Ω. Suppose that μ_Λ satisfies the Kolmogorov's consistency relation* (2.1.21), *and define*

$$\mathcal{A} = \left\{\pi^{-1}_{\Lambda J}(E) \subset \Omega^J \mid E \in \mathcal{B}(\Omega^{|\Lambda|}),\ \Lambda \subset J,\ |\Lambda| < \infty\right\},$$

where $\pi^{-1}_{\Lambda J}(E) = \{\omega \in \Omega^J \mid \omega\lceil_\Lambda \in E\}$. Then there exists a unique probability measure μ on $(\Omega^J, \sigma(\mathcal{A}))$ such that $\mu_\Lambda(E) = \mu(\pi_\Lambda^{-1}(E))$, for all $E \in \mathcal{B}(\Omega^\Lambda)$ with $|\Lambda| < \infty$.

Proof. Define the set function $\tilde{\mu} : \mathcal{A} \to [0, \infty)$ by

$$\tilde{\mu}(\pi^{-1}_{\Lambda J}(E)) = \mu_\Lambda(E), \quad E \in \mathcal{A}. \tag{2.1.22}$$

We show that (2.1.22) is well defined. When $\pi^{-1}_{\Lambda J}(E) = \pi^{-1}_{\Lambda' J}(E')$, it is seen for $\Lambda'' \supset \Lambda', \Lambda$ that

$$(\pi_{\Lambda\Lambda''} \circ \pi_{\Lambda'' J})^{-1}(E) = \pi^{-1}_{\Lambda J}(E) = \pi^{-1}_{\Lambda' J}(E') = (\pi_{\Lambda'\Lambda''} \circ \pi_{\Lambda'' J})^{-1}(E')$$

and thus $\pi^{-1}_{\Lambda'' J} \circ \pi^{-1}_{\Lambda\Lambda''}(E) = \pi^{-1}_{\Lambda'' J} \circ \pi^{-1}_{\Lambda'\Lambda''}(E')$. From this it follows that

$$\pi^{-1}_{\Lambda\Lambda''}(E) = \pi^{-1}_{\Lambda'\Lambda''}(E').$$

Hence

$$\mu_\Lambda(E) = \mu_{\Lambda''}(\pi^{-1}_{\Lambda\Lambda''}(E)) = \mu_{\Lambda''}(\pi^{-1}_{\Lambda'\Lambda''}(E')) = \mu_{\Lambda'}(E')$$

follows by Kolmogorov's consistency, and thus $\tilde{\mu}$ is well defined.

Next we extend the set function $\tilde{\mu}$ to a probability measure μ on $\sigma(\mathcal{A})$ by Hopf's extension theorem. In order to do that, it suffices to show that $\lim_{n\to\infty} \tilde{\mu}(A_n) = 0$ for every sequence $(A_n)_{n\in\mathbb{N}} \subset \mathcal{A}$ such that $A_1 \supset A_2 \supset \cdots$ and $\bigcap_{j=1}^{\infty} A_j = \emptyset$. Let $A_n = \pi_{\Lambda_n J}^{-1}(E_n)$ and suppose that it is decreasing, i. e.,

$$A_{n+1} = \left\{\omega \mid \omega\lceil_{\Lambda_{n+1}} \in E_{n+1}\right\} \subset A_n = \left\{\omega \mid \omega\lceil_{\Lambda_n} \in E_n\right\}.$$

We have $A_n \subset \{\omega \mid \omega\lceil_{\Lambda'_{n+1}} \in E'_{n+1}\}$, where $\Lambda'_{n+1} = \Lambda_{n+1} \cup \Lambda_n$ and $E'_{n+1} = E_{n+1} \cup E_n$. Hence we may assume that $E_n \subset E_{n+1}$ and $\Lambda_n \subset \Lambda_{n+1}$. Suppose that there exists $\varepsilon > 0$ such that $\tilde{\mu}(A_n) > \varepsilon$ for all n. Since Ω is a Polish space and $\mu_{\Lambda_n}(E_n) < \infty$, by Lemma 2.9 there exists a nonempty compact set $K_n \subset E_n$ such that $\mu_{\Lambda_n}(E_n \setminus K_n) < \varepsilon/2^{n+1}$. Let $B_n = \pi_{\Lambda_n}^{-1}(K_n) \subset A_n$. We obtain

$$\tilde{\mu}(A_n \setminus B_n) = \tilde{\mu}(\pi_{\Lambda_n}^{-1}(E_n \setminus K_n)) = \mu_{\Lambda_n}(E_n \setminus K_n) < \frac{\varepsilon}{2^{n+1}}.$$

Let $C_n = \bigcap_{j=1}^{n} B_j$, $C_n \subset A_n$, and

$$\tilde{\mu}(C_n) \geq \tilde{\mu}(A_n) - \tilde{\mu}\Big(\bigcup_{k=1}^{n}(A_k \setminus B_k)\Big) \geq \varepsilon - \sum_{k=1}^{n} \frac{\varepsilon}{2^{k+1}} > \frac{\varepsilon}{2}.$$

Hence $C_n \neq \emptyset$, and we will show in the next lemma that $D = \bigcap_{n=1}^{\infty} C_n \neq \emptyset$. Since $D \subset \bigcap_{j=1}^{\infty} A_j$, this contradicts $\bigcap_{j=1}^{\infty} A_j = \emptyset$. Thus $\lim_{n\to\infty} \tilde{\mu}(A_n) = 0$. □

Lemma 2.21. *It follows that* $D = \bigcap_{n=1}^{\infty} C_n \neq \emptyset$.

Proof. First, observe that

$$C_1 = B_1 = \pi_{\Lambda_1 J}^{-1}(\tilde{K}_1),$$

where $\tilde{K}_1 = K_1$ is compact, and

$$C_2 = B_1 \cap B_2 = \pi_{\Lambda_1 J}^{-1}(\tilde{K}_1) \cap \pi_{\Lambda_2 J}^{-1}(K_2) = \left\{\omega \mid \omega\lceil_{\Lambda_2} \in K_2 \cap \pi_{\Lambda_1\Lambda_2}^{-1}(\tilde{K}_1)\right\} = \pi_{\Lambda_2 J}^{-1}(\tilde{K}_2),$$

where $\tilde{K}_2 = K_2 \cap \pi_{\Lambda_1\Lambda_2}^{-1}(\tilde{K}_1)$ is a nonempty compact set. Repeating this procedure, we have

$$C_n = B_1 \cap \cdots \cap B_n = \pi_{\Lambda_n J}^{-1}(\tilde{K}_n)$$

and $\tilde{K}_n = K_n \cap \pi_{\Lambda_{n-1}\Lambda_n}^{-1}(\tilde{K}_{n-1})$ is a nonempty compact set. It is straightforward to check that $\pi_{\Lambda_m\Lambda_n}(\tilde{K}_n) \subset \tilde{K}_m$ for $m < n$. Thus

$$F_n = \pi_{\Lambda_1\Lambda_n}(\tilde{K}_n) \subset \tilde{K}_1$$

and F_n is closed and thus compact. Since

$$F_{n+1} = \pi_{\Lambda_1\Lambda_n}(\pi_{\Lambda_n\Lambda_{n+1}}(\tilde{K}_{n+1})) \subset \pi_{\Lambda_1\Lambda_n}(\tilde{K}_n) = F_n,$$

F_n, $n = 1, 2, \ldots$, is a nonincreasing sequence of compact sets in Ω. Hence $\bigcap_{n=1}^{\infty} F_n \neq \emptyset$ and there exists x_1 such that

$$x_1 \in \bigcap_{n=1}^{\infty} F_n \subset \tilde{K}_1.$$

Similarly,

$$F_n' = \pi_{\Lambda_2 \Lambda_n}(\tilde{K}_n) \cap \pi_{\Lambda_1 \Lambda_2}^{-1}(x_1) \subset \tilde{K}_2$$

is a nonincreasing sequence of compact sets. Note that

$$\pi_{\Lambda_1 \Lambda_2}(F_n') = \pi_{\Lambda_1 \Lambda_n}(\tilde{K}_n) \cap \{x_1\} = F_n \cap \{x_1\} = \{x_1\} \neq \emptyset.$$

Thus $F_n' \neq \emptyset$, which implies that $\tilde{K}_2 \supset \bigcap_{n=2}^{\infty} F_n' \neq \emptyset$ and there exists x_2 such that

$$x_2 \in \bigcap_{n=2}^{\infty} F_n' \subset \tilde{K}_2.$$

Furthermore, $F_n'' = \pi_{\Lambda_3 \Lambda_n}(\tilde{K}_n) \cap \pi_{\Lambda_2 \Lambda_3}^{-1}(x_2) \subset \tilde{K}_3$ is also a nonincreasing sequence of nonempty compact sets. Thus $\tilde{K}_3 \supset \bigcap_{n=3}^{\infty} F_n'' \neq \emptyset$ and there also exists x_3 such that

$$x_3 \in \bigcap_{n=3}^{\infty} F_n'' \subset \tilde{K}_3.$$

Repeating this procedure, we find a sequence x_m, $m = 1, 2, \ldots$, such that

$$x_m \in \pi_{\Lambda_m \Lambda_n}(\tilde{K}_n) \subset \tilde{K}_m$$

for all $n \geq m$. Define $\omega_* \in \Omega^J$ by

$$\omega_* = \begin{cases} \omega_* \lceil_{\Lambda_k} = x_k, & k = 1, 2, \ldots, \\ \omega_* \lceil_{J \setminus \bigcup_{k=1}^{\infty} \Lambda_k}, & \text{arbitrary.} \end{cases}$$

This ω_* is well defined by the construction of the sequence $(x_k)_{k=1}^{\infty}$. Hence $\pi_{\Lambda_n J}(\omega_*) = x_n \in \tilde{K}_n$ and $\omega_* \in \pi_{\Lambda_n J}^{-1}(\tilde{K}_n) = C_n$ for all $n \geq 1$, and we conclude that $\omega_* \in \bigcap_{n=1}^{\infty} C_n$. □

One application of Kolmogorov's extension theorem is in the construction of countable direct products of probability measures.

Example 2.22 (Countable product of probability measures). Let $(\mathbb{R}, \mathcal{B}(\mathbb{R}), \mu_k)$, $k \in \mathbb{N}$, be a family of probability measures. Take $\Lambda_1 \subset \Lambda_2 \subset \mathbb{N}$ and let $\pi_{\Lambda_1 \Lambda_2} : \mathbb{R}^{\Lambda_2} \to \mathbb{R}^{\Lambda_1}$ be the projection defined by $\pi_{\Lambda_1 \Lambda_2}(x) = x\lceil_{\Lambda_1}$ for $x \in \mathbb{R}^{\Lambda_2}$. Let $\Lambda_1 = \{k_1, \ldots, k_n\}$ and $\Lambda_2 = \{k_1, \ldots, k_n, k_{n+1}, \ldots, k_{n+m}\}$ for simplicity. Also, let $E = E_1 \times \cdots \times E_n \in \mathcal{B}(\mathbb{R}^n)$. Thus we see that

$$\pi_{\Lambda_1 \Lambda_2}^{-1}(E) = E \times \underbrace{\mathbb{R} \times \cdots \times \mathbb{R}}_{m}.$$

Hence it follows that $\mu_{\Lambda_1}(E) = \mu_{\Lambda_2}(\pi^{-1}_{\Lambda_1\Lambda_2}(E))$ and Kolmogorov's consistency holds. Thus by the Kolmogorov's extension theorem, there exists a unique probability measure μ on $(\mathbb{R}^{\mathbb{N}}, \mathcal{B}(\mathbb{R}^{\mathbb{N}}))$ such that

$$\mu\Big(\prod_{j\in\Lambda_1} E_j \times \prod_{k\in\mathbb{N}\setminus\Lambda_1} \mathbb{R}\Big) = \prod_{j\in\Lambda_1} \mu_j(E_j).$$

The measure μ can be denoted by

$$\mu = \prod_{k=1}^{\infty} \mu_k.$$

Countable direct products of Gaussian measures will be important in Chapters 2–4 in Volume 2.

We can also construct an uncountable product of probability measures.

Example 2.23 (Uncountable product of probability measures). Let $\mathbb{R}^{[0,\infty)}$ denote the set of all $\mathbb{R}$-valued functions on $[0,\infty)$. An n-dimensional *cylinder set* in $\mathbb{R}^{[0,\infty)}$ is defined by

$$C = \Big\{\omega \in \mathbb{R}^{[0,\infty)} \mid (\omega(t_1),\dots,\omega(t_n)) \in E_1 \times \cdots \times E_n,\ t_j \in [0,\infty),\ E_j \in \mathcal{B}(\mathbb{R}), j = 1,\dots,n\Big\}. \tag{2.1.23}$$

Let $\mathscr{A}$ denote the family of all finite dimensional cylinder sets, and $\mathscr{F} = \sigma(\mathscr{A})$. Also, let Λ be the set of finite sequences $t = (t_1,\dots,t_n)$ of distinct, nonnegative numbers, where the length n of these sequences ranges over the set of positive integers. For every $t = (t_1,\dots,t_n) \in \Lambda$ with length n, we assume that there exists a probability measure P_t on the measurable space $(\mathbb{R}^n, \mathcal{B}(\mathbb{R}^n))$ such that

(1) if $s = (t_{i_1},\dots,t_{i_n})$ is a permutation of t, then $P_t(E_1 \times \cdots \times E_n) = P_s(E_{i_1} \times \cdots \times E_{i_n})$;

(2) if $s = (t_1,\dots,t_m)$ for $m < n$, then $P_t(E_1 \times \cdots \times E_m \times \underbrace{\mathbb{R} \times \cdots \times \mathbb{R}}_{n-m}) = P_s(E_1 \times \cdots \times E_m)$.

By the Kolmogorov's extension theorem, there exists a probability measure P on the measurable space $(\mathbb{R}^{[0,\infty)}, \mathscr{F})$ such that $P(C) = P_t(E_1 \times \cdots \times E_n)$ for C in (2.1.23).

Finally, by introducing the locally uniform topology on $C(\mathbb{R}^+;\mathbb{R}) \subset \mathbb{R}^{[0,\infty)}$, with the standard notation $\mathbb{R}^+ = [0,\infty)$, we obtain the relationship between the σ-field generated by cylinder sets and the Borel σ-field. For later use, here we extend $C(\mathbb{R}^+;\mathbb{R})$ to

$$\mathscr{X} = C(\mathbb{R}^+;\mathbb{R}^d),$$

the space of $\mathbb{R}^d$-valued continuous functions on $\mathbb{R}^+$, equipped with the metric (2.1.3). As seen before, this metric induces the locally uniform topology in which $f_n \to f$ if and only if $f_n \to f$ uniformly on every compact set in $\mathbb{R}^+$. The so-obtained topology gives rise to the Borel σ-field $\mathcal{B}(\mathscr{X})$.

Lemma 2.24. *The Borel σ-field $\mathcal{B}(\mathscr{X})$ coincides with the σ-field generated by cylinder sets of the form*

$$A_{t_1,\dots,t_n}^{E_1,\dots,E_n} = \Big\{\omega \in \mathscr{X} \mid (\omega(t_1),\dots,\omega(t_n)) \in E_1 \times \cdots \times E_n\Big\} \tag{2.1.24}$$

with $E_1,\dots,E_n \in \mathcal{B}(\mathbb{R}^d)$, $0 \le t_1 < t_2 < \cdots < t_n$, *and* $n \in \mathbb{N}$.

Proof. Let $\mathscr{C}$ be the σ-field generated by cylinder sets of the form (2.1.24). We show that $\mathcal{B}(\mathscr{X}) = \mathscr{C}$. First, consider the inclusion $\mathscr{C} \subset \mathcal{B}(\mathscr{X})$. Note that $\mathscr{C}$ coincides with the σ-field generated by cylinder sets of the form (2.1.24) with open sets $E_1,\dots,E_n$. For $s \ge 0$ and open set $E \subset \mathbb{R}^d$, let

$$H = H(s,E) = \Big\{\omega \in \mathscr{X} \mid \omega(s) \in E\Big\}.$$

For $\omega_0 \in H$, define $B(\omega_0,\varepsilon) = \{\omega \in \mathscr{X} \mid \mathrm{dist}(\omega,\omega_0) < \varepsilon\}$, where $\mathrm{dist}(\cdot,\cdot)$ is the metric on $\mathscr{X}$, given by (2.1.3).

Let $\varepsilon > 0$ be sufficiently small. For every $\omega \in B(\omega_0,\varepsilon)$, we see that $\omega(s) \in E$. Hence $B(\omega_0,\varepsilon) \subset H$ and H is open. This implies that $H \in \mathcal{B}(\mathscr{X})$. Thus the cylinder set $A_{t_1,\dots,t_n}^{E_1,\dots,E_n} = \bigcap_{j=1}^n H(t_j,E_j)$ is also open for open sets E_j. Then $\mathscr{C} \subset \mathcal{B}(\mathscr{X})$ follows since $\mathscr{C}$ is the minimal σ-field containing all cylinder sets.

Next consider the converse inclusion $\mathscr{C} \supset \mathcal{B}(\mathscr{X})$. Let $A \in \mathcal{B}(\mathscr{X})$. Since $\mathscr{X}$ is separable, we have $A = \bigcup_{j=1}^\infty B(\omega_j,\varepsilon_j)$. Note that

$$B(\omega_0,\varepsilon) = \left\{\omega \in \mathscr{X} \;\middle|\; \sum_{n=1}^\infty \frac{(\sup_{0\le t\le n,\, t\in\mathbb{Q}} |\omega_0(t)-\omega(t)|) \wedge 1}{2^n} < \varepsilon\right\}.$$

We obtain $B(\omega_0,\varepsilon) \in \mathscr{C}$ by a simple limiting argument, and thus $\mathscr{C} \supset \mathcal{B}(\mathscr{X})$. □

2.1.3 Random variables

Let $(\Omega,\mathscr{F},P)$ be a probability space. The integral with respect to P of a Borel measurable function f is called *expectation* of f and will be denoted by any of the symbols

$$\int_\Omega f(\omega)dP(\omega) = \int_\Omega f dP = \mathbb{E}_P[f]. \tag{2.1.25}$$

Whenever possible, we omit the subscript P in $\mathbb{E}_P$ unless any confusion may arise, and write simply $\mathbb{E}$.

An $\mathscr{F}/\mathscr{S}$-measurable function X mapping a probability space $(\Omega,\mathscr{F},P)$ to a measurable space $(S,\mathscr{S})$ is called an S-valued *random variable*. The inverse map identifies the sample points in Ω on which the observation of event A depends. For an S-valued random variable X, the probability measure P_X on $(S,\mathscr{S})$ defined by

$$P_X(A) = P(X^{-1}(A)), \quad A \in \mathscr{S},$$

is called the *image measure* or the *distribution* of P under $X : \Omega \to S$. Using this, we have

$$\mathbb{E}[f(X)] = \int_S f(x) dP_X(x) \tag{2.1.26}$$

for a bounded Borel measurable function $f : S \to \mathbb{R}^d$. The *covariance* of two $\mathbb{R}^d$-valued random variables X and Y on the same probability space is defined by

$$\operatorname{cov}(X, Y) = \mathbb{E}[(X - m_X)(Y - m_Y)], \tag{2.1.27}$$

where $m_X = \mathbb{E}[X]$ and $m_Y = \mathbb{E}[Y]$ are the expectations of X and Y, respectively. We set $\operatorname{cov}(X,X) = \operatorname{cov}(X)$ for notational simplicity. For two S-valued random variables X and Y, not necessarily defined on the same probability space, we use the notation $X \stackrel{\mathrm{d}}{=} Y$ whenever X and Y are *identically distributed*, i. e., $P_X = P_Y$.

Example 2.25 (Gaussian random variable). Let $m \in \mathbb{R}$ and $\sigma > 0$. A random variable $X : \Omega \to \mathbb{R}$ with probability distribution

$$dP_X(x) = \frac{1}{\sqrt{2\pi\sigma^2}} e^{-\frac{(x-m)^2}{2\sigma^2}} dx$$

is called a *Gaussian random variable* and the notation $X \stackrel{\mathrm{d}}{=} N(m, \sigma^2)$ is used. A calculation gives that m is the mean, and σ^2 is the variance of X. An $N(0,1)$-distributed random variable is a *standard Gaussian random variable*. A random vector $X = (X_1, \dots, X_d) : \Omega \to \mathbb{R}^d$ is a *multivariate Gaussian random variable* if its probability distribution is given by

$$dP_X(x) = \frac{1}{(2\pi)^{d/2} (\det C)^{1/2}} e^{-\frac{1}{2}(x-m)\cdot C^{-1}(x-m)} dx,$$

where $x = (x_1, \dots, x_d) \in \mathbb{R}^d$, $C = (C_{ij})_{1 \le i,j \le d}$ is a $d \times d$ symmetric, strictly positive definite matrix, and $m = (m_1, \dots, m_d) \in \mathbb{R}^d$ is a vector. The matrix element C_{ij} gives the covariance $\operatorname{cov}(X_i, X_j)$ of the multivariate Gaussian random variable, and $m_j = \mathbb{E}[X_j]$.

Let $(S, \mathscr{S})$ be a measurable space, and X be an S-valued random variable on $(\Omega, \mathscr{F}, P)$. The notation $\sigma(X)$ will be used for the minimal σ-field such that X is measurable, i. e.,

$$\sigma(X) = \left\{ X^{-1}(U) \mid U \in \mathscr{S} \right\}. \tag{2.1.28}$$

Let Y be also a random variable on $(\Omega, \mathscr{F}, P)$. If the sub-σ-fields $\sigma(X)$ and $\sigma(Y)$ are independent, then X and Y are said to be *independent*. We consider the distribution of the sum of random variables. Let P and Q be probability measures on $(\mathbb{R}^d, \mathcal{B}(\mathbb{R}^d))$. The *convolution* of P and Q is a probability measure on $(\mathbb{R}^d, \mathcal{B}(\mathbb{R}^d))$ defined by

$$(P * Q)(A) = \int_{\mathbb{R}^d} Q(A - x)dP(x), \quad A \in \mathcal{B}(\mathbb{R}^d),$$

where $A-x = \{y-x \mid y \in A\}$. Note that $P*Q = Q*P$. Let X and Y be $\mathbb{R}^d$-valued independent random variables and P and Q be distributions on $\mathbb{R}^d$ of X and Y, respectively. It follows that

$$\mathbb{E}_P[f(X+Y)] = \int_{\mathbb{R}^d} dP(x) \int_{\mathbb{R}^d} f(x+y)dQ(y) = \int_{\mathbb{R}^d} f(z)(P * Q)(dz)$$

for all bounded Borel measurable functions f. Then the distribution of the sum of two independent random variables $X + Y$ is given by $P * Q$.

If X and Y are independent, then $\mathrm{cov}(f(X), g(Y)) = 0$ for all bounded Borel measurable functions f and g. In particular for two Gaussian random variables X and Y, it is well known that if X and Y are independent, then they are jointly Gaussian and their covariance is zero. Conversely, if X and Y are jointly Gaussian and their covariance is zero, then X and Y are independent. Here X and Y are said to be jointly Gaussian if and only if the linear combination $aX + bY$ is a Gaussian random variable for every $a, b \in \mathbb{R}$.

Lemma 2.26. *Let $(\Omega, \mathcal{F}, P)$ be a probability space and X_i, $i = 1, \dots, n$, real-valued random variables on it. Then X_i, $i = 1, \dots, n$, are independent if and only if*

$$\mathbb{E}[f_1(X_1) \cdots f(X_n)] = \prod_{i=1}^n \mathbb{E}[f_i(X_i)], \tag{2.1.29}$$

for all bounded Borel measurable functions f_i, $i = 1, \dots, n$.

Proof. Suppose that X_i, $i = 1, \dots, n$, are independent. Let $f_i(x) = 1_{E_i}(x)$ with $E_i \in \mathcal{B}(\mathbb{R})$. We see that

$$\mathbb{E}\left[\prod_{i=1}^n 1_{E_i}(X_i)\right] = P(X_1 \in E_1, \dots, X_n \in E_n) = \prod_{i=1}^n P(X_i \in E_i) = \prod_{i=1}^n \mathbb{E}[1_{E_i}(X_i)].$$

By a limiting argument, (2.1.29) follows. Conversely, suppose (2.1.29). Take $A_k \in \sigma(X_k)$, $k = 1, \dots, n$. We see that

$$P(A_1 \cap \cdots \cap A_n) = \mathbb{E}[\prod_{k=1}^n 1_{A_k}(X_k)] = \prod_{k=1}^n \mathbb{E}[1_{A_k}(X_k)] = \prod_{k=1}^n P(A_k).$$

Thus X_i, $i = 1, \dots, n$, are independent. □

We have a second useful criterion to show independence of random variables.

Lemma 2.27. *Let* $(\Omega, \mathscr{F}, P)$ *be a probability space and* X_i, $i = 1, \dots, n$, *real-valued random variables on it. Then* X_i, $i = 1, \dots, n$, *are independent if and only if for every* $(\xi_1, \dots, \xi_n) \in \mathbb{R}^n$,

$$\mathbb{E}[e^{i\sum_{j=1}^n \xi_j X_j}] = \prod_{j=1}^n \mathbb{E}[e^{i\xi_j X_j}]. \tag{2.1.30}$$

Proof. The necessity part is directly obtained from Lemma 2.26. To obtain the sufficiency part, suppose that (2.1.30) holds and let $f_j \in \mathscr{S}(\mathbb{R})$, for $j = 1, \dots, n$. Using that $f_j(x) = (2\pi)^{-1/2} \int_{\mathbb{R}} \hat{f}_j(k) e^{ikx} dk$, we have

$$\mathbb{E}\left[\prod_{j=1}^n f_j(X_j)\right] = \prod_{j=1}^n \mathbb{E}[f_j(X_j)]. \tag{2.1.31}$$

Let $C_0(\mathbb{R})$ be the set of continuous functions on $\mathbb{R}$ with compact support. For $f \in C_0(\mathbb{R})$, there exists a sequence $(g_n)_{n\in\mathbb{N}} \subset \mathscr{S}(\mathbb{R})$ such that $g_n \to f$, uniformly as $n \to \infty$. Equality (2.1.31) holds for every $f_j \in C_0(\mathbb{R})$ by a limiting argument. Fix $f_2, \dots, f_n \in C_0(\mathbb{R})$ and define

$$\mathscr{A} = \left\{A \in \mathcal{B}(\mathbb{R}) \,\middle|\, \mathbb{E}\left[1_A(X_1) \prod_{j=2}^n f_j(X_j)\right] = \mathbb{E}[1_A(X_1)]\mathbb{E}\left[\prod_{j=2}^n f_j(X_j)\right]\right\}.$$

Denote the family of finite unions of semiopen intervals in $\mathbb{R}$ by I. If $A = (a, b]$, then by approximating the indicator function 1_A pointwise by a function in $C_0(\mathbb{R})$ we see that $(a, b] \in \mathscr{A}$ by (2.1.31). Hence $I \subset \mathscr{A}$ follows. Since $\mathscr{A}$ is a monotone family, we get $\mathscr{A} = \mathcal{B}(\mathbb{R})$. Repeating this procedure, we see that $\mathbb{E}[\prod_{j=1}^n 1_{E_j}(X_j)] = \prod_{j=1}^n \mathbb{E}[1_{E_j}(X_j)]$ for arbitrary $E_j \in \mathcal{B}(\mathbb{R})$, and thus the independence of $X_1, \dots, X_n$ is proven. □

In what follows a real-valued random variable will be simply called a random variable. We summarize the notions of convergence of random variables next.

Definition 2.28 (Convergence of random variables). Let $(X_n)_{n\in\mathbb{N}}$ be a sequence of random variables and X another random variable, all on a given probability space $(\Omega, \mathscr{F}, P)$.

(1) We write $X_n \xrightarrow{\text{a.s.}} X$, i. e., X_n is convergent *almost surely* to X if

$$P\left(\lim_{n\to\infty} X_n = X\right) = 1.$$

(2) We write $X_n \xrightarrow{L^p} X$, i. e., X_n is convergent *in L^p-sense* to X for $1 \le p < \infty$, if

$$\lim_{n\to\infty} \mathbb{E}\left[|X_n - X|^p\right] = 0.$$

(3) We write $X_n \xrightarrow{P} X$, i. e., X_n is convergent *in probability* to X if for every $\varepsilon > 0$,

$$\lim_{n\to\infty} P(|X_n - X| \geq \varepsilon) = 0.$$

(4) We write $X_n \xrightarrow{\mathrm{d}} X$, i. e., X_n is convergent *in distribution* or *weakly* or *in law* to X if

$$\lim_{n\to\infty} \mathbb{E}[f(X_n)] = \mathbb{E}[f(X)]$$

for every bounded continuous function f.

Note that for any given sequence $(X_n)_{n\in\mathbb{N}}$ of random variables on $(\Omega, \mathscr{F}, P)$ with values in a metric space $(S, \mathcal{B}(S))$, the convergence $X_n \xrightarrow{\mathrm{d}} X$ is equivalent to weak convergence of $P \circ X_n^{-1}$ to $P \circ X^{-1}$.

To discuss the various types of convergence of measurable functions, first we study a criterion for relative sequential compactness in $L^1(\Omega) = L^1(\Omega, d\mu)$ in terms of the following concept.

Definition 2.29 (Uniform integrability). Let $(\Omega, \mathscr{F}, \mu)$ be a measure space. A set of functions $\mathscr{A} \subset L^1(\Omega, d\mu)$ is *uniformly integrable* if

$$\inf_{0\leq g\in L^1} \sup_{f\in\mathscr{A}} \int_\Omega (|f(x)| - g(x))^+ d\mu(x) = 0, \tag{2.1.32}$$

where $h^+ = \max\{h, 0\}$ denotes the nonnegative part of h.

Lemma 2.30. *Let $(\Omega, \mathscr{F}, \mu)$ be a finite measure space. Then $\mathscr{A} \subset L^1(\Omega)$ is uniformly integrable if and only if*

$$\inf_{a\in[0,\infty)} \sup_{f\in\mathscr{A}} \int_\Omega |f| 1_{\{|f|>a\}} d\mu = 0. \tag{2.1.33}$$

Proof. Assuming that (2.1.33) holds, we have

$$\inf_{a\in[0,\infty)} \sup_{f\in\mathscr{A}} \int_\Omega (|f| - a)^+ d\mu = 0. \tag{2.1.34}$$

Equality (2.1.34) furthermore implies the uniform integrability of $\mathscr{A}$ since the infimum is taken over the smaller set of constant functions.

Next suppose that $\mathscr{A}$ is uniformly integrable. There exists $0 \leq g_\varepsilon \in L^1(\Omega)$ such that $\sup_{f\in\mathscr{A}} \int_\Omega (|f| - g_\varepsilon)^+ d\mu \leq \varepsilon$. Define $\tilde{g}_\varepsilon = 2g_{\varepsilon/2}$. For $f \in \mathscr{A}$,

$$\int_\Omega |f| 1_{\{|f|>\tilde{g}_\varepsilon\}} d\mu \leq \int_\Omega (|f| - g_{\varepsilon/2})^+ 1_{\{|f|>\tilde{g}_\varepsilon\}} d\mu + \int_\Omega g_{\varepsilon/2} 1_{\{|f|>\tilde{g}_\varepsilon\}} d\mu.$$

We see that $\int_\Omega (|f| - g_{\varepsilon/2})^+ 1_{\{|f|>\tilde{g}_\varepsilon\}} d\mu \le \varepsilon/2$ and

$$g_{\varepsilon/2} 1_{\{|f|>\tilde{g}_\varepsilon\}} \le (|f| - g_{\varepsilon/2})^+ 1_{\{|f|>\tilde{g}_\varepsilon\}}.$$

Hence

$$\int_\Omega g_{\varepsilon/2} 1_{\{|f|>\tilde{g}_\varepsilon\}} d\mu \le \int_\Omega (|f| - g_{\varepsilon/2})^+ 1_{\{|f|>\tilde{g}_\varepsilon\}} d\mu \le \frac{\varepsilon}{2}.$$

Summing up, we have $\sup_{f\in\mathscr{A}} \int_\Omega |f| 1_{\{|f|>\tilde{g}_\varepsilon\}} d\mu \le \varepsilon$ and the same with ε replaced by $\varepsilon/2$. Choose a sufficiently large a_ε such that $\int_\Omega \tilde{g}_{\varepsilon/2} 1_{\{\tilde{g}_{\varepsilon/2}>a_\varepsilon\}} d\mu \le \varepsilon/2$.

We see that

$$\begin{aligned}\int_\Omega |f| 1_{\{|f|>a_\varepsilon\}} d\mu &\le \int_\Omega |f| 1_{\{|f|>a_\varepsilon\}} 1_{\{|f|>\tilde{g}_{\varepsilon/2}\}} d\mu + \int_\Omega |f| 1_{\{|f|>a_\varepsilon\}} 1_{\{|f|\le\tilde{g}_{\varepsilon/2}\}} d\mu \\ &\le \int_\Omega |f| 1_{\{|f|>\tilde{g}_{\varepsilon/2}\}} d\mu + \int_\Omega g_{\varepsilon/2} 1_{\{|f|>a_\varepsilon\}} d\mu \le \int_\Omega |f| 1_{\{|f|>\tilde{g}_{\varepsilon/2}\}} d\mu + \int_\Omega g_{\varepsilon/2} 1_{\{\tilde{g}_{\varepsilon/2}>a_\varepsilon\}} d\mu < \varepsilon. \qquad \square\end{aligned}$$

The relationships of the various types of convergence can be summarized as follows.

Theorem 2.31. *Let $(X_n)_{n\in\mathbb{N}}$ be a sequence of random variables and X another random variable, all on a probability space $(\Omega, \mathscr{F}, P)$.*

(1) *If $X_n \xrightarrow{\text{a.s.}} X$ or $X_n \xrightarrow{L^p} X$ for some $p \ge 1$, then $X_n \xrightarrow{P} X$ as $n \to \infty$.*
(2) *If $X_n \xrightarrow{P} X$, then $X_n \xrightarrow{\text{d}} X$ as $n \to \infty$.*
(3) *If $X_n \xrightarrow{P} X$, then there exists a subsequence X_{n_k}, $k \ge 1$, such that $X_{n_k} \xrightarrow{\text{a.s.}} X$ as $k \to \infty$.*
(4) (Dominated convergence theorem) *Let $X_n \xrightarrow{P} X$ as $n \to \infty$ and suppose that there is a random variable Y such that $|X_n| \le Y$ for all $n \in \mathbb{N}$, and $\mathbb{E}[Y] < \infty$. Then $X \in L^1(\Omega, dP)$ and $X_n \xrightarrow{L^1} X$.*
(5) (Monotone convergence theorem) *Let $X_n \xrightarrow{\text{a.s.}} X$ as $n \to \infty$. Suppose that the sequence $(X_n)_{n\in\mathbb{N}}$ is monotone increasing and there exists $M > 0$ such that $\mathbb{E}[X_n] < M$ for all n. Then $X \in L^1(\Omega, dP)$ and $\mathbb{E}[X_n] \to \mathbb{E}[X]$ as $n \to \infty$.*

Proof. (1) Let $\varepsilon > 0$. Suppose that $X_n \xrightarrow{\text{a.s.}} X$. Let

$$A_n = \left\{\omega \in \Omega \mid |X_n(\omega) - X(\omega)| > \varepsilon\right\},$$
$$E = \left\{\omega \in \Omega \mid \lim_{n\to\infty} X_n(\omega) = X(\omega)\right\}.$$

Thus $P(A_n) = P(A_n \cap E) + P(A_n \cap E^c) = P(A_n \cap E)$, since $P(E^c) = 0$. Thus $P(A_n) = P(A_n \cap E) \to 0$ as $n \to \infty$, and hence $X_n \xrightarrow{P} X$ as $n \to \infty$.

Next suppose that $X_n \xrightarrow{L^p} X$. We see that

$$P(A_n) \le \varepsilon^{-p} \|X_n - X\|_{L^p}^p$$

and $\lim_{n\to\infty} P(A_n) = 0$ follows. Thus $X_n \xrightarrow{P} X$ as $n \to \infty$.

(2) Let f be a bounded and uniformly continuous function, and choose $\varepsilon > 0$. We see that $|f(x) - f(y)| \le \varepsilon$ for all $x, y \in \mathbb{R}$ such that $|x - y| \le \delta$, with some $\delta > 0$. Let

$$K_n = \Big\{\omega \in \Omega \mid |X_n(\omega) - X(\omega)| \le \delta\Big\}.$$

Hence we see that

$$|\mathbb{E}[f(X_n) - f(X)]| \le \mathbb{E}[|f(X_n) - f(X)| 1_{K_n}] + \mathbb{E}[|f(X_n) - f(X)| 1_{K_n^c}] \le \varepsilon + M\mathbb{E}[1_{K_n^c}],$$

where $M = \sup_{x\in\mathbb{R}} |f(x)|$. Thus

$$\lim_{n\to\infty} |\mathbb{E}[f(X_n) - f(X)]| \le \varepsilon. \tag{2.1.35}$$

By a simple limiting argument, (2.1.35) holds for any bounded continuous function f.

(3) Since $X_n \xrightarrow{P} X$, there exists a subsequence n_k such that

$$P(|X - X_{n_k}| > 1/k) \le 2^{-k}.$$

Thus $\sum_{k=1}^{\infty} P(|X - X_{n_k}| > 1/k) < \infty$. By the Borel–Cantelli lemma, we then have

$$P\Big(\bigcup_{m=1}^{\infty} \bigcap_{k=m}^{\infty} \{|X - X_{n_k}| \le 1/k\}\Big) = 1.$$

Thus $X_{n_k} \to X$ as $n_k \to \infty$, almost surely.

(4) By Fatou's lemma, we have

$$\mathbb{E}[|X|] \le \liminf_{n\to\infty} \mathbb{E}[|X_n|] \le \mathbb{E}[|Y|] < \infty.$$

Thus $X \in L^1(\Omega, dP)$. Since $\mathbb{E}[|X_n| 1_{\{|X_n|>a\}}] \le \mathbb{E}[Y 1_{\{Y>a\}}]$, we have

$$\inf_{a\in[0,\infty)} \sup_{n\in\mathbb{N}} \mathbb{E}[|X_n| 1_{\{|X_n|>a\}}] \le \inf_{a\in[0,\infty)} \mathbb{E}[Y 1_{\{Y>a\}}] = 0,$$

and hence $(X_n)_{n\in\mathbb{N}}$ is uniformly integrable by Lemma 2.30. Suppose, to the contrary, that X_n does not converge to X in $L^1(\Omega, dP)$, i. e., there exist $\varepsilon > 0$ and a subsequence n_k, $k \in \mathbb{N}$, such that

$$\|X - X_{n_k}\|_{L^1} > 2\varepsilon. \tag{2.1.36}$$

We reset the counter and write X_{n_k} as X_n for notational simplicity. Note that $(X - X_n)_{n\in\mathbb{N}}$ is also uniformly integrable. There exists $0 \le g \in L^1(\Omega, dP)$ such that

$$\mathbb{E}[(|X - X_n| - g)^+] < \varepsilon.$$

This implies that

$$\limsup_{n\to\infty} \mathbb{E}[1_{\{|X-X_n|>g\}}|X - X_n|] < \varepsilon. \tag{2.1.37}$$

Define $g_n(\omega) = |X(\omega) - X_n(\omega)| \wedge g(\omega)$. It follows that $g_n \le g$ and $g_n \to 0$ almost surely as $n \to \infty$. Fatou's lemma again gives

$$\limsup_{n\to\infty} \mathbb{E}[g_n] = \mathbb{E}[g] - \liminf_{n\to\infty} \mathbb{E}[g - g_n] \le \mathbb{E}[g] - \mathbb{E}[\lim_{n\to\infty} (g - g_n)] = 0.$$

Since $\{|X - X_n| = g_n\} = \{|X - X_n| \le g\}$, together with (2.1.37) this implies that

$$\begin{aligned}\limsup_{n\to\infty} \|X - X_n\|_{L^1} &= \limsup_{n\to\infty} \left(\mathbb{E}[1_{\{|X-X_n|>g\}}|X - X_n|] + \mathbb{E}[1_{\{|X-X_n|\le g\}}|X - X_n|]\right) \\ &\le \limsup_{n\to\infty} \mathbb{E}[1_{\{|X-X_n|>g\}}|X - X_n|] + \limsup_{n\to\infty} \mathbb{E}[g_n] \le \varepsilon.\end{aligned}$$

This contradicts (2.1.36).

(5) This is a fundamental fact in measure theory. □

Corollary 2.32. *Suppose that $\{X_n\}_{n\in\mathbb{N}}$ is a uniformly integrable set of random variables on a probability space $(\Omega, \mathscr{F}, P)$. Then $\limsup_{n\to\infty} \mathbb{E}[|X_n|] < \infty$. If, in addition, $X_n \to X$ almost surely or in probability, then $X_n \xrightarrow{L^1} X$.*

Proof. Note that for every $\lambda > 0$ it follows that $\mathbb{E}[|X_n|] \le \lambda + \mathbb{E}[|X_n| 1_{\{|X_n|>\lambda\}}]$. Thus we have $\limsup_{n\to\infty} \mathbb{E}[|X_n|] < \infty$. For the second statement, in the same way as in the proof of (4) of Theorem 2.31, we see that $X_n \xrightarrow{L^1} X$ as $n \to \infty$. □

In probability theory we often have to deal with sums of independent, identically distributed random variables $\{X_n\}_{n\in\mathbb{N}}$ and related (e. g., Cesàro) means. One notable property of $\sum_{i=1}^n X_i/n$ is that it converges to the nonrandom number $\mathbb{E}[X_1]$, almost surely as $n \to \infty$. This is the celebrated *(strong) law of large numbers*. If almost sure convergence is replaced by convergence in probability, the proof of is more straightforward.

Lemma 2.33. *Let $(\Omega, \mathscr{F}, P)$ be a probability space and $(X_n)_{n\in\mathbb{N}}$ a sequence of independent random variables such that $\mathbb{E}[X_n^2] < \infty$ and $\mathbb{E}[X_n] = 0$, for all $n \in \mathbb{N}$. Let $a > 0$. Then*

$$P\left(\max_{1\le k\le n} \left|\sum_{i=1}^k X_i\right| > a\right) \le \frac{1}{a^2} \sum_{i=1}^n \sigma_i^2,$$

where $\sigma_i^2 = \mathbb{E}[X_i^2]$.

Proof. Let $A = \{\max_{1\leq k\leq n} |S_k| > a\} = \bigcup_{j=1}^n A_j$, where

$$S_k = \sum_{i=1}^k X_i, \quad A_j = \Big\{|S_i| < a,\ i = 1, \dots, j-1, \text{ and } |S_j| \geq a\Big\}.$$

Note that $A_i \cap A_j = \emptyset$ for $i \neq j$. We then have

$$\begin{aligned}\sum_{i=1}^n \sigma_i^2 &\geq \mathbb{E}[S_n^2 1_A] = \sum_{j=1}^n \mathbb{E}[S_n^2 1_{A_j}] = \sum_{j=1}^n \mathbb{E}[(S_n - S_j + S_j)^2 1_{A_j}] \\ &= \sum_{j=1}^n \mathbb{E}[((S_n - S_j)^2 - 2S_j(S_n - S_j) + S_j^2)1_{A_j}] \geq \sum_{j=1}^n \mathbb{E}[S_j^2 1_{A_j}] \geq \sum_{j=1}^n a^2 P(A_j) = a^2 P(A).\end{aligned}$$

Hence $P(A) \leq \sum_{i=1}^n \sigma_i^2 / a^2$ follows. □

Lemma 2.34. *Let $(\Omega, \mathscr{F}, P)$ be a probability space and $(X_n)_{n\in\mathbb{N}}$ a sequence of independent random variables such that $\sum_{n=1}^\infty \mathbb{E}[X_n^2] < \infty$. Then $S_n = \sum_{i=1}^n X_i$ converges almost surely as $n \to \infty$.*

Proof. It is sufficient to show that $(S_n)_{n\in\mathbb{N}}$ is a Cauchy sequence, almost surely. Let $\varepsilon > 0$ be fixed. Write $\mathbb{E}[X_n^2] = \sigma_n^2$. Define

$$A_N(\varepsilon) = \Big\{\omega \in \Omega \mid \exists i, j > N \text{ such that } |S_i(\omega) - S_j(\omega)| \geq 2\varepsilon\Big\}$$

and write $A = \bigcup_{n=m}^\infty \bigcap_{N=1}^\infty A_N(1/n)$, where $1/m < \varepsilon$; clearly, $\omega \in A$ implies that there exists $n = n(\omega) \in \mathbb{N}$ such that for all N there exist $i, j > N$ with $|S_i(\omega) - S_j(\omega)| \geq 2/n$. On the other hand, $\omega \in \Omega \setminus A$ implies that for every n there exists $N = N(\omega) \in \mathbb{N}$ such that for all $i, j > N$ it follows that $|S_i(\omega) - S_j(\omega)| < 2/n$. Thus it is sufficient to show $P(A) = 0$. Let $B_N = \{\omega \in \Omega \mid \exists i > N \text{such that} |S_i(\omega) - S_N(\omega)| \geq \varepsilon\}$. We have $A_N(\varepsilon) \subset B_N$, since $2\varepsilon \leq |S_i - S_j| \leq |S_i - S_N| + |S_N - S_j|$. Furthermore, we consider

$$\begin{aligned}B_{N,n} &= \Big\{\omega \in \Omega \mid \text{there exists } n \geq i > N \text{ such that } |S_i(\omega) - S_N(\omega)| \geq \varepsilon\Big\} \\ &= \Big\{\omega \in \Omega \mid \max_{N<i\leq n} |X_{N+1} + \cdots + X_i| \geq \varepsilon\Big\}.\end{aligned}$$

We see that $P(B_{N,n}) \leq \sum_{i=N+1}^n \sigma_i^2/\varepsilon^2$. By the dominated convergence theorem, we obtain $\lim_{n\to\infty} P(B_{N,n}) = P(B_N) \leq \sum_{i=N+1}^\infty \sigma_i^2/\varepsilon^2$. Thus for $1/n < \varepsilon$,

$$P\Big(\bigcap_{k=1}^\infty A_k(1/n)\Big) \leq P(A_N(1/n)) \leq P(B_N) \leq \sum_{i=N+1}^\infty \sigma_i^2/\varepsilon^2$$

for every N. Thus $P(\bigcap_{N=1}^\infty A_N(1/n)) = 0$ follows for $1/n < \varepsilon$. Hence we conclude that $P(A) = P(\bigcup_{n=m}^\infty \bigcap_{N=1}^\infty A_N(1/n)) \leq \sum_{n=m}^\infty P(\bigcap_{N=1}^\infty A_N(1/n)) = 0$. □

Lemma 2.35. *Let* $(\Omega, \mathscr{F}, P)$ *be a probability space and* $(X_n)_{n\in\mathbb{N}}$ *a sequence of independent random variables such that* $\mathbb{E}[X_n] = m_n$ *and* $\sum_{n=1}^{\infty} \mathbb{E}[|X_n - m_n|^2]/n^2 < \infty$. *Then*

$$\lim_{n\to\infty} \frac{1}{n} \sum_{i=1}^{n} (X_i - m_i) = 0 \quad \text{a.s.} \tag{2.1.38}$$

Proof. Note that $\sum_{j=1}^{\infty} a_j/j < \infty$ implies that $\lim_{n\to\infty} \sum_{i=1}^{n} a_i/n = 0$ as a general fact. Let $\hat{X}_j = (X_j - m_j)/j$. We see that $\sum_{j=1}^{\infty} \mathbb{E}[\hat{X}_j^2] < \infty$ and $\sum_{i=1}^{n} \hat{X}_i$ converges almost surely by Lemma 2.34. Thus (2.1.38) follows. □

Denote $\dot{\mathbb{N}} = \mathbb{N} \cup \{0\}$.

Theorem 2.36 (Strong law of large numbers). *Let* $(\Omega, \mathscr{F}, P)$ *be a probability space and* $(X_n)_{n\in\dot{\mathbb{N}}}$ *a sequence of independent and identically distributed random variables such that* $a = \mathbb{E}[X_0] < \infty$. *Then*

$$\lim_{n\to\infty} \frac{1}{n} \sum_{i=1}^{n} X_i = \mathbb{E}[X_0] \quad \text{a.s.}$$

Proof. Let $Y_n = 1_{[-n,n]}(X_n)$. Hence

$$\sigma_n^2 = \mathbb{E}[(Y_n - \mathbb{E}[Y_n])^2] \le \mathbb{E}[Y_n^2] = \int_{-n}^{n} x^2 dF(x),$$

where dF denotes the distribution of X_n. We show that

$$\sum_{n=1}^{\infty} \frac{\sigma_n^2}{n^2} = \sum_{n=1}^{\infty} \frac{1}{n^2} \int_{-n}^{n} x^2 dF(x) < \infty. \tag{2.1.39}$$

Indeed, we see that

$$\begin{aligned} \infty > C = \int_{\mathbb{R}} |x| dF(x) &\ge \sum_{n=0}^{N} \int_{n\le|x|<n+1} |x| dF(x) \\ &\ge \sum_{n=0}^{N} \frac{1}{n+1} \int_{n\le|x|<n+1} |x|^2 dF(x) = \sum_{n=0}^{N} \frac{a_{n+1} - a_n}{n+1}, \end{aligned}$$

where $a_n = \int_{-n}^{n} |x|^2 dF(x) = \sigma_n^2$. From this it follows that

$$\begin{aligned} C &> -a_0 + \frac{1}{1\cdot 2} a_1 + \frac{1}{2\cdot 3} a_2 + \cdots + \frac{1}{N(N+1)} a_N + \frac{a_{N+1}}{N+1} \\ &\ge -a_0 + \frac{1}{2^2} a_1 + \frac{1}{3^2} a_2 + \cdots + \frac{1}{(N+1)^2} a_N. \end{aligned}$$

Hence

$$\sum_{n=1}^{\infty} \frac{a_n}{n^2} = \sum_{n=1}^{\infty} \frac{a_n}{(n+1)^2} \frac{(n+1)^2}{n^2} \leq 4 \sum_{n=1}^{\infty} \frac{a_n}{(n+1)^2} < \infty, \tag{2.1.40}$$

and so (2.1.39) is proven. Thus $\lim_{n\to\infty} \sum_{j=0}^{n} (Y_j - m_j)/n = 0$ a. s. by Lemma 2.35. We also have $\lim_{n\to\infty} m_n = \int_{-n}^{n} x dF(x) = \int_{\mathbb{R}} x dF(x) = \mathbb{E}[X_0] = a$. Thus

$$\lim_{n\to\infty} \sum_{j=0}^{n} m_j/n = a$$

and we have

$$\lim_{n\to\infty} \frac{\sum_{j=0}^{n} Y_j}{n} = \lim_{n\to\infty} \frac{\sum_{j=0}^{n} (Y_j - m_j)}{n} + \lim_{n\to\infty} \frac{\sum_{j=0}^{n} m_j}{n} = a.$$

Let $A_n = \{X_n(\omega) \neq Y_n(\omega)\} = \{|X_n(\omega)| > n\}$. We have

$$\begin{aligned}\sum_{n=0}^{\infty} P(A_n) &= \sum_{n=0}^{\infty} P(|X_0| > n) = \sum_{n=0}^{\infty} \int_{|x|>n} dF(x) \\ &= \sum_{k=0}^{\infty} (k+1) \int_{k<|x|\leq k+1} dF(x) \leq \sum_{k=0}^{\infty} \int_{k<|x|\leq k+1} (1+|x|) dF(x) \leq \int_{\mathbb{R}} (1+|x|) dF(x) < \infty.\end{aligned}$$

By the Borel–Cantelli lemma, we see that $P(\bigcup_{n=1}^{\infty} \bigcap_{m=n}^{\infty} A_m^c) = 1$. Thus for every $\omega \in \bigcup_{n=1}^{\infty} \bigcap_{m=n}^{\infty} A_m^c$, there exists n such that for all $m > n$, $X_n(\omega) = X_m(\omega)$, and it follows that for $m > n$,

$$\begin{aligned}\frac{\sum_{i=0}^{m} X_i(\omega)}{m} &= \frac{\sum_{i=0}^{m} (X_i(\omega) - Y_i(\omega))}{m} + \frac{\sum_{i=0}^{m} Y_i(\omega)}{m} \\ &= \frac{\sum_{i=0}^{n} (X_i(\omega) - Y_i(\omega))}{m} + \frac{\sum_{i=0}^{m} Y_i(\omega)}{m} \to a\end{aligned}$$

as $m \to \infty$. Thus the statement follows. □

2.1.4 Conditional expectation and regular conditional probability measures

Let X be a random variable on a probability space $(\Omega, \mathcal{F}, P)$ and $\mathcal{G} \subset \mathcal{F}$ a sub-σ-field. Recall that the *conditional expectation*

$$Z = \mathbb{E}[X|\mathcal{G}]$$

of X with respect to $\mathcal{G}$ is defined as the unique $\mathcal{G}$-measurable random variable such that

$$\mathbb{E}[1_A X] = \mathbb{E}[1_A Z], \quad A \in \mathcal{G}. \tag{2.1.41}$$

If $X \geq 0$, the left-hand side of (2.1.41) defines a probability measure $P_X(A) = \mathbb{E}[1_A X]$ on $\mathcal{G}$ and $\mathbb{E}[X|\mathcal{G}]$ is in fact the *Radon–Nikodým derivative* dP_X/dP.

Let $P(A|B) = P(A \cap B)/P(B)$ for $A, B \in \mathcal{F}$ such that $P(B) \neq 0$. Then

$$\mathcal{F} \ni A \mapsto P_B(A) = P(A|B)$$

defines a probability measure on $(\Omega, \mathcal{F})$. Suppose that $A_n \in \mathcal{F}, n = 1, \ldots, N \leq \infty$, satisfy that $A_n \cap A_m = \emptyset$ for $n \neq m$ and $\bigcup_{n=1}^N A_n = \Omega$. Let $\mathcal{G} = \sigma(\{A_n\}_{n=1}^N)$ be the sub-σ-field generated by $\{A_n\}_{n=1}^N$. We have

$$\mathbb{E}[X|\mathcal{G}] = \sum_{n=1}^{N} \mathbb{E}_{P_{A_n}}[X] 1_{A_n}.$$

From this it can be said that the conditional expectation $\mathbb{E}[X|\mathcal{G}]$ gives an approximation of X in terms of $\mathcal{G}$-measurable functions.

We summarize some basic properties of conditional expectation below.

Proposition 2.37 (Properties of conditional expectation). *Let $(\Omega, \mathcal{F}, P)$ be a probability space, and X and Y be random variables on it such that $X, Y \in L^1(\Omega)$. Let $\mathcal{G} \subset \mathcal{H} \subset \mathcal{F}$ be sub-σ-fields. Then the following hold:*
(1) *(Linearity) $\mathbb{E}[aX + bY|\mathcal{G}] = a\mathbb{E}[X|\mathcal{G}] + b\mathbb{E}[Y|\mathcal{G}]$, for $a, b \in \mathbb{C}$.*
(2) *(Monotonicity) If $X \geq Y$ a. s., then $\mathbb{E}[X|\mathcal{G}] \geq \mathbb{E}[Y|\mathcal{G}]$ a. s.*
(3) *If Y is measurable with respect to $\mathcal{G}$ and $\mathbb{E}[|XY|] < \infty$, then $\mathbb{E}[XY|\mathcal{G}] = Y\mathbb{E}[X|\mathcal{G}]$.*
(4) *(Tower property) $\mathbb{E}[\mathbb{E}[X|\mathcal{H}]|\mathcal{G}] = \mathbb{E}[\mathbb{E}[X|\mathcal{G}]|\mathcal{H}] = \mathbb{E}[X|\mathcal{G}]$.*
(5) *(Triangle inequality) $|\mathbb{E}[X|\mathcal{G}]| \leq \mathbb{E}[|X||\mathcal{G}]$.*
(6) *(Independence) If $\sigma(X)$ and $\mathcal{G}$ are independent, then $\mathbb{E}[X|\mathcal{G}] = \mathbb{E}[X]$.*

Proof. (1) The right-hand side is $\mathcal{G}$-measurable, hence for $A \in \mathcal{G}$ we have

$$\mathbb{E}[1_A(a\mathbb{E}[X|\mathcal{G}] + b\mathbb{E}[Y|\mathcal{G}])] = a\mathbb{E}[1_A X] + b\mathbb{E}[1_A Y] = \mathbb{E}[1_A(aX + bY)].$$

Thus (1) follows.

(2) Let $A = \{\mathbb{E}[X|\mathcal{G}] < \mathbb{E}[Y|\mathcal{G}]\} \in \mathcal{G}$. Since from the assumption it follows that $0 \leq \mathbb{E}[1_A(X - Y)] = \mathbb{E}[1_A X|\mathcal{G}] - \mathbb{E}[1_A Y|\mathcal{G}]$, we have $P(A) = 0$. This gives (2).

(3) First, we assume that $X, Y \geq 0$. For every $n \in \mathbb{N}$, define $Y_n = 2^{-n}[2^n Y]$, where $[z]$ is the integer part of z. We have $Y_n = \sum_{k=1}^{\infty} 1_{\{Y_n = k2^{-n}\}} k2^{-n}$ and $Y_n \uparrow Y$ as $n \to \infty$. Let $A \in \mathcal{G}$. By the monotone convergence theorem, it can be seen that $\mathbb{E}[1_A Y_n \mathbb{E}[X|\mathcal{G}]] \to \mathbb{E}[1_A Y \mathbb{E}[X|\mathcal{G}]]$ as $n \to \infty$. On the other hand, since $\{Y_n = k2^{-n}\} \in \mathcal{G}$, we see that

$$\begin{aligned} \mathbb{E}[1_A Y_n \mathbb{E}[X|\mathcal{G}]] &= \sum_{k=1}^{\infty} \mathbb{E}[1_A 1_{\{Y_n = k2^{-n}\}} k2^{-n} \mathbb{E}[X|\mathcal{G}]] = \sum_{k=1}^{\infty} \mathbb{E}[1_A 1_{\{Y_n = k2^{-n}\}} k2^{-n} X] \\ &= \mathbb{E}[1_A Y_n X] \to \mathbb{E}[1_A Y X] \end{aligned}$$

as $n \to \infty$. Hence (3) follows for $X \geq 0$ and $Y \geq 0$. In the general case, write $X = X^+ - X^-$ and $Y = Y^+ - Y^-$ and use linearity of the conditional expectation. Then (3) follows.

(4) The second equality follows from (3) with $Y = \mathbb{E}[X|\mathscr{H}]$ and $X = 1$. Let $A \in \mathscr{G}$. Then $A \in \mathscr{H}$ and

$$\mathbb{E}[1_A \mathbb{E}[\mathbb{E}[X|\mathscr{G}]|\mathscr{H}]] = \mathbb{E}[1_A \mathbb{E}[X|\mathscr{G}]] = \mathbb{E}[1_A X] = \mathbb{E}[1_A \mathbb{E}[X|\mathscr{H}]].$$

Thus (4) follows.

(5) This follows from (1) and (2).

(6) Trivially, $\mathbb{E}[X]$ is $\mathscr{G}$-measurable. Let $A \in \mathscr{G}$. Since X and 1_A are independent, we see that $\mathbb{E}[\mathbb{E}[X|\mathscr{G}]1_A] = \mathbb{E}[X1_A] = \mathbb{E}[X]\mathbb{E}[1_A] = \mathbb{E}[\mathbb{E}[X]1_A]$. Then (6) is proven. □

The next lemma is of key importance.

Lemma 2.38. *Let $(\Omega, \mathscr{F}, P)$ be a probability space and $(S, \mathscr{S})$ a measurable space. Let, moreover, X be an S-valued random variable on Ω, and Y a real-valued $\sigma(X)$-measurable random variable on Ω. Then there exists an $\mathscr{S}$-measurable function f on S such that $Y = f \circ X$ a. s.*

Proof. It suffices to show the lemma for nonnegative Y. Let $\mathscr{L}$ be the set of real-valued $\sigma(X)$-measurable random variables on Ω satisfying the assumption of the lemma. If $A \in \sigma(X)$, then $1_A \in \mathscr{L}$, and all $\sigma(X)$-measurable step functions belong to $\mathscr{L}$. Since Y is $\sigma(X)$-measurable, there exists a sequence $(\phi_n)_{n\in\mathbb{N}}$ of nondecreasing step functions such that $\phi_n \to Y$ almost surely. Since $\phi_n \in \mathscr{L}$, there exists a sequence $(f_n)_{n\in\mathbb{N}}$ of $\mathscr{S}$-measurable functions such that $\phi_n = f_n \circ X$. Define the function f on S by $f(x) = \limsup_{n\to\infty} f_n(x)$ for $x \in S$. f is $\mathscr{S}$-measurable and

$$f(X(\omega)) = \limsup_{n\to\infty} f_n \circ X(\omega) = \lim_{n\to\infty} f_n \circ X(\omega) = Y(\omega),$$

and hence $Y \in \mathscr{L}$. □

Let X be a real-valued random variable on $(\Omega, \mathscr{F}, P)$ and F a measurable function on it. The random variable $\mathbb{E}[F|\sigma(X)]$ is a $\sigma(X)$-measurable function and, by Lemma 2.38, it can be expressed as $\mathbb{E}[F|\sigma(X)] = f(X)$, with a Borel measurable function f. This function is usually denoted by

$$f(x) = \mathbb{E}[F|X = x]. \tag{2.1.42}$$

Note that if $f(X) = k(X) = \mathbb{E}[F|\sigma(X)]$ a. s., then $f(x) = k(x)$ for $P \circ X^{-1}$-a. e. $x \in \mathbb{R}$, so the ambiguity in the expression (2.1.42) has measure zero under $P \circ X^{-1}$. Let Y be another random variable on $(\Omega, \mathscr{F}, P)$, and h, g Borel measurable functions on $\mathbb{R}$. Since $\mathbb{E}[h(Y)g(X)] = \mathbb{E}[h(Y)\mathbb{E}[g(X)|\sigma(Y)]]$ by the tower property of conditional expectations, using the notation (2.1.42) we have

$$\mathbb{E}[h(Y)g(X)] = \int_{\mathbb{R}^d} h(x)\mathbb{E}[g(X)|Y = x]\mu_Y(dx), \tag{2.1.43}$$

where $\mu_Y(dx)$ denotes the distribution of Y on $\mathbb{R}$.

The conditional expectation $\mathbb{E}[1_A|\mathscr{G}]$ is

$$P(A|\mathscr{G}) = \mathbb{E}[1_A|\mathscr{G}],$$

i. e., the *conditional probability* with respect to $\mathscr{G}$. It is in general not obvious that $P(\cdot|\mathscr{G})(\omega)$ defines a probability measure for almost every ω, since $P(A|\mathscr{G})(\cdot)$ is defined only almost surely for each A. Let $P(A|\mathscr{G})(\omega)$ be defined on $\omega \in \Omega \setminus N_A$ with $P(N_A) = 0$. Hence $P(\cdot|\mathscr{G})(\omega) : \mathscr{F} \to [0,1]$ is defined only for $\omega \in \Omega \setminus \bigcup_{A\in\mathscr{F}} N_A$. It is, however, in general not clear whether $P\left(\bigcup_{A\in\mathscr{F}} N_A\right) = 0$. This justifies the following definition.

Definition 2.39 (Regular conditional probability measure). Let $(\Omega, \mathscr{F}, P)$ be a probability space and $\mathscr{G}$ a sub-σ-field of $\mathscr{F}$. A function $Q : \Omega \times \mathscr{F} \to [0,1]$ is called a *regular conditional probability measure* for $\mathscr{F}$ given $\mathscr{G}$ if
(1) for every $\omega \in \Omega$, $Q(\omega, \cdot)$ is a probability measure on $(\Omega, \mathscr{F})$;
(2) for every $A \in \mathscr{F}$, $Q(\cdot, A)$ is $\mathscr{G}$-measurable;
(3) for every $A \in \mathscr{F}$, $Q(\omega, A) = P(A|\mathscr{G})(\omega)$, P-a. s.

For the purposes below, we will be particularly interested in the case where the sub-σ-field is generated by a random variable X, and give another variant of this definition. We write $\mathbb{E}[1_A|X = x] = P(A|X = x)$.

Definition 2.40 (Regular conditional probability measure). Let $(\Omega, \mathscr{F}, P)$ be a probability space and X a random variable on it with values in a measure space $(S, \mathscr{S})$. A function $Q : S \times \mathscr{F} \to [0,1]$ is called a *regular conditional probability measure* for $\mathscr{F}$ given X if
(1) for every $x \in S$, $Q(x, \cdot)$ is a probability measure on $(\Omega, \mathscr{F})$;
(2) for every $A \in \mathscr{F}$, $Q(\cdot, A)$ is $\mathscr{S}$-measurable;
(3) for every $A \in \mathscr{F}$, $Q(x, A) = P(A|X = x)$, $P \circ X^{-1}$-a. s.

Theorem 2.41 (Existence of regular conditional probability measures). *Let $(\Omega, \mathscr{F})$ be a standard measurable space and P a probability measure on $(\Omega, \mathscr{F})$.*
(1) *Let $\mathscr{G}$ be a sub-σ-field of $\mathscr{F}$. Then there exists a regular conditional probability measure $Q(\omega, A)$ for $\mathscr{F}$ given $\mathscr{G}$, and it is unique, i. e., if Q' is another regular conditional probability measure, then there exists $M \in \mathscr{G}$ with $P(M) = 0$, such that $Q(\omega, A) = Q'(\omega, A)$, for $A \in \mathscr{F}$ and $\omega \in \Omega \setminus M$.*
(2) *Let $Q(\omega, \cdot)$ be a regular conditional probability measure for $\mathscr{F}$ given $\mathscr{G}$. In addition to (1), suppose that $\mathscr{H}$ is a countably determined sub-σ-field of $\mathscr{G}$. Then there exists $N \in \mathscr{G}$ with $P(N) = 0$, such that $Q(\omega, A) = 1_A(\omega)$, for $A \in \mathscr{H}$ and $\omega \in \Omega \setminus N$.*
(3) *Let X be a random variable on $(\Omega, \mathscr{F})$ with values in a measurable space $(S, \mathscr{S})$. Then there exists a regular conditional probability measure $Q(x, A)$ for $\mathscr{F}$ given X, and it is*

unique, i. e., if Q' is another regular conditional probability measure, then there exists $M \in \mathcal{S}$ with $P \circ X^{-1}(M) = 0$, such that $Q(x,A) = Q'(x,A)$, for $A \in \mathcal{F}$ and $x \in S \setminus M$.

(4) *In addition to* (3), *suppose that $\mathcal{S}$ is countably determined and $\{x\} \in \mathcal{S}$ for every $x \in S$. Then there exists $N \in \mathcal{S}$ with $P \circ X^{-1}(N) = 0$, such that $Q(x, X^{-1}(B)) = 1_B(x)$ for $B \in \mathcal{S}$ and $x \in S \setminus N$. In particular, $Q(x, X^{-1}(x)) = 1$, for $x \in S \setminus N$.*

Proof. (1) We suppose that $(\Omega, \mathcal{F})$ is Borel isomorphic to $(\mathbb{M}, \mathcal{B}(\mathbb{M}))$; the other cases are more straightforward. Choose $\Omega = \mathbb{M}$ and $\mathcal{F} = \mathcal{B}(\mathbb{M})$. Let $\mathbb{M}_n = \{0,1\}^n$ and $\pi_n : \mathbb{M} \ni \omega = \{\omega_n\}_{n\in\mathbb{N}} \mapsto \{\omega_1, \dots, \omega_n\} \in \mathbb{M}_n$ be a projection. Define $\mathcal{F}_n = \pi_n^{-1}(\mathcal{B}(\mathbb{M}_n))$. We see that $\sigma(\bigcup_{n=1}^{\infty} \mathcal{F}_n) = \mathcal{B}(\mathbb{M})$. Notice that $\mathcal{F}_n$ is an increasing family of finite σ-fields. Note that

$$\int_B 1_A(\omega) dP(\omega) = \int_B P(A|\mathcal{G})(\omega) dP(\omega)$$

for every $B \in \mathcal{G}$. We set $P_n(\omega, A) = P(A|\mathcal{G})(\omega)$ for $A \in \mathcal{F}_n$. The $\mathcal{G}$-measurable function $\omega \mapsto P_n(\omega, A)$ satisfying above identity is uniquely determined for $\omega \notin N(A,n) \in \mathcal{G}$, where $P(N(A,n)) = 0$. Let $N_n = \bigcup_{A\in\mathcal{F}_n} N(A,n) \in \mathcal{G}$. We see that $P(N_n) = 0$ and $P_n(\omega, A)$ is uniquely determined for all $\omega \notin N_n$ and for all $A \in \mathcal{F}_n$. In particular, $P_n(\omega, \cdot)$ is a probability measure on $\mathcal{F}_n$ for $\omega \notin N_n$. Let $N = \bigcup_{n=1}^{\infty} N_n \in \mathcal{G}$; then $P(N) = 0$. We then construct the family of probability spaces $(\mathbb{M}_n, \mathcal{F}_n, P_n(\omega, \cdot))$, $n \in \mathbb{N}$, for every $\omega \notin N$, which satisfies the consistency relation

$$P_n(\omega, A) = P_{n-1}(\omega, A), \quad n \geq 2, \tag{2.1.44}$$

for $A \in \mathcal{F}_{n-1}$. We define the set function $P(\omega, \cdot)$ on $\bigcup_{k=1}^{\infty} \mathcal{F}_k$ by

$$P(\omega, A) = P_n(\omega, A), \quad A \in \mathcal{F}_n.$$

This $P(\omega, \cdot)$ is well defined by (2.1.44). Fix $\omega \notin N$. We extend $P(\omega, \cdot)$ to a measure on $\sigma(\bigcup_{k=1}^{\infty} \mathcal{F}_k)$. Let $A_n \in \bigcup_{k=1}^{\infty} \mathcal{F}_k$ be such that $A_n \supset A_{n+1} \supset \cdots$ and $\bigcap_{n=1}^{\infty} A_n = \emptyset$. By Hopf's extension theorem, it is sufficient to show that $\lim_{n\to\infty} P(\omega, A_n) = 0$ in order to see the existence of an extension of $P(\omega, \cdot)$. Since $\{A_n\}_{n\in\mathbb{N}}$ is a family of closed sets in a compact space $\mathbb{M}$, $\bigcap_{n=1}^{\infty} A_n = \emptyset$ implies that $A_m = \emptyset$ for all $m \geq N$ with some N. Thus $\lim_{n\to\infty} P(\omega, A_n) = 0$. By Hopf's extension theorem, there exists a measure $\bar{P}(\omega, \cdot)$ on $\sigma(\bigcup_{k=1}^{\infty} \mathcal{F}_k)$ such that $\bar{P}(\omega, A) = P(\omega, A)$ for any $A \in \bigcup_{k=1}^{\infty} \mathcal{F}_k$. Let μ be a probability measure on $(\mathbb{M}, \mathcal{B}(\mathbb{M}))$. Define

$$Q(\omega, A) = \begin{cases} \bar{P}(\omega, A), & \omega \notin N, \\ \mu(A), & \omega \in N. \end{cases}$$

This $Q(\omega, \cdot)$ is a probability measure on $(\mathbb{M}, \mathcal{B}(\mathbb{M}))$ for every $\omega \in \mathbb{M}$. For $A \in \bigcup_{k=1}^{\infty} \mathcal{F}_k$, $Q(\omega, A) = P(A|\mathcal{G})(\omega)$ for $\omega \in \mathbb{M} \setminus N$, and $Q(\cdot, A)$ is $\mathcal{G}$-measurable. These facts can be extended to $A \in \mathcal{B}(\mathbb{M})$ by a limiting argument. Then $Q(\cdot,\cdot)$ is a regular conditional probability measure for $\mathcal{B}(\mathbb{M})$ given $\mathcal{G}$. Next we show its uniqueness. Let $Q'(\cdot,\cdot)$ be a regular conditional probability measure for $\mathcal{B}(\mathbb{M})$ given $\mathcal{G}$, and consider

$$M = \Big\{\omega \mid Q(\omega, A) \neq Q'(\omega, A) \text{ for some } A \in \bigcup_{k=1}^{\infty} \mathcal{F}_k\Big\} \in \mathcal{G}.$$

We have $P(M) = 0$, since for $A \in \bigcup_{k=1}^{\infty} \mathcal{F}_k$ and $\omega \notin N$, $P(A|\mathcal{G})(\omega)$ is unique and it follows that $P(A|\mathcal{G})(\omega) = Q(\omega, A) = Q'(\omega, A)$. Hence we can conclude that $Q(\omega, A) = Q'(\omega, A)$ for $A \in \mathcal{B}(\mathbb{M})$ and for $\omega \notin M$ by a limiting argument, and uniqueness follows.

(2) Let $\mathcal{H}_0 \subset \mathcal{H}$ be a countable set. Since

$$\int_B 1_A(\omega) dP(\omega) = \int_B P(A|\mathcal{G})(\omega) dP(\omega) = \int_B Q(\omega, A) dP(\omega)$$

for $B \in \mathcal{G}$, for every $A \in \mathcal{H}_0$ there exists $N_A \in \mathcal{G}$ with $P(N_A) = 0$ such that $Q(\omega, A) = 1_A(\omega)$ for $\omega \notin N_A$. Both $Q(\omega, A)$ and $1_A(\omega)$ define probability measures on $\mathcal{H}$ and $Q(\omega, A) = 1_A(\omega)$ on $\mathcal{H}_0$ for $\omega \notin N_A$. We have $Q(\omega, A) = 1_A(\omega)$ on $\mathcal{H}$ for $\omega \notin N_A$. Let $N = \bigcup_{A \in \mathcal{H}_0} N_A$. Then $P(N) = 0$ and $P(\omega, A) = 1_A(\omega)$, for all $A \in \mathcal{H}$ and $\omega \notin N$.

(3) The proof is a minor modification of part (1). Choose $\Omega = \mathbb{M}$, $\mathcal{F} = \mathcal{B}(\mathbb{M})$, and note that

$$\int_B 1_A(\omega) dP(\omega) = \int_B P(A|\sigma(X))(\omega) dP(\omega) = \int_{X(B)} P(A|X = x) dP \circ X^{-1}(x)$$

for $B \in \sigma(X)$. Write $P_n(x, A) = P(A|X = x)$ for $A \in \mathcal{F}_n$. An $\mathcal{S}$-measurable function $x \mapsto P_n(x, A)$ is uniquely determined for $x \notin N(A, n) \in \mathcal{S}$, where $P \circ X^{-1}(N(A, n)) = 0$. Let $N_n = \bigcup_{A \in \mathcal{F}_n} N(A, n) \in \mathcal{S}$. We see that $P \circ X^{-1}(N_n) = 0$ and $P_n(x, A)$ is uniquely determined for all $x \notin N_n$ and $A \in \mathcal{F}_n$. In particular, $P_n(x, \cdot)$ is a probability measure on $\mathcal{F}_n$ for $x \notin N_n$. Let $N = \bigcup_{n=1}^{\infty} N_n \in \mathcal{S}$, and then $P \circ X^{-1}(N) = 0$. We then construct the family of probability spaces $(\mathbb{M}_n, \mathcal{F}_n, P_n(x, \cdot))$, $n \in \mathbb{N}$, for each $x \notin N$, which satisfies the consistency relation

$$P_n(x, A) = P_{n-1}(x, A), \quad n \geq 2, \tag{2.1.45}$$

for $A \in \mathcal{F}_{n-1}$. Define the set function $P(x, \cdot)$ on $\bigcup_{k=1}^{\infty} \mathcal{F}_k$ by

$$P(x, A) = P_n(x, A), \quad A \in \mathcal{F}_n.$$

Then $P(x, \cdot)$ is well defined by (2.1.45). Fix $x \notin N$. We can extend $P(x, \cdot)$ to a measure on $\sigma(\bigcup_{k=1}^{\infty} \mathcal{F}_k)$ by Hopf's extension theorem in a similar manner to (1). Let μ be a probability measure on $(\mathbb{M}, \mathcal{B}(\mathbb{M}))$. Define

$$Q(x,A) = \begin{cases} \bar{P}(x,A), & x \notin N, \\ \mu(A), & x \in N. \end{cases}$$

The set function $Q(x,\cdot)$ is a probability measure on $(\mathbb{M}, \mathcal{B}(\mathbb{M}))$ for every $x \in \mathbb{M}$. For $A \in \bigcup_{k=1}^{\infty} \mathcal{F}_k$, we have $Q(x,A) = P(A|X = x)$ for $x \notin N$ and $Q(\cdot,A)$ is $\mathcal{S}$-measurable. These facts can be extended for $A \in \mathcal{B}(\mathbb{M})$ by a limiting argument. Then $Q(\cdot,\cdot)$ is a regular conditional probability measure for $\mathcal{B}(\mathbb{M})$ given X, and we can show its uniqueness in the same way as in (1).

(4) The proof is similar to part (2). Note that

$$\int_{X^{-1}(B)} 1_A(X(\omega))dP(\omega) = \int_{X^{-1}(B)} 1_{X^{-1}(A)}(\omega)dP(\omega) = \int_B Q(x,X^{-1}(A))dP \circ X^{-1}(x)$$

for $B \in \mathcal{S}$. On the other hand, we have

$$\int_{X^{-1}(B)} 1_A(X(\omega))dP(\omega) = \int_B 1_A(x)dP \circ X^{-1}(x)$$

for $B \in \mathcal{S}$. Let $\mathcal{S}_0 \subset \mathcal{S}$ be a countable set. If $A \in \mathcal{S}_0$, then there exists $N_A \in \mathcal{S}$ with $P \circ X^{-1}(N_A) = 0$ such that $Q(x,X^{-1}(A)) = 1_A(x)$ for $x \notin N_A$. Both $Q(x,X^{-1}(A))$ and $1_A(x)$ define probability measures on $\mathcal{S}$, and $Q(x,X^{-1}(A)) = 1_A(x)$ on $\mathcal{S}_0$ for $x \notin N_A$. We see that $Q(x,X^{-1}(A)) = 1_A(x)$ on $\mathcal{H}$ for $x \notin N_A$. Set $N = \bigcup_{A \in \mathcal{H}_0} N_A$. Then $P \circ X^{-1}(N) = 0$ and $Q(x,X^{-1}(A)) = 1_A(x)$ for all $A \in \mathcal{S}$ and $x \notin N$. □

Intuitively, (2) and (4) of Theorem 2.41 suggest that the supports of $Q(\omega,\cdot)$ and $Q(x,\cdot)$ are $\{\omega\}$ and $X^{-1}(x)$, respectively. In what follows, the regular conditional probability measures $Q(\omega,A)$ and $Q(x,A)$ will be denoted by $P(A|\mathcal{G})$ and $P(A|X = x)$, respectively, unless any confusion may arise.

Remark 2.42.

(1) It is well known that a Polish space with a Borel σ-field is a standard measurable space, and every measurable subset of a standard measurable space with the induced σ-field is also a standard measurable space.

(2) If $(\Omega, \mathcal{F})$ is a standard measurable space, then $\mathcal{F}$ is countably determined, and $\{x\} \in \mathcal{F}$ for every $x \in \Omega$.

Corollary 2.43. *Let $(\Omega, \mathcal{F}, P)$ be a probability space such that Ω is a Polish space, and $\mathcal{F} = \mathcal{B}(\Omega)$. Let $\mathcal{G}$ be a sub-σ-field of $\mathcal{F}$, S also a Polish space, and $\mathcal{S} = \mathcal{B}(S)$. Consider the random variable $X : \Omega \to S$. The following hold:*

(1) (a) *There exists a regular conditional probability measure $Q(\omega,A)$ for $\mathcal{F}$ given $\mathcal{G}$, and it is unique.*

(b) *There exists $N \in \mathcal{G}$ with $P(N) = 0$, such that $Q(\omega,A) = 1_A(\omega)$ for $A \in \mathcal{G}$ and $\omega \in \Omega \setminus N$.*

(2) *There exists a unique regular conditional probability measure $Q(x, A)$ for $\mathscr{F}$ given X, such that $Q(x, X^{-1}(B)) = 1_B(x)$ for $B \in \mathscr{S}$ and $x \in S \setminus N$. In particular, $Q(x, X^{-1}(x)) = 1$, for $x \in S \setminus N$.*

Proof. Since $(\Omega, \mathscr{F})$ is a standard measurable space, $x \in S$ implies $\{x\} \in \mathscr{S}$, and both $\mathscr{F}$ and $\mathscr{S}$ are countably determined by Remark 2.42, thus the corollary follows directly from Theorem 2.41. □

Example 2.44. Let Ω and S be Polish spaces, and $(\Omega, \mathscr{F})$ and $(S, \mathscr{S})$ be measurable spaces, where both $\mathscr{F}$ and $\mathscr{S}$ are Borel σ-fields. Consider a probability measure P on $(\Omega, \mathscr{F})$ and a random variable $X : \Omega \to S$. Let $Q(x, \cdot)$, $x \in S$, be the related regular conditional probability measure, and dP_X the distribution of X on S. We see that

$$\int_\Omega 1_A(\omega) dP(\omega) = \int_S Q(x, A) dP_X(x) = \int_S dP_X(x) \int_\Omega 1_A(\omega) Q(x, d\omega).$$

Write $Q(x, d\omega) = P^X(d\omega)$, i. e., $\mathbb{E}_P^X[\dots] = \int_\Omega \dots Q(x, d\omega)$. Hence

$$\mathbb{E}_P[f] = \int_S \mathbb{E}_P^X[f] dP_X(x) \quad \text{and} \quad \mathbb{E}_P^X[f] = \mathbb{E}_P[f|X = x],$$

for bounded Borel measurable functions f on S.

2.2 Random processes

2.2.1 Basic concepts and facts

A random process is a mathematical model of a phenomenon evolving in time in a way which cannot be predicted with certainty. Let $(\Omega, \mathscr{F}, P)$ be a probability space and J be a given set. A family of S-valued random variables $(X_t)_{t \in J}$ is called a *random process with index set* J. When $S = \mathbb{R}^d$, we refer to it as a real-valued random process, no matter the dimension. If J is uncountable, then $(X_t)_{t \geq 0}$ is called a *continuous-time random process*, and if J is countable, then $(X_n)_{n \in \mathbb{N}}$ or $(X_n)_{n \in \dot{\mathbb{N}}}$ is called a *discrete-time random process*. Here we recall $\dot{\mathbb{N}} = \mathbb{N} \cup \{0\}$. For any fixed $t \in J$, the map $\omega \mapsto X_t(\omega)$ is an S-valued random variable on Ω, while for any fixed $\omega \in \Omega$ the map $t \mapsto X_t(\omega)$ is a function called *(sample) path*. This distinction justifies for any random process to address distributional properties, on the one hand, and sample path properties, on the other.

Let $(\Omega, \mathscr{F}, P)$ be a probability space and $X = (X_1, \dots, X_d)$ an $\mathbb{R}^d$-valued random variable on it. Define the measurable function $F : \mathbb{R}^d \to \mathbb{R}$ by

$$F(x_1, \dots, x_d) = P(X_1 \leq x_1, \dots, X_d \leq x_d) = P \circ X^{-1}((-\infty, x_1] \times \dots \times (-\infty, x_d]). \tag{2.2.1}$$

This is called the d-dimensional *distribution function* of X. Recall the more general definition.

Definition 2.45 (Distribution functions). The function $F : \mathbb{R}^d \to \mathbb{R}$ is called a *d-dimensional distribution function* if

(1) $F(x_1, \ldots, x_n)$ is right-continuous in each variable when the others are fixed;

(2) if $(a_1, \ldots, a_d)$ and $(b_1, \ldots, b_d)$ are points in $\mathbb{R}^d$ with $a_k < b_k$, then

$$\prod_{k=1}^{d} \Delta(a_k, b_k)F(x) \geq 0,$$

where $\Delta(a_k, b_k)F(x) = F(x_1, \ldots, \overset{k\text{th}}{b_k}, \ldots, x_d) - F(x_1, \ldots, \overset{k\text{th}}{a_k}, \ldots, x_d)$;

(3) if $x_j \downarrow -\infty$ for some j, then $F(x_1, \ldots, x_d) \downarrow 0$;

(4) if $x_j \uparrow +\infty$ for all $1 \leq j \leq d$, then $F(x_1, \ldots, x_d) \uparrow 1$.

By this definition, the function F of X defined by (2.2.1) is a distribution function. Conversely, for any d-dimensional distribution function F, we define

$$\tilde{P}_F((a_1, b_1] \times \cdots \times (a_d, b_d]) = \int_{\mathbb{R}^d} \prod_{k=1}^{d} \Delta(a_k, b_k)F(x)dx$$

and it is known that $\tilde{P}_F$ can be extended to a measure P_F on $(\mathbb{R}^d, \mathcal{B}(\mathbb{R}^d))$. The probability measure P_F on $\mathbb{R}^d$ is a *Lebesgue–Stieltjes measure*, denoted by $dF(x)$ instead of $P_F(dx)$. Let P_X be the distribution of X on $\mathbb{R}^d$ and F be defined by (2.2.1). We have

$$\mathbb{E}[f(X)] = \int_{\mathbb{R}^d} f(x)P_X(dx) = \int_{\mathbb{R}^d} f(x)dF(x). \tag{2.2.2}$$

Conversely, let F be a distribution function. It is known that there is a probability space and a d-dimensional random variable X such that its distribution function is F. A stronger statement is as follows.

A sequence $(F_n(x_1, \ldots, x_n)_n)_{n\in\mathbb{N}}$ of distribution functions is said to satisfy a *consistency relation* if

$$\lim_{x_n \uparrow \infty} F_n(x_1, \ldots, x_n) = F_{n-1}(x_1, \ldots, x_{n-1}), \quad n \geq 2. \tag{2.2.3}$$

Corollary 2.46. *Let $(F_n(x_1, \ldots, x_n))_{n\in\mathbb{N}}$ be a sequence of distribution functions satisfying the consistency relation (2.2.3). Then there exists a probability space $(\Omega, \mathscr{F}, P)$ and a discrete-time random process $(X_n)_{n\in\mathbb{N}}$ such that $P(X_1 \leq x_1, \ldots, X_n \leq x_n) = F_n(x_1, \ldots, x_n)$ for all $n \geq 1$.*

Proof. For $\Lambda_1 \subset \Lambda_2 \subset \mathbb{N}$, let $\pi_{\Lambda_1\Lambda_2} : \mathbb{R}^{|\Lambda_2|} \to \mathbb{R}^{|\Lambda_1|}$ be the projection defined as (2.1.20) with J replaced by the countable set $\mathbb{N}$. Let $\Omega = \mathbb{R}^{\mathbb{N}} = \{\omega : \mathbb{N} \to \mathbb{R}\}$ and

$$\mathscr{A} = \{\pi_{\Lambda\mathbb{N}}^{-1}(E) \subset \mathbb{R}^{\mathbb{N}} \mid E \in \mathcal{B}(\mathbb{R}^{|\Lambda|}), \Lambda \subset \mathbb{N}, |\Lambda| < \infty\}.$$

For a finite index set $\Lambda = \{j_1, \dots, j_n\}$, we define the probability measure

$$\mu_\Lambda(E) = \int_{\mathbb{R}^n} 1_E(x_1, \dots, x_n) dF_n$$

for $E \in \mathcal{B}(\mathbb{R}^n)$. By the consistency of $(F_n)_{n\in\mathbb{N}}$, it is straightforward to check that for $\Lambda_1 \subset \Lambda_2$,

$$\mu_{\Lambda_2}(\pi_{\Lambda_1\Lambda_2}^{-1}(E)) = \mu_{\Lambda_1}(E) \tag{2.2.4}$$

for $E \in \mathcal{B}(\mathbb{R}^{|\Lambda_1|})$. Define $\tilde{P} : \mathscr{A} \to [0, \infty)$ by $\tilde{P}(\pi_{\Lambda\mathbb{N}}^{-1}(E)) = \mu_\Lambda(E)$ for $E \in \mathcal{B}(\mathbb{R}^{|\Lambda|})$. Since (2.2.4) implies that the set of measures $\{\mu_\Lambda\}_{\Lambda\subset\mathbb{N},|\Lambda|<\infty}$ satisfies the Kolmogorov's consistency relation, there exists a probability measure P on $(\Omega, \sigma(\mathscr{A}))$ such that $P(\pi_{\Lambda\mathbb{N}}^{-1}(E)) = \mu_\Lambda(E)$ by the Kolmogorov's extension theorem. Define the random process $(X_n)_{n\in\mathbb{N}}$ on $(\Omega, \sigma(\mathscr{A}), P)$ by $X_n(\omega) = \omega(n)$ for $\omega = \{\omega(n)\}_{n\in\mathbb{N}} \in \mathbb{R}^{\mathbb{N}}$. We see that

$$\begin{aligned} P(X_1 \le a_1, \dots, X_n \le a_n) &= \mu_{\{1,\dots,n\}}(X_1^{-1}((-\infty, a_1]) \times \dots \times X_n^{-1}((-\infty, a_n])) \\ &= \int_{\mathbb{R}^n} \prod_{j=1}^n 1_{(-\infty,a_j]}(x_j) dF_n = F_n(a_1, \dots, a_n). \end{aligned}$$

Thus the corollary follows. □

We also have the continuous version of Corollary 2.46.

Corollary 2.47. *Let $(F_\Lambda)_{\Lambda\subset[0,\infty),|\Lambda|<\infty}$ be a sequence of distribution functions satisfying the consistency relation, i. e.,*

$$\lim_{x_n\uparrow\infty} F_{\{t_1,\dots,t_n\}}(x_1, \dots, x_n) = F_{\{t_1,\dots,t_{n-1}\}}(x_1, \dots, x_{n-1}) \tag{2.2.5}$$

for arbitrary $\Lambda = \{t_1, \dots, t_n\} \subset [0, \infty)$. There exist a probability space $(\Omega, \mathscr{F}, P)$ and a random process $(X_t)_{t\ge0}$ such that

$$P(X_{t_1} \le x_1, \dots, X_{t_n} \le x_n) = F_{\{t_1,\dots,t_n\}}(x_1, \dots, x_n)$$

for all $\Lambda = \{t_1, \dots, t_n\} \subset [0, \infty)$.

Proof. The proof is similar to that of Corollary 2.46. For $\Lambda_1 \subset \Lambda_2 \subset [0, \infty)$, let $\pi_{\Lambda_1\Lambda_2} : \mathbb{R}^{|\Lambda_2|} \to \mathbb{R}^{|\Lambda_1|}$ be the projection defined as (2.1.20) with J replaced by $[0, \infty)$. Let $\Omega = \mathbb{R}^{[0,\infty)} = \{\omega : [0, \infty) \to \mathbb{R}\}$ and

$$\mathscr{A} = \{\pi_{\Lambda[0,\infty)}^{-1}(E) \subset \mathbb{R}^{[0,\infty)} \mid E \in \mathcal{B}(\mathbb{R}^{|\Lambda|}), \Lambda \subset [0, \infty), |\Lambda| < \infty\}. \tag{2.2.6}$$

For the finite index set $\Lambda = \{t_1, \dots, t_n\}$, we define the set function $\tilde{\mu}_\Lambda$ on $\mathbb{R}^n$ by

$$\tilde{\mu}_\Lambda(E) = \int_{\mathbb{R}^n} 1_E dF_{\{t_1,\dots,t_n\}}.$$

The set function $\tilde{\mu}_\Lambda$ can be uniquely extended to a probability measure μ_Λ on $\mathcal{B}(\mathbb{R}^n)$. By the consistency of $\{F_\Lambda\}_{\Lambda\subset[0,\infty),|\Lambda|<\infty}$, we can also check that the Kolmogorov's consistency relation $\mu_{\Lambda_2}(\pi^{-1}_{\Lambda_1\Lambda_2}(E)) = \mu_{\Lambda_1}(E)$ holds for all $E \in \mathcal{B}(\mathbb{R}^{|\Lambda_1|})$. Define $\tilde{P} : \mathscr{A} \to [0,\infty)$ by $\tilde{P}(\pi^{-1}_{\Lambda[0,\infty)}(E)) = \mu_\Lambda(E)$ for $E \in \mathcal{B}(\mathbb{R}^{|\Lambda|})$. Since the set of measures $\{\mu_\Lambda\}_{\Lambda\subset[0,\infty),|\Lambda|<\infty}$ satisfies the Kolmogorov's consistency relation, there exists a probability measure P on $(\Omega, \sigma(\mathscr{A}))$ such that $P(\pi^{-1}_{\Lambda[0,\infty)}(E)) = \mu_\Lambda(E)$. Define the random process $(X_t)_{t\geq 0}$ on $(\Omega, \sigma(\mathscr{A}), P)$ by $X_t(\omega) = \omega(t)$ for $\omega \in \mathbb{R}^{[0,\infty)}$. We see that

$$P(X_{t_1} \leq a_1, \dots, X_{t_n} \leq a_n) = F_{\{t_1,\dots,t_n\}}(a_1, \dots, a_n)$$

and the corollary follows. □

The random processes $(X_t)_{t\in J}$ and $(Y_t)_{t\in J}$ on $(\Omega, \mathscr{F}, P)$ are called *versions* of each other if for every $t \geq 0$ there is a null set $N_t \in \mathscr{F}$ such that $X_t(\omega) = Y_t(\omega)$, for all $\omega \notin N_t$. When $t \mapsto Y_t(\omega)$ is continuous for almost every $\omega \in \Omega$, the process $(Y_t)_{t\in J}$ is called a *continuous version* of $(X_t)_{t\in J}$. In studying continuous versions, the following stronger concept will be useful.

Definition 2.48 (Hölder continuity). A function $f : [0,\infty) \to \mathbb{R}$ is called *locally γ-Hölder continuous* if for every compact interval $[S,T] \subset [0,\infty)$ there exists a constant $C = C(f, [S,T])$ such that

$$|f(s) - f(t)| \leq C|s-t|^\gamma$$

for all $s, t \in [S,T]$. The number γ is called the *Hölder exponent* and C is the *Hölder constant*.

Theorem 2.49 (Kolmogorov–Čentsov). *Let $(\Omega, \mathscr{F}, P)$ be a probability space and $(X_t)_{t\geq 0}$ a random process on it. Suppose that there exist $\alpha \geq 1$, $\beta > 0$, and $C > 0$ such that*

$$\mathbb{E}[|X_t - X_s|^\alpha] \leq C|t-s|^{1+\beta}. \tag{2.2.7}$$

Then there exists a version $(Y_t)_{0\leq t\leq T}$ of $(X_t)_{0\leq t\leq T}$ for every $T \geq 0$, with γ-Hölder continuous paths, where $\gamma \in (0, \beta/\alpha)$.

Proof. Suppose that $T = 1$ for simplicity, and let $\varepsilon > 0$ be arbitrary. Since

$$P(|X_t - X_s| \geq \varepsilon) \leq \varepsilon^{-\alpha}\mathbb{E}[|X_t - X_s|^\alpha] \leq C\varepsilon^{-\alpha}|t-s|^{1+\beta},$$

$X_s \to X_t$ as $s \to t$ in probability. We have

$$P(|X_{k/2^n} - X_{(k-1)/2^n}| \geq 2^{-\gamma n}) \leq C2^{\alpha\gamma n}(1/2^n)^{1+\beta} = C2^{-n(1+\beta-\alpha\gamma)}.$$

Hence

$$P\Big(\max_{1\leq k\leq 2^n} |X_{k/2^n} - X_{(k-1)/2^n}| \geq 2^{-\gamma n}\Big) = P\Big(\bigcup_{1\leq k\leq 2^n} \{|X_{k/2^n} - X_{(k-1)/2^n}| \geq 2^{-\gamma n}\}\Big) \leq C2^{-n(\beta-\alpha\gamma)}$$

and

$$\sum_{n=1}^{\infty} P\Big(\max_{1\leq k\leq 2^n} |X_{k/2^n} - X_{(k-1)/2^n}| \geq 2^{-\gamma n}\Big) \leq C\sum_{n=1}^{\infty} 2^{-n(\beta-\alpha\gamma)} < \infty. \tag{2.2.8}$$

Note that the exponent on the right-hand side of (2.2.8) is $\beta - \alpha\gamma > 0$ by the assumption. This implies by the Borel–Cantelli lemma that

$$P\Big(\bigcup_{m=1}^{\infty}\bigcap_{n=m}^{\infty}\{\max_{1\leq k\leq 2^n} |X_{k/2^n} - X_{(k-1)/2^n}| \geq 2^{-\gamma n}\}^c\Big) = 1.$$

There exists $\Omega_0 \in \mathscr{F}$ such that $P(\Omega_0) = 1$, and for all $\omega \in \Omega_0$ there exists $n_0 = n_0(\omega)$ such that

$$\max_{1\leq k\leq 2^n} |X_{k/2^n}(\omega) - X_{(k-1)/2^n}(\omega)| < 2^{-\gamma n} \tag{2.2.9}$$

for all $n > n_0$. For each n, let $D_n = \{k/2^n \mid k = 0, 1, \ldots, 2^n\}$ and $D = \bigcup_{n=1}^{\infty} D_n \subset [0,1]$ be the set of dyadic rationals in $[0,1]$. We fix $\omega \in \Omega_0$ and $n > n_0$, and show that for every $m > n$,

$$|X_t(\omega) - X_s(\omega)| \leq 2\sum_{j=n+1}^{m} 2^{-\gamma j} \tag{2.2.10}$$

for arbitrary $s, t \in D_m$ with $0 < t - s < 2^{-n}$. The estimate (2.2.10) can be proven as follows. For $m = n + 1$, we only have $t = k/2^m$ and $s = (k-1)/2^m$, and thus it follows from (2.2.9). For $m > n + 1$, it can be shown by induction. Suppose that (2.2.10) is valid for $m = n + 1, \ldots, M - 1$. Take $s, t \in D_M$ such that $s < t$ and consider the numbers $t^1 = \max\{u \in D_M \mid u \leq t\}$ and $s^1 = \min\{u \in D_{M-1} \mid u \geq s\}$. Note that $s \leq s^1 \leq t^1 \leq t$, $s^1 - s \leq 2^{-M}$ and $t - t^1 \leq 2^{-M}$. From (2.2.9) we have

$$|X_{s^1}(\omega) - X_s(\omega)| \leq 2^{-\gamma M}, \quad |X_t(\omega) - X_{t^1}(\omega)| \leq 2^{-\gamma M},$$

and from (2.2.10) with $m = M - 1$ we have

$$|X_{t^1}(\omega) - X_{s^1}(\omega)| \leq 2\sum_{j=n+1}^{M-1} 2^{-\gamma j}.$$

Thus

$$|X_t(\omega) - X_s(\omega)| \leq |X_t(\omega) - X_{t^1}(\omega)| + |X_{t^1}(\omega) - X_{s^1}(\omega)| + |X_{s^1}(\omega) - X_s(\omega)| \leq 2\sum_{j=n+1}^{M} 2^{-\gamma j},$$

leading to (2.2.10). For every $s, t \in D$ with $0 < t - s < 2^{-n_0(\omega)}$, choose $n \geq n_0(\omega)$ such that $2^{-(n+1)} \leq t - s \leq 2^{-n}$. We have from (2.2.10) that

$$|X_t(\omega) - X_s(\omega)| \leq \frac{2}{1-2^{-\gamma}} 2^{-\gamma(n+1)} \leq \frac{2}{1-2^{-\gamma}} |t-s|^\gamma \tag{2.2.11}$$

for $0 < t - s < 2^{-n_0(\omega)}$. This implies that $(X_t)_{t\in D}$ is uniformly Hölder continuous for every $\omega \in \Omega_0$. Define

$$\tilde{X}_t(\omega) = \begin{cases} 0, & \omega \notin \Omega_0, \\ X_t(\omega), & \omega \in \Omega_0, t \in D, \\ \lim_{n\to\infty} X_{s_n}(\omega), & \omega \in \Omega_0, t \notin D. \end{cases} \tag{2.2.12}$$

Here $(s_n)_{n\in\mathbb{N}} \subset D$ is a sequence of dyadic rationals such that $s_n \to t$ as $n \to \infty$. Note that $X_{s_n}(\omega)$ is a Cauchy sequence by (2.2.11), and there exists $\lim_{n\to\infty} X_{s_n}(\omega)$, which is independent of the choice of s_n. Hence we have

$$|\tilde{X}_t(\omega) - \tilde{X}_s(\omega)| \leq \frac{2}{1-2^{-\gamma}} |t-s|^\gamma$$

for $|t - s| < 2^{-n_0(\omega)}$. Thus $(\tilde{X}_t)_{t\geq 0}$ is a continuous random process. Note that $X_t(\omega) = \tilde{X}_t(\omega)$ for $t \in D$ and $\omega \in \Omega_0$, and $X_{s_n} \to X_t$ in probability, which implies that there exists a subsequence s'_n such that $X_{s'_n}$ converges almost surely to X_t. On the other hand, $X_{s'_n} \to \tilde{X}_t$ by the definition. Hence $X_t = \tilde{X}_t$ almost surely for every $t \in [0,1]$, and $(\tilde{X}_t)_{t\in[0,1]}$ is a continuous version of $(X_t)_{t\in[0,1]}$. □

2.2.2 Martingale properties

Martingales are a class of random processes with particularly convenient properties. In this section we briefly review some basic definitions and facts.

Definition 2.50 (Filtration). Let $(\Omega, \mathscr{F})$ be a measurable space and $(\mathscr{F}_t)_{t\geq 0}$ a family of sub-σ-fields of $\mathscr{F}$. The collection $(\mathscr{F}_t)_{t\geq 0}$ is called a *filtration* whenever $\mathscr{F}_s \subset \mathscr{F}_t$ for $s \leq t$. The given measurable space endowed with a filtration is called a *filtered space*. A filtration $(\mathscr{F}_t)_{t\geq 0}$ is called *right-continuous* if $\mathscr{F}_t = \bigcap_{s>t} \mathscr{F}_s$. A filtered space is denoted by $(\Omega, \mathscr{F}, (\mathscr{F}_t)_{t\geq 0})$.

Intuitively, $\mathscr{F}_t$ contains the information known to an observer at time t. Let $(X_t)_{t\geq 0}$ be an S-valued random process. A convenient choice of filtration is the *natural filtration* given by $\mathscr{F}_t^X = \sigma(X_s, 0 \leq s \leq t)$, containing the information accumulated by observing X_s up to time $s = t$.

Definition 2.51 (Adapted process). Let $(\Omega, \mathscr{F}, (\mathscr{F}_t)_{t\geq 0})$ be a filtered space and $(X_t)_{t\geq 0}$ a random process on this space. The process is called $(\mathscr{F}_t)_{t\geq 0}$-*adapted* whenever X_t is $\mathscr{F}_t$-measurable for every $t \geq 0$.

If a process $(X_t)_{t\geq 0}$ is $(\mathscr{F}_t)_{t\geq 0}$-adapted it means that the process does not depend on events at time t other than contained in $\mathscr{F}_t$, in particular, it does not depend on the future. Obviously, $(X_t)_{t\geq 0}$ is adapted to its natural filtration $(\mathscr{F}_t^X)_{t\geq 0}$, or equivalently, the natural filtration $(\mathscr{F}_t^X)_{t\geq 0}$ is the smallest filtration making $(X_t)_{t\geq 0}$ adapted.

Definition 2.52 (Martingale, supermartingale, and submartingale). Consider a filtered probability space $(\Omega, \mathscr{F}, (\mathscr{F}_t)_{t\geq 0}, P)$. The random process $(X_t)_{t\geq 0}$ is called an $(\mathscr{F}_t)_{t\geq 0}$-*martingale* (resp. *supermartingale, submartingale*) whenever
(1) $\mathbb{E}[|X_t|] < \infty$, for every $t \geq 0$;
(2) $(X_t)_{t\geq 0}$ is $(\mathscr{F}_t)_{t\geq 0}$-adapted;
(3) $\mathbb{E}[X_t|\mathscr{F}_s] = X_s$ (resp. $\leq, \geq$), almost surely for every $0 \leq s \leq t$.

The martingale property intuitively means that the best prediction of the future mean of a random process is the current value, i. e., there are no embedded biases or trends in the evolution of the process.

We recall a useful inequality next, which is an extension of the Schwarz inequality $\mathbb{E}[X]^2 \leq \mathbb{E}[X^2], X \in L^2(\Omega, dP)$, to arbitrary convex functions beyond the quadratic power function.

Proposition 2.53 (Jensen's inequality). *Let $(\Omega, \mathscr{F}, P)$ be a probability space and X a random variable on it. Let $\phi : \mathbb{R} \to \mathbb{R}$ be convex, and suppose that $X, \phi(X) \in L^1(\Omega)$. Then $\phi(\mathbb{E}[X|\mathscr{G}]) \leq \mathbb{E}[\phi(X)|\mathscr{G}]$, for every sub-σ-field $\mathscr{G} \subset \mathscr{F}$.*

Proof. By the definition of a convex function, the set $A = \{(x,y) \in \mathbb{R}^2 \mid y \geq \phi(x)\}$ is a convex subset of $\mathbb{R}^2$. Therefore we can find a straight line $\ell_x(y) = \phi(x) + a(y-x)$ through every boundary point $(x, \phi(x))$ of A such that $\{\ell_x(y) \mid y \in \mathbb{R}\}$ does not intersect the interior of A. This means that $\ell_x(y) \leq \phi(y)$, for all $y \in \mathbb{R}$. Choosing $x = \mathbb{E}[X|\mathscr{G}]$ and $y = X$, we obtain the pointwise inequality $\ell_{\mathbb{E}[X|\mathscr{G}]}(X) \leq \phi(X)$. Taking conditional expectations, we find

$$\mathbb{E}[\phi(X)|\mathscr{G}] \geq \mathbb{E}[\ell_{\mathbb{E}(X|\mathscr{G})}(X)|\mathscr{G}] = \mathbb{E}[\phi(\mathbb{E}[X|\mathscr{G}]) + a(X - \mathbb{E}[X|\mathscr{G}])|\mathscr{G}] = \phi(\mathbb{E}[X|\mathscr{G}]).$$

This shows the claim. □

Given a martingale $(X_t)_{t\geq 0}$, by Jensen's inequality it is immediate that $(\phi(X_t))_{t\geq 0}$ is a submartingale for any convex function ϕ if $\phi(X) \in L^1(\Omega, dP)$. In particular, $(X_t^2)_{t\geq 0}$ is a submartingale.

The following observation on conditional expectation will be useful below. Let $(\Omega, \mathscr{F}, P)$ be a probability space, $\mathscr{A} \subset \mathscr{F}$ a sub-σ-field, and $\mathscr{N} = \{A \in \mathscr{F} \mid P(A) = 0\}$. In the next lemma we show that $\mathbb{E}[X|\mathscr{A}]$ coincides with $\mathbb{E}[X|\sigma(\mathscr{A} \cup \mathscr{N})]$.

Lemma 2.54. *Let $(\Omega, \mathcal{F}, P)$ be a probability space. Suppose that $\mathcal{A}, \mathcal{B} \subset \mathcal{F}$ are independent sub-σ-fields. Let X be a random variable on it, and also suppose that $\sigma(\sigma(X) \cup \mathcal{A})$ and $\mathcal{B}$ are independent. Then $\mathbb{E}[X|\sigma(\mathcal{A} \cup \mathcal{B})] = \mathbb{E}[X|\mathcal{A}]$. In particular, $\mathbb{E}[X|\sigma(\mathcal{A} \cup \mathcal{N})] = \mathbb{E}[X|\mathcal{A}]$, where $\mathcal{N} = \{A \in \mathcal{F} \mid P(A) = 0\}$ denotes the family of all null sets.*

Proof. Let $Y = \mathbb{E}[X|\mathcal{A}]$. The random variable Y is $\mathcal{A}$-measurable, and hence $\sigma(\mathcal{A} \cup \mathcal{B})$-measurable. Let $\mathcal{C} = \{A \cap B \mid A \in \mathcal{A}, B \in \mathcal{B}\}$. Since $\mathcal{A}, \mathcal{B} \subset \mathcal{C}$, we can see $\sigma(\mathcal{A} \cup \mathcal{B}) \subset \sigma(\mathcal{C})$. On the other hand, $\mathcal{C}$ is a π-system. Hence the π–λ theorem yields $\sigma(\mathcal{C}) \subset \sigma(\mathcal{A} \cup \mathcal{B})$, and thus $\sigma(\mathcal{C}) = \sigma(\mathcal{A} \cup \mathcal{B})$. Since

$$\mathcal{D} = \{C \in \mathcal{F} \mid \mathbb{E}[1_C X] = \mathbb{E}[1_C Y]\}$$

is a σ-field, in order to show the lemma, it is sufficient to show that

$$\mathbb{E}[1_{A\cap B} X] = \mathbb{E}[1_{A\cap B} Y], \quad A \cap B \in \mathcal{C}. \tag{2.2.13}$$

Indeed, (2.2.13) implies $\mathcal{C} \subset \mathcal{D}$, which yields that $\sigma(\mathcal{C}) \subset \mathcal{D}$ and $\mathbb{E}[1_C X] = \mathbb{E}[1_C Y]$ for all $C \in \sigma(\mathcal{C})$. By the independence of $\mathcal{A}$ and $\mathcal{B}$, it follows that $\mathbb{E}[1_{A\cap B} Y] = \mathbb{E}[1_A 1_B Y] = \mathbb{E}[1_B]\mathbb{E}[1_A Y]$. Since $\mathbb{E}[1_A Y] = \mathbb{E}[1_A X]$, we see that $\mathbb{E}[1_{A\cap B} Y] = \mathbb{E}[1_B]\mathbb{E}[1_A X]$. From the independence of $\mathcal{B}$ and $\sigma(\sigma(X) \cup \mathcal{A})$, we also get $\mathbb{E}[1_B]\mathbb{E}[1_A X] = \mathbb{E}[1_B 1_A X] = \mathbb{E}[1_{A\cap B} X]$. Thus $\mathbb{E}[1_{A\cap B} X] = \mathbb{E}[1_{A\cap B} Y]$ follows. □

Proposition 2.55. *Let $(\Omega, \mathcal{F}, (\mathcal{F}_t)_{t\geq 0}, P)$ be a filtered probability space. Define*

$$\mathcal{F}_t^+ = \bigcap_{s>t} \mathcal{F}_s \quad \text{and} \quad \overline{\mathcal{F}}_t = \sigma(\mathcal{F}_t \cup \mathcal{N}),$$

where $\mathcal{N} = \{A \in \mathcal{F} \mid P(A) = 0\}$ is the family of all null sets. Then we have the following:
(1) *$\overline{\mathcal{F}}_t = \{A \in \mathcal{F} \mid \exists B \in \mathcal{F}_t$ such that $A \neq B, P(A \triangle B) = 0\}$;*
(2) *$\mathcal{F}_t \subset \mathcal{F}_t^+ \subset \overline{\mathcal{F}}_t$;*
(3) *$\overline{\mathcal{F}}_t$ is right-continuous;*
(4) *$\mathbb{E}[X|\mathcal{F}_t] = \mathbb{E}[X|\overline{\mathcal{F}}_t]$, for every random variable $X \in L^1(\Omega)$.*

Proof. Let $\mathcal{G}_t = \{A \in \mathcal{F} \mid$ there exists $B \in \mathcal{F}_t$ such that $A \neq B, P(A \triangle B) = 0\}$. We show that $\mathcal{G}_t = \overline{\mathcal{F}}_t$. Since $\mathcal{G}_t$ contains both $\mathcal{F}_t$ and $\mathcal{N}$, and is a σ-field, we have $\overline{\mathcal{F}}_t \subset \mathcal{G}_t$. To show the inverse inclusion, let $A \in \mathcal{G}_t$. There exists $B \in \mathcal{F}_t$ such that $P(A \triangle B) = 0$. Note that $N_1 = A \setminus B \in \mathcal{N}$ and $N_2 = B \setminus A \in \mathcal{N}$. As $A = (B \cup N_1) \setminus N_2$, we see that $A \in \sigma(\mathcal{F}_t \cup \mathcal{N}) = \overline{\mathcal{F}}_t$. Thus $\mathcal{G}_t \subset \overline{\mathcal{F}}_t$ and (1) is proven.

Next consider (2). The first inclusion is trivial, we thus show the second. Let $A \in \mathcal{F}_t^+$. It follows that

$$\mathbb{E}[1_A|\mathcal{F}_t] = \mathbb{E}[1_A|\mathcal{F}_t^+].$$

Consider $g = \mathbb{E}[1_A|\mathscr{F}_t]$, which is an $\mathscr{F}_t$-measurable function and $g = \mathbb{E}[1_A|\mathscr{F}_t^+] = 1_A$. The range of g is $\{0,1\}$. Let $B = g^{-1}(\{1\})$. Clearly, B is $\mathscr{F}_t$-measurable and $1_A = g = 1_B$ almost surely, which implies that $P(A \triangle B) = 0$. Then (2) is proven by (1).

Consider (3). It is sufficient to show $\bigcap_{r>s} \mathscr{F}_r \subset \mathscr{F}_s$. Take $A \in \bigcap_{r>s} \mathscr{F}_r$, and consider a sequence $(r_n)_{n\in\mathbb{N}}$ such that $r_n \downarrow s$. Since $A \in \mathscr{F}_{r_n}$, by part (1) there exists $B_n \in \mathscr{F}_{r_n}$ such that $P(A \triangle B_n) = 0$. Let $C = \bigcap_{n=1}^{\infty} \bigcup_{m=n}^{\infty} B_m$. Since $\bigcup_{m=n}^{\infty} B_m \in \mathscr{F}_{r_n}$, we can obtain that $C \in \bigcap_{n=1}^{\infty} \mathscr{F}_{r_n} = \mathscr{F}_s^+ \subset \overline{\mathscr{F}}_s$ by (2). Thus there exists $D \in \mathscr{F}_s$ such that

$$P(C \triangle D) = 0. \tag{2.2.14}$$

Since $A \triangle C \subset A \triangle (\bigcup_{n=1}^{\infty} B_n) = \bigcup_{n=1}^{\infty}(A \triangle B_n)$, we have

$$P(A \triangle C) \leq \sum_{n=1}^{\infty} P(A \triangle B_n) = 0. \tag{2.2.15}$$

Hence by (2.2.14) and (2.2.15), $P(A \triangle D) = 0$ and $D \in \mathscr{F}_s$ follows, and $A \in \overline{\mathscr{F}}_s$ is proven. Finally, (4) follows from Lemma 2.54. □

We have seen above that the filtration $(\mathscr{F}_t)_{t\geq 0}$ is not right-continuous in general, while the filtration $(\mathscr{F}_t^+)_{t\geq 0}$ is right-continuous by definition. However, we will see in the study of stopping times in the next section that it is desirable to work with a right-continuous filtration including all null sets. The family $(\overline{\mathscr{F}}_t)_{t\geq 0}$ defined below has all the required properties.

Definition 2.56 (Augmented filtration). Let $(\Omega, \mathscr{F}, (\mathscr{F}_t)_{t\geq 0}, P)$ be a filtered probability space. Define $\overline{\mathscr{F}}_t = \sigma(\mathscr{F}_t \cup \mathscr{N})$, where $\mathscr{N} = \{A \in \mathscr{F} \mid P(A) = 0\}$ denotes the family of all null sets. Then $(\overline{\mathscr{F}}_t)_{t\geq 0}$ is called an *augmented filtration.*

2.2.3 Stopping times and optional sampling

There is a useful weakening of the martingale property which removes the condition of integrability in Definition 2.52. To introduce and apply it below, first we need a concept of random times.

Definition 2.57 (Stopping time). Let $(\Omega, \mathscr{F}, (\mathscr{F}_t)_{t\geq 0}, P)$ be a filtered probability space. A random variable $\tau : \Omega \to [0, \infty]$ such that $\{\tau \leq t\} \in \mathscr{F}_t$ for all $t \geq 0$ is called a *stopping time.* For a stopping time τ, we consider

$$\mathscr{F}_\tau = \left\{A \in \mathscr{F} \mid A \cap \{\tau \leq t\} \in \mathscr{F}_t \text{ for all } t \geq 0\right\}.$$

Intuitively, $\mathscr{F}_\tau$ contains the events occurring before or at the given random time τ.

Lemma 2.58. *Suppose that the filtration $(\mathscr{F}_t)_{t\geq 0}$ is right-continuous. Then τ is a stopping time with respect to $(\mathscr{F}_t)_{t\geq 0}$ if and only if $\{\tau < t\} \in \mathscr{F}_t$, for every $t \geq 0$.*

Proof. Suppose that $\{\tau \le t\} \in \mathscr{F}_t$. It follows that $\{\tau < t\} = \bigcup_{n=1}^{\infty}\{\tau \le t - 1/n\} \in \mathscr{F}_t$. Next we suppose that $\{\tau < t\} \in \mathscr{F}_t$. Since $\{\tau \le t\} = \bigcap_{n=m}^{\infty}\{\tau < t + 1/n\} \in \mathscr{F}_{t+1/m}$, we have $\{\tau \le t\} \in \bigcap_{m=1}^{\infty}\mathscr{F}_{t+1/m} = \mathscr{F}_t$. Thus τ is a stopping time. □

Example 2.59. Let $(\mathscr{F}_t)_{t\ge 0}$ be right-continuous. If $(\tau_n)_{n\in\mathbb{N}}$ is a sequence of stopping times with respect to $(\mathscr{F}_n)_{n\in\mathbb{N}}$ and $\tau_n \downarrow \tau$ as $n \to \infty$, then τ is also a stopping time since $\{\tau < t\} = \bigcup_{n=1}^{\infty}\{\tau_n < t\} \in \mathscr{F}_t$. Similarly, if a sequence of stopping times $(\tau_n)_{n\in\mathbb{N}}$ satisfies that $\tau_n \uparrow \tau$ as $n \to \infty$, then again τ is a stopping time since $\{\tau \le t\} = \bigcap_{n=1}^{\infty}\{\tau_n < t\} \in \mathscr{F}_t$.

The first hitting time by an adapted process of an open or a closed set plays an important role in martingale theory. First hitting times are stopping times with respect to the augmented filtration.

Lemma 2.60 (First hitting time). *Let $(\Omega, \mathscr{F}, (\mathscr{F}_t)_{t\ge 0}, P)$ be a filtered probability space such that $(\mathscr{F}_t)_{t\ge 0}$ is right-continuous and contains all null sets. Also, let E be a Polish space and consider the measurable space $(E, \mathcal{B}(E))$. Let $(X_t)_{t\ge 0}$ be an E-valued continuous $(\mathscr{F}_t)_{t\ge 0}$-adapted process, and $A \subset E$ an open or closed set. The first hitting time of A by $(X_t)_{t\ge 0}$ defined by*

$$\tau_A = \inf\left\{t > 0 \mid X_t \in A\right\}$$

is a stopping time with respect to $(\mathscr{F}_t)_{t\ge 0}$.

Proof. There exists $N \in \mathscr{F}$ such that $P(N) = 0$ and the paths in $\Omega \setminus N \ni t \mapsto X_t(\omega) \in E$ are continuous. Let $Y_t(\omega) = X_t(\omega)$, $\omega \notin N$, and $Y_t(\omega) = e$, $\omega \in N$. Here $e \in E$ is arbitrary, and we fix it. The map $t \mapsto Y_t(\omega)$ is continuous for every $\omega \in \Omega$. Define $\tau'_A = \inf\{t > 0 \mid Y_t \in A\}$, and let $C = \{\tau_A \le t\}$ and $C' = \{\tau'_A \le t\}$. Since $\mathscr{F}_t \ni N$, we see that $\mathscr{F}_t \ni C$ if and only if $C \setminus N \in \mathscr{F}_t$, and, moreover, $C \setminus N \in \mathscr{F}_t$, $C' \setminus N \in \mathscr{F}_t$, and $C' \in \mathscr{F}_t$ are equivalent to each other. It follows that

$$\left\{\tau_A \le t\right\} \in \mathscr{F}_t \iff \left\{\tau'_A \le t\right\} \in \mathscr{F}_t.$$

Hence we may assume that $t \mapsto X_t(\omega)$ is continuous for every $\omega \in \Omega$ in the remainder of the proof. By the right-continuity of $(\mathscr{F}_t)_{t\ge 0}$, $\{\tau_A < t\} \in \mathscr{F}_t$ if and only if $\{\tau_A \le t\} \in \mathscr{F}_t$. This is due to Lemma 2.58.

Suppose that A is open. It is immediate to see that

$$\left\{\tau_A < t\right\} = \bigcup_{q\in\mathbb{Q}, q<t} \left\{X_q \in A\right\}.$$

Since $\{X_q \in A\} \in \mathscr{F}_t$ for $q < t$, it follows that $\bigcup_{q\in\mathbb{Q},q<t}\{X_q \in A\} \in \mathscr{F}_t$. Thus τ_A is a stopping time.

Next suppose that A is closed. Let $G_n = \{x \in E \mid \operatorname{dist}(x, A) < 1/n\}$, which is an open set containing A. Moreover, since A is closed, $\bigcap_{n=1}^{\infty} G_n = A$ follows. Hence τ_{G_n} is a stopping time. Since $A \subset G_n$, we have $\tau_{G_n} \le \tau_A$. Writing $\sup_{n\in\mathbb{N}}\tau_{G_n} = \tau$, we see that

$\tau_{G_n} \uparrow \tau$ as $n \uparrow \infty$. It then follows that $\tau \leq \tau_A$. We show $\tau \geq \tau_A$. If $\tau(\omega) = \infty$, then $\tau(\omega) = \tau_A(\omega) = \infty$. In case $\tau(\omega) < \infty$, fix $\omega \in \Omega$ and take an arbitrary r such that $\tau(\omega) < r$. Since $\tau_{G_n} < r$, there exists $t_n < r$ such that $X_{t_n} \in G_n$. Now $t_n \in (0, r]$ implies that there exists a convergent subsequence t'_n. We reset t'_n by t_n and let $t_n \to t$. By path continuity, it follows that $X_{t_n}(\omega) \to X_t(\omega)$. Therefore $\text{dist}(X_{t_n}(\omega), A) < 1/n$ and $\text{dist}(X_t(\omega), A) = 0$, giving $X_t(\omega) \in A$. This implies that $\tau_A(\omega) \leq t \leq r$. Since $r > \tau(\omega)$ is arbitrary, $\tau_A \leq \tau$ follows. Finally, we obtain $\tau = \tau_A$ and $\tau_{G_n} \uparrow \tau_A$ as $n \uparrow \infty$, and hence that τ_A is a stopping time. □

Lemma 2.61 (Approximation of stopping times). *Let τ be a stopping time with respect to a filtration $(\mathscr{F}_t)_{t\geq 0}$. Then there exists a sequence of stopping times $(\tau_n)_{n\in\mathbb{N}}$ such that* (1) *τ_n has a discrete range,* (2) *$\tau_n \geq \tau$,* (3) *$\tau_n \downarrow \tau$ as $n \uparrow \infty$.*

Proof. Define

$$\tau_n = \begin{cases} 0, & \tau = 0, \\ \frac{i+1}{n}, & \frac{i}{n} < \tau \leq \frac{i+1}{n}, \ i = 0, 1, 2, \ldots \end{cases} \tag{2.2.16}$$

Then (1)–(3) hold and it is sufficient to see that τ_n is a stopping time. We clearly have

$$\left\{\tau_n \leq t\right\} = \bigcup_{i,(i+1)/n \leq t} \left\{\tau_n = \frac{i+1}{n}\right\} = \bigcup_{i,(i+1)/n \leq t} \left\{\frac{i}{n} < \tau \leq \frac{i+1}{n}\right\} \in \mathscr{F}_t. \qquad \square$$

Now we consider a discrete-time martingale $(X_n)_{n\in\dot{\mathbb{N}}}$ with respect to $(\mathscr{F}_n)_{n\in\dot{\mathbb{N}}}$. Let $f = (f_n)_{n\in\dot{\mathbb{N}}}$ be a bounded and nonnegative random process such that f_n is $\mathscr{F}_{n-1}$-measurable for each n. We define the discrete time process $Y = (Y_n)_{n\in\dot{\mathbb{N}}}$ by

$$Y_n = \begin{cases} X_0, & n = 0, \\ Y_{n-1} + f_n \cdot (X_n - X_{n-1}), & n \geq 1. \end{cases} \tag{2.2.17}$$

It is seen that $(Y_n)_{n\in\dot{\mathbb{N}}}$ is also a martingale with respect to $(\mathscr{F}_n)_{n\in\dot{\mathbb{N}}}$; the process $(Y_n)_{n\in\dot{\mathbb{N}}}$ is called the *martingale transformation* of $(X_n)_{n\in\dot{\mathbb{N}}}$ by $(f_n)_{n\in\dot{\mathbb{N}}}$.

Theorem 2.62 (Discrete-time stopped process). *Let $\tau : \Omega \to \dot{\mathbb{N}}$ be a stopping time and $(X_n)_{n\in\dot{\mathbb{N}}}$ a martingale with respect to $(\mathscr{F}_n)_{n\in\dot{\mathbb{N}}}$. Then $(X_{n\wedge\tau})_{n\in\dot{\mathbb{N}}}$ is also a martingale with respect to $(\mathscr{F}_n)_{n\in\dot{\mathbb{N}}}$.*

Proof. Consider $f_n = 1_{\{n\leq\tau\}}$, which is $\mathscr{F}_{n-1}$-measurable since

$$\left\{n \leq \tau\right\} = \left(\bigcup_{k=0}^{n-1} \{\tau = k\}\right)^c. \tag{2.2.18}$$

Define the martingale $(Y_n)_{n\in\dot{\mathbb{N}}}$ by (2.2.17). Then $Y_n = X_{n\wedge\tau}$, and the result follows. □

The theorem above can be extended to a continuous-time version. In order to do this, we discuss the optional sampling theorem and martingale inequalities first.

Theorem 2.63 (Discrete optional sampling theorem: bounded case). *Let* $(X_n)_{n\in\dot{\mathbb{N}}}$ *be a discrete-time martingale, and* τ, ρ *be* $\dot{\mathbb{N}}$*-valued stopping times such that* $\tau(\omega) \leq \rho(\omega)$, *for all* $\omega \in \Omega$, *on a filtered probability space* $(\Omega, \mathscr{F}, (\mathscr{F}_n)_{n\in\dot{\mathbb{N}}}, P)$. *Suppose that there exists* $m < \infty$ *such that* $\rho(\omega) < m$ *for all* $\omega \in \Omega$. *Then*

$$\mathbb{E}[X_\rho | \mathscr{F}_\tau] = X_\tau. \tag{2.2.19}$$

Proof. It is sufficient to show that

$$\mathbb{E}[X_\rho 1_B] = \mathbb{E}[X_\tau 1_B] \tag{2.2.20}$$

for all $B \in \mathscr{F}_\tau$ to prove (2.2.19). Define $f_n = 1_{\{\tau<n\leq\rho\}} = 1_{\{n\leq\rho\}} - 1_{\{n\leq\tau\}}$, which is $\mathscr{F}_{n-1}$-measurable by (2.2.18), and $(Y_n)_{n\in\dot{\mathbb{N}}}$ defined by the martingale transform of $(X_n)_{n\in\dot{\mathbb{N}}}$ through $(f_n)_{n\in\dot{\mathbb{N}}}$ satisfies $Y_n - X_0 = X_{\rho\wedge n} - X_{\tau\wedge n}$. We have $\mathbb{E}[Y_n] = \mathbb{E}[X_0]$ for all n, since $(Y_n)_{n\in\dot{\mathbb{N}}}$ is a martingale. Hence

$$\mathbb{E}[Y_m - X_0] = \mathbb{E}[X_\rho] - \mathbb{E}[X_\tau] = 0. \tag{2.2.21}$$

Let $B \in \mathscr{F}_\tau$. Define

$$\tau_B(\omega) = \begin{cases} \tau(\omega), & \omega \in B, \\ m, & \omega \notin B, \end{cases} \quad \text{and} \quad \rho_B(\omega) = \begin{cases} \rho(\omega), & \omega \in B, \\ m, & \omega \notin B. \end{cases}$$

Both τ_B and ρ_B are stopping times, and it follows from (2.2.21) that $\mathbb{E}[X_{\rho_B}] = \mathbb{E}[X_{\tau_B}]$, which implies that $\mathbb{E}[X_\rho 1_B] + \mathbb{E}[X_m 1_{B^c}] = \mathbb{E}[X_\tau 1_B] + \mathbb{E}[X_m 1_{B^c}]$. Thus (2.2.20) follows. □

In Theorem 2.63, it was necessary to assume that both τ and ρ are finite stopping times, while in applications often stopping times are not necessarily finite. For a uniformly integrable martingale $(X_n)_{n\in\dot{\mathbb{N}}}$, we can remove this condition. To show this, we need several steps. Let $(X_n)_{n\in\dot{\mathbb{N}}}$ be $(\mathscr{F}_n)_{n\in\dot{\mathbb{N}}}$-adapted and $[a, b]$ an interval. Define

$$\begin{aligned} &\tau_1 = \min\{n \mid X_n \leq a\}, \quad \tau_2 = \min\{n \geq \tau_1 \mid X_n \geq b\}, \dots, \\ &\tau_{2k+1} = \min\{n \geq \tau_{2k} \mid X_n \leq a\}, \quad \tau_{2k+2} = \min\{n \geq \tau_{2k+1} \mid X_n \geq b\}, \dots \end{aligned} \tag{2.2.22}$$

Then $(\tau_n)_{n\in\mathbb{N}}$ is a sequence of stopping times. The number of upcrossings of a random process $(X_n)_{n\in\dot{\mathbb{N}}}$ for the interval $[a, b]$ is counted by $U_N^X(a, b) = \max\{k \mid \tau_{2k} \leq N\}$.

Lemma 2.64. *Let* $(X_n)_{n\in\dot{\mathbb{N}}}$ *be a submartingale on a filtered probability space* $(\Omega, \mathscr{F}, (\mathscr{F}_n)_{n\in\dot{\mathbb{N}}}, P)$. *Then*

$$\mathbb{E}[U_N^X(a, b)] \leq \frac{1}{b-a}\mathbb{E}[(X_N - a)^+ - (X_0 - a)^+], \tag{2.2.23}$$

where $h^+(\omega) = \max\{0, h(\omega)\}$.

Proof. Let $Y_n = (X_n - a)^+$. By Jensen's inequality, $(Y_n)_{n\in\dot{\mathbb{N}}}$ is a submartingale, and $U_N^X(a,b) = U_N^Y(0, b-a)$. Let τ_n be defined by (2.2.22) with $(X_t)_{t\geq 0}$ and a, b replaced by $(Y_t)_{t\geq 0}$ and $0, b-a$, respectively. Write $\tau_n' = \tau_n \wedge N$ for $n \geq 1$ and $\tau_0' = 0$. If $2k > N$, we have

$$Y_N - Y_0 = \sum_{n=1}^{2k}(Y_{\tau_n'} - Y_{\tau_{n-1}'}) = \sum_{n=1}^{k}(Y_{\tau_{2n}'} - Y_{\tau_{2n-1}'}) + \sum_{n=0}^{k-1}(Y_{\tau_{2n+1}'} - Y_{\tau_{2n}'}). \tag{2.2.24}$$

Note that

$$\mathbb{E}\left[\sum_{n=1}^{k}(Y_{\tau_{2n}'} - Y_{\tau_{2n-1}'})\right] \geq (b-a)U_N^Y(0, b-a) \quad \text{and} \quad \mathbb{E}\left[\sum_{n=0}^{k-1}(Y_{\tau_{2n+1}'} - Y_{\tau_{2n}'})\right] \geq 0$$

by the submartingale property. We obtain (2.2.23) by taking expectation on both sides of (2.2.24). □

Proposition 2.65 (Convergence of submartingales). *Let $(X_n)_{n\in\dot{\mathbb{N}}}$ be a submartingale on a filtered probability space $(\Omega, \mathscr{F}, (\mathscr{F}_n)_{n\in\dot{\mathbb{N}}}, P)$ such that $\sup_{n\in\dot{\mathbb{N}}} \mathbb{E}[X_n^+] < \infty$. Then $X_\infty = \lim_{n\to\infty} X_n$ exists almost surely and it is integrable.*

Proof. Since $\mathbb{E}[|X_n|] = 2\mathbb{E}[X_n^+] - \mathbb{E}[X_n] \leq 2\mathbb{E}[X_n^+] - \mathbb{E}[X_0]$, we have $\sup_{n\in\dot{\mathbb{N}}} \mathbb{E}[|X_n|] < \infty$. By Fatou's lemma, X_∞ is integrable whenever it exists. To show the existence of X_∞, we define $U_\infty^X(a,b)$ by $\lim_{N\to\infty} U_N^X(a,b)$. We have

$$\left\{\liminf_{n\in\dot{\mathbb{N}}} X_n < \limsup_{n\in\dot{\mathbb{N}}} X_n\right\} = \bigcup_{a,b\in\mathbb{Q}} \{U_\infty^X(a,b) = \infty\}.$$

Since $N \to U_N^X(a,b)$ is increasing, by the monotone convergence theorem, we have

$$\mathbb{E}[U_\infty^X(a,b)] = \lim_{N\to\infty} \mathbb{E}[U_N^X(a,b)] \leq \frac{1}{b-a} \lim_{N\to\infty} \mathbb{E}[(X_N - a)^+ - (X_0 - a)^+],$$

and this is finite due to $\sup_{n\in\dot{\mathbb{N}}} \mathbb{E}[X_n^+] < \infty$. Thus we have

$$P\left(\liminf_{n\in\dot{\mathbb{N}}} X_n < \limsup_{n\in\dot{\mathbb{N}}} X_n\right) = 0,$$

which implies $P(\liminf_{n\in\dot{\mathbb{N}}} X_n = \limsup_{n\in\dot{\mathbb{N}}} X_n) = 1$ and X_∞ exists almost surely. □

The convergence of submartingales can be extended to continuous submartingales.

Proposition 2.66 (Convergence of submartingales: continuous version). *Let $(X_t)_{t\geq 0}$ be a right-continuous submartingale on a filtered probability space $(\Omega, \mathscr{F}, (\mathscr{F}_t)_{t\geq}, P)$ such that $\sup_{t\geq 0} \mathbb{E}[X_t^+] < \infty$. Then $X_\infty = \lim_{t\to\infty} X_t$ exists almost surely and it is integrable.*

Proof. We give an outline of the proof. Since $(X_t)_{t\geq 0}$ is right-continuous, similarly to $U_N^X(a,b)$ we can define $U_t^X(a,b)$ for $t \geq 0$, and can show that

$$\mathbb{E}[U_\infty^X(a,b)] \leq \frac{C+|a|}{b-a}$$

with a constant C. The event $\{U_\infty^X(a,b)=\infty\}$ is negligible and

$$\left\{\liminf_{t\to\infty} X_t < \limsup_{t\to\infty} X_t\right\} = \bigcup_{a,b\in\mathbb{Q}} \{U_\infty^X(a,b)=\infty\}.$$

Therefore

$$P\Big(\liminf_{t\to\infty} X_t < \limsup_{t\to\infty} X_t\Big) = 0,$$

which implies that $P(\liminf_{t\to\infty} X_t = \limsup_{t\to\infty} X_t) = 1$ and X_∞ exists a. s. Moreover, $\mathbb{E}[|X_t|] = 2\mathbb{E}[X_t^+] - \mathbb{E}[X_t] \leq 2C - \mathbb{E}[X_0]$ yields $\mathbb{E}[|X_\infty|] < \infty$. □

Proposition 2.67. *Let $(X_n)_{n\in\dot{\mathbb{N}}}$ be a submartingale on a filtered probability space $(\Omega, \mathscr{F}, (\mathscr{F}_n)_{n\in\dot{\mathbb{N}}}, P)$ such that $\sup_{n\in\dot{\mathbb{N}}}\mathbb{E}[X_n^+] < \infty$. Define $X_\infty = \lim_{n\to\infty} X_n$, almost surely. Then $(X_n^+)_{n\in\dot{\mathbb{N}}}$ is uniformly integrable if and only if $\mathbb{E}[X_\infty|\mathscr{F}_n] \geq X_n$, for all $n \in \dot{\mathbb{N}}$, where equality holds if $(X_n)_{n\in\dot{\mathbb{N}}}$ is a martingale.*

Proof. Suppose that $X_n \leq \mathbb{E}[X_\infty|\mathscr{F}_n]$. Jensen's inequality yields

$$X_n^+ \leq (\mathbb{E}[X_\infty|\mathscr{F}_n])^+ \leq \mathbb{E}[X_\infty^+|\mathscr{F}_n]$$

and hence

$$\mathbb{E}[X_n^+ 1_{\{X_n^+>a\}}] \leq \mathbb{E}[X_\infty^+ 1_{\{X_n^+>a\}}]. \tag{2.2.25}$$

We truncate X_∞^+ by defining

$$(X_\infty^+)^N = \begin{cases} N, & X_\infty^+ > N, \\ X_\infty^+, & X_\infty^+ \leq N. \end{cases}$$

We see that $(X_\infty^+)^N \in L^2(\Omega)$ for each N. Take $\varepsilon > 0$. By the dominated convergence theorem, $|\mathbb{E}[(X_\infty^+)^N] - \mathbb{E}[X_\infty^+]| < \varepsilon$ for all $N > N_0$, with some N_0. Since $P(X_n^+ > a) \leq \mathbb{E}[X_n^+]/a \leq \mathbb{E}[X_\infty]/a$, it also follows that

$$\lim_{a\to\infty} \mathbb{E}[(X_\infty^+)^N 1_{\{X_n^+>a\}}] \leq \mathbb{E}[(X_\infty^+)^{2N}]^{1/2} P(X_n^+ > a)^{1/2} \leq \varepsilon$$

for all $a > a_0$ with some a_0. We see that

$$\mathbb{E}[X_\infty^+ 1_{\{X_n^+>a\}}] \leq \mathbb{E}[(X_\infty^+)^N 1_{\{X_n^+>a\}}] + |\mathbb{E}[(X_\infty^+)^N] - \mathbb{E}[X_\infty^+]| \leq 2\varepsilon$$

follows for all $a > a_0$ uniformly in n. This implies by (2.2.25) that

$$\sup_{n\in\dot{\mathbb{N}}} \mathbb{E}[X_n^+ 1_{\{X_n^+>a\}}] < 2\varepsilon$$

and thus $\inf_{a\geq 0}\sup_{n\in\dot{\mathbb{N}}} \mathbb{E}[X_n^+ 1_{\{X_n^+>a\}}] = 0$ is obtained, showing that $(X_n^+)_{n\in\dot{\mathbb{N}}}$ is uniformly integrable. Next we suppose that $(X_n^+)_{n\in\dot{\mathbb{N}}}$ is uniformly integrable. By Corollary 2.32, $X_n^+ \to X_\infty^+$ in $L^1(\Omega)$. Take $a > 0$. Then $X_n \vee (-a) \to X_\infty \vee (-a)$, as $n \to \infty$ in $L^1(\Omega)$, and $(X_n \vee (-a))_{n\in\dot{\mathbb{N}}}$ is a submartingale. We have

$$\mathbb{E}[X_\infty \vee (-a)|\mathscr{F}_n] = \lim_{m\to\infty} \mathbb{E}[X_m \vee (-a)|\mathscr{F}_n] \geq X_n \vee (-a),$$

and letting $a \uparrow \infty$ we obtain $\mathbb{E}[X_\infty|\mathscr{F}_n] \geq X_n$. □

Proposition 2.68 (Lévy's upward theorem). *Let $(\Omega, \mathscr{F}, (\mathscr{F}_n)_{n\in\dot{\mathbb{N}}}, P)$ be a filtered probability space, and $Y \in L^1(\Omega)$. Consider $X_n = \mathbb{E}[Y|\mathscr{F}_n]$. Then $(X_n)_{n\in\dot{\mathbb{N}}}$ is a uniformly integrable martingale and $X_\infty = \lim_{n\to\infty} X_n$ exists almost surely and $X_\infty \in L^1(\Omega)$. Furthermore,*

$$X_\infty = \mathbb{E}[Y|\mathscr{F}_\infty], \tag{2.2.26}$$

where $\mathscr{F}_\infty = \sigma(\bigcup_{n=0}^\infty \mathscr{F}_n)$.

Proof. Since $|X_n| \leq \mathbb{E}[|Y||\mathscr{F}_n]$, it can be proven that $(X_n)_{n\in\dot{\mathbb{N}}}$ is uniformly integrable in a similar manner to the proof of Proposition 2.67. Also, we see that $\mathbb{E}[X_m|\mathscr{F}_n] = \mathbb{E}[\mathbb{E}[Y|\mathscr{F}_m]|\mathscr{F}_n] = \mathbb{E}[Y|\mathscr{F}_n] = X_n$ for $m > n$, i. e., $(X_n)_{n\in\dot{\mathbb{N}}}$ is a martingale. We see thus that $X_\infty = \lim_{n\to\infty} X_n$ exists a. s., $X_\infty \in L^1(\Omega)$, and $\mathbb{E}[X_\infty|\mathscr{F}_n] = X_n$ by Proposition 2.67. Note that X_∞ is $\mathscr{F}_\infty$-measurable. Let

$$\mathscr{C} = \Big\{B \in \mathscr{F} \mid \mathbb{E}[X_\infty 1_B] = \mathbb{E}[Y 1_B]\Big\}.$$

If $B \in \mathscr{F}_n$, then $\mathbb{E}[Y1_B] = \mathbb{E}[X_n 1_B] = \mathbb{E}[X_\infty 1_B]$. Thus $\bigcup_{n=0}^\infty \mathscr{F}_n \subset \mathscr{C}$. Clearly, $\mathscr{C}$ is a λ-system, thus by the π–λ theorem it follows that $\mathscr{C} \supset \sigma(\bigcup_{n=0}^\infty \mathscr{F}_n)$. This proves (2.2.26). □

The martingale $(X_n)_{n\in\dot{\mathbb{N}}}$ defined by $X_n = \mathbb{E}[Y|\mathscr{F}_n]$ is called *Doob's martingale*.

Further assuming uniform integrability of $(X_n)_{n\in\dot{\mathbb{N}}}$, Theorem 2.63 can be extended to unbounded stopping times.

Theorem 2.69 (Discrete optional sampling theorem: unbounded case). *Let $(X_n)_{n\in\dot{\mathbb{N}}}$ be a uniformly integrable and discrete-time martingale on a filtered probability space $(\Omega, \mathscr{F}, (\mathscr{F}_n)_{n\in\dot{\mathbb{N}}}, P)$, and let τ, ρ be $\dot{\mathbb{N}}$-valued stopping times such that $\tau(\omega) \leq \rho(\omega)$, for all $\omega \in \Omega$. Then*

$$\mathbb{E}[X_\rho|\mathscr{F}_\tau] = X_\tau. \tag{2.2.27}$$

Proof. Note that $\sup_{n\in\dot{\mathbb{N}}} \mathbb{E}[|X_n|] < \infty$ by Corollary 2.32. In the same way as in the proof of Proposition 2.68, $X_\infty = \lim_{n\to\infty} X_n$ exists almost surely in $L^1(\Omega)$, and $X_n = \mathbb{E}[X_\infty|\mathscr{F}_n]$. Theorem 2.63 gives

$$\mathbb{E}[X_\infty|\mathscr{F}_{\tau\wedge n}] = X_{\tau\wedge n} = \mathbb{E}[X_n|\mathscr{F}_{\tau\wedge n}]. \tag{2.2.28}$$

Letting $n \to \infty$ on both sides above, by Proposition 2.68 we have

$$\mathbb{E}[X_\infty|\mathscr{G}] = X_\tau, \tag{2.2.29}$$

where $\mathscr{G} = \sigma(\bigcup_{n=0}^\infty \mathscr{F}_{\tau\wedge n})$. We need to show that $\mathscr{G}$ can be replaced by $\mathscr{F}_\tau$. Take $A \in \mathscr{F}_\tau$. We have $\mathbb{E}[1_A X_\infty] = \mathbb{E}[1_{A\cap\{\tau<\infty\}}X_\infty] + \mathbb{E}[1_{A\cap\{\tau=\infty\}}X_\infty]$. By (2.2.29), $\mathbb{E}[1_{A\cap\{\tau<\infty\}}X_\infty] = \mathbb{E}[1_{A\cap\{\tau<\infty\}}X_\tau]$, and we also see that $\mathbb{E}[1_{A\cap\{\tau=\infty\}}X_\infty] = \mathbb{E}[1_{A\cap\{\tau=\infty\}}X_\tau]$. Hence $\mathbb{E}[1_A X_\infty] = \mathbb{E}[1_A X_\tau]$ for $A \in \mathscr{F}_\tau$. Since X_τ is $\mathscr{F}_\tau$-measurable, $\mathbb{E}[X_\infty|\mathscr{F}_\tau] = X_\tau$ follows. Similarly, we have $\mathbb{E}[X_\infty|\mathscr{F}_\rho] = X_\rho$. Taking expectation with respect to $\mathscr{F}_\tau$, we have

$$\mathbb{E}[X_\rho|\mathscr{F}_\tau] = \mathbb{E}[\mathbb{E}[X_\infty|\mathscr{F}_\rho]|\mathscr{F}_\tau] = \mathbb{E}[X_\infty|\mathscr{F}_\tau] = X_\tau.$$

Here we used that $\mathscr{F}_\tau \subset \mathscr{F}_\rho$. □

We can furthermore extend the discrete optional sampling theorem to the continuous time setting.

Theorem 2.70 (Continuous optional sampling theorem). *Suppose that $(X_t)_{t\geq 0}$ is a right-continuous, uniformly integrable martingale on a filtered probability space $(\Omega, \mathscr{F}, (\mathscr{F}_t)_{t\geq 0}, P)$, and τ, ρ are stopping times such that $\tau(\omega) \leq \rho(\omega)$, for all $\omega \in \Omega$. Then $\mathbb{E}[X_\rho|\mathscr{F}_\tau] = X_\tau$.*

Proof. By Lemma 2.61, there exist stopping times τ_n and ρ_n such that both have discrete range, $\tau_n \leq \rho_n$, and $\tau_n \downarrow \tau$, $\rho_n \downarrow \rho$ as $n \uparrow \infty$. Since $(X_t)_{t\geq 0}$ is uniformly integrable, Theorem 2.63 provides $\mathbb{E}[X_{\rho_n}|\mathscr{F}_{\tau_n}] = X_{\tau_n}$. From $\mathscr{F}_\tau \subset \mathscr{F}_{\tau_n}$ it follows that

$$\mathbb{E}[X_{\rho_n}|\mathscr{F}_\tau] = \mathbb{E}[\mathbb{E}[X_{\rho_n}|\mathscr{F}_{\tau_n}]|\mathscr{F}_\tau] = \mathbb{E}[X_{\tau_n}|\mathscr{F}_\tau]. \tag{2.2.30}$$

We also see that

$$\mathbb{E}[X_{\rho_n}|\mathscr{F}_{\rho_{n+1}}] = X_{\rho_{n+1}}, \quad \mathbb{E}[X_{\tau_n}|\mathscr{F}_{\tau_{n+1}}] = X_{\tau_{n+1}}. \tag{2.2.31}$$

In particular, $\mathbb{E}[X_{\rho_n}]$ is independent of n, and $\sup_{n\in\mathbb{N}} \mathbb{E}[X_{\rho_n}] < \infty$. We show that both $\{X_{\rho_n}\}_{n\in\mathbb{N}}$ and $\{X_{\tau_n}\}_{n\in\mathbb{N}}$ are uniformly integrable. Let $M_n = X_{\tau_n}$, $\mathscr{F}_n = \mathscr{F}_{\tau_n}$, and take $a > 0$. By a straightforward computation, we arrive at the identity

$$\mathbb{E}[|M_n|1_{\{|M_n|>a\}}] = \mathbb{E}[M_n] - \mathbb{E}[M_n 1_{\{M_n\leq a\}}] - \mathbb{E}[M_n 1_{\{M_n<-a\}}]. \tag{2.2.32}$$

Since for $m \leq n$ we have, by (2.2.31),

$$\mathbb{E}[M_n 1_{\{M_n\leq a\}}] = \mathbb{E}[\mathbb{E}[M_m|\mathscr{F}_n]1_{\{M_n\leq a\}}] = \mathbb{E}[\mathbb{E}[M_m 1_{\{M_n\leq a\}}|\mathscr{F}_n]] = \mathbb{E}[X_m 1_{\{M_n\leq a\}}]$$

and similarly $\mathbb{E}[M_n 1_{\{M_n\leq -a\}}] = \mathbb{E}[X_m 1_{\{M_n\leq -a\}}]$, we obtain

$$\mathbb{E}[|M_n|1_{\{|M_n|>a\}}] = \mathbb{E}[|X_m|1_{\{|M_n|>a\}}], \quad m \le n. \tag{2.2.33}$$

Furthermore, let $M_n^- = \max\{-M_n, 0\}$. Since the function $x \mapsto \max\{x, 0\}$ is convex, it follows from Jensen's inequality that $\mathbb{E}[M_n^-|\mathscr{F}_m] \ge M_m^-$ for $n \le m$, and it is also true that $\mathbb{E}[M_n^-] \ge \mathbb{E}[M_m^-]$ for $n \le m$. Hence

$$\mathbb{E}[|M_n|] = \mathbb{E}[M_n] + 2\mathbb{E}[M_n^-] \le \sup_{n\in\dot{\mathbb{N}}} \mathbb{E}[M_n] + 2\mathbb{E}[M_0^-]$$

and $\sup_{n\in\dot{\mathbb{N}}} \mathbb{E}[|M_n|] < C$ follows with some C. We conclude that

$$P(|M_n| > a) \le \frac{1}{a} \sup_{n\in\dot{\mathbb{N}}} \mathbb{E}[|M_n|] = \frac{C}{a}. \tag{2.2.34}$$

We truncate M_m by considering

$$M_m^N = \begin{cases} N, & M_m > N, \\ M_m, & -N \le M_m \le N, \\ -N, & M_m < -N. \end{cases}$$

Note that $M_m^N \in L^2(\Omega)$ for each N. Let $\varepsilon > 0$ be given. By the dominated convergence theorem, $|\mathbb{E}[|M_m^N|] - \mathbb{E}[|M_m|]| < \varepsilon$ for all $N > N_0$ and some N_0. By (2.2.34), it also follows that

$$\mathbb{E}[|M_m^N|1_{\{|M_n|>a\}}] \le \mathbb{E}[|M_m^N|]^{1/2} P(|M_n| > a)^{1/2} \le \varepsilon$$

for all $a > a_0$ with some a_0, uniformly in n. It follows that

$$\mathbb{E}[|M_m|1_{\{|M_n|>a\}}] \le \mathbb{E}[|M_m^N|1_{\{|M_n|>a\}}] + |\mathbb{E}[|M_m^N|] - \mathbb{E}[|M_m|]| \le 2\varepsilon$$

for all $a > a_0$, uniformly in n. This implies by (2.2.33) that

$$\sup_{n\in\dot{\mathbb{N}}} \mathbb{E}[|X_n|1_{\{|M_n|>a\}}] = \sup_{n\in\dot{\mathbb{N}}} \mathbb{E}[|X_m|1_{\{|M_n|>a\}}] < 2\varepsilon$$

and then $\inf_{a\ge 0} \sup_{n\in\dot{\mathbb{N}}} \mathbb{E}[|M_n|1_{\{|M_n|>a\}}] = 0$ is obtained. We thus see that $\{X_{\tau_n}\}_{n\in\dot{\mathbb{N}}}$ is uniformly integrable. Similarly, we can show that $\{X_{\rho_n}\}_{n\in\dot{\mathbb{N}}}$ is uniformly integrable. Hence X_{τ_n} and X_{ρ_n} converge in $L^1(\Omega)$ by Corollary 2.32. By right-continuity, $X_{\tau_n} \to X_\tau$ and $X_{\rho_n} \to X_\rho$ follow almost surely, and we see that $X_{\tau_n} \to X_\tau$ and $X_{\rho_n} \to X_\rho$ in $L^1(\Omega)$. Now take $A \in \mathscr{F}_\tau$, by (2.2.30), $\mathbb{E}[X_{\tau_n}1_A] = \mathbb{E}[X_{\rho_n}1_A]$. By L^1-convergence, this yields $\mathbb{E}[X_\tau 1_A] = \mathbb{E}[X_\rho 1_A]$, if we let n tend to infinity. □

Corollary 2.71. *Let $(X_t)_{t\ge 0}$ be a right-continuous martingale on a filtered probability space $(\Omega, \mathscr{F}, (\mathscr{F}_t)_{t\ge 0}, P)$, and τ, ρ stopping times such that $\tau(\omega) \le \rho(\omega)$, for all $\omega \in \Omega$. Then*

$$\mathbb{E}[X_{t\wedge\rho}|\mathscr{F}_\tau] = X_{t\wedge\tau}.$$

Proof. Fix $t \geq 0$ and define the martingale by $(Y_T)_{T\geq 0} = (X_{t\wedge T})_{T\geq 0}$. Write $Y_\infty = X_t$. It is sufficient to show that $(Y_T)_{T\geq 0}$ is uniformly integrable by Theorem 2.70. Let $a > 0$. We estimate $\mathbb{E}[|Y_T| 1_{\{|Y_T|>a\}}]$. Note that $\mathbb{E}[Y_\infty|\mathscr{F}_T] = Y_T$. In the same way as when obtaining (2.2.32), we have the identity

$$\mathbb{E}[|Y_T|1_{\{|Y_T|>a\}}] = \mathbb{E}[Y_T] - \mathbb{E}[Y_T 1_{\{Y_T\leq a\}}] - \mathbb{E}[Y_T 1_{\{Y_T<-a\}}]. \tag{2.2.35}$$

Since

$$\mathbb{E}[Y_T 1_{\{Y_T\leq a\}}] = \mathbb{E}[\mathbb{E}[Y_\infty|\mathscr{F}_T]1_{\{Y_T\leq a\}}] = \mathbb{E}[\mathbb{E}[Y_\infty 1_{\{Y_T\leq a\}}|\mathscr{F}_T]] = \mathbb{E}[Y_\infty 1_{\{Y_T\leq a\}}]$$

and similarly $\mathbb{E}[Y_T 1_{\{Y_T\leq -a\}}] = \mathbb{E}[Y_\infty 1_{\{Y_T\leq -a\}}]$, we get

$$\mathbb{E}[|Y_T|1_{\{|Y_T|>a\}}] = \mathbb{E}[|Y_\infty|1_{\{|Y_T|>a\}}].$$

In the same way as in the proof of uniform integrability of $\{X_{\tau_n}\}_{n\in\mathbb{N}}$ in Theorem 2.70, we have $\sup_{T\geq 0}\mathbb{E}[|Y_T|] < \infty$ and $\inf_{a\geq 0}\sup_{T\geq 0}\mathbb{E}[|Y_T|1_{\{|Y_T|>a\}}] = 0$ follows. Thus $(Y_T)_{T\geq 0}$ is uniformly integrable. □

Theorem 2.72 (Martingale inequality). *Let $(X_t)_{t\geq 0}$ be a right-continuous martingale on a filtered probability space $(\Omega, \mathscr{F}, (\mathscr{F}_t)_{t\geq 0}, P)$.*

(1) *If $X_T \in L^p(\Omega)$ for some $p \geq 1$, then for all $\lambda > 0$,*

$$P\Big(\sup_{0\leq t\leq T}|X_t| \geq \lambda\Big) \leq \frac{1}{\lambda^p}\mathbb{E}[|X_T|^p]. \tag{2.2.36}$$

(2) *If $X_T \in L^p(\Omega)$ for some $p > 1$, then*

$$\mathbb{E}\Big[\sup_{0\leq t\leq T}|X_t|^p\Big] \leq \Big(\frac{p}{p-1}\Big)^p\mathbb{E}[|X_T|^p]. \tag{2.2.37}$$

Proof. Fix $\lambda > 0$. Let $(I_n)_{n\in\mathbb{N}}$ be an increasing sequence of finite subsets of $I = [0, T]$ such that $\bigcup_{n=1}^\infty I_n = [0, T] \cap \mathbb{Q} \cap \{T\}$. Let $(Y_t)_{t\geq 0}$ be a right-continuous submartingale with respect to $(\mathscr{F}_t)_{t\geq 0}$. Since $t \mapsto Y_t(\omega)$ is right-continuous, we see that

$$\Big\{\sup_{t\in I} Y_t \geq \lambda\Big\} = \bigcup_{n=1}^\infty\Big\{\max_{t\in I_n} Y_t \geq \lambda\Big\}.$$

Let $I_n = \{j_1, \dots, j_N\}$ be such that $j_1 < \dots < j_N$, and consider $(Y_{j_m})_{j_m\in I_n}$. Define

$$\tau = \begin{cases} \min\{j_n \mid Y_{j_n} \geq \lambda\} & \text{if } \{j_n \mid Y_{j_n} \geq \lambda\} \neq \emptyset, \\ j_N & \text{if } \{j_n \mid Y_{j_n} \geq \lambda\} = \emptyset. \end{cases}$$

Note that τ is a bounded stopping time such that $\tau \leq j_N$. Since $(Y_t)_{t\geq 0}$ is a submartingale, by Theorem 2.63 we have

$$\begin{aligned}\mathbb{E}[Y_{j_N}] &= \mathbb{E}[\mathbb{E}[Y_{j_N}|\mathcal{F}_\tau]] \geq \mathbb{E}[Y_\tau] = \mathbb{E}[Y_\tau 1_{\{\max_{1\leq n\leq N} Y_{j_n} \geq \lambda\}}] + \mathbb{E}[Y_{j_N} 1_{\{\max_{1\leq n\leq N} Y_{j_n} < \lambda\}}] \\ &\geq \lambda P\Big(\max_{n=1,\dots,N} Y_{j_n} \geq \lambda\Big) + \mathbb{E}[Y_{j_N} 1_{\{\max_{1\leq n\leq N} Y_{j_n} < \lambda\}}].\end{aligned}$$

We furthermore have

$$\lambda P\Big(\max_{n=1,\dots,N} Y_{j_n} \geq \lambda\Big) \leq \mathbb{E}[Y_{j_N}] - \mathbb{E}[Y_{j_N} 1_{\{\max_{1\leq n\leq N} Y_{j_n} < \lambda\}}] = \mathbb{E}[Y_{j_N} 1_{\{\max_{1\leq n\leq N} Y_{j_n} \geq \lambda\}}] \leq \mathbb{E}[|Y_{j_N}|]. \tag{2.2.38}$$

Hence for arbitrary I_n, we get $\lambda P(\mathcal{I}_n) \leq \mathbb{E}[|Y_{\max I_n}|]$, where $\mathcal{I}_n = \{\max_{t\in I_n} Y_t \geq \lambda\}$. Since I_n is increasing and $\bigcup_{n=1}^{\infty} \mathcal{I}_n = \{\sup_{t\in I} Y_t \geq \lambda\}$, we see that

$$\lim_{n\to\infty} \lambda P(\mathcal{I}_n) = \lambda P\Big(\sup_{t\in I} Y_t \geq \lambda\Big) \leq \lim_{n\to\infty} \mathbb{E}[|Y_{\max I_n}|] = \mathbb{E}[|Y_T|].$$

Take $|X_t|^p = Y_t$, which is a submartingale. Then

$$\lambda^p P\Big(\sup_{t\in I} |X_t| \geq \lambda\Big) = \lambda^p P\Big(\sup_{t\in I} |X_t|^p \geq \lambda^p\Big) \leq \mathbb{E}[|X_T|^p].$$

Hence (2.2.36) follows. Next we show (2.2.37). For $I_n = \{j_1, \dots, j_N\}$ with $j_1 < \dots < j_N$, let $Y = \max_{0\leq n\leq N} |X_{j_n}|$. By (2.2.38), we have $\lambda P(Y \geq \lambda) \leq \int_\Omega 1_{\{Y\geq\lambda\}} |X_{j_N}| dP$ and obtain that

$$\begin{aligned}\mathbb{E}[Y^p] &= \int_\Omega dP \int_0^Y p\lambda^{p-1} d\lambda = \int_\Omega dP \int_0^\infty 1_{\{Y\geq\lambda\}} p\lambda^{p-1} d\lambda \\ &= p\int_0^\infty \lambda^{p-1} P(Y \geq \lambda) d\lambda \leq p \int_0^\infty d\lambda \int_\Omega \lambda^{p-2} 1_{\{Y\geq\lambda\}} |X_{j_N}| dP \\ &= \frac{p}{p-1} \int_\Omega Y^{p-1} |X_{j_N}| dP \leq \frac{p}{p-1} \mathbb{E}[Y^{p-1}|X_{j_N}|] \\ &\leq \frac{p}{p-1} (\mathbb{E}[|X_{j_N}|^p])^{1/p} (\mathbb{E}[Y^p])^{(p-1)/p}.\end{aligned}$$

This gives $\mathbb{E}[\max_{0\leq n\leq N} |X_{j_n}|^p] \leq \left(\frac{p}{p-1}\right)^p \mathbb{E}[|X_{j_N}|^p]$ and implies

$$\mathbb{E}\Big[\sup_{t\in\bigcup_{n=1}^\infty I_n} |X_t|^p\Big] \leq \left(\frac{p}{p-1}\right)^p \mathbb{E}[|X_{\sup\bigcup_{n=1}^\infty I_n}|^p]$$

by the monotone convergence theorem. By the right-continuity of $X_\cdot$, we have

$$\sup_{t\in\bigcup_{n=1}^\infty I_n} |X_t|^p = \sup_{t\in I} |X_t|^p, \quad |X_{\sup\bigcup_{n=1}^\infty I_n}| = X_T.$$

Hence (2.2.37) follows. □

For $p = 2$, the inequality (2.2.37) is called *Doob–Kolmogorov inequality*.

Theorem 2.73 (Continuous-time stopped process). *Let $(X_t)_{t\geq 0}$ be a right-continuous martingale on a probability space $(\Omega, \mathscr{F}, P)$, and τ a stopping time with respect to the natural filtration of $(X_t)_{t\geq 0}$. Then $(X_{t\wedge\tau})_{t\geq 0}$ is also a right-continuous martingale with respect to the same filtration.*

Proof. Let $\mathscr{M}_t = \sigma(X_r, 0 \leq r \leq t)$. Choose $0 \leq t_1 < t_2 < \cdots < t_m = s < t$, and consider

$$A = \bigcap_{i=1}^{m} \left\{X_{t_i\wedge\tau} \in B_i\right\}, \tag{2.2.39}$$

where $B_i \in \mathcal{B}(\mathbb{R})$. Let τ_n be the stopping time constructed by (2.2.16) in the proof of Lemma 2.61.

First, assume that there exists an integer j such that $s = (j+1)/n$. We see that $X_{s\wedge\tau_n} = X_{\tau_n} = X_{t\wedge\tau_n}$ on $\{\tau \leq (j+1)/n\} = \{\tau_n \leq (j+1)/n\}$, and obtain

$$\mathbb{E}[1_{A\cap\{\tau\leq(j+1)/n\}} X_{s\wedge\tau_n}] = \mathbb{E}[1_{A\cap\{\tau\leq(j+1)/n\}} X_{t\wedge\tau_n}]. \tag{2.2.40}$$

We also see that

$$\mathbb{E}[1_{A\cap\{\tau>(j+1)/n\}} X_{t\wedge\tau_n}] = \sum_{l=j+1}^{M} \mathbb{E}[1_{A\cap\{l/n<\tau\leq(l+1)/n\}} X_{(l+1)/n}] + \mathbb{E}[1_{A\cap\{\tau>(M+1)/n\}} X_t], \tag{2.2.41}$$

where M is such that $(M+1)/n < t \leq (M+2)/n$. Notice that

$$A \cap \{\tau > (M+1)/n\} = \bigcap_{i=1}^{m} \{X_{t_i} \in B_i\} \cap \{\tau > (M+1)/n\} \in \mathscr{M}_{(M+1)/n}.$$

It then follows that

$$\mathbb{E}[1_{A\cap\{\tau>(M+1)/n\}} X_t] = \mathbb{E}[\mathbb{E}[1_{A\cap\{\tau>(M+1)/n\}} X_t | \mathscr{M}_{(M+1)/n}]] = \mathbb{E}[1_{A\cap\{\tau>(M+1)/n\}} X_{(M+1)/n}],$$

since $(X_t)_{t\geq 0}$ is a martingale. By (2.2.41), we have

$$\mathbb{E}[1_{A\cap\{\tau>(j+1)/n\}} X_{t\wedge\tau_n}] = \sum_{l=j+1}^{M-1} \mathbb{E}[1_{A\cap\{l/n<\tau\leq(l+1)/n\}} X_{(l+1)/n}] + \mathbb{E}[1_{A\cap\{\tau>M/n\}} X_{(M+1)/n}]. \tag{2.2.42}$$

Proceeding in this manner step by step, we arrive at

$$\begin{aligned}\mathbb{E}[1_{A\cap\{\tau>(j+1)/n\}} X_{t\wedge\tau_n}] &= \mathbb{E}[1_{A\cap\{\tau>(j+1)/n\}} X_{(j+2)/n}] = \mathbb{E}[1_{A\cap\{\tau>(j+1)/n\}} X_{(j+1)/n}] \\ &= \mathbb{E}[1_{A\cap\{\tau>(j+1)/n\}} X_s] = \mathbb{E}[1_{A\cap\{\tau>(j+1)/n\}} X_{s\wedge\tau_n}],\end{aligned}$$

since $s = (j+1)/n$, $A \cap \{\tau > (j+1)/n\} = \bigcap_{i=1}^{m} \{X_{t_i} \in B_i\} \cap \{\tau > (j+1)/n\}$ is in $\mathscr{M}_{(j+1)/n}$ and $s \wedge \tau_n = s$ on this set. Recalling (2.2.40), we have

$$\mathbb{E}[1_A X_{s\wedge\tau_n}] = \mathbb{E}[1_A X_{t\wedge\tau_n}]. \tag{2.2.43}$$

Next we consider the case of $j/n < s < (j+1)/n$ for some j. We modify τ_n by taking

$$\tau_n = \begin{cases} 0, & \tau = 0, \\ \frac{i+1}{n}, & \frac{i}{n} < \tau \leq \frac{i+1}{n}, i < j, \\ s, & \frac{j}{n} < \tau \leq s, \\ \frac{j+1}{n}, & s < \tau \leq \frac{j+1}{n}. \end{cases} \tag{2.2.44}$$

Then τ_n satisfies the properties in Lemma 2.61, and (2.2.43) follows as before.

We complete the proof of the theorem first in the case when τ is bounded, i. e., $\tau \leq N$ with some $N < \infty$. The range of τ_n is $\{s_i^n\}_{i\in\dot{\mathbb{N}}} \subset [0, N+1)$. Applying the proof of (2.2.43) to $|X_t|$ instead of X_t, we obtain

$$\mathbb{E}[1_A |X_{s\wedge\tau_n}|] \leq \mathbb{E}[1_A |X_{t\wedge\tau_n}|] \tag{2.2.45}$$

whenever $t > s$, since $|X_t|$ is a submartingale. Taking

$$A = \{|X_{s\wedge\tau_n}| > \lambda\} = \{X_{s\wedge\tau_n} \in [-\lambda, \lambda]^c\}, \quad t = N+1,$$

we get from (2.2.45) that

$$\mathbb{E}[1_{\{|X_{s\wedge\tau_n}|>\lambda\}} |X_{s\wedge\tau_n}|] \leq \mathbb{E}[1_{\{|X_{s\wedge\tau_n}|>\lambda\}} |X_{\tau_n}|]. \tag{2.2.46}$$

Since $|X_t|$ is a submartingale, we see that

$$\begin{aligned} \mathbb{E}[1_{\{|X_{s\wedge\tau_n}|>\lambda\}\cap\{\tau_n=s_i^n\}} |X_{\tau_n}|] &= \mathbb{E}[1_{\{|X_{s\wedge\tau_n}|>\lambda\}\cap\{\tau_n=s_i^n\}} |X_{s_i^n}|] \\ &\leq \mathbb{E}[\mathbb{E}[1_{\{|X_{s\wedge\tau_n}|>\lambda\}\cap\{\tau_n=s_i^n\}} |X_{N+1}| | \mathscr{M}_{s_i^n}]] = \mathbb{E}[1_{\{|X_{s\wedge\tau_n}|>\lambda\}\cap\{\tau_n=s_i^n\}} |X_{N+1}|]. \end{aligned} \tag{2.2.47}$$

Summing over i, we have

$$\mathbb{E}[1_{\{|X_{s\wedge\tau_n}|>\lambda\}} |X_{\tau_n}|] \leq \mathbb{E}[1_{\{|X_{s\wedge\tau_n}|>\lambda\}} |X_{N+1}|].$$

From (2.2.46), it follows that

$$\mathbb{E}[1_{\{|X_{s\wedge\tau_n}|>\lambda\}} |X_{s\wedge\tau_n}|] \leq \mathbb{E}[1_{\{|X_{s\wedge\tau_n}|>\lambda\}} |X_{N+1}|]. \tag{2.2.48}$$

Consider $\sup_{n\in\mathbb{N}} \sup_{s\geq 0} |X_{s\wedge\tau_n}|$. Since τ is bounded and $\tau_n \downarrow \tau$ as $n \uparrow \infty$, we have

$$\sup_{n\in\mathbb{N}} \sup_{s\geq 0} |X_{s\wedge\tau_n}| = \sup_{n\in\mathbb{N}} \sup_{0\leq r\leq\tau_n} |X_r| \leq \sup_{0\leq r\leq\tau_1} |X_r| \leq \sup_{0\leq r\leq N} |X_r|.$$

Since $(X_t)_{t\geq 0}$ is a right-continuous martingale, we have

$$P\Big(\sup_{n\in\mathbb{N}}\sup_{s\geq 0}|X_{s\wedge\tau_n}|>\lambda\Big)\leq P\Big(\sup_{0\leq r\leq N}|X_r|>\lambda\Big)\leq\frac{\mathbb{E}[|X_N|]}{\lambda}$$

by the martingale inequality quoted in Theorem 2.72. Hence

$$P\Big(\sup_{n\in\mathbb{N}}\sup_{s\geq 0}|X_{s\wedge\tau_n}|>\lambda\Big)\to 0$$

as $\lambda\to\infty$. The right-hand side of (2.2.48) can be estimated as

$$\mathbb{E}[1_{\{|X_{s\wedge\tau_n}|>\lambda\}}|X_{N+1}|]\leq\mathbb{E}[1_{\{\sup_{0\leq r\leq N}|X_r|>\lambda\}}|X_{N+1}|]\to 0$$

as $\lambda\to\infty$. Thus

$$\inf_{\lambda\geq 0}\sup_{n\in\mathbb{N}}\mathbb{E}[1_{\{|X_{s\wedge\tau_n}|>\lambda\}}|X_{s\wedge\tau_n}|]\leq\inf_{\lambda\geq 0}\mathbb{E}[1_{\{\sup_{0\leq r\leq N}|X_r|>\lambda\}}|X_{N+1}|]=0,$$

$\{X_{s\wedge\tau_n}\}_{n\in\mathbb{N}}$ is uniformly integrable, and $X_{s\wedge\tau_n}$ converges in $L^1(\Omega)$ as $n\to\infty$. Similarly, we can show that $\{X_{t\wedge\tau_n}\}_{n\in\mathbb{N}}$ is also uniformly integrable. Since $t\mapsto X_t$ is right-continuous and $\tau_n\downarrow\tau$ as $n\uparrow\infty$, it can be seen that $\lim_{n\to\infty}X_{t\wedge\tau_n}=X_{t\wedge\tau}$ in $L^1(\Omega)$. It follows that $\infty>\limsup_{n\to\infty}\mathbb{E}[|X_{t\wedge\tau_n}|]=\mathbb{E}[|X_{t\wedge\tau}|]$ from Corollary 2.32. Moreover, from (2.2.43) it follows that

$$\mathbb{E}[1_A X_{s\wedge\tau}]=\mathbb{E}[1_A X_{t\wedge\tau}].\tag{2.2.49}$$

Since this is true for any set A of the form (2.2.39), $(X_{t\wedge\tau})_{t\geq 0}$ is a martingale and right-continuous in t.

Finally, suppose that τ is unbounded. The stopping time $\tau\wedge N$ is bounded. Thus with $N>t$ we get $\mathbb{E}[|X_{t\wedge(\tau\wedge N)}|]=\mathbb{E}[|X_{t\wedge\tau}|]<\infty$, and

$$\mathbb{E}[1_A X_{s\wedge\tau\wedge N}]=\mathbb{E}[1_A X_{t\wedge\tau\wedge N}]\tag{2.2.50}$$

is obtained. Taking $N>t$, the relation (2.2.49) follows. □

To conclude this section, we discuss the Doob–Meyer decomposition of submartingales. Let $(X_n)_{n\in\dot{\mathbb{N}}}$ be a discrete submartingale with respect to $(\mathcal{F}_n)_{n\in\dot{\mathbb{N}}}$. Define $A_n=A_{n-1}+\mathbb{E}[X_n-X_{n-1}|\mathcal{F}_{n-1}]$ for $n\geq 1$ with $A_0=0$. We have

$$X_n=M_n+A_n,$$

where $M_n=X_n-A_n$. It is direct to see that A_n is adapted, $A_n\leq A_{n+1}$, and M_n is a martingale. We can extend this to the continuous case.

Definition 2.74 (Increasing process). An adapted process $(A_t)_{t\geq 0}$ on a filtered probability space $(\Omega,\mathcal{F},(\mathcal{F}_t)_{t\geq 0},P)$ is called an *increasing process* if for a. e. $\omega\in\Omega$,

(1) $A_0(\omega)=0$;

(2) $t \mapsto A_t(\omega)$ is a nondecreasing, right-continuous function;
(3) $\mathbb{E}[A_t] < \infty$, for every $t \geq 0$

hold. Furthermore, if $\mathbb{E}[A_\infty] < \infty$ with $A_\infty = \lim_{t\to\infty} A_t$, an increasing process $(A_t)_{t\geq 0}$ is called an *integrable increasing process.*

Definition 2.75 (Natural process). An increasing process $(A_t)_{t\geq 0}$ is called a *natural process* if for every bounded martingale $(X_t)_{t\geq 0}$,

$$\mathbb{E}\Big[\int_{(0,t]} X_s dA_s\Big] = \mathbb{E}\Big[\int_{(0,t]} X_{s-} dA_s\Big]$$

holds, for every $t \geq 0$.

Remark 2.76. It is known that an integrable increasing process is natural if and only if it is predictable (see Definition 3.64). If an integrable increasing process is continuous, then it is natural.

Definition 2.77 (Classes D and DL). Let $(\Omega, \mathscr{F}, (\mathscr{F}_t)_{t\geq 0}, P)$ be a filtered probability space. Let $\mathscr{S}$ (resp. $\mathscr{S}_a$) be the set of all stopping times τ satisfying $P(\tau < \infty) = 1$ (respectively, $P(\tau \leq a) = 1$ for a given finite number $a > 0$). An adapted process $(X_t)_{t\geq 0}$ with respect to $(\mathscr{F}_t)_{t\geq 0}$ is said to be of *class D* if the family $\{X_\tau\}_{\tau\in\mathscr{S}}$ is uniformly integrable, and of *class DL* if the family $\{X_\tau\}_{\tau\in\mathscr{S}_a}$ is uniformly integrable for every $0 < a < \infty$.

Every right-continuous martingale $(X_t)_{t\geq 0}$ is of class DL, since

$$\mathbb{E}[|X_\sigma| 1_{\{|X_\sigma|>c\}}] \leq \mathbb{E}[|X_a| 1_{\{|X_\sigma|>c\}}]$$

for $\sigma \in \mathscr{S}_a$ by Theorem 2.70. Here we do not need to assume that $(X_t)_{t\geq 0}$ is uniformly integrable since σ is bounded. From the inequality

$$\sup_{\sigma\in\mathscr{S}_a} P(|X_\sigma| > c) \leq \sup_{\sigma\in\mathscr{S}_a} \frac{\mathbb{E}[|X_\sigma|]}{c} \leq \frac{\mathbb{E}[|X_a|]}{c},$$

it is seen that the family of random variables $\{X_\sigma\}_{\sigma\in\mathscr{S}_a}$ is uniformly integrable. Similarly, we can show that every right-continuous uniformly integrable martingale $(X_t)_{t\geq 0}$ is of class D.

If a right-continuous submartingale $X = (X_t)_{t\geq 0}$ is represented as $X_t = M_t + A_t$, where $(M_t)_{t\geq 0}$ is a martingale and $A = (A_t)_{t\geq 0}$ an increasing process, then $(X_t)_{t\geq 0}$ is of class DL. The converse statement is known as *Doob–Meyer decomposition.* We state this without proof.

Theorem 2.78 (Doob–Meyer decomposition). *Let $(\Omega, \mathscr{F}, (\mathscr{F}_t)_{t\geq 0}, P)$ be a filtered probability space. Suppose that $(\mathscr{F}_t)_{t\geq 0}$ is a right-continuous and complete filtration, and $(X_t)_{t\geq 0}$ is a right-continuous submartingale of class DL. Then it has the decomposition*

$$X_t = M_t + A_t, \quad 0 \le t < \infty, \tag{2.2.51}$$

where $(M_t)_{t\ge0}$ is a right-continuous martingale and $(A_t)_{t\ge0}$ an increasing process. The random process $(A_t)_{t\ge0}$ can be chosen to be a natural process, and with this extra condition $(M_t)_{t\ge0}$ and $(A_t)_{t\ge0}$ are uniquely determined. Furthermore, if $(X_t)_{t\ge0}$ is of class D, then $(X_t)_{t\ge0}$ is uniformly integrable and $(A_t)_{t\ge0}$ is integrable.

2.2.4 Markov properties

The Markov property of a random process is a formalized expression of independence of the future of the process from its past. In this subsection we discuss some equivalent statements of the Markov property of a random process.

Definition 2.79 (Markov process). Let $(X_t)_{t\ge0}$ be an adapted process on a filtered probability space $(\Omega, \mathcal{F}, (\mathcal{F}_t)_{t\ge0}, P)$. The process $(X_t)_{t\ge0}$ is said to be a *Markov process* with respect to $(\mathcal{F}_t)_{t\ge0}$ whenever

$$\mathbb{E}\left[f(X_t)|\mathcal{F}_s\right] = \mathbb{E}\left[f(X_t)|\sigma(X_s)\right], \quad 0 \le s \le t, \tag{2.2.52}$$

for all bounded Borel measurable functions f.

Markov processes can be characterized by their probability transition kernels.

Definition 2.80 (Probability transition kernel). A map $p : \mathbb{R}^+ \times \mathbb{R}^+ \times \mathbb{R}^d \times \mathcal{B}(\mathbb{R}^d) \to \mathbb{R}$, $(s, t, x, A) \mapsto p(s, t, x, A)$, is called a *probability transition kernel* if
(1) $p(s, t, x, \cdot)$ is a probability measure on $\mathcal{B}(\mathbb{R}^d)$;
(2) $p(s, t, \cdot, A)$ is Borel measurable;
(3) for every $0 \le r \le s \le t$, the *Chapman–Kolmogorov identity*

$$\int_{\mathbb{R}^d} p(s, t, y, A) p(r, s, x, dy) = p(r, t, x, A) \tag{2.2.53}$$

holds.

Let $(X_t)_{t\ge0}$ be a Markov process. We define

$$p_X(s, t, x, A) = \mathbb{E}[1_A(X_t)|X_s = x] \tag{2.2.54}$$

for $A \in \mathcal{B}(\mathbb{R}^d)$, $0 \le s \le t < \infty$, and $x \in \mathbb{R}^d$. Recall the notation at the right-hand side of (2.2.54). The random variable $\mathbb{E}[1_A(X_t)|\sigma(X_s)]$ is $\sigma(X_s)$-measurable and by Lemma 2.38 it can be expressed as $\mathbb{E}[1_A(X_t)|\sigma(X_s)] = f(X_s)$ with a Borel measurable function f, and we use the notation $f(x) = \mathbb{E}[1_A(X_t)|X_s = x]$. By (2.2.54), we then see that $p_X(s, t, X_s, A) = \mathbb{E}[1_A(X_t)|\sigma(X_s)]$.

Lemma 2.81. *Suppose that* $(X_t)_{t\geq 0}$ *is a Markov process on a filtered probability space* $(\Omega, \mathscr{F}, (\mathscr{F}_t)_{t\geq 0}, P)$. *Then* p_X *is a probability transition kernel.*

Proof. Properties (1)–(2) of Definition 2.79 are straightforward to check, therefore we only show (3). Since $\int_{\mathbb{R}^d} 1_A(y) p_X(r,s,x,dy) = \mathbb{E}[1_A(X_s)|X_r = x]$, we have the equality $\int_{\mathbb{R}^d} f(y) p_X(r,s,x,dy) = \mathbb{E}[f(X_s)|X_r = x]$ for all bounded Borel measurable functions f. Hence

$$\int_{\mathbb{R}^d} p_X(s,t,y,A) p_X(r,s,x,dy) = \mathbb{E}[p_X(s,t,X_s,A)|X_r = x]. \tag{2.2.55}$$

Here $p_X(s,t,X_s,A)$ is given by $p_X(s,t,y,A)$ evaluated at $y = X_s$. As $p_X(s,t,X_s,A) = \mathbb{E}[1_A(X_t)|\sigma(X_s)]$, we have $p_X(s,t,X_s,A) = \mathbb{E}[1_A(X_t)|\mathscr{F}_s]$ by the Markov property. Inserting this into (2.2.55), we obtain

$$\int_{\mathbb{R}^d} p_X(s,t,y,A) p_X(r,s,x,dy) = \mathbb{E}[\mathbb{E}[1_A(X_t)|\mathscr{F}_s]|X_r = x]. \tag{2.2.56}$$

Since $\sigma(X_r) \subset \mathscr{F}_s$, we have $\mathbb{E}[\mathbb{E}[1_A(X_t)|\mathscr{F}_s]|\sigma(X_r)] = \mathbb{E}[1_A(X_t)|\sigma(X_r)]$, and therefore

$$\mathbb{E}[\mathbb{E}[1_A(X_t)|\mathscr{F}_s]|X_r = x] = p_X(r,t,x,A).$$

Combining this with (2.2.56), $\int_{\mathbb{R}^d} p_X(s,t,y,A) p_X(r,s,x,dy) = p_X(r,t,x,A)$ is obtained. □

The probability measure

$$P_0(A) = P(X_0 \in A) \tag{2.2.57}$$

is the *initial distribution* of the Markov process describing the random variable at $t = 0$. The family of finite-dimensional distributions of $(X_t)_{t\geq 0}$ is determined by its probability transition kernel and initial distribution through the formula

$$P(X_{t_0} \in A_0, \dots, X_{t_n} \in A_n) = \int_{(\mathbb{R}^d)^{(n+1)}} \left(\prod_{i=0}^n 1_{A_i}(x_i)\right) \left(\prod_{i=1}^n p(t_{i-1}, t_i, x_{i-1}, dx_i)\right) 1_{A_0}(x_0) P_0(dx_0), \tag{2.2.58}$$

for $n = 1, 2, \dots$, and where $t_0 = 0$.

Proposition 2.82. *Let* $(X_t)_{t\geq 0}$ *be an* $\mathbb{R}^d$*-valued random process on a probability space* $(\Omega, \mathscr{F}, P)$, P_0 *a probability measure on* $\mathbb{R}^d$, *and* $p(s,t,x,A)$ *a probability transition kernel. Suppose that the finite-dimensional distributions of the process are given by* (2.2.58). *Then* $(X_t)_{t\geq 0}$ *is a Markov process under the natural filtration with probability transition kernel* $p(s,t,x,A)$.

Proof. It suffices to show that

$$\mathbb{E}\Big[1_A(X_r)\prod_{j=0}^{n}1_{A_j}(X_{t_j})\Big] = \mathbb{E}\Big[\mathbb{E}\,[1_A(X_r)|\sigma(X_t)]\prod_{j=0}^{n}1_{A_j}(X_{t_j})\Big] \tag{2.2.59}$$

for every $0 = t_0 < t_1 \le \cdots \le t_n \le t \le r$, and every $A, A_j \in \mathcal{B}(\mathbb{R}^d), j = 0, 1, \ldots, n$. From the identity $\mathbb{E}[1_A(X_t)\mathbb{E}[f(X_r)|\sigma(X_t)]] = \mathbb{E}[1_A(X_t)f(X_r)]$, it follows that

$$\int_{\mathbb{R}^d} 1_A(y)\mathbb{E}[f(X_r)|X_t = y]P_t(dy) = \int_{\mathbb{R}^d} 1_A(y)f(y')p(t,r,y,dy')P_t(dy),$$

where $P_t(dy) = \int_{\mathbb{R}^d} P_0(dy_0)p(0,t,y_0,dy)$ and $P_t(dy)$ is the distribution of X_t on $\mathbb{R}^d$. Thus it is seen that

$$\mathbb{E}[f(X_r)|X_t = y] = \int_{\mathbb{R}^d} f(y')p(t,r,y,dy'). \tag{2.2.60}$$

By the definition of $\mathbb{E}[f(X_r)|X_t = y]$, we have

$$\mathbb{E}[f(X_r)|\sigma(X_t)] = \int_{\mathbb{R}^d} f(y)p(t,r,X_t,dy) \tag{2.2.61}$$

and thus

$$\mathbb{E}[1_A(X_r)|\sigma(X_t)] = p(t,r,X_t,A). \tag{2.2.62}$$

Using (2.2.62), (2.2.58), and the Chapman–Kolmogorov identity, (2.2.59) follows. □

Furthermore, we have the following strong description of the converse situation.

Proposition 2.83. *Suppose that a map p satisfying the properties in Definition* 2.80 *and a probability distribution P_0 on $\mathbb{R}^d$ are given. Then there exist a filtered probability space $(\Omega, \mathcal{F}, (\mathcal{F}_t)_{t\ge 0}, P)$ and a Markov process $(X_t)_{t\ge s}$ on it such that p is its probability transition kernel and*

$$\mathbb{E}[1_A(X_t)|X_s = x] = p(s,t,x,A) \tag{2.2.63}$$

holds; moreover, $P \circ X_0^{-1} = P_0$.

Proof. Let $\Omega = (\mathbb{R}^d)^{[0,\infty)}$ be the set of all functions $\omega : [0,\infty) \to \mathbb{R}^d$, let $\mathcal{G} = \sigma(\mathcal{A})$ be the σ-field generated by cylinder sets $\mathcal{A}$ defined by (2.2.6) with $\mathbb{R}$ replaced by $\mathbb{R}^d$. Consider the measurable space $(\Omega, \mathcal{G})$. For $I \subset [0,\infty)$, we define a sub-σ-field $\mathcal{G}_I$ by $\mathcal{G}_I = \sigma(\{\omega \in \Omega \mid \omega(u), u \in I\})$, and $\mathcal{G}_s = \mathcal{G}_{[0,s]}$. We also define a family of finite-dimensional distributions by choosing arbitrary $0 = t_0 < t_1 < \cdots < t_n$, and $x_i \in \mathbb{R}^d$ for $0 \le i \le n$, and every $n \in \mathbb{N}$, by the expressions

$$F_{\{t_0,t_1,\dots,t_n\}}(x_0,x_1,\dots,x_n) = \int_{(\mathbb{R}^d)^{n+1}} \left(\prod_{i=0}^{n} 1_{(-\infty,x_i]}(y_i)\right) \prod_{i=1}^{n} p(t_{i-1},t_i,y_{i-1},dy_i)P_0(dy_0), \quad n \in \mathbb{N},$$

$$F_{\{t_0\}}(x_0) = \int_{\mathbb{R}^d} 1_{(-\infty,x_0]}(y_0)P_0(dy_0).$$

The Chapman–Kolmogorov identity gives for $0 \le k \le n$,

$$\lim_{x_k \uparrow \infty} F_{\{t_0,t_1,\dots,t_n\}}(x_0,x_1,\dots,x_n) = F_{\{t_0,\dots,t_{k-1},t_{k+1},\dots,t_n\}}(x_0,\dots,x_{k-1},x_{k+1},\dots,x_n).$$

Thus the family of distribution functions $(F_\Lambda)_{\Lambda\subset[0,\infty),|\Lambda|<\infty}$ is consistent in the sense of (2.2.5). Hence by the Kolmogorov's extension theorem, there exists a probability measure P on $(\Omega,\mathscr{G})$ such that

$$P(X_{t_0} \le x_0,\dots,X_{t_n} \le x_n) = F_{\{t_0,t_1,\dots,t_n\}}(x_0,x_1,\dots,x_n), \tag{2.2.64}$$

and thus

$$P(X_{t_0} \in E_0,\dots,X_{t_n} \in E_n) = \int_{(\mathbb{R}^d)^{n+1}} \left(\prod_{i=0}^{n} 1_{E_i}(x_i)\right) dF_{\{t_0,t_1,\dots,t_n\}}(x_0,x_1,\dots,x_n)$$

for $E_i \in \mathcal{B}(\mathbb{R}^d)$, $i = 0,1,\dots,n$, $n \in \mathbb{N}$. Note that $X_t(\omega) = \omega(t)$ is the coordinate process. Taking $n = 0$ in (2.2.64), we get $P_0 = P \circ X_0^{-1}$. Equality (2.2.63) is equivalent to

$$\mathbb{E}[1_A(X_t)|\sigma(X_s)] = p(s,t,X_s,A). \tag{2.2.65}$$

In order to see this, it is sufficient to prove that

$$\mathbb{E}[1_A(X_t)|\mathscr{G}_s] = \mathbb{E}[1_A(X_t)|\sigma(X_s)], \quad s \le t, \tag{2.2.66}$$

$$\mathbb{E}[1_A(X_t)|\mathscr{G}_s] = p(s,t,X_s,A). \tag{2.2.67}$$

Note that (2.2.67) implies (2.2.66), which means that $(X_t)_{t\ge0}$ is a Markov process. To show (2.2.67), it suffices to see that

$$\mathbb{E}[1_A(X_t)1_D] = \mathbb{E}[p(s,t,X_s,A)1_D], \tag{2.2.68}$$

for every set of the form $D = \{X_{t_0} \in A_0,\dots,X_{t_n} \in A_n\}$, with $0 = t_0 < t_1 < \dots < t_n = s \le t$ and $A, A_j \in \mathcal{B}(\mathbb{R}^d)$ for $j = 0,\dots,n$. Once this is shown, (2.2.68) follows for all $D \in \mathscr{G}_{[0,s]}$ by a limiting argument. Write $t_{n+1} = t$ and $A_{n+1} = A$. The left-hand side of (2.2.68) becomes

$$\begin{aligned}\mathbb{E}[1_A(X_t)1_D] &= P(D \cap \{X_t \in A\}) \\ &= \int_{(\mathbb{R}^d)^{n+2}} \left(\prod_{i=0}^{n+1} 1_{A_i}(y_i)\right) \prod_{i=1}^{n+1} p(t_{i-1},t_i,y_{i-1},dy_i)P_0(dy_0)\end{aligned}$$

$$= \int_{(\mathbb{R}^d)^{n+1}} p(t_n, t_{n+1}, y_n, A_{n+1}) \left(\prod_{i=0}^{n} 1_{A_i}(y_i) \right) \prod_{i=1}^{n} p(t_{i-1}, t_i, y_{i-1}, dy_i) P_0(dy_0). \tag{2.2.69}$$

We prove next that

$$\int_{(\mathbb{R}^d)^{n+1}} f(y_n) \left(\prod_{i=0}^{n} 1_{A_i}(y_i) \right) \prod_{i=1}^{n} p(t_{i-1}, y_{i-1}, t_i, dy_i) P_0(dy_0) = \mathbb{E}\left[\left(\prod_{i=0}^{n} 1_{\{X_{t_i} \in A_i\}} \right) f(X_{t_n}) \right], \tag{2.2.70}$$

where f is an arbitrary Borel measurable function on $\mathbb{R}^d$. Then substituting the map

$$y_n \mapsto \int_{\mathbb{R}^d} 1_{A_{n+1}}(y_{n+1}) p(t_n, t_{n+1}, y_n, dy_{n+1})$$

into $f(y_n)$ in (2.2.70), we have by (2.2.69)

$$\mathbb{E}[1_A(X_t) 1_D] = \mathbb{E}\left[\left(\prod_{i=0}^{n} 1_{\{X_{t_i} \in A_i\}} \right) p(t_n, t_{n+1}, X_{t_n}, A_{n+1}) \right] = \mathbb{E}\left[\left(\prod_{i=0}^{n} 1_{\{X_{t_i} \in A_i\}} \right) p(s, t, X_s, A) \right],$$

and hence (2.2.68) follows. Let f be the indicator function 1_F with $F \in \mathcal{B}(\mathbb{R}^d)$, and $A'_n = A_n \cap F$. We have

$$\begin{aligned} &\int_{(\mathbb{R}^d)^{n+1}} 1_F(y_n) \left(\prod_{i=0}^{n} 1_{A_i}(y_i) \right) \prod_{i=1}^{n} p(t_{i-1}, t_i, y_{i-1}, dy_i) P_0(dy_0) \\ &= \int_{(\mathbb{R}^d)^{n+1}} 1_{A'_n}(y_n) \left(\prod_{i=0}^{n-1} 1_{A_i}(y_i) \right) \prod_{i=1}^{n} p(t_{i-1}, t_i, y_{i-1}, dy_i) P_0(dy_0) \\ &= \mathbb{E}\left[\left(\prod_{i=0}^{n} 1_{\{X_{t_i} \in A_i\}} \right) 1_F(X_{t_n}) \right] \end{aligned}$$

by the Chapman–Kolmogorov identity. Hence (2.2.70) follows by approximation. We see that $\mathbb{E}[f(X_t)|\mathcal{G}_s] = \mathbb{E}[f(X_t)|\sigma(X_s)]$ for every bounded Borel measurable function f, and conclude that $(X_t)_{t\geq 0}$ is a Markov process. □

When the probability transition kernel $p(s, t, X, A)$ of a Markov process satisfies $p(s, t, x, A) = p(0, t-s, x, A)$, the process is said to be a *stationary Markov process*. In this case, by a slight abuse of notation, we write $p(t, x, A)$ for $p(0, t, x, A)$, and the Chapman–Kolmogorov identity reduces to

$$\int_{\mathbb{R}^d} p(s, y, A) p(t, x, dy) = p(s + t, x, A). \tag{2.2.71}$$

In case $(X_t)_{t\in\mathbb{R}}$ is a Markov process indexed by the full time-line $\mathbb{R}$, it is called a *reversible Markov process* if the joint distributions of $X_{t_1},\dots,X_{t_n}$ and $X_{s-t_1},\dots,X_{s-t_n}$ are the same for all $t_1,\dots,t_n,s\in\mathbb{R}$, and all $n\in\mathbb{N}$. A reversible Markov process is necessarily stationary, since both $X_{t_1},\dots,X_{t_n}$ and $X_{s+t_1},\dots,X_{s+t_n}$ have the same joint distribution as $X_{-t_1},\dots,X_{-t_n}$.

2.2.5 Feller transition kernels and generators

Markov processes are intimately related to operator semigroups, which intuitively describe the time evolution of the related probability distributions. For essential background material on operators, see Section 3.1.5 below.

Let $C_\infty(\mathbb{R}^d)$ be the set of real-valued continuous functions on $\mathbb{R}^d$ such that $\lim_{|x|\to\infty} f(x) = 0$. With the norm $\|f\|_\infty = \sup_{x\in\mathbb{R}^d} |f(x)|$ on $C_\infty(\mathbb{R}^d)$, the normed space $(C_\infty(\mathbb{R}^d), \|\cdot\|_\infty)$ is a Banach space.

Definition 2.84 (Feller transition kernel). A probability transition kernel $p(t,x,A)$ is said to be a *Feller transition kernel* whenever the following conditions hold:
(1) $p(t,x,\mathbb{R}^d) = 1$, for all $t\geq 0$ and $x\in\mathbb{R}^d$;
(2) $\int_{\mathbb{R}^d} p(t,\cdot,dy)f(y)\in C_\infty(\mathbb{R}^d)$ for all $t\geq 0$ and $f\in C_\infty(\mathbb{R}^d)$;
(3) $\lim_{t\downarrow 0}\int_{\mathbb{R}^d} p(t,x,dy)f(y) = f(x)$, for every $f\in C_\infty(\mathbb{R}^d)$ and $x\in\mathbb{R}^d$.

Consider

$$(P_t f)(x) = \int_{\mathbb{R}^d} p(t,x,dy)f(y), \quad f\in C_\infty(\mathbb{R}^d). \tag{2.2.72}$$

When $P_t f\geq 0$ for $f\geq 0$, for all $t\geq 0$, the semigroup $\{P_t : t\geq 0\}$ is called *nonnegative.*

Definition 2.85 (C_0-semigroup). The one-parameter family of bounded operators $\{S_t : t\geq 0\}$ on a Banach space $\mathcal{B}$ is said to be a *strongly continuous one-parameter semigroup* or *C_0-semigroup* whenever
(1) $S_0 = 1$;
(2) $S_s S_t = S_{s+t}$;
(3) S_t is strongly continuous in t, i. e., $\text{s-}\lim_{t\to 0} S_t = S_0 = 1$.

The next result gives the relationship between Markov processes and C_0-semigroups.

Proposition 2.86. *Let $p(t,x,A)$ be a Feller transition kernel. Then $\{P_t : t\geq 0\}$ is a nonnegative C_0-semigroup on $C_\infty(\mathbb{R}^d)$.*

Proof. It is easy to see that $P_0 = 1$ and P_t is nonnegative, while $P_tP_s = P_{t+s}$ follows by the Chapman–Kolmogorov identity. Moreover, $\|P_tf\|_\infty \leq \|f\|_\infty$ can also easily be seen. The nontrivial part is to show strong continuity, i. e.,

$$\lim_{t\downarrow 0} P_tf = f \tag{2.2.73}$$

in the sup-norm. Let $\alpha > 0$ and define $f_\alpha(x) = \int_0^\infty \alpha e^{-\alpha t} P_tf(x)dt$. Then $f_\alpha \in C_\infty(\mathbb{R}^d)$, since $f \in C_\infty(\mathbb{R}^d)$. Let $\mathscr{M}$ be the closed subspace of $C_\infty(\mathbb{R}^d)$ spanned by f such that (2.2.73) holds. Take $f \geq 0$ and define $f_{\alpha,h}(x) = \int_h^\infty \alpha e^{-\alpha t} P_tf(x)dx$; then $f_{\alpha,h}(x) \uparrow f_\alpha(x)$ as $h \downarrow 0$. This convergence is uniform as the sequence of continuous monotone functions $f_{\alpha,h}$, $h \geq 0$, converges to f_α pointwise on the compact set $\mathbb{R}^d \cup \{\infty\}$. Since $\mathscr{M}$ is closed, $f_\alpha \in \mathscr{M}$ follows for any nonnegative $f \in C_\infty(\mathbb{R}^d)$ and therefore for all $f \in C_\infty(\mathbb{R}^d)$. Let $l \in C_\infty(\mathbb{R}^d)^*$ be a linear functional on $C_\infty(\mathbb{R}^d)$. Since $\|f_\alpha\|_\infty \leq \|f\|_\infty$ and, for every x, $\lim_{\alpha\to\infty} f_\alpha(x) = f(x)$, it follows that $f_\alpha \to f$ weakly as $\alpha \to \infty$ and hence $l(f_\alpha) \to l(f)$. Suppose $l(g) = 0$ for all $g \in \mathscr{M}$. We see that $l(f) = \lim_{\alpha\to\infty} l(f_\alpha) = 0$ for every $f \in C_\infty(\mathbb{R}^d)$. Thus the Hahn–Banach theorem gives $C_\infty(\mathbb{R}^d) = \mathscr{M}$. □

The converse statement of Proposition 2.86 also holds.

Proposition 2.87. *For every nonnegative C_0-semigroup on $C_\infty(\mathbb{R}^d)$, there exists a unique Feller transition kernel such that* (2.2.72) *holds.*

Proof. Let $\{P_t : t \geq 0\}$ be a nonnegative C_0-semigroup. It will be shown in Lemma 2.88 below that there exists a unique kernel $p(t, x, dy)$ such that

$$P_tf(x) = \int_{\mathbb{R}^d} p(t, x, dy)f(y), \quad f \in C_\infty(\mathbb{R}^d).$$

From $\|P_t\| \leq 1$ it follows that $p(t, x, \mathbb{R}^d) \leq 1$. By the continuity of $x \mapsto P_tf(x)$, we get the measurability of $x \mapsto p(t, x, B)$. The Chapman–Kolmogorov identity can be shown by the semigroup property $P_tP_s = P_{s+t}$. Finally, the strong continuity of $t \mapsto P_tf$ yields (3) of Definition 2.84. □

Lemma 2.88. *Let l be a continuous linear functional on $C_\infty(\mathbb{R}^d)$. Then there exists a finite signed measure ρ such that $l(f) = \int_{\mathbb{R}^d} f(x)d\rho(x)$ for all $f \in C_\infty(\mathbb{R}^d)$.*

Proof. Let $K = (\dot{\mathbb{R}^d}) = \mathbb{R}^d \cup \{\infty\}$ be the one-point compactification of $\mathbb{R}^d$. For every $f \in C_\infty(\mathbb{R}^d)$, we define $\tilde{f} \in C(K)$ by $\tilde{f}(x) = f(x)$ for $x \in \mathbb{R}^d$, and $\tilde{f}(\infty) = 0$. Any $g \in C(K)$ can be represented by $g = c + \tilde{f}$, where c is a constant and $f \in C_\infty(\mathbb{R}^d)$. The map i : $C(K) \ni g \mapsto f \in C_\infty(\mathbb{R}^d)$ is linear and $\|ig\| \leq 2\|g\|$. We define a linear functional $\tilde{l}$ on $C(K)$ by $\tilde{l}(g) = l(ig)$. Hence $\tilde{l}$ is a continuous linear functional on $C(K)$ and there exists a measure on $(K, \mathcal{B}(K))$ such that

$$\tilde{l}(g) = \int_K g(x) d\rho(x)$$

by the *Riesz theorem*. It follows that $l(f) = \int_K \tilde{f}(x)\rho(dx) = \int_{\mathbb{R}^d} f(x)\rho(dx)$. □

Let $(X_t)_{t\geq 0}$ be a Markov process and consider its Feller transition kernel. Then $\{P_t : t \geq 0\}$ in (2.2.72) is a C_0-semigroup by Proposition 2.86 and there exists a closed operator L on $C_\infty(\mathbb{R}^d)$ by the Hille–Yoshida theorem (Proposition 4.80 below) such that

$$\lim_{t\to 0} \frac{1}{t}(P_t f - f) = Lf,$$

where L is called the *generator* of the Markov process and is formally written as $P_t = e^{tL}$. The domain of L is defined by

$$D(L) = \left\{ f \in C_\infty(\mathbb{R}^d) \,\middle|\, \exists g \text{ such that } \lim_{t\downarrow 0} \left\| \frac{P_t f - f}{t} - g \right\|_\infty = 0 \right\}.$$

We will see in Section 3.1.5 below that Feller transition kernels form a rich class and further examples of generators will be given.

2.2.6 Invariant measures

Finally, we discuss time-invariant measures μ on $(\mathbb{R}^d, \mathcal{B}(\mathbb{R}^d))$ related to Markov processes. As seen below, formally an invariant measure is the eigenfunction associated with the zero-eigenvalue for the adjoint of the generator L of the Markov process. Furthermore, by using the invariant measure, we can extend the semigroup $\{P_t : t \geq 0\}$ on $C_\infty(\mathbb{R}^d)$ to $L^p(\mathbb{R}^d, d\mu)$ with suitable $p > 0$.

Definition 2.89 (Invariant measure). Let $(X_t)_{t\geq 0}$ be an $\mathbb{R}^d$-valued stationary Markov process on a filtered probability space $(\Omega, \mathcal{F}, (\mathcal{F}_t)_{t\geq 0}, P)$, with probability transition kernel $p(t, x, A)$. A probability measure μ on $(\mathbb{R}^d, \mathcal{B}(\mathbb{R}^d))$ is an *invariant measure* for $(X_t)_{t\geq 0}$ if it satisfies

$$\int_{\mathbb{R}^d} p(t, x, A)\mu(dx) = \mu(A)$$

for every $t \geq 0$ and $A \in \mathcal{B}(\mathbb{R}^d)$.

Lemma 2.90. *Let $(X_t)_{t\geq 0}$ be a d-dimensional stationary Markov process on a filtered probability space $(\Omega, \mathcal{F}, (\mathcal{F}_t)_{t\geq 0}, P)$ with a Feller transition kernel $p(t, x, A)$. Define the semigroup $\{P_t : t \geq 0\}$ on $C_\infty(\mathbb{R}^d)$ by (2.2.72). Then a measure μ is invariant for $(X_t)_{t\geq}$ if and only if for all $f \in C_\infty(\mathbb{R}^d)$ and all $t \geq 0$,*

$$\int_{\mathbb{R}^d} (P_t f)(x) d\mu(x) = \int_{\mathbb{R}^d} f(x) d\mu(x). \tag{2.2.74}$$

Proof. By the definition of the invariant measure, μ is invariant if and only if (2.2.74) holds for all $f(x) = 1_A(x)$ with $A \in \mathcal{B}(\mathbb{R}^d)$. By a limiting argument, the claim follows. □

Verifying directly that a given μ is an invariant measure is often difficult. Therefore the following criterion is useful.

Theorem 2.91. *Let the assumptions of Lemma 2.90 hold, and L be the generator of the semigroup $\{P_t : t \geq 0\}$. Then μ is an invariant measure for $(X_t)_{t\geq 0}$ if and only if for all $f \in D(L)$,*

$$\int_{\mathbb{R}^d} (Lf)(x)d\mu(x) = 0. \tag{2.2.75}$$

Proof. Let μ be an invariant measure and $f \in D(L)$. We have

$$P_tf - f = \int_0^t \frac{d}{ds}P_sfds = \int_0^t P_sLfds.$$

It follows that the family $\{\frac{1}{t}(P_tf - f)\}_{t>0}$ is uniformly bounded, and thus by the dominated convergence theorem,

$$\int_{\mathbb{R}^d} (Lf)(x)d\mu(x) = \lim_{t\to 0} \frac{1}{t} \int_{\mathbb{R}^d} (P_tf - f)d\mu(x) = 0.$$

Conversely, let (2.2.75) hold for all $f \in D(L)$. It follows from general theory that for every $t > 0$ and $g \in C_\infty(\mathbb{R}^d)$ there exists $f \in D(L)$ with $f - tLf = g$. It follows that $\int_{\mathbb{R}^d} f(x)d\mu(x) = \int_{\mathbb{R}^d} g(x)d\mu(x)$, and by iteration we get

$$\int_{\mathbb{R}^d} (1 - tL)^{-n}g(x)d\mu(x) = \int_{\mathbb{R}^d} f(x)d\mu(x)$$

for all $n \in \mathbb{N}$. The integrand at the left-hand side is uniformly bounded and converges to P_tg; thus, by the dominated convergence theorem, the claim follows. □

Theorem 2.92. *Choose $p \geq 1$. Let the assumptions of Lemma 2.90 hold, and suppose that $C_\infty(\mathbb{R}^d)$ is dense in $L^p(\mathbb{R}^d, d\mu)$. Then $\{P_t : t \geq 0\}$ can be extended to a C_0-semigroup on $L^p(\mathbb{R}^d, d\mu)$.*

Proof. Let $f \in C_\infty(\mathbb{R}^d)$. We have $P_tf(x) = \mathbb{E}[f(X_t)|X_0 = x]$ and

$$|P_tf(x)|^p \leq \mathbb{E}[|f(X_t)|^p|X_0 = x] = P_t(|f|^p)(x).$$

Then

$$\int_{\mathbb{R}^d} |P_tf(x)|^p d\mu(x) \leq \int_{\mathbb{R}^d} P_t(|f|^p)(x)d\mu(x) = \int_{\mathbb{R}^d} |f(x)|^p d\mu(x).$$

Here we used that μ is an invariant measure. Thus $\|P_t f\|_{L^p(\mathbb{R}^d,d\mu)} \le \|f\|_{L^p(\mathbb{R}^d,d\mu)}$ and $C_\infty(\mathbb{R}^d)$ is dense in $L^p(\mathbb{R}^d, d\mu)$. We can extend P_t to a contraction operator on L^p, which we denote by $\bar{P}_t$. It is straightforward to see that $\bar{P}_t\bar{P}_s = \bar{P}_{t+s}$ for $s, t \ge 0$, and $\bar{P}_0 = I$. We show strong continuity of $t \mapsto \bar{P}_t$. For this, it is sufficient to show the claim at $t = 0$. Since $p(t, x, dy)$ is a Feller transition kernel, $P_t f(x)$ is continuous at $t = 0$ for every $x \in \mathbb{R}^d$ and all $f \in C_\infty(\mathbb{R}^d)$. As $C_\infty(\mathbb{R}^d)$ is dense in $L^p(\mathbb{R}^d, d\mu)$, the L^p-continuity follows from the dominated convergence theorem. □

Finally, we introduce a martingale associated with the Markov process $(X_t)_{t\ge0}$ and its generator.

Theorem 2.93. *Let $(X_t)_{t\ge0}$ be a d-dimensional stationary Markov process on a filtered probability space $(\Omega, \mathscr{F}, (\mathscr{F}_t)_{t\ge0}, P)$ with a Feller transition kernel $p(t, x, A)$. Define the semigroup $\{P_t : t \ge 0\}$ on $C_\infty(\mathbb{R}^d)$ by (2.2.72), and let L be the generator of P_t and $f \in D(L)$. Then*

$$M_t = f(X_t) - f(X_0) - \int_0^t (Lf)(X_s)ds, \quad t \ge 0,$$

is a martingale on $(\Omega, \mathscr{F}, (\mathscr{F}_t)_{t\ge0}, P)$.

Proof. It is straightforward to see that $\mathbb{E}[|M_t|] < \infty$, $t \ge 0$. By the Markov property,

$$\mathbb{E}[f(X_t)|\mathscr{F}_s] = P_{t-s}f(X_s).$$

It is trivial to see that

$$\mathbb{E}[M_t|\mathscr{F}_s] = M_s + \mathbb{E}\left[f(X_t) - f(X_s) - \int_s^t Lf(X_r)dr \,\middle|\, \mathscr{F}_s\right]. \tag{2.2.76}$$

Since $f \in D(L)$, we have $\frac{d}{dt}P_t f = LP_t f = P_t Lf$ and $P_t f - f = \int_0^t LP_s f ds$. It follows that

$$\begin{aligned}\mathbb{E}\left[f(X_t) - f(X_s) - \int_s^t Lf(X_r)dr \,\middle|\, \mathscr{F}_s\right] &= P_{t-s}f(X_s) - f(X_s) - \int_s^t LP_{r-s}f(X_s)dr \\ &= P_{t-s}f(X_s) - f(X_s) - \int_0^{t-s} LP_r f(X_s)dr = 0.\end{aligned}$$

Hence the second term on the right-hand side of (2.2.76) is zero and the martingale property is proven. □

The asymptotic behavior of $\frac{1}{t}\int_0^t (Lf)(X_s)ds$ for given (classes of) functions, as $t \to \infty$, is an interesting object to study, leading to *functional central limit theorems*, extending the standard central limit theorem.

2.3 Brownian motion and Wiener measure

2.3.1 Construction of Brownian motion

After this extensive preparation of background material, we turn now to discussing Brownian motion, which is a random process of fundamental importance in pure and applied probability, and will play a central role throughout below. While Brownian motion is one of the best studied random processes, we offer here a brief summary of the aspects that we will need for our purposes in this book, and refer to the literature for many interesting properties which we do not even mention. Since Brownian motion lies at the intersection of many desirable properties a random process may have, by being a process with independent increments, a Gaussian process, a (strong) Markov process, a martingale, and the only Lévy process with almost surely continuous paths, many of the general facts discussed before can now be applied by particularizing the underlying distribution.

Definition 2.94 (Brownian motion). A real-valued random process $(B_t)_{t\geq 0}$ on a probability space $(\Omega, \mathscr{F}, P^x)$, $x \in \mathbb{R}$, is called *Brownian motion*, or *Wiener process*, starting at x, whenever the following conditions hold:

(1) $P^x(B_0 = x) = 1$;
(2) the increments $(B_{t_i} - B_{t_{i-1}})_{1\leq i\leq n}$ are independent Gaussian random variables for every collection $0 = t_0 < t_1 < \cdots < t_n$, $n \in \mathbb{N}$, of time-points, with

$$B_{t_i} - B_{t_{i-1}} \stackrel{\mathrm{d}}{=} N(0, t_i - t_{i-1});$$

(3) the function $t \mapsto B_t(\omega)$ is continuous for a. e. $\omega \in \Omega$.

In particular, $(B_t)_{t\geq 0}$ is called a *standard Brownian motion* when $x = 0$. We write simply $P = P^0$ when no confusion arises. A d-component vector of independent Brownian motions $(B_t^1, \ldots, B_t^d)_{t\geq 0}$ is called a *d-dimensional Brownian motion*.

Brownian motion is thus a real-valued continuous-time random process with index set $\mathbb{R}^+ = [0, \infty)$. Consider

$$\mathscr{X} = C(\mathbb{R}^+; \mathbb{R}^d) \tag{2.3.1}$$

equipped with the metric given by (2.1.3). First, we discuss the canonical representation of Brownian motion on the probability space $(\mathscr{X}, \mathcal{B}(\mathscr{X}), \mathcal{W}^x)$, where $\mathcal{W}^x$ is *Wiener measure* to be defined below. In this representation, a realization of the process $(B_t)_{t\geq 0}$ is given by

$$B_t(\omega) = \omega(t) \in \mathbb{R}^d, \quad \omega \in \mathscr{X}, \tag{2.3.2}$$

called the *coordinate process*. For each fixed $\omega \in \mathscr{X}$, we think of $\mathbb{R}^+ \ni t \mapsto B_t(\omega) \in \mathbb{R}^d$ as a *Brownian path*. We write

$$\Pi_t(x) = \frac{1}{(2\pi t)^{d/2}} \exp\left(-\frac{|x|^2}{2t}\right), \quad t > 0,\ x \in \mathbb{R}^d, \tag{2.3.3}$$

called the d-dimensional *heat kernel*.

Definition 2.95 (Wiener measure). Let $n \in \mathbb{N}$ be an arbitrary natural number. Consider an arbitrary collection of time-points $0 = t_0 < t_1 < \cdots < t_n$, and an arbitrary collection of Borel sets $E_j \in \mathcal{B}(\mathbb{R}^d)$, $j = 1, \dots, n$. The unique probability measure $\mathcal{W}^x$ on $(\mathscr{X}, \mathcal{B}(\mathscr{X}))$ whose family of finite-dimensional distributions is given by

$$\mathcal{W}^x(B_0 = x) = 1 \tag{2.3.4}$$

and

$$\mathcal{W}^x(B_{t_1} \in E_1, \dots, B_{t_n} \in E_n) = \int_{\mathbb{R}^{dn}} \left(\prod_{j=1}^n 1_{E_j}(x_j)\right)\left(\prod_{j=1}^n \Pi_{t_j - t_{j-1}}(x_{j-1} - x_j)\right) \prod_{j=1}^n dx_j \tag{2.3.5}$$

where $x_0 = x$ and $B_t(\omega) = \omega(t)$ for $\omega \in \mathscr{X}$, is called *Wiener measure* starting at $x \in \mathbb{R}^d$.

We write $\mathcal{W} = \mathcal{W}^0$, and for expectation with respect to Wiener measure we will use the notations

$$\mathbb{E}^x[F] = \mathbb{E}^x_{\mathcal{W}}[F] = \mathbb{E}_{\mathcal{W}^x}[F], \quad \mathbb{E}^0[F] = \mathbb{E}^0_{\mathcal{W}}[F] = \mathbb{E}_{\mathcal{W}^0}[F] = \mathbb{E}[F]. \tag{2.3.6}$$

By the definition above, we have

$$\mathbb{E}^x\left[\prod_{j=1}^n f_j(B_{t_j})\right] = \int_{\mathbb{R}^{dn}} \left(\prod_{j=1}^n f_j(x_j)\right)\left(\prod_{j=1}^n \Pi_{t_j - t_{j-1}}(x_{j-1} - x_j)\right) \prod_{j=1}^n dx_j \tag{2.3.7}$$

for all bounded Borel measurable functions $f_1, \dots, f_n$. In particular,

$$\mathbb{E}^x[f(B_t)] = \int_{\mathbb{R}^d} f(y)\Pi_t(x-y)dy. \tag{2.3.8}$$

A prime application of this formula is to obtain the moments of Brownian motion.

Proposition 2.96 (Moments). *Let $(B_t)_{t\geq 0}$ be a one-dimensional standard Brownian motion, and $s, t \geq 0$. Then we have the following expressions:*

(1) *(Characteristic function)*

$$\phi(u) = \mathbb{E}[e^{iu(B_t - B_s)}] = e^{-\frac{|t-s|u^2}{2}}, \quad u \in \mathbb{R};$$

(2) *(Moment generating function)*

$$\psi(u) = \mathbb{E}[e^{u(B_t - B_s)}] = e^{\frac{|t-s|u^2}{2}}, \quad u \in \mathbb{R};$$

(3) *(Moments)*

$$\mathbb{E}[(B_t - B_s)^n] = \begin{cases} \frac{(2k)!}{2^k k!}|t-s|^k & \text{if } n = 2k,\ k \in \dot{\mathbb{N}}; \\ 0 & \text{if } n = 2k+1,\ k \in \dot{\mathbb{N}}. \end{cases}$$

Proof. The characteristic and moment generating functions are obtained by completion of the square in the exponent and Gaussian integration, i. e., we have

$$\phi(u) = \int_{\mathbb{R}} e^{iux}\Pi_{t-s}(x)dx = e^{-\frac{|t-s|u^2}{2}}.$$

In the same way, we can see (2). The moments can be obtained from either of these functions. Write $X = B_t - B_s$. By (1), we see that $\phi \in C^\infty(\mathbb{R})$ and

$$\phi(u) = \sum_{k=0}^{2n} \frac{1}{k!}\frac{d^k\phi}{dx^k}(0)u^k + R_{2n+1} = \sum_{k=0}^{\infty} \frac{1}{k!}|t-s|^k u^{2k},$$

where R_{2n+1} is the remainder term and it gives $R_{2n+1} = \sum_{k=2n+1}^{\infty} \frac{1}{k!}|t-s|^k u^{2k}$ by the uniqueness of the Taylor expansion. Hence $R_{2n+1} \to 0$ as $n \to \infty$ follows. Since $\mathbb{E}[|\frac{d^k}{dx^k}e^{iuX}|] \le \mathbb{E}[|u^k|] < \infty$, we can exchange $\mathbb{E}$ and $\frac{d^k}{dx^k}$, and $\frac{d^k\phi}{dx^k}(u) = \mathbb{E}[(iu)^k e^{iuX}]$. Notice that the odd moments vanish for symmetry reasons. We have

$$\mathbb{E}[e^{iuX}] = \sum_{k=0}^{\infty} \frac{1}{k!} i^k \mathbb{E}[X^k]u^k = \sum_{n=0}^{\infty} \frac{(-1)^n}{(2n)!}\mathbb{E}[X^{2n}]u^{2n}.$$

Using the right-hand side of the expression of the characteristic function, we obtain

$$e^{-\frac{1}{2}|t-s|u^2} = \sum_{n=0}^{\infty} \frac{1}{n!}\left(-\frac{1}{2}|t-s|u^2\right)^n = \sum_{n=0}^{\infty} \frac{1}{n!}\frac{(-1)^n}{2^n}|t-s|^n u^{2n}.$$

Comparing the coefficients on the two sides yields the required result. □

Our next goal is to discuss the existence of Wiener measure. The main challenge is to show that there is no conflict between the Gaussian increments and path continuity. In Chapter 3 below we will see that path continuity is exceptional, and in a large canonical class of processes it only occurs for Brownian motion.

Theorem 2.97 (Existence of Wiener measure). *Wiener measure $\mathcal{W}^x$ exists on the measurable space $(\mathscr{X}, \mathcal{B}(\mathscr{X}))$, and the coordinate process $B_t(\omega) = \omega(t)$, $t \ge 0$, on the probability space $(\mathscr{X}, \mathcal{B}(\mathscr{X}), \mathcal{W}^x)$ is a Brownian motion starting at $x \in \mathbb{R}^d$.*

Proof. We assume $d = 1$, for $d \geq 2$ the proof is similar. The idea of the proof is first to construct a probability measure on all real-valued functions on the positive semiaxis, and next show that this measure is actually supported on the subspace of continuous functions.

Let $\Lambda_1 \subset \Lambda_2 \subset [0, \infty)$ with $|\Lambda_1| < \infty$ and $|\Lambda_2| < \infty$. Without restricting generality, we chose $\Lambda_1 = \{t_1, \ldots, t_n\}$ and $\Lambda_2 = \{t_1, \ldots, t_n, s_1, \ldots, s_m\}$, and define the projections $\pi_{\Lambda_1\Lambda_2} : \mathbb{R}^{|\Lambda_2|} \to \mathbb{R}^{|\Lambda_1|}$ by the map

$$(x_{t_1}, \ldots, x_{t_n}, x_{s_1}, \ldots, x_{s_m}) \mapsto (x_{t_1}, \ldots, x_{t_n}).$$

We define the projection $\pi_\Lambda : \mathbb{R}^{[0,\infty)} \to \mathbb{R}^{|\Lambda|}$ similarly. Also, define a probability measure μ_Λ on $\mathbb{R}^{|\Lambda|}$ for $\Lambda = \{t_1, \ldots, t_n\}$ by

$$\mu_\Lambda(E_1 \times \cdots \times E_n) = \int_{\mathbb{R}^n} \left(\prod_{j=1}^n 1_{E_j}(x_j) \right) \left(\prod_{j=1}^n \Pi_{t_j - t_{j-1}}(x_{j-1} - x_j) \right) \prod_{j=1}^n dx_j$$

with $x_0 = 0$. It is direct to check that $\pi^{-1}_{\Lambda_2\Lambda_1}(E_1 \times \cdots \times E_n) = E_1 \times \cdots \times E_n \times \underbrace{\mathbb{R} \times \cdots \times \mathbb{R}}_{m}$ and

$$\mu_{\Lambda_1}(E_1 \times \cdots \times E_n) = \mu_{\Lambda_2}(E_1 \times \cdots \times E_n \times \underbrace{\mathbb{R} \times \cdots \times \mathbb{R}}_{m}). \tag{2.3.9}$$

Equality (2.3.9) is the Kolmogorov's consistency relation (2.1.21) for μ_Λ above. Define the set function μ on

$$\mathscr{A} = \left\{ \pi^{-1}_{\Lambda[0,\infty)}(E) \subset \mathbb{R}^{[0,\infty)} \mid E \in \mathcal{B}(\mathbb{R}^{|\Lambda|}, \Lambda \subset [0, \infty), |\Lambda| < \infty \right\}$$

by $\tilde{\mu}(\pi^{-1}_{\Lambda[0,\infty)}(E)) = \mu_\Lambda(E)$. In a first step of the proof, we show that $\tilde{\mu}$ is well defined and can be uniquely extended to a measure μ on $(\mathbb{R}^{[0,\infty)}, \mathscr{G})$ satisfying (2.3.4) and (2.3.5), where $\mathscr{G}$ is the σ-field generated by the cylinder sets in $\mathbb{R}^{[0,\infty)}$, i. e., $\mathscr{G} = \sigma(\mathscr{A})$. This results by an application of the Kolmogorov's extension theorem (Theorem 2.20). Let $(Y_t)_{t\geq 0}$ be the coordinate process on $(\mathbb{R}^{[0,\infty)}, \mathscr{G}, \mu)$, i. e., $Y_t(\omega) = \omega(t)$ for $\omega \in \mathbb{R}^{[0,\infty)}$. We then see that

(1) $\mu(Y_0 = 0) = 1$;
(2) the increments $\{Y_{t_j} - Y_{t_{j-1}}\}_{j=1}^n$ are independent for every division $0 = t_0 < t_1 < \cdots < t_n$, and $Y_{t_j} - Y_{t_{j-1}} \stackrel{\mathrm{d}}{=} N(0, t_j - t_{j-1})$.

Next, in a second step we prove that there exists a continuous version $(\tilde{Y}_t)_{t\geq 0}$ of $(Y_t)_{t\geq 0}$ by using

$$\mathbb{E}_\mu[|Y_t - Y_s|^4] \leq C|t - s|^2 \tag{2.3.10}$$

and the Kolmogorov–Čentsov theorem (Theorem 2.49). Note that $(\tilde{Y}_t)_{t\geq 0}$ is not the coordinate process. Although $\tilde{Y}_t(\omega) \neq Y_t(\omega)$ in general, it is true that $\tilde{Y}_t(\omega) = Y_t(\omega)$ for

$\omega \notin N_t$, where the null set N_t depends on t. The process $(\tilde{Y}_t)_{t\geq 0}$ satisfies (1)–(3) holding for $(Y_t)_{t\geq 0}$.

With the so-constructed process $(\tilde{Y}_t)_{t\geq 0}$ on $(\mathbb{R}^{[0,\infty)}, \mathscr{G}, \mu)$, let

$$X_\cdot : (\mathbb{R}^{[0,\infty)}, \mathscr{G}, \mu) \to (\mathscr{X}, \mathcal{B}(\mathscr{X}))$$

be defined by $X_\cdot(\omega) = \tilde{Y}_\cdot(\omega) \in \mathscr{X}$ for $\omega \in \mathbb{R}^{[0,\infty)}$. It can be checked directly that $X^{-1}(E) \in \mathscr{G}$ for any cylinder set E, and for $E = \{\omega \in \mathscr{X} \mid \omega(t_j) \in E_j, E_j \in \mathcal{B}(\mathbb{R}), j = 1, \ldots, n\}$ we can indeed see that

$$\begin{aligned}
X^{-1}(E) &= \left\{\omega \in \mathbb{R}^{[0,\infty)} \mid X_{t_j}(\omega) \in E_j, j = 1, \ldots, n\right\} \\
&= \left\{\omega \in \mathbb{R}^{[0,\infty)} \mid \tilde{Y}_{t_j}(\omega) \in E_j, j = 1, \ldots, n\right\} \\
&\subset \left\{\omega \in \mathbb{R}^{[0,\infty)} \mid Y_{t_j}(\omega) \in E_j, j = 1, \ldots, n\right\} \bigcup_{j=1}^{N} N_{t_j} \\
&= \left(\bigcap_{j=1}^{n} \left\{\omega \in \mathbb{R}^{[0,\infty)} \mid \omega(t_j) \in E_j\right\}\right) \bigcup_{j=1}^{N} N_{t_j} \in \mathscr{G}.
\end{aligned}$$

Here $N_{t_j}, j = 1, \ldots, n$, are null sets such that $Y_{t_j} = \tilde{Y}_{t_j}$ on $\mathbb{R}^{[0,\infty)} \setminus N_{t_j}$. Thus $X_\cdot \in \mathscr{G}/\mathcal{B}(\mathscr{X})$, since $\mathcal{B}(\mathscr{X})$ coincides with the σ-field generated by the cylinder sets. The image measure μ_X of μ under X on $(\mathscr{X}, \mathcal{B}(\mathscr{X}))$ is denoted by $\mathcal{W}^0$. It is seen that $\mathcal{W}^0$ is Wiener measure with $x = 0$, and the coordinate process $B_t : \omega \in \mathscr{X} \mapsto \omega(t)$ is a Brownian motion on $(\mathscr{X}, \mathcal{B}(\mathscr{X}), \mathcal{W}^0)$ starting at 0. Indeed, the finite-dimensional distributions are

$$\mathbb{E}^0\Big[\prod_{j=1}^{n} 1_{E_j}(B_{t_j})\Big] = \mu \circ X^{-1}(\{\bar{\omega} \in \mathscr{X} \mid B_{t_j}(\bar{\omega}) \in E_j, j = 1, \ldots, n\}), \quad n \in \mathbb{N}.$$

Note that $\bar{\mathscr{X}} = X(\mathbb{R}^{[0,\infty)}) \subset \mathscr{X}$ satisfies $\mathcal{W}^0(\bar{\mathscr{X}}) = 1$. The paths $\bar{\omega} \in \bar{\mathscr{X}} \setminus \bar{N}$ can be described as $\bar{\omega}(t) = X_t(\omega)$ with some $\omega \in \mathbb{R}^{[0,\infty)}$, where $\mathcal{W}^0(\bar{N}) = 0$. Hence we have

$$\begin{aligned}
\mathbb{E}^0\Big[\prod_{j=1}^{n} 1_{E_j}(B_{t_j})\Big] &= \mu \circ X^{-1}(\{\bar{\omega} \in \mathscr{X} \mid B_{t_j}(\bar{\omega}) = \bar{\omega}(t_j) \in E_j, j = 1, \ldots, n\}) \\
&= \mu \circ X^{-1}(\{X_\cdot(\omega) \in \mathscr{X} \mid X_{t_j}(\omega) \in E_j, j = 1, \ldots, n\}) \\
&= \mu(\{\omega \in \mathbb{R}^{[0,\infty)} \mid X_{t_j}(\omega) \in E_j, j = 1, \ldots, n\} \\
&= \mu(\{\omega \in \mathbb{R}^{[0,\infty)} \mid \tilde{Y}_{t_j}(\omega) \in E_j, j = 1, \ldots, n\}) \\
&= \mu(\{\omega \in \mathbb{R}^{[0,\infty)} \mid Y_{t_j}(\omega) \in E_j, j = 1, \ldots, n\}).
\end{aligned}$$

Here we used that $\tilde{Y}_t = Y_t$ almost surely. We see that $B_t \stackrel{\mathrm{d}}{=} Y_t$, and thus

$$\mathcal{W}^0(B_0 = 0) = \mu(Y_0 = 0) = 1, \quad \mathbb{E}^0[B_t B_s] = \mathbb{E}_\mu[Y_t Y_s] = t \wedge s,$$

$\{B_{t_j} - B_{t_{j-1}}\}, j = 1, \ldots, n$, are independent, and $B_{t_j} - B_{t_{j_1}} \stackrel{d}{=} N(0, t_j - t_{j-1})$ follows. Hence $(B_t)_{t\geq 0}$ is a Brownian motion. Define

$$Z_\cdot : (\mathscr{X}, \mathcal{B}(\mathscr{X}), \mathcal{W}^0) \to (\mathscr{X}, \mathcal{B}(\mathscr{X}))$$

by $Z_t(\omega) = B_t(\omega) + x$. It can be seen that $Z \in \mathcal{B}(\mathscr{X})/\mathcal{B}(\mathscr{X})$. We denote the image measure of $\mathcal{W}^0$ under Z on $(\mathscr{X}, \mathcal{B}(\mathscr{X}))$ by $\mathcal{W}^x$. This is then Wiener measure starting at x, and the coordinate process $B_t(\omega) = \omega(t), t \geq 0$, under $\mathcal{W}^x$ is a Brownian motion starting at x. □

Now we turn to studying some fundamental distributional properties of Brownian motion. By the definition of the standard Brownian motion, we see that $B_t = B_t - B_0$ a. s., and hence

$$B_t \stackrel{d}{=} N(0, t), \quad t \geq 0.$$

Thus B_t is a Gaussian random variable for every $t > 0$, with $\mathbb{E}[B_t] = 0$ and variance $\mathbb{E}[B_t^2] = t$. In the next result, we show that this structure goes over also to process level.

Theorem 2.98 (Gaussian property). *Let $s, t \geq 0$. Standard Brownian motion $(B_t)_{t\geq 0}$, $B_t = (B_t^1, \ldots, B_t^d)$, is a Gaussian process with mean and covariance*

$$\begin{aligned} \mathbb{E}[B_t^j] &= 0, \quad j = 1, \ldots, d, \\ \mathbb{E}[B_s^j B_t^k] &= (s \wedge t)\, \delta_{jk}, \quad j, k = 1, \ldots, d, \end{aligned}$$

where δ_{jk} is the Kronecker symbol.

Notice that for the standard Brownian motion the last expression coincides with the covariance. For a one-dimensional Brownian motion starting at x, the mean is $\mathbb{E}^x[B_t] = x$ and the covariance becomes

$$\operatorname{cov}(B_s, B_t) = \mathbb{E}^x[(B_s - \mathbb{E}^x[B_s])(B_t - \mathbb{E}^x[B_t])] = \mathbb{E}^x[B_s B_t] - x^2.$$

Proof. For simplicity, we discuss the proof for the one-dimensional standard Brownian motion only, the general case causes no special difficulty. Recall that an $\mathbb{R}$-valued random process $(X_t)_{t\geq 0}$ is called a Gaussian process if for all $0 \leq t_1 \leq \cdots \leq t_n$, every linear combination $\sum_{i=1}^n c_i X_{t_i}$, $c_1, \ldots, c_n \in \mathbb{R}$, is a Gaussian random variable, for all $n \in \mathbb{N}$.

Let $n \in \mathbb{N}$ be arbitrary, $c_1, \ldots, c_n \in \mathbb{R}$, and consider any $0 \leq t_1 \leq \cdots \leq t_n$. Then notice that we can write by telescoping

$$\sum_{i=1}^n c_i B_{t_i} = \sum_{i=1}^n (c_i + \cdots + c_n)(B_{t_i} - B_{t_{i-1}}).$$

Since the increments $B_{t_i} - B_{t_{i-1}}$ are all independent, and constant multiples of Gaussian random variables are again Gaussian random variables, the right-hand side above is the sum of independent Gaussian random variables, hence another Gaussian random variable. Next we calculate the mean and covariance. The first equality is immediate. For the second, choose first $s < t$. Since $B_t - B_s$ and B_s are independent, we have for the covariance

$$\mathbb{E}[(B_t - \mathbb{E}[B_t])(B_s - \mathbb{E}[B_s])] = \mathbb{E}[B_t B_s] = \mathbb{E}[(B_t - B_s)B_s] + \mathbb{E}[B_s^2] = s,$$

and similarly obtain t in case $t < s$. □

This property allows a characterization of Brownian motion and we have the following converse statement, which we state for convenience for the one-dimensional case only.

Theorem 2.99. *Let $(X_t)_{t\geq 0}$ be a Gaussian process such that $X_0 = 0$ a. s. and*

(1) *$\mathbb{E}[X_t] = 0$ and $\mathbb{E}[X_s X_t] = s \wedge t$, for all $s, t \geq 0$;*
(2) *$t \mapsto X_t$ is a. s. continuous.*

Then $(X_t)_{t\geq 0}$ is a standard Brownian motion.

Proof. We only need to prove that the increments $(X_{t_i} - X_{t_{i-1}})_{1\leq i\leq n}$ are independent Gaussian random variables distributed by $N(0, t_i - t_{i-1})$, for any partition $0 = t_0 < t_1 < \cdots < t_n$, $n \in \mathbb{N}$. Let $0 < r < s < t$. Then

$$\mathbb{E}[(X_r - X_0)(X_t - X_s)] = \mathbb{E}[X_r(X_t - X_s)] = \mathbb{E}[X_r X_t] - \mathbb{E}[X_r X_s] = r \wedge t - r \wedge s = 0.$$

On the other hand, also $\mathbb{E}[X_r - X_0]\mathbb{E}[X_t - X_s] = 0$, and since they are Gaussian random variables, this implies that $X_r - X_0$ and $X_t - X_s$ are independent. Furthermore, we have $\mathbb{E}[X_{t_i} - X_{t_{i-1}}] = 0$ and

$$\begin{aligned} \operatorname{var}(X_{t_i} - X_{t_{i-1}}) &= \mathbb{E}[(X_{t_i} - X_{t_{i-1}})^2] = \mathbb{E}[X_{t_i}^2] - 2\mathbb{E}[X_{t_{i-1}} X_{t_i}] + \mathbb{E}[X_{t_{i-1}}^2] \\ &= t_i - 2t_{i-1} \wedge t_i + t_{i-1} = t_i - 2t_{i-1} + t_{i-1} = t_i - t_{i-1}. \end{aligned}$$

Since all other conditions are satisfied, we have that $(X_t)_{t\geq 0}$ is a standard Brownian motion. □

The following are basic invariance properties of Brownian motion.

Theorem 2.100 (Symmetry properties). *Let $(B_t)_{t\geq 0}$ be a standard Brownian motion. Then the following symmetry properties hold:*

(1) (spatial homogeneity) *If $(B_t^x)_{t\geq 0}$ is a Brownian motion starting at x, then $B_t + x \overset{\mathrm{d}}{=} B_t^x$;*
(2) (reflection symmetry) $-B_t \overset{\mathrm{d}}{=} B_t$;
(3) (self-similarity) $\sqrt{c} B_{t/c} \overset{\mathrm{d}}{=} B_t$ *for all $c > 0$;*
(4) (time inversion) *Let* $Z_t(\omega) = \begin{cases} 0, & \text{if } t = 0, \\ tB_{1/t}(\omega), & \text{if } t > 0, \end{cases}$ *then* $Z_t \overset{\mathrm{d}}{=} B_t$;
(5) (time reversibility) $B_{t-s} - B_t \overset{\mathrm{d}}{=} B_s$ *for all* $0 \leq s \leq t$.

Proof. These equalities follow easily from (2.3.7). The fact that $Z_0 = 0$ a. s. in (4) is an application of the strong law of large numbers discussed in Section 2.3.7. □

2.3.2 Two-sided Brownian motion

For our purposes below, it will be sometimes of interest to have Brownian motion extended over the full time-line $\mathbb{R}$ instead of running it on $[0, \infty)$ only. This can be done on the measurable space $(\mathfrak{X}, \mathcal{B}(\mathfrak{X}))$ with path space

$$\mathfrak{X} = C(\mathbb{R}; \mathbb{R}^d). \tag{2.3.11}$$

Consider $\tilde{\mathscr{X}} = \mathscr{X} \times \mathscr{X}$, with $\mathscr{X}$ as given by (2.3.1), and $\mu^x = \mathcal{W}^x \times \mathcal{W}^x$. Let $\omega = (\omega_1, \omega_2) \in \tilde{\mathscr{X}}$ and define

$$\tilde{B}_t(\omega) = \begin{cases} \omega_1(t) & \text{if } t \geq 0, \\ \omega_2(-t) & \text{if } t < 0. \end{cases}$$

Since $\tilde{B}_t(\omega)$ is continuous in $t \in \mathbb{R}$ under μ^x, $X : (\tilde{\mathscr{X}}, \mathcal{B}(\tilde{\mathscr{X}})) \to (\mathfrak{X}, \mathcal{B}(\mathfrak{X}))$ can be defined by $X_\cdot(\omega) = \tilde{B}_\cdot(\omega)$. It can be seen that $X \in \mathcal{B}(\tilde{\mathscr{X}})/\mathcal{B}(\mathfrak{X})$ by showing that $X^{-1}(E) \in \mathcal{B}(\tilde{\mathscr{X}})$ for any cylinder set $E \in \mathcal{B}(\mathfrak{X})$. Thus X is an $\mathfrak{X}$-valued random variable on $\tilde{\mathscr{X}}$. Denote the image measure of μ^x under X on $(\mathfrak{X}, \mathcal{B}(\mathfrak{X}))$ by $\tilde{\mathcal{W}}^x$. The coordinate process denoted by

$$\tilde{B}_t : \mathfrak{X} \ni \omega \mapsto \omega(t) \in \mathbb{R}^d \tag{2.3.12}$$

is a *Brownian motion over* $\mathbb{R}$ on $(\mathfrak{X}, \mathcal{B}(\mathfrak{X}), \tilde{\mathcal{W}}^x)$, with $\tilde{B}_0 = x$ almost surely. The properties of Brownian motion indexed by the real line $\mathbb{R}$ can be summarized as follows.

Proposition 2.101 (Two-sided Brownian motion). *The process $(\tilde{B}_t)_{t\in\mathbb{R}}$ satisfies the following:*

(1) $\tilde{\mathcal{W}}^x(\tilde{B}_0 = x) = 1$;
(2) *The increments $(\tilde{B}_{t_i} - \tilde{B}_{t_{i-1}})_{1\leq i\leq n}$ are independent Gaussian random variables for every collection $0 = t_0 < t_1 < \cdots < t_n$, $n \in \mathbb{N}$, with $\tilde{B}_t - \tilde{B}_s \stackrel{\mathrm{d}}{=} N(0, t-s)$ for $t > s \geq 0$;*
(3) *The increments $(\tilde{B}_{-t_{i-1}} - \tilde{B}_{-t_i})_{1\leq i\leq n}$ are independent Gaussian random variables for every collection $0 = -t_0 > -t_1 > \cdots > -t_n$, $n \in \mathbb{N}$, with $\tilde{B}_{-t} - \tilde{B}_{-s} \stackrel{\mathrm{d}}{=} N(0, s-t)$ for $0 \geq -t > -s$;*
(4) *The function $\mathbb{R} \ni t \mapsto \tilde{B}_t(\omega) \in \mathbb{R}$ is continuous for a. e. ω;*
(5) *$B_{-t} \stackrel{d}{=} B_t$ for $t \in \mathbb{R}$, and B_t, B_s for $t > 0 > s$ are independent.*

We use the previous notation B_t and $\mathcal{W}^x$ for $\tilde{B}_t$ and $\tilde{\mathcal{W}}^x$, respectively, unless any confusion may arise.

Next we consider time-shifts for the two-sided Brownian motion. It can be checked directly through the finite-dimensional distributions (2.3.7) that all joint distributions of $\tilde{B}_{t_0}, \ldots, \tilde{B}_{t_n}$ for arbitrary time-points $-\infty < t_0 < t_1 < \cdots < t_n < \infty$, $n \in \mathbb{N}$, with respect to $dx \otimes d\tilde{\mathcal{W}}^x$ are invariant under time-shift under some conditions. We highlight this through some simple cases. Let $0 < s \leq t$. We see that

$$\int_{\mathbb{R}^d} \mathbb{E}^x_{\mathcal{W}}[f(B_s)g(B_t)]dx = \int_{\mathbb{R}^d \times \mathbb{R}^d} f(x)g(y)\Pi_s(x)\Pi_{t-s}(y-x)dxdy.$$

On the other hand,

$$\int_{\mathbb{R}^d} \mathbb{E}^x_{\mathcal{W}}[f(B_0)g(B_{t-s})]dx = \int_{\mathbb{R}^d \times \mathbb{R}^d} f(x)g(y)\Pi_{t-s}(y-x)dxdy,$$

and thus we conclude that in general

$$\int_{\mathbb{R}^d} \mathbb{E}^x_{\mathcal{W}}[f(B_s)g(B_t)]dx \neq \int_{\mathbb{R}^d} \mathbb{E}^x_{\mathcal{W}}[f(B_0)g(B_{t-s})]dx.$$

It follows that $\int_{\mathbb{R}^d} \mathbb{E}^x_{\mathcal{W}}[f(B_s)g(B_t)]dx$ for $0 < s \leq t$ is not time-shift invariant. Now let $-s \leq 0 \leq t$. We see that

$$\int_{\mathbb{R}^d} \mathbb{E}^x_{\tilde{\mathcal{W}}}[f(\tilde{B}_{-s})g(\tilde{B}_t)]dx = \int_{\mathbb{R}^d} \mathbb{E}^x_{\tilde{\mathcal{W}}}[f(\tilde{B}_{-s})]\mathbb{E}^x_{\mathcal{W}}[g(B_t)]dx$$

by the independence of $\tilde{B}_{-s}$ and $\tilde{B}_t$, and further from the reflection symmetry we have

$$\begin{aligned} &= \int_{\mathbb{R}^d} \mathbb{E}^x_{\mathcal{W}}[f(B_s)]\mathbb{E}^x_{\mathcal{W}}[g(B_t)]dx = \int_{\mathbb{R}^d \times \mathbb{R}^d \times \mathbb{R}^d} f(z)g(y)\Pi_s(z-x)\Pi_t(z-y)dxdydz \\ &= \int_{\mathbb{R}^d \times \mathbb{R}^d} f(x)g(y)\Pi_{s+t}(x-y)dxdy = \int_{\mathbb{R}^d} \mathbb{E}^x_{\mathcal{W}}[f(B_0)g(B_{s+t})]dx. \end{aligned}$$

Hence

$$\int_{\mathbb{R}^d} \mathbb{E}^x_{\tilde{\mathcal{W}}}[f(\tilde{B}_{-s})g(\tilde{B}_t)]dx = \int_{\mathbb{R}^d} \mathbb{E}^x_{\mathcal{W}}[f(B_0)g(B_{s+t})]dx$$

follows. Furthermore, by a similar computation we see that

$$\int_{\mathbb{R}^d} \mathbb{E}^x_{\tilde{\mathcal{W}}}[f(\tilde{B}_{-s})g(\tilde{B}_t)]dx = \int_{\mathbb{R}^d} \mathbb{E}^x_{\tilde{\mathcal{W}}}[f(\tilde{B}_{-s+r})g(\tilde{B}_{t+r})]dx$$

for $-t \leq r \leq s$. Thus $\int_{\mathbb{R}^d} \mathbb{E}^x_{\tilde{\mathcal{W}}}[f(\tilde{B}_{-s})g(\tilde{B}_t)]dx$ is r-shift invariant for $-t \leq r \leq s$. We emphasize again that for $0 \leq s \leq t$,

$$\int_{\mathbb{R}^d} \mathbb{E}^x_{\tilde{\mathcal{W}}}[f(\tilde{B}_0)g(\tilde{B}_t)]dx \neq \int_{\mathbb{R}^d} \mathbb{E}^x_{\tilde{\mathcal{W}}}[f(\tilde{B}_s)g(\tilde{B}_{s+t})]dx$$

while

$$\int_{\mathbb{R}^d} \mathbb{E}^x_{\tilde{\mathcal{W}}}[f(\tilde{B}_0)g(\tilde{B}_t)]dx = \int_{\mathbb{R}^d} \mathbb{E}^x_{\tilde{\mathcal{W}}}[f(\tilde{B}_{-s})g(\tilde{B}_{t-s})]dx.$$

Remark 2.102. In Section 4.2.1 below, we will see that $\mathbb{E}^x[f(B_t)] = (e^{-th}f)(x)$, where $h = -(1/2)\Delta$, and Δ is the Laplacian in d-dimensions. Thus we have

$$(f, e^{-th}g)_{L^2(\mathbb{R}^d)} = \int_{\mathbb{R}^d} \mathbb{E}^x_{\mathcal{W}}[\bar{f}(B_0)g(B_t)]dx.$$

The left-hand side is equal to

$$(f, e^{-th}g)_{L^2(\mathbb{R}^d)} = (e^{-sh}f, e^{-(t-s)h}g)_{L^2(\mathbb{R}^d)}, \quad 0 \le s \le t, \tag{2.3.13}$$

and the right-hand side above is given by

$$(e^{-sh}f, e^{-(t-s)h}g)_{L^2(\mathbb{R}^d)} = \int_{\mathbb{R}^d} \mathbb{E}^x_{\mathcal{W}}[\bar{f}(B_s)]\mathbb{E}^x_{\mathcal{W}}[g(B_{t-s})]dx$$
$$= \int_{\mathbb{R}^d} \mathbb{E}^x_{\tilde{\mathcal{W}}}[\bar{f}(\tilde{B}_{-s})]\mathbb{E}^x_{\mathcal{W}}[g(B_{t-s})]dx = \int_{\mathbb{R}^d} \mathbb{E}^x_{\tilde{\mathcal{W}}}[\bar{f}(\tilde{B}_{-s})g(\tilde{B}_{t-s})]dx.$$

We have thus identically

$$\int_{\mathbb{R}^d} \mathbb{E}^x_{\tilde{\mathcal{W}}}[\bar{f}(B_0)g(B_t)]dx = \int_{\mathbb{R}^d} \mathbb{E}^x_{\tilde{\mathcal{W}}}[\bar{f}(\tilde{B}_{-s})g(\tilde{B}_{t-s})]dx, \quad 0 \le s \le t.$$

This allows then to view time-shift invariance as a semigroup property given by (2.3.13).

We conclude by the following result.

Proposition 2.103 (Time-shift invariance). *Let $m, n \in \mathbb{N}$, $t_1 \le \cdots \le t_n \le 0 \le t_{n+1} \le \cdots \le t_{n+m}$ be arbitrary time-points, $f_1, f_{n+m} \in L^2(\mathbb{R}^d)$, and f_i, $2 \le i \le n+m-1$, be bounded Borel measurable functions. Then*

$$\int_{\mathbb{R}^d} \mathbb{E}^x_{\tilde{\mathcal{W}}}\left[\prod_{i=1}^{n+m} f_i(\tilde{B}_{t_i})\right]dx = \int_{\mathbb{R}^d} \mathbb{E}^x_{\mathcal{W}}\left[\prod_{i=1}^{n+m} f_i(B_{t_i - t_1})\right]dx. \tag{2.3.14}$$

In particular,

$$\int_{\mathbb{R}^d} \mathbb{E}^x_{\tilde{\mathcal{W}}}\left[\prod_{i=1}^{n+m} f_i(\tilde{B}_{t_i})\right]dx = \int_{\mathbb{R}^d} \mathbb{E}^x_{\tilde{\mathcal{W}}}\left[\prod_{i=1}^{n+m} f_i(\tilde{B}_{t_i+s})\right]dx \tag{2.3.15}$$

for all $-t_{n+m} \le s \le -t_1$.

Proof. Both sides of (2.3.14) are finite by the assumptions on the functions f_j. We prove (2.3.14). Let $-t_1 \le \cdots \le -t_n \le 0 \le t_{n+1} \le \cdots \le t_{n+m}$. By the independence of $\tilde{B}_{-t}$ and $\tilde{B}_s$ for $-t \le 0 \le s$ and reflection symmetry, we have

$$\int_{\mathbb{R}^d} \mathbb{E}^x_{\tilde{\mathcal{W}}}\left[\prod_{i=1}^n f_i(\tilde{B}_{-t_i}) \prod_{j=n+1}^{n+m} f_j(\tilde{B}_{t_j})\right] dx = \int_{\mathbb{R}^d} \mathbb{E}^x_{\tilde{\mathcal{W}}}\left[\prod_{i=1}^n f_i(\tilde{B}_{-t_i})\right] \mathbb{E}^x_{\tilde{\mathcal{W}}}\left[\prod_{j=n+1}^{n+m} f_j(\tilde{B}_{t_j})\right] dx$$
$$= \int_{\mathbb{R}^d} \mathbb{E}^x_{\mathcal{W}}\left[\prod_{i=1}^n f_i(B_{t_i})\right] \mathbb{E}^x_{\mathcal{W}}\left[\prod_{j=n+1}^{n+m} f_j(B_{t_j})\right] dx.$$

Hence it is seen that

$$= \int_{\mathbb{R}^d} dx \int_{\mathbb{R}^{d(n+m)}} \prod_{i=1}^{n+m} dx_i \left(\prod_{i=1}^n f_i(x_i)\right) \Pi_{t_n}(x - x_n) \left(\prod_{i=n}^{2} \Pi_{t_{i-1}-t_i}(x_{i-1} - x_i)\right)$$
$$\times \left(\prod_{j=n+1}^{n+m} f_j(x_j)\right) \Pi_{t_{n+1}}(x_{n+1} - x) \prod_{j=n+2}^{n+m} \Pi_{t_j - t_{j-1}}(x_j - x_{j-1}).$$

Since $\int_{\mathbb{R}^d} \Pi_{t_n}(x - x_n)\Pi_{t_{n+1}}(x_{n+1} - x)dx = \Pi_{t_n + t_{n+1}}(x_{n+1} - x_n)$, we have

$$= \int_{\mathbb{R}^{d(n+m)}} \prod_{i=1}^{n+m} dx_i \left(\prod_{i=1}^{n+m} f_i(x_i)\right)\left(\prod_{i=2}^{n} \Pi_{t_{i-1}-t_i}(x_{i-1} - x_i)\right)$$
$$\times \Pi_{t_n + t_{n+1}}(x_{n+1} - x_n) \prod_{j=n+2}^{n+m} \Pi_{t_j - t_{j-1}}(x_j - x_{j-1}). \tag{2.3.16}$$

On the other hand, we have

$$\int_{\mathbb{R}^d} \mathbb{E}^x_{\mathcal{W}}\left[\prod_{i=1}^n f_i(B_{t_1 - t_i}) \prod_{i=n+1}^{n+m} f_i(B_{t_1 + t_j})\right] dx$$
$$= \int_{\mathbb{R}^d} dx \int_{\mathbb{R}^{d(n+m)}} \prod_{i=1}^{n+m} dx_i \left(\prod_{i=1}^{n+m} f_i(x_i)\right) \Pi_0(x, x_1) \left(\prod_{i=2}^{n} \Pi_{t_{i-1}-t_i}(x_{i-1} - x_i)\right)$$
$$\times \Pi_{t_n + t_{n+1}}(x_{n+1} - x_n) \prod_{j=n+2}^{n+m} \Pi_{t_j - t_{j-1}}(x_j - x_{j-1})$$
$$= \int_{\mathbb{R}^{d(n+m)}} \prod_{i=1}^{n+m} dx_i \left(\prod_{i=1}^{n+m} f_i(x_i)\right)\left(\prod_{i=2}^{n} \Pi_{t_{i-1}-t_i}(x_{i-1} - x_i)\right)$$
$$\times \Pi_{t_n + t_{n+1}}(x_{n+1} - x_n) \prod_{j=n+2}^{n+m} \Pi_{t_j - t_{j-1}}(x_j - x_{j-1}). \tag{2.3.17}$$

Here we used $\int_{\mathbb{R}^d} \Pi_0(x, x_1)dx = 1$. Comparing (2.3.16) and (2.3.17), we obtain (2.3.14), and hence (2.3.15) follows. □

2.3.3 Conditional Wiener measure

Conditionals of the Wiener path measure and related conditioned processes have important applications, and we will often use them below. We discuss next Brownian bridge measure obtained by pinning down paths at a starting and an ending position.

Definition 2.104 (Brownian bridge measure). Let $(B_t)_{t\geq 0}$ be a Brownian motion. Let $0 < T_1 < T_2$ and $x, y \in \mathbb{R}^d$. Then the probability measure $\mathcal{W}^{x,y}_{[T_1,T_2]}$ on $(\mathscr{X}, \mathcal{B}(\mathscr{X}))$ is defined by

$$\mathcal{W}^{x,y}_{[T_1,T_2]}(A) = \mathbb{E}[1_A | B_{T_1} = x, B_{T_2} = y] \quad A \in \mathcal{B}(\mathscr{X}). \tag{2.3.18}$$

This is referred to as the *Brownian bridge measure* starting at x at $t = T_1$ and ending at y at $t = T_2$. Furthermore, the measure

$$\Pi_{T_2-T_1}(x-y)\mathcal{W}^{x,y}_{[T_1,T_2]} \tag{2.3.19}$$

is called the *conditional Wiener measure.*

In Definition 2.104, $\mathbb{E}[\ldots | B_{T_1} = x, B_{T_2} = y]$ is defined by (2.1.42). That is, for the $\mathbb{R}^{2d}$-valued random variable $X = (B_{T_1}, B_{T_2})$, the conditional expectation $\mathbb{E}[F|\sigma(X)]$ can be expressed as $\mathbb{E}[F|\sigma(X)] = f(X)$ with a Borel measurable function f on $\mathbb{R}^{2d}$. The function is denoted by $f(x,y) = \mathbb{E}[F|B_{T_1} = x, B_{T_2} = y]$. Note that the Brownian bridge measure is a probability measure, while the conditional Wiener measure is not. Let $A_1, \ldots, A_n$ be Borel sets in $\mathbb{R}^d$, and $n \in \mathbb{N}$ arbitrary. Given a partition $0 < T_1 < t_1 < \cdots < t_n < T_2$ and $x, y \in \mathbb{R}^d$, we see that

$$\begin{aligned}
&\mathcal{W}^{x,y}_{[T_1,T_2]}(B_{t_1} \in A_1, \ldots, B_{t_n} \in A_n) \\
&= \int_{\mathbb{R}^{nd}} \frac{\Pi_{t_1-T_1}(x-x_1)\Pi_{t_2-t_1}(x_1-x_2)\cdots\Pi_{T_2-t_n}(x_n-y)}{\Pi_{T_2-T_1}(x-y)} \left(\prod_{i=1}^n 1_{A_i}(x_i)\right) \prod_{i=1}^n dx_i.
\end{aligned} \tag{2.3.20}$$

We write $\mathbb{E}^{x,y}_{[T_1,T_2]}$ for the expectation with respect to $\mathcal{W}^{x,y}_{[T_1,T_2]}$. From the finite-dimensional distribution of the Brownian bridge measure, for any bounded measurable function f,

$$\mathcal{W}^{x,y}_{[T_1,T_2]}[f(B_t)] = \int_{\mathbb{R}^d} f(z) \frac{\Pi_a(x-z)\Pi_b(z-y)}{\Pi_{a+b}(x-y)} dz.$$

Here $a = t - T_1$ and $b = T_2 - t$. Then $a + b = T_2 - T_1$. Now

$$\Pi_a(x-z)\Pi_b(z-y) = \frac{1}{(2\pi a)^{d/2}(2\pi b)^{d/2}} \exp\left(-\frac{|x-z|^2}{2a} - \frac{|z-y|^2}{2b}\right).$$

Completing the square in z, we obtain

$$\frac{|x-z|^2}{2a} + \frac{|z-y|^2}{2b} = \frac{a+b}{2ab}\left|z - \frac{bx+ay}{a+b}\right|^2 + \frac{|x-y|^2}{2(a+b)}.$$

Therefore

$$\frac{\Pi_a(x-z)\Pi_b(z-y)}{\Pi_{a+b}(x-y)} = N(m,\sigma^2 I) = \frac{1}{(2\pi\sigma^2)^{d/2}} \exp\left(-\frac{(z-m)^2}{2\sigma^2}\right),$$

where

$$m = \frac{bx+ay}{a+b}, \quad \sigma^2 = \frac{ab}{a+b}.$$

For example, it follows that

$$\mathbb{E}_{\mathcal{W}^{x,y}_{[T_1,T_2]}}[B_t] = m = x + \frac{t-T_1}{T_2-T_1}(y-x).$$

From the definition, it is directly seen that

$$\mathbb{E}\left[\prod_{j=1}^n f_j(B_{t_j})\right] = \int_{\mathbb{R}^d\times\mathbb{R}^d} \Pi_{T_1}(x)\Pi_{T_2-T_1}(x-y)\mathbb{E}^{x,y}_{[T_1,T_2]}\left[\prod_{j=1}^n f_j(B_{t_j})\right] dxdy$$

for all Borel measurable bounded functions $f_1,\dots,f_n$ and for $0 < T_1 < t_1 < \cdots < t_n < T_2$. In particular $\mathbb{E}^x[\dots]$ can be expressed in terms of the conditional Wiener measure. For $0 < t_1 < \cdots < t_n < T$,

$$\mathbb{E}^x\left[\prod_{j=1}^n f_j(B_{t_j})\right] = \mathbb{E}\left[\prod_{j=1}^n f_j(B_{t_j}+x)\right] = \int_{\mathbb{R}^d} \Pi_T(x-y)\mathbb{E}^{x,y}_{[0,T]}\left[\prod_{j=1}^n f_j(B_{t_j})\right] dy \tag{2.3.21}$$

and

$$\mathbb{E}^x\left[f(B_0)\left(\prod_{j=1}^n f_j(B_{t_j})\right)g(B_T)\right] = \int_{\mathbb{R}^d} f(x)g(y)\Pi_T(x-y)\mathbb{E}^{x,y}_{[0,T]}\left[\prod_{j=1}^n f_j(B_{t_j})\right] dy.$$

Definition 2.105 (Full Wiener measure). We refer to the measure

$$d\bar{\mathcal{W}} = d\mathcal{W}^x \otimes dx \quad \text{on } \mathscr{X} \times \mathbb{R}^d \tag{2.3.22}$$

as *full Wiener measure*, carrying infinite mass.

Although full Wiener measure is not a probability measure on $\mathscr{X} \times \mathbb{R}^d$, we use the notation $\mathbb{E}_{\bar{\mathcal{W}}}[\dots]$ for $\int_{\mathscr{X}\times\mathbb{R}^d} \dots d\bar{\mathcal{W}}$. Full Wiener measure $d\bar{\mathcal{W}}$ is expressed in terms of the conditional Wiener measure as

$$d\bar{\mathcal{W}} = \left(\int_{\mathbb{R}^d} \Pi_T(x-y)d\mathcal{W}^{x,y}_{[0,T]}dy\right)dx,$$

or in other words,

$$\mathbb{E}_{\bar{\mathcal{W}}}\left[\prod_{i=1}^{n} f_i(B_{t_i})\right] = \int_{\mathbb{R}^d\times\mathbb{R}^d} \Pi_T(x-y)\mathbb{E}^{x,y}_{[0,T]}\left[\prod_{i=1}^{n} f_i(B_{t_i})\right] dxdy. \tag{2.3.23}$$

Thus for $0 < t_1 < \cdots < t_n < T$, we have

$$\mathbb{E}_{\bar{\mathcal{W}}}\left[f(B_0)\left(\prod_{j=1}^{n} f_j(B_{t_j})\right)g(B_T)\right] = \int_{\mathbb{R}^d\times\mathbb{R}^d} f(x)g(y)\Pi_T(x-y)\mathbb{E}^{x,y}_{[0,T]}\left[\prod_{i=1}^{n} f_i(B_{t_i})\right] dxdy. \tag{2.3.24}$$

2.3.4 Martingale properties of Brownian motion

Brownian motion is a martingale of primary importance in the class of continuous-time random processes. In fact, the martingale property of a process with almost surely continuous paths, and of a related quadratic process, forces it to be a Brownian motion.

Proposition 2.106. *Let $(B_t)_{t\geq 0}$ be a Brownian motion on a filtered probability space $(\Omega, \mathcal{F}, (\mathcal{F}_t)_{t\geq 0}, P^x)$, where $(\mathcal{F}_t)_{t\geq 0}$ is the natural filtration of $(B_t)_{t\geq 0}$. Then both $(B_t)_{t\geq 0}$ and $(B_t^2 - t)_{t\geq 0}$ are $(\mathcal{F}_t)$-martingales.*

Proof. Adaptedness of the two processes and absolute integrability of B_t and $B_t^2 - t$ are straightforward. Let $t \geq s$. We have

$$\begin{aligned}\mathbb{E}^x[B_t|\mathcal{F}_s] &= \mathbb{E}^x[(B_t - B_s) + B_s|\mathcal{F}_s] = \mathbb{E}^x[B_t - B_s|\mathcal{F}_s] + \mathbb{E}^x[B_s|\mathcal{F}_s]\\ &= \mathbb{E}^x[B_t - B_s] + B_s = B_s\end{aligned} \tag{2.3.25}$$

by the independence of $B_t - B_s$ and B_r for $r \leq s$. By a similar calculation, it is easy to show also the second claim. □

Remarkably, there is a strong converse of this proposition.

Proposition 2.107 (Lévy martingale characterization theorem). *Let $(X_t)_{t\geq 0}$ be a random process on a probability space $(\Omega, \mathcal{F}, P)$, and $(\mathcal{F}_t)_{t\geq 0}$ the natural filtration of $(X_t)_{t\geq 0}$. Suppose that* (1) *$X_0 = 0$ almost surely,* (2) *$t \mapsto X_t$ is almost surely continuous, and* (3) *both $(X_t)_{t\geq 0}$ and $(X_t^2 - t)_{t\geq 0}$ are $(\mathcal{F}_t)$-martingales. Then $(X_t)_{t\geq 0}$ is a standard Brownian motion.*

Proof. Let $0 \leq s \leq t$. It is sufficient to show that $X_t - X_s$ is independent of $\mathcal{F}_s$ and $X_t - X_s \stackrel{\mathrm{d}}{=} N(0, t-s)$. This follows from

$$\mathbb{E}[e^{i\xi(X_t - X_s)}|\mathcal{F}_s] = e^{-(t-s)\xi^2/2}, \quad \xi \in \mathbb{R}. \tag{2.3.26}$$

Indeed, (2.3.26) yields that $\mathbb{E}[e^{i\xi(X_t-X_s)}] = e^{-(t-s)\xi^2/2}$ and $\mathbb{E}[e^{i\xi(X_t-X_s)}1_A] = \mathbb{E}[e^{i\xi(X_t-X_s)}]P(A)$ for $A \in \mathcal{F}_s$. This proves that $X_t - X_s$ is independent of $\mathcal{F}_s$ by Lemma 2.27. In order to prove (2.3.26), we use Taylor expansion. Set $\Delta X = X_t - X_s$ and $\delta = t - s$. We have

$$e^{i\xi\Delta X} - e^{-\xi^2\delta/2} = i\xi\Delta X + \frac{\xi^2}{2}\delta - \frac{\xi^2}{2}(\Delta X)^2 - \frac{\xi^4}{8}\delta^2 - \frac{i\xi^3}{6}(\eta\Delta X)^3 + \frac{\xi^6}{48}(\theta\delta)^3,$$

where $\eta \in (0,1)$ and $\theta \in (0,1)$. Taking conditional expectation with respect to $\mathscr{F}_s$, we have

$$\mathbb{E}[e^{i\xi\Delta X} - e^{-\xi^2\delta/2}|\mathscr{F}_s] = -\frac{\xi^4}{8}\delta^2 - \frac{i\xi^3}{6}\mathbb{E}[(\eta\Delta X)^3|\mathscr{F}_s] + \frac{\xi^6}{48}(\theta\delta)^3.$$

Here we used $\mathbb{E}[i\xi\Delta X + \frac{\xi^2}{2}\delta - \frac{\xi^2}{2}(\Delta X)^2|\mathscr{F}_s] = 0$, which is obtained from the fact that $X_t^2 - t$ is a martingale. Using Hölder inequality, we find

$$\mathbb{E}[(\eta\Delta X)^3|\mathscr{F}_s] \le \mathbb{E}[|\Delta X|^3|\mathscr{F}_s] \le (\mathbb{E}[(\Delta X)^4|\mathscr{F}_s])^{3/4} \le 3^{3/4}\delta^{3/2}.$$

The latter estimate is obtained in Lemma 2.108 below. Therefore we have

$$\mathbb{E}[e^{i\xi\Delta X} - e^{-\xi^2\delta/2}|\mathscr{F}_s] \le c_\xi(\delta^2 + \delta^3 + \delta^{3/2})$$

with constant c_ξ. Suppose that $\delta < 1$. Then

$$\mathbb{E}[e^{i\xi\Delta X} - e^{-\xi^2\delta/2}|\mathscr{F}_s] \le c'_\xi\delta^{3/2}. \tag{2.3.27}$$

Let $t_j = s + (t-s)j/n$. Using the bound $\left|\prod_{j=1}^n a_j - \prod_{j=1}^n b_j\right| \le \sum_{j=1}^n |a_j - b_j|$ for $a_j, b_j \in \mathbb{C}$ and $|a_j|, |b_j| \le 1$, we find

$$\begin{aligned}\left|\mathbb{E}[e^{i\xi(X_t-X_s)} - e^{-\frac{1}{2}\xi^2(t-s)}|\mathscr{F}_s]\right| &\le \mathbb{E}\left[\left|\prod_{j=1}^n e^{i\xi(X_{t_j}-X_{t_{j-1}})} - \prod_{j=1}^n e^{-\frac{1}{2}\xi^2(t_j-t_{j-1})}\right| |\mathscr{F}_s\right] \\ &\le \sum_{j=1}^n \mathbb{E}\left[\left|e^{i\xi(X_{t_j}-X_{t_{j-1}})} - e^{-\frac{1}{2}\xi^2(t_j-t_{j-1})}\right| |\mathscr{F}_s\right].\end{aligned}$$

If n is sufficiently large, we get $t_j - t_{j-1} < 1$. Hence we have by (2.3.27) that

$$\left|\mathbb{E}[e^{i\xi(X_t-X_s)} - e^{-\frac{1}{2}\xi^2(t-s)}|\mathscr{F}_s]\right| \le c'_\xi \sum_{j=1}^n (t_j - t_{j-1})^{3/2} = \frac{c'_\xi(t-s)}{\sqrt{n}}.$$

The right-hand side converges to zero as $n \to \infty$. □

Lemma 2.108. *We have that* $\mathbb{E}[(X_t - X_s)^4|\mathscr{F}_s] \le 3|t-s|^2$.

Proof. Let $t_j = s + (t-s)j/n$ and $\Delta X_j = X_{t_j} - X_{t_{j-1}}$, $j = 1, \dots, n$. By Fatou's lemma, we have

$$\mathbb{E}\left[\liminf_{n\to\infty}\sum_{j=1}^n \Delta X_j^2\right] \le \liminf_{n\to\infty}\mathbb{E}\left[\sum_{j=1}^n \Delta X_j^2\right] = \liminf_{n\to\infty}\sum_{j=1}^n (t_j - t_{j-1}) = t - s,$$

where we used $\mathbb{E}[(X_t - X_s)|\mathscr{F}_s] = t - s$ for $s \le t$. Therefore $\liminf_{n\to\infty} \sum_{j=1}^n (\Delta X_j)^2$ is almost surely finite, and from path continuity we obtain

$$\liminf_{n\to\infty} \sum_{j=1}^{n} |\Delta X_j|^{2+\varepsilon} \le \Big(\lim_{n\to\infty} \max_{1\le k\le n} |\Delta X_k|^{\varepsilon}\Big) \liminf_{n\to\infty} \sum_{j=1}^{n} (\Delta X_j)^2 = 0.$$

Thus

$$\begin{aligned}(X_t - X_s)^4 = \Big(\sum_{j=1}^{n} \Delta X_j\Big)^4 &= \sum_{j=1}^{n} (\Delta X_j)^4 + 4 \sum_{j\neq k} (\Delta X_j)^3 \Delta X_k + 6 \sum_{j<k} (\Delta X_j)^2 (\Delta X_k)^2 \\ &+ 12 \sum_{i<j<k} \big((\Delta X_i)^2 \Delta X_j \Delta X_k + (\Delta X_j)^2 \Delta X_i \Delta X_k + (\Delta X_k)^2 \Delta X_i \Delta X_j\big) \\ &+ 24 \sum_{i<j<k<\ell} \Delta X_i \Delta X_j \Delta X_k \Delta X_\ell .\end{aligned}$$

We find

$$\mathbb{E}[(\Delta X_k)^2 (\Delta X_j)^2 | \mathscr{F}_s] = \mathbb{E}[\mathbb{E}[(\Delta X_k)^2 (\Delta X_j)^2 | \mathscr{F}_{t_j}] | \mathscr{F}_s] = \mathbb{E}[\mathbb{E}[(\Delta X_k)^2 | \mathscr{F}_{t_j}] (\Delta X_j)^2 | \mathscr{F}_s] = \delta_k \delta_j,$$

where $\delta_i = t_i - t_{i-1} = (t-s)/n$. Since $\mathbb{E}[X_t - X_s | \mathscr{F}_s] = 0$, we also see that

$$\mathbb{E}[(\Delta X_i)^3 \Delta X_j | \mathscr{F}_s] = \mathbb{E}[(\Delta X_i)^2 \Delta X_j \Delta X_k | \mathscr{F}_s] = \mathbb{E}[\Delta X_i \Delta X_j \Delta X_k \Delta X_\ell | \mathscr{F}_s] = 0$$

if the indices i, j, k, and ℓ are pairwise different. Hence we conclude that

$$\mathbb{E}[(X_t - X_s)^4 | \mathscr{F}_s] \le \liminf_{n\to\infty} 6 \sum_{j<k} \delta_j \delta_k = \liminf_{n\to\infty} 6 \frac{(t-s)^2}{n^2} \frac{n(n-1)}{2} = 3(t-s)^2. \qquad \square$$

Definition 2.109 (Augmented filtration of Brownian motion). We define

$$(\mathscr{F}_t^{\mathrm{BM}})_{t\ge 0} = \sigma(\mathscr{F}_t \cup \mathscr{N})$$

to be the augmented filtration of the natural filtration of Brownian motion $(B_t)_{t\ge 0}$, where $\mathscr{F}_t = \sigma(B_r, 0 \le r \le t)$ and $\mathscr{N}$ is the family of all null sets.

Brownian motion is again a martingale with respect to the augmented filtration.

Proposition 2.110. *Let $(B_t)_{t\ge 0}$ be a Brownian motion. Then both $(B_t)_{t\ge 0}$ and $(B_t^2 - t)_{t\ge 0}$ are $(\mathscr{F}_t^{\mathrm{BM}})_{t\ge 0}$-martingales.*

Proof. This is obtained by a combination of Proposition 2.106 and Lemma 2.54. □

Finally, we discuss Blumenthal's zero–one law, which will be applied below in studying some global path properties of Brownian motion.

Proposition 2.111 (Blumenthal's zero–one law). *Let $(B_t)_{t\geq 0}$ be a Brownian motion, and $\mathcal{F}_t = \sigma(B_r, 0 \leq r \leq t)$. Consider the germ field at zero given by $\mathcal{F}_0^+ = \bigcap_{t>0} \mathcal{F}_t$. Then $\mathcal{F}_0^+$ is a trivial σ-field.*

Proof. Define the random variables $Y_n = (B_{2^{-n}+t} - B_{2^{-n}})_{t\in[0,2^{-n}]}$, $n \in \mathbb{N}$, with values in $C([0, 2^{-n}])$. Note that $\{\sigma(Y_n)\}_{n\in\mathbb{N}}$ is a family of independent sub-σ-fields. By Kolmogorov's zero–one law, the tail σ-field $\mathcal{T} = \bigcap_{n=1}^{\infty} \sigma(\bigcup_{m=n}^{\infty} \sigma(Y_m))$ is trivial. On the other hand, we have $\sigma(\bigcup_{m=n}^{\infty} \sigma(Y_m)) = \mathcal{F}_{2^{-n+1}}$ and hence $\mathcal{F}_0^+ = \bigcap_{t>0} \mathcal{F}_t = \bigcap_{n=1}^{\infty} \mathcal{F}_{2^{-n+1}} = \mathcal{T}$ is trivial. □

2.3.5 Markov properties of Brownian motion

In addition to the martingale property, Brownian motion has the equally important Markov property. We will discuss here the usual Markov and semigroup properties it satisfies, and a stronger variant as it will be explained below.

Theorem 2.112. *Let $(B_t)_{t\geq 0}$ be a d-dimensional Brownian motion starting at $x \in \mathbb{R}^d$ on a filtered probability space $(\Omega, \mathcal{F}, (\mathcal{F}_t^{\mathrm{BM}})_{t\geq 0}, P^x)$. Then $(B_t)_{t\geq 0}$ is a stationary Markov process and*

$$\mathbb{E}^x[f(B_{s+t_0}, \ldots, B_{s+t_n})|\mathcal{F}_s] = \mathbb{E}^{B_s}[f(B_{t_0}, \ldots, B_{t_n})], \quad x \in \mathbb{R}^d, \tag{2.3.28}$$

holds for every bounded Borel function f, and all $t_0 < \cdots < t_n$, $s > 0$, and $n \in \mathbb{N}$. Moreover, its initial distribution $P_0 = \delta_x$ is a Dirac measure with mass at x, while its probability transition kernel p_{BM} is given by

$$p_{\mathrm{BM}}(t, x, A) = \int_{\mathbb{R}^d} 1_A(z)\Pi_t(x - z)dz \tag{2.3.29}$$

or, equivalently,

$$p_{\mathrm{BM}}(t, x, dy) = \Pi_t(x - y)dy. \tag{2.3.30}$$

Proof. The Markov property with probability transition kernel (2.3.30) follows by the definition of the finite-dimensional distributions of Brownian motion and Proposition 2.82. Note that $p_{\mathrm{BM}}(s, t, x, A) = \int_{\mathbb{R}^d} 1_A(z)\Pi_{t-s}(x - z)dz$. We show (2.3.28). Let $h(y) = \mathbb{E}^x[\prod_{j=1}^n f(B_{s+t_j})|B_s = y]$. From the identity

$$\mathbb{E}^x\left[1_A(B_s)\prod_{j=1}^n f_j(B_{s+t_j})\right] = \mathbb{E}^x\left[1_A(B_s)\mathbb{E}^x\left[\prod_{j=1}^n f_j(B_{s+t_j})\,\middle|\,\sigma(B_s)\right]\right],$$

we have

$$\int_{(\mathbb{R}^d)^{n+1}} 1_A(y)\Pi_s(y-x)\left(\prod_{j=1}^n f_j(x_j)\right)\Pi_{t_1}(x_1-y)\left(\prod_{j=2}^n \Pi_{t_j-t_{j-1}}(x_j-x_{j-1})\right)dy\prod_{j=1}^n dx_j$$
$$= \int_{\mathbb{R}^d} 1_A(y)h(y)\Pi_s(y-x)dy.$$

Comparing both sides above,

$$h(y) = \int_{(\mathbb{R}^d)^{n+1}} \left(\prod_{j=1}^n f_j(x_j)\right)\Pi_{t_1}(x_1-y)\left(\prod_{j=2}^n \Pi_{t_j-t_{j-1}}(x_j-x_{j-1})\right)\prod_{j=1}^n dx_j$$

is obtained, while the right-hand side equals $\mathbb{E}^y[\prod_{j=1}^n f_j(B_{t_j})]$. Hence

$$\mathbb{E}^x\left[\prod_{j=1}^n f(B_{s+t_j})\,\Big|\,\sigma(B_s)\right] = \mathbb{E}^{B_s}\left[\prod_{j=1}^n f_j(B_{t_j})\right],$$

and (2.3.28) follows. □

From the proof of Theorem 2.112, it follows that the Markov property (2.2.52) can be reformulated as

$$\mathbb{E}^x[f(B_t)|\mathscr{F}_s] = \mathbb{E}^{B_s}[f(B_{t-s})], \quad t \geq s. \tag{2.3.31}$$

In what follows, the Markov property will be considered with respect to the natural filtration, unless specified otherwise.

Corollary 2.113. *Let $(B_t)_{t\geq 0}$ be a d-dimensional Brownian motion starting at $x \in \mathbb{R}^d$ on the filtered probability space $(\Omega, \mathscr{F}, (\mathscr{F}_t^{\mathrm{BM}})_{t\geq 0}, P^x)$. Let $\xi \in \mathbb{R}^d$. Then the random process $\left(e^{\xi\cdot B_t - \frac{1}{2}|\xi|^2 t}\right)_{t\geq 0}$ is an $(\mathscr{F}_t^{\mathrm{BM}})_{t\geq 0}$-martingale.*

Proof. By (2.3.31), we have

$$\mathbb{E}^x\left[e^{\xi\cdot B_t-\frac{1}{2}|\xi|^2 t}\,\Big|\,\mathscr{F}_s\right] = \mathbb{E}^{B_s}\left[e^{\xi\cdot B_{t-s}-\frac{1}{2}|\xi|^2 t}\right].$$

The right-hand side above can be computed as

$$\mathbb{E}^{B_s}\left[e^{\xi\cdot B_{t-s}-\frac{1}{2}|\xi|^2 t}\right] = e^{-\frac{1}{2}|\xi|^2 t}\int_{\mathbb{R}^d} e^{\xi\cdot y}\Pi_{t-s}(B_s-y)dy = e^{\xi\cdot B_s-\frac{1}{2}|\xi|^2 s},$$

and the corollary follows. □

Brownian motion has a yet stronger Markov property than (2.3.28) since the same equality holds even for random times.

Theorem 2.114 (Strong Markov property). *Let $(B_t)_{t\geq 0}$ be a d-dimensional Brownian motion on a filtered probability space $(\Omega, \mathscr{F}, (\mathscr{F}_t^{\mathrm{BM}})_{t\geq 0}, P^x)$. Let τ be a stopping time such that $P^x(\tau < \infty) = 1$. Then $W_t = B_{t+\tau} - B_\tau$, $t \geq 0$, is also a Brownian motion, which is independent of $\mathscr{F}_\tau^{\mathrm{BM}}$. Moreover,*

$$\mathbb{E}^x[f(B_{t+\tau})|\mathscr{F}_\tau] = \mathbb{E}^{B_\tau}[f(B_t)],$$

for every bounded continuous function f.

Proof. Fix $0 = t_0 \leq t_1 \leq \cdots \leq t_n$ and $\xi \in \mathbb{R}^d$. Let τ_n be an approximation of τ, i. e.,

$$\tau_n = \begin{cases} 0, & \tau = 0, \\ \frac{i+1}{n}, & \frac{i}{n} < \tau \leq \frac{i+1}{n}. \end{cases}$$

For all $0 \leq s < t$ and all $F \in \mathscr{F}_\tau^{\mathrm{BM}}$, we have by the definition of τ_n that

$$\begin{aligned}\mathbb{E}^x[e^{i\xi\cdot(B_{\tau+t}-B_{\tau+s})}1_F] &= \lim_{n\to\infty} \mathbb{E}^x[e^{i\xi\cdot(B_{\tau_n+t}-B_{\tau_n+s})}1_F] = \lim_{n\to\infty} \sum_{k=0}^{\infty} \mathbb{E}^x[e^{i\xi\cdot(B_{\frac{k+1}{n}+t}-B_{\frac{k+1}{n}+s})}1_{\{\tau_n=\frac{k+1}{n}\}}1_F] \\ &= \lim_{n\to\infty} \sum_{k=0}^{\infty} \mathbb{E}^x[e^{i\xi\cdot(B_{\frac{k+1}{n}+t}-B_{\frac{k+1}{n}+s})}1_{\{\frac{k}{n}<\tau\leq\frac{k+1}{n}\}}1_F].\end{aligned}$$

We notice that $B_{\frac{k+1}{n}+t} - B_{\frac{k+1}{n}+s}$ is independent from $\mathscr{F}_{\frac{k+1}{n}}^{\mathrm{BM}}$ and $\left\{\frac{k}{n} < \tau \leq \frac{k+1}{n}\right\} \cap F \in \mathscr{F}_{\frac{k+1}{n}}^{\mathrm{BM}}$. We see that this is, furthermore,

$$\begin{aligned}&= \lim_{n\to\infty} \sum_{k=0}^{\infty} \mathbb{E}^x[e^{i\xi\cdot(B_{\frac{k+1}{n}+t}-B_{\frac{k+1}{n}+s})}]\mathbb{E}^x[1_{\{\frac{k}{n}<\tau\leq\frac{k+1}{n}\}}1_F] \\ &= \lim_{n\to\infty} \sum_{k=0}^{\infty} \mathbb{E}^x[e^{i\xi\cdot B_{t-s}}]P^x\left(\left\{\frac{k}{n} < \tau \leq \frac{k+1}{n}\right\} \cap F\right) = \mathbb{E}^x[e^{i\xi\cdot B_{t-s}}]P^x(F).\end{aligned}$$

In the last equality, we used $\bigcup_{k=0}^{\infty}\left\{\frac{k}{n} < \tau \leq \frac{k+1}{n}\right\} \cap F = \{\tau < \infty\}$ for all n, and $P^x(\tau < \infty) = 1$. The same calculation applies to finitely many increments, and we obtain that

$$\mathbb{E}^x[e^{i\sum_{j=1}^n \xi_j\cdot(B_{\tau+t_j}-B_{\tau+t_{j-1}})}1_F] = \prod_{j=1}^{n} \mathbb{E}^x[e^{i\xi_j\cdot B_{t_j-t_{j-1}}}]P^x(F).$$

This means that the increments $\{B_{\tau+t_j} - B_{\tau+t_{j-1}}\}_{j=1}^n = \{W_{t_j} - W_{t_{j-1}}\}_{j=1}^n$ are independent by Lemma 2.27 and distributed as

$$W_{t_j} - W_{t_{j-1}} \stackrel{\mathrm{d}}{=} B_{t_j-t_{j-1}} \stackrel{\mathrm{d}}{=} N(0, t_j - t_{j-1}).$$

Hence $(W_t)_{t\geq 0}$ is a Brownian motion. Moreover, $B_{\tau+t_j} - B_{\tau+t_{j-1}}$, $j = 1, \ldots, n$, are independent of $\mathscr{F}_\tau^{\mathrm{BM}}$, and hence $\sigma(B_{\tau+t} - B_\tau, t \geq 0)$ are independent of $\mathscr{F}_\tau^{\mathrm{BM}}$.

In the same way, we have for $F \in \mathscr{F}_\tau^{\mathrm{BM}}$ that

$$\begin{aligned}\mathbb{E}^x[f(B_{t+\tau})1_F] &= \lim_{n\to\infty}\mathbb{E}^x[f(B_{t+\tau_n})1_F] = \lim_{n\to\infty}\sum_{k=0}^{\infty}\mathbb{E}^x[f(B_{t+\tau_n})1_{\{\tau_n=\frac{k+1}{n}\}}1_F]\\ &= \lim_{n\to\infty}\sum_{k=0}^{\infty}\mathbb{E}^x[f(B_{t+\frac{k+1}{n}})1_{\{\tau_n=\frac{k+1}{n}\}}1_F]\\ &= \lim_{n\to\infty}\sum_{k=0}^{\infty}\mathbb{E}^x\left[\mathbb{E}^x\left[f(B_{t+\frac{k+1}{n}})1_{\{\tau_n=\frac{k+1}{n}\}}1_F|\mathscr{F}_{\frac{k+1}{n}}^{\mathrm{BM}}\right]\right].\end{aligned}$$

Since $F \in \mathscr{F}_\tau^{\mathrm{BM}}$ and $\tau \le \tau_n$, we have $F \in \mathscr{F}_{\tau_n}^{\mathrm{BM}}$. Hence $F \cap \left\{\tau_n = \frac{k+1}{n}\right\} \in \mathscr{F}_{\frac{k+1}{n}}^{\mathrm{BM}}$. By the Markov property of Brownian motion, this is furthermore

$$\begin{aligned}&= \lim_{n\to\infty}\sum_{k=0}^{\infty}\mathbb{E}^x[1_{\{\tau_n=\frac{k+1}{n}\}}1_F\mathbb{E}^x[f(B_{t+\frac{k+1}{n}})\mid\mathscr{F}_{\frac{k+1}{n}}^{\mathrm{BM}}]]\\ &= \lim_{n\to\infty}\sum_{k=0}^{\infty}\mathbb{E}^x\left[1_{\{\tau_n=\frac{k+1}{n}\}}1_F\mathbb{E}^{B_{\frac{k+1}{n}}}[f(B_t)]\right]\\ &= \lim_{n\to\infty}\sum_{k=0}^{\infty}\mathbb{E}^x[1_{\{\tau_n=\frac{k+1}{n}\}}1_F\mathbb{E}^{B_{\tau_n}}[f(B_t)]] = \mathbb{E}^x[1_F\mathbb{E}^{B_\tau}[f(B_t)]].\end{aligned}$$

We can derive $\mathbb{E}^x[f(B_{\tau+t})1_F] = \mathbb{E}^x[1_F\mathbb{E}^{B_\tau}[f(B_t)]]$ for every $F \in \mathscr{F}_\tau^{\mathrm{BM}}$, which implies that $\mathbb{E}^x[f(B_{\tau+t})\mid\mathscr{F}_\tau^{\mathrm{BM}}] = \mathbb{E}^{B_\tau}[f(B_t)]$. □

The theorem is valid also for unbounded stopping times but we stated it in this form for simplicity. A notable application of the strong Markov property of Brownian motion is the so-called reflection principle.

Corollary 2.115. *Let $(B_t)_{t\ge0}$ be a standard Brownian motion on a filtered probability space $(\Omega, \mathscr{F}, (\mathscr{F}_t^{\mathrm{BM}})_{t\ge0}, P)$. Let $\tau_a = \inf\{t > 0 \mid B_t = a\}$ be the first hitting time of level a. Then the process $(\tilde{B}_t)_{t\ge0}$ defined by*

$$\tilde{B}_t = \begin{cases} B_t, & t < \tau_a,\\ 2a - B_t, & t \ge \tau_a \end{cases} \tag{2.3.32}$$

is also a Brownian motion.

Proof. Note that τ_a is a stopping time with respect to $(\mathscr{F}_t^{\mathrm{BM}})_{t\ge0}$. Consider the processes $(Y_t)_{0\le t\le\tau_a}$, $Y_t = B_t$, and $(Z_t)_{t\ge0}$, $Z_t = B_{t+\tau_a} - a$. By the strong Markov property, $(Z_t)_{t\ge0}$ is a Brownian motion, independent of $(Y_t)_{0\le t\le\tau_a}$. By symmetry, $(-Z_t)_{t\ge0}$ is also a Brownian motion, independent of $(Y_t)_{0\le t\le\tau_a}$. Thus $(Y_t, Z_t) \stackrel{\mathrm{d}}{=} (Y_t, -Z_t)$. Let

$$\phi(Y,Z)_t = Y_t1_{\{t\le\tau_a\}} + (a + Z_{t-\tau_a})1_{\{t>\tau_a\}}, \quad t \ge 0,$$

which is a continuous process, and has the same law as $\phi(Y,-Z)_t$. However, since $\phi(Y,Z)_t = B_t$ and $\phi(Y,-Z)_t = \tilde{B}_t$, the corollary follows. □

Corollary 2.116 (Reflection principle). *Let $a > 0$, $(B_t)_{t\geq 0}$ be a standard Brownian motion on a filtered probability space $(\Omega, \mathcal{F}, (\mathcal{F}_t^{\mathrm{BM}})_{t\geq 0}, P)$, and $S_t = \sup_{0\leq s\leq t} B_s$, the running maximum of Brownian motion. Then*

$$P(S_t \geq a,\, B_t \leq a - y) = P(B_t \geq a + y). \tag{2.3.33}$$

In particular, $P(S_t \geq a) = 2P(B_t \geq a)$.

Proof. Let $\tau_a = \inf\{t > 0 \mid B_t = a\}$ be the first hitting time of level a, and $(\tilde{B})_{t\geq 0}$ be defined by (2.3.32). Denote $\tilde{S}_t = \sup_{0\leq s\leq t} \tilde{B}_s$. We have

$$P(S_t \geq a,\, B_t \leq a - y) = P(\tilde{S}_t \geq a,\, \tilde{B}_t \leq a - y) = P(t \geq \tau_a,\, B_t \geq a + y).$$

Since

$$P(B_t \geq a + y) = P(t \geq \tau_a,\, B_t \geq a + y) + P(t < \tau_a,\, B_t \geq a + y) = P(t \geq \tau_a,\, B_t \geq a + y),$$

equality (2.3.33) follows. By (2.3.33), we also see that

$$\begin{aligned} P(S_t \geq a) &= P(S_t \geq a,\, B_t \leq a) + P(S_t \geq a,\, B_t > a) \\ &= 2P(S_t \geq a,\, B_t \geq a) = 2P(B_t \geq a). \end{aligned}$$ □

We conclude by briefly discussing the transition semigroup of Brownian motion, and its generator.

Proposition 2.117. *The probability transition kernel p_{BM} of Brownian motion is a Feller transition kernel, and its generator is $\frac{1}{2}\Delta$ on $C_0^\infty(\mathbb{R}^d)$.*

Proof. We have already shown that $p_{\mathrm{BM}}(t,x,A) = \int_{\mathbb{R}^d} 1_A(z)\Pi_t(x-z)dz$ in (2.3.29). Thus $p_{\mathrm{BM}}(t,x,\mathbb{R}^d) = 1$ holds identically. As $P_t^{\mathrm{BM}}f(x) = (2\pi t)^{-d/2}\int_{\mathbb{R}^d} e^{-|z|^2/2t}f(x+z)dz$ for $f \in C_\infty(\mathbb{R}^d)$, by dominated convergence it is clear that $P_t^{\mathrm{BM}}f(x)$ is continuous in x and $\lim_{|x|\to\infty} P_t^{\mathrm{BM}}f(x) = 0$. Thus $P_t^{\mathrm{BM}}f \in C_\infty(\mathbb{R}^d)$. Next we check that

$$\lim_{t\to 0} P_t^{\mathrm{BM}}f(x) = f(x). \tag{2.3.34}$$

Notice that f is uniformly continuous on the compact set $\mathbb{R}^d \cup \{\infty\}$. Thus for every $\varepsilon > 0$ there exists $\delta > 0$ such that $|f(x+y) - f(x)| \leq \varepsilon$, for all $x \in \mathbb{R}^d$ and $|y| < \delta$. Hence

$$|P_t^{\mathrm{BM}}f(x) - f(x)| \leq \int\limits_{|y|<\delta} |f(x+y) - f(x)|\Pi_t(y)dy + \int\limits_{|y|\geq\delta} |f(x+y) - f(x)|\Pi_t(y)dy$$

$$\leq \varepsilon \int_{|y|<\delta} \Pi_t(y)dy + M \int_{|y|\geq\delta} \Pi_t(y)dy,$$

where $M = \sup_{x\in\mathbb{R}^d\cup\{\infty\}} |f(x)|$. Thus $\lim_{t\to 0} |P_t^{\mathrm{BM}} f(x) - f(x)| \leq \varepsilon$, whence (2.3.34) follows.

Let L_{BM} denote the generator of the transition semigroup P_t^{BM}, $t \geq 0$, and take $f \in C_0^\infty(\mathbb{R}^d)$. The Fourier transform of $P_t^{\mathrm{BM}} f$ is equal to $e^{-t|k|^2/2}\hat{f}(k)$. This gives

$$\lim_{t\downarrow 0} \left\| \frac{1}{t}\left(\widehat{P_t^{\mathrm{BM}} f} - \hat{f}\right) + \frac{1}{2}|k|^2 \hat{f} \right\| = 0, \tag{2.3.35}$$

and thus

$$\lim_{t\downarrow 0} \frac{1}{t}\left(P_t^{\mathrm{BM}} f - f\right) = \frac{1}{2}\Delta f \tag{2.3.36}$$

in $L^2(\mathbb{R}^d)$. By taking a subsequence t', (2.3.36) stays valid for almost every $x \in \mathbb{R}^d$, while the right-hand side (2.3.36) is equal to $L_{\mathrm{BM}} f$. Hence $L_{\mathrm{BM}} f = \frac{1}{2}\Delta f$ Lebesgue almost surely. □

2.3.6 Local path properties of Brownian motion

A second aspect in the analysis of a random process are its sample path properties. We begin by discussing the behavior of Brownian paths $[0,\infty) \ni t \mapsto B_t(\omega) \in \mathbb{R}^d$, for fixed but arbitrary $\omega \in \mathscr{X}$, at short range.

Proposition 2.118 (Hölder continuity). *Let $(B_t)_{t\geq 0}$ be a standard Brownian motion on a probability space $(\Omega, \mathscr{F}, P)$. Then the following hold:*

(1) *Let $\gamma \in (0, 1/2)$. Then*

$$P\left(\lim_{h\to 0} \sup_{\substack{0\leq s<t\leq T \\ |t-s|<h}} \frac{|B_t - B_s|}{|t-s|^\gamma} = 0 \right) = 1. \tag{2.3.37}$$

In particular, for every $T > 0$, $\gamma \in (0,1/2)$, and for a. e. $\omega \in \mathscr{X}$, there is a constant $\kappa = \kappa(T,\gamma,\omega) > 0$ such that

$$|B_t(\omega) - B_s(\omega)| \leq \kappa |t-s|^\gamma, \quad s,t \in [0,T]. \tag{2.3.38}$$

(2) *Let $\gamma \in [1/2, \infty)$. Then*

$$P\left(\lim_{h\to 0} \sup_{\substack{0\leq s<t\leq T \\ |t-s|<h}} \frac{|B_t - B_s|}{|t-s|^\gamma} = \infty \right) = 1. \tag{2.3.39}$$

In particular, the function $t \mapsto B_t$ is almost surely nowhere Hölder continuous of order $\gamma \geq 1/2$, i. e., for a. e. $\omega \in \mathscr{X}$, there is no $\kappa = \kappa(\omega)$ such that

$$|B_t(\omega) - B_s(\omega)| \leq \kappa |t-s|^\gamma, \quad s,t \in [0,T]. \tag{2.3.40}$$

Proof. (1) First we show that (2.3.38) and (2.3.37) are equivalent. Suppose (2.3.38) holds, and let $\gamma \in (0,1/2)$ be given. Take $\gamma' \in (0,1/2)$ such that $\gamma < \gamma'$. By the assumption, there exists a null set N such that for all $\omega \in \mathscr{X} \setminus N$ there is a $\kappa = \kappa(\omega)$ such that $|B_t(\omega) - B_s(\omega)| \leq \kappa|t-s|^{\gamma'}$ for $t, s \in [0,T]$. This implies that

$$\lim_{h\to 0} \sup_{\substack{0\leq s<t\leq T\\ |t-s|<h}} \frac{|B_t(\omega) - B_s(\omega)|}{|t-s|^{\gamma}} \leq \lim_{h\to 0} \kappa h^{\gamma'-\gamma} = 0$$

for $\omega \in \mathscr{X} \setminus N$, and (2.3.37) follows. Next let (2.3.37) hold. There exists a null set M such that for $\omega \in \mathscr{X} \setminus M$,

$$\lim_{h\to 0} \sup_{\substack{0\leq s<t\leq T\\ |t-s|<h}} \frac{|B_t(\omega) - B_s(\omega)|}{|t-s|^{\gamma}} = 0.$$

Fix $\omega \in \mathscr{X} \setminus M$. For an arbitrary $\varepsilon > 0$, there exists $0 < h_0$ such that for every $h < h_0$,

$$\sup_{\substack{0\leq s<t\leq T\\ |t-s|<h}} \frac{|B_t(\omega) - B_s(\omega)|}{|t-s|^{\gamma}} < \varepsilon$$

and $|B_t(\omega) - B_s(\omega)| \leq \varepsilon|t-s|^{\gamma}$ follows for all $t, s \in [0,T]$ with $|t-s| < h$. Take $n \in \mathbb{N}$ such that $n \geq T/h$. We have

$$|B_t(\omega) - B_s(\omega)| \leq n\varepsilon|t-s|^{\gamma}$$

for every $t, s \in [0,T]$. This implies (2.3.38). Next we prove (2.3.38). Note that by part (3) of Proposition 2.96 and the Kolmogorov–Čentsov theorem, there exists a γ-Hölder continuous version $(\tilde{B}_t)_{t\geq 0}$ of $(B_t)_{t\geq 0}$, given by

$$\tilde{B}_t(\omega) = \begin{cases} 0, & \omega \notin \Omega_0, \\ B_t(\omega), & \omega \in \Omega_0,\ t \in D, \\ \lim_{n\to\infty} B_{s_n}(\omega), & \omega \in \Omega_0,\ t \notin D, \end{cases}$$

where D denotes the dyadic rationals in $[0,T]$, and $(s_n)_{n\in\mathbb{N}} \subset D$ is a sequence such that $\lim_{n\to\infty} s_n = t$. Here $P(\Omega_0) = 1$. Note that $\lim_{n\to\infty} B_{s_n}(\omega) = B_t(\omega)$ almost surely since $t \mapsto B_t$ is almost surely continuous. Hence $B_t(\omega) = \tilde{B}_t(\omega)$ almost surely for all t. Since $\gamma \in (0, \frac{n-1}{2n})$ for every $n \geq 2$, the expression (2.3.38) follows.

(2) It suffices to show the claim for $\gamma = 1/2$. For $c > 0$, consider

$$A_n = \bigcap_{k=1}^{n} \{|B_{k/n} - B_{(k-1)/n}| \leq c/n^{1/2}\}.$$

By the independence of increments of Brownian motion, we have

$$P(A_n) = \prod_{k=1}^{n} P(\{|B_{k/n} - B_{(k-1)/n}| \leq c/n^{1/2}\}) = a^n,$$

where

$$a = \int_{-c/n^{1/2}}^{c/n^{1/2}} \frac{1}{\sqrt{2\pi/n}} e^{-nx^2/2} dx = \int_{-c}^{c} \frac{1}{\sqrt{2\pi}} e^{-x^2/2} dx < 1.$$

This implies that $\sum_{n=1}^{\infty} P(A_n) = \sum_{n=1}^{\infty} a^n < \infty$, and thus the Borel–Cantelli lemma yields that $P(\bigcup_{n=1}^{\infty} \bigcap_{k=n}^{\infty} A_n^c) = 1$. There is a null set N such that for every $\omega \in \mathscr{X} \setminus N$ there exists $n_0 \in \mathbb{N}$ such that for every $n \geq n_0$, $|B_{k/n}(\omega) - B_{(k-1)/n}(\omega)| > c/n^{1/2}$ for some $k \leq n$. This gives

$$P\left(\lim_{h \to 0} \sup_{\substack{0 \leq s < t \leq T \\ |t-s|<h}} \frac{|B_t - B_s|}{|t-s|^\gamma} \geq c \right) = 1.$$

Here c is arbitrary and (2.3.39) follows. Also, (2.3.40) follows for $\omega \in \mathscr{X} \setminus N$ from (2.3.39). □

An important consequence of the lack of Hölder continuity for exponents $\gamma \geq \frac{1}{2}$ is that Brownian paths are in particular not Lipschitz continuous, and thus nondifferentiable at any point.

Proposition 2.119 (Nowhere differentiability). *Let $(B_t)_{t \geq 0}$ be a standard Brownian motion on a probability space $(\Omega, \mathscr{F}, P)$. The function $t \mapsto B_t$ is almost surely nowhere differentiable.*

Proof. Fix $c > 0$ and $\gamma > 1/2$. Consider

$$\mathcal{A}_n = \left\{ \omega \in \mathscr{X} \,\middle|\, \exists s \in [0,1] \text{ such that } |B_t(\omega) - B_s(\omega)| \leq c|t-s|^\gamma \text{ if } |t-s| \leq \frac{3}{n} \right\}.$$

Note that $\mathcal{A}_n$ is nondecreasing. Define

$$\begin{aligned} Y_{k,n} &= \max \left\{ |B_{k/n} - B_{(k-1)/n}|, |B_{(k+1)/n} - B_{k/n}|, |B_{(k+2)/n} - B_{(k+1)/n}| \right\}, \\ \mathcal{B}_n &= \left\{ \max_{1 \leq k \leq n-2} Y_{k,n} \leq c(1+2^\gamma)/n^\gamma \right\}. \end{aligned}$$

Let $\omega \in \mathcal{A}_n$ and $s \in [k/n, (k+1)/n)$. We see that

$$\begin{aligned} &|B_{(k-1)/n} - B_s| \leq c(2/n)^\gamma, \quad |B_{k/n} - B_s| \leq c(1/n)^\gamma, \\ &|B_{(k+1)/n} - B_s| \leq c(1/n)^\gamma, \quad |B_{(k+2)/n} - B_s| \leq c(2/n)^\gamma. \end{aligned}$$

By the triangle inequality, it follows that $\omega \in \mathcal{B}_n$, and so we have $\mathcal{A}_n \subset \mathcal{B}_n$. Setting $c(1+2^\gamma) = \xi$, we have

$$P(\mathcal{A}_n) \leq P(\mathcal{B}_n) \leq \sum_{k=1}^{n-2} P(Y_{k,n} \leq \xi/n^\gamma) = (n-1)P(Y_{1,n} \leq \xi/n^\gamma) = (n-1)P(B_{1/n} \leq \xi/n^\gamma)^3$$

$$= (n-1)\frac{\sqrt{n}}{\sqrt{2\pi}}\Bigg(\int_{|x|\le\xi/n^{\gamma}} e^{-nx^2/2}dx\Bigg)^3 = \frac{n-1}{n}\Bigg(\int_{|y|\le\xi n^{1/2-\gamma}} e^{-|y|^2/2}dy\Bigg)^3.$$

Due to $1/2-\gamma<0$, we see that $\lim_{n\to\infty}P(\mathcal{A}_n)=0$. Since $P(\mathcal{A}_m)\le P(\mathcal{A}_n)$ for all $n\ge m$, it follows that $P(\mathcal{A}_m)=0$ for all m. Take $\gamma=1$. Since c is arbitrary, it implies that the function $t\mapsto B_t$ is almost surely nowhere diferentiable. □

The last local property that we discuss is the variation of paths over arbitrary intervals of time. This will have an important role in Section 2.4 below, where we discuss Brownian motion as an integrator. Let $p>0$, $I\subset\mathbb{R}$ be a bounded interval, and $\Delta_n=\{t_0<t_1<\cdots<t_n\}$ be a partition of I such that t_0 and t_n coincide with the two endpoints of I. The *p-variation* over I and Δ_n of Brownian motion is defined by

$$V^p_{I,\Delta_n}(\omega)=\sum_{k=0}^{n-1}|B_{t_{k+1}}(\omega)-B_{t_k}(\omega)|^p.$$

For $p=1$, this gives the *total variation* and for $p=2$ the *quadratic variation.*

Proposition 2.120 (*p*-variation). *Let $p>0$ and $(B_t)_{t\ge0}$ be a d-dimensional standard Brownian motion on a probability space $(\Omega,\mathcal{F},P)$. Consider a bounded interval $I\subset\mathbb{R}$, and let $\Delta_n=\{t_0<t_1<\cdots<t_n\}$ be a partition of I such that t_0 and t_n coincide with the two endpoints of I, with mesh $m(\Delta_n)=\max_{k=0,\dots,n-1}|t_{k+1}-t_k|$. Then the following hold:*

(1) $\limsup_{m(\Delta_n)\to0}V^p_{I,\Delta_n}(\omega)=\begin{cases}\infty & p<2,\\ d|I| & p=2,\\ 0 & p>2\end{cases}$ *holds almost surely;*

(2) $\limsup_{m(\Delta_n)\to0}V^p_{I,\Delta_n}(\omega)=\begin{cases}d|I|, & p=2,\\ 0, & p>2\end{cases}$ *in $L^q(\Omega,dP)$, $q\ge1$, where $|I|$ denotes the length of the interval I.*

Proof. We only show a special case for simplicity, a generalization is left to the reader. Suppose that $d=1$, $I=[0,1]$, and $t_k=k/2^n$. We have $\Delta_n=\{0,1/2^n,2/2^n,\dots,2^n/2^n\}$ and $m(\Delta_n)=1/2^n$. Write $V^p_{I,\Delta_n}=V^p_n$ and let

$$d(n,p)=\mathbb{E}[|B_{k/2^n}-B_{(k-1)/2^n}|^p]=2^{-np/2}c_p,$$

with constant $c_p=(2\pi)^{-1/2}\int_{\mathbb{R}}|x|^pe^{-|x|^2/2}dx$. Note that $c_2=1$. We obtain

$$\begin{aligned}\mathbb{E}\Bigg[\Bigg|\sum_{k=1}^{2^n}(|B_{k/2^n}-B_{(k-1)/2^n}|^p-d(n,p))\Bigg|^2\Bigg]&=\sum_{k=1}^{2^n}\mathbb{E}\Big[\big||B_{k/2^n}-B_{(k-1)/2^n}|^p-d(n,p)\big|^2\Big]\\&\le\sum_{k=1}^{2^n}\mathbb{E}\Big[|B_{k/2^n}-B_{(k-1)/2^n}|^{2p}\Big]=2^nd(n,2p),\end{aligned}$$

where we used that $\mathbb{E}[(X - \mathbb{E}[X])^2] \le \mathbb{E}[X^2]$. This gives

$$\mathbb{E}[|V_n^p - 2^{n(1-(p/2))}c_p|^2] \le 2^{n(1-p)}c_{2p}. \tag{2.3.41}$$

(1) Suppose $p < 2$. For a given $k > 0$, taking sufficiently large n, we have that $\frac{1}{2}c_p 2^{n(1-(1/2)p)} > k$. For such n, it follows that

$$\begin{aligned} P(V_n^p < k) &\le P\left(V_n^p < \frac{1}{2}c_p 2^{n(1-(1/2)p)}\right) \\ &= P\left(V_n^p - c_p 2^{n(1-(1/2)p)} < -\frac{1}{2}c_p 2^{n(1-(1/2)p)}\right) \\ &\le \frac{1}{(\frac{1}{2}c_p 2^{n(1-(1/2)p)})^2}\mathbb{E}[|V_n^p - c_p 2^{n(1-(1/2)p)}|^2] \le 4c_p^{-2}c_{2p}2^{-n}. \end{aligned}$$

Thus $\sum_{n=1}^\infty P(V_n^p < k) < \infty$ follows. By the Borel–Cantelli lemma, this implies that $P(\bigcup_{m=1}^\infty \bigcap_{n=m}^\infty \{V_n^p \ge k\}) = 1$, and $V_n^p \to \infty$ follows almost surely.

Next suppose $p = 2$. By (2.3.41), we have $\mathbb{E}[|V_n^2 - 1|^2] \le 2^{-n}c_4$. In the same way as for $p < 2$, it follows that

$$P(|V_n^2 - 1| \ge \varepsilon) \le \frac{1}{\varepsilon^2}2^{-n}c_4 \tag{2.3.42}$$

and, by the Borel–Cantelli lemma, we have $P(\bigcup_{m=1}^\infty \bigcap_{n=m}^\infty \{|V_n^2 - 1| \le \varepsilon\}) = 1$, giving $V_n^2 \to 1$ as $n \to \infty$, almost surely.

Suppose that $p > 2$. By (2.3.41), we have $\mathbb{E}[|V_n^p - 2^{n(1-(p/2))}c_p|^2] \le 2^{n(1-p)}c_{2p}$. For every $\varepsilon > 0$, it follows that

$$P(V_n^p > \varepsilon) \le \frac{1}{\varepsilon^2}\mathbb{E}[|V_n^p|^2] \le \frac{2}{\varepsilon^2}\left(\mathbb{E}[|V_n^p - 2^{n(1-(p/2))}|^2] + 2^{2n(1-(p/2))}\right). \tag{2.3.43}$$

We have $\sum_{n=1}^\infty P(V_n^p > \varepsilon) < \infty$, and from the Borel–Cantelli lemma it follows again that $P(\bigcup_{m=1}^\infty \bigcap_{n=m}^\infty \{V_n^p \le \varepsilon\}) = 1$. Hence we see that $V_n^p \to 0$ as $n \to \infty$, almost surely. By (2.3.42) and (2.3.43), both convergences hold also in probability.

(2) Next we show L^q-convergence. It can be seen that

$$\mathbb{E}[|V_n^p|^q] \le 2^{nq}\mathbb{E}[|B_{1/2^n}|^{qp}] \le 2^{nq(1-(1/2)p)}.$$

Thus in the case of $p > 2$, $V_n^p \to 0$ in L^q. Suppose next $p = 2$. Write for simplicity $X_j = |B_{j/2^n} - B_{(j-1)/2^n}|$ and choose $q = 2$. We see that

$$\mathbb{E}[|V_n^2 - 1|^2] = \mathbb{E}\left[\left(\sum_{j=1}^{2^n} X_j^2\right)^2 - 2\sum_{j=1}^{2^n} X_j^2 + 1\right].$$

We note that $\{X_n\}_{n\in\mathbb{N}}$ are independent random variables and $\mathbb{E}[X_n] = 0$. We obtain

$$\begin{aligned}\mathbb{E}[|V_n^2-1|^2] &= \sum_{j'\neq j}^{2^n}\mathbb{E}[X_{j'}^2X_j^2]+\sum_{j=1}^{2^n}\mathbb{E}[X_j^4]-2\sum_{j=1}^{2^n}\mathbb{E}[X_j^2]+1\\ &= \sum_{j'\neq j}^{2^n}2^{-2n}+\sum_{j=1}^{2^n}2^{-2n}c_4-2\sum_{j=1}^{2^n}2^{-n}+1\\ &= 2^n(2^n-1)2^{-2n}+2^n\cdot 2^{-2n}c_4-2\cdot 2^n\cdot 2^{-n}+1.\end{aligned}$$

Thus $V_n^2 \to 1$ as $n \to \infty$ in L^2. Next suppose $q = 4$. In the same way as for L^2, we see that

$$\mathbb{E}[|V_n^2-1|^4]=\mathbb{E}\left[\left(\sum_{j=1}^{2^n}X_j^2\right)^4-4\left(\sum_{j=1}^{2^n}X_j^2\right)^3+6\left(\sum_{j=1}^{2^n}X_j^2\right)^2-4\left(\sum_{j=1}^{2^n}X_j^2\right)+1\right].$$

We compute the fourth-order term and get

$$\begin{aligned}\mathbb{E}\left[\left(\sum_{j=1}^{2^n}X_j^2\right)^4\right] &= \sum_{i\neq j\neq k\neq l}^{2^n}\mathbb{E}[X_i^2X_j^2X_k^2X_l^2]+\sum_{i\neq j\neq k}^{2^n}\mathbb{E}[X_i^2X_j^2X_k^4]\\ &\quad+\sum_{i\neq j}^{2^n}\mathbb{E}[X_i^4X_j^4]+\sum_{i\neq j}^{2^n}\mathbb{E}[X_i^6X_j^2]+\sum_{i=1}^{2^n}\mathbb{E}[X_i^8],\end{aligned}$$

where $\sum_{i\neq j\neq k\neq l}^{2^n}$ denotes summation over all $1 \le i,j,k,l \le 2^n$ such that i, j, k, l are all different from each other (the other sums are similarly understood). We get

$$\sum_{i\neq j\neq k\neq l}^{2^n}\mathbb{E}[X_i^2X_j^2X_k^2X_l^2]=2^{-4n}2^n(2^n-1)(2^n-2)(2^n-3)\to 1$$

as $n \to \infty$, and

$$\sum_{i\neq j\neq k}^{2^n}\mathbb{E}[X_i^2X_j^2X_k^4]=2^{-4n}c_4 2^n(2^n-1)(2^n-2)\to 0$$

as $n \to \infty$. Similarly,

$$\begin{aligned}&\sum_{i\neq j}^{2^n}\mathbb{E}[X_i^4X_j^4]=2^{-4n}c_4^2 2^n(2^n-1)\to 0,\\ &\sum_{i\neq j}^{2^n}\mathbb{E}[X_i^6X_j^2]=2^{-4n}2^n(2^n-1)\to 0,\\ &\sum_{i=1}^{2^n}\mathbb{E}[X_i^8]=2^{-4n}c_8 2^n\to 0.\end{aligned}$$

A further computation of the same kind yields that

$$\mathbb{E}\left[\left(\sum_{j=1}^{2^n} X_j^2\right)^3\right] \to 1, \quad \mathbb{E}\left[\left(\sum_{j=1}^{2^n} X_j^2\right)^2\right] \to 1, \quad \text{and} \quad \mathbb{E}\left[\left(\sum_{j=1}^{2^n} X_j^2\right)\right] \to 1.$$

Putting together these results, we obtain that $V_n^2 \to 1$ in L^4. Take now $q = 2m$. In this case,

$$\mathbb{E}[|V_n^2 - 1|^{2m}] = \sum_{m'=0}^{2m} \binom{2m}{m'} (-1)^{2m-m'} \mathbb{E}\left[\left(\sum_{j=1}^{2^n} X_j^2\right)^{m'}\right],$$

and we see that $\lim_{n\to\infty} \mathbb{E}[(\sum_{j=1}^{2^n} X_j^2)^{m'}] = 1$ similarly as above. Hence

$$\lim_{n\to\infty} \mathbb{E}[|V_n^2 - 1|^{2m}] = \sum_{m'=0}^{2m} \binom{2m}{m'} (-1)^{2m-m'} = 0,$$

which implies $V_n^2 \to 1$ as $n \to \infty$ in L^{2m}. □

2.3.7 Global path properties of Brownian motion

Next we discuss the typical long time behavior of Brownian motion.

Proposition 2.121. *Let $(B_t)_{t\geq 0}$ be a standard Brownian motion on a probability space $(\Omega, \mathscr{F}, P)$. Then*

$$\liminf_{t\to\infty} B_t = -\infty \quad \textit{and} \quad \limsup_{t\to\infty} B_t = \infty$$

hold almost surely.

Proof. Fix a sequence $(t_n)_{n\geq 1}$ such that $t_n \to \infty$ as $n \to \infty$. Due to the self-similarity of Brownian motion, we have $B_{t_n} \stackrel{d}{=} \sqrt{t_n} B_1$, and thus $P(B_{t_n} > K) = P(B_1 > K/\sqrt{t_n}) \to (2\pi)^{-1/2} \int_0^\infty e^{-|x|^2/2} dx = 1/2$ as $n \to \infty$, for every $K > 0$. By Fatou's lemma, we get

$$P\left(\bigcup_{m=1}^{\infty} \bigcap_{n=m}^{\infty} \{B_{t_n} > K\}\right) = \mathbb{E}\left[\limsup_{n\to\infty} 1_{\{B_{t_n} > K\}}\right] \geq \limsup_{n\to\infty} \mathbb{E}[1_{\{B_{t_n} > K\}}] = \frac{1}{2}.$$

Hence $P(\limsup_{n\to\infty} B_{t_n} > K) \geq 1/2$. Therefore

$$P\left(\limsup_{n\to\infty} B_{t_n} = \infty\right) \geq \frac{1}{2}.$$

Let $t_0 = 0$ and $Z_n = B_{t_n} - B_{t_{n-1}}$. The random variables $(Z_n)_{n\in\mathbb{N}}$ are independent and $B_{t_n} = \sum_{j=1}^n Z_j$. We have

$$I = \left\{\limsup_{n\to\infty} B_{t_n} = \infty\right\} = \left\{\limsup_{n\to\infty}(B_{t_n} - B_{t_m}) = \infty\right\} \in \sigma\Big(\bigcup_{n=m+1}^{\infty} \sigma(Z_n)\Big)$$

for each m. Since $I \in \bigcap_{m=1}^{\infty} \sigma(\bigcup_{n=m+1}^{\infty} \sigma(Z_n))$ and $\bigcap_{m=1}^{\infty} \sigma(\bigcup_{n=m+1}^{\infty} \sigma(Z_n))$ is a tail-σ-field, Kolmogorov's zero–one law yields $P(I) \in \{0,1\}$. Since this probability is not less than 1/2 by the above, it must then be 1. Hence it follows that $\limsup_{n\to\infty} B_{t_n} = \infty$ almost surely. By reflection symmetry of Brownian motion, we immediately obtain $\liminf_{n\in\mathbb{N}} B_{t_n} = -\infty$ almost surely. □

Proposition 2.122. *Let $(B_t)_{t\geq 0}$ be a standard Brownian motion on a probability space $(\Omega, \mathscr{F}, P)$. Then*

$$\liminf_{t\to\infty} \frac{B_t}{\sqrt{t}} = -\infty \quad \textit{and} \quad \limsup_{t\to\infty} \frac{B_t}{\sqrt{t}} = \infty$$

hold almost surely.

Proof. Let $K > 0$ be arbitrary and consider the stopping time $\tau = \inf\{t > 0 \mid B_t \geq K\sqrt{t}\}$. Take the events $A_s = \{\tau < s\}$ and $A = \{\tau = 0\}$. It is directly seen that

$$A = \bigcap_{s>0} A_s \in \bigcap_{t>0} \sigma(B_r, 0 \leq r \leq t).$$

By Blumenthal's zero–one law, it follows that $P(A) \in \{0,1\}$. Using $B_t/\sqrt{t} \overset{\mathrm{d}}{=} B_1$, we obtain that

$$P(A) = \inf_{s>0} P(A_s) \geq \inf_{s>0} P(B_s \geq K\sqrt{s}) = P(B_1 \geq K) > 0$$

and thus $P(A) = 1$. This means, in particular, that for every $\varepsilon > 0$ there exists $0 < t < \varepsilon$ such that $B_t/\sqrt{t} \geq K$ almost surely, for every $K > 0$. Thus $\limsup_{t\downarrow 0} B_t/\sqrt{t} = \infty$ almost surely. Since $tB_{1/t} \overset{\mathrm{d}}{=} B_t$, we also see that $\limsup_{t\to\infty} B_t/\sqrt{t} = \infty$ almost surely. It follows similarly that $\liminf_{t\to\infty} B_t/\sqrt{t} = -\infty$. □

Proposition 2.123 (Strong law of large numbers). *Let $(B_t)_{t\geq 0}$ be a standard Brownian motion on a probability space $(\Omega, \mathscr{F}, P)$. Then*

$$\lim_{t\to\infty} \frac{B_t}{t} = 0$$

almost surely.

Proof. By the martingale inequality shown in Theorem 2.72, we see that

$$\mathbb{E}\Big[\sup_{S\leq t\leq T} |B_t/t|^2\Big] \leq \frac{1}{S^2}\mathbb{E}\Big[\sup_{S\leq t\leq T} |B_t|^2\Big] \leq \frac{4}{S^2}\mathbb{E}[|B_T|^2] = \frac{4}{S^2}T.$$

This implies that

$$P\Big(\sup_{S\le t\le T}|B_t/t|>\varepsilon\Big)\le\frac{1}{\varepsilon^2}\mathbb{E}\Big[\sup_{S\le t\le T}|B_t/t|^2\Big]\le\frac{1}{\varepsilon^2}\frac{4T}{S^2}.$$

Put $S=2^n$ and $T=2^{n+1}$. We have $P(\sup_{2^n\le t\le 2^{n+1}}|B_t/t|>\varepsilon)\le\frac{8}{\varepsilon^2}2^{-n}$ and

$$\sum_{n=1}^{\infty}P(\sup_{2^n\le t\le 2^{n+1}}|B_t/t|>\varepsilon)<\infty.$$

Hence the Borel–Cantelli lemma yields

$$P\Big(\bigcup_{m=1}^{\infty}\bigcap_{n=m}^{\infty}\{\sup_{2^n\le t\le 2^{n+1}}|B_t/t|\le\varepsilon\}\Big)=1,$$

which implies the statement. □

The last two statements above over- and underestimated the typical growth profile of Brownian motion. The following result identifies the almost sure growth rate precisely.

Proposition 2.124 (Law of iterated logarithm). *Let $(B_t)_{t\ge0}$ be a standard Brownian motion on a probability space $(\Omega,\mathscr{F},P)$. Then*
(1) $\limsup_{t\to\infty}B_t/\sqrt{2t\log\log t}=1$;
(2) $\liminf_{t\to\infty}B_t/\sqrt{2t\log\log t}=-1$;
(3) $\limsup_{t\downarrow0}B_t/\sqrt{2t\log\log(1/t)}=1$;
(4) $\liminf_{t\downarrow0}B_t/\sqrt{2t\log\log(1/t)}=-1$

hold almost surely.

Proof. First, we consider (1) and divide the proof into two steps.
Step 1: (Upper bound) Fix $\varepsilon>0$ and $r>1$. Let

$$\mathcal{A}_n=\Big\{\sup_{0\le t\le r^n}B_t\ge(1+\varepsilon)\sqrt{2r^n\log\log r^n}\Big\}.$$

By the reflection principle obtained in Corollary 2.116, we have

$$P(\mathcal{A}_n)\le 2P\left(B_{r^n}\ge(1+\varepsilon)\sqrt{2r^n\log\log r^n}\right)=2P\left(B_{r^n}/\sqrt{r^n}\ge(1+\varepsilon)\sqrt{2\log\log r^n}\right).$$

Since $B_{r^n}/\sqrt{r^n}\overset{\mathrm{d}}{=}B_1$, we obtain

$$P(\mathcal{A}_n)\le 2P\left(B_1\ge(1+\varepsilon)\sqrt{2\log\log r^n}\right).\tag{2.3.44}$$

It can be directly seen that

$$(2\pi)^{-1/2}\frac{x}{1+x^2}e^{-x^2/2}\le P(B_1>x)\le(2\pi)^{-1/2}\frac{1}{x}e^{-x^2/2},\quad x>0.\tag{2.3.45}$$

By (2.3.44) and (2.3.45), we have

$$P(\mathcal{A}_n) \le \frac{1}{(1+\varepsilon)\sqrt{\pi}\sqrt{\log\log r^n}} e^{-(1+\varepsilon)^2 \log\log r^n} \le c(n\log r)^{-(1+\varepsilon)^2}.$$

Thus $\sum_{n=1}^{\infty} P(\mathcal{A}_n) < \infty$, and hence $P(\bigcup_{m=1}^{\infty}\bigcap_{n=m}^{\infty}\mathcal{A}_n^c) = 1$ follows from the Borel–Cantelli lemma, which implies that

$$\limsup_{n\to\infty} \frac{\sup_{0\le t\le r^n} B_t}{\sqrt{2r^n \log\log r^n}} \le 1+\varepsilon$$

almost surely. For every $t > 1$, there exists n such that $t \in [r^{n-1}, r^n]$. Note that $t \mapsto \sqrt{2t\log\log t}$ is increasing. For a sufficiently large t, we have

$$\frac{B_t}{\sqrt{2t\log\log t}} \le \frac{\sup_{0\le s\le r^n} B_s}{\sqrt{2r^n\log\log r^n}} \frac{\sqrt{2r^n\log\log r^n}}{\sqrt{2r^{n-1}\log\log r^{n-1}}}.$$

Thus we conclude that

$$\limsup_{t\to\infty} \frac{B_t}{\sqrt{2t\log\log t}} \le (1+\varepsilon)\sqrt{r}.$$

Since $\varepsilon > 0$ and $r > 1$ are arbitrary, we obtain

$$\limsup_{t\to\infty} \frac{B_t}{\sqrt{2t\log\log t}} \le 1. \tag{2.3.46}$$

Step 2: (Lower bound) Let

$$\mathcal{B}_n = \{B_{r^n} - B_{r^{n-1}} \ge \sqrt{2(r^n - r^{n-1})\log\log r^n}\}$$

for $n \ge 1$. The events $\mathcal{B}_n$, $n = 1, 2, \ldots$, are independent due to the independence of the increments of $(B_t)_{t\ge 0}$. Moreover, it follows that

$$P(\mathcal{B}_n) = P\left(\frac{B_{r^n} - B_{r^{n-1}}}{\sqrt{r^n - r^{n-1}}} \ge \sqrt{2\log\log r^n}\right) = P\left(\frac{B_{r^n - r^{n-1}}}{\sqrt{r^n - r^{n-1}}} \ge \sqrt{2\log\log r^n}\right).$$

Since $\frac{B_{r^n - r^{n-1}}}{\sqrt{r^n - r^{n-1}}} \overset{\mathrm{d}}{=} B_1$, we have

$$P\left(\frac{B_{r^n - r^{n-1}}}{\sqrt{r^n - r^{n-1}}} \ge \sqrt{2\log\log r^n}\right) \ge \frac{1}{\sqrt{8\pi}} \frac{1}{\sqrt{2\log\log r^n}} e^{-\log\log r^n} \ge \frac{c}{n\sqrt{\log n}}.$$

Here we used the lower bound in (2.3.45), and $2x/(1+x^2) \ge 1/x$ for all $x > 1$. Hence $\sum_{n=1}^{\infty} P(\mathcal{B}_n) = \infty$, and thus it follows by the Borel–Cantelli lemma that $P(\bigcap_{m=1}^{\infty}\bigcup_{n=m}^{\infty}\mathcal{B}_n) = 1$. Hence for infinitely many $n \ge 1$,

$$B_{r^n} \geq \sqrt{2(r^n - r^{n-1}) \log \log r^n} + B_{r^{n-1}}.$$

Using the estimate in Step 1 for $(-B_t)_{t\geq 0}$, we get that

$$-B_{r^{n-1}}(\omega) \leq 2\sqrt{2r^{n-1} \log \log r^{n-1}} \leq \frac{2}{\sqrt{r}} \sqrt{2r^n \log \log r^n}, \tag{2.3.47}$$

almost surely for every $n \geq n_0(\omega)$, with some $n_0(\omega)$. We then obtain

$$B_{r^n} \geq \sqrt{2(r^n - r^{n-1}) \log \log r^n} - \frac{2}{\sqrt{r}} \sqrt{2r^n \log \log r^n}$$

and conclude that

$$\limsup_{t\to\infty} \frac{B_t}{\sqrt{2t \log \log t}} \geq \limsup_{n\to\infty} \frac{B_{r^n}}{\sqrt{2r^n \log \log r^n}} \geq \sqrt{1 - \frac{1}{r}} - \frac{2}{\sqrt{r}}. \tag{2.3.48}$$

Letting $r \to \infty$ along a countable sequence completes the proof of (1). Since $(-B_t)_{t\geq 0}$ and $(tB_{1/t})_{t\geq 0}$ are again Brownian motions, parts (2), (3), and (4) follow from (1). □

2.4 Stochastic calculus based on Brownian motion

2.4.1 The classical integral and its extensions

Classical analysis provides a working notion of integral called after Riemann. In this concept, a real-valued function f on an interval $I = [a, b]$ is considered, I is divided into subintervals $\Delta_n = \{a = x_0 < \cdots < x_n = b\}$, and limits of sums of the type

$$\sum_{j=0}^{n-1} f(\xi_j)(x_{j+1} - x_j), \quad \xi_j \in [x_j, x_{j+1}),$$

are considered when $\lim_{n\to\infty} m(\Delta_n) = 0$, where $m(\Delta_n)$ is the mesh of the partition. When such a limit exists independent of the choice of the partition and the values ξ_j, we call it the *Riemann integral* $\int_a^b f(x)dx$ of f. A sufficient condition for the Riemann integral of f to exist is that f is continuous.

If for some purpose one is interested in integrating a function against the increments of another, it is possible to extend the concept of Riemann integral to cover this case. This extension is achieved by the *Stieltjes integral* $\int_a^b f(x)dg(x)$, obtained as the limit of sums of the type

$$\sum_{j=0}^{n-1} f(\xi_j)(g(x_{j+1}) - g(x_j)), \quad \xi_j \in [x_j, x_{j+1}).$$

The Riemann integral then follows when the integrator is the identity function, $g(x) = x$. Two sufficient conditions for the Stieltjes integral of f with respect to g to exist are (1) f and g have no discontinuities at the same point x, (2) f is continuous and g is of bounded variation. However, (2) is not a necessary condition.

Surprisingly, there are few results on weaker conditions on the existence of the Riemann–Stieltjes integral. Young has shown as early as in 1936 that if f is ρ-Hölder continuous and g is γ-Hölder continuous such that $\gamma + \rho > 1$, then $\int_a^b f(x)dg(x)$ exists. A nearly optimal (sufficient and "very nearly" also necessary) condition is that

(1) f and g are not both discontinuous at the same point x;
(2) f has bounded p-variation and g has bounded q-variation for some $p, q > 0$ such that $1/p + 1/q = 1$.

Our interest is to define integrals of the type $\int_a^b X_t dB_t$ where we would use Brownian motion as integrator and a random process X_t as integrand. Clearly, because Brownian motion is nowhere differentiable, the usual Riemann–Stieltjes integral will not work. However, since Brownian motion is of bounded p-variation for $p > 2$, see Proposition 2.120, provided the integrand is of bounded q-variation with suitable $q < 2$, Young's criterion can be applied. This is possible, in particular, when $(X_t)_{t\geq 0}$ has bounded variation, i. e., $q = 1$. A sufficient condition for this is that $t \mapsto X_t$ is differentiable and has a bounded derivative. However, the approach offered by Young or the conditions above quickly show their limitations as they would fail to cover the simple integral $\int_a^b B_t dB_t$. In order to include this and more interesting cases, a genuinely new type of integral is needed.

2.4.2 Stochastic integrals

Recall the basic theorem of calculus: If $f \in C^1(\mathbb{R}^d; \mathbb{R})$ and $g \in C^1(\mathbb{R}; \mathbb{R}^d)$, then

$$f(g(t)) - f(g(0)) = \int_0^t \nabla f(g(s)) \cdot (dg(s)/ds)\, ds.$$

Since Brownian paths are nowhere differentiable, we cannot substitute B_s for $g(s)$ in the formula above. In spite of that, a similar formula still exists with an extra term appearing as a correction. Indeed, we will see that almost surely we have

$$f(B_t) - f(B_0) = \int_0^t \nabla f(B_s) \cdot dB_s + \frac{1}{2} \int_0^t \Delta f(B_s)\, ds,$$

with an expression formally written as dB_s "=" $(dB_s/ds)ds$, which will be explained below. The first term on the right-hand side above leads to a new concept of integral.

Following the conventional concept of Riemann–Stieltjes integrals, in order to construct the integral with respect to Brownian motion, we divide the interval into n subintervals and look at approximating sums

$$S_n = \sum_{j=1}^{n} f(B_{s_j})(B_{t_{j+1}} - B_{t_j}), \quad s_j \in [t_j, t_{j+1}).$$

The pathwise (that is, in $\mathscr{X}$ pointwise) limit of the random variable S_n does not exist due to the unbounded variation of Brownian paths, thus it cannot be defined as a Stieltjes integral. However, as will be seen below, when S_n is regarded as an element of $L^2(\mathscr{X}, dW)$, it does have a limit. But even in this situation, we face the peculiar feature that, unlike in the case of the Riemann–Stieltjes integral, the limit depends on how the point s_j is chosen and a different object is obtained when $s_j = t_j$ or when $s_j = t_{j+1}$.

First, we consider a one-dimensional standard Brownian motion.

Example 2.125. Take a division of the interval $[0, T]$ into subintervals of equal length, i. e., choose $t_j = jT/n$. Then a direct calculation gives $\lim_{n\to\infty} \sum_{j=0}^{n-1} B_{t_j}(B_{t_{j+1}} - B_{t_j}) = \frac{1}{2}(B_T^2 - T)$ and $\lim_{n\to\infty} \sum_{j=0}^{n-1} B_{t_{j+1}}(B_{t_{j+1}} - B_{t_j}) = \frac{1}{2}(B_T^2 + T)$ in L^2-sense. Indeed, a rearrangement of the terms gives for the first sum

$$\sum_{j=0}^{n-1} B_{t_j}(B_{t_{j+1}} - B_{t_j}) = \frac{1}{2}B_T^2 - \frac{1}{2}\sum_{j=0}^{n-1}(B_{t_{j+1}} - B_{t_j})^2.$$

By Proposition 2.120, the latter sum converges in L^2 sense to the quadratic variation $V_T^2 = T$, thus the first claim follows. The other limit can be computed similarly.

To remove the problem of dependence on the intermediary point, we restrict here to the choice $s_j = t_j$, i. e., we choose the integrand to be adapted to the natural filtration $(\mathscr{F}_t^{\mathrm{BM}})_{t\geq 0}$ of the standard Brownian motion. We call $(\phi(t))_{t\geq 0}$ a *random step process* if there is a sequence $0 \leq t_0 < \cdots < t_n$ and random variables $f_0, \ldots, f_{n-1} \in L^2(\mathscr{X}, dW)$, f_j are $\mathcal{F}_{t_j}^{\mathrm{BM}}$-measurable, and

$$\phi(t, \omega) = \sum_{j=0}^{n-1} f_j(\omega) 1_{[t_j, t_{j+1})}(t). \tag{2.4.1}$$

The notation M_{step}^2 will be used for the space of random step processes.

Definition 2.126. We denote by $M^2(S, T)$ the class of functions $f : \mathbb{R}^+ \times \mathscr{X} \to \mathbb{R}$ such that

(1) $f \in \mathcal{B}(\mathbb{R}^+) \times \mathcal{B}(\mathscr{X})/\mathcal{B}(\mathbb{R})$;
(2) $(f(t, \cdot))_{t\geq 0}$ is $(\mathscr{F}_t^{\mathrm{BM}})_{t\geq 0}$-adapted;
(3) $\mathbb{E}[\int_S^T |f(t, \omega)|^2 dt] < \infty$.

We note that the filtration $(\mathcal{F}_t^{\mathrm{BM}})_{t\geq 0}$ is right-continuous and contains all the null sets, while the natural filtration $(\mathcal{F}_t)_{t\geq 0}$ of Brownian motion is not. There are general reasons in stochastic analysis how the filtration is selected. For instance, first hitting times of an open or a closed set are stopping times under $(\mathcal{F}_t^{\mathrm{BM}})_{t\geq 0}$, see Proposition 2.60, and thus makes this filtration more convenient to use.

Let $\phi(t,\omega)$ be as in (2.4.1) with $S = t_0 < t_1 < \cdots < t_n = T$, and $(B_t)_{t\geq 0}$ one-dimensional Brownian motion. Define

$$\int_S^T \phi(t,\omega)dB_t = \sum_{j=0}^{n-1} f_j(\omega)(B_{t_{j+1}}(\omega) - B_{t_j}(\omega)). \tag{2.4.2}$$

Lemma 2.127. *Let $\phi(t,\omega)$ be as in* (2.4.2). *Then*

$$\mathbb{E}^x\Big[\Big|\int_S^T \phi(t,\omega)dB_t\Big|^2\Big] = \mathbb{E}^x\Big[\int_S^T |\phi(t,\omega)|^2 dt\Big]. \tag{2.4.3}$$

Proof. Denote in shorthand $I(\phi) = \int_S^T \phi(t,\omega)dB_t$, $\Delta B_j = B_{t_{j+1}} - B_{t_j}$, and $\Delta t_j = t_{j+1} - t_j$. Clearly, $\mathbb{E}^x[\Delta B_j] = 0$, $\mathbb{E}^x[(\Delta B_j)^2] = \Delta t_j$. We have

$$|I(\phi)|^2 = \sum_{j=0}^{n-1} f_j^2(\Delta B_j)^2 + 2\sum_{0\leq j<k\leq n-1} f_j f_k \Delta B_j \Delta B_k.$$

By independence, if $j < k$, then $\mathbb{E}^x[f_j f_k \Delta B_j \Delta B_k] = 0$. On the other hand, for $j = k$, $\mathbb{E}^x[f_j^2(\Delta B_j)^2] = \mathbb{E}^x[f_j^2]\Delta t_j$. Hence

$$\mathbb{E}^x[|I(\phi)|^2] = \sum_{j=0}^{n-1} \mathbb{E}^x[f_j^2]\Delta t_j. \tag{2.4.4}$$

Moreover, $|\phi(t)|^2 = \sum_{j=0}^{n-1} f_j^2 1_{[t_j,t_{j+1})}(t)$, i. e.,

$$\mathbb{E}^x\Big[\Big|\int_S^T |\phi(t)|^2 dt\Big|\Big] = \sum_{j=0}^{n-1} \mathbb{E}^x[f_j^2]\Delta t_j \tag{2.4.5}$$

follows. A comparison of the right-hand sides of (2.4.4) and (2.4.5) above completes the proof. □

Equality (2.4.3) is called *Itô isometry*. Using Itô isometry, we can extend the definition of the stochastic integral $\int_S^T f(t,\omega)dB_t$ to $f \in M^2(S,T)$. First, we construct in the lemma below a sequence of random step functions $(\phi_n)_{n\in\mathbb{N}} \subset M^2_{\mathrm{step}}$ convergent to a given $f \in M^2(S,T)$, that is,

$$\mathbb{E}^x\Big[\int_S^T |f(t,\omega)-\phi_n(t,\omega)|^2 dt\Big] \to 0 \quad \text{as } n \to \infty. \tag{2.4.6}$$

Lemma 2.128. *Let $f \in M^2(S,T)$. Then there exists a sequence $(\phi_n)_{n\in\mathbb{N}} \subset M^2_{\text{step}}$ satisfying* (2.4.6).

Proof. We divide the proof into four steps.

Step 1. Let $g \in M^2(S,T)$ be bounded and $g(\cdot,\omega)$ be continuous for every $\omega \in \mathscr{X}$. There exists a sequence of random step functions $(\phi_n)_{n\in\mathbb{N}} \subset M^2_{\text{step}}$ such that

$$\int_S^T |g(t,\omega)-\phi_n(t,\omega)|^2 dt \to 0 \tag{2.4.7}$$

as $n \to \infty$. In fact, we can take $\phi_n = \sum_{j=0}^{n-1} g(t_j,\omega)1_{[t_j,t_{j+1})}(t)$.

Step 2. Let $h \in M^2(S,T)$ be bounded. There exists a bounded sequence $(g_n(\cdot,\omega))_{n\in\mathbb{N}} \subset M^2(S,T)$ continuous for every ω, with

$$\int_S^T |h(t,\omega)-g_n(t,\omega)|^2 dt \to 0 \tag{2.4.8}$$

as $n \to \infty$. We can take $g_n(t,\omega) = \int_0^t \psi_n(s-t)h(s,\omega)ds$, where (1) ψ_n is a nonnegative and continuous function, (2) $\operatorname{supp}\psi_n \subset [-1/n,0]$, and (3) $\int_{\mathbb{R}} \psi_n(x)dx = 1$.

Step 3. Let $f \in M^2(S,T)$. There exists a sequence $(h_n)_{n\in\mathbb{N}} \subset M^2(S,T)$ of bounded functions and

$$\int_S^T |f(t,\omega)-h_n(t,\omega)|^2 dt \to 0 \tag{2.4.9}$$

as $n \to \infty$. Here we can take

$$h_n(t,\omega) = \begin{cases} -n, & f(t,\omega) < -n, \\ f(t,\omega), & -n \le f(t,\omega) \le n, \\ n, & f(t,\omega) > n. \end{cases}$$

Step 4. If $f \in M^2(S,T)$, by (2.4.7)–(2.4.9) we can choose a sequence of random step functions $(\phi_n)_{n\in\mathbb{N}} \subset M^2_{\text{step}}$ such that $\mathbb{E}^x[\int_S^T |f(t,\omega)-\phi_n(t,\omega)|^2 dt] \to 0$ as $n \to \infty$. □

By (2.4.6) and the Itô isometry, we see that $(\int_S^T \phi_n dB_t)_{n\in\mathbb{N}}$ is a Cauchy sequence in $L^2(\mathscr{X}, d\mathcal{W}^x)$. Thus $\int_S^T \phi_n dB_t$ converges to a random variable as $n \to \infty$ in $L^2(\mathscr{X}, d\mathcal{W}^x)$.

Now we are in the position to define the Itô integral.

Definition 2.129 (Itô integral/stochastic integral). Let $f \in M^2(S,T)$ and a sequence $(\phi_n)_{n\in\mathbb{N}} \subset M^2_{\text{step}}$ be such that $\mathbb{E}^x[\int_S^T |f(t,\omega) - \phi_n(t,\omega)|^2 dt] \to 0$ as $n \to \infty$. We define the *Itô integral* of f by

$$\int_S^T f(t,\omega)dB_t = \lim_{n\to\infty} \int_S^T \phi_n(t,\omega)dB_t. \tag{2.4.10}$$

Note that the definition of $\int_S^T f(t,\omega)dB_s$ is independent of the choice of random step functions ϕ_n.

Example 2.130. We consider a one-dimensional standard Brownian motion. We compute $\int_0^t B_s dB_s$ as follows. Take a division $0 = s_0 < \cdots < s_n = t$ of the interval, with $s_j = jt/n$. Write $\phi_n(t) = \sum_{j=0}^{n-1} B_{s_j} 1_{[s_j,s_{j+1})}$. The integrals $\int_0^t \phi_n(s,\omega)dB_s$ approximate $\int_0^t B_s dB_s$ in L^2-sense, and we see by Example 2.125 that

$$\int_0^t B_s dB_s = \lim_{n\to\infty} \int_0^t \phi_n(t)dB_s = \lim_{n\to\infty} \sum_{j=0}^{n-1} B_{s_j}(B_{s_{j+1}} - B_{s_j}) = \frac{1}{2}B_t^2 - \frac{t}{2}.$$

Next we present some basic properties of the Itô integral.

Theorem 2.131 (Itô isometry). *Let $f \in M^2(S,T)$. The expectation of any Itô integral is*

$$\mathbb{E}^x\Big[\int_S^T f(s,\omega)dB_s\Big] = 0,$$

and its covariance is given by

$$\mathbb{E}^x\Big[\Big|\int_S^T f(s,\omega)dB_s\Big|^2\Big] = \mathbb{E}^x\Big[\int_S^T |f(s,\omega)|^2 ds\Big].$$

The latter equality is called Itô isometry for this class of functions.

Proof. This follows from the definition of $\int_S^T f(t,\omega)dB_t$ and (2.4.3). □

Theorem 2.132 (Martingale property). *Let $f \in M^2(0,t)$ for all $t \geq 0$. The random process $(I_t)_{t\geq 0}$, $I_t = \int_0^t f(s,\omega)dB_s$, is an $(\mathscr{F}_t^{\text{BM}})$-martingale, i. e.,* (1) *I_t is $\mathscr{F}_t^{\text{BM}}$-measurable,* (2) *$\mathbb{E}^x[|I_t|] < \infty$,* (3) *$\mathbb{E}^x[I_t|\mathscr{F}_s^{\text{BM}}] = I_s$ for $s \leq t$.*

Proof. Statements (1) and (2) are immediate; we show (3). Let $I_t(n) = \int_0^t \phi_n(s,\omega)dB_s$, where $\phi_n = \sum_{j=0}^{n-1} f_j^{(n)}(\omega)1_{[t_{j+1}^{(n)}-t_j^{(n)})}(t)$; $I_t(n)$ is an approximating sequence of I_t. We have

$$\mathbb{E}^x[I_t(n)|\mathscr{F}_s^{\mathrm{BM}}] = \int_0^s \phi_n(u,\omega)dB_u + \mathbb{E}^x\Big[\int_s^t \phi_n(r,\omega)dB_r\big|\mathscr{F}_s^{\mathrm{BM}}\Big].$$

The claim follows by showing that the second term on the right-hand side above vanishes. We have

$$\mathbb{E}^x\Big[\int_s^t \phi_n(r,\omega)dB_r\big|\mathscr{F}_s^{\mathrm{BM}}\Big] = \sum_{s\le t_j^{(n)}\le t_{j+1}^{(n)}\le t} \mathbb{E}^x\Big[f_j^{(n)}(B_{t_{j+1}^{(n)}} - B_{t_j^{(n)}})\Big|\mathscr{F}_s^{\mathrm{BM}}\Big],$$

where $f_j^{(n)}(B_{t_{j+1}^{(n)}} - B_{t_j^{(n)}})$ is independent of $\mathscr{F}_s^{\mathrm{BM}}$ and thus $\mathbb{E}[f_j^{(n)}(B_{t_{j+1}^{(n)}} - B_{t_j^{(n)}})|\mathscr{F}_s^{\mathrm{BM}}] = \mathbb{E}[f_j^{(n)}(B_{t_{j+1}^{(n)}} - B_{t_j^{(n)}})] = 0$. Therefore,

$$\mathbb{E}^x\Big[\int_s^t \phi_n(r,\omega)dB_r\big|\mathscr{F}_s^{\mathrm{BM}}\Big] = 0.$$ □

Theorem 2.133 (Continuous version). *Let $f \in M^2(0,T)$. The random process $(I_t)_{0\le t\le T}$, $I_t = \int_0^t f(s,\omega)dB_s$, has a continuous version $(\tilde{I}_t)_{0\le t\le T}$, i. e., $\tilde{I}_t$ is almost surely continuous in t and $\mathcal{W}^x(\tilde{I}_t = I_t) = 1$ for all $0 \le t \le T$.*

Proof. Without restricting generality, we set $x = 0$. Let $(\phi_n(t))_{n\in\mathbb{N}} \subset M^2_{\mathrm{step}}$ be such that $\mathbb{E}[\int_0^t |f(s,\omega) - \phi_n(s,\omega)|^2 ds] \to 0$ as $n \to \infty$. Write $I_t(n) = \int_0^t \phi_n(s,\omega)dB_s$. Note that $I_t(n)$ is continuous in t for all n and is an $\mathscr{F}_t^{\mathrm{BM}}$-martingale. This implies that $I_t(n) - I_t(m)$ is also an $\mathscr{F}_t^{\mathrm{BM}}$-martingale, thus by the martingale inequality (see Theorem 2.72) we have

$$\begin{aligned}\mathcal{W}\Big(\sup_{0\le t\le T} |I_t(n) - I_t(m)| > \varepsilon\Big) &\le \frac{1}{\varepsilon^2}\mathbb{E}[|I_T(n) - I_T(m)|^2] \\ &= \frac{1}{\varepsilon^2}\mathbb{E}\Big[\int_0^T |\phi_n(s) - \phi_m(s)|^2 ds\Big] \to 0\end{aligned}$$

as $m,n \to \infty$. Hence we can choose a subsequence $n_k \to \infty$ such that $P(A_k) \le 2^{-k}$, where

$$A_k = \Big\{\omega \in \mathscr{X} \mid \sup_{0\le t\le T} |I_t(n_{k+1},\omega) - I_t(n_k,\omega)| > 2^{-k}\Big\}.$$

Since $\sum_{k=1}^\infty \mathcal{W}(A_k) < \infty$, we get $\mathcal{W}(\bigcap_{N=1}^\infty \bigcup_{k=N}^\infty A_k) = 0$ by the Borel–Cantelli lemma. Thus for almost every $\omega \in \mathscr{X}$ it follows that

$$\sup_{0\le t\le T} |I_t(n_{k+1},\omega) - I_t(n_k,\omega)| \le 2^{-k}$$

for all $k > N(\omega)$, with some $N(\omega)$. Therefore $(I_t(n_k, \omega))_{k\in\mathbb{N}}$ is a Cauchy sequence and so it converges uniformly in $t \in [0, T]$ as $k \to \infty$ for almost every $\omega \in \mathscr{X}$, which we denote by $\tilde{I}_t(\omega)$. Note that $\tilde{I}_t(\omega)$ is continuous in t for almost every $\omega \in \mathscr{X}$ since the convergence is uniform in t. Since $I_t(n_k) \to I_t$ in $L^2(\mathscr{X})$, we conclude that $\tilde{I}_t = I_t$ almost surely. □

From now on we will consider the continuous version of the Itô integral of a random process without explicitly stating it.

Next we consider a d-dimensional Brownian motion $(B_t)_{t\geq 0}$ and $f = (f_1, \ldots, f_d)$ such that $f_\mu \in M^2(0, t)$, $\mu = 1, \ldots, d$. We define the Itô integral

$$\int_0^t f(s,\omega)\cdot dB_s = \sum_{\mu=1}^d \int_0^t f_\mu(s,\omega)dB_s^\mu. \tag{2.4.11}$$

It is direct to see that the basic properties $\mathbb{E}^x\left[\int_0^t f(s,\omega)\cdot dB_s\right] = 0$ and

$$\mathbb{E}^x\Big[\int_0^t f_\mu(s,\omega)dB_s^\mu \int_0^t f_\nu(s,\omega)dB_s^\nu\Big] = \delta_{\mu\nu}\mathbb{E}^x\Big[\int_0^t f_\mu(s,\omega)f_\nu(s,\omega)ds\Big]$$

similarly hold.

Let $C_b(\mathbb{R}\times\mathbb{R}^d)$ denote the set of bounded continuous functions on $\mathbb{R}\times\mathbb{R}^d$ and $C_b^n(\mathbb{R}\times\mathbb{R}^d)$ the set of bounded and n times continuously differentiable functions on $\mathbb{R}\times\mathbb{R}^d$. Consider the following specific cases.

Corollary 2.134. *Let $(B_t)_{t\geq 0}$ be a d-dimensional Brownian motion. Write $t_j = tj/2^m$ and $\Delta B_j^\mu = B_{t_j}^\mu - B_{t_{j-1}}^\mu$, $1 \leq j \leq 2^m$.*

(1) *Let $f \in C_b(\mathbb{R}^d)$. Then*

$$\lim_{n\to\infty}\sum_{j=1}^{2^n} f(B_{t_{j-1}})\Delta B_j^\mu = \int_0^t f(B_s)dB_s^\mu. \tag{2.4.12}$$

(2) *Let $f \in C_b^2(\mathbb{R}^d)$. Then*

$$\lim_{n\to\infty}\sum_{j=1}^{2^n}\frac{1}{2}\left(f(B_{t_{j-1}})+f(B_{t_j})\right)\Delta B_j^\mu = \int_0^t f(B_s)dB_s^\mu + \frac{1}{2}\int_0^t \partial_\mu f(B_s)ds. \tag{2.4.13}$$

The limits in both expressions above are understood in $L^2(\mathscr{X}, d\mathcal{W}^x)$ sense.

Proof. Statement (1) is directly obtained from the definition of Itô integral. For (2), we write

$$f(B_{t_j}) - f(B_{t_{j-1}}) = \sum_{\nu=1}^d \partial_\nu f(B_{t_{j-1}})\Delta B_j^\nu + \frac{1}{2}\sum_{\rho,\nu=1}^d \partial_\rho\partial_\nu f(\xi)\Delta B_j^\rho \Delta B_j^\nu$$

with some $\xi = \xi(\omega)$ by the Taylor expansion. We see that

$$\sum_{j=1}^{2^n} \Big| \big(f(B_{t_j}) - f(B_{t_{j-1}})\big) \Delta B_j^\mu - \sum_{\nu=1}^{d} \partial_\nu f(B_{t_{j-1}}) \Delta B_j^\nu \Delta B_j^\mu \Big| \le C \sum_{j=1}^{2^n} |B_{t_j} - B_{t_{j-1}}|^3 \stackrel{n\to\infty}{\longrightarrow} 0$$

in $L^2(\mathscr{X}, d\mathcal{W}^x)$, where C is a constant independent of ω. Let μ be fixed. By writing $f(B_{t_{j-1}}) + f(B_{t_j}) = 2f(B_{t_{j-1}}) + (f(B_{t_j}) - f(B_{t_{j-1}}))$ and making a Taylor expansion we have

$$\begin{aligned} \text{l. h. s. (2.4.13)} &= \lim_{n\to\infty} \sum_{j=1}^{2^n} f(B_{t_{j-1}}) \Delta B_j^\mu + \frac{1}{2} \lim_{n\to\infty} \sum_{j=1}^{2^n} \sum_{\nu=1}^{d} \partial_\nu f(B_{t_{j-1}}) \Delta B_j^\nu \Delta B_j^\mu \\ &= \int_0^t f(B_s) dB_s^\mu + \frac{1}{2} \int_0^t \partial_\mu f(B_s) ds, \end{aligned}$$

yielding (2.4.13). □

Definition 2.135 (Stratonovich integral). The right-hand side of (2.4.13) defines the *Stratonovich integral*, i. e., for $f = (f_1, \dots, f_d) \in (C_b^2(\mathbb{R}^d))^d$,

$$\int_0^t f(B_s) \circ dB_s = \int_0^t f(B_s) \cdot dB_s + \frac{1}{2} \int_0^t \nabla \cdot f(B_s) ds. \tag{2.4.14}$$

2.4.3 Extension of stochastic integrals

Finally, we define a more general stochastic integral for a wider class of integrands by weakening the conditions of $M^2(0,T)$. Condition $\mathbb{E}[\int_0^T |f(s,\omega)|^2 ds] < \infty$ is too strong to derive the so-called Feynman–Kac–Itô formula in Chapter 4. We then want to define the stochastic integral $\int_0^T f(s,\omega) dB_s$ for f such that it does not necessarily satisfy $\mathbb{E}[\int_0^T |f(s,\omega)|^2 ds] < \infty$.

Lemma 2.136. *Let $f, g \in M^2(0,t)$, and let τ and ρ be stopping times with respect to $(\mathscr{F}_t^{\mathrm{BM}})_{t\ge0}$ such that $\tau \le \rho$ a. s. Then we have the following:*

(1) $\mathbb{E}[\int_0^{t\wedge\rho} f dB_s | \mathscr{F}_\tau^{\mathrm{BM}}] = \int_0^{t\wedge\tau} f dB_s$;

(2) $X_t = (\int_0^t f dB_s)^2 - \int_0^t |f|^2 ds$, $t \ge 0$, *is an* $(\mathscr{F}_t^{\mathrm{BM}})_{t\ge0}$*-martingale;*

(3) *It follows that*

$$\mathbb{E}\Big[\Big(\int_0^{t\wedge\rho} f dB_s - \int_0^{t\wedge\tau} f dB_s\Big)\Big(\int_0^{t\wedge\rho} g dB_s - \int_0^{t\wedge\tau} g dB_s\Big)\Big| \mathscr{F}_\tau^{\mathrm{BM}}\Big] = \mathbb{E}\Big[\int_{t\wedge\tau}^{t\wedge\rho} fg ds \Big| \mathscr{F}_\tau^{\mathrm{BM}}\Big];$$

in particular, we have

$$\mathbb{E}\left[\left(\int_0^{t\wedge\rho} f dB_s - \int_0^{t\wedge\tau} f dB_s\right)^2 \middle| \mathscr{F}_\tau^{\mathrm{BM}}\right] = \mathbb{E}\left[\int_{t\wedge\tau}^{t\wedge\rho} |f|^2 ds \middle| \mathscr{F}_\tau^{\mathrm{BM}}\right];$$

(4) $\int_0^{t\wedge\tau} f dB_s = \int_0^t f 1_{\{s\le\tau\}} dB_s$.

Proof. Since $\left(\int_0^t f dB_s\right)_{t\ge 0}$ is a continuous martingale, statement (1) follows from Corollary 2.71. For (2), we see that

$$X_t = \left(\int_0^s f dB_r\right)^2 + 2\left(\int_0^s f dB_r\right)\left(\int_s^t f dB_r\right) + \left(\int_s^t f dB_r\right)^2 - \int_0^t |f|^2 dr.$$

It is straightforward to show $\mathbb{E}[X_t|\mathscr{F}_s^{\mathrm{BM}}] = X_s$. We show (3). Write $I_t(f) = \int_0^t f dB_s$. We have

$$\mathbb{E}[(I_{t\wedge\rho}(f) - I_{t\wedge\tau}(f))^2|\mathscr{F}_\tau^{\mathrm{BM}}] = I_{t\wedge\tau}(f)^2 - 2I_{t\wedge\tau}(f)\mathbb{E}[I_{t\wedge\rho}(f)|\mathscr{F}_\tau^{\mathrm{BM}}] + \mathbb{E}[I_{t\wedge\rho}(f)^2|\mathscr{F}_\tau^{\mathrm{BM}}]. \tag{2.4.15}$$

Since $(I_t)_{t\ge 0}$ and $(X_t)_{t\ge 0}$ are continuous martingales by (2), we have

$$\mathbb{E}[X_{t\wedge\rho}|\mathscr{F}_\tau^{\mathrm{BM}}] = X_{t\wedge\tau}, \quad \mathbb{E}[I_{t\wedge\rho}|\mathscr{F}_\tau^{\mathrm{BM}}] = I_{t\wedge\tau}$$

by Corollary 2.71. Inserting this into the right-hand side of (2.4.15), we have

$$\mathbb{E}[(I_{t\wedge\rho}(f) - I_{t\wedge\tau}(f))^2|\mathscr{F}_\tau^{\mathrm{BM}}] = \mathbb{E}\left[\int_0^{t\wedge\rho} |f|^2 ds \middle| \mathscr{F}_\tau^{\mathrm{BM}}\right] - \int_0^{t\wedge\tau} |f|^2 ds = \mathbb{E}\left[\int_{t\wedge\tau}^{t\wedge\rho} |f|^2 ds \middle| \mathscr{F}_\tau^{\mathrm{BM}}\right].$$

Replacing f in this equality first by $f+g$ and then by $f-g$, and subtracting the resulting equation, we obtain (3). Finally, we prove (4). Let $\tilde{f}(s,\omega) = f(s,\omega)1_{\{s\le\tau(\omega)\}}$. We have

$$I_{t\wedge\tau}(f) - I_t(\tilde{f}) = I_{t\wedge\tau}(f-\tilde{f}) - (I_t(\tilde{f}) - I_{t\wedge\tau}(\tilde{f})).$$

By (3), we see that

$$\mathbb{E}\left[(I_{t\wedge\tau}(f-\tilde{f}) - I_{s\wedge\tau}(f-\tilde{f}))^2|\mathscr{F}_{s\wedge\tau}^{\mathrm{BM}}\right] = \mathbb{E}\left[\int_{s\wedge\tau}^{t\wedge\tau} |f-\tilde{f}|^2 ds \middle| \mathscr{F}_{s\wedge\tau}^{\mathrm{BM}}\right] = 0 \tag{2.4.16}$$

for arbitrary $s \ge 0$, and then $I_{t\wedge\tau}(f-\tilde{f}) = 0$ follows from taking the expectation on both sides of (2.4.16). Moreover,

$$\mathbb{E}[(I_t(\tilde{f}) - I_{t\wedge\tau}(\tilde{f}))^2|\mathscr{F}_{t\wedge\tau}^{\mathrm{BM}}] = \mathbb{E}\left[\int_{t\wedge\tau}^{t} |\tilde{f}|^2 ds \middle| \mathscr{F}_{t\wedge\tau}^{\mathrm{BM}}\right] = 0$$

is derived. Then $\mathbb{E}[(I_{t\wedge\tau}(f) - I_t(\tilde{f}))^2] = 0$ follows, which gives the proof of (4). □

Definition 2.137. Let $\mathbb{M}^2(S,T)$ be the class of $f : \mathbb{R}^+ \times \mathscr{X} \to \mathbb{R}$ such that
(1) $f \in \mathcal{B}(\mathbb{R}^+) \times \mathcal{B}(\mathscr{X})/\mathcal{B}(\mathbb{R})$;
(2) $(f(t,\cdot))_{t\geq 0}$ is $(\mathscr{F}_t^{\mathrm{BM}})_{t\geq 0}$-adapted;
(3) $\mathcal{W}(\int_S^T |f(s,\omega)|^2 ds < \infty) = 1$.

Let $f \in \mathbb{M}^2(0,T)$ and define a stopping time with respect to $(\mathscr{F}_t^{\mathrm{BM}})_{t\geq 0}$ by

$$\tau_n = \begin{cases} \inf\left\{t \geq 0 \,\middle|\, \int_0^t f(s,\omega)^2 ds \geq n\right\} & \text{if } \left\{t \geq 0 \,\middle|\, \int_0^t f(s,\omega)^2 ds \geq n\right\} \neq \emptyset, \\ \infty & \text{if } \left\{t \geq 0 \,\middle|\, \int_0^t f(s,\omega)^2 ds \geq n\right\} = \emptyset. \end{cases} \tag{2.4.17}$$

Note that $(\int_0^t f(s,\cdot)^2 ds)_{t\geq 0}$ is a continuous adapted process with respect to the filtration $(\mathscr{F}_t^{\mathrm{BM}})_{t\geq 0}$. The first hitting time τ_n of $(\int_0^t f(s,\cdot)^2 ds)_{t\geq 0}$ to the closed set $[n,\infty)$ is a stopping time by Proposition 2.60. It follows that $\tau_n \uparrow \infty$ as $n \uparrow \infty$, since we assume $\mathcal{W}(\int_0^T |f(s,\omega)|^2 ds < \infty) = 1$. Let $f \in \mathbb{M}^2(0,t)$. Although $f \in M^2(0,t)$ is not necessarily satisfied, we see that $\tilde{f}(s,\omega) = f(s,\omega)1_{\{s\leq\tau_n(\omega)\}} \in M^2(0,t)$ for each n. We can define the stochastic integral $\int_0^t \tilde{f}(s,\omega)dB_s$. Define

$$I_t^n(f) = \int_0^{t\wedge\tau_n} \tilde{f}(s,\omega)dB_s.$$

More precisely, $I_t^n(f)$ is defined by $\int_0^y \tilde{f}(s,\omega)dB_s$ evaluated as $y = t\wedge\tau_n$.

Lemma 2.138. *Let $f \in \mathbb{M}^2(0,T)$, and the stopping time τ_n be defined by* (2.4.17). *Then for $m < n$,*

$$I_t^n(f) = I_t^m(f), \quad 0 \leq t \leq \tau_m.$$

Proof. By (3) of Lemma 2.136, we see that

$$\mathbb{E}[(I_t^n(f) - I_t^m(f))^2 | \mathscr{F}_{\tau_m}^{\mathrm{BM}}] = \mathbb{E}\left[\int_{t\wedge\tau_m}^{t\wedge\tau_n} |f(s,\omega)|^2 ds \,\middle|\, \mathscr{F}_{\tau_m}^{\mathrm{BM}}\right].$$

Thus

$$\mathbb{E}[(I_t^n(f) - I_t^m(f))^2 1_A] = \mathbb{E}\left[\int_{t\wedge\tau_m}^{t\wedge\tau_n} |f(s,\omega)|^2 ds 1_A\right] \tag{2.4.18}$$

for any $A \in \mathscr{F}_{\tau_m}^{\mathrm{BM}}$. Take $A = \{t \leq \tau_m\} \in \mathscr{F}_m^{\mathrm{BM}}$. From (2.4.18) we have

$$\mathbb{E}[(I_t^n(f) - I_t^m(f))^2 1_A] = \mathbb{E}\left[\int_t^t |f(s,\omega)|^2 ds 1_A\right] = 0,$$

which implies that $I_t^n(f) = I_t^m(f)$ if $t \le \tau_m$. □

Definition 2.139 (Extension of Itô/stochastic integral). Let $f \in \mathbb{M}^2(0,t)$ and define the stopping time τ_n by (2.4.17). Then the stochastic integral $\int_0^t f(s,\omega)dB_s$ is defined by

$$\int_0^t f(s,\omega)dB_s = \int_0^{t\wedge\tau_n} f(s,\omega)1_{\{s\le\tau_n\}}dB_s, \quad 0 \le t \le \tau_n. \tag{2.4.19}$$

Remark 2.140.

(1) The extension of the stochastic integral (2.4.19) is well defined by Lemma 2.138. Let $f \in \mathbb{M}^2(0,t)$. Then by (4) of Lemma 2.136, we have

$$\int_0^{t\wedge\tau_n} f(s,\omega)1_{\{s\le\tau_n\}}dB_s = \int_0^t f(s,\omega)1_{\{s\le\tau_n\}}dB_s. \tag{2.4.20}$$

(2) We can define the stochastic integral not only for Brownian motion but also for a continuous *local martingale* $(X_t)_{t\ge0}$. A random process $(X_t)_{t\ge0}$ on a filtered probability space $(\Omega, \mathscr{F}, (\mathscr{F}_t)_{t\ge0}, P)$ is said to be a local martingale whenever there exists a nondecreasing sequence $(\tau_n)_{n\in\mathbb{N}}$ of stopping times such that $(X_{t\wedge\tau_n})_{t\ge0}$ is a martingale and $P(\lim_{n\to\infty}\tau_n = \infty) = 1$. The resulting integral is denoted by $\int_0^T f(s,\omega)dX_s$.

(3) Let $f \in \mathbb{M}^2(0,t)$ for any $t > 0$. Note that the integral in Definition 2.139 is not an $(\mathscr{F}_t^{\mathrm{BM}})_{t\ge0}$-martingale and the Itô isometry is also not automatically satisfied. By (1) of Lemma 2.136, it follows that

$$\mathbb{E}\left[\int_0^{t\wedge\tau_n} f1_{\{r\le\tau_n\}}dB_r \,\middle|\, \mathscr{F}_s^{\mathrm{BM}}\right] = \int_0^{s\wedge\tau_n} f1_{\{r\le\tau_n\}}dB_r$$

for all n with $0 \le t \le \tau_n$ and $0 \le s \le t$. Then $\left(\int_0^t f dB_r\right)_{t\ge0}$ is a local martingale.

Example 2.141. Let $f \in L^2_{\mathrm{loc}}(\mathbb{R})$. It is not necessarily true that $f \in M^2(0,t)$; however, it is true that $f \in \mathbb{M}^2(0,t)$. Therefore $\int_0^t f(B_s)dB_s$ is well defined in the sense of (2.4.19).

Using random processes given by Itô integrals, we have the following important class.

Definition 2.142 (Itô process). A random process $(X_t)_{t\ge0}$ on the probability space $(\mathscr{X}, \mathcal{B}(\mathscr{X}), \mathcal{W})$ is called an *Itô process* if it has the form

$$X_t = X_0 + \int_0^t b_s ds + \int_0^t \sigma_s \cdot dB_s, \quad t \geq 0, \tag{2.4.21}$$

where

(1) $(b_t)_{t\geq 0}$ is an $(\mathscr{F}_t^{\mathrm{BM}})_{t\geq 0}$-adapted process with $\int_0^t |b_s| ds < \infty$;

(2) $\sigma_s = (\sigma_s^1, \ldots, \sigma_s^d)$ and $\sigma_\cdot^i \in \mathbb{M}^2(0,t)$, $i = 1, \ldots, d$.

The stochastic integral $\int_0^t \sigma_s \cdot dB_s$ is understood in the sense of Definition 2.139, i. e.,

$$\int_0^t \sigma_s \cdot dB_s = \int_0^{t\wedge\tau_n} 1_{\{s\leq\tau_n\}} \sigma_s \cdot dB_s, \quad 0 \leq t \leq \tau_n.$$

In concise differential notation, the above integral equality is written as

$$dX_t = b_t dt + \sigma_t \cdot dB_t. \tag{2.4.22}$$

The process $(b_t)_{t\geq 0}$ is called the *drift term* and $(\sigma_t)_{t\geq 0}$ the *diffusion term*. It should be stressed that the *stochastic differentials* dX_t and dB_t have no meaning other than the shorthand notation explained above.

2.4.4 Itô formula

Another landmark result of stochastic analysis is the Itô formula which is the counterpart of the fundamental change of variable rule in classical analysis. In particular, this provides a tool for computing Itô integrals without directly having to use the definition. Recall that $C^{1,2}(\mathbb{R}^+ \times \mathbb{R})$ denotes the space of functions on $\mathbb{R}^+ \times \mathbb{R}$, which are C^1 with respect to the first variable and C^2 with respect to the second.

Theorem 2.143 (Itô formula for Brownian motion). *Let $h \in C^{1,2}(\mathbb{R}^+ \times \mathbb{R})$ and define $X_t = h(t, B_t)$. Then $(X_t)_{t\geq 0}$ is an Itô process and we have*

$$X_t - X_0 = \int_0^t \left(\dot{h}(s, B_s) + \frac{1}{2}\partial_x^2 h(s, B_s) \right) ds + \int_0^t \partial_x h(s, B_s) dB_s, \tag{2.4.23}$$

with $\dot{h} = \partial_s h(s,x)$ or, in equivalent differential notation,

$$dX_t = \left(\dot{h}(t, B_t) + \frac{1}{2}\partial_x^2 h(t, B_t) \right) dt + \partial_x h(t, B_t) dB_t. \tag{2.4.24}$$

The following more general formula is readily obtained.

Theorem 2.144 (Itô formula for Itô process). *Let $(B_t)_{t\geq 0}$ be a d-dimensional Brownian motion and $(X_t)_{t\geq 0} = (X_t^i)_{t\geq 0, 1\leq i\leq D}$ an $\mathbb{R}^D$-valued Itô process on a filtered probability space $(\mathscr{X}, \mathcal{B}(\mathscr{X}), (\mathscr{F}_t^{\mathrm{BM}})_{t\geq 0}, \mathcal{W})$ such that*

$$dX_t^i = b_t^i dt + \sigma_t^i \cdot dB_t, \quad i = 1, \ldots, D, \tag{2.4.25}$$

where $(b_t^i)_{t\geq 0}$ is $(\mathscr{F}_t^{\mathrm{BM}})_{t\geq 0}$-adapted and

$$\begin{aligned} &b_t^i(\omega) \in L^1_{\mathrm{loc}}(\mathbb{R}^+, dt), \quad \forall \omega \in \mathscr{X}, \\ &\sigma_t^i = (\sigma_t^{i\mu})_{1\leq \mu \leq d} \subset \bigcup_{T\geq 0} \mathbb{M}^2(0, T). \end{aligned}$$

Let $h = (h^1, \ldots, h^m) \in C^{1,2}(\mathbb{R}^+ \times \mathbb{R}^D; \mathbb{R}^m)$. Then $Y_t^k = h^k(t, X_t)$ is also an Itô process with

$$Y_t^k - Y_0^k = \int_0^t \left(\dot{h}^k + \sum_{i=1}^D h_i^k b_s^i + \frac{1}{2} \sum_{i,j=1}^D h_{ij}^k (\sigma_s \sigma_s^T)_{ij} \right) ds + \int_0^t \sum_{i=1}^D h_i^k \sigma_s^i \cdot dB_s, \quad 1 \leq k \leq m. \tag{2.4.26}$$

Here $\dot{h}^k = \partial_t h^k$, $h_i^k = \partial_{x_i} h^k$ and $h_{ij}^k = \partial_{x_i}\partial_{x_j} h^k$, evaluated at (s, X_s). In differential notation,

$$dY_t^k = \left(\dot{h}^k + \sum_{i=1}^D h_i^k b_t^i + \frac{1}{2} \sum_{i,j=1}^D h_{ij}^k (\sigma_t \sigma_t^T)_{ij} \right) dt + \sum_{i=1}^D h_i^k \sigma_t^i \cdot dB_t. \tag{2.4.27}$$

Proof. Step 1. Suppose that $h \in C^\infty(\mathbb{R}^+ \times \mathbb{R}^D; \mathbb{R}^m)$ in Steps 1–6 below. We define a $(D+1)$-dimensional stochastic process by

$$\hat{X}_t^i = \begin{cases} X_t^i, & 1 \leq i \leq D, \\ t, & i = D+1, \end{cases}$$

which satisfies

$$d\hat{X}_t^i = \hat{b}_t^i dt + \hat{\sigma}_t^i \cdot dB_t, \quad i = 1, \ldots, D+1, \tag{2.4.28}$$

where

$$\hat{b}_t^i = \begin{cases} b_t^i, & 1 \leq i \leq D, \\ 1, & i = D+1, \end{cases} \qquad \hat{\sigma}_t^{i\mu} = \begin{cases} \sigma_t^{i\mu}, & 1 \leq i \leq D, \\ 0, & i = D+1. \end{cases}$$

Let $\hat{h} = (\hat{h}^1, \ldots, \hat{h}^m) \in C^\infty(\mathbb{R}^{D+1}; \mathbb{R}^m)$ and define $\hat{Y}_t = \hat{h}(\hat{X}_t)$. Equation (2.4.27) is then reduced to

$$d\hat{Y}_t^k = \left(\sum_{i=1}^{D+1} \hat{h}_i^k \hat{b}_t^i + \frac{1}{2} \sum_{i,j=1}^{D+1} \hat{h}_{ij}^k (\hat{\sigma}_t \hat{\sigma}_t^T)_{ij} \right) dt + \sum_{i=1}^{D+1} \hat{h}_i^k \hat{\sigma}_t^i \cdot dB_t. \tag{2.4.29}$$

We show (2.4.29). For this, reset $D+1$ to D and for simplicity write h, b_t^i, and $\sigma_t^{i\mu}$ for $\hat{h}$, $\hat{b}_t^i$, and $\hat{\sigma}_t^{i\mu}$, respectively. Also, we assume $h \in C^\infty(\mathbb{R}^D;\mathbb{R}^m)$.

Step 2. Let $M_t^i = \int_0^t \sigma_s^i \cdot dB_s$, $A_t^i = \int_0^t b_s^i ds$, $M_t = (M_t^1,\dots,M_t^D)$, and $A_t = (A_t^1,\dots,A_t^D)$. For every $t \geq 0$ and n, we define the set $I_n(t)$ by

$$I_n(t) = \Big\{|X_0(\omega)| > n\} \cup \{|M_t(\omega)| > n\} \cup \{|A_t(\omega)| > n\Big\}.$$

Define the hitting time

$$\tau_n = \begin{cases} \inf\{t > 0 \mid I_n(t) \neq \emptyset\} & \text{if } \{t > 0 \mid I_n(t) \neq \emptyset\} \neq \emptyset, \\ \infty & \text{if } \{t > 0 \mid I_n(t) \neq \emptyset\} = \emptyset. \end{cases} \tag{2.4.30}$$

Then τ_n is a stopping time with respect to $(\mathscr{F}_t^{\mathrm{BM}})_{t\geq 0}$. We define the stopped process by $X_t^{(n)} = X_{t\wedge\tau_n}$. We see that $X_t^{(n)}(\omega) \to X_t(\omega)$ as $n \to \infty$. For a continuous function f, we define the stopping time $\rho_m = \inf\{t \geq 0 \mid \int_0^t |f(X_s)|^2 ds \geq m\}$. Since we have

$$\mathbb{E}\left[\left|\int_0^{\tau_n\wedge\rho_m} f(X_s^{(n)})dB_s - \int_0^{\tau_n\wedge\rho_m} f(X_s)dB_s\right|^2\right] = \mathbb{E}\left[\int_0^{\tau_n\wedge\rho_m} |f(X_s^{(n)}) - f(X_s)|^2 ds\right] = 0,$$

we see that $\int_0^t f(X_s^{(n)})dB_s = \int_0^t f(X_s)dB_s$ for $0 \leq t \leq \tau_n\wedge\rho_m$. We also see that $\int_0^t f(X_s^{(n)})ds \to \int_0^t f(X_s)ds$ as $n \to \infty$ almost surely. Hence it suffices to show (2.4.29) with X_t replaced by $X_t^{(n)}$. Furthermore, since $h^k(X_t^{(n)})$, $h_i^k(X_t^{(n)})$, and $h_{ij}^k(X_t^{(n)})$ are bounded, we can assume that $h^k \in C_b^\infty(\mathbb{R}^D)$, i. e., h^k, h_i^k, and h_{ij}^k are continuous and bounded for all i,j in what follows.

Step 3. We reset $X_t^{(n)}$ to X_t. It is sufficient to prove that

$$dY_t^k = \left(\sum_{i=1}^D h_i^k b_t^i + \frac{1}{2}\sum_{i,j=1}^D h_{ij}^k(\sigma_t\sigma_t^T)_{ij}\right)dt + \sum_{i=1}^D h_i^k \sigma_t^i \cdot dB_t \tag{2.4.31}$$

for $h^k \in C_b^\infty(\mathbb{R}^D)$. Let $0 = s_0 < s_1 < \cdots < s_J = t$. First, consider the case when $\sigma_s^{i\mu}$ is a step function of the form

$$\sigma_s^{i\mu}(\omega) = \sum_{j=1}^J \sigma_j^{i\mu}(\omega)1_{[s_{j-1},s_j)}(s),$$

where $\sigma_j^{i\mu}$ is measurable with respect to $\mathscr{F}_{s_j}^{\mathrm{BM}}$. Notice that

$$h^k(X_t) - h^k(X_0) = \sum_{m=1}^J \left(h^k(X_{s_m\wedge t}) - h^k(X_{s_{m-1}\wedge t})\right).$$

In order to show (2.4.31), it is sufficient to prove

$$h^k(X_u) - h^k(X_s) = \int_s^u h_i^k(X_r)b_r^i dr + \frac{1}{2}\int_s^u h_{ij}^k(X_r)\sigma_r^{i\mu}\sigma_r^{j\mu} dr + \int_s^u h_i^k(X_r)\sigma_r^{i\mu} dB_r^\mu. \tag{2.4.32}$$

Here $s_{p-1} \le s < u < s_p$ with some p, and the summation over repeated indices is understood. In what follows, we fix k and write h for h^k for notational simplicity. Since $s, u \in [s_{p-1}, s_p)$, X_u^i is of the form

$$X_u^i = X_s^i + \sigma_p^{i\mu}(B_u^\mu - B_s^\mu) + \int_s^u b_r^i dr, \quad i = 1, \dots, D. \tag{2.4.33}$$

Hence $u \mapsto X_u$ is continuous.

Step 4. We divide the interval $[s, u]$ into n intervals and set $t_l = s + (u - s)l/n$, $0 \le l \le n$. Let

$$\Delta X_{l-1}^i = X_{t_l}^i - X_{t_{l-1}}^i, \quad l = 1, \dots, n.$$

By Taylor expansion, we have

$$h(X_u) - h(X_s) = \sum_{l=1}^n h_i(X_{t_{l-1}})\Delta X_{l-1}^i + \frac{1}{2}\sum_{l=1}^n h_{ij}(X_{t_{l-1}})\Delta X_{l-1}^i \Delta X_{l-1}^j + R, \tag{2.4.34}$$

where

$$R = \sum_{l=1}^n h_{ijk}(\xi_{l-1})\Delta X_{l-1}^i \Delta X_{l-1}^j \Delta X_{l-1}^k$$

with some $\xi_{l-1} = \theta_l X_{t_{l-1}} + (1 - \theta_l)X_{t_l}$ and $\theta_l = \theta_l(\omega) \in (0, 1)$. Let

$$\Delta_{l-1}^i B = \sigma_p^{i\mu}(B_{t_l}^\mu - B_{t_{l-1}}^\mu), \quad \Delta_{l-1}^i b = \int_{t_{l-1}}^{t_l} b_r^i dr, \quad 1 \le l \le n.$$

We have $\lim_{n\to\infty} \mathbb{E}[|\sum_{l=1}^n \Delta_{l-1}^i B \Delta_{l-1}^j B \Delta_{l-1}^k B|^2] = 0$. Taking a subsequence n', we also see that $\lim_{n'\to\infty} \mathbb{E}[|\sum_{l=1}^{n'} \Delta_{l-1}^i b \Delta_{l-1}^j b \Delta_{l-1}^k b|^2] = 0$. Hence $\lim_{n'\to\infty} \mathbb{E}[|R|^2] = 0$ follows. In what follows, we reset n' to n. Inserting (2.4.33) into (2.4.34), we can rewrite (2.4.34) as

$$h(X_t) - h(X_s) = \sum_{j=1}^5 I_n^{(j)} + R, \tag{2.4.35}$$

where

$$I_n^{(1)} = \sum_{l=1}^{n} h_i(X_{t_{l-1}})\Delta_{l-1}^i B, \quad I_n^{(2)} = \sum_{l=1}^{n} h_i(X_{t_{l-1}})\Delta_{l-1}^i b,$$
$$I_n^{(3)} = \frac{1}{2}\sum_{l=1}^{n} h_{ij}(X_{t_{l-1}})\Delta_{l-1}^i B\Delta_{l-1}^j B, \quad I_n^{(4)} = \sum_{l=1}^{n} h_{ij}(X_{t_{l-1}})\Delta_{l-1}^i B\Delta_{l-1}^j b,$$
$$I_n^{(5)} = \frac{1}{2}\sum_{l=1}^{n} h_{ij}(X_{t_{l-1}})\Delta_{l-1}^i b\Delta_{l-1}^j b.$$

Step 5. We prove the following:

$$I_n^{(1)} \overset{L^2}{\to} \sigma_p^{i\mu} \int_s^u h_i(X_r)dB_r^\mu, \tag{2.4.36}$$

$$I_n^{(2)} \overset{a.s.}{\to} \int_s^u h_i(X_r)b_r^i dr, \tag{2.4.37}$$

$$I_n^{(3)} \overset{L^2}{\to} \frac{1}{2}\sigma_p^{i\mu}\sigma_p^{j\mu} \int_s^u h_{ij}(X_r)dr, \tag{2.4.38}$$

$$I_n^{(4)} \overset{a.s.}{\to} 0, \tag{2.4.39}$$

$$I_n^{(5)} \overset{a.s.}{\to} 0. \tag{2.4.40}$$

We have

$$I_n^{(1)} = \sigma_p^{i\mu} \int_s^u \Phi_{i,n}(r)dB_r^\mu,$$

where $\Phi_{i,n}(r) = \sum_{l=1}^n h_i(X_{t_{l-1}})1_{[t_{l-1},t_l)}(r)$. Since h_i is bounded, using the dominated convergence theorem, we see that

$$\lim_{n\to\infty} \mathbb{E}\left[\int_s^u |\Phi_{i,n}(r) - h_i(X_r)|^2 dr\right] = 0.$$

Hence (2.4.36) follows; (2.4.37) is straightforward. Next we show (2.4.38). Fix μ and μ', and denote $\Psi(X_{t_{l-1}}) = \Psi^{\mu\mu'}(X_{t_{l-1}}) = h_{ij}(X_{t_{l-1}})\sigma_p^{i\mu}\sigma_p^{j\mu'}$. It is sufficient to show that

$$\sum_{l=1}^{n} \Psi(X_{t_{l-1}})(B_{t_l}^\mu - B_{t_{l-1}}^\mu)(B_{t_l}^{\mu'} - B_{t_{l-1}}^{\mu'}) \overset{L^2}{\to} \delta_{\mu\mu'} \int_s^u \Psi(X_r)dr.$$

We write $Z_l = Z_l^{\mu\mu'} = (B_{t_l}^\mu - B_{t_{l-1}}^\mu)(B_{t_l}^{\mu'} - B_{t_{l-1}}^{\mu'}) - (t_l - t_{l-1})\delta_{\mu\mu'}$. It follows that

$$\mathbb{E}\left[\left|\sum_{l=1}^{n}\Psi(X_{t_{l-1}})(B_{t_l}^{\mu}-B_{t_{l-1}}^{\mu})(B_{t_l}^{\mu'}-B_{t_{l-1}}^{\mu'})-\delta_{\mu\mu'}\sum_{l=1}^{n}\Psi(X_{t_{l-1}})(t_l-t_{l-1})\right|^2\right]$$
$$=\mathbb{E}\left[\left|\sum_{l=1}^{n}\Psi(X_{t_{l-1}})Z_l\right|^2\right]=J_n^1+J_n^2.$$

Here $J_n^1=\sum_{l=1}^n\mathbb{E}[|\Psi(X_{t_{l-1}})Z_l|^2]$ and $J_n^2=\sum_{l\neq l'}^n\mathbb{E}[\Psi(X_{t_{l-1}})Z_l\Psi(X_{t_{l'-1}})Z_{l'}]$. It can be directly seen that $0\le J_n^1\le\sup_{x\in\mathbb{R}^d}\Psi(x)^2\sum_{l=1}^n\mathbb{E}[Z_l^2]$ and $\lim_{n\to\infty}\sum_{l=1}^n\mathbb{E}[Z_l^2]=0$. Hence it follows that $\lim_{n\to\infty}J_n^1=0$. On the other hand, we see that for $l<l'$,

$$\mathbb{E}[\Psi(X_{t_{l-1}})Z_l\Psi(X_{t_{l'-1}})Z_{l'}]=\mathbb{E}[\Psi(X_{t_{l-1}})\Psi(X_{t_{l'-1}})Z_{l'}]\mathbb{E}[Z_l]=0.$$

Hence $J_n^2=0$. Together with $\sum_{l=1}^n\Psi(X_{t_{l-1}})(t_l-t_{l-1})\xrightarrow{L^2}\int_s^u\Psi(X_r)dr$, (2.4.38) follows. We see that

$$|I_n^{(4)}|\le\sup_{x\in\mathbb{R}^D}|h_{ij}(x)||\sigma_p^{i\mu}|\max_{1\le l\le n}|B_{t_l}^{\mu}-B_{t_{l-1}}^{\mu}|\int_s^u|b_r^j|dr.$$

Since $t\mapsto B_t$ is almost surely continuous, the right-hand side above almost surely converges to zero as $n\to\infty$. In a similar way, we see that (2.4.40) converges to zero as $n\to\infty$.

Step 6. Suppose that $\sigma_\cdot^{i\mu}\in\mathbb{M}^2(0,T)$. The integral $\int_0^t\sigma_r^{i\mu}dB_r$ is defined by

$$\int_0^t\sigma_r^{i\mu}dB_r=\int_0^t\sigma_r^{i\mu}1_{\{r\le\rho_m\}}dB_r,\quad t\le\rho_m,$$

where ρ_m is the stopping time defined by

$$\rho_m=\begin{cases}\inf\left\{t\ge0\mid\int_0^t|\sigma_s^{i\mu}|^2ds\ge m\right\} & \text{if }\left\{t\ge0\mid\int_0^t|\sigma_s^{i\mu}|^2ds\ge m\right\}\neq\emptyset,\\ \infty & \text{if }\left\{t\ge0\mid\int_0^t|\sigma_s^{i\mu}|^2ds\ge m\right\}=\emptyset.\end{cases}$$

There exists a sequence of step functions $\sigma_t^{i\mu}(n)$ such that

$$\mathbb{E}\left[\int_0^t|\sigma_r^{i\mu}(n)-\sigma_r^{i\mu}1_{\{r\le\rho_m\}}|^2dr\right]\to0\quad\text{as }n\to\infty.$$

Set $\tilde{\sigma}_r^{i\mu}=\sigma_r^{i\mu}1_{\{r\le\rho_m\}}$. Define

$$X_t^i(n)=X_0^i(n)+\int_0^t b_r^i dr+\int_0^t\sigma_s^{i\mu}(n)dB_s^{\mu}.$$

By Steps 1–5, (2.4.32) with $s = 0$ and $u = t$ is valid with X_t^i and $\sigma_t^{i\mu}$ replaced by $X_t^i(n)$ and $\sigma_t^{i\mu}(n)$, respectively, i. e.,

$$h(X_t(n)) - h(X_0(n)) = \int_0^t h_i(X_r(n))b_r^i dr + \frac{1}{2}\int_0^t h_{ij}(X_r(n))\sigma_r^{i\mu}(n)\sigma_r^{j\mu}(n)dr + \int_0^t h_i(X_r(n))\sigma_r^{i\mu}(n)dB_r^\mu. \quad (2.4.41)$$

We see that

$$\begin{aligned}\mathbb{E}\left[\left|\int_0^t h_i(X_r(n))\sigma_r^{i\mu}(n)dB_r^\mu - \int_0^t h_i(X_r)\tilde{\sigma}_r^{i\mu}dB_r^\mu\right|^2\right] &\le 2\mathbb{E}\left[\left|\int_0^t h_i(X_r(n))(\sigma_r^{i\mu}(n) - \tilde{\sigma}_r^{i\mu})dB_r^\mu\right|^2\right] + 2\mathbb{E}\left[\left|\int_0^t (h_i(X_r(n)) - h_i(X_r))\tilde{\sigma}_r^{i\mu}dB_r^\mu\right|^2\right] \\ &\le 2\sup_{x\in\mathbb{R}^d}|h_i(x)|^2\mathbb{E}\left[\int_0^t |\sigma_r^{i\mu}(n) - \tilde{\sigma}_r^{i\mu}|^2 dr\right] + 2\mathbb{E}\left[\int_0^t |(h_i(X_r(n)) - h_i(X_r))\tilde{\sigma}_r^{i\mu}|^2 dr\right].\end{aligned}$$

The right-hand side above converges to zero as $n \to \infty$. There exists a subsequence n' such that $\int_0^t h_i(X_r(n'))\sigma_r^{i\mu}(n')dB_r^\mu$ almost surely converges to $\int_0^t h_i(X_r)\tilde{\sigma}_r^{i\mu}dB_r^\mu$. Note that $X_t(n) \to X_t$ uniformly in $t \in [0, T]$ almost surely. Thus $h(X_t(n)) - h(X_0(n)) \to h(X_t) - h(X_0)$ almost surely as $n \to \infty$. It follows that

$$\lim_{n\to\infty}\mathbb{E}\left[\left|\int_0^t (h_i(X_r(n)) - h_i(X_r))b_r^i dr\right|^2\right] = 0$$

by the dominated convergence theorem. Similarly, we see that

$$\mathbb{E}\left[\left|\int_0^t h_{ij}(X_r(n))\sigma_r^{i\mu}(n)\sigma_r^{j\mu}(n)dr - \int_0^t h_{ij}(X_r)\tilde{\sigma}_r^{i\mu}\tilde{\sigma}_r^{j\mu}dr\right|^2\right] \to 0$$

as $n \to \infty$. Then the theorem follows by taking the limit as $n \to \infty$ on both sides of (2.4.41).

Step 7. Finally, we show (2.4.26) for $h \in C^{1,2}(\mathbb{R}^+ \times \mathbb{R}^d; \mathbb{R}^m)$. Take a sequence $(h_n)_{n\in\mathbb{N}} \subset C^\infty(\mathbb{R}^+ \times \mathbb{R}^d; \mathbb{R}^m)$ such that h^n, $\dot{h}^n$, h_i^n, h_{ij}^n locally uniformly converges to h, $\dot{h}$, h_i, h_{ij}, respectively, as $n \to \infty$. Equation (2.4.26) is valid for h^n for a. e. $\omega \in \mathscr{X} \setminus N_n$, where N_n is a null set and depends on h_n. Let $N = \bigcup_{n=1}^\infty N_n$. By a simple limiting argument (if necessary, taking a subsequence), it is seen that (2.4.26) is valid for h for a. e. $\omega \in \mathscr{X} \setminus N$. □

Remark 2.145 (Rules of Itô differential calculus). For practical purposes, the following formal rules of Itô differential calculus are useful:

$$dtdt = 0, \quad dtdB_t^\mu = 0, \quad dB_t^\mu dt = 0, \quad dB_t^\mu \cdot dB_t^\nu = \delta_{\mu\nu} dt \tag{2.4.42}$$

for all $\mu, \nu = 1, \dots, d$. The last property intuitively corresponds to the fact that the quadratic variation of Brownian motion grows linearly in time, that is, B_t has large fluctuations on a small scale, $dB_t = O(\sqrt{dt})$.

Using the Itô rules, we can express the Itô formula in Theorem 2.144 in differential form as

$$dY_t^k = \dot{h}^k dt + \sum_{i=1}^{D} h_i^k dX_t^i + \frac{1}{2} \sum_{i,j=1}^{D} h_{ij}^k dX_t^i dX_t^j, \quad 1 \le k \le m. \tag{2.4.43}$$

An application of the Itô formula gives the following useful result.

Corollary 2.146 (Product formula). *Let $(X_t)_{t\ge 0}$ and $(Y_t)_{t\ge 0}$ be two Itô processes. Then the product formula*

$$d(X_t Y_t) = dX_t \cdot Y_t + X_t \cdot dY_t + dX_t \cdot dY_t \tag{2.4.44}$$

holds.

Proof. Choose $g(x,y) = xy$, $x = X_t$, and $y = Y_t$. Then the formula follows directly from the Itô formula. □

Example 2.147. When the integrand $f(s)$ is continuous and of bounded variation but nonrandom, i. e., independent of ω, an application of the Itô formula gives

$$\int_0^t f(s) dB_s = f(t) B_t - \int_0^t B_s df(s).$$

Example 2.148. Let $Z_t = g(t, X_t)$ with $g(t,x) \in C^{1,2}(\mathbb{R}^+ \times \mathbb{R})$, where

$$dX_t = V(B_t)dt + a(B_t) \cdot dB_t$$

and $a = (a_1, \dots, a_d)$. Then

$$dZ_t = \left(\dot{g} + \frac{1}{2}(a \cdot a)\Delta_x g + V \partial_x g\right) dt + (\partial_x g) a \cdot dB_t. \tag{2.4.45}$$

This formula shows the relationship between Stratonovich and Itô integrals. In the particular case $g(t,x) = g(x)$ and $X_t = B_t$,

$$g(B_t) = g(B_0) + \int_0^t (\partial_x g)(B_s) \circ dB_s. \tag{2.4.46}$$

We conclude this section by two further applications of the Itô formula.

Example 2.149 (Wick product). Put $h(t,x) = e^{ax-\frac{1}{2}a^2t}$; notice that $\frac{1}{2}\Delta h + \dot{h} = 0$ is satisfied. The Itô formula yields $dh(t, B_t) = ahdB_t$. Hence,

$$\int_0^t h(s, B_s)dB_s = \frac{1}{a}(h(t, B_t) - 1). \tag{2.4.47}$$

Define the *Wick product* of e^{aB_t} by

$$:\exp(aB_t): = \exp\left(aB_t - \frac{1}{2}a^2 t\right), \tag{2.4.48}$$

giving

$$:\exp\left(\int_0^t f(s, B_s)dB_s\right): = \exp\left(\int_0^t f(s, B_s)dB_s - \frac{1}{2}\mathbb{E}^x\left[\int_0^t |f(s, B_s)|^2 ds\right]\right).$$

This definition will be useful in the discussion of the Cameron–Martin formula (2.4.97) below. The Wick product of B_t^n can then be written as

$$:B_t^n: = \left(\frac{d^n h}{da^n}\right)(t, B_t)\restriction_{a=0},$$

and as a result we obtain by (2.4.47)

$$\int_0^t :B_s^n:dB_s = \frac{1}{n+1}:B_t^{n+1}:.$$

The following inequality offers useful bounds on the moments of the martingale $\int_0^T f(s, B_s) \cdot dB_s$.

Proposition 2.150 (Burkholder–Davis–Gundy (BDG) inequality). *Let $m \in \mathbb{N}$ and suppose that $\mathbb{E}[\int_0^T |f(t, B_t)|^{2m} dt] < \infty$. Then*

$$\mathbb{E}\left[\left|\int_0^T f(t, B_t) \cdot dB_t\right|^{2m}\right] \leq (m(2m-1))^m T^{m-1}\mathbb{E}\left[\int_0^T |f(t, B_t)|^{2m} dt\right]. \tag{2.4.49}$$

Proof. We apply the Itô formula to $|X_t|^{2m}$, where $dX_t = f(t, B_t) \cdot dB_t$, and obtain

$$d|X_t|^{2m} = m(2m-1)|X_t|^{2m-2}|f(t, B_t)|^2 dt + 2m|X_t|^{2m-1}f(t, B_t) \cdot dB_t.$$

Hence

$$\mathbb{E}\left[|X_t|^{2m}\right] \leq m(2m-1)\int_0^T \mathbb{E}\left[|X_s|^{2m-2}|f(s, B_s)|^2\right] ds. \tag{2.4.50}$$

By Jensen's inequality, we have $\mathbb{E}[|X_t|^{2m}|\mathscr{F}_s] \geq |X_s|^{2m}$ for $t \geq s$. Thus

$$\mathbb{E}\left[|X_s|^{2m}\right] \leq \mathbb{E}\left[|X_T|^{2m}\right].$$

By (2.4.50) and the Hölder inequality, it follows that

$$\mathbb{E}\left[|X_T|^{2m}\right] \leq m(2m-1)\left(\mathbb{E}[|X_T|^{(2m-2)p}]\right)^{1/p}\left(\mathbb{E}\left[\left(\int_0^T |f(s,B_s)|^2 ds\right)^q\right]\right)^{1/q}$$

for $1/p + 1/q = 1$. Putting $p = m/(m-1)$ and $q = m$, we obtain

$$(\mathbb{E}[|X_T|^{2m}])^{1/m} \leq m(2m-1)\mathbb{E}\left[\left(\int_0^T |f(s,B_s)|^2 ds\right)^m\right]^{1/m}.$$

Taking the mth power of both sides, by the Hölder inequality

$$\mathbb{E}[|X_T|^{2m}] \leq [m(2m-1)]^m T^{m-1}\mathbb{E}\left[\int_0^T |f(s,B_s)|^{2m} ds\right]. \qquad \square$$

Since $\left(\int_0^t f(s,B_s)\cdot dB_s\right)_{t\geq 0}$ is a continuous martingale provided that $f \in M^2(0,t)$ for all $t \geq 0$, combining the BDG inequality and the martingale inequalities (2.2.36) and (2.2.37), further useful inequalities can be derived.

2.4.5 Stochastic differential equations

In this section we consider a class of stochastic differential equations (SDEs) and their solutions. An SDE can be thought of as arising from an ordinary differential equation $\dot{X}(t) = a(t)X(t)$ in which $a(t)$ is replaced by a factor consisting of a nonrandom and a noise term. The solution of such an equation is a random process. The theory of SDEs makes it possible to implement the noise term mathematically and allows showing that under sufficient regularity a solution exists, which is continuous in t and has the Markov property.

Consider the system of SDEs

$$X_t^i = x^i + \int_0^t b_s^i(X_s)ds + \int_0^t \sigma_s^i(X_s)\cdot dB_s, \quad X_0^i = x^i, \quad i = 1,\dots,n, \tag{2.4.51}$$

or in differential form,

$$dX_t^i = b_t^i(X_t)dt + \sigma_t^i(X_t)\cdot dB_t, \quad X_0 = x, \quad i = 1,\dots,n. \tag{2.4.52}$$

Here $(B_t)_{t\geq 0}$ is a d-dimensional Brownian motion, $b_t = (b_t^1, \dots, b_t^n) : \mathbb{R}^n \to \mathbb{R}^n$, and $\sigma_t = (\sigma_t^1, \dots, \sigma_t^n) : \mathbb{R}^n \to \mathbb{R}^{d\times n}$.

Definition 2.151 (Strong solution). A strong solution of SDE (2.4.51) or (2.4.52) on $(\mathscr{X}, \mathcal{B}(\mathscr{X}), \mathcal{W})$ is a process $(X_t^x)_{t\geq 0}$ with the following properties:
(1) $t \mapsto X_t^x$ is almost surely continuous;
(2) $(X_t^x)_{t\geq 0}$ is adapted with respect to $(\mathscr{F}_t^{\mathrm{BM}})_{t\geq 0}$;
(3) $\mathcal{W}(X_0^x = x) = 1$;
(4) $\mathcal{W}(\int_0^t (|b_s^i(X_s)| + |\sigma^{ij}(X_s)|^2) ds < \infty) = 1$ holds for $1 \leq i \leq n$, $1 \leq j \leq d$, and $0 \leq t < \infty$;
(5) (2.4.51) holds almost surely.

Theorem 2.152 (Strong solution of SDEs). *Suppose that $b_t = (b_t^1, \dots, b_t^n) : \mathbb{R}^n \to \mathbb{R}^n$ and $\sigma_t = (\sigma_t^1, \dots, \sigma_t^n) : \mathbb{R}^n \to \mathbb{R}^{d\times n}$ satisfy, for $t \in [0, T]$,*

$$|b_t(x)| + |\sigma_t(x)| \leq C(1 + |x|), \tag{2.4.53}$$

$$|b_t(x) - b_t(y)| + |\sigma_t(x) - \sigma_t(y)| \leq D|x - y|, \tag{2.4.54}$$

with some constants C and D independent of $x, y \in \mathbb{R}^n$ and $t \in [0, T]$, where we write $|\sigma_t(x)| = (\sum_{i=1}^n \sum_{j=1}^d |\sigma_t^{ij}(x)|^2)^{1/2}$ and $|b_t(x)| = (\sum_{i=1}^n |b_t^i(x)|^2)^{1/2}$. Then there exists a unique strong solution $(X_t^x)_{t\in[0,T]}$ of

$$dX_t^i = b_t^i(X_t)dt + \sigma_t^i(X_t) \cdot dB_t, \quad X_0^i = x^i \in \mathbb{R}, \quad i = 1, \dots, n, \tag{2.4.55}$$

as an n-dimensional random process on a probability space $(\mathscr{X}, \mathcal{B}(\mathscr{X}), \mathcal{W})$ such that
(1) $(X_t^x)_{t\geq 0}$ is $(\mathscr{F}_t^{\mathrm{BM}})_{t\geq 0}$-adapted;
(2) $\mathbb{E}\left[\int_0^T |X_t^x|^2 dt\right] < \infty$;
(3) X_t^x is almost surely continuous in $t \in [0, T]$.

Proof. Let $X_t = (X_t^1, \dots, X_t^n)$, $b_t = (b_t^1, \dots, b_t^n)$, and $\sigma_t = (\sigma_t^1, \dots, \sigma_t^n)$. The SDE is represented as

$$X_t - X_0 = \int_0^t b_s(X_s)ds + \int_0^t \sigma_s(X_s) \cdot dB_s. \tag{2.4.56}$$

We first show the uniqueness of the solution. Suppose that X_t and Z_t are solutions of (2.4.56). Put $a_s = b_s(X_s) - b_s(Z_s)$ and $\gamma_s = \sigma_s(X_s) - \sigma_s(Z_s)$. We have

$$\begin{aligned}\mathbb{E}[|X_t - Z_t|^2] &= \mathbb{E}\left[\left|\int_0^t a_s ds + \int_0^t \gamma_s dB_s\right|^2\right] \leq 2t\mathbb{E}\left[\int_0^t |a_s|^2 ds\right] + 2\mathbb{E}\left[\int_0^t |\gamma_s|^2 ds\right] \\ &\leq 2(1 + t)D^2 \int_0^t \mathbb{E}[|X_s - Z_s|^2] ds.\end{aligned} \tag{2.4.57}$$

Writing $v(t) = \mathbb{E}[|X_t - Z_t|^2]$, we have $v(t) \le A \int_0^t v(s)ds$, where $A = 2(1+T)D^2$. By the *Gronwall inequality* quoted in Proposition 2.153, we conclude that $v(t) = 0$ for all $t \ge 0$. Thus $\mathcal{W}(X_t - Z_t = 0 \text{ for all } t \ge 0) = 1$.

Next we prove existence of a solution. Define $Y_t^{(0)} = X_0 = x$ and $Y_t^{(k)}$ inductively by

$$Y_t^{(k+1)} = X_0 + \int_0^t b_s(Y_s^{(k)})ds + \int_0^t \sigma_s(Y_s^{(k)}) \cdot dB_s. \tag{2.4.58}$$

In the same way as in (2.4.57), we have

$$\mathbb{E}[|Y_t^{(k+1)} - Y_t^{(k)}|^2] \le (1+T)2D^2 \int_0^t \mathbb{E}[|Y_s^{(k)} - Y_s^{(k-1)}|^2]ds$$

for $k \ge 1$ and $t \le T$, and

$$\mathbb{E}[|Y_t^{(1)} - Y_t^{(0)}|^2] \le 2C^2 t(1 + \mathbb{E}[|X_0|^2]) + 2C^2 t^2(1 + \mathbb{E}[|X_0|^2]) \le A_1 t,$$

where $A_1 = 2C^2(1+T)(1+\mathbb{E}[|X_0|^2])$ only depends on C, T, and $\mathbb{E}[|X_0|^2]$. By induction, we obtain

$$\mathbb{E}[|Y_t^{(k+1)} - Y_t^{(k)}|^2] \le \frac{(A_2 t)^{k+1}}{(k+1)!} \tag{2.4.59}$$

for $k \ge 0$ and $t \le T$. Here A_2 only depends on C, T, and $\mathbb{E}[|X_0|^2]$. We have

$$\sup_{t\in[0,T]} |Y_t^{(k+1)} - Y_t^{(k)}| \le \int_0^T |b_s(Y_s^{(k)}) - b_s(Y_s^{(k-1)})|ds + \sup_{t\in[0,T]} \left| \int_0^t (\sigma_s(Y_s^{(k)}) - \sigma_s(Y_s^{(k-1)})) \cdot dB_s \right|.$$

By

$$\left\{ \sup_{t\in[0,T]} |Y_t^{(k+1)} - Y_t^{(k)}| > 2^{-k} \right\} \subset \left\{ \int_0^T |b_s(Y_s^{(k)}) - b_s(Y_s^{(k-1)})|ds > \frac{2^{-k}}{2} \right\}$$
$$\cup \left\{ \sup_{t\in[0,T]} \left| \int_0^t (\sigma_s(Y_s^{(k)}) - \sigma_s(Y_s^{(k-1)})) \cdot dB_s \right| > \frac{2^{-k}}{2} \right\},$$

we have

$$\mathcal{W}\Big(\sup_{t\in[0,T]} |Y_t^{(k+1)} - Y_t^{(k)}| > 2^{-k} \Big) \le \mathcal{W}\Big(\left| \int_0^T |b_s(Y_s^{(k)}) - b_s(Y_s^{(k-1)})|ds \right|^2 > 2^{-2k-2} \Big)$$

$$+\mathcal{W}\Big(\sup_{t\in[0,T]}\Big|\int_0^t(\sigma_s(Y_s^{(k)})-\sigma_s(Y_s^{(k-1)}))\cdot dB_s\Big|>2^{-k-1}\Big).$$

By the martingale inequality we also have

$$\mathcal{W}\Big(\sup_{t\in[0,T]}|Y_t^{(k+1)}-Y_t^{(k)}|>2^{-k}\Big)$$
$$\le 2^{2k+2}T\int_0^T\mathbb{E}[|b_s(Y_s^{(k)})-b_s(Y_s^{(k-1)})|^2]ds+2^{2k+2}\int_0^T\mathbb{E}[|\sigma_s(Y_s^{(k)})-\sigma_s(Y_s^{(k-1)})|^2]ds$$
$$\le 2^{2k+2}D^2(T+1)\int_0^T\frac{(A_2t)^k}{k!}dt\le\frac{(4A_2T)^{k+1}}{(k+1)!},$$

if $A_2\ge D^2(T+1)$. Since $\sum_{k=0}^\infty\mathcal{W}(\sup_{t\in[0,T]}|Y_t^{(k+1)}-Y_t^{(k)}|>2^{-k})<\infty$, by the Borel–Cantelli lemma, we see that

$$\mathcal{W}\Big(\bigcup_{m=1}^\infty\bigcap_{k=m}^\infty\{\sup_{t\in[0,T]}|Y_t^{(k+1)}-Y_t^{(k)}|\le 2^{-k}\}\Big)=0.$$

For a. e. ω, there exists $k_0=k_0(\omega)$ such that

$$\sup_{t\in[0,T]}|Y_t^{(k+1)}(\omega)-Y_t^{(k)}(\omega)|\le 2^{-k}$$

for all $k\ge k_0$. Therefore,

$$Y_t^{(n)}(\omega)=Y_t^{(0)}(\omega)+\sum_{k=0}^{n-1}(Y_t^{(k+1)}(\omega)-Y_t^{(k)}(\omega))$$

uniformly converges in $[0,T]$ almost surely. Denote the limit by $X_t(\omega)=\lim_{n\to\infty}Y_t^{(n)}(\omega)$. It is seen that $t\to X_t$ is continuous, since $Y_t^{(n)}$ is continuous and the convergence is uniform in t. Moreover, X_t is $\mathcal{F}_t$-measurable, since $Y_t^{(n)}$ is $\mathcal{F}_t$-measurable for all n. Next note that by (2.4.59),

$$\big(\mathbb{E}[|Y_t^{(m)}-Y_t^{(n)}|^2]\big)^{1/2}\le\sum_{k=n}^{m-1}\big(\mathbb{E}[|Y_t^{(k+1)}-Y_t^{(k)}|^2]\big)^{1/2}\le\sum_{k=n}^\infty\Big(\frac{(A_2t)^{k+1}}{(k+1)!}\Big)^{1/2}\to 0$$

as $n\to\infty$ and $(Y_t^{(n)})_{n\in\mathbb{N}}$ is a Cauchy sequence in the Hilbert space $L^2(\mathcal{X},d\mathcal{W})$. Thus a subsequence $Y_t^{(n')}$ almost surely converges to some Y_t. We reset n' by n. We have $Y_t=X_t$ a. s. In particular, $\infty>\mathbb{E}[|Y_t|^2]=\mathbb{E}[|X_t|^2]$ follows. It remains to show that X_t satisfies (2.4.56). For all n, we have

$$Y_t^{(n+1)} = X_0 + \int_0^t b_s(Y_s^{(n)})ds + \int_0^t \sigma_s(Y_s^{(n)}) \cdot dB_s. \tag{2.4.60}$$

It follows that $\lim_{n\to\infty} Y_t^{(n+1)} = X_t$ a. s. We note that

$$\mathbb{E}\left[\int_0^T |X_t - Y_t^{(n)}|^2 dt\right] \le \limsup_{m\to\infty} \mathbb{E}\left[\int_0^T |Y_t^{(m)} - Y_t^{(n)}|^2 dt\right] \le T\left(\sum_{k=n}^{\infty}\left(\frac{(A_2 t)^{k+1}}{(k+1)!}\right)^{1/2}\right)^2 \tag{2.4.61}$$

by Fatou's lemma. By (2.4.61) and the Itô isometry, taking a subsequence n', we have

$$\lim_{n'\to\infty} \int_0^t \sigma_s(Y_s^{(n')}) \cdot dB_s = \int_0^t \sigma_s(X_s) \cdot dB_s, \tag{2.4.62}$$

and by the Hölder inequality,

$$\lim_{n'\to\infty} \int_0^t b_s(Y_s^{(n')})ds = \int_0^t b_s(X_s)ds \tag{2.4.63}$$

in $L^2(\mathscr{X}, d\mathcal{W})$. Therefore, taking the limit of both sides of (2.4.60), we can obtain the desired results. □

Proposition 2.153 (Gronwall inequality). *Let $v(t)$ be a nonnegative function such that $v(t) \le C + A\int_0^t v(s)ds$ for $t \in [0,T]$ with some constants $A, C \ge 0$. Then it follows that $v(t) \le Ce^{At}$ for $t \in [0,T]$.*

Proof. We may suppose $A \ne 0$. Let $f(t) = w(t)e^{-At}$ and $w(t) = \int_0^t v(s)ds$. Since $f(t) = \int_0^t f'(s)ds \le -\frac{C}{A}(e^{-At} - 1)$, we have $w(t) \le \frac{C}{A}(e^{At} - 1)$. Hence $v(t) \le C + aw(t) \le Ce^{At}$ follows. □

An important class of solutions of SDEs is the following.

Definition 2.154 (Diffusion process). A random process $(X_t)_{t\ge0}$ on a filtered probability space $(\Omega, \mathscr{F}, (\mathscr{F}_t)_{t\ge0}, P)$ is a *diffusion process* whenever
(1) $(X_t)_{t\ge0}$ is a Markov process;
(2) paths $t \mapsto X_t(\omega)$ are continuous for a. e. $\omega \in \mathscr{X}$.

We now consider (2.4.55) with b_t and σ_t replaced by time-independent b and σ, respectively. We present a result without a proof.

Theorem 2.155 (Itô diffusion). *Suppose that the functions $b = (b^1, \dots, b^n) : \mathbb{R}^n \to \mathbb{R}^n$ and $\sigma = (\sigma^1, \dots, \sigma^n) : \mathbb{R}^n \to \mathbb{R}^{d\times n}$ satisfy*

$$|b(x) - b(y)| + |\sigma(x) - \sigma(y)| \leq D|x - y|$$

with some constant D independent of $x, y \in \mathbb{R}^n$. Then the unique solution $(X_t^x)_{t\geq 0}$, $X_t^x = (X_t^{1,x}, \dots, X_t^{n,x})$, of

$$dX_t^i = b^i(X_t)dt + \sigma^i(X_t) \cdot dB_t, \quad X_0^i = x^i \in \mathbb{R}, \quad i = 1, \dots, n, \tag{2.4.64}$$

satisfies the Markov property

$$\mathbb{E}[f(X_{s+t}^x) | \mathscr{F}_s^{\mathrm{BM}}] = \mathbb{E}[f(X_t^{X_s^x})]$$

for every bounded Borel function f, where $\mathbb{E}[f(X_t^{X_s^x})] = \mathbb{E}[f(X_t^y)]$ evaluated at $y = X_s^x$. In particular, $\mathbb{E}[f(X_{s+t}^x) | \mathscr{F}_s^{\mathrm{BM}}] = \mathbb{E}[f(X_{s+t}^x) | \sigma(X_s^x)]$ holds.

The diffusion process in Theorem 2.155 is called an *Itô diffusion*. Let $(X_t^x)_{t\geq 0}$, $X_t^x = (X_t^{1,x}, \dots, X_t^{n,x})$, be the solution of SDE (2.4.51) and let $\mathscr{U}_t : L^\infty(\mathbb{R}^d) \to L^\infty(\mathbb{R}^d)$ be given by

$$\mathscr{U}_t f(x) = \mathbb{E}[f(X_t^x)]. \tag{2.4.65}$$

Definition 2.156 (Infinitesimal generator). The *infinitesimal generator H_X* of the solution $(X_t^x)_{t\geq 0}$ is defined by

$$H_X f(x) = \lim_{t \downarrow 0} \frac{1}{t}(\mathscr{U}_t f(x) - f(x)). \tag{2.4.66}$$

Theorem 2.157. *Let $(X_t^x)_{t\geq 0}$ be the solution of*

$$dX_t^i = b^i(X_t)dt + \sigma^i(X_t) \cdot dB_t, \quad X_0 = x, \quad i = 1, \dots, n, \tag{2.4.67}$$

where $b = (b^1, \dots, b^n)$ and $\sigma = (\sigma^1, \dots, \sigma^n)$ are independent of t and suppose (2.4.53) *and* (2.4.54). *Then its generator is given by the second-order partial differential operator*

$$H_X = \frac{1}{2} \sum_{i,j=1}^n (\sigma\sigma^T)_{ij}(x) \partial_i \partial_j + \sum_{i=1}^n b^i(x) \partial_i \tag{2.4.68}$$

acting on $C_0^2(\mathbb{R}^d)$.

Proof. Let $(Y_t^i)_{t\geq 0}$ be an Itô process, $dY_t^i = u_t^i dt + v_t^i \cdot dB_t$, with $Y_0^i = x^i$. The Itô formula (2.4.31) yields

$$f(Y_t) - f(Y_0) = \int_0^t \left(\frac{1}{2} \sum_{i,j=1}^n (v_s v_s^T)_{ij} \partial_i \partial_j f(Y_s) + \sum_{i=1}^n u_s^i \partial_i f(Y_s) \right) ds + \int_0^t \sum_{i=1}^n (\partial_i f(Y_s)) v_s^i \cdot dB_s.$$

Taking expectations, the martingale part $\int_0^t \sum_{i=1}^n (\partial_i f) v_s^i \cdot dB_s$ vanishes, and we are left with

$$\mathbb{E}[f(Y_t)] = f(x) + \mathbb{E}\left[\int_0^t \left(\frac{1}{2}\sum_{i,j=1}^n (v_s v_s^T)_{ij}\partial_i\partial_j f(Y_s) + \sum_{i=1}^n u_s^i \partial_i f(Y_s)\right) ds\right].$$

Hence

$$\mathscr{U}_t f(x) - f(x) = \mathbb{E}\left[\int_0^t \left(\frac{1}{2}\sum_{i,j=1}^n (\sigma\sigma^T)_{ij}(X_s^x)\partial_i\partial_j f(X_s^x) + \sum_{i=1}^n b^i(X_s^x)\partial_i f(X_s^x)\right) ds\right].$$

Since X_s^x is continuous in s, the theorem now follows directly. □

The last equality is, in fact, a simple version of *Dynkin's formula.*

Example 2.158 (Linear SDE). Assume that the $n \times n$, $n \times 1$, and $n \times d$ matrices A_t, a_t, and σ_t, respectively, are nonrandom and locally bounded. Let Φ_t be the $n \times n$ matrix function solving the nonrandom equation $\dot{\Phi}_t = A_t \Phi_t$. Then the solution of the linear SDE

$$dX_t = (A_t \cdot X_t + a_t)\, dt + \sigma_t \cdot dB_t \tag{2.4.69}$$

is given by

$$X_t = \Phi_t \left(X_0 + \int_0^t \Phi_s^{-1} a_s ds + \int_0^t \Phi_s^{-1}\sigma_s \cdot dB_s \right), \tag{2.4.70}$$

where Φ_t^{-1} is the inverse matrix of Φ_t. This is easily checked by an application of the Itô formula.

Example 2.159 (Langevin equation). Let $d = 1$ and $a, \sigma > 0$. Consider the SDE

$$dX_t = -aX_t dt + \sigma dB_t, \quad X_0 = x \in \mathbb{R}. \tag{2.4.71}$$

By (2.4.70), the solution of (2.4.71) is given by

$$X_t^x = x\mathrm{e}^{-at} + \sigma \int_0^t \mathrm{e}^{-a(t-s)} dB_s. \tag{2.4.72}$$

The random process $(X_t^x)_{t\geq 0}$ is called an *Ornstein–Uhlenbeck process.* We have by (2.4.72) that

$$\mathbb{E}[X_t^x] = x\mathrm{e}^{-at} \quad \text{and} \quad \mathrm{cov}[X_t^x X_s^x] = \frac{\sigma^2}{2a} e^{-a(s+t)}(\mathrm{e}^{2a(s\wedge t)} - 1). \tag{2.4.73}$$

Moreover, the generator of X_t^x is given by

$$L = \frac{\sigma^2}{2}\Delta - ax \cdot \nabla. \tag{2.4.74}$$

2.4.6 Brownian bridge

Brownian bridge is a continuous Gaussian process starting in a given position a at a time $t = T_1$ and ending in a given position b at a time $t = T_2$.

Definition 2.160 (Brownian bridge). A *Brownian bridge* on the interval $[T_1, T_2]$ starting in $a \in \mathbb{R}^d$ and ending in $b \in \mathbb{R}^d$ is a continuous multivariate Gaussian random process $(X_t)_{T_1 \le t \le T_2}$ on a probability space $(\Omega, \mathscr{F}, P)$ such that
(1) $X_{T_1} = a$ and $X_{T_2} = b$ a. s.;
(2) $\mathbb{E}[X_t] = a(1 - \frac{t-T_1}{T_2-T_1}) + b\frac{t-T_1}{T_2-T_1}$;
(3) $\operatorname{cov}(X_s, X_t) = s \wedge t - \frac{st}{T_2-T_1}$.

There are several representations of Brownian bridge of which now we present three.

Example 2.161 (SDE for Brownian bridge). The Brownian bridge starting in a at $t = 0$ and ending in b at $t = T$ solves the SDE

$$dX_t = \frac{b - X_t}{T - t} dt + dB_t, \quad 0 \le t < T, \quad X_0 = a. \tag{2.4.75}$$

As a linear SDE, it can be solved exactly through (2.4.70), giving

$$X_t = a\left(1 - \frac{t}{T}\right) + b\frac{t}{T} + (T - t) \int_0^t \frac{1}{T - s} dB_s, \quad 0 \le t < T. \tag{2.4.76}$$

Let

$$\alpha_t = (T - t) \int_0^t \frac{1}{T - r} dB_r, \quad 0 \le t < T. \tag{2.4.77}$$

Note that α_t is a Gaussian random variable with $\mathbb{E}[\alpha_t] = 0$ and

$$\mathbb{E}[\alpha_s \alpha_t] = (T - t)(T - s) \int_0^{t \wedge s} \frac{1}{(T - r)^2} dr = s \wedge t - \frac{st}{T}.$$

Thus

$$\mathbb{E}[X_t] = a\left(1 - \frac{t}{T}\right) + b\frac{t}{T} \quad \text{and} \quad \operatorname{cov}(X_t, X_s) = s \wedge t - \frac{st}{T}.$$

It is easily seen that $X_t \to b$ as $t \to T$ in $L^2(\mathscr{X}, d\mathcal{W})$. Furthermore, it can also be shown that $X_t(\omega) \to b$ as $t \to T$ for a. e. $\omega \in \mathscr{X}$. Define

$$\tilde{X}_t = \begin{cases} X_t, & 0 \le t < T, \\ b, & t = T. \end{cases} \tag{2.4.78}$$

Hence $(\tilde{X}_t)_{0\le t\le T}$ is a Brownian bridge starting in a at $t = 0$ and ending b at $t = T$.

Example 2.162 (Brownian bridge by time change). Let $(B_t)_{t\ge0}$ be one-dimensional Brownian motion and define

$$\beta_t = (T - t)B_{\frac{t}{(T-t)T}}, \quad 0 \le t < T.$$

It is seen that β_t is a Gaussian random variable with $\mathbb{E}[\beta_t] = 0$ and $\mathbb{E}[\beta_s\beta_t] = s \wedge t - \frac{st}{T}$. Let

$$X_t = a\left(1 - \frac{t}{T}\right) + b\frac{t}{T} + \beta_t, \quad 0 \le t < T.$$

It can also be shown that $X_t \to b$ as $t \to T$ a. s. Define $\tilde{X}_t$ by (2.4.78) with α_t replaced by β_t. Hence $(\tilde{X}_t)_{0\le t\le T}$ is a Brownian bridge starting in a at $t = 0$ and ending b at $t = T$.

Example 2.163. Let $(B_t)_{t\ge0}$ be a one-dimensional Brownian motion starting from 0, and define

$$\gamma_t = B_t - \frac{t}{T}B_T, \quad 0 \le t \le T.$$

Clearly, γ_t is a Gaussian random variable with $\mathbb{E}[\gamma_t] = 0$ and $\mathbb{E}[\gamma_s\gamma_t] = s \wedge t - \frac{st}{T}$. Let

$$X_t = a\left(1 - \frac{t}{T}\right) + b\frac{t}{T} + \gamma_t, \quad 0 \le t \le T.$$

Hence $(X_t)_{0\le t\le T}$ is a Brownian bridge starting in a at $t = 0$ and ending in b at $t = T$.

We finally state the relationship between the Brownian bridge measure $\mathcal{W}^{x,y}_{[0,T]}$ and the Brownian bridge process.

Theorem 2.164. *Let $(X_t)_{0\le t\le T}$ be a Brownian bridge starting in a at $t = 0$ and ending in b at $t = T$ on a probability space $(\Omega, \mathscr{F}, P)$. Let $0 = t_0 < t_1 < \cdots < t_n \le T$ and*

$$Z_j = \frac{X_{t_j}}{T - t_j} - \frac{X_{t_{j-1}}}{T - t_{j-1}}.$$

Then we have the following:
(1) *$\{Z_j\}_{j=1,\dots,n}$ are independent;*
(2) *$\{Z_j\}_{j=1,\dots,n}$ is a collection of Gaussian random variables such that*

$$\mathbb{E}[Z_j] = \frac{b(t_j - t_{j-1})}{(T - t_j)(T - t_{j-1})} \quad \text{and} \quad \operatorname{cov}(Z_j, Z_j) = \frac{(t_j - t_{j-1})}{(T - t_j)(T - t_{j-1})}.$$

Proof. Both statements follow directly from above. □

Using this theorem, we can compute the finite-dimensional distributions of a Brownian bridge.

Corollary 2.165. *Let $(X_t^{a,b})_{0\le t\le T}$ be a Brownian bridge starting in a at $t = 0$ and ending in b at $t = T$ on a probability space $(\Omega, \mathscr{F}, P)$. Let $0 = t_0 < t_1 < \cdots < t_n \le T$ and $A_j \in \mathcal{B}(\mathbb{R})$, $j = 1, \ldots, n$. Then*

$$\mathbb{E}_P\left[\prod_{j=1}^n 1_{A_j}(X_{t_j}^{a,b})\right] = \int_{\mathbb{R}^n} \frac{\Pi_{t_1}(a - x_1)\Pi_{t_2-t_1}(x_1 - x_2)\cdots\Pi_{T-t_n}(x_n - b)}{\Pi_T(b - a)}\left(\prod_{j=1}^n 1_{A_j}(x_j)\right)\prod_{j=1}^n dx_j. \tag{2.4.79}$$

We recall that

$$\mathbb{E}^x\left[\prod_{j=1}^n f_j(B_{t_j})\right] = \int_{\mathbb{R}} \Pi_T(x-y)\mathbb{E}_{[0,T]}^{x,y}\left[\prod_{j=1}^n f_j(B_{t_j})\right] dy$$

for $0 < t_1 < \cdots < t_n < T$. Combining this with Corollary 2.165, we obtain the following corollary.

Corollary 2.166 (Brownian bridge and Brownian bridge measure). *Let $(X_t^{x,y})_{0\le t\le T}$ be a Brownian bridge starting in x at $t = 0$ and ending in y at $t = T$ on a probability space $(\mathscr{X}, \mathcal{B}(\mathscr{X}), \mathcal{W}^0)$. Then*

$$\mathbb{E}^x\left[\prod_{j=1}^n f_j(B_{t_j})\right] = \int_{\mathbb{R}} \Pi_T(x-y)\mathbb{E}^0\left[\prod_{j=1}^n f_j(X_{t_j}^{x,y})\right] dy \tag{2.4.80}$$

for $0 < t_1 < \cdots < t_n < T$.

Proof. By comparing Corollary 2.165 with (2.3.20), we obtain

$$\mathbb{E}^0\left[\prod_{j=1}^n f_j(X_{t_j}^{a,b})\right] = \mathbb{E}_{\mathcal{W}_{[0,T]}^{a,b}}\left[\prod_{j=1}^n f_j(B_{t_j})\right].$$

Then the corollary is proven. □

2.4.7 Weak solution and time change

In the previous section we studied strong solutions of SDEs. In this section we briefly study the notion of weak solutions.

Definition 2.167 (Weak solution). Let $(B_t)_{t\ge 0}$ be a d-dimensional Brownian motion, $b_t = (b_t^1, \ldots, b_t^n) : \mathbb{R}^n \to \mathbb{R}^n$, and $\sigma_t = (\sigma_t^1, \ldots, \sigma_t^n) : \mathbb{R}^n \to \mathbb{R}^{nd}$. A weak solution of

$$X_t^i = \int_0^t b_s^i(X_s)ds + \int_0^t \sigma_s^i(X_s) \cdot dB_s, \quad X_0 = x, \quad i = 1, \ldots, n, \tag{2.4.81}$$

is a triplet

$$((\tilde{X}_t)_{t\geq 0}, (\tilde{B}_t)_{t\geq 0}), \quad (\Omega, \mathscr{F}, P), \quad (\mathscr{F}_t)_{t\geq 0}, \tag{2.4.82}$$

where
(1) $(\Omega, \mathscr{F}, P)$ is a probability space;
(2) $(\mathscr{F}_t)_{t\geq 0}$ is an augmented filtration on $(\Omega, \mathscr{F}, P)$;
(3) $(\tilde{B}_t)_{t\geq 0}$ is an n-dimensional Brownian motion on $(\Omega, \mathscr{F}, P)$;
(4) $(\tilde{X}_t)_{t\geq 0}$ is an $\mathbb{R}^n$-valued continuous and adapted process with respect to $(\mathscr{F}_t)_{t\geq 0}$;
(5) $P(\tilde{X}_0 = x) = 1$;
(6) $((\tilde{X}_t)_{t\geq 0}, (\tilde{B}_t)_{t\geq 0})$ satisfies

$$\tilde{X}_t^i = \int_0^t b_s^i(\tilde{X}_s)ds + \int_0^t \sigma_s^i(\tilde{X}_s) \cdot d\tilde{B}_s, \quad i = 1, \ldots, n. \tag{2.4.83}$$

The concept of weak solution of SDEs is convenient for mathematical reasons, since there exist SDEs which have no strong solutions but still have a weak solution.

Example 2.168 (Tanaka equation). Consider the one-dimensional SDE with initial condition $X_0 = 0$,

$$dX_t = \operatorname{sgn}(X_t)dB_t, \tag{2.4.84}$$

where

$$\operatorname{sgn}(x) = \begin{cases} 1, & x > 0, \\ -1, & x \leq 0. \end{cases}$$

Equation (2.4.84) is called the *Tanaka equation*. Note that $\sigma_t(x) = \sigma(x) = \operatorname{sgn}(x)$ does not satisfy the Lipschitz condition given by (2.4.54) and thus Theorem 2.152 cannot be applied. Indeed, it is known that (2.4.84) has no strong solution. A weak solution of (2.4.84) can, however, be obtained as follows. Let $(\tilde{X}_t)_{t\geq 0}$ be a Brownian motion on a probability space $(\Omega, \mathscr{F}, P)$ and $(\mathscr{F}_t)_{t\geq 0}$ the natural filtration of $(\tilde{X}_t)_{t\geq 0}$. Define $(W_t)_{t\geq 0}$ by

$$W_t = \int_0^t \operatorname{sgn}(\tilde{X}_s)d\tilde{X}_s. \tag{2.4.85}$$

Note that we take the continuous version of $(W_t)_{t\geq 0}$. Thus $(W_t)_{t\geq 0}$ is a continuous $(\mathscr{F}_t)$-martingale with covariance $\mathbb{E}_P[|W_t|^2] = \int_0^t \mathbb{E}_P[|\operatorname{sgn}(\tilde{X}_s)|^2]ds = t$. Hence $(W_t^2 - t)_{t\geq 0}$ is a martingale, which implies that $(W_t)_{t\geq 0}$ is a Brownian motion by Proposition 2.107. Hence the triplet $((\tilde{X}_t)_{t\geq 0}, (W_t)_{t\geq 0}), (\Omega, \mathscr{F}, P), (\mathscr{F}_t)_{t\geq 0}$ is a weak solution of (2.4.84).

Example 2.169 (Bessel process). Let $(B_t)_{t\geq 0}$ be a d-dimensional Brownian motion on a probability space $(\Omega, \mathscr{F}, P)$. Consider the Euclidean distance from the origin

$$R_t = |B_t| = \sqrt{(B_t^1)^2 + \cdots + (B_t^d)^2}. \tag{2.4.86}$$

The random process $(R_t)_{t\geq 0}$ is called a *Bessel process*. The Bessel process starting at $r \geq 0$ is a weak solution of the SDE

$$R_t = r + \int_0^t \frac{d-1}{2R_s} ds + \hat{B}_t, \tag{2.4.87}$$

where $(\hat{B}_t)_{t\geq 0}$ is a one-dimensional Brownian motion. Note that as $\mathbb{E}[\int_0^t |B_s^i/R_s|^2 ds] \leq t < \infty$, the random process $(\int_0^t B_s^i/R_s dB_s^i)_{t\geq 0}$ is a continuous martingale for every i, moreover, we have $\mathbb{E}[(\sum_{i=1}^d \int_0^t B_s^i/R_s dB_s^i)^2] = t$. Hence

$$\tilde{B}_t = \sum_{i=1}^d \int_0^t \frac{B_s^i}{R_s} dB_s^i \tag{2.4.88}$$

is a one-dimensional Brownian motion by Theorem 2.107. Heuristically, one can apply the Itô formula to derive (2.4.87) for $R_t = f(B_t^1, \ldots, B_t^d)$, where $f(x_1, \ldots, x_d) = \sqrt{\sum_{i=1}^d x_i^2}$. The difficulty here is that f is not differentiable at the origin, and the Itô formula cannot be applied directly to f. Let $Y_t = R_t^2$. We have

$$Y_t = r^2 + 2\sum_{i=1}^d \int_0^t B_s^i dB_s^i + td.$$

Let $g(y) = \sqrt{y}$ and define

$$g_\varepsilon(y) = \begin{cases} \frac{3}{8}\sqrt{\varepsilon} + \frac{3}{4\sqrt{\varepsilon}}y - \frac{1}{8\varepsilon\sqrt{\varepsilon}}y^2, & y < \varepsilon, \\ \sqrt{y}, & y \geq \varepsilon. \end{cases}$$

It follows that $g_\varepsilon \in C^2(\mathbb{R})$ and $\lim_{\varepsilon\downarrow 0} g_\varepsilon(y) = g(y)$. The Itô formula yields

$$g_\varepsilon(Y_t) = g_\varepsilon(r^2) + \sum_{i=1}^d I_t^i(\varepsilon) + J_t(\varepsilon) + K_t(\varepsilon),$$

where

$$I_t^i(\varepsilon) = \int_0^t \left(1_{\{Y_s \geq \varepsilon\}} \frac{1}{R_s} + 1_{\{Y_s < \varepsilon\}} \frac{1}{2\sqrt{\varepsilon}}(3 - \tfrac{Y_s}{\varepsilon})\right) B_s^i dB_s^i, \quad J_t(\varepsilon) = \int_0^t 1_{\{Y_s \geq \varepsilon\}} \frac{d-1}{2R_s} ds,$$

$$K_t(\varepsilon) = \int_0^t 1_{\{Y_s < \varepsilon\}} \frac{1}{4\sqrt{\varepsilon}} \left(3d - (d+2)\frac{Y_s}{\varepsilon}\right) ds.$$

It is directly seen that $\lim_{\varepsilon \downarrow 0} J_t(\varepsilon) = \int_0^t \frac{d-1}{2R_s} ds$ a. s. and $\mathbb{E}[K_t(\varepsilon)] \to 0$ as $\varepsilon \downarrow 0$. Finally, using (2.4.88), we have

$$\begin{aligned} \mathbb{E}[(\tilde{B}_t^i - I_t^i(\varepsilon))^2] &= \mathbb{E}\left[\int_0^t 1_{\{Y_s < \varepsilon\}} \left(\frac{1}{R_s} - \frac{1}{2\sqrt{\varepsilon}}\left(3 - \frac{Y_s}{\varepsilon}\right)\right)^2 (B_s^i)^2 ds\right] \\ &= \mathbb{E}\left[\int_0^t 1_{\{Y_s < \varepsilon\}} \left(1 - \frac{1}{2}\sqrt{\frac{Y_s}{\varepsilon}}\left(3 - \frac{Y_s}{\varepsilon}\right)\right)^2 \left(\frac{B_s^i}{R_s}\right)^2 ds\right] \leq \int_0^t P(Y_s < \varepsilon) ds \to 0 \end{aligned}$$

as $\varepsilon \downarrow 0$. Thus R_t satisfies

$$R_t = r + \int_0^t \frac{d-1}{2R_s} ds + \tilde{B}_t,$$

and the Bessel process is a weak solution of (2.4.87). By (2.4.87), the generator of the Bessel process $(R_t)_{t \geq 0}$ is given by

$$\frac{1}{2}\frac{d^2}{dx^2} + \frac{d-1}{2x}\frac{d}{dx}.$$

In the remainder of this section, we briefly discuss time changes of random processes. In the special case of one-dimensional SDEs, it is possible to construct weak solutions by a combined drift and time change. Consider the one-dimensional SDE

$$dX_t = b(X_t)dt + \sigma(X_t)dB_t, \quad X_0 = x_0. \tag{2.4.89}$$

The function

$$S(x) = \int_{x_0}^x e^{-\int_{x_0}^y \frac{2b(z)}{\sigma^2(z)} dz} dy \tag{2.4.90}$$

is called a *scale function*. It is directly seen that S satisfies the differential equation

$$\frac{1}{2}\sigma^2(x)S''(x) + b(x)S'(x) = 0. \tag{2.4.91}$$

By using the Itô formula for $Y_t = S(X_t)$, we obtain

$$dY_t = \left((\sigma S') \circ S^{-1}\right)(Y_t)dB_t, \quad Y_0 = S(x_0), \tag{2.4.92}$$

that is, the process on its natural scale has no drift term. This equation can be solved by a time change via an application of the following result.

Theorem 2.170. *Let $(\Omega, \mathscr{F}, P)$ be a probability space, $I \subset \mathbb{R}$ be a bounded interval, and consider the one-dimensional SDE* (2.4.89) *such that $b, \sigma : I \to \mathbb{R}$ are continuous functions with the property that $\sigma(x) > 0$ for all $x \in I$. Let τ be a stopping time with respect to $(\mathscr{F}_t^{\mathrm{BM}})_{t\geq 0}$. Then the following statements are equivalent:*
(1) *The random process $(X_t)_{0\leq t<\tau}$ on $(\Omega, \mathscr{F}, P)$ is a weak solution of* (2.4.89) *up to time τ.*
(2) *The random process $(Y_t)_{0\leq t<\tau}$, $Y_t = S(X_t)$, on $(\Omega, \mathscr{F}, P)$ is a weak solution of* (2.4.92) *up to time τ.*
(3) *The random process $(Y_t)_{0\leq t<\tau}$ can be represented in the form $Y_t = B^1_{h^{-1}(t)}$, where*

$$h(t) = \int_0^t \frac{1}{((\sigma S') \circ S^{-1})^2 (B_s^1)} ds$$

and $(B_t^1)_{t\geq 0}$ is an independent one-dimensional Brownian motion such that $B_0^1 = S(x_0)$ a. s.

2.4.8 Girsanov theorem and Cameron–Martin formula

In this section we show that adding a pathwise shift to Brownian motion results in a new Brownian motion on a suitable probability space.

Fix $T > 0$ and suppose that $a_t = a_t(\omega) = (a_t^1, \dots, a_t^d)$ is adapted to the filtration of Brownian motion $(B_t)_{t\geq 0}$ and satisfies

$$\mathbb{E}\left[\int_0^T |a_t|^2 dt\right] < \infty.$$

Define the random process $(Z_t)_{0\leq t\leq T}$ by

$$dZ_t = a_t \cdot dB_t - \frac{1}{2} a_t^2 dt. \tag{2.4.93}$$

Next, define the probability measure $\hat{P}_T$ on $(\mathscr{X}, \mathscr{F}_T)$, with $\mathscr{F}_T = \sigma(B_s, 0 \leq s \leq T)$, and the d-dimensional random process $(\hat{B}_t)_{0\leq t\leq T}$ by putting

$$\hat{P}_T(A) = \mathbb{E}[e^{Z_T} 1_A], \quad A \in \mathscr{F}_T,$$

$$\hat{B}_t(\omega) = B_t(\omega) - \int_0^t a_s(\omega) ds, \quad t \in [0, T].$$

Note that $\hat{P}_T(\mathscr{X}) = 1$.

Proposition 2.171 (Girsanov theorem). *The random process $(e^{Z_t})_{0\le t\le T}$ defined in (2.4.93) is an $(\mathscr{F}_t)_{0\le t\le T}$-martingale and $(\hat{B}_t)_{0\le t\le T}$ is a d-dimensional Brownian motion on the probability space $(\mathscr{X}, \mathscr{F}_T, \hat{P}_T)$.*

Proof. Since $de^{Z_t} = e^{Z_t} a_t \cdot dB_t$, for $\phi \in C^2(\mathbb{R}^d)$ a combination of the Itô formula and the product formula (2.4.44) gives

$$d\left(e^{Z_t}\phi(\hat{B}_t)\right) = e^{Z_t}(\nabla\phi(\hat{B}_t) + \phi(\hat{B}_t)a_t)\cdot dB_t + \frac{1}{2}e^{Z_t}\Delta\phi(\hat{B}_t)dt.$$

From this and the martingale property of e^{Z_t} (as seen in Corollary 2.113), it follows that for $A \in \mathscr{F}_s$, $s \le t$,

$$\begin{aligned}
\mathbb{E}_{\hat{P}_T}[\phi(\hat{B}_t)1_A] &= \mathbb{E}[\phi(\hat{B}_t)e^{Z_T}1_A] = \mathbb{E}[\phi(\hat{B}_t)e^{Z_t}1_A] \\
&= \mathbb{E}\left[\left(\phi(\hat{B}_0) + \int_0^s e^{Z_r}(\nabla\phi(\hat{B}_r) + \phi(\hat{B}_r)a_r)\cdot dB_r\right)1_A\right] + \frac{1}{2}\int_0^t \mathbb{E}\left[1_A e^{Z_r}\Delta\phi(\hat{B}_r)\right]dr \\
&= \mathbb{E}[1_A e^{Z_s}\phi(\hat{B}_s)] + \frac{1}{2}\int_s^t \mathbb{E}\left[1_A e^{Z_r}\Delta\phi(\hat{B}_r)\right]dr = \mathbb{E}_{\hat{P}_T}\left[\left(\phi(\hat{B}_s) + \frac{1}{2}\int_s^t \Delta\phi(\hat{B}_r)dr\right)1_A\right].
\end{aligned}$$

Hence it can be seen that $Y_t = \phi(\hat{B}_t) - \phi(\hat{B}_0) - \frac{1}{2}\int_0^t \Delta\phi(\hat{B}_s)ds$ is an $(\mathscr{F}_t)_{0\le t\le T}$-martingale on $(\mathscr{X}, \mathscr{F}_T, \hat{P}_T)$. Set $\phi(x) = e^{i\xi\cdot x}$, $\xi \in \mathbb{R}^d$, and substitute this into Y_t. The random process

$$e^{i\xi\cdot\hat{B}_t} + \frac{|\xi|^2}{2}\int_0^t e^{i\xi\cdot\hat{B}_s}ds, \quad 0 \le t \le T,$$

is an $(\mathscr{F}_t)_{0\le t\le T}$-martingale on $(\mathscr{X}, \mathscr{F}_T, \hat{P}_T)$. Since $\frac{d}{dt}\mathbb{E}_{\hat{P}_T}[Y_t|\mathscr{F}_s] = \frac{d}{dt}Y_s = 0$, we have

$$\frac{d}{dt}\mathbb{E}_{\hat{P}_T}[e^{i\xi\cdot\hat{B}_t}|\mathscr{F}_s] = -\frac{|\xi|^2}{2}\mathbb{E}_{\hat{P}_T}[e^{i\xi\cdot\hat{B}_t}|\mathscr{F}_s].$$

Solving this equation yields

$$\mathbb{E}_{\hat{P}_T}[e^{i\xi(\hat{B}_t-\hat{B}_s)}|\mathscr{F}_s] = e^{-|\xi|^2(t-s)/2}. \tag{2.4.94}$$

Thus $(\hat{B}_t^2 - t)_{0\le t\le T}$ is an $(\mathscr{F}_t)_{0\le t\le T}$-martingale, therefore $(\hat{B}_t)_{0\le t\le T}$ is a Brownian motion on $(\mathscr{X}, \mathscr{F}_T, \hat{P}_T)$ by Proposition 2.107, as claimed. □

We conclude by two applications of Girsanov's theorem.

Example 2.172 (Drift transformation). Consider the process $(X_t)_{t\ge 0}$ given by the solution of

$$dX_t = dB_t + a_t(X_t)dt. \tag{2.4.95}$$

By the Girsanov theorem, $\tilde{B}_t = B_t - \int_0^t a_s(B_s)ds$, $t \in [0,T]$, is a Brownian motion on $(\mathcal{X}, \mathcal{F}_T, e^{Z_T}\mathcal{W})$, where Z_T is given by (2.4.93). From the definition of $\tilde{B}_t$, we have

$$dB_t = d\tilde{B}_t + a_t(B_t)dt. \tag{2.4.96}$$

Put

$$d\tilde{Z}_t = a_t(B_t) \cdot dB_t - \frac{1}{2}a_t(B_t)^2 dt.$$

We regard B_t as a solution of the SDE (2.4.96) on the probability space $(\mathcal{X}, \mathcal{F}_T, e^{\tilde{Z}_T}\mathcal{W})$. Its generator is $\frac{1}{2}\Delta + a \cdot \nabla$, which coincides with that of the solution of (2.4.95) on $(\mathcal{X}, \mathcal{B}(\mathcal{X}), \mathcal{W})$. Therefore,

$$\mathbb{E}[f(X_t)] = \mathbb{E}[f(B_t)e^{\tilde{Z}_T}].$$

Let Q be the distribution of $\mathcal{W}$ under the random variable $\mathcal{X} \ni \omega \to X_\cdot(\omega) \in \mathcal{X}$. Then

$$\frac{dQ}{d\mathcal{W}} = \exp\left(\int_0^T a_s(B_s) \cdot dB_s - \frac{1}{2}\int_0^T a_s(B_s)^2 ds\right).$$

Example 2.173 (Cameron–Martin formula). Here we consider the effect of pathwise shifts applied to Brownian motion. Let

$$dY_t = dB_t + a_t dt, \quad d\bar{Z}_t = a_t dB_t - \frac{1}{2}a_t^2 dt.$$

Note that a_t is independent of the path $\omega \in \mathcal{X}$ and

$$Y_t(\omega) = B_t(\omega) + \int_0^t a_s ds = \omega(t) + \int_0^t a_s ds$$

is a shift of ω by the path $\int_0^\cdot a_s ds$. Since $\bar{B}_t = B_t - \int_0^t a_s ds$ is a Brownian motion on $(\mathcal{X}, \mathcal{F}_T, e^{\bar{Z}_T}\mathcal{W})$ by the Girsanov theorem, we have

$$\mathbb{E}[f(Y_t)] = \mathbb{E}\left[f\left(\bar{B}_t + \int_0^t a_s ds\right)e^{\bar{Z}_T}\right] = \mathbb{E}[f(B_t)e^{\bar{Z}_T}].$$

Here we used that $\int_0^t a_s ds$ is independent of path ω. Hence

$$\mathbb{E}[f(Y_t)] = \mathbb{E}[f(B_t)e^{\bar{Z}_T}] \tag{2.4.97}$$

is obtained. Let Q be the distribution of $\mathcal{W}$ under the random variable $\mathscr{X} \in \omega \to Y_{\cdot}(\omega) \in \mathscr{X}$ on $(\mathscr{X}, \mathcal{B}(\mathscr{X}))$. By (2.4.97), we then have

$$\frac{dQ}{d\mathcal{W}} = \exp\left(\int_0^t a_s \cdot dB_s - \frac{1}{2} \int_0^t a_s^2 ds \right), \tag{2.4.98}$$

known as the *Cameron–Martin formula.*

3 Lévy processes

3.1 Lévy processes and the Lévy–Khintchine formula

3.1.1 Infinitely divisible random variables

Brownian motion is a specific case of a wider class of random processes which itself is of interest for our purposes. It is a simple fact of elementary probability that the distributions of sums of independent, identically distributed random variables are obtained as convolutions of their distributions. A natural question is then whether one can break up a given random process into sums of others, simpler or canonical ones. This leads to considering the following concept.

Definition 3.1 (Infinite divisibility). An $\mathbb{R}^d$-valued random variable X on a probability space $(\Omega, \mathcal{F}, P)$ is *infinitely divisible* if for every $n \in \mathbb{N}$ there exist independent, identically distributed random variables $X_1^{(n)}, \dots, X_n^{(n)}$ such that

$$X \stackrel{\mathrm{d}}{=} X_1^{(n)} + \cdots + X_n^{(n)}. \tag{3.1.1}$$

Equivalently, the probability distribution μ on $\mathbb{R}^d$ of X is infinitely divisible if there exists a probability distribution μ_n common to all $X_j^{(n)}$ such that

$$\mu = \underbrace{\mu_n * \cdots * \mu_n}_{n\text{-fold}}. \tag{3.1.2}$$

Recall that the convolution of distributions μ_1 and μ_2 on $\mathbb{R}^d$ is defined by the formula

$$(\mu_1 * \mu_2)(A) = \int_{\mathbb{R}^d \times \mathbb{R}^d} 1_A(x+y)\mu_1(dx)\mu_2(dy)$$

for $A \in \mathcal{B}(\mathbb{R}^d)$. In terms of the characteristic function $\hat{\mu}$ of μ, i. e.,

$$\hat{\mu}(u) = \mathbb{E}[e^{iu\cdot X}] = \int_{\mathbb{R}^d} e^{iu\cdot x}\mu(dx), \quad u \in \mathbb{R}^d,$$

we can alternatively say that a random variable is infinitely divisible if and only if for every $n \in \mathbb{N}$ there is a characteristic function $\hat{\mu}_n$ satisfying

$$\hat{\mu}(u) = (\hat{\mu}_n(u))^n. \tag{3.1.3}$$

Example 3.2. Examples of infinitely divisible random variables include the following:

(1) *Gaussian distribution.* Take a Gaussian random variable $X \stackrel{\mathrm{d}}{=} N(m, \sigma^2)$ over $\mathbb{R}$ with probability distribution

https://doi.org/10.1515/9783112231463-003

$$\mu(dx) = \frac{1}{(2\pi\sigma^2)^{1/2}} \exp\left(-\frac{(x-m)^2}{2\sigma^2}\right) dx.$$

Since by direct computation

$$\hat{\mu}(u) = e^{ium-(1/2)u^2\sigma^2} = \left(\exp\left(iu\frac{m}{n} - \frac{1}{2}u^2\left(\frac{\sigma}{\sqrt{n}}\right)^2\right)\right)^n,$$

we have $X_1^{(n)}, \ldots, X_n^{(n)} \stackrel{d}{=} N\left(m/n, (\sigma/\sqrt{n})^2\right)$.

(2) *Poisson distribution.* For a Poisson random variable $X \stackrel{d}{=} \mathrm{Poi}(\lambda)$ with distribution

$$\mu(\{k\}) = e^{-\lambda}\frac{\lambda^k}{k!}, \quad k \in \mathbb{N} \cup \{0\},$$

we have

$$\hat{\mu}(u) = \exp\left(\lambda(e^{iu} - 1)\right) = \left(\exp\left(\frac{\lambda}{n}\left(e^{iu} - 1\right)\right)\right)^n,$$

and we can identify $X_1^{(n)}, \ldots, X_n^{(n)} \stackrel{d}{=} \mathrm{Poi}\,(\lambda/n)$.

(3) *Compound Poisson distribution.* Let $(Z_n)_{n\in\mathbb{N}}$ be independent and identically distributed random variables, and $N \stackrel{d}{=} \mathrm{Poi}(\lambda)$ a Poisson random variable. Then

$$Y = \sum_{n=1}^{N} Z_n$$

is called a *compound Poisson random variable.* Its characteristic function is

$$\hat{\mu}(u) = \exp(\lambda(\hat{\sigma}(u) - 1)),$$

where $\hat{\sigma}(u)$ denotes the characteristic function of Z_1. Since we have

$$\hat{\mu}(u) = \left(\exp\left(\frac{\lambda}{n}(\hat{\sigma}(u) - 1)\right)\right)^n,$$

Y is an infinitely divisible random variable.

(4) *Cauchy distribution.* Let $X \stackrel{d}{=} C(\alpha, \beta)$, $\alpha > 0$, $\beta \in \mathbb{R}^d$, be a random variable with the Cauchy distribution on $\mathbb{R}^d$,

$$\mu(dx) = \frac{\Gamma((d+1)/2)}{\pi^{(d+1)/2}} \frac{\alpha}{(|x-\beta|^2 + \alpha^2)^{(d+1)/2}} dx.$$

Then

$$\hat{\mu}(u) = \exp\left(-\alpha|u| + i\beta \cdot u\right) = \left(\exp\left(-\frac{\alpha}{n}|u| + i\frac{\beta}{n} \cdot u\right)\right)^n.$$

Thus we can identify $X_1^{(n)}, \ldots, X_n^{(n)} \overset{\mathrm{d}}{=} C(\alpha/n, \beta/n)$.

Two important counterexamples are the uniform and binomial distributions.

Let μ be a probability measure on $\mathbb{R}^d$. The characteristic function of μ has the properties (1) $\hat{\mu}(0) = 1$ and $|\hat{\mu}(u)| \leq 1$; (2) $\hat{\mu}$ is uniformly continuous; (3) $\hat{\mu}$ is nonnegative definite, i. e.,

$$\sum_{k,l=1}^{n} \hat{\mu}(u_k - u_l) z_k \bar{z}_l \geq 0 \tag{3.1.4}$$

for $u_j \in \mathbb{R}^d$, $z_i \in \mathbb{C}$, $i, j = 1, \ldots, n$, $n \geq 1$. A remarkable fact is that the converse of this statement is also true: If a function $\varphi : \mathbb{R}^d \to \mathbb{C}$ satisfies properties (1)–(3) above, then there exists a probability measure μ on $\mathbb{R}^d$ such that the characteristic function of μ is $\hat{\mu} = \varphi$. This is *Bochner's theorem.*

In order to see the relationship between a measure and its characteristic function, next we discuss the Stone–Weierstrass theorem. Let $\mathbb{K} = \mathbb{R}$ or $\mathbb{K} = \mathbb{C}$, and (E, d) be a metric space. Also, let $C_b(E; \mathbb{K})$ denote the Banach space of bounded continuous, $\mathbb{K}$-valued functions on E, equipped with the supremum norm.

A subset $\mathscr{C} \subset C_b(E; \mathbb{K})$ is called an *algebra* if
(1) $1 \in \mathscr{C}$;
(2) $f, g \in \mathscr{C}$ implies that $f + g, fg \in \mathscr{C}$;
(3) if $a \in \mathbb{K}$ and $f \in \mathscr{C}$, then $af \in \mathscr{C}$.

We say that $\mathscr{C}$ *separates points* if for any two points $x, y \in E$ with $x \neq y$, there is an $f \in \mathscr{C}$ such that $f(x) \neq f(y)$.

Proposition 3.3 (Stone–Weierstrass theorem). *Let E be a compact Hausdorff space, and $\mathscr{C} \subset C_b(E; \mathbb{K})$ an algebra that separates points. In addition, in the case of $\mathbb{K} = \mathbb{C}$ we suppose that $\mathscr{C}$ is closed under complex conjugation, i. e., $f \in \mathscr{C}$ implies $\bar{f} \in \mathscr{C}$. Then $\mathscr{C}$ is dense in $C_b(E; \mathbb{K})$ with respect to the supremum norm $\|\cdot\|_\infty$.*

Proof. We proceed through several steps.

Step 1. Let $\mathbb{K} = \mathbb{R}$. Take a sequence of polynomials $p_n(t)$ approximating the function $\sqrt{t}$ on $[0, 1]$, uniformly as $n \to \infty$. If $f \in \mathscr{C}$, then we have

$$|f| = \|f\|_\infty \times \lim_{n\to\infty} p_n(f^2/\|f\|_\infty^2).$$

Hence $|f| \in \bar{\mathscr{C}}$.

Step 2. By Step 1, we also have $f \wedge g = \frac{1}{2}(f + g + |f - g|) \in \bar{\mathscr{C}}$.

Step 3. For every $f \in C_b(E; \mathbb{R})$, $x \in E$, and $\varepsilon > 0$, there exists $g_x \in \bar{\mathscr{C}}$ such that

$$g_x(x) = f(x), \quad g_x(y) \leq f(y) + \varepsilon \quad \text{for all } y \in E. \tag{3.1.5}$$

The function g_x above is constructed as follows. Since $\mathscr{C}$ separates points, for every $z \in E \setminus \{x\}$ there exists $H_z \in \mathscr{C}$ with $H_z(z) \neq H_z(x) = 0$. For such z, define $h_z \in \mathscr{C}$ by

$$h_z(y) = f(x) + \frac{f(z) - f(x)}{H_z(z)} H_z(y)$$

for all $y \in E$. It follows that $h_z(x) = f(x)$ and $h_z(z) = f(z)$, for all $z \in E$. Since f and h_z are continuous, for every z there exists an open set $U_z \ni z$ such that $h_z(y) \leq f(y) + \varepsilon$, for all $y \in U_z$. Since E is compact, there exists a finite covering $\{U_{z_1}, \dots, U_{z_n}\}$ of E consisting of such neighborhoods and we define $g_x = \min\{h_{z_1}, \dots, h_{z_n}\}$. By Step 2, we have $g_x \in \bar{\mathscr{C}}$ and (3.1.5) is satisfied.

Step 4. Let $f \in C_b(E; \mathbb{R})$, $\varepsilon > 0$, $x \in E$, and g_x be as in Step 3. Since f and g_x are continuous, for every $x \in E$ there exists an open set $V_x \ni x$ such that $g_x(y) > f(y) - \varepsilon$ for $y \in V_x$. Similarly to Step 3, we take a finite covering $\{V_{x_1}, \dots, V_{x_n}\}$ of E and define $g = \max\{g_{x_1}, \dots, g_{x_n}\}$. We have $g \in \mathscr{C}$ and $\|g - f\|_\infty < \varepsilon$ by construction. Thus we obtain $\bar{\mathscr{C}} = C_b(E; \mathbb{R})$.

Step 5. Let $\mathbb{K} = \mathbb{C}$. If $f \in \mathscr{C}$, then $\Re f$ and $\Im f$ are in $\mathscr{C}$ by closedness of complex conjugation. In particular, $\mathscr{C}_0 = \{\Re f \mid f \in \mathscr{C}\} \subset \mathscr{C}$ is a real algebra that separates points. Hence $\mathscr{C}_0$ is dense in $C_b(E; \mathbb{R})$ by Step 4. Since $\mathscr{C} = \mathscr{C}_0 + i\mathscr{C}_0$, the theorem follows. □

Lemma 3.4. *A finite measure μ on $(\mathbb{R}^d, \mathcal{B}(\mathbb{R}^d))$ is uniquely described by its characteristic function $\hat{\mu}$, i. e., $\hat{\mu}_1 = \hat{\mu}_2$ if and only if $\mu_1 = \mu_2$.*

Proof. For $u \in \mathbb{R}^d$, we define $f_u : \mathbb{R}^d \to \mathbb{C}$ by $f_u(y) = e^{iu \cdot y}$. It is direct to see that $\mathscr{C} =$ linear hull of $\{f_u \mid u \in \mathbb{R}^d\}$ is an algebra that separates points and that is closed under complex conjugation. Note, however, that $\mathbb{R}^d$ is a Hausdorff space but it is not compact, and therefore the Stone–Weierstrass theorem cannot be applied directly. Let μ_1, μ_2 be finite measures such that $\hat{\mu}_1(u) = \hat{\mu}_2(u)$, for all $u \in \mathbb{R}^d$. Choose $f \in C_b(\mathbb{R}^d)$ and $\varepsilon > 0$ be fixed, and $N \in \mathbb{N}$. Suppose that

$$(1 + 2\|f\|_\infty)(\mu_1 + \mu_2)(\mathbb{R}^d \setminus [-N, N]^d) < \frac{\varepsilon}{2}.$$

Let $\mathscr{C}'$ be the algebra of the linear hull of $\{f_{2\pi m} \mid m \in \mathbb{Z}^d\}$, and consider the subset of functions $\mathscr{C}_N = \{g\lceil_{[-N,N]^d} \mid g \in \mathscr{C}'\}$. The algebra $\mathscr{C}_N$ is closed under complex conjugation. Hence the Stone–Weierstrass theorem yields that $\mathscr{C}_N$ is dense in $C_b([-N, N]^d; \mathbb{C})$. Thus there exists $g \in \mathscr{C}'$ such that

$$\delta = \sup_{x \in [-N,N]^d} |g(x) - f(x)| < 1 \wedge \frac{\varepsilon}{2(\mu_1 + \mu_2)(\mathbb{R}^d)}.$$

Since $g(x) = g(x - kN)$ for all $k \in \mathbb{Z}^d$, we have $\|g\|_\infty = \sup_{x \in [-N,N]^d} |g(x)| \leq 1 + \|f\|_\infty$ and

$$\|g - f\|_\infty \leq \|f\|_\infty + \|g\|_\infty \leq 1 + 2\|f\|_\infty.$$

Thus we have

$$\left| \int_{\mathbb{R}^d} f d\mu_1 - \int_{\mathbb{R}^d} f d\mu_2 \right| \leq \|f - g\|_\infty (\mu_1 + \mu_2) \left(\mathbb{R}^d \setminus [-N, N]^d\right) + \delta (\mu_1 + \mu_2) (\mathbb{R}^d) \leq \frac{\varepsilon}{2} + \frac{\varepsilon}{2} = \varepsilon$$

and $\int_{\mathbb{R}^d} f d\mu_1 = \int_{\mathbb{R}^d} f d\mu_2$ for $f \in C_b(\mathbb{R}^d)$. Let $A \subset \mathbb{R}^d$ be closed. Define $\rho_\varepsilon : \mathbb{R}^d \to [0,1]$ by $\rho_\varepsilon(x) = 1 - \varphi(d(x,A)/\varepsilon)$, where $d(x,A)$ denotes the distance between x and A, and $\varphi : \mathbb{R} \to [0,1]$ by $\varphi(t) = (t \vee 0) \wedge 1$. Thus $\rho_\varepsilon(x) = 1$ for $x \in A$, and $\rho_\varepsilon(x) = 0$ for x such that $d(x,A) \geq \varepsilon$. Hence we have $\int_{\mathbb{R}^d} \rho_\varepsilon(x)\mu_1(dx) = \int_{\mathbb{R}^d} \rho_\varepsilon(x)\mu_2(dx)$. Taking the limit as $\varepsilon \downarrow 0$ on both sides, we have $\mu_1(A) = \mu_2(A)$, and thus $\mu_1 = \mu_2$ follows. □

Next we discuss some technicalities concerning characteristic functions.

Lemma 3.5. *Let μ and $(\mu_n)_{n\in\mathbb{N}}$ be probability measures on $\mathbb{R}^d$, and write $\hat{\mu}$ and $(\hat{\mu}_n)_{n\in\mathbb{N}}$ for their characteristic functions, respectively. Then we have the following:*

(1) *We have $|\hat{\mu}(u)|^2 = \widehat{\mu * \bar{\mu}}(u)$, where $\hat{\bar{\mu}}(u) = \hat{\mu}(-u)$. In particular, $|\hat{\mu}(u)|^2$ is also a characteristic function of the probability measure $\mu * \bar{\mu}$.*
(2) *If $\mu_n \to \mu$ weakly as $n \to \infty$, then $\hat{\mu}_n(u) \to \hat{\mu}(u)$ uniformly on any compact set as $n \to \infty$.*
(3) *Suppose that $\hat{\mu}_n(u) \to \phi(u)$ as $n \to \infty$ for every u, and $\phi(u)$ is continuous at $u = 0$. Then $\phi(u)$ is a characteristic function of a probability measure μ, and μ_n weakly converges to μ as $n \to \infty$.*

Proof. Statements (1) and (2) are straightforward. We prove (3). By Bochner's theorem, it is sufficient to show that $\phi(u)$ is uniformly continuous. For every $s, t \in \mathbb{R}$, we have

$$|\phi(t) - \phi(s)|^2 \leq 4\phi(0)|\phi(0) - \phi(t - s)|. \tag{3.1.6}$$

To see this, note that $(\phi(t_i - t_j))_{1\leq i,j\leq n}$ is a Hermitian positive definite matrix. In particular, choosing $(t_1, t_2, t_3) = (t, s, 0)$ gives that the matrix

$$M = \begin{pmatrix} \phi(0) & \phi(t-s) & \phi(t) \\ \overline{\phi(t-s)} & \phi(0) & \phi(s) \\ \overline{\phi(t)} & \overline{\phi(s)} & \phi(0) \end{pmatrix}$$

has a nonnegative determinant. From $|\phi(t)| \leq \phi(0)$ and $|\phi(s)| \leq \phi(0)$, we have

$$\det M \leq \phi(0)^3 - \phi(0)(|\phi(t) - \phi(s)|^2 + |\phi(t-s)|^2) + 2\phi(0)^2|\phi(0) - \phi(t-s)|.$$

Rearranging terms, we then obtain

$$|\phi(t) - \phi(s)|^2 \leq \phi(0)^2 - |\phi(t-s)|^2 + 2\phi(0)|\phi(0) - \phi(t-s)| \leq 4\phi(0)|\phi(0) - \phi(t-s)|.$$

This proves (3.1.6). Thus continuity at $u = 0$ implies uniform continuity of $\phi(u)$. Hence there exists a probability measure μ such that $\hat{\mu}(u) = \phi(u)$.

Next we show that $\mu_n \to \mu$. Let $\varepsilon > 0$. We claim that the sequence $(\mu_n)_{n\in\mathbb{N}}$ is tight, i. e., there exists a compact set $K \subset \mathbb{R}^d$ such that $\mu_n(K) > 1-\varepsilon$ for sufficiently large n. We see that

$$\begin{aligned}\int_{\mathbb{R}^d} 1_{\{|y|\geq 2/t\}}\mu_n(dy) &= 2\int_{\mathbb{R}^d} 1_{\{|y|\geq 2/t\}}\frac{1}{2}\mu_n(dy) \leq 2\int_{\mathbb{R}^d} 1_{\{|y|\geq 2/t\}}\left(1-\frac{1}{t|y|}\right)\mu_n(dy)\\ &\leq 2\int_{\mathbb{R}^d} 1_{\{|y|\geq 2/t\}}\left(1-\frac{\sin y}{t|y|}\right)\mu_n(dy) \leq 2\int_{\mathbb{R}^d}\left(1-\frac{\sin y}{t|y|}\right)\mu_n(dy)\\ &= \int_{\mathbb{R}^d}\mu_n(dy)\int_{-t}^{t}\frac{1-e^{i\xi y}}{t}d\xi = \frac{1}{t}\int_{-t}^{t}(1-\phi_n(\xi))d\xi.\end{aligned}$$

Thus $\limsup_{n\to\infty}\int_{\mathbb{R}^d} 1_{\{|y|\geq 2/t\}}\mu_n(dy) \leq \frac{1}{t}\int_{-t}^{t}(1-\phi(\xi))d\xi \leq \varepsilon/2$ for sufficiently small $t > 0$. This gives $\mu_n(\{y \in \mathbb{R}^d \mid |y| > 2/t\}) < \varepsilon$, for small enough t and sufficiently large n, implying that $\mu_n(\{y \in \mathbb{R}^d \mid |y| \leq 2/t\}) > 1-\varepsilon$. Thus $(\mu_n)_{n\in\mathbb{N}}$ is tight, and thus there exists a subsequence $(\mu'_n)_{n\in\mathbb{N}}$ weakly convergent to a probability measure μ', and $\mu' = \mu$. Thus every subsequence $(\mu'_n)_{n\in\mathbb{N}}$ of $(\mu_n)_{n\in\mathbb{N}}$ contains a subsequence $(\mu''_n)_{n\in\mathbb{N}}$ such that $\mu_{n''}$ weakly converges to μ. Hence μ_n weakly converges to μ. □

Lemma 3.6. *If μ is an infinitely divisible probability measure, then $\hat{\mu}$ has no zeroes.*

Proof. For every $n \in \mathbb{N}$, there exists a characteristic function $\hat{\mu}_n$ such that $\hat{\mu}(u) = \hat{\mu}_n(u)^n$. We have $|\hat{\mu}_n(u)|^2 = |\hat{\mu}(u)|^{2/n}$, which implies that

$$\lim_{n\to\infty}|\hat{\mu}_n(n)|^2 = \phi(u) = \begin{cases}1, & \hat{\mu}(u) \neq 0,\\ 0, & \hat{\mu}(u) = 0.\end{cases}$$

Since $\hat{\mu}(0) = 1$ and $\hat{\mu}(u)$ is continuous, $\phi(u) = 1$ in a neighborhood of $u = 0$. We see that $|\hat{\mu}_n(u)|^2$ is a characteristic function and the limit $\phi(u)$ is also a characteristic function by Lemma 3.5. Hence ϕ is continuous and $\phi(u) = 1$, for all $u \in \mathbb{R}^d$. Hence there is no point u such that $\hat{\mu}(u) = 0$. □

More generally, it can be seen that for a continuous function $c : \mathbb{R}^d \to \mathbb{C}$ such that $c(0) = 1$ and c has no zero, there exists a unique continuous function $f : \mathbb{R}^d \to \mathbb{C}$ such that $c(u) = \exp(f(u))$ and $f(0) = 0$. Moreover, for every positive integer k there exists a unique continuous function g_k such that $g_k(0) = 1$ and $c(u) = g_k(u)^k$. Also, it can be shown that $g_k(u) = \exp(f(u)/k)$. In particular, if $c(u)$ is a characteristic function and so it has no zero, by Bochner's theorem it follows that $g_k(u)$ is also a characteristic function. Let μ be an infinitely divisible probability measure. Then $\hat{\mu}$ has no zero and it has the representation $\hat{\mu}(u) = e^{\eta(u)}$. Thus $\log\hat{\mu}$ and $\hat{\mu}(u)^{1/k}$ are defined by $\eta(u)$ and $e^{\eta(u)/k}$, respectively. Furthermore, the probability measure with characteristic function $\hat{\mu}(u)^{1/k}$ will be denoted by $\mu^{1/k*}$, i. e., $\widehat{\mu^{1/k*}} = \hat{\mu}^{1/k}$.

Lemma 3.7. *Let $(\mu_n)_{n\in\mathbb{N}}$ be a sequence of infinitely divisible probability measures, and $\mu_n \to \mu$ weakly as $n \to \infty$. Then μ is also an infinitely divisible probability measure.*

Proof. Since $|\hat{\mu}_n(u)|^{2/k} \to |\hat{\mu}(u)|^{2/k}$ and $|\hat{\mu}_n(u)|^{2/k} = |\hat{\mu}_n(u)^{1/k}|^2$, both $|\hat{\mu}_n(u)|^{2/k}$ and $|\hat{\mu}(u)|^{2/k}$ are characteristic functions. Thus $\hat{\mu}(u) \neq 0$ and $\hat{\mu}(u)^{1/k}$ can be defined. Furthermore, $\hat{\mu}_n(u)^{1/k} \to \hat{\mu}(u)^{1/k}$ uniformly on every compact set. In particular, $\hat{\mu}(u)^{1/k}$ is continuous in a neighborhood of $u = 0$. Thus $\hat{\mu}(u)^{1/k}$ is also a characteristic function and $\hat{\mu}$ is infinitely divisible. □

Theorem 3.8. *Let μ be an infinitely divisible probability measure and $t \geq 0$. Then μ^{t*} is well defined and is an infinitely divisible probability measure.*

Proof. Let $t = 1/n$ with $n \in \mathbb{N}$. It is trivial to see that $\hat{\mu}(u)^{1/n} = (\hat{\mu}(u)^{1/kn})^k$ and $\hat{\mu}^{1/kn}$ is a characteristic function. Thus $\mu^{1/n*}$ is infinitely divisible, and so is $\mu^{m/n*} = \underbrace{\mu^{1/n*} * \cdots * \mu^{1/n*}}_{m}$, for all $n, m \in \mathbb{N}$. There exists $r_n \in \mathbb{Q}$ such that $r_n \to t$ as $n \to \infty$. Hence $\hat{\mu}(u)^{r_n} \to \hat{\mu}(u)^t$ uniformly on every compact set and $\hat{\mu}(u)^t$ is continuous in $u = 0$. Thus $\hat{\mu}(u)^t$ is a characteristic function of μ^{t*} and $\mu^{r_n*} \to \mu^{t*}$ weakly by (3) of Lemma 3.5. By Lemma 3.7, it follows that μ^{t*} is infinitely divisible. □

3.1.2 Lévy–Khintchine formula

Next we will see that there is a general formula describing the characteristic function of any infinitely divisible random variable.

Definition 3.9 (Lévy measure). A σ-finite Borel measure ν on $\mathcal{B}(\mathbb{R}^d \setminus \{0\})$ satisfying

$$\int_{\mathbb{R}^d\setminus\{0\}} (|y|^2 \wedge 1)\nu(dy) < \infty \tag{3.1.7}$$

is called a *Lévy measure*.

An infinitely divisible distribution is uniquely characterized by three parameters called *Lévy triplet*, as given by the following fundamental result.

Theorem 3.10 (Lévy–Khintchine formula).
(1) *If a probability distribution μ on the measurable space $(\mathbb{R}^d, \mathcal{B}(\mathbb{R}^d))$ is infinitely divisible, then there exist a vector $b \in \mathbb{R}^d$, a nonnegative definite symmetric $d \times d$ matrix A, and a Lévy measure ν such that $\hat{\mu}(u) = e^{\eta(u)}$ for all $u \in \mathbb{R}^d$, where*

$$\eta(u) = ib \cdot u - \frac{1}{2}u \cdot Au + \int_{\mathbb{R}^d\setminus\{0\}} (e^{iu\cdot y} - 1 - iu \cdot y 1_{\{|y|<1\}})\, \nu(dy). \tag{3.1.8}$$

(2) *The triplet* (b, A, ν) *is uniquely given.*
(3) *Conversely, given a triplet* (b, A, ν) *with the above properties, there exists an infinitely divisible probability distribution* μ *on* $\mathbb{R}^d$*, whose characteristic function is* $\hat{\mu} = e^{\eta(u)}$.

Remark 3.11. The integrand in (3.1.8) satisfies

$$|e^{iu\cdot y} - 1 - iu \cdot y 1_{\{|y|<1\}}| \le \begin{cases} \frac{1}{2}|u|^2|y|^2 + O(|y|^3), & |y| < 1, \\ 2, & |y| \ge 1, \end{cases}$$

ensuring that the integral is well defined. Furthermore, if $\int_{\mathbb{R}^d\setminus\{0\}} |y| 1_{\{|y|<1\}} \nu(dy)$ is bounded, then (3.1.8) becomes

$$\eta(u) = ib' \cdot u - \frac{1}{2} u \cdot Au + \int_{\mathbb{R}^d\setminus\{0\}} (e^{iu\cdot y} - 1)\nu(dy)$$

with $b' = b - \int_{\{|y|<1\}} y\nu(dy)$.

Remark 3.12. L et $c : \mathbb{R}^d \to \mathbb{R}$ be a bounded continuous function with the properties $\lim_{|y|\to 0}(c(y) - 1)/|y| = 0$ and $\lim_{|y|\to\infty} |y|c(y) < \infty$. Then (3.1.8) is represented as

$$\eta(u) = ib_c \cdot u - \frac{1}{2} u \cdot Au + \int_{\mathbb{R}^d\setminus\{0\}} (e^{iu\cdot y} - 1 - iu \cdot y\, c(y))\, \nu(dy), \tag{3.1.9}$$

where $b_c = b + \int_{\mathbb{R}^d} y(c(y) - 1_{\{|y|<1\}})\nu(dy)$.

We carry on now with the proof of Theorem 3.10.

Proof. We prove (2) and (3). Suppose that $\eta(u)$ is given by (3.1.8). Let $s \in \mathbb{R}$. It is direct to see that $\lim_{s\to\infty} s^{-2}\eta(su) = -u \cdot Au/2$. Thus A is uniquely determined by η. Let

$$\psi(u) = ib \cdot u + \int_{\mathbb{R}^d\setminus\{0\}} (e^{iu\cdot y} - 1 - iu \cdot y 1_{\{|y|<1\}})\, \nu(dy).$$

Hence we have

$$\psi(u) - \psi(u + w) = \int_{\mathbb{R}^d\setminus\{0\}} (e^{iu\cdot y} - e^{i(u+w)\cdot y} + iw \cdot y 1_{\{|y|<1\}})\, \nu(dy) - ib \cdot w.$$

By Fubini's theorem, we have

$$\int_{[-1,1]^d} (\psi(u) - \psi(u + w))dw = \int_{\mathbb{R}^d\setminus\{0\}} e^{iu\cdot y}\rho(dy),$$

where

$$\rho(dy) = 2^d \left(1 - \prod_{i=1}^{d} \frac{\sin y_i}{y_i}\right) \nu(dy).$$

Here it is understood that $(\sin y_i)/y_i = 1$ for $y_i = 0$. Since $\rho(dy) \le a(|y|^2 \wedge 1)\nu(dy)$ with a constant a, we have that ρ is a finite measure on $\mathbb{R}^d$. We note that $\int_{\mathbb{R}^d\setminus\{0\}} e^{iu\cdot y}\rho(dy)$ is the Fourier transform of $\rho(dy)$. Thus ρ is uniquely determined by η due to Lemma 3.4, and also ν is uniquely determined. Uniqueness of b follows from the uniqueness of A and ν. Hence (2) is proven. To show (3), suppose that there exists a triplet (b, A, ν). We denote the right-hand side of (3.1.8) by $\phi(u)$. For $n > 0$, we define

$$\phi_n(u) = ib \cdot u - \frac{1}{2} u \cdot Au + \int_{1/n \le |y| < 1} (e^{iu\cdot y} - 1 - iu \cdot y 1_{\{|y|<1\}})\, \nu(dy).$$

There exists b_n such that

$$\phi_n(u) = ib_n \cdot u - \frac{1}{2} u \cdot Au + \int_{1/n \le |y|} (e^{iu\cdot y} - 1)\, \nu(dy)$$

and ν is a finite measure on $|y| \ge 1/n$. Note that it is not necessary that ν is a finite measure on $(0, 1)$. We have

$$\int_{1/n \le |y|} (e^{iu\cdot y} - 1)\, \nu(dy) = c(\hat{m}(u) - 1),$$

where $c = \int_{1/n \le |y|} \nu(dy)$ and $\hat{m}(u) = \int_{1/n \le |y|} e^{iu\cdot y}\, \nu(dy)/c$. Then $e^{c(\hat{m}(u)-1)}$ is the characteristic function of a compound Poisson process, and it is infinitely divisible. Since a Gaussian distribution is infinitely divisible, $e^{\phi_n(u)}$ is also the characteristic function of an infinitely divisible probability measure. We see that $e^{\phi_n(u)} \to e^{\eta(u)}$ as $n \to \infty$, and $e^{\eta(u)}$ is continuous at $u = 0$. Hence $e^{\eta(u)}$ is also the characteristic function of an infinitely divisible probability measure. Finally, we prove (1). Let μ be an infinitely divisible probability measure on $\mathbb{R}^d$. By Lemma 3.8, there exists a probability measure $\mu^{1/n*}$ such that $\widehat{\mu^{1/n*}}(u) = \hat{\mu}(u)^{1/n}$. Define μ_n by

$$\hat{\mu}_n(u) = \exp(n(\hat{\mu}(u)^{1/n} - 1)) = \exp\left(n \int_{\mathbb{R}^d\setminus\{0\}} (e^{iu\cdot y} - 1)\mu^{1/n*}(dy) \right).$$

This $\hat{\mu}_n(u)$ is the characteristic function of a compound Poisson process, thus infinitely divisible. Note that

$$\hat{\mu}_n(u) = \exp(n(e^{(1/n)\log \hat{\mu}(u)} - 1)) = \exp(n(n^{-1} \log \hat{\mu}(u) + o(n^{-2}))).$$

It follows that $\lim_{n\to\infty}\hat{\mu}_n(u) = \lim_{n\to\infty}\exp(n(n^{-1}\log\hat{\mu}(u) + o(n^{-2}))) = \hat{\mu}(u)$. To conclude, (1) follows from Lemma 3.13. □

Lemma 3.13. *Suppose that $(\mu_n)_{n\in\mathbb{N}}$ are infinitely divisible probability measures on $\mathbb{R}^d$ and $\hat{\mu}_n$ has a Lévy–Khintchine representation with triplet (b_n, A_n, ν_n) for every $n \in \mathbb{N}$. Let μ be a probability measure on $\mathbb{R}^d$. Then $\mu_n \to \mu$ weakly as $n \to \infty$ if and only if μ is infinitely divisible and the triplet (b, A, ν) associated with $\hat{\mu}$ satisfies the following:*

(1) *If $f \in C_\infty(\mathbb{R}^d)$, then $\lim_{n\to\infty}\int_{\mathbb{R}^d} f(y)\nu_n(dy) = \int_{\mathbb{R}^d} f(y)\nu(dy)$.*
(2) *Let the symmetric nonnegative definite matrices $A_{n,\varepsilon}$ be defined through*

$$u\cdot A_{n,\varepsilon}u = u\cdot A_n u + \int_{|y|\le\varepsilon}(u\cdot y)^2\nu_n(dy).$$

Then $\lim_{\varepsilon\downarrow 0}\limsup_{n\to\infty}|u\cdot A_{n,\varepsilon}u - u\cdot Au| = 0$ for all $u \in \mathbb{R}^d$.
(3) *The limit $\lim_{n\to\infty} b_n = b$ holds.*

Proof. We use representation (3.1.9). Suppose that $\mu_n \to \mu$ weakly. μ is also infinitely divisible and $\log\hat{\mu}_n(u) \to \log\hat{\mu}(u)$ uniformly on every compact set. Let

$$g(u,y) = e^{iu\cdot y} - 1 - i(u\cdot y)c(y).$$

By the assumption, we have

$$\log\hat{\mu}_n(u) = ib_n\cdot u - \frac{1}{2}u\cdot A_n u + \int_{\mathbb{R}^d\setminus\{0\}} g(u,y)\nu_n(dy). \tag{3.1.10}$$

Let $\rho_n(dy) = (|y|^2\wedge 1)\nu_n(dy)$. We show that

$$\sup_{n\in\mathbb{N}}\rho_n(\mathbb{R}^d) < \infty, \tag{3.1.11}$$

$$\lim_{l\to\infty}\sup_{n\in\mathbb{N}}\int_{|y|>l}\rho_n(dy) = 0. \tag{3.1.12}$$

Consider $C(h) = [-h,h]^d$. We have

$$-\int_{C(h)}\log\hat{\mu}_n(u)du = \frac{2^{d-1}h^{d+2}}{3}\sum_{i=1}^d A_{ii}^{(n)} + (2h)^d\int_{\mathbb{R}^d}\left(1 - \prod_{i=1}^d\frac{\sin hx_i}{hx_i}\right)\nu_n(dx). \tag{3.1.13}$$

Here $A_n = (A_{ij}^{(n)})_{1\le i,j\le d}$. Taking the limit as $n\to\infty$ on both sides gives

$$-\int_{C(h)}\log\hat{\mu}(u)du \ge \lim_{n\to\infty}\frac{2^{d-1}h^{d+2}}{3}\sum_{i=1}^d A_{ii}^{(n)}$$

$$+ (2h)^d \inf_{x\in\mathbb{R}^d} \left\{ \left(1 - \prod_{i=1}^{d} \frac{\sin hx_i}{hx_i}\right) (|x|^2 \wedge 1)^{-1} \right\} \lim_{n\to\infty} \rho_n(\mathbb{R}^d). \qquad (3.1.14)$$

Hence (3.1.11) follows. We can also see that $\lim_{n\to\infty} \sum_{i=1}^{d} A_{ii}^{(n)} < \infty$.

Next take the limit as $h \downarrow 0$ in (3.1.14). Since $\log \hat{\mu}(0) = \log 1 = 0$ and $u \mapsto \hat{\mu}(u)$ is continuous at $u = 0$, for sufficiently small ε there exists $\delta > 0$ such that $|\log \hat{\mu}(u)| < \varepsilon$ for any u with $|u| < \delta$. We see that for $0 < \sqrt{d}h < \delta$, $\frac{-1}{(2h)^d} \int_{C(h)} \log \hat{\mu}(u)du < \varepsilon$. It follows that

$$\lim_{h\downarrow 0} \frac{-1}{(2h)^d} \int_{C(h)} \log \hat{\mu}(u)du = 0$$

and hence

$$\lim_{h\downarrow 0} \lim_{n\to\infty} \int_{\mathbb{R}^d} \left(1 - \prod_{i=1}^{d} \frac{\sin hx_i}{hx_i}\right) \nu_n(dx) = 0.$$

By (3.1.13), for every $\varepsilon > 0$ there exist n_0 and h_0 such that for all $n \geq n_0$,

$$\int_{\mathbb{R}^d} \left(1 - \prod_{i=1}^{d} \frac{\sin h_0 x_i}{h_0 x_i}\right) \nu_n(dx) < \varepsilon.$$

Noting that $1 - \prod_{i=1}^{d} \frac{\sin h_0 x_i}{h_0 x_i} > 1/2$ for $|x| > 2\sqrt{d}/h_0$, we get $\frac{1}{2} \int_{|x|>2\sqrt{d}/h_0} \nu_n(dx) < \varepsilon$ for all $n > n_0$, which implies $\frac{1}{2} \int_{|x|>2\sqrt{d}/h_0} \rho_n(dx) < \varepsilon$ for all $n > n_0$. Thus (3.1.12) follows. Together with Helly's second theorem, (3.1.11) and (3.1.12) imply that there exists a subsequence ρ_{n_k} weakly convergent to a probability measure ρ. Define the probability measure ν by $\nu(dx) = (|x|^2 \wedge 1)^{-1} \rho(dx)$ for $|x| > 0$ and $\nu(\{0\}) = 0$. We have

$$\log \hat{\mu}_n(u) = ib_n \cdot u - \frac{1}{2} u \cdot A_{n,\varepsilon} u + I_1 + I_2, \qquad (3.1.15)$$

where

$$I_1 = \int_{|y|\geq\varepsilon} g(u,y) \frac{\rho_n(dy)}{|y|^2 \wedge 1} \quad \text{and} \quad I_2 = \int_{|y|<\varepsilon} (g(u,y) + (u\cdot y)^2/2) \frac{\rho_n(dy)}{|y|^2 \wedge 1}.$$

We replace n with n_k and denote ρ_{n_k} by ρ_n. Define $E = \left\{\varepsilon > 0 \mid \int_{|x|=\varepsilon} \rho(dx) = 0\right\}$. We have

$$\lim_{n\to\infty} I_1 = \int_{|y|>\varepsilon} g(u,y) \frac{\rho(dy)}{|y|^2 \wedge 1} + \int_{|y|=\varepsilon} g(u,y) \frac{\rho(dy)}{|y|^2 \wedge 1} = \int_{|y|>\varepsilon} g(u,y) \frac{\rho(dy)}{|y|^2 \wedge 1} \qquad (3.1.16)$$

for $\varepsilon \in E$, and

$$\lim_{E\ni\varepsilon\downarrow 0} \lim_{n\to\infty} I_1 = \int_{\mathbb{R}^d\setminus\{0\}} g(u,y) \frac{\rho(dy)}{|y|^2 \wedge 1}. \qquad (3.1.17)$$

By (3.1.12), we also see that

$$\lim_{\varepsilon\downarrow 0}\sup_{n\in\mathbb{N}} |I_2| = 0. \tag{3.1.18}$$

By considering the real and the imaginary parts of (3.1.15), we have

$$\lim_{E\ni\varepsilon\downarrow 0}\limsup_{n\to\infty} u\cdot A_{n,\varepsilon}u = \lim_{E\ni\varepsilon\downarrow 0}\liminf_{n\to\infty} u\cdot A_{n,\varepsilon}u < \infty, \tag{3.1.19}$$

$$\limsup_{n\to\infty} b_n\cdot u = \liminf_{n\to\infty} b_n\cdot u < \infty. \tag{3.1.20}$$

Thus (3.1.19) is represented by $u\cdot Au$ with a nonnegative definite symmetric matrix A, and (3.1.20) by $b\cdot u$ with a suitable b. In (3.1.19), we may drop the restriction ε since $u\cdot A_{n,\varepsilon}u$ is monotone. Thus

$$\Re \log\hat{\mu}(u) = -u\cdot Au/2 + \Re\int_{\mathbb{R}^d\setminus\{0\}} g(u,y)\nu(dy),$$

$$\Im \log\hat{\mu}(u) = b\cdot u + \Im\int_{\mathbb{R}^d\setminus\{0\}} g(u,y)\nu(dy).$$

This means that $\hat{\mu}(u)$ is represented by (3.1.10). By monotonicity, we can remove $\lim_{E\ni\varepsilon\downarrow 0}$ in (3.1.19). Furthermore, by the uniqueness of the representation of $\hat{\mu}$, we can choose a subsequence $(\rho_n'')_{n\in\mathbb{N}}$ from any subsequence $(\rho_n')_{n\in\mathbb{N}}$ of $(\rho_n)_{n\in\mathbb{N}}$ such that ρ_n'' converges to ρ. Hence $\rho_n\to\rho$ without choosing a subsequence. Then (1) follows, and from this (2) and (3) follow without going to a subsequence.

Next we assume (1)–(3). Define $\rho_n(dy) = (|y|^2\wedge 1)\nu_n(dy)$ and $\rho(dy) = (|y|^2\wedge 1)\nu(dy)$. From (1), we can derive (3.1.16) and (3.1.17). Using (1) and (2), we have (3.1.11), implying (3.1.18). Thus we have

$$\lim_{n\to\infty}\hat{\mu}_n(u) = ib\cdot u - \frac{1}{2}u\cdot Au + \int_{\mathbb{R}^d\setminus\{0\}} g(u,y)\nu(dy),$$

whose right-hand side equals η. □

3.1.3 Lévy processes

We consider the following large class of random processes.

Definition 3.14 (Lévy process). A random process $(X_t)_{t\geq 0}$ on a probability space $(\Omega,\mathscr{F},P)$ is called a *Lévy process* if

(1) $X_0(\omega) = 0$ almost surely,
(2) the increments $(X_{t_i} - X_{t_{i-1}})_{1\leq i\leq n}$ are independent random variables for every choice of time-points $0 = t_0 < t_1 < \cdots < t_n$, and all $n\in\mathbb{N}$,

(3) $X_t - X_s \overset{d}{=} X_{t-s}$, for every $0 \le s \le t$,
(4) the function $t \mapsto X_t$ is stochastically continuous, i. e., for all $t \ge 0$ and $\varepsilon > 0$

$$\lim_{s\to t} P(|X_s - X_t| > \varepsilon) = 0.$$

A Lévy process $(X_t)_{t\ge 0}$ can also be defined to start at an arbitrary $x \in \mathbb{R}$ by replacing (1) of Definition 3.14 with $X_0(\omega) = x$ almost surely.

There is a close relationship between infinitely divisible distributions and Lévy processes.

Theorem 3.15.
(1) *Let $(X_t)_{t\ge 0}$ be a Lévy process. Then for every $t > 0$, the distribution μ_t of X_t is infinitely divisible and $\mu_t = \mu^{t*}$, i. e., $\widehat{\mu_t}(u) = \hat{\mu}(u)^t$, where μ is the distribution of X_1.*
(2) *Conversely, for every infinitely divisible distribution μ on $\mathbb{R}^d$ with a triplet (b, A, ν), where $b \in \mathbb{R}^d$, A is a $d \times d$ nonnegative definite symmetric matrix, and ν is a Lévy measure, there exists a Lévy process $(X_t)_{t\ge 0}$ on $(\Omega, \mathcal{F}, P)$ such that the distribution of X_1 is μ and the characteristic triplet of X_t is $(tb, tA, t\nu)$, i. e., $\mathbb{E}[e^{iuX_t}] = e^{t\eta(u)}$, where η is given by the Lévy–Khintchine formula* (3.1.8).

Proof. (1) Let $t_k = tk/n$. By telescoping

$$X_t = (X_{t_1} - X_{t_0}) + \cdots + (X_{t_n} - X_{t_{n-1}}),$$

the random variable X_t is obtained as a sum of independent and identically distributed random variables. Thus X_t is an infinitely divisible random variable for every $t \ge 0$. By $X_1 = (X_{1/n} - X_0) + \cdots + (X_{n/n} - X_{n-1/n})$ we have $\hat{\mu}_{1/n}(u)^n = \hat{\mu}(u)$, i. e., $\hat{\mu}_{1/n}(u) = \hat{\mu}(u)^{1/n}$. We also have

$$X_{m/n} = (X_{1/n} - X_0) + \cdots + (X_{m/n} - X_{(m-1)/n})$$

and hence $\hat{\mu}_{m/n}(u) = \hat{\mu}(u)^{m/n}$ also follows. Thus the distribution of $X_{m/n}$ is $\mu^{m/n*}$. In general, if a random processes Y_n converges to Y in probability as $n \to \infty$, then the distribution μ_n of Y_n weakly converges to the distribution μ of Y. Let $r_n \to t$ with $r_n \in \mathbb{Q}$. Since $(X_t)_{t\ge 0}$ is a Lévy process, $X_{r_n} \to X_t$ in probability, and thus $\mu^{r_n*} \to \mu_t$ weakly. On the other hand, $\widehat{\mu^{r_n*}} = \hat{\mu}^{r_n} \to \hat{\mu}^t$. Thus $\hat{\mu}_t = \hat{\mu}^t$, which implies $\mu_t = \mu^{t*}$.

(2) Suppose that μ is an infinitely divisible probability measure. Note that $\mu^{t*} * \mu^{s*} = \mu^{(t+s)*}$. Let $\Omega = (\mathbb{R}^d)^{[0,\infty)}$, $\mathcal{A}$ the set of cylinder sets, and $X_t(\omega) = \omega(t)$, $\omega \in \Omega$, be the coordinate process. For $E_j \in \mathcal{B}(\mathbb{R}^d)$, $j = 0, 1, \ldots, n$, define

$$\mu_{\{t_0,\ldots,t_n\}}(E_0 \times \cdots \times E_n) = \int_{(\mathbb{R}^d)^{(n+1)}} 1_{E_0}(y_0) 1_{E_1}(y_0 + y_1) \cdots 1_{E_n}(y_0 + \cdots + y_n) \prod_{i=0}^{n} \mu^{(t_i - t_{i-1})*}(dy_i),$$

where we understand $t_{-1} = 0$. In the same way as in the proof of Theorem 2.97, the existence of a probability measure on $(\Omega, \sigma(\mathscr{A}))$ follows. In fact, the family of measures $\{\mu_\Lambda\}_{\Lambda\subset[0,\infty),|\Lambda|<\infty}$ satisfies the Kolmogorov's consistency relation. Thus there exists a unique probability measure P on $\sigma(\mathscr{A})$ such that

$$P(X_{t_0} \in E_0, \dots, X_{t_n} \in E_n) = \mu_{\{t_0,\dots,t_n\}}(E_0 \times \cdots \times E_n).$$

The remaining part is to show that $(X_t)_{t\geq 0}$ is a Lévy process. We have

$$\mathbb{E}\left[f(X_{t_0}, \dots, X_{t_n})\right] = \int_{(\mathbb{R}^d)^{(n+1)}} f(y_0, y_0 + y_1, \dots, y_0 + \cdots + y_n) \prod_{i=0}^{n} \mu^{(t_i - t_{i-1})*}(dy_i) \tag{3.1.21}$$

for every Borel measurable bounded function f. Choose $e^{i\sum_{j=1}^n u_j\cdot(x_j - x_{j-1})}$, with $u_j \in \mathbb{R}$, for $f(x_0, \dots, x_n)$ in (3.1.21). We have

$$\mathbb{E}\left[e^{i\sum_{j=1}^n u_j\cdot(X_{t_j} - X_{t_{j-1}})}\right] = \prod_{j=0}^{n} \int_{\mathbb{R}^d} e^{iu_j\cdot y_j} \mu^{(t_j - t_{j-1})*}(dy_j).$$

In particular, putting $u_i = 0$ for $i \neq j$, we have

$$\mathbb{E}\left[e^{iu_j\cdot(X_{t_j} - X_{t_{j-1}})}\right] = \int_{\mathbb{R}^d} e^{iu_j\cdot y_j} \mu^{(t_j - t_{j-1})*}(dy_j). \tag{3.1.22}$$

Thus the distribution of $X_{t_j} - X_{t_{j-1}}$ is $\mu^{(t_j - t_{j-1})*}(dy)$. Hence

$$\mathbb{E}\left[e^{i\sum_{j=1}^n u_j\cdot(X_{t_j} - X_{t_{j-1}})}\right] = \prod_{j=1}^{n} \mathbb{E}[e^{iu_j\cdot(X_{t_j} - X_{t_{j-1}})}].$$

This implies that the increments $\{X_{t_j} - X_{t_{j-1}}\}_{j=0}^n$ are independent by Lemma 2.27. Similarly, we have

$$\mathbb{E}[e^{iu_j\cdot(X_{t_j} - X_{t_{j-1}})}] = \int_{\mathbb{R}^d} e^{iu_j\cdot y_j} \mu^{(t_j - t_{j-1})*}(dy_j) = \mathbb{E}[e^{iu_j\cdot(X_{t_j+s} - X_{t_{j-1}+s})}].$$

It follows that $X_t \stackrel{d}{=} X_{t+s} - X_s$. Finally, to show the stochastic continuity of $t \mapsto X_t$, it suffices to prove that $\lim_{t\downarrow 0} P(|X_t| > \varepsilon) = 0$ for every $\varepsilon > 0$. We have

$$P(|X_t| > \varepsilon) = \int_{\mathbb{R}^d} 1_{\{|y|>\varepsilon\}} \mu^{t*}(dy).$$

Taking $t \to 0$, we get $\widehat{\mu^{t*}}(u) = \hat{\mu}(u)^t \to 1$. Thus we have $\int_{\mathbb{R}^d} f(y)\mu^{t*}(dy) \to f(0)$ as $t \to 0$, for every bounded function f, continuous at $u = 0$. So $\lim_{\varepsilon\to 0} P(|X_t| > \varepsilon) = 0$, and $(X_t)_{t\geq 0}$

is a Lévy process. By (3.1.22), the distribution of X_t under P is μ^{t*}, and $\widehat{\mu^{t*}}(u) = \hat{\mu}(u)^t$, where μ is the distribution of X_1. Since $\hat{\mu}(u) = e^{\eta(u)}$, where $\eta(u)$ is given by (3.1.8), we have $\widehat{\mu^{t*}}(u) = \hat{\mu}(u)^t = e^{t\eta(u)}$. Thus the characteristic triplet for X_t is $(tb, tA, t\nu)$. □

The vector b is called *drift term*, the matrix A *diffusion matrix*, the measure ν *Lévy (jump) measure*, and the triplet (b, A, ν) the *characteristics* of the Lévy process $(X_t)_{t\geq 0}$, while η given by (3.1.8) is called its *Lévy symbol*. By Theorem 3.15, the processes in Example 3.2 are all Lévy processes.

Example 3.16 (Brownian motion). The Lévy triplet of Brownian motion with constant drift is $(b, A, 0)$ and its Lévy symbol is $\eta(u) = ib \cdot u - \frac{1}{2}u \cdot Au$.

Example 3.17 (Poisson process). A nonnegative integer-valued Lévy process $(N_t)_{t\geq 0}$ is called a *Poisson process* with intensity $\lambda > 0$ if it has Lévy triplet $(0, 0, \lambda\delta_1)$, where δ_1 is a one-dimensional Dirac point measure, and Lévy symbol $\eta(u) = \lambda(e^{iu} - 1)$. It has independent, identically distributed increments $(N_{t_i} - N_{t_{i-1}})_{0\leq i\leq n}$, for every choice of time-points $0 = t_0 < t_1 < \cdots < t_n$, and every $n \in \mathbb{N}$, with Poisson distribution $N_{t_i} - N_{t_{i-1}} \stackrel{\mathrm{d}}{=} \mathrm{Poi}(\lambda(t_i - t_{i-1}))$, $i = 1, \ldots, n$. In particular, we have

$$\mathbb{E}[1_{\{n\}}(N_t)] = \frac{(\lambda t)^n}{n!}e^{-\lambda t}, \quad n \geq 0$$

and characteristic function

$$\mathbb{E}[e^{iuX_t}] = e^{\lambda t(e^{iu}-1)}, \quad u \in \mathbb{R}.$$

Example 3.18 (Compound Poisson processes). Let $(Z_n)_{n\in\mathbb{N}}$ be independent and identically distributed random variables with characteristic function $\hat{\sigma}(u)$, and $(N_t)_{t\geq 0}$ a Poisson process such that $N_t \stackrel{\mathrm{d}}{=} \mathrm{Poi}(\lambda t)$. Then

$$Y_t = \sum_{n=1}^{N_t} Z_n$$

is called a *compound Poisson process*. Its characteristic function is

$$\hat{\mu}_t(u) = \exp(\lambda(\hat{\sigma}(u) - 1)t),$$

and $\eta(u) = \lambda(\hat{\sigma}(u) - 1)$.

The Cauchy process is a specific case of a class of special interest, which we describe in the next example.

Example 3.19 (Stable processes). A random variable X is called *stable* if for every $n \in \mathbb{N}$ there exist $a_n \geq 0$ and $b_n \in \mathbb{R}^d$ such that $X_1 + \cdots + X_n \stackrel{\mathrm{d}}{=} a_nX + b_n$, and *strictly stable* if $b_n = 0$, where $X_1, \ldots, X_n$ are independent, identically distributed copies of X. (Stability under taking sums explains the name of this class.) A Lévy process $(X_t)_{t\geq 0}$ is called a *stable process* if all random variables X_t, $t > 0$, are stable. For a stable process, we have that for every $a > 0$ there exist $b > 0$ and $c \in \mathbb{R}^d$ such that $X_{at} \stackrel{d}{=} bX_t + ct$, which holds for a strictly stable process with $c = 0$. This is an extension of the scaling property of Brownian motion; see Proposition 2.100 (5) below. From the definition, it can be seen that for a stable process there exist $0 < \alpha \leq 2$ and $b_t \in \mathbb{R}^d$ such that $X_t \stackrel{d}{=} t^{1/\alpha}X_1 + b_t$; for a strictly stable process, $b_t = 0$. Here α is called the *stability index* of the stable process. For a stable process, either $\alpha = 2$ and its Lévy triplet is $(b, A, 0)$, i. e., it is a Gaussian process with mean b and covariance matrix A, or $0 < \alpha < 2$ and its Lévy triplet is $(b, 0, \nu)$ with Lévy measure

$$\nu(E) = \int_{S^{d-1}} \lambda(d\theta) \int_0^\infty 1_E(r\theta)\frac{dr}{r^{1+\alpha}}, \tag{3.1.23}$$

where S^{d-1} is the $(d-1)$-dimensional unit sphere centered in the origin. The above integral can be computed; for $d = 1$, the calculation gives

$$\nu(dy) = \left(\frac{c_1}{y^{1+\alpha/2}}1_{(0,\infty)} + \frac{c_2}{(-y)^{1+\alpha/2}}1_{(-\infty,0)}\right)dy, \quad c_1, c_2 \geq 0. \tag{3.1.24}$$

Stable processes with index $1 \leq \alpha \leq 2$ are type-C Lévy processes, while those with $0 < \alpha < 1$ are type-B. A stable process with index $0 < \alpha \leq 2$ is rotationally invariant if $\mathscr{R}X_t \stackrel{d}{=} X_t$, for all $\mathscr{R} \in O(\mathbb{R}^d)$, i. e., under the action of the orthogonal group. In this case the Lévy symbol of the process becomes $\eta(u) = -c|u|^\alpha$, with some $c > 0$, and for $0 < \alpha < 2$ the Lévy measure has a uniform distribution on S_{d-1}. In particular, the rotationally invariant 1-stable process is the Cauchy process.

Let $(X_t)_{t\geq 0}$ be a d-dimensional rotationally invariant α-stable process on a probability space $(\Omega, \mathscr{F}, P)$. The characteristic function of $(X_t)_{t\geq 0}$ is given by

$$\mathbb{E}[e^{i\xi\cdot X_t}] = e^{-t|\xi|^\alpha}, \quad \xi \in \mathbb{R}^d,\ t \geq 0. \tag{3.1.25}$$

The probability transition (or heat) kernel $p_t^X(x)$ of $(X_t)_{t\geq 0}$ is a smooth real-valued function on $\mathbb{R}^d$ given by

$$p_t^X(x) = \frac{1}{(2\pi)^d}\int_{\mathbb{R}^d} e^{-t|\xi|^\alpha}e^{-ix\cdot\xi}d\xi, \quad x \in \mathbb{R}^d,\ t > 0.$$

We have the scaling property

$$p_t^X(x) = t^{-d/\alpha}p_1^X(t^{-1/\alpha}x),$$

and thus the generic bound

$$p_t^X(x) \le t^{-d/\alpha} \frac{1}{(2\pi)^d} \int_{\mathbb{R}^d} e^{-|\xi|^\alpha} d\xi. \tag{3.1.26}$$

The asymptotic behavior of $p_1^X(x)$ can be computed as

$$\lim_{|x|\to\infty} |x|^{d+\alpha} p_1^X(x) = \alpha 2^{\alpha-1} \frac{1}{\pi^{(d+2)/2}} \sin\left(\frac{\alpha\pi}{2}\right) \Gamma\left(\frac{d+\alpha}{2}\right) \Gamma\left(\frac{\alpha}{2}\right). \tag{3.1.27}$$

Hence for every $\varepsilon > 0$ there exist $a_1(\varepsilon), a_2(\varepsilon) > 0$ such that

$$\frac{a_1(\varepsilon)t}{|x|^{d+\alpha}} \le p_t^X(x) \le \frac{a_2(\varepsilon)t}{|x|^{d+\alpha}}, \quad |x| > \varepsilon.$$

For every $\varepsilon > 0$, together with (3.1.26) we have the kernel estimates

$$b_1(\varepsilon)\left(\frac{t}{|x|^{d+\alpha}} \wedge t^{-d/\alpha}\right) \le p_t^X(x) \le b_2(\varepsilon)\left(\frac{t}{|x|^{d+\alpha}} \wedge t^{-d/\alpha}\right), \quad |x| > \varepsilon$$

with positive constants $b_1(\varepsilon)$ and $b_2(\varepsilon)$. By a more precise calculation, these estimates improve to

$$\frac{1}{C}\left(\frac{t}{|x|^{d+\alpha}} \wedge t^{-d/\alpha}\right) \le p_t^X(x) \le C\left(\frac{t}{|x|^{d+\alpha}} \wedge t^{-d/\alpha}\right) \tag{3.1.28}$$

with a constant $C > 0$. The Lévy measure of $(X_t)_{t\ge0}$ is given by

$$\nu(dx) = \mathcal{A}_{d,-\alpha}|x|^{-d-\alpha}dx \quad \text{with} \quad \mathcal{A}_{d,\gamma} = 2^{-\gamma}\pi^{-d/2}\frac{\Gamma((d-\gamma)/2)}{|\Gamma(\gamma/2)|}.$$

Next we discuss some properties of Lévy processes. In general, their sums are not necessarily Lévy processes again, but independent sums give rise to Lévy processes.

Theorem 3.20 (Sums of Lévy processes). *Let $(X_t)_{t\ge0}$ and $(Y_t)_{t\ge0}$ be Lévy processes on a probability space $(\Omega, \mathcal{F}, P)$. If $(X_t)_{t\ge0}$ and $(Y_t)_{t\ge0}$ are independent, then $(Z_t)_{t\ge0}$, $Z_t = X_t + Y_t$, is also a Lévy process.*

Proof. By independence

$$\mathbb{E}[e^{iu\cdot(Z_t-Z_s)}] = \mathbb{E}[e^{iu\cdot(X_t-X_s)}]\mathbb{E}[e^{iu\cdot(Y_t-Y_s)}] = \mathbb{E}[e^{iu\cdot X_{t-s}}]\mathbb{E}[e^{iu\cdot Y_{t-s}}] = \mathbb{E}[e^{iu\cdot Z_{t-s}}].$$

Thus $Z_t - Z_s \overset{\mathrm{d}}{=} Z_{t-s}$ follows, and the independent increments property of the process $(Z_t)_{t\ge0}$ is similarly shown. We have

$$P(|Z_t - Z_s| > \varepsilon) \le P(|X_t - X_s| > \varepsilon/2) + P(|Y_t - Y_s| > \varepsilon/2),$$

showing stochastic continuity. □

Theorem 3.21 (Convergence of Lévy processes). *Let $(X_t)_{t\geq 0}$ be a random process on a probability space $(\Omega, \mathscr{F}, P)$ and $(X_t^{(n)})_{t\geq 0}$, $n \in \mathbb{N}$, be a sequence of Lévy processes on it. Suppose that $X_t^{(n)}$ converges to X_t in probability as $n \to \infty$ for every $t \geq 0$, and $\lim_{n\to\infty} \limsup_{t\to\infty} P(|X_t^{(n)} - X_t| > a) = 0$, for all $a > 0$. Then $(X_t)_{t\geq 0}$ is also a Lévy process.*

Proof. It is direct to see that $X_0 = 0$. Since convergence in probability implies convergence of the characteristic functions, we have

$$\mathbb{E}[e^{iu\cdot(X_t - X_s)}] = \lim_{n\to\infty} \mathbb{E}[e^{iu\cdot(X_t^{(n)} - X_t^{(n)})}] = \lim_{n\to\infty} \mathbb{E}[e^{iu\cdot X_{t-s}^{(n)}}] = \mathbb{E}[e^{iu\cdot X_{t-s}}],$$

thus $X_t - X_s \overset{\mathrm{d}}{=} X_{t-s}$. Similarly, we can show that

$$\mathbb{E}\Big[\prod_{j=1}^{n} e^{iu\cdot(X_{t_j} - X_{t_{j-1}})}\Big] = \prod_{j=1}^{n} \mathbb{E}[e^{iu\cdot(X_{t_j} - X_{t_{j-1}})}],$$

thus the increments of $(X_t)_{t\geq 0}$ are independent. Furthermore, we have

$$P(|X_t| > a) \leq P(|X_t - X_t^{(n)}| > a/2) + P(|X_t^{(n)}| > a/2)$$

and hence

$$\limsup_{t\to 0} P(|X_t| > a) \leq \limsup_{t\to 0} P(|X_t - X_t^{(n)}| > a/2) + \limsup_{t\to 0} P(|X_t^{(n)}| > a/2).$$

Note that $\limsup_{t\to 0} P(|X_t^{(n)}| > a/2) = \lim_{t\to 0} P(|X_t^{(n)}| > a/2) = 0$, which also means that $\lim_{t\to 0} P(|X_t| > \varepsilon) \leq \limsup_{t\to 0} P(|X_t| > \varepsilon) = 0$, giving stochastic continuity. □

Theorem 3.22 (Versions of Lévy processes). *Let $(X_t)_{t\geq 0}$ be a Lévy process on a probability space $(\Omega, \mathscr{F}, P)$ and $(Y_t)_{t\geq 0}$ a version of $(X_t)_{t\geq 0}$. Then $(Y_t)_{t\geq 0}$ is also a Lévy process.*

Proof. The fact $Y_0 = 0$ is obvious. Fix $0 \leq s < t$. Let

$$D_{s,t} = \{\omega \in \Omega \mid X_s(\omega) = Y_s(\omega) \text{ and } X_t(\omega) = Y_t(\omega)\}.$$

Note that $P(D_{s,t}^c) \leq P(X_s \neq Y_s) + P(X_t \neq Y_t) = 0$, implying $P(D_{s,t}) = 1$. Also, we have

$$\begin{aligned} P(Y_t - Y_s \in A) &= P(Y_t - Y_s \in A, D_{s,t}) + P(Y_t - Y_s \in A, D_{s,t}^c) \\ &= P(Y_t - Y_s \in A, D_{s,t}) = P(X_t - X_s \in A, D_{s,t}) \\ &\leq P(X_t - X_s \in A) = P(X_{t-s} \in A) = P(Y_{t-s} \in A). \end{aligned}$$

The reverse inequality is proven in a similar fashion, and thus $Y_t - Y_s \overset{\mathrm{d}}{=} Y_{t-s}$ follows. It follows similarly that the increments of $(Y_t)_{t\geq 0}$ are independent. Stochastic continuity is implied by $P(|Y_t - Y_s| > \varepsilon) \leq P(|X_t - X_s| > \varepsilon)$. □

3.1.4 Martingale properties of Lévy processes

As in the case of Brownian motion, more general Lévy processes also have martingale properties, which are very helpful whenever available.

Proposition 3.23. *Let $(X_t)_{t\geq 0}$ be a Lévy process on a probability space $(\Omega, \mathscr{F}, P)$ with Lévy symbol η, and take a fixed $u \in \mathbb{R}^d$. Then $(Y_t)_{t\geq 0}$, $Y_t = e^{iu\cdot X_t - t\eta(u)}$, is a martingale with respect to the natural filtration $(\mathscr{F}_t)_{t\geq 0}$ of $(X_t)_{t\geq 0}$.*

Proof. $\mathbb{E}[|Y_t|] < \infty$ is straightforward. Also, $\mathbb{E}[Y_t|\mathscr{F}_s] = Y_s\mathbb{E}[e^{iu\cdot(X_t - X_s)-(t-s)\eta(u)}] = Y_s$, for all $0 \leq s \leq t$. □

Below the following martingales constructed by subtracting a term called *compensator* will be important. Let $(N_t)_{t\geq 0}$ be a Poisson process with intensity λ. The process $(\tilde{N}_t)_{t\geq 0} = (N_t - \lambda t)_{t\geq 0}$ is called a *compensated Poisson process.*

Lemma 3.24 (Martingales for Poisson processes). *Let $(N_t)_{t\geq 0}$ be a Poisson process with intensity λ on a probability space $(\Omega, \mathscr{F}, P)$. Then $(\tilde{N}_t)_{t\geq 0}$ and $(\tilde{N}_t^2 - \lambda t)_{t\geq 0}$ are martingales with respect to the natural filtration $(\mathscr{F}_t)_{t\geq 0}$ of $(N_t)_{t\geq 0}$.*

Proof. The first statement follows by $\mathbb{E}[\tilde{N}_t|\mathscr{F}_s] = \mathbb{E}[N_t - N_s + N_s - \lambda t|\mathscr{F}_s] = \mathbb{E}[N_t - N_s] + \mathbb{E}[N_s - \lambda t|\mathscr{F}_s] = \lambda(t-s) - \lambda t + N_s = \tilde{N}_s$. The proof of the second statement is completely similar. □

3.1.5 Markov properties of Lévy processes

In this section we show that Lévy processes are strong Markov processes possessing Feller transition kernels, and derive their generators.

Let $(X_t)_{t\geq 0}$ be a Lévy process on $(\Omega, \mathscr{F}, P)$. Let μ_t be the infinitely divisible distribution of X_t and define $p(s,t,x,A)$ by

$$p(s,t,x,A) = \mu_{t-s}(A - x) \tag{3.1.29}$$

for $0 \leq s \leq t$, $x \in \mathbb{R}^d$, and $A \in \mathcal{B}(\mathbb{R}^d)$, where $A - x = \{y - x \mid y \in A\}$. Note that $\mu_t(A - x) = P(X_t + x \in A)$. By the definition of a Lévy process we have $p(s,t,x,A) = P(X_t - X_s \in A - x)$ and $p(s,t,x,A)$ is a probability transition kernel. The Chapman–Kolmogorov identity results from the semigroup property

$$\mu_{t-s} * \mu_{u-t} = \mu_{u-s}, \quad s < t < u, \tag{3.1.30}$$

i. e.,

$$\int_{\mathbb{R}^d} \mu_{t-s}(A - y)\mu_{u-t}(dy) = \mu_{u-s}(A), \quad A \in \mathcal{B}(\mathbb{R}^d). \tag{3.1.31}$$

Write

$$p(s,t,x,A) = p(t-s,x,A). \tag{3.1.32}$$

Proposition 3.25. *A Lévy process $(X_t)_{t\geq 0}$ is a stationary Markov process and its probability transition kernel is given by* (3.1.32).

Proof. By independence of the increments, the finite-dimensional distributions of the process are

$$P(X_0 \in A_0,\ldots,X_n \in A_n) = \int\limits_{(\mathbb{R}^d)^{(n+1)}} \left(\prod_{i=0}^{n} 1_{A_i}(x_i)\right) \prod_{i=1}^{n} p(t_i - t_{i-1}, x_{i-1}, dx_i) P^0(dx_0), \tag{3.1.33}$$

where $P^0(dx) = \delta(x)$ is Dirac point measure with mass at $x = 0$. Thus by Proposition 2.82, $(X_t)_{t\geq 0}$ is a Markov process with probability transition kernel $p(t,x,A)$. □

Theorem 3.26 (C_0-semigroup on $C_\infty(\mathbb{R}^d)$). *The probability transition kernel $p(t,x,A)$ of a Lévy process is a Feller transition kernel. In particular,*

$$f \mapsto P_t f(x) = \int\limits_{\mathbb{R}^d} f(y) p(t,x,dy)$$

defines a C_0-semigroup on $C_\infty(\mathbb{R}^d)$.

Proof. The fact $p(t,x,A) \leq 1$ is trivial. We show that $P_t f \in C_\infty(\mathbb{R}^d)$ for $f \in C_\infty(\mathbb{R}^d)$. Let $x_n \to x$ as $n \to \infty$. By the dominated convergence theorem, we have

$$\lim_{n\to\infty} P_t f(x_n) = \int\limits_{\mathbb{R}^d} \lim_{n\to\infty} f(x_n + y) p(t,0,dy) = P_t f(x)$$

and

$$\lim_{|x|\to\infty} P_t f(x) = \int\limits_{\mathbb{R}^d} \lim_{|x|\to\infty} f(x+y) p(t,0,dy) = 0.$$

Thus $P_t f \in C_\infty(\mathbb{R}^d)$. To conclude, we show that $\lim_{t\to 0} P_t f = f$ for $f \in C_\infty(\mathbb{R}^d)$ in the sup-norm. Since f is uniformly continuous, for every $\varepsilon > 0$ there exists an $r > 0$ such that $\sup_{x\in\mathbb{R}^d} |f(x+y) - f(x)| < \varepsilon$ for all $|y| < r$. Moreover, by stochastic continuity of the Lévy process, $\int_{|y|>r} p(t,0,dy) = P(|X_t| > r) \to 0$ as $t \to 0$. Thus there exists t_0 such that $\int_{|y|>r} p(t,0,dy) < \varepsilon$ for $0 < t < t_0$. Hence

$$\sup_{x\in\mathbb{R}^d} |P_t f(x) - f(x)| = \sup_{x\in\mathbb{R}^d} \left| \int\limits_{\mathbb{R}^d} (f(x+y) - f(x)) p(t,0,dy) \right|$$

$$= \sup_{x\in\mathbb{R}^d}\left|\int_{|y|\le r}(f(x+y)-f(x))p(t,0,dy)+\int_{|y|>r}(f(x+y)-f(x))p(t,0,dy)\right|$$

$$\le \varepsilon\int_{|y|<r}p(t,0,dy)+\varepsilon\|f\|_\infty$$

for $0 < t < t_0$. Thereby $\lim_{t\to 0} P_t f = f$ follows. □

If the infinitely divisible distribution μ_t of $(X_t)_{t\ge 0}$ satisfies

$$\mu_t(-A) = \mu_t(A), \tag{3.1.34}$$

for all Borel measurable A, then P_t can be extended to a C_0-semigroup on $L^2(\mathbb{R}^d)$.

Theorem 3.27 (C_0-semigroup on $L^2(\mathbb{R}^d)$). *Let property* (3.1.34) *hold. Then* $\{P_t : t \ge 0\}$ *can be extended to a symmetric* C_0*-semigroup on* $L^2(\mathbb{R}^d)$.

Proof. Suppose that $f, g \ge 0$ and $\int_{\mathbb{R}^d} g(x)(P_t f)(x)dx < \infty$. Since

$$\int_{\mathbb{R}^d} g(x)(P_t f)(x)dx = \int_{\mathbb{R}^d} g(x)dx\int_{\mathbb{R}^d} f(x+y)p(t,0,dy)$$
$$= \int_{\mathbb{R}^d} g(x-y)dx\int_{\mathbb{R}^d} f(x)p(t,0,dy),$$

the symmetry property (3.1.34) implies that

$$\int_{\mathbb{R}^d} g(x)(P_t f)(x)dx = \int_{\mathbb{R}^d} g(x+y)dx\int_{\mathbb{R}^d} f(x)p(t,0,dy) = \int_{\mathbb{R}^d}(P_t g)(x)f(x)dx. \tag{3.1.35}$$

Let $f \in L^2(\mathbb{R}^d) \cap C_\infty(\mathbb{R}^d)$. Thus by Schwarz inequality and the symmetry in the equality (3.1.35), we have

$$\int_{\mathbb{R}^d}|(P_t f)(x)|^2dx \le \int_{\mathbb{R}^d}P_t(|f|^2)(x)dx = \int_{\mathbb{R}^d}1\cdot P_t(|f|^2)(x)dx = \int_{\mathbb{R}^d}P_t 1\cdot|f|^2(x)dx.$$

If $P_t 1 = 1$, this implies that $\|P_t f\| \le \|f\|$. Since $L^2(\mathbb{R}^d) \cap C_\infty(\mathbb{R}^d)$ is dense in $L^2(\mathbb{R}^d)$, the estimate $\|P_t f\| \le \|f\|$ can be extended to $f \in L^2(\mathbb{R}^d)$, and the theorem follows. □

By the Hille–Yosida theorem (see Proposition 4.80), the semigroup $\{P_t : t \ge 0\}$ associated with a Lévy process is of the form $P_t = e^{tL}$ with a generator L on $C_\infty(\mathbb{R}^d)$. The operator L is closed on $C_\infty(\mathbb{R}^d)$ in the sup-norm. The semigroup $\{P_t : t \ge 0\}$ can be also viewed as a pseudodifferential operator with symbol $e^{t\eta(k)}$. Denote by $\mathscr{S}(\mathbb{R}^d)$ the Schwartz space over $\mathbb{R}^d$, i. e., the space of infinitely differentiable, rapidly decreasing functions on $\mathbb{R}^d$.

Proposition 3.28 (Generator of a Lévy process). *Let $(X_t)_{t\geq 0}$ be a Lévy process with characteristics (b, A, ν) and symbol η. Let $\{P_t : t \geq 0\}$ be the associated C_0-semigroup on $C_\infty(\mathbb{R}^d)$ and L its generator, i. e., $P_t = e^{tL}$. Then the following hold:*

(1) *For $t \geq 0, f \in \mathscr{S}(\mathbb{R}^d)$, and $x \in \mathbb{R}^d$,*

$$(P_t f)(x) = \frac{1}{(2\pi)^{d/2}} \int_{\mathbb{R}^d} e^{ik\cdot x} e^{t\eta(k)} \hat{f}(k) dk. \tag{3.1.36}$$

(2) *For $f \in \mathscr{S}(\mathbb{R}^d)$ and $x \in \mathbb{R}^d$,*

$$Lf(x) = \frac{1}{(2\pi)^{d/2}} \int_{\mathbb{R}^d} e^{ik\cdot x} \eta(k) \hat{f}(k) dk. \tag{3.1.37}$$

In particular,

$$L = \frac{1}{2} \nabla \cdot A\nabla + b \cdot \nabla + L_\nu, \tag{3.1.38}$$

where

$$L_\nu f(x) = \int_{\mathbb{R}^d \setminus \{0\}} \left(f(x+y) - f(x) - y \cdot \nabla f(x) 1_{\{|y|<1\}} \right) \nu(dy).$$

Proof. (1) By the definition of P_t, we have

$$(P_t f)(x) = \frac{1}{(2\pi)^{d/2}} \mathbb{E} \left[\int_{\mathbb{R}^d} e^{ik\cdot(x+X_t)} \hat{f}(k) dk \right].$$

Since $\mathbb{E}[e^{ik\cdot X_t}] = e^{t\eta(k)}$, we have (1). To prove (2), consider the generator L given by

$$(Lf)(x) = \frac{1}{(2\pi)^{d/2}} \lim_{t\downarrow 0} \int_{\mathbb{R}^d} e^{ik\cdot x} \hat{f}(k) \left(\frac{e^{t\eta(k)} - 1}{t} \right) dk.$$

Since $|e^{ik\cdot x} \hat{f}(k) (\frac{e^{t\eta(k)}-1}{t})| \leq (1 + |k|^2) \hat{f} \in L^2(\mathbb{R}^d)$, the result follows by the dominated convergence theorem. □

Example 3.29. The generators of some specific Lévy processes are as follows:

(1) For a Brownian motion with constant drift,

$$L = \frac{1}{2} \nabla \cdot A\nabla + b \cdot \nabla. \tag{3.1.39}$$

(2) For a Poisson process with intensity λ,

$$Lf(x) = \lambda(f(x+1) - f(x)). \tag{3.1.40}$$

(3) The infinitesimal generator of a rotationally symmetric α-stable process is

$$L_{0,\alpha} = -(-\Delta)^{\alpha/2}, \tag{3.1.41}$$

i. e., the *fractional Laplacian* of order $\alpha \in (0,2)$. This will be further discussed in Example 4.267 below.

Finally, we show the strong Markov property of Lévy processes.

Theorem 3.30 (Strong Markov property). *Let $(X_t)_{t\geq 0}$ be a Lévy process on a filtered probability space $(\Omega, \mathscr{F}, (\mathscr{F}_t)_{t\geq 0}, P)$, where $(\mathscr{F}_t)_{t\geq 0}$ is the natural filtration of $(X_t)_{t\geq 0}$. Suppose that τ is a stopping time such that $P(\tau < \infty) = 1$. Define $X_t^\tau = X_{t+\tau} - X_\tau$. Then $(X_t^\tau)_{t\geq 0}$ is also a Lévy process, which is independent of $\mathscr{F}_\tau$, and $X_t^\tau \stackrel{d}{=} X_t$. Furthermore, $(X_t^\tau)_{t\geq 0}$ is $(\mathscr{F}_{\tau+t})$-adapted.*

Proof. In a similar way to the proof of the strong Markov property for Brownian motion (see Theorem 2.114), we can show

$$\mathbb{E}[e^{i\sum_{j=1}^n \xi_j\cdot(X_{\tau+t_j}-X_{\tau+t_{j-1}})}1_F] = \prod_{j=1}^n \mathbb{E}[e^{i\xi_j\cdot(X_{t_j}-X_{t_{j-1}})}]P(F) \tag{3.1.42}$$

for $F \in \mathscr{F}_\tau$ and $\xi_j \in \mathbb{R}^d, j = 1,\dots,n$, for all n. Taking $F = \Omega$, $n = 1$, $\xi_1 = \xi$, and $t_1 = t$ in (3.1.42), we have $\mathbb{E}[e^{i\xi\cdot X_t^\tau}] = \mathbb{E}[e^{i\xi\cdot X_t}]$. This gives $X_t^\tau \stackrel{d}{=} X_t$. From (3.1.42) it also follows that $\{X_{\tau+t_j}-X_{\tau+t_{j-1}}\}_{j=1}^n = \{X_{t_j}^\tau - X_{t_{j-1}}^\tau\}_{j=1}^n$ are independent. Furthermore, the stochastic continuity of X_t^τ follows from that of $(X_t)_{t\geq 0}$, proving that $(X_t^\tau)_{t\geq 0}$ is a Lévy process. By (3.1.42) again, we have $\mathbb{E}[e^{i\xi\cdot X_t^\tau}1_F] = \mathbb{E}[e^{i\xi\cdot X_t}]P(F) = \mathbb{E}[e^{i\xi\cdot X_t^\tau}]P(F)$ for $F \in \mathscr{F}_\tau$, which implies that X_t^τ is independent of $\mathscr{F}_\tau$. □

3.2 Sample path properties of Lévy processes

3.2.1 Càdlàg version

It can be shown that every Lévy process has a version with paths $t \mapsto X_t$ which are right-continuous and have left limits; functions, and thereby random processes, of this type are called *càdlàg* retaining the original French terminology "continue à droite, limite à gauche." By Theorem 3.22, a càdlàg version is also a Lévy process. Left-continuous paths having right limits can also be considered, similarly called *càglàd* paths.

Let $(X_t)_{t\geq 0}$ be a random process on a probability space $(\Omega, \mathscr{F}, P)$, $M \subset [0,\infty)$, $\varepsilon > 0$, and fix $\omega \in \Omega$. The function $t \mapsto X_t(\omega)$ is said to have *ε-oscillation n times in M* if there exist $t_0 < t_1 < \cdots < t_n$ in M such that $|X_{t_j}(\omega) - X_{t_{j-1}}(\omega)| > \varepsilon$ for $j = 1,\dots,n$. Define

$$B(p,\varepsilon,M) = \{\omega \in \Omega \mid X_t(\omega) \text{ has } \varepsilon\text{-oscillation } p \text{ times in } M\}.$$

Also, $X_t(\omega)$ has ε-oscillation infinitely often in M if for every n, $X_t(\omega)$ has ε-oscillation n times in M. Put

$$B(\infty, \varepsilon, M) = \{\omega \in \Omega \mid X_t(\omega) \text{ has } \varepsilon\text{-oscillation infinitely often in } M\}.$$

Let

$$K = \left\{\omega \in \Omega \,\middle|\, \exists \lim_{s \in \mathbb{Q}, s \downarrow t} X_s(\omega) \text{ and } \exists \lim_{s \in \mathbb{Q}, s \uparrow t} X_s(\omega) \text{ for all } t > 0\right\}.$$

Define $A_{N,k} = \Omega \setminus B(\infty, 1/k, [0, N] \cap \mathbb{Q})$ and $K' = \bigcap_{N=1}^{\infty} \bigcap_{k=1}^{\infty} A_{N,k}$. We have $K' \in \mathscr{F}$, since $\mathbb{Q}$ is countable.

Lemma 3.31. *It follows that $K' = K$.*

Proof. It is trivial to see that $K' \supset K$. We show that $K' \subset K$. Let $\omega \in K'$ and $t_n \in \mathbb{Q}$ be such that $t_n \downarrow t$ as $n \to \infty$. For every k, there exists n_0 such that $|X_{t_n}(\omega) - X_{t_0}(\omega)| \leq 1/k$ for all $n \geq n_0$. Hence $\lim_{n\to\infty} X_{t_n}(\omega)$ exists and thus so does $\lim_{s \in \mathbb{Q}, s\downarrow t} X_s(\omega)$. Similarly, we can show that $\lim_{s \in \mathbb{Q}, s\uparrow t} X_s(\omega)$ exists. □

Next we show that both càdlàg and càglàd versions exist.

Lemma 3.32. *Suppose that a random process $(X_t)_{t\geq 0}$ is stochastically continuous on a probability space $(\Omega, \mathscr{F}, P)$ and $P(K') = 1$. Then there exist $(Y_t)_{t\geq 0}$ and $(Z_t)_{t\geq 0}$ such that (1) $P(X_t = Y_t) = 1$ and $t \mapsto Y_t(\omega)$ is càdlàg, (2) $P(X_t = Z_t) = 1$ and $t \mapsto Z_t(\omega)$ is càglàd, for every $\omega \in \Omega$.*

Proof. For $\omega \in K'$, we can define $Y_t(\omega) = \lim_{s \in \mathbb{Q}, s\downarrow t} X_s(\omega)$ by Lemma 3.31. For $\omega \notin K'$, we set $Y_t(\omega) = 0$. Thus $t \to Y_t(\omega)$ is càdlàg for every $\omega \in \Omega$. By the stochastic continuity of $t \mapsto X_t$, there exists a sequence $\mathbb{Q} \supset (s_n)_{n\in\mathbb{N}}$ such that $X_{s_n} \to X_t$ in probability as $\mathbb{Q} \ni s_n \downarrow t$, while $X_{s_n} \to Y_t$ almost surely due to $P(K') = 1$. Hence $P(X_t = Y_t) = 1$ follows for $t \geq 0$. In the same way as with $(Y_t)_{t\geq 0}$, for $\omega \in K'$ we define $Z_t(\omega) = \lim_{s\in\mathbb{Q}, s\uparrow t} X_s(\omega)$, and for $\omega \notin K'$ we set $Z_t(\omega) = 0$. Thus $t \to Z_t(\omega)$ is càglàd for every $\omega \in \Omega$ and $P(X_t = Z_t) = 1$ follows for $t \geq 0$. □

Denote by $B_\varepsilon(x)$ the open ball centered at $x \in \mathbb{R}^d$ with radius ε. Let $(X_t)_{t\geq 0}$ be a Markov process on a probability space $(\Omega, \mathscr{F}, P)$ and $p^X(s, t, x, A)$ the probability transition kernel of $(X_t)_{t\geq 0}$. Define

$$\alpha_{\varepsilon,T}(u) = \sup\left\{p^X(s, t, x, B_\varepsilon(x)^c) \mid x \in \mathbb{R}^d, s, t \in [0, T] \text{ with } t - s \in [0, u]\right\}, \tag{3.2.1}$$

where

$$\begin{aligned} p^X(s, t, x, B_\varepsilon(x)^c) &= P(X_t - X_s \in B_\varepsilon(x)^c \setminus \{x\}) \\ &= P(X_t - X_s \in B_\varepsilon(0)^c \setminus \{0\}) = P(|X_t - X_s| \geq \varepsilon), \end{aligned}$$

and then $\alpha_{\varepsilon,T}(u) = \sup_{s,t\in[0,T], t-s\in[0,u]} P(|X_t - X_s| \geq \varepsilon)$.

Lemma 3.33. *Let $(X_t)_{t\geq 0}$ be a Markov process on a probability space $(\Omega, \mathscr{F}, P)$. Let p be a positive integer and $0 \leq u \leq t_1 < \cdots < t_n \leq v \leq T$ and $M = \{t_1, \ldots, t_n\}$. Then*

$$\mathbb{E}[1_{B(p,4\varepsilon,M)}] \leq (2\alpha_{\varepsilon,T}(v-u))^p.$$

Proof. We prove this by induction on p. First, let $p = 1$ and consider

$$C_k = \Big\{\omega \in \Omega \mid \max_{1\leq j\leq k-1} |X_{t_j}(\omega) - X_u(\omega)| \leq 2\varepsilon \text{ and } |X_{t_k}(\omega) - X_u(\omega)| > 2\varepsilon\Big\},$$
$$D_k = \Big\{\omega \in \Omega | \, |X_v(\omega) - X_{t_k}(\omega)| > \varepsilon\Big\}.$$

The sets $C_1, \ldots, C_n$ are pairwise disjoint and

$$B(1, 4\varepsilon, M) \subset \bigcup_{k=1}^{n} C_k \subset \Big\{\omega \in \Omega \mid |X_v(\omega) - X_u(\omega)| \geq \varepsilon\Big\} \cup \bigcup_{k=1}^{n}(C_k \cap D_k).$$

Hence

$$\mathbb{E}[1_{B(1,4\varepsilon,M)}] \leq \mathbb{E}[1_{\{|X_v - X_u|\geq\varepsilon\}}] + \sum_{k=1}^{n} \mathbb{E}[1_{C_k}1_{D_k}] \leq \alpha_{\varepsilon,T}(v-u) + \sum_{k=1}^{n} \mathbb{E}[1_{C_k}]\alpha_{\varepsilon,T}(v-u)$$

by the Markov property. Thus

$$\mathbb{E}[1_{B(1,4\varepsilon,M)}] \leq 2\alpha_{\varepsilon,T}(v-u) \tag{3.2.2}$$

follows, and the statement holds for $p = 1$. Next assume that it is true for $p - 1$. Let

$$F_k = B(p-1, 4\varepsilon, \{t_1, \ldots, t_k\}) \setminus B(p-1, 4\varepsilon, \{t_1, \ldots, t_{k-1}\}),$$
$$G_k = B(1, 4\varepsilon, \{t_k, \ldots, t_n\}).$$

Note that $B(p-1, 4\varepsilon, \{t_1, \ldots, t_k\}) = \emptyset$ if $p - 1 < k$. The sets $F_1, \ldots, F_n$ are disjoint and $B(p-1, 4\varepsilon, M) = \bigcup_{k=1}^{n} F_k$; moreover, $B(p, 4\varepsilon, M) \subset \bigcup_{k=1}^{n}(F_k \cap G_k)$. Hence we have

$$\mathbb{E}[1_{B(p,4\varepsilon,M)}] \leq \sum_{k=1}^{n} \mathbb{E}[1_{F_k}1_{G_k}] \leq 2\bigcup_{k=1}^{n} \mathbb{E}[1_{F_k}]\alpha_{\varepsilon,T}(v-u)$$

by the Markov property and (3.2.2). Thus

$$\mathbb{E}[1_{B(p,4\varepsilon,M)}] \leq \mathbb{E}[1_{B(p-1,4\varepsilon,M)}]2\alpha_{\varepsilon,T}(v-u) \leq (2\alpha_{\varepsilon,T}(v-u))^p$$

and the lemma follows. □

The following fundamental result has a similar role for general Lévy processes than the Kolmogorov–Čentsov theorem for Brownian motion. The *Dynkin–Kinney condition* given by (3.2.3) below replaces the sufficient condition in Theorem 2.49 for path continuity, and establishes a càdlàg and a càglàd property of sample paths instead.

Theorem 3.34 (Dynkin–Kinney condition). *Let $(X_t)_{t\geq 0}$ be a Markov process on a probability space $(\Omega, \mathscr{F}, P)$ such that*

$$\lim_{u\downarrow 0} \alpha_{\varepsilon,T}(u) = 0 \tag{3.2.3}$$

for every $\varepsilon > 0$ and $T > 0$, where $\alpha_{\varepsilon,T}$ is given by (3.2.1). Then there exist $(Y_t)_{t\geq 0}$ and $(Z_t)_{t\geq 0}$ such that (1) $P(X_t = Y_t) = 1$ and $t \mapsto Y_t(\omega)$ is càdlàg, (2) $P(X_t = Z_t) = 1$ and $t \mapsto Z_t(\omega)$ is càglàd, for every $\omega \in \Omega$.

Proof. Condition (3.2.3) implies stochastic continuity of $(X_t)_{t\geq 0}$. Thus it is sufficient to prove that $P(K') = 1$ by Lemma 3.32, and actually it suffices to show that $P(A^c_{N,k}) = 0$ for every fixed N and k. Choose l large enough so that $2\alpha_{1/(4k)}(N/l) < 1$. We have

$$P(A^c_{N,k}) = P\left(B\left(\infty, \frac{1}{k}, [0,N] \cap \mathbb{Q}\right)\right) \leq \sum_{j=1}^{l} \lim_{p\to\infty} P\left(B\left(p, \frac{1}{k}, \left[\frac{j-1}{l}N, \frac{j}{l}N\right] \cap \mathbb{Q}\right)\right).$$

Enumerating the elements of $\left[\frac{j-1}{l}N, \frac{j}{l}N\right] \cap \mathbb{Q}$ as $t_1, t_2, \ldots$, we have

$$P\left(B\left(p, \frac{1}{k}, \left[\frac{j-1}{l}N, \frac{j}{l}N\right] \cap \mathbb{Q}\right)\right) = \lim_{n\to\infty} P\left(B\left(p, \frac{1}{k}, \{t_1, \ldots, t_n\}\right)\right).$$

Since

$$P\left(B\left(p, \frac{1}{k}, \{t_1, \ldots, t_n\}\right)\right) \leq (2\alpha_{1/(4k),N}(N/l))^p$$

by Lemma 3.33, we get on letting $n \to \infty$ that

$$P\left(B\left(p, \frac{1}{k}, \left[\frac{j-1}{l}N, \frac{j}{l}N\right] \cap \mathbb{Q}\right)\right) \leq (2\alpha_{1/(4k),N}(N/l))^p.$$

The right-hand side goes to zero as $p \to \infty$. Hence $P(A^c_{N,k}) = 0$ follows. □

Corollary 3.35 (Càdlàg and càglàd versions). *Let $(X_t)_{t\geq 0}$ be a Lévy process on a probability space $(\Omega, \mathscr{F}, P)$. Then a càdlàg version and a càglàd version of $(X_t)_{t\geq 0}$ exist.*

Proof. Let p^X be the probability transition kernel of $(X_t)_{t\geq 0}$ and $B_\varepsilon(x)$ be the ball centered at x with radius $\varepsilon > 0$. We have $p^X(s,t,x,B_\varepsilon(x)^c) = P(|X_t - X_s| \geq \varepsilon)$. We can show that $P(|X_t - X_s| \geq \varepsilon)$ uniformly converges to zero as $|s-t| \to 0$ for $s, t \in [0,T]$ in Lemma 3.36. Thus $\lim_{u\downarrow 0} \alpha_{\varepsilon,T}(u) = 0$ and the statement follows by Lemma 3.32. □

The following gives a strengthened version of stochastic continuity.

Lemma 3.36 (Uniform stochastic continuity). *Let $(X_t)_{t\geq 0}$ be a Lévy process on a probability space $(\Omega, \mathscr{F}, P)$. Then for every $T > 0$ we have that $P(|X_t - X_s| \geq \varepsilon)$ uniformly converges to zero as $|s-t| \to 0$, for $s, t \in [0,T]$.*

Proof. Fix $\eta > 0$. By stochastic continuity, for every $t > 0$ there exists $\delta_t > 0$ such that $P(|X_t - X_s| \geq \varepsilon/2) \leq \eta/2$, for $|s - t| < \delta_t$. Let $I_t = (t - \delta_t/2, t + \delta_t/2)$. There is a finite covering $I_{t_j}, j = 1, \ldots, n$, such that $\bigcup_{j=1}^n I_{t_j} \supset [0, T]$. Take $\delta = \min_{1\leq j\leq n} \delta_{t_j}$. If $|s - t| < \delta$ and $s, t \in [0, T]$, then $t \in I_{t_j}$ for some j; hence $|s - t_j| < \delta_{t_j}$ and

$$P(|X_t - X_s| > \varepsilon) \leq P(|X_t - X_{t_j}| > \varepsilon/2) + P(|X_s - X_{t_j}| > \varepsilon/2) < \eta. \qquad \square$$

To conclude the discussion of sample path properties, we finally point out the topological structure of the space of càdlàg paths. We will denote the set of $\mathbb{R}^d$-valued càdlàg paths on $[0, \infty)$ by

$$\mathscr{X}_D = D([0, \infty); \mathbb{R}^d). \tag{3.2.4}$$

To introduce a topology on $\mathscr{X}_D$, let $\mathscr{I}'$ be the collection of strictly increasing functions $h : [0, \infty) \to [0, \infty)$. In particular, we have $h(0) = 0$, $\lim_{t\to\infty} h(t) = \infty$, and $t \mapsto h(t)$ is continuous. Let $\mathscr{I}$ be the set of Lipschitz continuous functions $h \in \mathscr{I}'$ such that

$$\gamma(h) = \sup_{\substack{s,t\geq 0\\ s\neq t}} \left|\log \frac{h(t) - h(s)}{t - s}\right| < \infty.$$

For $X, Y \in \mathscr{X}_D$, we define the metric

$$d_D(X, Y) = \inf_{h\in\mathscr{I}} \left(\gamma(h) \vee \int_0^\infty e^{-u} (\sup_{t\geq 0} |X_{t\wedge u} - Y_{h(u)\wedge u}| \wedge 1) du \right). \tag{3.2.5}$$

The properties of the so-induced topological space are summarized below without proof.

Proposition 3.37 (Skorokhod topology).
(1) *The space $(\mathscr{X}_D, d_D)$ is a separable and complete metric (i. e., a Polish) space.*
(2) *Let $\mathscr{B}$ be the smallest σ-field containing all finite-dimensional cylinder sets, and $\mathscr{B}(\mathscr{X}_D)$ be the Borel σ-field with the topology induced by d_D. Then $\mathscr{B} = \mathscr{B}(\mathscr{X}_D)$.*

3.2.2 Two-sided Lévy processes

Next we describe how Lévy processes can be extended to run over the whole real line. Let $(X_t)_{t\geq 0}$ be the coordinate process of a Lévy process on a probability space $(\mathscr{X}_D, \mathcal{B}(\mathscr{X}_D), P^X)$, and $(\bar{X}_t)_{t\geq 0}$ the coordinate process of a Lévy process on a probability space $(\mathscr{X}_{\bar{D}}, \mathcal{B}(\mathscr{X}_{\bar{D}}), \bar{P}^X)$, such that $t \mapsto X_t$ is càdlàg and $t \mapsto \bar{X}_t$ is càglàd. Consider the product probability space

$$(\mathscr{X}_D \times \mathscr{X}_{\bar{D}}, \mathcal{B}(\mathscr{X}_D) \times \mathcal{B}(\mathscr{X}_{\bar{D}}), P^X \otimes \bar{P}^X)$$

and define for $\omega = (\omega_1, \omega_2) \in \mathscr{X}_D \times \mathscr{X}_{\bar{D}}$,

$$\widehat{X}_t(\omega) = \begin{cases} \omega_1(t), & t \geq 0, \\ \omega_2((-t)-), & t < 0. \end{cases}$$

Then $t \mapsto \widehat{X}_t(\omega)$ is càdlàg for $t \in \mathbb{R}$. Let

$$\mathfrak{X}_D = D(\mathbb{R}; \mathbb{R}^d) \tag{3.2.6}$$

be the space of càdlàg paths on $\mathbb{R}$, and $\mathcal{B}(\mathfrak{X}_D)$ be the family of cylinder sets on $\mathfrak{X}_D$. Denote the image measure of $P^x \otimes \bar{P}^x$ on $(\mathfrak{X}_D, \mathcal{B}(\mathfrak{X}_D))$ with respect to $\widehat{X}_.$ by

$$Q^x = (P^x \otimes \bar{P}^x) \circ \widehat{X}_.^{-1},$$

and the coordinate process on $(\mathfrak{X}_D, \mathcal{B}(\mathfrak{X}_D), Q^x)$ by $(Y_t)_{t\in\mathbb{R}}$, which is a Lévy process indexed by $\mathbb{R}$ on this space. The properties of the so obtained process can be summarized as follows.

Proposition 3.38 (Two-sided Lévy process). *The following hold:*
(1) $Q^x(Y_0 = x) = 1$;
(2) *The increments* $\{Y_{t_i} - Y_{t_{i-1}}\}_{1\leq i\leq n}$ *are independent for every* $0 = t_0 < t_1 < \cdots < t_n$ *with* $Y_t - Y_s \stackrel{\mathrm{d}}{=} Y_{t-s}$ *for* $t > s$;
(3) *The increments* $\{Y_{-t_{i-1}} - Y_{-t_i}\}_{1\leq i\leq n}$ *are independent for every* $0 = -t_0 > -t_1 > \cdots > -t_n$ *with* $Y_{-t} - Y_{-s} \stackrel{\mathrm{d}}{=} Y_{s-t}$ *for* $-t > -s$;
(4) *The function* $\mathbb{R} \ni t \mapsto Y_t(\omega) \in \mathbb{R}$ *is almost surely càdlàg;*
(5) $Y_{-t} \stackrel{\mathrm{d}}{=} Y_t$ *for* $t \in \mathbb{R}$, *and* Y_t *and* Y_s *for* $t > 0$ *and* $s < 0$ *are independent.*

It can be checked through the finite-dimensional distributions that the joint distributions of $Y_{t_0}, \ldots, Y_{t_n}$, $-\infty < t_0 < t_1 < \cdots < t_n < \infty$, $n \in \mathbb{N}$, are invariant with respect to time-shift under some conditions.

Remark 3.39. As seen in Remark 2.102 for Brownian motion, time-shift invariance of an $\mathbb{R}^d$-valued two-sided Lévy process $(Y_t)_{t\in\mathbb{R}}$ is also related to the symmetry of the semigroup generated by process. Consider $P_t f(x) = \mathbb{E}^x[f(Y_t)]$, satisfying the semigroup property $P_{t+s} = P_t P_s$ for $t, s \geq 0$, by the Markov property of the process. Then we have

$$\int_{\mathbb{R}^d} \mathbb{E}^x[\bar{f}(Y_0) g(Y_t)] dx = (f, P_t g)_{L^2(\mathbb{R}^d)}.$$

When P_t is symmetric, i. e., $(f, P_t g)_{L^2(\mathbb{R}^d)} = (P_t f, g)_{L^2(\mathbb{R}^d)}$ holds, we also have $(f, P_t g)_{L^2(\mathbb{R}^d)} = (P_s f, P_{t-s} g)_{L^2(\mathbb{R}^d)}$ for $0 \leq s \leq t$, which can equally be represented as

$$\int_{\mathbb{R}^d} \mathbb{E}^x[\bar{f}(Y_0) g(Y_t)] dx = \int_{\mathbb{R}^d} \mathbb{E}^x[\bar{f}(Y_s)] \mathbb{E}^x[g(Y_{t-s})] dx.$$

The right-hand side above can be computed as

$$\int_{\mathbb{R}^d} \mathbb{E}^x[\bar{f}(Y_0)g(Y_t)]dx = \int_{\mathbb{R}^d} \mathbb{E}^x[\bar{f}(Y_{-s})]\mathbb{E}^x[g(Y_{t-s})]dx = \int_{\mathbb{R}^d} \mathbb{E}^x[\bar{f}(Y_{-s})g(Y_{t-s})]dx.$$

Hence $\int_{\mathbb{R}^d} \mathbb{E}^x[\bar{f}(Y_0)g(Y_t)]dx$ is time-shift invariant for all $0 \le s \le t$. The crucial assumption under which shift invariance is obtained is that P_t is symmetric. Let μ_t be the distribution of Y_t for $t \ge 0$. If $\mu_t(A) = \mu_t(-A)$ for $A \in \mathcal{B}(\mathbb{R}^d)$, then as seen in Theorem 3.27, it follows that

$$(f, P_t g)_{L^2(\mathbb{R}^d)} = \int_{\mathbb{R}^d} \bar{f}(x)g(x+y)\mu_t(dy)dx = \int_{\mathbb{R}^d} \bar{f}(x-y)g(x)\mu_t(dy)dx.$$

Since $\mu_t(A) = \mu_t(-A)$, replacing y for $-y$, we see that

$$(f, P_t g)_{L^2(\mathbb{R}^d)} = \int_{\mathbb{R}^d} \bar{f}(x+y)g(x)\mu_t(dy)dx = (P_t f, g)_{L^2(\mathbb{R}^d)}.$$

Hence P_t is symmetric. Thus the condition $\mu_t(A) = \mu_t(-A)$ ensures that P_t is symmetric.

Proposition 3.40 (Time-shift invariance). *Let $(Y_t)_{t\in\mathbb{R}}$ be an $\mathbb{R}^d$-valued two-sided Lévy process, with $Y_0 = 0$ almost surely, and μ_t be the distribution of Y_t for $t \ge 0$. Suppose that $\mu_t(A) = \mu_t(-A)$, for every $t \ge 0$ and $A \in \mathcal{B}(\mathbb{R}^d)$. Let λ be a shift invariant measure on $\mathbb{R}^d$, i. e., $\lambda(A + a) = \lambda(A)$ for every $A \in \mathcal{B}(\mathbb{R}^d)$ and $a \in \mathscr{A}$, where $\mathscr{A} = \bigcup_{t\ge 0} \operatorname{Ran} Y_t$. Let $t_1 \le \cdots \le t_n \le 0 \le t_{n+1} \le \cdots \le t_{n+m}$. For bounded functions f_i, $i = 2, \ldots, n+m-1$ and $f_1, f_{n+m} \in L^2(\mathbb{R}^d, \lambda(dx))$, it follows that*

$$\int_{\mathbb{R}^d} \mathbb{E}^x\left[\prod_{i=1}^{n+m} f_i(Y_{t_i})\right]\lambda(dx) = \int_{\mathbb{R}^d} \mathbb{E}^x\left[\prod_{i=1}^{n+m} f_i(Y_{t_i+s})\right]\lambda(dx) \tag{3.2.7}$$

for $-t_{n+m} \le s \le -t_0$. Here

$$\int_{\mathbb{R}^d} \mathbb{E}^x\left[\prod_{i=1}^{n} f_i(Y_{t_i})\right]\lambda(dx) = \int_{\mathbb{R}^d} \mathbb{E}\left[\prod_{i=1}^{n} f_i(Y_{t_i}+x)\right]\lambda(dx).$$

Proof. Let $P_t(x, A) = \mathbb{E}[1_A(Y_t + x)]$, for $A \in \mathcal{B}(\mathbb{R}^d)$. We have

$$\int_{\mathbb{R}^d} \mathbb{E}^x\left[\prod_{j=1}^{n} f_j(Y_{t_j})\right]\lambda(dx) = \int_{\mathbb{R}^d\times\mathbb{R}^{dn}} \left(\prod_{j=1}^{n} f_j(x_j)\right)\prod_{j=1}^{n} P_{t_j-t_{j-1}}(x_{j-1}, dx_j)\lambda(dx),$$

where $t_0 = 0$ and $x_0 = x$. Let $-t_1 \le \cdots \le -t_n \le 0 \le t_{n+1} \le \cdots \le t_{n+m}$. We show that

$$\int_{\mathbb{R}^d} \mathbb{E}^x\left[\prod_{i=1}^{n} f_i(Y_{-t_i}) \prod_{j=n+1}^{n+m} f_j(Y_{t_j})\right] \lambda(dx) = \int_{\mathbb{R}^d} \mathbb{E}^x\left[\prod_{i=1}^{n} f_i(Y_{t_1-t_i}) \prod_{j=n+1}^{n+m} f_j(Y_{t_1+t_j})\right] \lambda(dx). \quad (3.2.8)$$

Thus (3.2.7) follows from (3.2.8). By independence of Y_{-t} and Y_s for $-t \le 0 \le s$ and reflection symmetry, we have

$$\begin{aligned}\int_{\mathbb{R}^d} \mathbb{E}^x\left[\prod_{i=1}^{n} f_i(Y_{-t_i}) \prod_{j=n+1}^{n+m} f_j(Y_{t_j})\right] \lambda(dx) &= \int_{\mathbb{R}^d} \mathbb{E}^x\left[\prod_{i=1}^{n} f_i(Y_{-t_i})\right] \mathbb{E}^x\left[\prod_{j=n+1}^{n+m} f_j(Y_{t_j})\right] \lambda(dx) \\ &= \int_{\mathbb{R}^d} \mathbb{E}^x\left[\prod_{i=1}^{n} f_i(Y_{t_i})\right] \mathbb{E}^x\left[\prod_{j=n+1}^{n+m} f_j(Y_{t_j})\right] \lambda(dx). \end{aligned} \quad (3.2.9)$$

Hence we can compute the left-hand side of (3.2.9) as

$$\begin{aligned}\text{l. h. s. (3.2.9)} = \int_{\mathbb{R}^d} \lambda(dx) \int_{\mathbb{R}^{d(n+m)}} \left(\prod_{i=1}^{n+m} f_i(x_i)\right) P_{t_n}(x, dx_n) \left(\prod_{i=n}^{2} P_{t_{i-1}-t_i}(x_i, dx_{i-1})\right) \\ \times P_{t_{n+1}}(x, dx_{n+1}) \prod_{j=n+2}^{n+m} P_{t_j - t_{j-1}}(x_{j-1}, dx_j).\end{aligned}$$

By Lemma 3.41 we see the equality of measures on $\mathbb{R}^d \times \mathbb{R}^d$, i. e.,

$$\int_{\mathbb{R}^d} P_{t_n}(x, dx_n) P_{t_{n+1}}(x, dx_{n+1}) \lambda(dx) = P_{t_n + t_{n+1}}(x_n, dx_{n+1}) \lambda(dx_n).$$

Thus we have

$$\begin{aligned}\text{l. h. s. (3.2.9)} = \int_{\mathbb{R}^{d(n+m)}} \left(\prod_{i=1}^{n+m} f_i(x_i)\right) \left(\prod_{i=2}^{n} P_{t_{i-1}-t_i}(x_i, dx_{i-1})\right) \lambda(dx_n) \\ \times P_{t_n + t_{n+1}}(x_n, dx_{n+1}) \prod_{j=n+2}^{n+m} P_{t_j - t_{j-1}}(x_{j-1}, dx_j).\end{aligned} \quad (3.2.10)$$

On the other hand, the right-hand side of (3.2.9) can be computed as

$$\begin{aligned}\text{r. h. s. (3.2.9)} &= \int_{\mathbb{R}^d} \lambda(dx) \int_{\mathbb{R}^{d(n+m)}} \left(\prod_{i=1}^{n+m} f_i(x_i)\right) P_0(x, dx_1) \left(\prod_{i=2}^{n} P_{t_{i-1}-t_i}(x_{i-1}, dx_i)\right) \\ &\quad \times P_{t_n + t_{n+1}}(x_n, dx_{n+1}) \prod_{j=n+2}^{n+m} P_{t_j - t_{j-1}}(x_{j-1}, dx_j) \\ &= \int_{\mathbb{R}^{d(n+m)}} \left(\prod_{i=1}^{n+m} f_i(x_i)\right) \lambda(dx_1) \left(\prod_{i=2}^{n} P_{t_{i-1}-t_i}(x_{i-1}, dx_i)\right)\end{aligned}$$

$$\times P_{t_n+t_{n+1}}(x_n, dx_{n+1}) \prod_{j=n+2}^{n+m} P_{t_j-t_{j-1}}(x_{j-1}, dx_j). \tag{3.2.11}$$

Here we used $\int_{\mathbb{R}^d} P_0(x, dx_1)\lambda(dx) = \lambda(dx_1)$ as measures, due to $\int_A \int_{\mathbb{R}^d} P_0(x, dx_1)\lambda(dx) = \int_A P_0(x, \mathbb{R}^d)\lambda(dx) = \int_A \lambda(dx) = \lambda(A)$. By Lemma 3.42, we also see that

$$\left(\prod_{i=2}^{n} P_{t_{i-1}-t_i}(x_i, dx_{i-1})\right)\lambda(dx_n) = \lambda(dx_1)\left(\prod_{i=2}^{n} P_{t_{i-1}-t_i}(x_{i-1}, dx_i)\right) \tag{3.2.12}$$

as measures. Comparing (3.2.10) and (3.2.11) with (3.2.12), we obtain (3.2.8), and hence (3.2.7) follows for every $s \in \mathbb{R}$. □

To complete the proof of Proposition 3.40, it remains to show Lemmas 3.41–3.42.

Lemma 3.41. *Let the assumptions of Proposition* 3.40 *hold, and consider* $P_t(x, A) = \mathbb{E}[1_A(Y_t + x)]$. *Then for* $s, t \geq 0$,

$$\int_{\mathbb{R}^d} P_s(x, dy)P_t(x, dz)\lambda(dx) = P_{s+t}(y, dz)\lambda(dy). \tag{3.2.13}$$

Proof. For short, we denote the left-hand side of (3.2.13) by m, and the right-hand side by n. As $\lambda(A + a) = \lambda(A)$ for $a \in \mathscr{A}$, we have

$$\begin{aligned} m(A \times B) &= \int_{\mathbb{R}^d} P_s(x, A)P_t(x, B)\lambda(dx) = \int_{\mathbb{R}^d} \lambda(dx) \int_{\mathbb{R}^d\times\mathbb{R}^d} 1_A(y + x)1_B(z + x)\mu_s(dy)\mu_t(dz) \\ &= \int_{\mathbb{R}^d\times\mathbb{R}^d} \rho(-y + z)\mu_s(dy)\mu_t(dz), \end{aligned}$$

where $\rho(X) = \int_{\mathbb{R}^d} 1_A(x)1_B(x + X)\lambda(dx)$. We may think of y, z as elements of $\mathscr{A}$. By the assumption $\mu_t(A) = \mu_t(-A)$, we can replace $-y$ by y. Hence by the definition of the convolution of measures, we have

$$m(A \times B) = \int_{\mathbb{R}^d\times\mathbb{R}^d} \rho(y + z)\mu_s(dy)\mu_t(dz) = \int_{\mathbb{R}^d} \rho(p)(\mu_s * \mu_t)(dp).$$

By the Markov property $\mu_s * \mu_t = \mu_{s+t}$, we also have

$$\begin{aligned} m(A \times B) &= \int_{\mathbb{R}^d} \rho(p)\mu_{s+t}(dp) = \int_{\mathbb{R}^d\times\mathbb{R}^d} 1_A(x)1_B(x + p)\mu_{s+t}(dp)\lambda(dx) \\ &= \int_{\mathbb{R}^d} 1_A(x)P_{s+t}(x, B)\lambda(dx) = n(A \times B), \end{aligned}$$

which implies (3.2.13). □

Lemma 3.42. *Let the assumptions of Proposition* 3.40 *hold. Then* (3.2.12) *also holds.*

Proof. We prove the proposition by induction. Take $n = 2$. For simplicity, write $P_{j-1} = P_{t_{j-1}-t_j}$, and for the distribution of Y_{t_j} write μ_j. Let $A = A_1 \times A_2$. We see that

$$\int_A P_1(x_1, dx_2)\lambda(dx_1) = \int_{A_1} P_1(x_1, A_2)\lambda(dx_1) = \int_{\mathbb{R}^d} \lambda(dx_1) \int_{\mathbb{R}^d} \mu_1(dy)1_{A_1}(x_1)1_{A_2}(x_1 + y).$$

Since $\mu_j(-A) = \mu_j(A)$ and $\lambda(A + y) = \lambda(A)$ for $y \in \mathscr{A}$, we see that

$$\begin{aligned}\int_A P_1(x_1, dx_2)\lambda(dx_1) &= \int_{\mathbb{R}^d} \lambda(dx_1) \int_{\mathbb{R}^d} \mu_1(dy)1_{A_1}(x_1)1_{A_2}(x_1 - y)\\ &= \int_{\mathbb{R}^d} \lambda(dx_1) \int_{\mathbb{R}^d} \mu_1(dy)1_{A_1}(x_1 + y)1_{A_2}(x_1) = \int_A P_1(x_2, dx_1)\lambda(dx_2).\end{aligned}$$

Hence $\lambda(dx_1)P_1(x_1, dx_2) = P_1(x_2, dx_1)\lambda(dx_2)$ follows. From this equality, formally we can see that

$$\begin{aligned}&\lambda(dx_1)P_1(x_1, dx_2)\cdots P_{n-1}(x_{n-1}, dx_n)\\ &= P_1(x_2, dx_1)\lambda(dx_2)P_1(x_2, dx_3)\cdots P_{n-1}(x_{n-1}, dx_n)\\ &= P_1(x_2, dx_1)P_1(x_3, dx_2)\lambda(dx_3)\cdots P_{n-1}(x_{n-1}, dx_n)\\ &= \ldots = P_1(x_2, dx_1)P_1(x_3, dx_2)\cdots P_{n-1}(x_n, dx_{n-1})\lambda(dx_n).\end{aligned}$$

To verify this rigorously, assume that (3.2.12) holds for $n = 2, \ldots, k$. Let $A = A_1 \times \cdots \times A_{k+1}$. We have

$$\begin{aligned}\int_A \left(\prod_{i=1}^{k} P_i(x_i, dx_{i+1})\right)\lambda(dx_1) &= \int_{A_1\times\cdots\times A_k} \left(\prod_{i=1}^{k-1} P_i(x_i, dx_{i+1})\right) P_k(x_k, A_{k+1})\lambda(dx_1)\\ &= \int_{A_1\times\cdots\times A_k} \left(\prod_{i=1}^{k-1} P_i(x_{i+1}, dx_i)\right) P_k(x_k, A_{k+1})\lambda(dx_k) = \int_{A_k} \xi(x_k)P_k(x_k, A_{k+1})\lambda(dx_k).\end{aligned}$$

The second equality follows from the induction hypothesis, and ξ is given by

$$\xi(x_k) = \int_{A_1\times\cdots\times A_{k-1}} \prod_{i=1}^{k-1} P_i(x_{i+1}, dx_i).$$

Since $\lambda(dx_k)P_k(x_k, dx_{k+1}) = P_k(x_{k+1}, dx_k)\lambda(dx_{k+1})$ as measures, we see that

$$\int_A \left(\prod_{i=1}^{k} P_i(x_i, dx_{i+1})\right)\lambda(dx_1) = \int_{A_k\times A_{k+1}} \xi(x_k)1_{A_{k+1}}(x_{k+1})P_k(x_k, dx_{k+1})\lambda(dx_k)$$

$$= \int_{A_k \times A_{k+1}} \xi(x_k) 1_{A_{k+1}}(x_{k+1}) P_k(x_{k+1}, dx_k) \lambda(dx_{k+1}) = \int_A \left(\prod_{i=1}^{k} P_i(x_{i+1}, dx_i) \right) \lambda(dx_{k+1})$$

follows. This implies that (3.2.12) also holds for $n = k + 1$. □

Example 3.43 (Two-sided stable process). Let $(X_t)_{t\geq 0}$ be a d-dimensional, rotationally symmetric α-stable process, and consider its extension to $(X_t)_{t\in\mathbb{R}}$ defined on the whole real line $\mathbb{R}$. The distribution μ_t of X_t, $t \geq 0$, is rotationally invariant, which implies that $\mu_t(A) = \mu_t(-A)$ for $A \in \mathcal{B}(\mathbb{R}^d)$. Thus $(X_t)_{t\geq 0}$ is time-shift invariant.

Next we further consider time-shift invariance of two-sided Lévy processes, without assuming the reflection symmetry $\mu_t(A) = \mu_t(-A)$. A crucial example is the so-called *spin* (in probability theory better known as the *telegraph*) *process* $(\sigma_t)_{t\in\mathbb{R}}$ defined by $\sigma_t = (-1)^{N_t}$, where $(N_t)_{t\in\mathbb{R}}$ is a Poisson process on $\mathbb{R}$. The Poisson process is a nonnegative integer-valued random process, not having reflection symmetry, and the generator A of $(N_t)_{t\geq 0}$ given by $Af(z) = f(z+1) - f(z)$, $z \in \mathbb{Z}$, is not symmetric. Thus

$$(f, Ag)_{L^2(\mathbb{Z})} = \sum_{z\in\mathbb{Z}} \bar{f}(z)(g(z+1) - g(z)) = \sum_{z\in\mathbb{Z}} (\bar{f}(z-1) - \bar{f}(z))g(z) \neq (Af, g)_{L^2(\mathbb{Z})}.$$

However, σ_t is shift invariant. We consider this situation more generally next.

Proposition 3.44 (Time-shift invariance). *Let $(Y_t)_{t\in\mathbb{R}}$ be an $\mathbb{R}^d$-valued two-sided Lévy process with $Y_0 = 0$ almost surely, and ρ a function on $\mathbb{R}^d$ such that $\rho(-x) = \rho(x)$ for $x \in \mathbb{R}^d$. Define*

$$Z_t = \rho(Y_t), \quad t \in \mathbb{R}.$$

Let λ be a shift invariant measure on $\mathbb{R}^d$, i. e., $\lambda(A + a) = \lambda(A)$ for every $A \in \mathcal{B}(\mathbb{R}^d)$ and $a \in \mathscr{A}$. Let $t_1 \leq \cdots \leq t_n \leq 0 \leq t_{n+1} \leq \cdots \leq t_{n+m}$. Then for all bounded functions f_i, $i = 2, \ldots, n+m$, and $f_1, f_{n+m} \in L^2(\mathbb{R}^d, \lambda(dx))$, it follows that

$$\int_{\mathbb{R}^d} \mathbb{E}^x \left[\prod_{i=1}^{n+m} f_i(Z_{t_i}) \right] \lambda(dx) = \int_{\mathbb{R}^d} \mathbb{E}^x \left[\prod_{i=1}^{n+m} f_i(Z_{t_i+s}) \right] \lambda(dx) \tag{3.2.14}$$

for all $-t_{n+m} \leq s \leq -t_0$. Here

$$\int_{\mathbb{R}^d} \mathbb{E}^x \left[\prod_{i=1}^{n} f_i(Z_{t_i}) \right] \lambda(dx) = \int_{\mathbb{R}^d} \mathbb{E} \left[\prod_{i=1}^{n} f_i(\rho(Y_{t_i} + x)) \right] \lambda(dx).$$

Proof. Let μ_t be the distribution of Y_t on $\mathbb{R}^d$. The distribution of $-Y_t$ is given by $\lambda_t = \mu_t \circ r$, where $r : \mathbb{R}^d \to \mathbb{R}^d$ is defined by $r(a) = -a$. Let $P_t(x, A) = \mathbb{E}[1_A(Y_t + x)]$ and $Q_t(x, A) = \mathbb{E}[1_A(-Y_t + x)]$ for $A \in \mathcal{B}(\mathbb{R}^d)$. Let $g_j = f_j \circ \rho$. Thus $f_j(Z_t) = g_j(Y_t) = g_j(-Y_t)$ follows. We have for $0 \leq s_1 \leq \cdots \leq s_n$,

$$\int_{\mathbb{R}^d} \mathbb{E}^x \left[\prod_{j=1}^{n} g_j(Y_{s_j}) \right] \lambda(dx) = \int_{\mathbb{R}^d \times \mathbb{R}^{dn}} \left(\prod_{j=1}^{n} g_j(x_j) \right) \prod_{j=1}^{n} P_{s_j - s_{j-1}}(x_{j-1}, dx_j) \lambda(dx)$$
$$= \int_{\mathbb{R}^d \times \mathbb{R}^{dn}} \left(\prod_{j=1}^{n} g_j(x_j) \right) \prod_{j=1}^{n} Q_{s_j - s_{j-1}}(x_{j-1}, dx_j) \lambda(dx),$$

where $s_0 = 0$ and $x_0 = x$. Although $t_1 \leq \cdots \leq t_n \leq 0 \leq t_{n+1} \leq \cdots \leq t_{n+m}$, for the reader's convenience, we assume that $t_j \geq 0$ for $j = 1, \ldots, n+m$ and $-t_1 \leq \cdots \leq -t_n \leq 0 \leq t_{n+1} \leq \cdots \leq t_{n+m}$ in the proof below, and show that

$$\int_{\mathbb{R}^d} \mathbb{E}^x \left[\prod_{i=1}^{n} g_i(Y_{-t_i}) \prod_{j=n+1}^{n+m} g_j(Y_{t_j}) \right] \lambda(dx) = \int_{\mathbb{R}^d} \mathbb{E}^x \left[\prod_{i=1}^{n} g_i(Y_{t_1 - t_i}) \prod_{j=n+1}^{n+m} g_j(Y_{t_1 + t_j}) \right] \lambda(dx). \tag{3.2.15}$$

Thus (3.2.14) follows from (3.2.15). We have

$$\int_{\mathbb{R}^d} \mathbb{E}^x \left[\prod_{i=1}^{n} g_i(Y_{-t_i}) \prod_{j=n+1}^{n+m} g_j(Y_{t_j}) \right] \lambda(dx) = \int_{\mathbb{R}^d} \mathbb{E}^x \left[\prod_{i=1}^{n} g_i(Y_{t_i}) \right] \mathbb{E}^x \left[\prod_{j=n+1}^{n+m} g_j(Y_{t_j}) \right] \lambda(dx).$$

Hence we compute the left-hand side above as

$$\text{l. h. s. (3.2.15)} = \int_{\mathbb{R}^d} \lambda(dx) \int_{\mathbb{R}^{d(n+m)}} \left(\prod_{i=1}^{n+m} g_i(x_i) \right) Q_{t_n}(x, dx_n) \left(\prod_{i=n}^{2} Q_{t_{i-1} - t_i}(x_i, dx_{i-1}) \right)$$
$$\times P_{t_{n+1}}(x, dx_{n+1}) \prod_{j=n+2}^{n+m} P_{t_j - t_{j-1}}(x_{j-1}, dx_j).$$

By Lemma 3.45 below, we have the equality of measures on $\mathbb{R}^d \times \mathbb{R}^d$, i. e.,

$$\int_{\mathbb{R}^d} Q_{t_n}(x, dx_n) P_{t_{n+1}}(x, dx_{n+1}) \lambda(dx) = P_{t_n + t_{n+1}}(x_n, dx_{n+1}) \lambda(dx_n).$$

Thus we obtain

$$\text{l. h. s. (3.2.15)} = \int_{\mathbb{R}^{d(n+m)}} \left(\prod_{i=1}^{n+m} g_i(x_i) \right) \left(\prod_{i=2}^{n} Q_{t_{i-1} - t_i}(x_i, dx_{i-1}) \right) \lambda(dx_n)$$
$$\times P_{t_n + t_{n+1}}(x_n, dx_{n+1}) \prod_{j=n+2}^{n+m} P_{t_j - t_{j-1}}(x_{j-1}, dx_j). \tag{3.2.16}$$

On the other hand,

$$\text{r. h. s. (3.2.15)} = \int_{\mathbb{R}^{d(n+m)}} \left(\prod_{i=1}^{n+m} g_i(x_i)\right) \lambda(dx_1) \left(\prod_{i=2}^{n} P_{t_{i-1}-t_i}(x_{i-1}, dx_i)\right)$$
$$\times P_{t_n+t_{n+1}}(x_n, dx_{n+1}) \prod_{j=n+2}^{n+m} P_{t_j-t_{j-1}}(x_{j-1}, dx_j). \tag{3.2.17}$$

By Lemma 3.45 again, we also see that

$$\left(\prod_{i=2}^{n} Q_{t_{i-1}-t_i}(x_i, dx_{i-1})\right) \lambda(dx_n) = \lambda(dx_1) \left(\prod_{i=2}^{n} P_{t_{i-1}-t_i}(x_{i-1}, dx_i)\right). \tag{3.2.18}$$

Comparing (3.2.16) and (3.2.17) with (3.2.18), we can conclude (3.2.15) and hence (3.2.14) follows for any $s \in \mathbb{R}$. □

To complete the proof of Proposition 3.44, we show Lemma 3.45.

Lemma 3.45. *Let the assumptions of Proposition* 3.44 *hold.*
(1) *Let* $P_t(x, A) = \mathbb{E}[1_A(Y_t + x)]$ *and* $Q_t(x, A) = \mathbb{E}[1_A(-Y_t + x)]$. *Then*

$$\int_{\mathbb{R}^d} Q_s(x, dy) P_t(x, dz) \lambda(dx) = P_{s+t}(y, dz) \lambda(dy). \tag{3.2.19}$$

(2) *Equality* (3.2.18) *holds.*

Proof. (1) We denote the left-hand side of (3.2.19) by m and the right-hand side by n. We have

$$m(A \times B) = \int_{\mathbb{R}^d} Q_s(x, A) P_t(x, B) \lambda(dx) = \int_{\mathbb{R}^d \times \mathbb{R}^d} \rho(-y + z) \lambda_s(dy) \mu_t(dz),$$

where $\rho(X) = \int_{\mathbb{R}^d} 1_A(x) 1_B(x + X) \lambda(dx)$. As $\lambda_t(A) = \mu_t(-A)$, we see that

$$m(A \times B) = \int_{\mathbb{R}^d \times \mathbb{R}^d} \rho(y + z) \mu_s(dy) \mu_t(dz) = \int_{\mathbb{R}^d} \rho(p) (\mu_s * \mu_t)(dp)$$
$$= \int_{\mathbb{R}^d} \rho(p) \mu_{s+t}(dp) = \int_{\mathbb{R}^d \times \mathbb{R}^d} 1_A(x) 1_B(x + p) \mu_{s+t}(dp) \lambda(dx) = n(A \times B),$$

which implies (3.2.19).

(2) We proceed by induction. For ease of notation, we write $P_{j-1} = P_{t_{j-1}-t_j}$, $Q_{j-1} = Q_{t_{j-1}-t_j}$, and denote the distributions of Y_{t_j} and $-Y_{t_j}$ by λ_j and μ_j, respectively. Assume $n = 2$, Let $A = A_1 \times A_2$. Since $\lambda_j(-A) = \mu_j(A)$ and $\lambda(A + a) = \lambda(A)$ for $a \in \mathscr{A}$, we see that

$$\int_A P_1(x_1, dx_2) \lambda(dx_1) = \int_{\mathbb{R}^d} \lambda(dx_1) \int_{\mathbb{R}^d} 1_{A_1}(x_1) 1_{A_2}(x_1 + y) \lambda_1(dy)$$

$$= \int_{\mathbb{R}^d} \lambda(dx_1) \int_{\mathbb{R}^d} 1_{A_1}(x_1+y)1_{A_2}(x_1)\mu_1(dy)$$
$$= \int_{\mathbb{R}^d} \lambda(dx_2) \int_{\mathbb{R}^d} 1_{A_1}(x_2+y)1_{A_2}(x_2)\mu_1(dy) = \int_A Q_1(x_2, dx_1)\lambda(dx_2).$$

Hence $\lambda(dx_1)P_1(x_1, dx_2) = Q_1(x_2, dx_1)\lambda(dx_2)$ follows. Suppose that (3.2.18) holds for $n = 2, \dots, k$. Let $A = A_1 \times \dots \times A_{k+1}$. We have

$$\int_A \prod_{i=1}^{k} P_i(x_i, dx_{i+1})\lambda(dx_1) = \int_{A_1\times\dots\times A_k} \left(\prod_{i=1}^{k-1} P_i(x_i, dx_{i+1})\right) P_k(x_k, A_{k+1})\lambda(dx_1)$$
$$= \int_{A_1\times\dots\times A_k} \left(\prod_{i=1}^{k-1} Q_i(x_{i+1}, dx_i)\right) P_k(x_k, A_{k+1})\lambda(dx_k) = \int_{A_k} \xi(x_k)P_k(x_k, A_{k+1})\lambda(dx_k).$$

The second equality follows by the induction hypothesis, and ξ is given by

$$\xi(x_k) = \int_{A_1\times\dots\times A_{k-1}} \prod_{i=1}^{k-1} Q_i(x_{i+1}, dx_i).$$

Since $\lambda(dx_k)P_k(x_k, dx_{k+1}) = Q_k(x_{k+1}, dx_k)\lambda(dx_{k+1})$ as measures, we see that

$$\int_A \lambda(dx_1)\left(\prod_{i=1}^{k} P_i(x_i, dx_{i+1})\right) = \int_{A_k\times A_{k+1}} \xi(x_k)1_{A_{k+1}}(x_{k+1})\lambda(dx_k)P_k(x_k, dx_{k+1})$$
$$= \int_{A_k\times A_{k+1}} \xi(x_k)1_{A_{k+1}}(x_{k+1})Q_k(x_{k+1}, dx_k)\lambda(dx_{k+1}) = \int_A \left(\prod_{i=1}^{k} Q_i(x_{i+1}, dx_i)\right)\lambda(dx_{k+1})$$

follows. This implies that (3.2.18) also holds for $n = k + 1$. □

Finally, we consider bridge measures for Lévy processes. Let $(X_t)_{t\geq 0}$ be a Lévy process on a probability space $(\Omega, \mathcal{F}, P^X)$, and recall that p_t^X denotes the probability transition kernel of $(X_t)_{t\geq 0}$.

Definition 3.46 (Lévy bridge measure). Given a partition $0 \leq T_1 \leq t_1 < \dots < t_n \leq T_2$ and $x, y \in \mathbb{R}^d$, the probability measure on $(\mathscr{X}_D, \mathcal{B}(\mathscr{X}_D))$ defined by

$$P^{x,y}_{[T_1,T_2]}(X_{t_1} \in A_1, \dots, X_{t_n} \in A_n)$$
$$= \int_{\mathbb{R}^{nd}} \frac{p^X_{t_1-T_1}(x - x_1)p^X_{t_2-t_1}(x_1 - x_2)\cdots p^X_{T_2-t_n}(x_n - y)}{p^X_{T_2-T_1}(x - y)} \left(\prod_{i=1}^{n} 1_{A_i}(x_i)\right)\prod_{i=1}^{n} dx_i \qquad (3.2.20)$$

is called a *Lévy bridge measure* starting in x at $t = T_1$ and ending in y at $t = T_2$. Furthermore, the nonnormalized measure given by $p^X_{T_2-T_1}(x-y)P^{x,y}_{[T_1,T_2]}$ is called a *conditional Lévy measure.*

Write $\mathbb{E}^{x,y}_{[T_1,T_2]}$ for the expectation with respect to $P^{x,y}_{[T_1,T_2]}$. From the definition, it is directly seen that P^x and $P^{x,y}_{[T_1,T_2]}$ are related through

$$\mathbb{E}^x\left[\prod_{j=1}^n f_j(X_{t_j})\right] = \int_{\mathbb{R}^d} p^X_{T_2-T_1}(x-y)\mathbb{E}^{x,y}_{[T_1,T_2]}\left[\prod_{j=1}^n f_j(X_{t_j})\right] dy$$

with $T_1 \leq t_1 \leq \cdots \leq t_n \leq T_2$ and for every Borel measurable bounded function f_j. In particular,

$$\mathbb{E}^x\left[f(X_{T_1})\left(\prod_{j=1}^n f_j(X_{t_j})\right)g(X_{T_2})\right] = \int_{\mathbb{R}^d} p^X_{T_2-T_1}(x-y)f(x)g(y)\mathbb{E}^{x,y}_{[T_1,T_2]}\left[\prod_{j=1}^n f_j(X_{t_j})\right] dy.$$

3.3 Random measures and Lévy–Itô decomposition

3.3.1 Poisson random measures

Definition 3.47 (Random measure). Let $(S, \mathscr{S})$ be a measurable space and $(\Omega, \mathscr{F}, P)$ a probability space. Furthermore, let $M = (M(A))_{A\in\mathscr{S}}$ be a collection of random variables $M(A)(\cdot) : \Omega \to \mathbb{R}$ indexed by $A \in \mathscr{S}$; M is a *random measure* whenever
(1) $M(\emptyset) = 0$;
(2) $M(\bigcup_{n=1}^\infty A_n) = \sum_{n=1}^\infty M(A_n)$ almost surely, provided that $A_n \cap A_m = \emptyset$ for $n \neq m$;
(3) $M(A)$ and $M(B)$ are independent whenever $A \cap B = \emptyset$.

The following is a random measure of fundamental interest.

Definition 3.48 (Poisson random measure). If, in addition to (1)–(3) of Definition 3.47, every $M(A)$ has a Poisson distribution with intensity $\lambda(A)$, i. e., if $\lambda(A) < \infty$,

$$P(M(A) = n) = e^{-\lambda(A)}\frac{\lambda(A)^n}{n!} \tag{3.3.1}$$

holds, with $\lambda(A) = \infty$ giving $M(A) = \infty$, then M is called a *Poisson random measure.*

From (3.3.1) it follows that $\lambda(A) = \mathbb{E}[M(A)]$, and it is a measure on $\mathscr{S}$ since M is a random measure. Remarkably, a converse property also holds.

Theorem 3.49. *Given a σ-finite measure λ on a measurable space $(S, \mathscr{S})$, there exist a probability space $(\Omega, \mathscr{F}, P)$ and a Poisson random measure $M = (M(A))_{A\in\mathscr{S}}$ such that $\mathbb{E}[M(A)] = \lambda(A)$ for $A \in \mathscr{S}$.*

Proof. Let $U_n \in \mathscr{S}, n \in \mathbb{N}$, such that $0 < \lambda(U_n) < \infty$ and $\bigcup_{m=1}^{\infty} U_m = S$. We take random variables $\{X_n(m)\}_{n\in\mathbb{N}}, m \in \mathbb{N}$, and $\{N_n\}_{n\in\mathbb{N}}$ on $(\Omega, \mathscr{F}, P)$ such that

(1) for all n, m, $X_n(m)$ is a U_n-valued random variable with $P(X_n(m) \in A) = \lambda(A)/\lambda(U_n)$, for $A \in \mathscr{S}$;
(2) for all n, we have that N_n is an integer-valued random variable such that

$$P(N_n = k) = \frac{1}{k!}\lambda(U_n)^k e^{-\lambda(U_n)}, \quad k = 0, 1, 2, \ldots$$

(3) $X_n(m)$ and N_n, $n, m \in \mathbb{N}$, are mutually independent.

Write $M(A) = \sum_{m=1}^{\infty} M_m(A)$, where $M_m(A) = \sum_{n=1}^{N_m} 1_{A\cap U_m}(X_n(m))1_{\{N_m\geq 1\}}$, for $A \in \mathscr{S}$. Let $A_1, \ldots, A_k$ be pairwise disjoint sets and $\bigcup_{j=1}^{k} A_j = \Omega$, such that $n_1 + \cdots + n_k = n$. We have

$$\begin{aligned}
&P(M_m(A_1) = n_1, \ldots, M_m(A_k) = n_k) \\
&\quad = P(M_m(A_1) = n_1, \ldots, M_m(A_k) = n_k | M_m(\Omega) = n)P(M_m(\Omega) = n).
\end{aligned}$$

Note that $M_m(\Omega) = N_m$, and we arrive at the multinomial distribution

$$\begin{aligned}
&P(M_m(A_1) = n_1, \ldots, M_m(A_k) = n_k | M_m(\Omega) = n) \\
&\quad = P\left(\sum_{j=1}^{n} 1_{A_1\cap U_m}(X_j(m)) = n_1, \ldots, \sum_{j=1}^{n} 1_{A_k\cap U_m}(X_j(m)) = n_k\right) \\
&\quad = \frac{n!}{n_1!\cdots n_k!}\left(\frac{\lambda(A_1\cap U_m)}{\lambda(U_m)}\right)^{n_1}\cdots\left(\frac{\lambda(A_k\cap U_m)}{\lambda(U_m)}\right)^{n_k}.
\end{aligned}$$

We furthermore obtain

$$\begin{aligned}
&P(M_m(A_1) = n_1, \ldots, M_m(A_k) = n_k) \\
&\quad = \frac{n!}{n_1!\cdots n_k!}\left(\frac{\lambda(A_1\cap U_m)}{\lambda(U_m)}\right)^{n_1}\cdots\left(\frac{\lambda(A_k\cap U_m)}{\lambda(U_m)}\right)^{n_k}\frac{\lambda(U_m)^n}{n!}e^{-\lambda(U_m)} \\
&\quad = \prod_{j=1}^{k} e^{-\lambda(A_j\cap U_m)}\frac{\lambda(A_j\cap U_m)^{n_j}}{n_j!}.
\end{aligned}$$

Summing over $n_1, \ldots, n_k$ except for n_j, we have

$$P(M_m(A_j) = n_j) = e^{-\lambda(A_j\cap U_m)}\frac{\lambda(A_j\cap U_m)^{n_j}}{n_j!}.$$

Thus $M_m(A)$ is a Poisson random variable with intensity $\lambda(A \cap U_m)$ for every A. Since $M(A) = \sum_{m=1}^{\infty} M_m(A)$ and $(M_m(A))_{m\geq 1}$ are independent of each other, and

$$\mathbb{E}[M(A)] = \sum_{m=1}^{\infty}\mathbb{E}[M_m(A)] = \sum_{m=1}^{\infty}\lambda(A\cap U_m) = \lambda(A),$$

it follows that $M(A)$ is a Poisson process whenever $\lambda(A) < \infty$. In case $\lambda(A) = \infty$, we have

$$\sum_{m=1}^{\infty} P(M_m(A) \geq 1) = \sum_{m=1}^{\infty} (1 - e^{-\lambda(A \cap U_m)}) \geq \sum_{m=1}^{\infty} \frac{1}{2}\lambda(A \cap U_m) \wedge a = \infty$$

with $a > 0$. Hence $P(M_m(A) \geq 1$ for infinitely many $m) = 1$ by the Borel–Cantelli lemma. Thus $M(A) = \infty$ almost surely. This proves that $(M(A))_{A \in \mathscr{S}}$ is a Poisson random measure with $\mathbb{E}[M(A)] = \lambda(A)$, for all $A \in \mathscr{S}$. □

Next we define the *counting measure* associated with a Lévy process $(X_t)_{t\geq 0}$ on a probability space $(\Omega, \mathscr{F}, P)$. Every càdlàg function is bounded on compact intervals and attains its extrema there; moreover, every such function is Borel measurable. Denote $X_{t-} = \lim_{s \uparrow t} X_s$, $t > 0$, and define the jump at t by

$$\Delta X_t = X_t - X_{t-}. \tag{3.3.2}$$

A càdlàg path can only have jump discontinuities, and the set

$$\mathcal{J} = \{t \in [0, \infty) \mid \Delta X_t \neq 0\} \tag{3.3.3}$$

of time points where a jump occurs is at most countable but does not have accumulation points. The Lévy processes with symbol (A, b, ν) can be classified by their jump properties:

(type I) $A = 0$ and $\nu(\mathbb{R}^d \setminus \{0\}) < \infty$;

(type II) $A = 0$, $\nu(\mathbb{R}^d \setminus \{0\}) = \infty$ and $\int_{0<|y|\leq 1} |y| \nu(dy) < \infty$;

(type III) $A \neq 0$ or $\int_{0<|y|\leq 1} |y| \nu(dy) = \infty$.

A type III process has a finite number of jumps in every bounded interval. If $\nu(\mathbb{R}^d \setminus \{0\}) = \infty$ holds, then $\mathcal{J}$ is a countable dense subset of $[0, \infty)$. If a process is of type I or II, then the paths are of bounded variation on every bounded interval, while if it is of type III, then paths are of unbounded variation on any interval.

The *counting measure* $N(t, \cdot) : \mathcal{B}(\mathbb{R}^d \setminus \{0\}) \to \mathbb{N} \cup \{0\}$ adds up jumps of specific size, i. e., for $0 \leq t$ and $A \in \mathcal{B}(\mathbb{R}^d \setminus \{0\})$,

$$N(t, A) = |\{0 \leq s \leq t \mid \Delta X_s \in A\}|\,. \tag{3.3.4}$$

It is seen that $N(t, \cdot)(\omega)$ is a measure on $\mathcal{B}(\mathbb{R}^d \setminus \{0\})$ and $\mathbb{E}[N(t, \cdot)]$ is a Borel measure on $\mathcal{B}(\mathbb{R}^d \setminus \{0\})$, while

$$\mu(\cdot) = \mathbb{E}[N(1, \cdot)] \tag{3.3.5}$$

is called *intensity measure*. In Theorem 3.58 below, we will see that μ coincides with the Lévy measure ν of a Lévy process $(X_t)_{t\geq 0}$. When the expectation of the number of small jumps at $t = 1$ of a Lévy process is zero, the Lévy symbol is simply

$$\eta(u) = ib \cdot u - \frac{1}{2} u \cdot Au + \int_{|y| \geq 1} (e^{iu \cdot y} - 1)\nu(dy).$$

Next we discuss the random jump times of Lévy processes.

Lemma 3.50. *Let $(X_t)_{t \geq 0}$ be a Lévy process on a probability space $(\Omega, \mathscr{F}, P)$ such that it is almost surely increasing. Define a sequence of stopping times $(\tau_n)_{n \in \mathbb{N}}$ with respect to the natural filtration $(\mathscr{F}_t)_{t \geq 0}$ of $(X_t)_{t \geq 0}$ recursively by*

$$\tau_0 = 0 \quad \text{and} \quad \tau_n = \inf\left\{t > \tau_{n-1} \mid X_t - X_{\tau_{n-1}} \neq 0\right\}, \qquad n \in \mathbb{N}. \tag{3.3.6}$$

Then $\tau_1, \tau_2 - \tau_1, \dots, \tau_n - \tau_{n-1}, \dots$ are independent and identically distributed random variables.

Proof. We have $\{\tau_n - \tau_{n-1} > t\} = \{X_t - X_{\tau_{n-1}} = 0\}$. Thus $\{\tau_n - \tau_{n-1} > t\}$ is independent of $\mathscr{F}_{\tau_{n-1}}$ by the strong Markov property of $(X_t)_{t \geq 0}$, and so $\tau_n - \tau_{n-1}$ is independent of $\mathscr{F}_{\tau_{n-1}}$. On the other hand, $\tau_n - \tau_{n-1}$ is $\mathscr{F}_{\tau_n}$-measurable. In particular, $\tau_n - \tau_{n-1}$ and $\tau_m - \tau_{m-1}$ are independent for $n \neq m$. By the strong Markov property of $(X_t)_{t \geq 0}$, we see that

$$\tau_n - \tau_{n-1} = \inf\{t > 0 \mid X_{t+\tau_{n-1}} - X_{\tau_{n-1}} \neq 0\} \stackrel{\mathrm{d}}{=} \inf\{t > 0 \mid X_t - X_0 \neq 0\} = \tau_1.$$

Thus $\tau_1, \tau_2 - \tau_1, \dots, \tau_n - \tau_{n-1}, \dots$ are independent and identically distributed. □

The following result gives a sufficient condition for a Lévy process to be a Poisson process.

Lemma 3.51. *Let $(X_t)_{t \geq 0}$ be a Lévy process on a probability space $(\Omega, \mathscr{F}, P)$ such that it is almost surely increasing and $\Delta X_t \in \{0, 1\}$, for all $t \geq 0$. Then $(X_t)_{t \geq 0}$ is a Poisson process.*

Proof. Define a sequence of stopping times $(\tau_n)_{n \in \mathbb{N}}$ by (3.3.6). Note that if $X_{t+s} = 0$, then $X_s = 0$. We have

$$\begin{aligned} P(\tau_1 > t + s) &= P(X_{t+s} = 0) = P(X_s = 0, X_{t+s} - X_s = 0) \\ &= P(X_s = 0)P(X_{t+s} - X_s = 0) = P(X_s = 0)P(X_t = 0) = P(\tau_1 > s)P(\tau_1 > t). \end{aligned} \tag{3.3.7}$$

Note that $t \mapsto P(\tau_1 > t)$ is decreasing, since $t \mapsto N_t$ is increasing, and $t \mapsto P(\tau_1 > t)$ is continuous at $t = 0$ since $t \mapsto N_t$ is stochastically continuous. Thus a solution to the functional equation (3.3.7) is of the form $P(\tau_1 > t) = e^{-\lambda t}$ with a suitable $\lambda > 0$. Thus

$$P(N_t = 0) = P(\tau_1 > t) = e^{-\lambda t}.$$

Hence $e^{-\lambda t} = \int_t^\infty P(s)ds$ for the distribution function $P(s)$ of τ_1, and then $P(s)ds = \lambda e^{-\lambda s} 1_{[0,\infty)}(s)ds$. By Lemma 3.50, $\tau_1, \tau_2 - \tau_1, \dots, \tau_n - \tau_{n-1}, \dots$ are independent and identically distributed. Noting that $\tau_n = (\tau_n - \tau_{n-1}) + \cdots + (\tau_1 - \tau_0)$, the distribution of τ_n is given by $e^{-\lambda s} \frac{\lambda^n s^{n-1}}{(n-1)!} 1_{[0,\infty)}(s)ds$. Then for $k \geq 0$,

$$
\begin{aligned}
P(X_t = n) &= P(\tau_{n+1} > t, \tau_n \le t) = P(\tau_{n+1} > t) - P(\tau_n > t)\\
&= \int_t^\infty e^{-\lambda s} \frac{\lambda^{n+1} s^n}{n!} ds - \int_t^\infty e^{-\lambda s} \frac{\lambda^n s^{n-1}}{(n-1)!} ds = \frac{(\lambda t)^n}{n!} e^{-\lambda t},
\end{aligned}
$$

showing that $(X_t)_{t\ge0}$ is a Poisson process. □

Lemma 3.52. *Let $(X_t)_{t\ge0}$ be a Lévy process on a probability space $(\Omega, \mathcal{F}, P)$. Define $N(t, A)$ by (3.3.4). Then $N(t, A) < \infty$ almost surely, for all $A \in \mathcal{B}(\mathbb{R}^d \setminus \{0\})$ such that $0 \notin \bar{A}$.*

Proof. Define a sequence of stopping times $(\tau_n)_{n\in\mathbb{N}}$ by $\tau_1 = \inf\{t > 0 \mid \Delta X_t \in A\}$ and $\tau_{n+1} = \inf\{t > \tau_n \mid \Delta X_t \in A\}$, for $n \ge 1$. Since the set of jumps in A has no accumulation point as $0 \notin \bar{A}$, we have $\tau_1 > 0$ and $\tau_n \to \infty$ as $n \to \infty$. Therefore we have $N(t, A) = \sum_{n=1}^{\infty} 1_{\{\tau_n \le t\}} < \infty$. □

Theorem 3.53. *Let $(X_t)_{t\ge0}$ be a Lévy process on a probability space $(\Omega, \mathcal{F}, P)$. Define $N(t, A)$ by (3.3.4). The following properties hold:*
(1) *Let $A \in \mathcal{B}(\mathbb{R}^d \setminus \{0\})$ such that $0 \notin \bar{A}$. Then $(N(t, A))_{t\ge0}$ is a Poisson process with intensity $\mu(A)$.*
(2) *The random processes $N(t, A_1), \dots, N(t, A_n)$ are independent for all pairwise disjoint sets $A_j \in \mathcal{B}(\mathbb{R}^d \setminus \{0\})$ such that $0 \notin \bar{A}_j$, $j = 1, \dots, n$.*

Proof. Note that $\Delta N(t, A) = N(t, A) - N(t-, A)$ takes values in $\{0, 1\}$ and $t \mapsto N(t, A)$ is increasing. By Lemma 3.51, it is sufficient to show that $(N(t, A))_{t\ge0}$ is a Lévy process. We have $N(t, A) - N(s, A) = n$ if and only if there exists $s < t_1 < \dots < t_n \le t$ such that $\Delta X_{t_j} \in A$, $j = 1, \dots, n$. Furthermore, $\Delta X_u \in A$ if and only if there exists $a \in A$ for which, given $\varepsilon > 0$, there exists $\delta > 0$ such that $|u - v| < \delta$ implies that $|X_v - X_u - a| < \varepsilon$. Thus we deduce that

$$\{N(t, A) - N(s, A) = n\} \in \sigma(X_v - X_u, s \le u < v \le t), \tag{3.3.8}$$

giving $\sigma(N(t, A) - N(s, A)) \in \sigma(X_v - X_u \mid s \le u < v \le t)$. This implies that the increments of $(N(t, A))_{t\ge0}$ are independent. Since

$$
\begin{aligned}
N(t, A) - N(s, A) &= |\{s < r \le t \mid \Delta X_r \in A\}| = |\{0 < r \le t - s \mid \Delta X_r \in A\}|\\
&= N(t - s, A) - N(0, A),
\end{aligned}
$$

the process $(N(t, A))_{t\ge0}$ is stationary. Next we show stochastic continuity. Using the independence of increments, we see that

$$
\begin{aligned}
&P(N(t, A) = 0)\\
&= P\left(N\left(\frac{t}{n}, A\right) = 0, N\left(\frac{2t}{n}, A\right) - N\left(\frac{t}{n}, A\right) = 0, \dots, N(t, A) - N\left(\frac{(n-1)t}{n}, A\right) = 0\right)\\
&= P\left(N\left(\frac{t}{n}, A\right) = 0\right)^n.
\end{aligned}
$$

This gives

$$\limsup_{t\to 0} P(N(t,A)=0) = \lim_{n\to\infty} \limsup_{t\to 0} P(N(t/n,A)=0)^n = \lim_{n\to\infty} \limsup_{s\to 0} P(N(s,A)=0)^n,$$
$$\liminf_{t\to 0} P(N(t,A)=0) = \lim_{n\to\infty} \liminf_{t\to 0} P(N(t/n,A)=0)^n = \lim_{n\to\infty} \liminf_{s\to 0} P(N(s,A)=0)^n.$$

We consider the following two cases:
(1) $\lim_{t\to 0} P(N(t,A)=0)$ exists, i. e., $\lim_{t\to 0} P(N(t,A)=0) = 1$ or $= 0$.
(2) $\lim_{t\to 0} P(N(t,A)=0)$ does not exist, i. e., $\limsup_{t\to 0} P(N(t,A)=0) = 1$ and $\liminf_{t\to 0} P(N(t,A)=0) = 0$.

Suppose case (2) holds. Since $t \mapsto P(N(t,A)=0)$ is decreasing, if $P(N(t,A)=0) > \varepsilon$ for some t, then $\liminf_{t\to 0} P(N(t,A)=0) \geq \varepsilon$. Hence if $\liminf_{t\to 0} P(N(t,A)=0) = 0$, $P(N(t,A)=0) = 0$ for all t and so $\limsup_{t\to 0} P(N(t,A)=0) = 0$, which yields a contradiction. Thus (2) cannot hold. Assume case (1) and $\lim_{t\to 0} P(N(t,A)=0) = 0$. Thus $\lim_{t\to 0} P(N(t,A)\neq 0) = 1$. Let A and B be disjoint. We have

$$\lim_{t\to 0} P(N(t,A\cup B)\neq 0) = \lim_{t\to 0}(P(N(t,A)\neq 0) + P(N(t,B)\neq 0)).$$

Hence we deduce that $\lim_{t\to 0} P(N(t,A)=0) = 1$ and $\lim_{t\to 0} P(N(t,A)\neq 0) = 0$, which implies that $\lim_{t\to 0} P(N(t,A) > \varepsilon) = 0$ for arbitrary $\varepsilon > 0$. Thus part (1) of the proposition is proven. For part (2), in a similar way as (3.3.8) was obtained, we see that the processes $\{N(t,A_1)=n_1\},\dots,\{N(t,A_m)=n_m\}$ are elements of an independent σ-field. □

The difference between the counting measure $N(t,\cdot)$ and the intensity measure μ will be denoted by

$$\tilde N(t,A) = N(t,A) - t\mu(A), \quad A \in \mathcal{B}(\mathbb{R}^d\setminus\{0\}). \tag{3.3.9}$$

If $0 \notin \bar A$, $(N(t,A))_{t\geq 0}$ is a Poisson process and then $(\tilde N(t,A))_{t\geq 0}$ is called the associated *compensated Poisson process*. Here are some basic properties of $N(t,A)$ and $\tilde N(t,A)$.
(1) For every $t > 0$ and $\omega \in \Omega$, $N(t,\cdot)(\omega)$ is a counting measure on the measurable space $(\mathbb{R}^d\setminus\{0\}, \mathcal{B}(\mathbb{R}^d\setminus\{0\}))$.
(2) For every $A \in \mathcal{B}(\mathbb{R}^d\setminus\{0\})$ such that $0 \notin \bar A$, $(N(t,A))_{t\geq 0}$ is a Poisson process with intensity $\mu(A) = \mathbb{E}[N(1,A)]$, and $(\tilde N(t,A))_{t\geq 0}$ is a martingale.

Next consider the integral $\int_A f(x)N(t,dx)(\omega)$, which can be defined for every $\omega \in \Omega$, $t \geq 0$, $A \in \mathcal{B}(\mathbb{R}^d\setminus\{0\})$, and suitable f. Suppose that $0 \notin \bar A$. For a step function $f(x) = \sum_{j=1}^n a_j 1_{A_j}(x)$, we have

$$\int_A f(x)N(t,dx)(\omega) = \sum_{j=1}^n a_j N(t, A\cap A_j)(\omega) = \sum_{x\in A}\sum_{j=1}^n a_j 1_{A_j}(x) N(t,\{x\})(\omega),$$

where $\sum_{x\in A}$ is the sum over jumps $x \in A$ such that $x = \Delta X_s$ with some $0 \le s \le t$. In particular, $\sum_{x\in A}$ is a finite sum for each path $\omega \in \Omega$. For a nonnegative $\mathcal{B}(\mathbb{R}^d \setminus \{0\})$-measurable function f, there exists a sequence of step functions such that $f_n(x) \uparrow f(x)$ for every x. It follows that

$$\int_A f(x)N(t,dx) = \sum_{x\in A} f(x)N(t,\{x\}) \quad \text{a.s.,} \tag{3.3.10}$$

and (3.3.10) holds for general $\mathcal{B}(\mathbb{R}^d \setminus \{0\})$-measurable functions.

We have the following on related characteristic functions.

Theorem 3.54. *Let $(X_t)_{t\ge 0}$ be a Lévy process on a probability space $(\Omega, \mathscr{F}, P)$ and $N(t,\cdot)$ the Poisson random measure associated with $(X_t)_{t\ge 0}$. Let μ be the intensity measure of $N(t,\cdot)$. Suppose that $A \in \mathcal{B}(\mathbb{R}^d \setminus \{0\})$ and $0 \notin \bar{A}$.*

(1) *Let $f : \mathbb{R}^d \to \mathbb{R}^d$ be Borel measurable and $u \in \mathbb{R}^d$. We have*

$$\mathbb{E}\left[e^{i\int_A u\cdot f(x)N(t,dx)}\right] = e^{t\int_A (e^{iu\cdot f(x)}-1)\mu(dx)}.$$

(2) *If $f \in L^1(A,\mu)$, then $\mathbb{E}[\int_A f(x)N(t,dx)] = t\int_A f(x)\mu(dx)$.*

(3) *If $f \in L^2(A,\mu)$, then $\operatorname{cov}(\int_A f(x)N(t,dx)) = t\int_A |f(x)|^2\mu(dx)$.*

Proof. Let $f : \mathbb{R}^d \to \mathbb{R}^d$ be Borel measurable. For $y = (y_1,\dots,y_d) \in \mathbb{Z}^d$, define

$$C_y^n = \{x = (x_1,\dots,x_d) \in \mathbb{R}^d \mid 2^{-n}(y_j - 1) < x_j \le 2^{-n}y_j, 1 \le j \le d\}.$$

We have $\mathbb{R}^d = \bigcup_{y\in\mathbb{Z}^d} C_y^n$, where $C_y^n \cap C_z^n = \emptyset$ for $y \ne z$. Choose a point x_y^n in each C_y^n and define a function $f_n : A \to \mathbb{R}^d$ by $f_n(x) = x_y^n$ for $x \in f^{-1}(C_y^n)$. Thus $|f(x) - f_n(x)| \le \sqrt{d}/2^n$. Let $Y_n = \int_A f_n(x)N(t,dx)$ and $Y = \int_A f(x)N(t,dx)$. We have $|Y_n(\omega) - Y(\omega)| \le N(t,A)(\omega)\sqrt{d}/2^n$. Hence $|Y_n(\omega) - Y(\omega)| \to 0$ as $n \to \infty$, for every $\omega \in \Omega$. Since $Y_n = \sum_{y\in\mathbb{Z}^d} x_y^n N(t, f^{-1}(C_y^n))$, we compute

$$\mathbb{E}[e^{iu\cdot\int_A f_n(x)N(t,dx)}] = \prod_{y\in\mathbb{Z}^d} \mathbb{E}[e^{i(u\cdot x_y^n)N(t,f^{-1}(C_y^n))}] = \prod_{y\in\mathbb{Z}^d} e^{t(e^{i(u\cdot x_y^n)}-1)\mu(f^{-1}(C_y^n))} = e^{t\int_A (e^{iu\cdot f_n(x)}-1)\mu(dx)}.$$

We have $\int_A f_n(x)N(t,dx) \to \int_A f(x)N(t,dx)$ as $n \to \infty$ by (3.3.10). By the dominated convergence theorem, (1) holds. Replacing f by sf with $s \in \mathbb{R}$ in (1) and differentiating both sides above at $s = 0$, we obtain (2) and (3). □

We can extend Theorem 3.54 to Poisson random measures. Let $(S, \mathscr{S})$ be a measurable space and $M = (M(A))_{A\in\mathscr{S}}$ a Poisson random measure with intensity $\lambda(\cdot)$. Similarly to Theorem 3.54, we have the following result.

Theorem 3.55. *Let $A \in \mathcal{S}$.*
(1) *Suppose that $f : \mathbb{R}^d \to \mathbb{R}^d$ is Borel measurable and $u \in \mathbb{R}^d$. Then*

$$\mathbb{E}\left[e^{i\int_A u\cdot f(x)M(dx)}\right] = e^{\int_A (e^{iu\cdot f(x)}-1)\lambda(dx)}.$$

(2) *If $f \in L^1(A,\lambda)$, then $\mathbb{E}[\int_A f(x)M(dx)] = \int_A f(x)\lambda(dx)$.*
(3) *If $f \in L^2(A,\lambda)$, then $\operatorname{cov}(\int_A f(x)M(dx)) = \int_A |f(x)|^2\lambda(dx)$.*

A similar result holds also for the compensated Poisson random measure $\tilde{N}(t,A)$. Let $f \in L^1(A,\mu)$ and define

$$\int_A f(x)\tilde{N}(t,dx) = \int_A f(x)N(t,dx) - t\int_A f(x)\mu(dx). \tag{3.3.11}$$

Theorem 3.56. *Let $(X_t)_{t\geq 0}$ be a Lévy process on a probability space $(\Omega,\mathcal{F},P)$ and $N(t,\cdot)$ the Poisson random measure associated with $(X_t)_{t\geq 0}$. Let $A \in \mathcal{B}(\mathbb{R}^d \setminus \{0\})$ such that $0 \notin \bar{A}$, and $f \in L^1(A,\mu)$. Then we have the following:*
(1) $\mathbb{E}[e^{i\int_A f(x)\tilde{N}(t,dx)}] = e^{t\int_A (e^{if(x)}-1-if(x))\mu(dx)}$.
(2) $\mathbb{E}[\int_A f(x)\tilde{N}(t,dx)] = 0$.
(3) *In addition, suppose that $f \in L^2(A,\mu)$, $g \in L^2(B,\mu)$, and $0 \notin \bar{B} \in \mathcal{B}(\mathbb{R}^d \setminus \{0\})$. Then*

$$\mathbb{E}\left[\int_A f(x)\tilde{N}(t,dx)\int_B g(x)\tilde{N}(t,dx)\right] = t\int_{A\cap B} f(x)g(x)\mu(dx). \tag{3.3.12}$$

In particular, the random variables $\int_A f(x)\tilde{N}(t,dx)$ and $\int_B g(x)\tilde{N}(t,dx)$ are independent whenever $A \cap B = \emptyset$.
(4) *The random process $\left(\int_A f(x)\tilde{N}(t,dx)\right)_{t\geq 0}$ is a martingale with respect to the natural filtration $(\mathcal{F}_t)_{t\geq 0}$ of $(X_t)_{t\geq 0}$.*

Proof. The proofs of (1)–(3) are similar to those of Theorem 3.54. Let $f(x) = 1_B(x)$ for $B \in \mathcal{B}(\mathbb{R}^d \setminus \{0\})$. We have

$$Y_t = \int_A f(x)\tilde{N}(t,dx) = N(t,A\cap B) - t\mu(A\cap B).$$

Since the increments of $(N(t,A\cap B))_{t\geq 0}$ are independent and we have $\mathbb{E}[N(t,A\cap B)] = t\mu(A\cap B)$, it is direct that $\mathbb{E}[Y_t|\mathcal{F}_s] = Y_s$. We obtain (4) by a limiting argument. □

We have seen above that $(N(t,A))_{t\geq 0}$ is a Poisson process for all $A \in \mathcal{B}(\mathbb{R}^d \setminus \{0\})$ such that $0 \notin \bar{A}$. Now we will see also that $(\int_A f(x)N(t,dx))_{t\geq 0}$ is a compound Poisson process.

Proposition 3.57 (Compound Poisson process). *Let $(X_t)_{t\geq 0}$ be a Lévy process on a probability space $(\Omega,\mathcal{F},P)$ and $N(t,\cdot)$ the Poisson random measure associated with $(X_t)_{t\geq 0}$. Let*

$f : \mathbb{R}^d \to \mathbb{R}$ be measurable, and $A \in \mathcal{B}(\mathbb{R}^d \setminus \{0\})$ such that $0 \notin \bar{A}$. Define τ_n and Y_n by $\tau_n = \tau_n^A = \inf\{t > 0 \mid \Delta X_{t+\tau_{n-1}} \in A\}$ and

$$Y_n = \int_A f(x)N(\tau_n, dx) - \int_A f(x)N(\tau_{n-1}, dx).$$

Then we have the following:

(1) *$(\Delta X_{\tau_n})_{n\in\mathbb{N}}$ and $\{\tau_n\}_{n\in\mathbb{N}}$ are independent;*

(2) *Y_n, $n \in \mathbb{N}$, are independent;*

(3) *The random process $\left(\int_A f(x)N(t, dx)\right)_{t\geq 0}$ is a compound Poisson process, in particular, a Lévy process.*

3.3.2 Lévy–Itô decomposition

Lévy processes are a class large enough to contain random processes with jumps. A natural question is if there is a canonical way of separating the continuous part from the part with discontinuous paths. As we will see now, any Lévy process can be decomposed into a sum of four independent random processes accounting for these components. To obtain this for a Lévy process $(X_t)_{t\geq 0}$, we define the total sum of jumps pathwise. For $A \in \mathcal{B}(\mathbb{R}^d \setminus \{0\})$,

$$\int_0^t \int_A zN(dsdz) = \int_A zN(tdz) = \sum_{0\leq u\leq t} \Delta X_u 1_A(\Delta X_u)$$

is the sum of all the jumps taking values in set A up to time t. Since the paths of $(X_t)_{t\geq 0}$ are càdlàg, this is a finite sum of random variables.

Note that the random processes $(\int_A z\tilde{N}(t, dz))_{t\geq 0}$ and $(\int_B z\tilde{N}(t, dz))_{t\geq 0}$ are independent as soon as $A \cap B = \emptyset$, and $(\int_A z\tilde{N}(t, dz))_{t\geq 0}$ and $(\int_B zN(t, dz))_{t\geq 0}$ are independent whenever $A \cap B = \emptyset$ and $\int_{|z|\leq 1} |z|\mu(dz) < \infty$. Lemma 3.54 allows computing the characteristic functions of

$$Z_t^{(1)} = \int_{|z|\geq 1} zN(t, dz) \quad \text{and} \quad Z_t^{(2)}(\varepsilon) = \int_{\varepsilon<|z|<1} z\tilde{N}(t, dz),$$

giving $\mathbb{E}[e^{iuZ_t^{(1)}}] = e^{t\eta^{(1)}(u)}$ and $\mathbb{E}[e^{iuZ_t^{(2)}(\varepsilon)}] = e^{t\eta^{(2)}(u,\varepsilon)}$, respectively, where

$$\eta^{(1)}(u) = \int_{|\lambda|\geq 1} (e^{iu\cdot y} - 1)\mu(dy), \tag{3.3.13}$$

$$\eta^{(2)}(u, \varepsilon) = \int_{\varepsilon<|\lambda|<1} (e^{iu\cdot y} - 1 - u\cdot y1_{\{|\lambda|<1\}})\mu(dy). \tag{3.3.14}$$

Thus $(Z_t^{(1)})_{t\geq 0}$ and $(Z_t^{(2)}(\varepsilon))_{t\geq 0}$ are compound Poisson processes. Let $(B_t)_{t\geq 0}$ be a Brownian motion independent of $(Z_t^{(1)})_{t\geq 0}$ and $(Z_t^{(2)}(\varepsilon))_{t\geq 0}$. Thus the characteristic function of the process

$$X_t(\varepsilon) = bt + \sqrt{A}\cdot B_t + Z_t^{(1)} + Z_t^{(2)}(\varepsilon), \quad t \geq 0, \tag{3.3.15}$$

with $b \in \mathbb{R}^d$, a $d\times d$ nonnegative matrix A such that $A = \sqrt{A}\sqrt{A}^T$, is given by $\mathbb{E}[e^{iu\cdot X_t(\varepsilon)}] = e^{t\eta(u,\varepsilon)}$, where

$$\eta(u,\varepsilon) = iu\cdot b - \frac{1}{2}u\cdot Au + \int\limits_{\varepsilon\leq|\lambda|} (e^{iu\lambda} - 1 - u\cdot\lambda 1_{\{|\lambda|<1\}})\mu(d\lambda). \tag{3.3.16}$$

Finally, we give the Lévy–Itô decomposition theorem saying that every Lévy process can be represented as the sum of four independent components of which a part has almost surely continuous paths and the other part can be expressed as a compensated sum of jumps. We state the theorem without proof.

Theorem 3.58 (Lévy–Itô decomposition). *Let $(X_t)_{t\geq 0}$ be a Lévy process with characteristics (b,A,ν), $N(\cdot,\cdot)$ the counting measure associated with $(X_t)_{t\geq 0}$ on $\mathcal{B}(\mathbb{R}^+)\times\mathcal{B}(\mathbb{R}^d\setminus\{0\})$ as in* (3.3.4), *and $\mu(A) = \mathbb{E}[N(1,A)]$. Then $\mu = \nu$ and the following hold:*

(1) *X_t is obtained as*

$$X_t = X_t^{(1)} + X_t^{(2)} + X_t^{(3)} + X_t^{(4)}, \quad t \geq 0, \tag{3.3.17}$$

where

$$X_t^{(1)} = bt, \quad X_t^{(3)} = \int\limits_0^t \int\limits_{|z|\geq 1} zN(dsdz),$$

$$X_t^{(2)} = \sqrt{A}\cdot B_t, \quad X_t^{(4)} = \lim_{\varepsilon\downarrow 0}\left(\int\limits_0^t \int\limits_{\varepsilon<|z|<1} zN(dsdz) - t\int\limits_{\varepsilon<|z|<1} z\nu(dz)\right).$$

Here the limit $\lim_{\varepsilon\downarrow 0}\left(\int_0^t\int_{\varepsilon<|z|<1} zN(dsdz) - t\int_{\varepsilon<|z|<1} z\nu(dz)\right)$ is taken in $L^2(\Omega)$ sense, and is uniform with respect to t in every compact interval.

(2) *$(X_t^{(1)})_{t\geq 0}$, $(X_t^{(2)})_{t\geq 0}$, and $(X_t^{(3)} + X_t^{(4)})_{t\geq 0}$ are independent.*

(3) *If $\int_{|\lambda|\leq 1}|\lambda|\nu(d\lambda) < \infty$, then $(X_t^{(1)})_{t\geq 0}$, $(X_t^{(2)})_{t\geq 0}$, $(X_t^{(3)})_{t\geq 0}$, and $(X_t^{(4)})_{t\geq 0}$ are independent.*

From the above decomposition, we see that $X_t^{(1)}$ is a constant drift with Lévy symbol $\eta_1(u) = iu\cdot b$, $X_t^{(2)}$ is a Brownian motion with $\eta_2(u) = -(1/2)u\cdot Au$, where $A = \sqrt{A}\sqrt{A}^T$, $X_t^{(3)}$ is a compound Poisson process with $\eta_3(u) = \int_{|y|\geq 1}(e^{iu\cdot y} - 1)\nu(dy)$, and $X_t^{(4)}$ is a square integrable martingale with $\eta_4(u) = \int_{0<|y|<1}(e^{iu\cdot y} - 1 - iu\cdot y)\nu(dy)$.

3.4 Itô formula for semimartingales

3.4.1 Point processes

In this section first we discuss point processes adapted to a filtration, and then their stochastic calculus.

Definition 3.59 (Point function). Let $(S, \mathscr{S})$ be a measurable space and D_p a countable subset of $(0, \infty)$. A *point function* on S is a map $p : D_p \subset (0, \infty) \to S$.

Given a point function, we define a counting measure N_p on $\mathcal{B}((0, \infty)) \times \mathscr{S}$ by

$$N_p(A \times U) = \left|\{s \in D_p \mid s \in A,\ p(s) \in U\}\right|, \quad A \in \mathcal{B}((0, \infty)),\ U \in \mathscr{S}. \tag{3.4.1}$$

Let P_S be the set of all point functions on S, and $\mathcal{B}(P_S)$ the smallest σ-field on P_S with respect to which $P_S \ni p \mapsto N_p((0, t] \times U)$, $t > 0$, $U \in \mathscr{S}$, are measurable. In what follows, we write $N_p(t, U)$ instead of $N_p((0, t] \times U)$.

Definition 3.60 (Point process). Let $(S, \mathscr{S})$ be a measurable space. A *point process* p on S is a $(P_S, \mathcal{B}(P_S))$-valued random variable; $(S, \mathscr{S})$ is called its *state space.*

A point process p is called *stationary*, if for every $t > 0$, $\theta_t p \stackrel{d}{=} p$, where $\theta_t p$ is defined by $D_{\theta_t p} = \{s \in (0, \infty) \mid s + t \in D_p\}$ and $\theta_t p(s) = p(t + s)$. Moreover, a point process p is called a *Poisson point process* if $N_p(dtdx)$ is a Poisson random measure on $\mathcal{B}((0, \infty)) \times \mathscr{S}$. It is known that a Poisson point process is stationary if and only if its intensity measure $\mathbb{E}[N_p(dtdx)] = dtn(dx)$ with a measure $n(dx)$ on $(S, \mathscr{S})$.

Let $(\Omega, \mathscr{F}, (\mathscr{F}_t)_{t\geq 0}, P)$ be a filtered probability space and p be a point process on $(S, \mathscr{S})$ defined on Ω, i. e., $p(\omega) \in P_S$ for every $\omega \in \Omega$. We say that p is $(\mathscr{F}_t)_{t\geq 0}$*-adapted* whenever for every $t > 0$ and $U \in \mathscr{S}$, $N_p(t, U) = \sum_{s\in D_p, s\leq t} 1_U(p_s)$ is $\mathscr{F}_t$-measurable. A point process p is called *σ-finite* if there exists $U_n \in \mathscr{S}$ such that $U_n \uparrow S$ as $n \to \infty$ and $\mathbb{E}[N_p(t, U_n)] < \infty$, for all $t > 0$ and $n \in \mathbb{N}$. For a σ-finite and $(\mathscr{F}_t)_{t\geq 0}$-adapted point process p, let

$$\Gamma_p = \{U \in \mathscr{S} \mid \mathbb{E}[N_p(t, U)] < \infty \text{ for all } t > 0\}.$$

Definition 3.61 (Quasi-left-continuity). Let p be an $(\mathscr{F}_t)_{t\geq 0}$-adapted point process on $(S, \mathscr{S})$ defined on a probability space $(\Omega, \mathscr{F}, P)$. It is called *quasi-left-continuous* if it is σ-finite and there exists a natural integrable increasing process $\hat{N}_p(t, U)$ such that
(1) for $U \in \Gamma_p$, $t \mapsto \hat{N}_p(t, U)$ is a continuous $(\mathscr{F}_t)_{t\geq 0}$-adapted increasing process;
(2) for t, $\mathscr{S} \ni U \mapsto \hat{N}_p(t, U)$ is σ-finite measure on $(S, \mathscr{S})$ almost surely;
(3) for $U \in \Gamma_p$, $t \mapsto \tilde{N}_p(t, U) = N_p(t, U) - \hat{N}_p(t, U)$ is an $(\mathscr{F}_t)_{t\geq 0}$-martingale.

The random measure $\hat{N}_p(t, \cdot)$ is called the *compensator* of the point process p.

Note that if $U \in \Gamma_p$, then $t \mapsto N_p(t, U)$ is an adapted and integrable increasing process, and hence by the Doob–Meyer decomposition there exists a natural integrable increasing process $\hat{N}_p(t, U)$ such that

$$\tilde{N}_p(t, U) = N_p(t, U) - \hat{N}_p(t, U), \quad t \geq 0, \tag{3.4.2}$$

is a martingale. It is not necessary, however, that $t \mapsto \hat{N}_p(t, U)$ is continuous and $\hat{N}_p(t, \cdot)$ is a measure.

Definition 3.62 ($(\mathscr{F}_t)_{t\geq 0}$-Poisson point process). A point process p is called an $(\mathscr{F}_t)_{t\geq 0}$-*Poisson point process* if it is an $(\mathscr{F}_t)_{t\geq 0}$-adapted, σ-finite Poisson point process such that $(N_p(t+s, U) - N_p(t, U))_{s>0, U\in\mathscr{S}}$ is independent of $\mathscr{F}_t$, for all $t \geq 0$.

Remark 3.63. It is known that an $(\mathscr{F}_t)_{t\geq 0}$-Poisson point process is quasi-left-continuous if and only if $t \mapsto \mathbb{E}[N_p(t, U)]$ is continuous for $U \in \Gamma_p$. In this case the compensator $\hat{N}_p(t, U)$ is given by $\hat{N}_p(t, U) = \mathbb{E}[N_p(t, U)]$. In particular, a stationary $(\mathscr{F}_t)_{t\geq 0}$-Poisson point process is quasi-left-continuous and $\hat{N}_p(t, U)$ is given by $\hat{N}_p(t, U) = \mathbb{E}[N_p(t, U)] = tn(U)$.

Generally, a random process $(X_t)_{t\geq 0}$ on a probability space $(\Omega, \mathscr{F}, P)$ is called measurable if the map $[0, \infty) \times \Omega \ni (t, \omega) \mapsto X_t(\omega) \in \mathbb{R}^d$ is $\mathcal{B}([0, \infty)) \times \mathscr{F} / \mathcal{B}(\mathbb{R}^d)$-measurable. Let $\mathscr{L}$ be the smallest σ-field on the set $[0, \infty) \times \Omega$ such that every left-continuous $(\mathscr{F}_t)_{t\geq 0}$-adapted process $Y : [0, \infty) \times \Omega \ni (t, \omega) \mapsto Y_t(\omega) \in \mathbb{R}^d$ is measurable.

Definition 3.64 (Predictable process). A random process $(X_t)_{t\geq 0}$ on a filtered probability space $(\Omega, \mathscr{F}, (\mathscr{F}_t)_{t\geq 0}, P)$ is called *predictable* with respect to $(\mathscr{F}_t)_{t\geq 0}$ if the map $[0, \infty) \times \Omega \ni (t, \omega) \mapsto X_t(\omega) \in \mathbb{R}^d$ is $\mathscr{L} / \mathcal{B}(\mathbb{R}^d)$-measurable.

Lemma 3.65. *Let $(\Omega, \mathscr{F}, (\mathscr{F}_t)_{t\geq 0}, P)$ be a filtered probability space and $(f_t)_{t\geq 0}$ a bounded predictable process with respect to $(\mathscr{F}_t)_{t\geq 0}$. Let p be quasi-left-continuous and $U \in \Gamma_p$. Define $X_t = \int_0^t f_s \tilde{N}_p(ds, U)$, where*

$$\int_0^t f_s \tilde{N}_p(ds, U) = \sum_{\substack{s\leq t \\ s\in D_p}} f_s \mathbb{1}_U(p(s)) - \int_0^t f_s \hat{N}_p(ds, U). \tag{3.4.3}$$

Then $(X_t)_{t\geq 0}$ is an $(\mathscr{F}_t)_{t\geq 0}$-martingale.

Proof. By Lemma 3.66, it is sufficient to assume that $(f_t)_{t\geq 0}$ is a bounded and left-continuous adapted process. Let

$$f_{s,n} = f_0 \mathbb{1}_{\{s=0\}}(s) + \sum_{k=0}^{\infty} f_{k/2^n} \mathbb{1}_{(k/2^n, (k+1)/2^n]}(s).$$

We have $f_{s,n} \to f_s$ as $n \to \infty$ for every s. Then it suffices to prove the claim for $f_{s,n}$. It is clear that

$$\int_0^t f_{s,n}\tilde{N}_p(ds, U) = \sum_{k=0}^{\infty} f_{k/2^n}\left(\tilde{N}_p\left(\frac{k+1}{2^n} \wedge t, U\right) - \tilde{N}_p\left(\frac{k}{2^n} \wedge t, U\right)\right)$$

is an $(\mathscr{F}_t)_{t\geq 0}$-martingale, since $U \in \Gamma_p$. □

Lemma 3.66. *Let L be the linear space of real-valued and bounded measurable processes satisfying the following:*

(1) *L contains all bounded, left-continuous $(\mathscr{F}_t)_{t\geq 0}$-adapted processes;*
(2) *if $(X_n)_{n\in\mathbb{N}}$ is an increasing sequence of random processes in L such that $X = \sup_{n\in\mathbb{N}} X_n$ is bounded, then $X \in L$.*

Then L contains all bounded predictable processes.

Proof. Note that a bounded predictable process is a monotone increasing limit of $\mathscr{L}$-measurable step functions on $[0,\infty) \times \Omega$, so it is sufficient to show $1_A \in L$ for $A \in \mathscr{L}$. Let $\mathscr{L}' = \{A \subset [0,\infty) \times \Omega \mid 1_A \in L\}$. This $\mathscr{L}'$ has the following properties: (1) $[0,\infty) \times \Omega \in \mathscr{L}'$, (2) $A, B \in \mathscr{L}'$ and $A \subset B$ implies $B \setminus A \in \mathscr{L}'$, (3) $A_n \in \mathscr{L}'$, $A_n \subset A_{n+1}$, $n = 1, 2, \ldots$, implies $\sup_{n\in\mathbb{N}} A_n \in \mathscr{L}'$. Then $\mathscr{L}'$ is a λ-system. We denote the smallest λ-system containing $\mathscr{L}'$ by $\lambda(\mathscr{L}')$. Let Y_i, $i = 1, \ldots, k$, be left-continuous $(\mathscr{F}_t)_{t\geq 0}$-adapted processes and E_i, $i = 1, \ldots, k$, open sets in $\mathbb{R}$. We have $\bigcap_{i=1}^k Y_i^{-1}(E_i) \in \mathscr{L}'$. Indeed, $1_{\bigcap_{i=1}^k Y_i^{-1}(E_i)}(t,\omega) = \prod_{i=1}^k 1_{E_i}(Y_i(t,\omega))$ and there exists a sequence of bounded continuous functions $(\phi_n^i)_{n\in\mathbb{N}}$ on $\mathbb{R}$ such that $\phi_n^i(x) \uparrow 1_{E_i}(x)$ as $n \to \infty$,

$$\prod_{i=1}^k \phi_n^i(Y_i(t,\omega)) \uparrow \prod_{i=1}^k 1_{E_i}(Y_i(t,\omega)).$$

Here $\prod_{i=1}^k \phi_n^i(Y_i(t,\omega))$ is a bounded, left-continuous, and $(\mathscr{F}_t)_{t\geq 0}$-adapted process contained in L, and thus $\prod_{i=1}^k 1_{E_i}(Y_i(t,\omega))$ is in L by (2). Let $\mathscr{C}$ be the collection of all the sets of the form $\bigcap_{i=1}^k Y_i^{-1}(E_i) \in \mathscr{L}'$. By definition, $\sigma(\mathscr{C}) = \mathscr{L}$. On the other hand, we see that $\mathscr{C} \subset \mathscr{L}'$, and by the $\pi - \lambda$ theorem we have $\mathscr{L} = \sigma(\mathscr{C}) = \lambda(\mathscr{C}) \subset \lambda(\mathscr{L}') = \mathscr{L}'$. □

Lemma 3.67. *Let p be a quasi-left-continuous point process on a filtered probability space $(\Omega, \mathscr{F}, (\mathscr{F}_t)_{t\geq 0}, P)$. Then for $U \in \Gamma_p$, the random process $\tilde{N}_p(t, U)$ is a square integrable martingale and $\tilde{N}_p(t, U_1)\tilde{N}_p(t, U_2) - \hat{N}_p(t, U_1 \cap U_2)$ is also a martingale for $U_1, U_2 \in \Gamma_p$.*

Proof. Let $\tau_n = \inf\{t \geq 0 \mid \tilde{N}_p(t, U_1) \wedge \tilde{N}_p(t, U_1) \geq n\}$. This is a stopping time, and $\tau_n \uparrow \infty$ as $n \to \infty$; also, $\tilde{N}_p^n(t, U) = \tilde{N}_p(t \wedge \tau_n, U)$ are bounded in t. We directly see that

$$\tilde{N}_p^n(t, U_1)\tilde{N}_p^n(t, U_2) = \int_0^t \tilde{N}_p^n(s-, U_1)\tilde{N}_p^n(ds, U_2) + \int_0^t \tilde{N}_p^n(s, U_2)\tilde{N}_p^n(ds, U_1).$$

Thus

$$\tilde{N}_p^n(t, U_1)\tilde{N}_p^n(t, U_2) = \int_0^t \tilde{N}_p^n(s-, U_1)\tilde{N}_p^n(ds, U_2) + \int_0^t \tilde{N}_p^n(s-, U_2)\tilde{N}_p^n(ds, U_1)$$
$$+ \int_0^t (\tilde{N}_p^n(s, U_2) - \tilde{N}_p^n(s-, U_2))\tilde{N}_p^n(ds, U_1).$$

The first two terms on the right-hand side are martingales by Lemma 3.65 and the last term is equal to

$$\sum_{\substack{s \le t\wedge\tau_n \\ s\in D_p}} 1_{U_1\cap U_2}(p(s)) = N_p(t\wedge\tau_n, U_1\cap U_2) = \hat{N}_p(t\wedge\tau_n, U_1\cap U_2) + \tilde{N}_p(t\wedge\tau_n, U_1\cap U_2).$$

Since $\tilde{N}_p(t\wedge\tau_n, U_1\cap U_2)$ is a martingale, the lemma follows. □

Next we define a stochastic integral with respect to a given quasi-left-continuous point process, with $N_p(dsdx)$, $\tilde{N}_p(dsdx)$, and $\hat{N}_p(dsdx)$ defined as above.

Definition 3.68 ($(\mathcal{F}_t)_{t\ge0}$-predictable process). Let $(\Omega, \mathcal{F}, (\mathcal{F}_t)_{t\ge0}, P)$ be a filtered probability space and $(S, \mathcal{S})$ a measurable space. An $\mathbb{R}$-valued function $g(t, z, \omega)$ on $[0,\infty)\times S\times\Omega$ is called $(\mathcal{F}_t)_{t\ge0}$-*predictable* if the map $(t, z, \omega)\mapsto g(t, z, \omega)$ is $\mathcal{P}/\mathcal{B}(\mathbb{R})$-measurable, where $\mathcal{P}$ is the smallest σ-field on $[0,\infty)\times S\times\Omega$ with respect to which all g having the following properties are measurable:
(1) for every $t\ge 0$, the map $(z,\omega)\mapsto g(t,z,\omega)$ is $\mathcal{S}\times\mathcal{F}_t/\mathcal{B}(\mathbb{R})$-measurable;
(2) for every $(z,\omega)\in S\times\Omega$, the map $t\mapsto g(t,z,\omega)$ is left-continuous.

Note that if g is $(\mathcal{F}_t)_{t\ge0}$-predictable, then the process $(g(t,x,\cdot))_{t\ge0}$ is $(\mathcal{F}_t)_{t\ge0}$-adapted for every x.

Definition 3.69 (Classes $\mathbb{P}$, $\mathbb{P}_1$, $\mathbb{P}_2$, $\mathbb{P}_{2,\mathrm{loc}}$). Let $(\Omega, \mathcal{F}, (\mathcal{F}_t)_{t\ge0}, P)$ be a filtered probability space and $(S,\mathcal{S})$ a measurable space. Let p be a quasi-left-continuous point process on S. Define the class of functions on $[0,\infty)\times S\times\Omega$ by the following conditions:
(1) $f\in\mathbb{P}$ if f is $(\mathcal{F}_t)_{t\ge0}$-predictable and

$$\int_0^{t+}\int_S |f(s,x,\omega)| N_p(dsdx) < \infty \quad \text{a.s.}, t>0;$$

(2) $f\in\mathbb{P}_1$ if f is $(\mathcal{F}_t)_{t\ge0}$-predictable and

$$\mathbb{E}\left[\int_0^t\int_S |f(s,x,\cdot)|\hat{N}_p(dsdx)\right] < \infty, \quad t>0;$$

(3) $f \in \mathbb{P}_2$ if f is $(\mathscr{F}_t)_{t\geq 0}$-predictable and

$$\mathbb{E}\left[\int_0^t\int_S |f(s,x,\cdot)|^2\hat{N}_p(dsdx)\right] < \infty, \quad t > 0;$$

(4) $f \in \mathbb{P}_{2,\mathrm{loc}}$ if f is $(\mathscr{F}_t)_{t\geq 0}$-predictable and there exists a sequence of stopping times $(\tau_n)_{n\in\mathbb{N}}$ such that $\tau_n \uparrow \infty$ almost surely and $1_{[0,\tau_n]}(t)f(t,x,\omega) \in \mathbb{P}_2$ for $n \in \mathbb{N}$.

For $f \in \mathbb{P}$, we write $f(s,x)$ for $f(s,x,\omega)$ unless any confusion may arise. It follows from the definition above that for $f \in \mathbb{P}$, $\int_0^{t+}\int_S f(s,x)N_p(dsdx)$ is well defined and

$$\int_0^t\int_S f(s,x)N_p(dsdx) = \sum_{\substack{s\leq t\\ s\in D_p}} f(s,p(s)).$$

Lemma 3.70. *Let $(\Omega,\mathscr{F},(\mathscr{F}_t)_{t\geq 0},P)$ be a filtered probability space and $(S,\mathscr{S})$ a measurable space. Let p be a quasi-left-continuous point process on S defined on $(\Omega,\mathscr{F},P)$. Then we have the following:*

(1) *Let $f \in \mathbb{P}_1$. Then*

$$\mathbb{E}\left[\int_0^{t+}\int_S |f(s,x)|N_p(dsdx)\right] = \mathbb{E}\left[\int_0^{t+}\int_S |f(s,x)|\hat{N}_p(dsdx)\right].$$

(2) *Let $f \in \mathbb{P}_1$. Then $\left(\int_0^{t+}\int_S f(s,x)\tilde{N}_p(dsdx)\right)_{t\geq 0}$ is an $(\mathscr{F}_t)_{t\geq 0}$-martingale.*

(3) *Let $f \in \mathbb{P}_1 \cap \mathbb{P}_2$. Then*

$$\int_0^{t+}\int_S f(s,x)\tilde{N}_p(dsdx) = \int_0^{t+}\int_S f(s,x)N_p(dsdx) - \int_0^{t+}\int_S f(s,x)\hat{N}_p(dsdx) \tag{3.4.4}$$

is a square integrable $(\mathscr{F}_t)_{t\geq 0}$-martingale and

$$\left(\int_0^{t+}\int_S f(s,x)\tilde{N}_p(dsdx)\right)^2 - \int_0^{t+}\int_S |f(s,x)|^2\hat{N}_p(dsdx)$$

is a martingale.

Proof. The proofs are similar to those of Lemmas 3.65 and 3.67. □

Consider the case when $f \in \mathbb{P}_2$ but $f \notin \mathbb{P}_1 \cap \mathbb{P}_2$. Write

$$f_n(s,x) = 1_{(-n,n)}(f(s,x))1_{U_n}(x)f(s,x).$$

We have $f_n \in \mathbb{P}_1 \cap \mathbb{P}_2$, and $\int_0^{t+} \int_S f_n(s,x)\tilde{N}_p(dsdx)$ is well defined for every n. Furthermore,

$$\mathbb{E}\left[\left|\int_0^{t+}\int_S f_n(s,x)\tilde{N}_p(dsdx) - \int_0^{t+}\int_S f_m(s,x)\tilde{N}_p(dsdx)\right|^2\right]$$

$$= \mathbb{E}\left[\int_0^{t+}\int_S |f_n(s,x) - f_m(s,x)|^2 \hat{N}_p(dsdx)\right].$$

Thus $\left(\int_0^{t+}\int_S f_n(s,x)\tilde{N}_p(dsdx)\right)_{n\in\mathbb{N}}$ is a Cauchy sequence in $L^2(\Omega)$. We will use the same notation

$$\int_0^{t+}\int_S f(s,x)\tilde{N}_p(dsdx) = \lim_{n\to\infty}\int_0^{t+}\int_S f_n(s,x)\tilde{N}_p(dsdx)$$

for the strong limit in $L^2(\Omega)$. Note that in this case (3.4.4) no longer holds.

Example 3.71. The following is an example of a Poisson point process related to the jumps of Lévy processes. Let $(X_t)_{t\geq 0}$ be a Lévy process on a probability space $(\Omega, \mathscr{F}, P)$. Define $p : \omega \mapsto \Delta X_\cdot(\omega)$ with domain $D_p = \{t > 0 \mid \Delta X_t \neq 0\}$ and $p(\omega)(t) = \Delta X_t(\omega)$ for $t \in D_p$. Note that $p(\omega)(t)$ measures the size of jumps at t. The counting measure is

$$N(t,A) = \left|\{s \in D_p \mid p(s) \in A, s \leq t\}\right|, \quad A \in \mathcal{B}(\mathbb{R}^d \setminus \{0\}).$$

If $A \in \mathcal{B}(\mathbb{R}^d \setminus \{0\})$ such that $0 \notin \bar{A}$, then, as shown in Theorem 3.53 and Theorem 3.58, $(N(t,A))_{t\geq 0}$ is a Poisson process and $\mathbb{E}[N(t,A)] = t\nu(A)$, where ν denotes the Lévy measure of $(X_t)_{t\geq 0}$. Although $N(t,A)$ is a Poisson process for A with $0 \notin \bar{A}$, the set function $\mathcal{B}(\mathbb{R}^d \setminus \{0\}) \ni A \mapsto N(t,A)$ is not a Poisson random measure. In order to construct a Poisson point process with characteristic measure ν, we can apply Theorem 3.72 below. Due to this theorem, there exists a stationary Poisson point process $\bar{p}$ on $(\mathbb{R}^d \setminus \{0\}, \mathcal{B}(\mathbb{R}^d \setminus \{0\}))$ defined on a probability space $(\bar{\Omega}, \bar{\mathscr{F}}, \bar{P})$ such that $\mathbb{E}[N_{\bar{p}}(t,A)] = t\nu(A)$, for all $A \in \mathcal{B}(\mathbb{R}^d \setminus \{0\})$. Let $\mathscr{G}_t = \sigma(N_{\bar{p}}(s,A), 0 \leq s \leq t, A \in \mathcal{B}(\mathbb{R}^d \setminus \{0\}))$. We see that $\bar{p}$ is a stationary $(\mathscr{G}_t)_{t\geq 0}$-Poisson point process. In particular, $\bar{p}$ is quasi-left-continuous by Remark 3.63. As $N_p(t,A) \overset{\mathrm{d}}{=} N_{\bar{p}}(t,A)$ for A with $0 \notin \bar{A}$, it follows that

$$\int_0^{t+}\int_A f(s,x)N_p(dsdx) \overset{\mathrm{d}}{=} \int_0^{t+}\int_A f(s,x)N_{\bar{p}}(dsdx)$$

for A with $0 \notin \bar{A}$.

Theorem 3.72. *Let λ be a σ-finite measure on a measurable space $(S, \mathscr{S})$. Then there exists a stationary Poisson point process on $(S, \mathscr{S})$ defined on a probability space $(\Omega, \mathscr{F}, P)$ with characteristic measure λ.*

Proof. The proof is similar to Theorem 3.49. Let $U_n \in \mathscr{S}$ be such that $0 < \lambda(U_m) < \infty$ and $\bigcup_{m=1}^{\infty} U_m = S$. We take random variables $(X_n(m))_{n\in\mathbb{N}}$ and $(Y_n(m))_{n\in\mathbb{N}}$, $m = 1, 2, \ldots$, on a probability space $(\Omega, \mathscr{F}, P)$ such that

(1) for each n and m, $X_n(m)$ is an U_n-valued random variable with

$$P(X_n(m) \in A) = \frac{\lambda(A)}{\lambda(U_n)}$$

for $A \in \mathscr{S}$;

(2) for each $n, m \geq 1$, $Y_n(m)$ is nonnegative and $P(Y_n(m) > t) = e^{-t\lambda(U_m)}$ for $t > 0$;

(3) $X_n(m)$ and $Y_n(m)$, $(n, m \geq 1)$, are mutually independent.

Write for the domain

$$D_p = \bigcup_{k=1}^{\infty} \{Y_1(k), Y_1(k) + Y_2(k), \ldots, Y_1(k) + \cdots + Y_m(k), \ldots\}, \tag{3.4.5}$$

which is a disjoint union. Define $p : \Omega \to P_S$ by $p(\omega)(s) = X_n(m)(\omega)$ for $s = Y_1(k) + \cdots + Y_m(k) \in D_p$. As in Theorem 3.49, it can be directly checked that p is a stationary point process such that $\mathbb{E}[N_p((0,t] \times A)] = tn(A)$. □

3.4.2 Itô formula for semimartingales

In view of the Lévy–Itô decomposition a Lévy process can be split into four integrals with respect to ds, dB_s, $N(dsdz)$, and $\tilde{N}(dsdz)$. This justifies to consider a general stochastic integral of the form

$$\begin{aligned} X_t = X_0 &+ \int_0^t a(s,\omega)ds + \int_0^t b(s,\omega) \cdot dB_s \\ &+ \int_0^t \int_S c(s,z,\omega)N(dsdz) + \int_0^t \int_S d(s,z,\omega)\tilde{N}(dsdz), \end{aligned} \tag{3.4.6}$$

where the integrands satisfy suitable conditions for the integrals to exist. Furthermore, we want to consider the process $Y_t = F(X_t)$ for a function F and are interested whether Y_t is again of the form as in (3.4.6).

First, we introduce semimartingales which are obtained as extensions of the Itô processes seen before.

Definition 3.73 (D-dimensional semimartingale). Let $(\Omega, \mathscr{F}, (\mathscr{F}_t)_{t\geq 0}, P)$ be a filtered probability space and $(S, \mathscr{S})$ a measurable space. Let p be an $(\mathscr{F}_t)_{t\geq 0}$-adapted quasi-left-continuous point process on $(S, \mathscr{S})$, on a probability space $(\Omega, \mathscr{F}, P)$. Assume that for every $i = 1, \ldots, D$,

(1) X_0^i is $\mathscr{F}_0$-measurable;
(2) $M = (M_t)_{t\geq 0} = (M_t^i)_{t\geq 0, 1\leq i\leq D}$ is a D-dimensional locally square integrable continuous martingale, i. e., $t \mapsto M_t$ is a. s. continuous, and there exists a sequence of stopping times $(\tau_n)_{n\in\mathbb{N}}$ such that $\tau_n \uparrow \infty$ as $n \to \infty$ and $M_{t\wedge\tau_n}$ is a square integrable martingale for each n;
(3) $A = (A_t)_{t\geq 0} = (A_t^i)_{t\geq 0, 1\leq i\leq D}$ is a D-dimensional continuous $(\mathscr{F}_t)_{t\geq 0}$-adapted process such that $A_0^i = 0$ and $t \mapsto A_t^i$ has a. s. bounded variation on every bounded interval;
(4) $h_1^i \in \mathbb{P}$, $h_2^i \in \mathbb{P}_{2,\mathrm{loc}}$, and $h_1^i h_2^i = 0$.

The D-dimensional random process $(X_t)_{t\geq 0} = (X_t^i)_{t\geq 0, 1\leq i\leq D}$ defined by

$$X_t^i = X_0^i + M_t^i + A_t^i + \int_0^{t+}\int_S h_1^i(s,z,\omega)N_p(dsdz) + \int_0^{t+}\int_S h_2^i(s,z,\omega)\tilde{N}_p(dsdz) \tag{3.4.7}$$

on $(\Omega, \mathscr{F}, P)$ is called a *semimartingale.*

For our purposes below, we discuss a special case of semimartingales. A typical example of a local martingale is given by a stochastic integral and adapted process defined as an integral of an adapted process.

(M) Let $1 \leq \mu \leq d$ and $1 \leq i \leq D$. Let $f_\mu^i \in \mathbb{M}^2(0,t)$ for all $t \geq 0$ for each μ and i, i. e., $(f_\mu^i(t))_{t\geq 0}$ is $(\mathscr{F}_t)_{t\geq 0}$-adapted and for all $t \geq 0$,

$$P\left(\int_0^t |f_\mu^i(s,\omega)|^2 ds < \infty\right) = 1.$$

(A) Let $1 \leq i \leq D$. Let $(g^i(t))_{t\geq 0}$ be $(\mathscr{F}_t)_{t\geq 0}$-adapted and $g^i(t) \in L^1_{\mathrm{loc}}(\mathbb{R}, dt)$ almost surely.

An example of $(M_t)_{t\geq 0}$ and $(A_t)_{t\geq 0}$ is of the form

$$M_t^i = \int_0^t f^i(s,\omega)\cdot dB_s, \quad A_t^i = \int_0^t g^i(s,\omega)ds.$$

Proposition 3.74 (Itô formula for semimartingales). *Let $(B_t)_{t\geq 0}$ be d-dimensional Brownian motion on a filtered probability space $(\Omega, \mathscr{F}, (\mathscr{F}_t)_{t\geq 0}, P)$, and p an $(\mathscr{F}_t)_{t\geq 0}$-adapted quasi-left-continuous point process on a measurable space $(S, \mathscr{S})$ defined on $(\Omega, \mathscr{F}, P)$. Let $F \in C^2(\mathbb{R}^D)$ and $(X_t)_{t\geq 0} = (X_t^i)_{t\geq 0, 1\leq i\leq D}$ be a D-dimensional semimartingale given by*

$$X_t^i = X_0^i + \int_0^t f^i(s,\omega)\cdot dB_s + \int_0^t g^i(s,\omega)ds$$

$$+\int_0^{t+}\int_S h_1^i(s,z,\omega)N_p(dsdz)+\int_0^{t+}\int_S h_2^i(s,z,\omega)\tilde{N}_p(dsdz) \tag{3.4.8}$$

under (M), (A), *and* (1) *and* (4) *in Definition* 3.73. *Suppose*

$$\sup_{1\le i\le D,0\le t,\omega\in\Omega,x\in\mathbb{R}^d}|h_2^i(t,x,\omega)|\le a$$

with some $a>0$. *Then* $(F(X_t))_{t\ge 0}$ *is also a semimartingale and the following formula holds:*

$$\begin{aligned}F(X_t)=F(X_0)&+\sum_{i=1}^D\int_0^t F_i(X_s)f^i\cdot dB_s+\int_0^t\left(\sum_{i=1}^D F_i(X_s)g^i+\frac{1}{2}\sum_{i,j=1}^D\int_0^t F_{ij}(X_s)f^i\cdot f^j\right)ds\\&+\int_0^{t+}\int_S\Big(F(X_{s-}+h_1)-F(X_{s-})\Big)N_p(dsdz)\\&+\int_0^{t+}\int_S\Big(F(X_{s-}+h_2)-F(X_{s-})\Big)\tilde{N}_p(dsdz)\\&+\int_0^{t}\int_S\left(F(X_s+h_2)-F(X_s)-\sum_{i=1}^D h_2^iF_i(X_s)\right)\hat{N}_p(dsdz),\end{aligned} \tag{3.4.9}$$

where $F_i=\partial_{x_i}F$ *and* $F_{ij}=\partial_{x_i}\partial_{x_j}F$.

Proof. We give an outline of the proof. To avoid notational complexity, we assume that $D=1$. We show that

$$\begin{aligned}F(X_t)-F(X_0)=&\int_0^t F'(X_s)f\cdot dB_s+\int_0^t F'(X_s)gds+\frac{1}{2}\int_0^t F''(X_s)|f|^2ds\\&+\int_0^{t+}\int_S(F(X_{s-}+h_1(s,x))-F(X_{s-}))N_p(dsdx)\\&+\int_0^{t+}\int_S(F(X_{s-}+h_2(s,x))-F(X_{s-}))\tilde{N}_p(dsdx)\\&+\int_0^{t+}\int_S(F(X_{s-}+h_2(s,x))-F(X_{s-})-h_2(s,x)F'(X_s))\hat{N}_p(dsdx).\end{aligned}$$

In Theorem 2.144, we have obtained formula (3.4.9) for

$$X_t=X_0+\int_0^t f(s,\omega)dB_s+\int_0^t g(s,\omega)ds.$$

Let

$$h_1^{(m)}(s,x,\omega)=\begin{cases} m, & h_1(s,x,\omega)\geq m,\\ h_1(s,x,\omega), & |h_1(s,x,\omega)|\leq m,\\ -m, & h_1(s,x,\omega)\leq -m.\end{cases}$$

By approximation, it is sufficient to prove the claim for $h_1^{(m)}$ for h_1. We reset $h_1^{(m)}$ to h_1. We can assume that $h_2 \in \mathbb{P}_2$ and F is such that F, F', and F'' are bounded. There exists $U_n \in \mathscr{S}$ such that $U_n \subset U_{n+1} \subset \ldots, \bigcup_{n=1}^{\infty} U_n = S$, and $\mathbb{E}[N_p((0,t]\times U_n)) < \infty$ for all $t \geq 0$. For each n, we define

$$X_t(n) = X_0(n) + \int_0^t f_s dB_s + \int_0^t g_s ds + \int_0^{t+}\int_S h_1^n(s,z)N(dsdz) + \int_0^{t+}\int_S h_2^n(s,z)\tilde{N}(dsdz),$$

where $h_j^n(s,x,\omega) = h_j(s,x,\omega)1_{U_n}(x)$. Also, we define the point process p_n with domain $D_{p_n} = \{s \in D_p \mid p(s) \in U_n\}$ and $p_n(s) = p(s)$ for $s \in D_{p_n}$. Let $D_{p_n} = \{\tau_1, \tau_2, \ldots\}$, where $0 < \tau_1 < \tau_2 < \cdots$. Thus τ_m is an $(\mathscr{F}_t)_{t\geq 0}$-stopping time, and it is seen that

$$\begin{aligned} X_t(n) - X_0(n) &= \int_0^t f_s dB_s + \int_0^t g_s ds \\ &\quad + \sum_{\tau_m\leq t} h_1(\tau_m, p(\tau_m)) + \sum_{\tau_m\leq t} h_2(\tau_m, p(\tau_m)) + \int_0^{t+}\int_S h_2^n(s,x)\hat{N}(dsdz). \end{aligned}$$

Setting $\tau_0 = 0$, we have

$$\begin{aligned} F(X_t(n)) - F(X_0) &= \sum_{m=1}^{\infty} \{F(X_{\tau_m\wedge t}(n)) - F(X_{\tau_{m-1}\wedge t}(n))\} \\ &= \sum_{m=1}^{\infty} \{F(X_{\tau_m\wedge t}(n)) - F(X_{\tau_m\wedge t-}(n))\} + \sum_{m=1}^{\infty} \{F(X_{\tau_m\wedge t-}(n)) - F(X_{\tau_{m-1}\wedge t}(n))\} \\ &= I_1(t) + I_2(t). \end{aligned}$$

Here

$$F(X_{\tau_m\wedge t-}(n)) = \begin{cases} F(X_{\tau_m-}(n)), & \tau_m \leq t,\\ F(X_t(n)), & t < \tau_m. \end{cases}$$

Note that $p(\tau_m-) = p(\tau_{m-1}), p(\tau_m-) \neq p(\tau_m)$, and

$$\sum_{\tau_l\leq\tau_m-} f(\tau_l, p(\tau_l)) = \lim_{t\to\tau_m-} \sum_{\tau_l\leq t} f(\tau_l, p(\tau_l)) = \sum_{\tau_l\leq\tau_{m-1}} f(\tau_l, p(\tau_l)).$$

Hence, intuitively, I_1 describes the jump part and I_2 the continuous part. By Theorem 2.144, we have

$$F(X_{\tau_m\wedge t-}(n)) - F(X_{\tau_{m-1}\wedge t}(n)) = \int_{\tau_{m-1}\wedge t}^{\tau_m\wedge t} F'(X_s(n))f\cdot dB_s + \int_{\tau_{m-1}\wedge t}^{\tau_m\wedge t} F'(X_s(n))g ds$$
$$+ \frac{1}{2}\int_{\tau_{m-1}\wedge t}^{\tau_m\wedge t} F''(X_s(n))|f|^2 ds - \int_{\tau_{m-1}\wedge t}^{\tau_m\wedge t} F'(X_s(n))dA_n(s),$$

where $A_n(t) = \int_0^t \int_S h_2^n(s,x)\hat{N}_p(dsdx)$ and

$$\int_{\tau_{m-1}\wedge t}^{\tau_m\wedge t} F'(X_s(n))dA_n(s) = \int_{\tau_{m-1}\wedge t}^{\tau_m\wedge t} F'(X_s(n))h_2^n(s,x)\hat{N}_p(dsdx).$$

Thus we obtain

$$I_2(t) = \int_0^t F'(X_s(n))f\cdot dB_s + \int_0^t F'(X_s(n))g ds + \frac{1}{2}\int_0^t F''(X_s(n))|f|^2 ds - \int_0^t F'(X_s(n))dA_n(s). \tag{3.4.10}$$

Next consider the jump part $I_1(t)$. We decompose $I_1(t)$ in three parts. By the definition of the integral $\int_0^{t+}\int_S f N_p(dsdx)$, $h_1(s,x)h_2(s,x) = 0$, and since

$$X_{\tau_m} = X_{\tau_m-} + h_1(\tau_m, p(\tau_m)) + h_2(\tau_m, p(\tau_m)),$$

we have

$$\begin{aligned}
I_1(t) &= \sum_{m=1}^{\infty}(F(X_{\tau_m}(n)) - F(X_{\tau_m-}(n)))1_{\{\tau_m\le t, h_1(\tau_m,p(\tau_m))\neq 0\}} \\
&\quad + \sum_{m=1}^{\infty}(F(X_{\tau_m}(n)) - F(X_{\tau_m-}(n)))1_{\{\tau_m\le t, h_2(\tau_m,p(\tau_m))\neq 0\}} \\
&= \int_0^{t+}\int_S (F(X_{s-}(n) + h_1^n(s,x)) - F(X_{s-}(n)))N_p(dsdx) \\
&\quad + \int_0^{t+}\int_S (F(X_{s-}(n) + h_2^n(s,x)) - F(X_{s-}(n)))\tilde{N}_p(dsdx) \\
&\quad + \int_0^{t}\int_S (F(X_s(n) + h_2^n(s,x)) - F(X_s(n)))\hat{N}_p(dsdx).
\end{aligned} \tag{3.4.11}$$

From (3.4.10)–(3.4.11), we can derive (3.4.9) for $(X_t(n))_{t\geq 0}$, and (3.4.9) for $(X_t)_{t\geq 0}$ is obtained by letting $n \to \infty$. First, by Lemma 3.70,

$$\int_0^{t+}\int_S h_2^n(s,x)\tilde{N}_p(dsdx) \to \int_0^{t+}\int_S h_2(s,x)\tilde{N}_p(dsdx)$$

as $n \to \infty$ in $L^2(\Omega)$, and hence taking a subsequence if necessary, we may assume that this convergence is uniform in t on each bounded interval almost surely. Also it can be seen that

$$\int_0^{t+}\int_S h_1^n(s,x)\tilde{N}_p(dsdx) \to \int_0^{t+}\int_S h_1(s,x)\tilde{N}_p(dsdx)$$

as $n \to \infty$ in $L^2(\Omega)$ uniformly in t on each finite interval almost surely. Consequently, $X_t(n)$ converges to X_t uniformly in t on each bounded interval almost surely and hence $F(X_t(n)) - F(X_0(n)) \to F(X_t) - F(X_0)$ as $n \to \infty$, almost surely. We also see, by the dominated convergence theorem, that $\int_0^t F'(X_s(n))f \cdot dB_s \to \int_0^t F'(X_s)f \cdot dB_s$ as $n \to \infty$ in $L^2(\Omega)$, and

$$\int_0^t F'(X_s(n))gds \to \int_0^t F'(X_s)gds \quad \text{and} \quad \int_0^t F''(X_s(n))|f|^2ds \to \int_0^t F''(X_s)|f|^2ds,$$

as $n \to \infty$, almost surely. The convergence of the jump part follows by

$$\begin{aligned}
&\int_0^t\int_S \left(F(X_s(n) + h_2(s,x)) - F(X_s(n)) - h_2(s,x)F'(X_s(n))\right)\hat{N}_p(dsdx)\\
&\qquad \to \int_0^t\int_S \left(F(X_s + h_2(s,x)) - F(X_s) - h_2(s,x)F'(X_s)\right)\hat{N}_p(dsdx) \quad \text{a. s.},\\
&\int_0^{t+}\int_S \left(F(X_{s-}(n) + h_1^n(s,x)) - F(X_{s-}(n))\right)N_p(dsdx)\\
&\qquad \to \int_0^{t+}\int_S \left(F(X_{s-} + h_1(s,x)) - F(X_{s-})\right)N_p(dsdx) \quad \text{a. s.},\\
&\int_0^{t+}\int_S \left(F(X_{s-}(n) + h_2^n(s,x)) - F(X_{s-}(n))\right)\tilde{N}_p(dsdx)\\
&\qquad \to \int_0^{t+}\int_S \left(F(X_{s-} + h_2(s,x)) - F(X_{s-})\right)\tilde{N}_p(dsdx) \quad \text{in } L^2(\Omega)
\end{aligned}$$

as $n \to \infty$. □

Example 3.75. An application of the Itô formula above is the following variation formula for Lévy processes. Consider the Lévy process $(X_t)_{t\geq 0}$, $X_t = \lim_{\varepsilon\downarrow 0} X_t^\varepsilon$, where X_t^ε is given by the Lévy–Itô decomposition, i. e.,

$$X_t^\varepsilon = bt + A\cdot B_t + \int_0^t \int_{|z|\geq 1} zN(dsdz) + \int_0^t \int_{\varepsilon<|z|<1} z\tilde{N}(dsdz).$$

Since $1_{\{|z|\geq 1\}}\cdot 1_{\{\varepsilon<|z|<1\}} = 0$, for $F\in C_b^2(\mathbb{R}^n)$, $Y_t^\varepsilon = F(X_t^\varepsilon)$ is a semimartingale of the form

$$\begin{aligned} Y_t^\varepsilon = Y_0^\varepsilon &+ \sum_{\mu=1}^d \sum_{i=1}^n \int_0^t F_i(X_s^\varepsilon)A_{i\mu}dB_s^\mu + \int_0^t \left(\sum_{i=1}^n b_iF_i(X_s^\varepsilon) + \frac{1}{2}\sum_{i,j=1}^n \int F_{ij}(X_s^\varepsilon)(AA^T)_{ij}\right)ds \\ &+ \int_0^{t+}\int_{|z|\geq 1} (F(X_{s-}^\varepsilon + z1_{\{|z|\geq 1\}}) - F(X_{s-}^\varepsilon))\,N(dsdz) \\ &+ \int_0^{t+}\int_{\varepsilon<|z|<1} (F(X_{s-}^\varepsilon + z1_{\{\varepsilon<|z|<1\}}) - F(X_{s-}^\varepsilon))\,\tilde{N}(dsdz) \\ &+ \int_0^t \int_{\mathbb{R}^d\setminus\{0\}} \left(F(X_s^\varepsilon + z1_{\{\varepsilon<|z|<1\}}) - F(X_s^\varepsilon) - \frac{1}{2}\sum_{i=1}^n z_i 1_{\{\varepsilon<|z|<1\}}F_i(X_s^\varepsilon)\right)ds\nu(dz). \end{aligned}$$

Since $\sup_{t\in[0,T]}\mathbb{E}[|X_t - X_t^\varepsilon|^2] \to 0$ as $\varepsilon\downarrow 0$, there exists a subsequence ε' such that $X_t^{\varepsilon'}(\omega)\to X_t(\omega)$ for a. e. $(t,\omega)\in[0,T]\times\Omega$. By a simple limiting argument we have

$$\begin{aligned} Y_t = Y_0 &+ \sum_{\mu=1}^d \sum_{i=1}^n \int_0^t F_i(X_s)A_{i\mu}dB_s^\mu + \int_0^t \left(\sum_{i=1}^n b_iF_i(X_s) + \frac{1}{2}\sum_{i,j=1}^n \int F_{ij}(X_s)(AA^T)_{ij}\right)ds \\ &+ \int_0^{t+}\int_{|z|\geq 1} (F(X_{s-} + z1_{\{|z|\geq 1\}}) - F(X_{s-}))\,N(dsdz) \\ &+ \lim_{\varepsilon\downarrow 0}\int_0^{t+}\int_{\varepsilon<|z|<1} (F(X_{s-} + z1_{\{\varepsilon<|z|<1\}}) - F(X_{s-}))\,\tilde{N}(dsdz) \\ &+ \int_0^t \int_{\mathbb{R}^d\setminus\{0\}} \left(F(X_s + z1_{\{0<|z|<1\}}) - F(X_s) - \frac{1}{2}\sum_{i=1}^n z_i 1_{\{0<|z|<1\}}F_i(X_s)\right)ds\nu(dz). \end{aligned}$$

Finally, we show the product Itô formula for semimartingales. Write (3.4.8) as

$$dX_t^i = g^i(t)dt + f^i(t)\cdot dB_t + \int_S h_1^i dN_t + \int_S h_2^i d\tilde{N}_t$$

in differential notation. Take $n = 1$ and let $(Z_t)_{t\geq 0}$, $(Y_t)_{t\geq 0}$ be semimartingales given by

$$dZ_t = u_t^{(1)} dt + v_t^{(1)} \cdot dB_t + \int_S f_t^{(1)} dN_t + \int_S g_t^{(1)} d\tilde{N}_t,$$
$$dY_t = u_t^{(2)} dt + v_t^{(2)} \cdot dB_t + \int_S f_t^{(2)} dN_t + \int_S g_t^{(2)} d\tilde{N}_t.$$

By Proposition 3.74, we have the following product Itô formula.

Corollary 3.76. *Let $f_t^{(1)} g_t^{(1)} = 0$, $f_t^{(1)} g_t^{(2)} = 0$, $f_t^{(2)} g_t^{(2)} = 0$, and $f_t^{(2)} g_t^{(1)} = 0$. Then*

$$\begin{aligned} d(Z_t Y_t) = {} & Z_t u_t^{(2)} dt + Z_t v_t^{(2)} \cdot dB_t + \int_S Z_{t-} f_t^{(2)} dN_t + \int_S Z_{t-} g_t^{(2)} d\tilde{N}_t \\ & + Y_t u_t^{(1)} dt + Y_t v_t^{(1)} \cdot dB_t + \int_S Y_{t-} f_t^{(1)} dN_t + \int_S Y_{t-} g_t^{(1)} d\tilde{N}_t \\ & + v_t^{(1)} \cdot v_t^{(2)} dt + \int_S (f_t^{(1)} f_t^{(2)} + g_t^{(1)} g_t^{(2)}) dN_t. \end{aligned}$$

In concise differential notation, this is written as

$$d(Z_t Y_t) = dZ_t \cdot Y_t + Z_t \cdot dY_t + dZ_t \cdot dY_t.$$

3.5 Exponentials of Lévy processes and recurrence properties

3.5.1 Exponential functionals of Lévy processes

In this section we study some exponential functionals of Lévy processes $(X_t)_{t\geq 0}$, which will be useful in investigating the decay properties of bound states in quantum theory applications below. The first result is an estimate on the running maximum of the process.

Lemma 3.77. *Let $(X_t)_{t\geq 0}$ be a Lévy process on a probability space $(\Omega, \mathscr{F}, P)$, with jumps of size ≤ 1. Then there exist constants $c_0, D > 0$ such that*

$$\mathbb{E}[e^{c \sup_{s\in[0,t]} |X_s|}] \leq De^t, \tag{3.5.1}$$

for all $c \in (0, c_0)$.

Proof. Write $Y_t = \sup_{s\in[0,t]} |X_s|$ for notational simplicity, and define the family of hitting times $\tau_1 = \inf\{t > 0 \mid |X_t| > 1\}$ and $\tau_k = \inf\{t > \tau_{k-1} \mid |X_t - X_{\tau_{k-1}}| > 1\}$ for $k \geq 2$. It can be shown in the same way as in the proof of Lemma 3.50 that $\{\tau_n - \tau_{n-1}\}_{n=1}^k$ are independent and identically distributed random variables, by using the strong Markov property of $(X_t)_{t\geq 0}$. Hence we have

$$\mathbb{E}[e^{-\tau_k}] = \mathbb{E}[e^{-(\tau_k-\tau_{k-1}+\tau_{k-1}-\tau_{k-2}+\cdots+\tau_1-\tau_0)}] = \prod_{j=1}^{k} \mathbb{E}[e^{-(\tau_j-\tau_{j-1})}] = a^k,$$

where $\tau_0 = 0$ and $a = \mathbb{E}[e^{-\tau_1}] < 1$. We also see that

$$P(Y_t \geq 2k) \leq P(t \geq \tau_k) \leq e^t \mathbb{E}[e^{-\tau_k}] = e^t a^k.$$

Hence it follows that

$$\begin{aligned}\mathbb{E}[e^{cY_t}] &= \mathbb{E}[e^{cY_t}1_{\{Y_t<2\}}] + \sum_{k=1}^{\infty} \mathbb{E}[e^{cY_t}1_{\{2k\leq Y_t<2(k+1)\}}] \\ &\leq e^{2c} + \sum_{k=1}^{\infty} e^{2(k+1)c} P(Y_t \geq 2k) \leq e^{2c} e^t \left(\sum_{k=0}^{\infty} e^{(2c+\log a)k} \right) = \frac{e^{2c}}{1-ae^{2c}} e^t.\end{aligned}$$

Thus the desired result follows provided that $c > 0$ is small enough so that $2c + \log a < 0$. □

Theorem 3.78. *Let $(X_t)_{t\geq 0}$ be a Lévy process on a probability space $(\Omega, \mathscr{F}, P)$ such that the Lévy measure ν associated with $(X_t)_{t\geq 0}$ is exponentially localized, i. e.,*

$$\int_{|z|>1} e^{c|z|} \nu(dz) < \infty$$

for all $c < \delta$ with some $\delta > 0$. Then there exist constants $\delta > c_0 > 0$ and $D > 0$ such that

$$\mathbb{E}[e^{c \sup_{s\in[0,t]} |X_s|}] \leq D \exp\left(t \left(1 + \int_{|z|>1} (e^{c|z|} - 1)\nu(dz) \right) \right)$$

for all $c \in (0, c_0)$. In particular, it follows that

$$\mathbb{E}[e^{c \sup_{s\in[0,t]} |X_s|}] \leq De^{bt}$$

with a constant $b > 0$.

Proof. By using the Lévy–Itô decomposition, $(X_t)_{t\geq 0}$ can be represented as

$$X_t = \int_0^t \int_{|z|>1} zN(dsdz) + Z_t, \quad t \geq 0,$$

where the Lévy process $(Z_t)_{t\geq 0}$ and $(\int_0^t \int_{|z|>1} zN(dsdz))_{t\geq 0}$ are independent. Denote $Y_t = \int_0^t \int_{|z|>1} zN(dsdz)$. Consequently, we have

$$\mathbb{E}[e^{c \sup_{s\in[0,t]} |X_s|}] \leq \mathbb{E}[e^{c \sup_{s\in[0,t]} |Z_s|}]\mathbb{E}[e^{c \sup_{s\in[0,t]} |Y_s|}]. \tag{3.5.2}$$

Note that the jumps of $(Z_t)_{t\geq 0}$ are of size up to 1 only, and, according to Lemma 3.77, we have

$$\mathbb{E}[e^{c\sup_{s\in[0,t]}|Z_s|}] \leq De^t,$$

for all $t \in (0, c_0)$ with some c_0. The second factor in (3.5.2) is directly estimated as

$$\mathbb{E}[e^{c\sup_{s\in[0,t]}|Y_s|}] \leq \mathbb{E}[e^{c\int_{|z|>1}|z|N(tdz)}] = e^{t\int_{|z|>1}(e^{c|z|}-1)\nu(dz)}. \qquad \square$$

3.5.2 Capacitary measures

We summarize some concepts and facts related to capacitary measures associated with Lévy processes without proof. Consider the family of linear operators U^q, $q > 0$, associated with Lévy process $(X_t)_{t\geq 0}$, called *q-resolvent operator*. For every measurable function $f \geq 0$, these operators act by

$$U^q f(x) = \mathbb{E}^x\left[\int_0^\infty e^{-qt} f(X_t)dt\right]. \tag{3.5.3}$$

Formally, $U^q f(x) = (q - L)^{-1} f(x)$ with the generator L of $(X_t)_{t\geq 0}$. Consider the family of finite measures $(U^q(x, dy))_{x\in\mathbb{R}^d}$ on $(\mathbb{R}^d, \mathcal{B}(\mathbb{R}^d))$ associated with the q-resolvent operators

$$U^q f(x) = \int_{\mathbb{R}^d} f(y) U^q(x, dy), \tag{3.5.4}$$

called *q-resolvent kernel*. It is known that if the resolvent kernel is absolutely continuous with respect to Lebesgue measure, then there exists a unique measurable function u^q : $\mathbb{R}^d \to [0, \infty]$ such that $U^q f(x) = \int_{\mathbb{R}^d} f(y) u^q(y - x)dy$, for all $f \geq 0$.

Definition 3.79 (Capacitary measure and capacity). Let $q > 0$, and $B \subset \mathbb{R}^d$ be an open or a closed set. Let $(X_t)_{t\geq 0}$ be a Lévy process on a probability space $(\Omega, \mathcal{F}, P)$, and consider the stopping time τ_B given by the first hitting time $\tau_B = \inf\{t \geq 0 \mid X_t \in B\}$. The measure μ_B^q on $(\mathbb{R}^d, \mathcal{B}(\mathbb{R}^d))$ defined by

$$\mu_B^q(A) = q \int_{\mathbb{R}^d} \mathbb{E}^x[e^{-q\tau_B} 1_{\{X_{\tau_B}\in A\}}]dx \tag{3.5.5}$$

is called the *q-capacitary measure of B*. The total mass $C^q(B) = \mu_B^q(\mathbb{R}^d)$ of the q-capacitary measure of B is called the *q-capacity of set B*.

Let $q > 0$. The capacity C^q has the following properties:

(1) If $B \subset A$, then $C^q(B) \leq C^q(A)$.

(2) We have $C^q(A \cup B) + C^q(A \cap B) \leq C^q(A) + C^q(B)$.
(3) Let $(B_n)_{n\in\mathbb{N}}$ be an increasing sequence of open sets with $\bigcup_{n=1}^{\infty} B_n = B$. Then $\lim_{n\to\infty} C^q(B_n) = C^q(B)$.
(4) Let B be closed. Then $C^q(B) = \inf\{C^q(A) \mid B \subset A,\ A \text{ is open}\}$.

Lemma 3.80. *The q-capacitary measure of $B \subset \mathbb{R}^d$ is a Radon measure, and its support is contained in the closure $\bar{B}$.*

The following result gives an expression of the expectation of $e^{-\lambda\tau_B}$ in terms of the related capacitary measure.

Lemma 3.81. *Let $q > 0$, $(X_t)_{t\geq 0}$ be a Lévy process, suppose that its q-resolvent kernel is absolutely continuous with respect to Lebesgue measure, and let u^q be its density function. Let $B \subset \mathbb{R}^d$ be either open or closed, and $\tau_B = \inf\{t \geq 0 \mid X_t \in B\}$. Then it follows that*

$$\mathbb{E}^x[e^{-q\tau_B}] = \int_{\mathbb{R}^d} u^q(x-y)\mu_B^q(dy), \quad a.\,e.\ x \in \mathbb{R}^d. \tag{3.5.6}$$

By Lemma 3.81, taken together with Lemma 3.80, we may replace $\mathbb{R}^d$ in (3.5.6) with the support $\bar{B}$ of the q-capacitary measure, i. e., we alternatively have

$$\mathbb{E}^x[e^{-q\tau_B}] = \int_{\bar{B}} u^q(x-y)\mu_B^q(dy). \tag{3.5.7}$$

3.5.3 Recurrence properties of Lévy processes

Now we explore some fundamental properties of Lévy processes relating to how often they (re)visit regions of space.

Definition 3.82 (Recurrence and transience). A random process $(X_t)_{t\geq 0}$ defined on a probability space $(\Omega, \mathcal{F}, P)$ is said to be *recurrent* if

$$P\left(\liminf_{t\to\infty} |X_t| = 0\right) = 1$$

and it is called *transient* if

$$P\left(\lim_{t\to\infty} |X_t| = \infty\right) = 1.$$

It can be proven that a Lévy process is either recurrent or transient. Consider a ball $B_a(0) \subset \mathbb{R}^d$ centered in the origin and of radius a, and consider the functional

$$U_t(a) = \int_0^t 1_{\{X_s \in B_a(0)\}} ds \tag{3.5.8}$$

giving the total time spent by the random process in the ball within the time interval $[0, t]$, called *occupation time* up to t in $B_a(0)$.

Theorem 3.83.
(1) *A random process $(X_t)_{t\geq 0}$ is recurrent if and only if for all $a > 0$ it follows that $\mathbb{E}[U_\infty(a)] = \infty$, alternatively $U_\infty(a) = \infty$ a. s.*
(2) *A random process $(X_t)_{t\geq 0}$ is transient if and only if for all $a > 0$ it follows that $\mathbb{E}[U_\infty(a)] < \infty$, alternatively $U_\infty(a) < \infty$ a. s.*

The functional $U_\infty(a)$ is called the *potential measure* of $(X_t)_{t\geq 0}$ for the ball $B_a(0)$.

In general, these conditions are difficult to check, and the recurrence properties can be characterized by the following rule.

Theorem 3.84 (Chung–Fuchs criterion). *A Lévy process $(X_t)_{t\geq 0}$ is transient if and only if for some $a > 0$,*

$$\int_{B_a(0)} \mathbb{R}\left(\frac{1}{\eta(u)}\right) du < \infty, \tag{3.5.9}$$

where $\eta(u)$ is the Lévy symbol of $(X_t)_{t\geq 0}$.

Example 3.85. Using the characterization given by the Chung–Fuchs criterion, we have the following situations in some specific cases.
(1) Since, by using the Lévy–Khintchine representation, $|\eta(u)| \leq u \cdot Au$ follows for all $u \in \mathbb{R}^d$, where A is the diffusion matrix in the Lévy triplet, every Lévy process with nonzero A is transient if $d \geq 3$.
(2) Brownian motion is recurrent if $d \leq 2$ and transient if $d \geq 3$.
(3) A rotationally symmetric non-Gaussian α-stable process in $d = 1$ is recurrent if $1 \leq \alpha < 2$ and transient if $0 < \alpha < 1$, and if $d = 2$ it is transient for all $0 < \alpha < 2$.

Example 3.86. Let $(X_t)_{t\geq 0}$ be a d-dimensional rotationally symmetric α-stable process, $\alpha \in (0, 2)$. We will investigate Riesz potentials G_0^α in Section 4.9.4, and use them here to characterize recurrence properties in terms of the related Green functions.
(1) Let $\alpha < d$. Then the process $(X_t)_{t\geq 0}$ is transient with Riesz potential

$$G_0^\alpha(y - x) = \int_0^\infty p_t^X(y - x)dt = \mathcal{A}_{d,\alpha} \frac{1}{|y - x|^{d-\alpha}}, \quad x, y \in \mathbb{R}^d,$$

where

$$\mathcal{A}_{d,\alpha} = \frac{\Gamma((d - \alpha)/2)}{2^\alpha \pi^{d/2} \Gamma(\alpha/2)}.$$

(2) Let $\alpha \geq d$. Then the process is recurrent (pointwise recurrent when $\alpha > d = 1$). In this case we can consider the compensated kernel, i. e., for $\alpha \geq d$, we put

$$\tilde{G}_0^\alpha(y-x) = \int_0^\infty \left(p_t^X(y-x) - p_t^X(x_0)\right) dt,$$

where $x_0 = 0$ for $\alpha > d = 1$ and $x_0 = 1$ for $\alpha = d = 1$. In this case

$$\tilde{G}_0^\alpha(x) = \begin{cases} -\frac{1}{\pi}\log|x| & \text{if } \alpha = d = 1, \\ \frac{1}{2\Gamma(\alpha)\cos(\pi\alpha/2)}\frac{1}{|x|^{1-\alpha}} & \text{if } \alpha > d = 1. \end{cases}$$

It is known that in one-dimension the occupation time of Brownian motion is absolutely continuous with respect to Lebesgue measure and its density is the random process called *Brownian local time* $(L_t^x)_{t\geq 0}$, i. e., we have

$$U_t(a) = \int_{-a}^{a} L_t^y dy.$$

This makes it possible to define the local time process of Brownian motion as the *mesure de voisinage*, introduced by P. Lévy and given by

$$L_t^x = \lim_{\varepsilon \downarrow 0} \frac{1}{2\varepsilon} \int_0^t 1_{\{B_s \in (x-\varepsilon, x+\varepsilon)\}} ds, \tag{3.5.10}$$

which then measures the amount of time over the interval $[0, t]$ spent by Brownian motion near the point x and also gives an intuitively clear definition of the local time process. Local times can be defined also to more general Lévy processes but we will not discuss them here.

3.6 Subordinators and Bernstein functions

3.6.1 Subordinators and subordinate Brownian motion

We conclude this chapter by discussing a special class of Lévy processes.

Definition 3.87 (Subordinator). A one-dimensional Lévy process $(T_t)_{t\geq 0}$ is called a *subordinator* whenever $s \leq t$ implies $T_s \leq T_t$ almost surely.

Since $T_{t+s} - T_s \stackrel{\mathrm{d}}{=} T_t$, for all $s, t \geq 0$, the definition readily implies that $T_t \geq 0$ almost surely, for every $t > 0$. A subordinator can be regarded as a random time since it is nondecreasing and nonnegative.

Theorem 3.88 (Characterization of subordinators). *The Lévy symbol of a subordinator is*

$$\eta(u) = ibu + \int_0^\infty (e^{iuy} - 1)\lambda(dy), \tag{3.6.1}$$

where $b \geq 0$ and the Lévy measure satisfies

$$\lambda((-\infty, 0)) = 0 \quad \text{and} \quad \int_0^\infty (y \wedge 1)\lambda(dy) < \infty. \tag{3.6.2}$$

Conversely, for every $\eta(u)$ given by (3.6.1), *there exists a subordinator $(T_t)_{t\geq 0}$ whose Lévy symbol is of the form $\eta(u)$.*

Proof. Suppose that $(T_t)_{t\geq 0}$ is a subordinator with Lévy triplet (b, A, λ). Then $(T_t)_{t\geq 0}$ has no negative jumps and is positive, thus $\lambda((-\infty, 0)) = 0$ and $A = 0$ by the Lévy–Itô decomposition. Hence

$$T_t = tb + \lim_{\varepsilon \downarrow 0} \int_{(0,t]\times(\varepsilon,\infty)} xN(dsdx).$$

Let $T_{t,\varepsilon} = \int_{(0,t]\times(\varepsilon,\infty)} xN(dsdx)$. We see that $Y_t = \lim_{\varepsilon\downarrow 0} T_{t,\varepsilon} = \int_{(0,t]\times(0,\infty)} xN(dsdx)$ exists in L^2 sense and is bounded by $T_t - bt$. We have

$$\begin{aligned}
\mathbb{E}[e^{-uT_{t,\varepsilon}}] &= \exp\left(t \int_{(\varepsilon,\infty)} (e^{-ux} - 1)\lambda(dx)\right) \\
&= \exp\left(t \int_{(\varepsilon,\infty)} (e^{-ux} - 1 + ux1_{\{|x|\leq 1\}}(x))\lambda(dx) - tu \int_{(\varepsilon,1]} x\lambda(dx)\right).
\end{aligned} \tag{3.6.3}$$

As $\varepsilon \downarrow 0$ we have $\mathbb{E}[e^{-uT_{t,\varepsilon}}] \to \mathbb{E}[e^{-uY_t}] > 0$, and $\int_{(\varepsilon,\infty)} (e^{-ux} - 1 - ux1_{\{|x|\leq 1\}}(x))\lambda(dx)$ tends to the integral over $(0, \infty)$, which is finite. Hence $\int_{(0,1]} x\lambda(dx) < \infty$ by (3.6.3). Thus $b \geq 0$, and (3.6.2)–(3.6.1) follow.

Conversely, assume that (3.6.1)–(3.6.2) hold. Let $(T_t)_{t\geq 0}$ be a Lévy process with Lévy symbol (3.6.1). We see that $N((0, t] \times (-\infty, 0)) = t\lambda((-\infty, 0)) = 0$ almost surely, i. e., $(X_t)_{t\geq 0}$ has no negative jumps. Hence by the Lévy–Itô decomposition, we have

$$T_t = tb + \int_{(0,t]\times(0,\infty)} xN(dsdx).$$

This shows that $(X_t)_{t\geq 0}$ is increasing. □

By Theorem 3.88, it is seen that the map $u \mapsto \mathbb{E}[e^{iuT_t}] = e^{t\eta(u)}$ can be analytically continued into the region $\{iu \in \mathbb{C} : u > 0\}$. This implies that the Laplace transform of T_t is well defined and given by

$$\mathbb{E}[e^{-uT_t}] = e^{-t\Psi(u)}, \tag{3.6.4}$$

where

$$\Psi(u) = bu + \int_0^\infty (1 - e^{-uy})\lambda(dy) \tag{3.6.5}$$

for every $u > 0$.

Example 3.89 ($\alpha/2$-stable subordinator). Let $b = 0$, $0 < \alpha < 2$ and using (3.1.24) take

$$\nu(dx) = \frac{\alpha}{2\Gamma(1-\alpha/2)} \frac{1_{(0,\infty)}(x)}{x^{1+\alpha/2}} dx, \tag{3.6.6}$$

where Γ denotes the Gamma function. This is called $\frac{\alpha}{2}$*-stable subordinator* and

$$\mathbb{E}^0[e^{-uT_t}] = e^{-tu^{\alpha/2}}$$

holds.

Example 3.90 (Inverse Gaussian subordinator). Let $(\Omega, \mathscr{F}, P)$ be a probability space, $m \geq 0$ and $\delta > 0$ given constants, and consider the first hitting time of one-dimensional standard Brownian motion with constant drift m,

$$T_t(m, \delta) = \inf\left\{s > 0 \mid B_s + ms = \delta t\right\}. \tag{3.6.7}$$

The process $(T_t(m, \delta))_{t \geq 0}$ is called the *inverse Gaussian subordinator* for $m > 0$, and

$$\mathbb{E}[e^{-uT_t(m,\delta)}] = \exp\left(-t\delta\left(\sqrt{2u + m^2} - m\right)\right) \tag{3.6.8}$$

holds. The distribution of $T_t(m, \delta)$ can be explicitly determined to be

$$\rho(s) = \frac{\delta t}{\sqrt{2\pi}} e^{\delta m t} \frac{1}{s^{3/2}} \exp\left(-\frac{1}{2}\left(t^2\delta^2\frac{1}{s} + m^2 s\right)\right).$$

To see this, recall the exponential martingale of Brownian motion $(e^{\alpha B_t - \frac{1}{2}\alpha^2 t})_{t \geq 0}$, see Corollary 2.113. Consider the stopped process, $(e^{\alpha B_{T_t \wedge n} - \frac{1}{2}\alpha^2 T_t \wedge n})_{t \geq 0}$, which is also a martingale. Thus

$$\mathbb{E}[e^{\alpha B_{T_t \wedge n} - \frac{1}{2}\alpha^2 T_t \wedge n}] = \mathbb{E}[e^{\alpha B_0}] = 1, \quad t \geq 0,\ n \in \mathbb{N}.$$

Let $A_n = \{T_t \le n\}$ and

$$\mathbb{E}[e^{aB_{T_t\wedge n}-\frac{1}{2}a^2T_t\wedge n}] = \mathbb{E}[e^{aB_{T_t\wedge n}-\frac{1}{2}a^2T_t\wedge n}1_{A_n}] + \mathbb{E}[e^{aB_{T_t\wedge n}-\frac{1}{2}a^2T_t\wedge n}1_{A_n^c}].$$

It is immediate to see that

$$\mathbb{E}\left[e^{aB_{T_t\wedge n}-\frac{1}{2}a^2T_t\wedge n}1_{A_n^c}\right] \le e^{-\frac{1}{2}a^2n}\mathbb{E}\left[1_{A_n^c}e^{aB_n}\right]$$

and, since $B_n < \delta t - mn$ on A_n^c, it follows that $\lim_{n\to\infty}\mathbb{E}[e^{aB_{T_t\wedge n}-\frac{1}{2}a^2T_t\wedge n}1_{A_n^c}] = 0$. Furthermore,

$$\lim_{n\to\infty}\mathbb{E}\left[e^{aB_{T_t\wedge n}-\frac{1}{2}a^2T_t\wedge n}1_{A_n}\right] = \mathbb{E}\left[e^{a(\delta t-mT_t)}e^{-\frac{1}{2}a^2T_t}\right].$$

This implies that $1 = \mathbb{E}[e^{a(\delta t-mT_t)}e^{-\frac{1}{2}a^2T_t}]$, and hence

$$e^{-a\delta t} = \mathbb{E}\left[e^{-\frac{1}{2}a(a+2m)T_t}\right]$$

follows. Setting $a = \sqrt{2u+m^2} - m$ proves the claim.

Definition 3.91 (Subordinate Brownian motion). Let $(B_t)_{t\ge0}$ be a Brownian motion and $(T_t^\Psi)_{t\ge0}$ a subordinator with Laplace exponent Ψ given by (3.6.4). The random process defined by

$$(X_t^\Psi)_{t\ge0} = (B_{T_t^\Psi})_{t\ge0} \tag{3.6.9}$$

is called a *subordinate Brownian motion* with subordinator $(T_t^\Psi)_{t\ge0}$.

Let $(X_t^\Psi)_{t\ge0}$ be a subordinate Brownian motion as defined above. It follows that

$$\mathbb{E}[e^{i\xi\cdot X_t^\Psi}] = \int_0^\infty e^{-|\xi|^2y/2}p_t^\Psi(y)dy,$$

where p_t^Ψ denotes the distribution of T_t^Ψ on $[0,\infty)$. A subordinate Brownian motion is by definition the composition of a Brownian motion and a subordinator, and it is a Lévy process. Hence the generator of $(X_t^\Psi)_{t\ge0}$ is given by $-\Psi(-\frac{1}{2}\Delta)$. For instance, when $\Psi(u) = (2u)^{\alpha/2}$, $0 < \alpha < 2$, we obtain that $(X_t^\Psi)_{t\ge0}$ is a rotationally symmetric α-stable process with generator given by the fractional Laplacian $-(-\Delta)^{\alpha/2}$.

3.6.2 Bernstein functions

Subordinators have a deep relationship with a specific class of functions. We will make use of this in the path integral representations of operators semigroups below.

Definition 3.92 (Bernstein function). Let $f \in C^{\infty}((0,\infty))$ with $f \geq 0$. A function f is called a *Bernstein function* whenever $(-1)^n \frac{d^n f}{dx^n}(x) \leq 0$, for all $n \in \mathbb{N}$. We denote the set of Bernstein functions by $\mathcal{B}$.

It is possible to characterize Bernstein functions through completely monotone functions.

Definition 3.93 (Completely monotone function). Let $f \in C^{\infty}((0,\infty))$. The function f is called a *completely monotone function* whenever $(-1)^n \frac{d^n f}{dx^n}(x) \geq 0$, for all $n \in \dot{\mathbb{N}}$.

By the definition, any completely monotone function f is positive and not increasing, and hence $\lim_{x\to\infty} f(x) \geq 0$ exists.

Proposition 3.94 (Characterization of completely monotone functions). *Let f be a completely monotone function. Then there exists a unique measure μ on $[0,\infty)$ such that*

$$f(u) = \int_{[0,\infty)} e^{-uy}\mu(dy). \tag{3.6.10}$$

Conversely, whenever the right-hand side of (3.6.10) *is finite, $f(u)$ is a completely monotone function.*

Proof. Suppose that f is completely monotone. We prove (3.6.10) in four steps.

Step 1. Suppose that $f(0+) = 1$ and $f(\infty) = 0$. By Taylor expansion, we have

$$f(u) = \sum_{k=0}^{n-1} \frac{(-1)^k f^{(k)}(a)}{k!}(a-u)^k + \int_u^a \frac{(-1)^n f^{(n)}(s)}{(n-1)!}(s-u)^{n-1}ds. \tag{3.6.11}$$

Let $a \to \infty$. Since the first term on the right-hand side of (3.6.11) is positive, we have

$$\lim_{a\to\infty} \int_u^a \frac{(-1)^n f^{(n)}(s)}{(n-1)!}(s-u)^{n-1}ds = \int_u^\infty \frac{(-1)^n f^{(n)}(s)}{(n-1)!}(s-u)^{n-1}ds \leq f(u).$$

Hence (3.6.11) converges for every n as $a \to \infty$. Thus every term converges to a nonnegative limit as $a \to \infty$. For $n \geq 0$, let

$$\rho_n(u) = \lim_{a\to\infty} \frac{(-1)^n f^{(n)}(a)}{n!}(a-u)^n.$$

We note that this limit is independent of $u > 0$. Let $c_n = \sum_{k=0}^{n-1} \rho_k(u)$. We have

$$f(u) = c_n + \int_{[u,\infty)} \frac{(-1)^n f^{(n)}(s)}{(n-1)!}(s-u)^{n-1}ds.$$

Clearly, $f(u) \geq c_n \geq 0$. Since $f(\infty) = 0$, it follows that $c_n = 0$ for every n. Thus we have

$$f(u) = \int_{[u,\infty)} \frac{(-1)^n f^{(n)}(s)}{(n-1)!}(s-u)^{n-1}ds. \tag{3.6.12}$$

By the monotone convergence theorem, we obtain

$$1 = \lim_{u\to 0} f(u) = \int_{[0,\infty)} \frac{(-1)^n f^{(n)}(s)}{(n-1)!} s^{n-1}ds. \tag{3.6.13}$$

Substitution into (3.6.12) gives

$$f(u) = \int_{[0,\infty)} \left(1 - \frac{ut}{n}\right)_+^{n-1} f_n(t)dt, \tag{3.6.14}$$

where

$$f_n(s) = \begin{cases} \frac{(-1)^n}{n!} f^{(n)}(n/s)(n/s)^{n+1}, & s > 0, \\ 0, & s = 0. \end{cases}$$

Note that $\int_{[0,\infty)} f_n(t)dt = 1$ by (3.6.13). By Helly's second theorem, there exist a subsequence n_k and a measure μ on $(0,\infty)$ such that $f_{n_k}(t)dt$ vaguely converges to $\mu(dt)$. Since $f_{n_k}(t)dt$, $k \in \mathbb{N}$, are probability measures, μ is a finite Radon measure. Let $\mu((0,\infty)) = m$. We reset n_k by n. Note that $(1 - \frac{ut}{n})_+^{n-1} \to e^{-ut}$ as $n \to \infty$ uniformly in $t \in [0,\infty)$. For every $\varepsilon > 0$, there exist $K_0 > 0$ and $n_0 \in \mathbb{N}$ such that $\mu((K,\infty)) < \varepsilon$ for every $K > K_0$, and $\sup_{t>0} \left|(1 - \frac{ut}{n})_+^{n-1} - e^{-ut}\right| < \epsilon$ for every $n > n_0$. Let $K > K_0$ and $n > n_0$. We have

$$\begin{aligned} &\int_{[0,\infty)} \left(1 - \frac{ut}{n}\right)_+^{n-1} f_n(t)dt - \int_{[0,\infty)} e^{-ut}\mu(dt) \\ &= \int_{[0,K]} \left(1 - \frac{ut}{n}\right)_+^{n-1} f_n(t)dt - \int_{[0,K]} e^{-ut}\mu(dt) + \int_{(K,\infty)} \left(1 - \frac{ut}{n}\right)_+^{n-1} f_n(t)dt - \int_{(K,\infty)} e^{-ut}\mu(dt). \end{aligned}$$

We estimate as

$$\begin{aligned} &\left| \int_{[0,K]} \left(1 - \frac{ut}{n}\right)_+^{n-1} \mu(dt) - \int_{[0,K]} e^{-ut}\mu(dt) \right| \\ &\leq \left| \int_{[0,K]} \left(1 - \frac{ut}{n}\right)_+^{n-1} f_n(t)dt - \int_{[0,K]} e^{-ut} f_n(t)dt \right| + \left| \int_{[0,K]} e^{-ut} f_n(t)dt - \int_{[0,K]} e^{-ut}\mu(dt) \right| \\ &\leq \epsilon \left| \int_{[0,K]} f_n(t)dt \right| + \left| \int_{[0,K]} e^{-ut} f_n(t)dt - \int_{[0,K]} e^{-ut}\mu(dt) \right|. \end{aligned}$$

On the other hand, we see that

$$\left| \int_{(K,\infty)} \left(1 - \frac{ut}{n}\right)_+^{n-1} \mu(dt) - \int_{(K,\infty)} e^{-ut}\mu(dt) \right|$$
$$\leq \left| \int_{(K,\infty)} \left(1 - \frac{ut}{n}\right)_+^{n-1} f_n(t)dt - \int_{(K,\infty)} e^{-ut} f_n(t)dt \right| + \left| \int_{(K,\infty)} e^{-ut} f_n(t)dt - \int_{(K,\infty)} e^{-ut}\mu(dt) \right|$$
$$\leq \varepsilon + e^{-uK}(1+m).$$

Since $f_n(t)dt$ is vaguely convergent to $\mu(dt)$, we have

$$\lim_{n\to\infty} \left| \int_0^\infty \left(1 - \frac{ut}{n}\right)_+^{n-1} f_n(t)dt - \int_0^\infty e^{-ut}\mu(dt) \right|$$
$$\leq 2\varepsilon + e^{-uK}(1+m) + \lim_{n\to\infty} \left| \int_{[0,K]} e^{-ut} f_n(t)dt - \int_{[0,K]} e^{-ut}\mu(dt) \right| = 2\varepsilon + e^{-uK}(1+m).$$

Here we can take an arbitrarily large K. Taking the limit of (3.6.14) along a subsequence n_k, it follows that

$$f(u) = \int_{[0,\infty)} e^{-ut}\mu(dt). \tag{3.6.15}$$

Set $u = 0$. We have $1 = f(0) = \mu([0,\infty))$. Then μ is a probability measure. Uniqueness is further proven by the expression

$$\sum_{0\leq k\leq ux} (-1)^k \frac{u^k}{k!} \frac{d^k}{du^k} \int_{[0,\infty)} e^{-ut}\mu(dt) = \int_{[0,\infty)} \sum_{0\leq k\leq ux} \frac{(ut)^k}{k!} e^{-ut}\mu(dt).$$

From this, we have

$$\lim_{u\to\infty} \sum_{0\leq k\leq ux} (-1)^k \frac{u^k}{k!} \frac{d^k}{du^k} \int_{[0,\infty)} e^{-ut}\mu(dt) = \mu([0,\infty)).$$

This shows that μ can be recovered from its Laplace transform, and thus uniqueness follows.

Step 2. Suppose that $f(0+) < \infty$ and $f(\infty) = 0$. By looking at $g(u) = f(u)/f(0+)$, we have $g(u) = \int_{[0,\infty)} e^{-ut}\mu(dt)$ with a probability measure μ. We see that f is represented by the measure $f(0+)\mu$, i. e.,

$$f(u) = \int_{[0,\infty)} e^{-ut} f(0+)\mu(dt).$$

Step 3. Suppose that $f(\infty) = 0$. Let $f_a(u) = f(u + a)$. The function f_a is also completely monotone and $f_a(0+) = f(a) < \infty$ and $f_a(\infty) = 0$. Thus there exists a unique finite measure μ_a on $[0, \infty)$ such that $f_a(u) = \int_{[0,\infty)} e^{-ut}\mu_a(dt)$. Since $e^{at}\mu_a(dt) = e^{bt}\mu_b(dt)$, we can consistently define $\mu(dt) = e^{at}\mu_a(dt)$, for all $a > 0$. We have

$$f(u) = f_{u/2}(u/2) = \int_{[0,\infty)} e^{(-u/2)t}\mu_{u/2}(dt) = \int_{[0,\infty)} e^{-ut}e^{(u/2)t}\mu_{u/2}(dt) = \int_{[0,\infty)} e^{-ut}\mu(dt).$$

Step 4. In this step, we prove (3.6.10) for arbitrary completely monotone functions. Suppose that $f(\infty) = c > 0$. Let $g(u) = f(u) - c$. The function g is completely monotone and $g(\infty) = 0$. By Step 3, we have the representation $g(u) = \int_{[0,\infty)} e^{-ut}\mu(dt)$ with a finite measure μ. Thus

$$f(u) = \int_{[0,\infty)} e^{-ut}\mu(dt) + c\int_{[0,\infty)} e^{-ut}d\delta_0,$$

where δ_0 is the Dirac measure with mass at 0. Thus $\mu + c\delta_0$ represents f.

Next we prove the converse. Write $f(u) = \int_{[0,\infty)} e^{-ut}\mu(dt)$. Fix $u > 0$ and pick $\varepsilon \in (0, u)$. Since $t^n \le n!\varepsilon^{-n}e^{\varepsilon t}$ for all $t > 0$, we find

$$\int_{[0,\infty)} t^n e^{-ut}\mu(dt) \le \frac{n!}{\varepsilon^n}\int_{[0,\infty)} t^n e^{-(u-\varepsilon)t}\mu(dt) = \frac{n!}{\varepsilon^n}f(u - \varepsilon),$$

and this shows that we may use the differentiation lemma for parameter-dependent integrals to get

$$(-1)^n f^{(n)}(u) = (-1)^n \int_{[0,\infty)} t^n \frac{d^n}{du^n}e^{-ut}\mu(dt) \ge 0. \qquad \square$$

It follows from the definition that Bernstein functions are positive, increasing, and concave. Define a subset of $\mathcal{B}$ by

$$\mathcal{B}_0 = \left\{\Psi \in \mathcal{B} \,\middle|\, \lim_{u\to 0+} \Psi(u) = 0\right\}. \tag{3.6.16}$$

Examples of functions in $\mathcal{B}_0$ include $\Psi(u) = cu^{\alpha/2}$ with $c > 0$ and $0 < \alpha \le 2$, and $\Psi(u) = 1 - e^{-au}$ with $a \ge 0$. Furthermore, it can be seen that sums $\sum_{j=1}^n (1 - e^{-a_j u})$ of Bernstein functions are also Bernstein functions, and u^α can be expressed as $u^\alpha = \int_0^\infty (1-e^{-yu})\nu(dy)$ by using the measure ν in (3.6.6). The right-hand side is derived from the analytic continuation of the Lévy symbol of subordinator (3.6.1).

We will see next that $\mathcal{B}_0$ can be characterized by subordinators. We define the following subset of Lévy measures.

Definition 3.95 (Lévy measures and Bernstein functions). Let $\mathscr{L}$ be the set of Borel measures λ on $\mathbb{R} \setminus \{0\}$ such that $\lambda((-\infty, 0)) = 0$ and $\int_0^\infty (y \wedge 1)\lambda(dy) < \infty$.

Note that $\lambda \in \mathscr{L}$ satisfies $\int_0^\infty (y^2 \wedge 1)\lambda(dy) < \infty$, thus it is a Lévy measure. The important fact connecting $\mathscr{L}$ and $\mathcal{B}_0$ is given by the next proposition.

Proposition 3.96 (Characterization of Bernstein functions). *For every* $\Psi \in \mathcal{B}_0$, *there exists* $(b, \lambda) \in \mathbb{R}_+ \times \mathscr{L}$ *such that*

$$\Psi(u) = bu + \int_{(0,\infty)} (1 - e^{-uy})\lambda(dy). \tag{3.6.17}$$

Conversely, the right-hand side of (3.6.17) *is in* $\mathcal{B}_0$ *for every* $(b, \lambda) \in \mathbb{R}_+ \times \mathscr{L}$.

Proof. Let Ψ be a Bernstein function. Then Ψ' is a completely monotone function. According to Proposition 3.94, there exists a measure μ on $[0, \infty)$ such that

$$\Psi'(u) = \int_{[0,\infty)} e^{-uy}\mu(dy).$$

Let $\mu(\{0\}) = b$. We have

$$\Psi(u) - \Psi(\varepsilon) = \int_\varepsilon^u \Psi'(s)ds = (u - \varepsilon)b + \int_\varepsilon^u ds \int_{(0,\infty)} e^{-ys}\mu(dy)$$

and

$$\Psi(u) = \Psi(u) - \Psi(0+) = bu + \int_{(0,u)} ds \int_{(0,\infty)} e^{-ys}\mu(dy) = bu + \int_{(0,\infty)} (1 - e^{-ys})\frac{1}{y}\mu(dy).$$

Set $\lambda(dy) = \frac{1}{y}\mu(dy)\restriction_{(0,\infty)}$. Then (3.6.17) follows. Conversely, suppose that Ψ is given by (3.6.17). Since $(1 - e^{-1})(1 \wedge t) \leq 1 - e^{-t}$ for $t \geq 0$, we infer

$$\int_{(0,\infty)} (1 \wedge y)\lambda(dy) \leq \frac{e}{e-1} \int_{(0,\infty)} (1 - e^{-y})\lambda(dy) = \frac{e}{e-1}(\Psi(1) - b) < \infty.$$

Since $ye^{-uy} \leq y \wedge (eu)^{-1}$ and $y \wedge (eu)^{-1}$ is integrable with respect to $\lambda(dy)$, $\Psi'(u) = \int_{[0,\infty)} e^{-uy}\nu(dy)$, where $\nu(dy) = y\lambda(dy) + b\delta_0(dy)$. This shows that Ψ' is completely monotone. □

In the examples below, we give some examples of Bernstein functions of basic interest, from which the defining parameters can be read off directly.

Example 3.97. The following are Bernstein functions with right limit vanishing at zero:

(1) Let $\Psi(u) = u^{\alpha/2}$ for $0 < \alpha < 2$. Since

$$u^{\alpha/2} \equiv \frac{\alpha}{2\Gamma(1-\alpha/2)} \int_{(0,\infty)} (1-e^{-uy}) \frac{1}{y^{1+\alpha/2}} dy, \quad u \geq 0,$$

it follows that

$$b = 0 \quad \text{and} \quad \lambda(dy) = \frac{\alpha}{2\Gamma(1-\alpha/2)} \frac{1_{\{y \geq 0\}}}{y^{1+\alpha/2}} dy.$$

(2) Let $\Psi(u) = (u + m^{2/\alpha})^{\alpha/2} - m$ for $0 < \alpha < 2$ and $m > 0$. Since

$$(u + m^{2/\alpha})^{\alpha/2} - m \equiv \frac{\alpha}{2\Gamma(1-\alpha/2)} \int_{(0,\infty)} (1-e^{-uy}) \frac{e^{-m^{2/\alpha}y}}{y^{1+\alpha/2}} dy, \quad u \geq 0,$$

it follows that

$$b = 0 \quad \text{and} \quad \lambda(dy) = \frac{\alpha}{2\Gamma(1-\alpha/2)} \frac{e^{-m^{2/\alpha}y}}{y^{1+\alpha/2}} 1_{\{y \geq 0\}} dy.$$

(3) Let $\Psi(u) = \log(1+u)$. Since

$$\log(1+u) \equiv \int_{(0,\infty)} (1-e^{-uy}) \frac{e^{-y}}{y} dy, \quad u \geq 0,$$

it follows that

$$b = 0 \quad \text{and} \quad \lambda(dy) = \frac{\alpha}{2\Gamma(1-\alpha/2)} \frac{e^{-y}}{y} 1_{\{y \geq 0\}} dy.$$

Due to Proposition 3.96, for $\Psi \in \mathcal{B}_0$ there exists $(b, \lambda) \in (\mathbb{R}_+, \mathcal{L})$ such that (3.6.17) is satisfied.

Proposition 3.98. *The map $\mathcal{B}_0 \ni \Psi \mapsto (b, \lambda) \in (\mathbb{R}_+, \mathcal{L})$ is a bijection.*

Proof. By Proposition 3.96, it is clear that the map is surjective. For a given $\Psi \in \mathcal{B}_0$, the constant b is uniquely determined by

$$b = \lim_{u \to \infty} \frac{\Psi(u)}{u}.$$

Since $f = d\Psi/du$ is a completely monotone function, there exists a unique measure μ on $[0, \infty)$ such that $f(u) = \int_0^\infty e^{-us} d\mu(s)$, for $u > 0$. On the other hand, we have

$$\frac{d\Psi}{du} = b + \int_0^\infty e^{-uy} y \lambda(dy) = \int_0^\infty e^{-uy} \tilde{\lambda}(dy),$$

where $\tilde{\lambda}(dy) = b\delta_0 + y\lambda(dy)$. Thus $b\delta_0 + y\lambda(dy) = \mu(dy)$. Since b is unique, the measure $\lambda \in \mathscr{L}$ for the Bernstein function Ψ is also uniquely determined. □

Denote the set of subordinators by $\mathscr{S}$. By the one-to-one correspondence between $\mathcal{B}_0$ and $(\mathbb{R}_+, \mathscr{L})$ we also have a bijection between $\mathscr{S}$ and $\mathcal{B}_0$.

Proposition 3.99. *If $\Psi \in \mathcal{B}_0$, then there exists a unique subordinator $(T_t)_{t\geq 0} \in \mathscr{S}$ such that*

$$\mathbb{E}[e^{-uT_t}] = e^{-t\Psi(u)}. \tag{3.6.18}$$

Conversely, if $(T_t)_{t\geq 0} \in \mathscr{S}$, then there exists $\Psi \in \mathcal{B}_0$ such that (3.6.18) *holds.*

Proof. Let $\Psi \in \mathcal{B}_0$ be given. There exists $(b, \lambda) \in \mathbb{R}_+ \times \mathscr{L}$ such that (3.6.17) holds. This means that there exists a subordinator $(T_t)_{t\geq 0}$ with Lévy triplet $(b, 0, \lambda)$. Moreover, (3.6.18) follows by analytic continuation of $\mathbb{E}[e^{iuT_t}] = e^{t\eta(u)}$ with exponent $\eta(u) = ibu + \int_0^\infty (e^{iuy} - 1)\lambda(dy)$. Conversely, let $(T_t)_{t\geq 0} \in \mathscr{S}$ be given. For $(b, \lambda) \in \mathbb{R}_+ \times \mathscr{L}$ associated with $(T_t)_{t\geq 0}$, we define the Bernstein function by (3.6.17), and the result follows. □

Example 3.100. In Example 3.97 above, the Bernstein functions Ψ correspond respectively to a (1) $\frac{\alpha}{2}$-stable subordinator, (2) $\frac{\alpha}{2}$-relativistic stable subordinator, (3) Gamma-subordinator, and the corresponding subordinate Brownian motions are (1) rotationally symmetric α-stable process, (2) rotationally symmetric relativistic α-stable process, (3) variance gamma (VG) process.

By Propositions 3.6.17–3.6.18, it can be seen that there is a pairwise bijective relation among subordinators $\mathscr{S}$, Bernstein functions $\mathcal{B}_0$, and $(\mathbb{R}_+, \mathscr{L})$. This will be useful in constructing path integral representations of semigroups generated by Bernstein functions of the Laplacian, discussed in the next chapter.

4 Feynman–Kac formulae

4.1 Schrödinger semigroups

4.1.1 Schrödinger equation and path integral solutions

As we have explained in Chapter 1, a basic reason behind the interest in Feynman–Kac formulae is the analysis of the solutions of the Schrödinger equation

$$i\partial_t \varphi = -\frac{1}{2}\Delta\varphi + V\varphi \tag{4.1.1}$$

with potential V, or its imaginary-time counterpart, the heat equation with spatially nonhomogeneous dissipation

$$\partial_t \varphi = \frac{1}{2}\Delta\varphi - V\varphi. \tag{4.1.2}$$

In fact, the scope of Feynman–Kac formulae in the context of partial differential equations goes far beyond these two cases; however, in this book we have no space to discuss the PDE aspect separately.

Let $f : \mathbb{R} \times \mathbb{R}^d \to \mathbb{R}$ be a solution of the heat equation, i. e.,

$$\partial_t f(t,x) = \frac{1}{2}\Delta f(t,x), \quad f(0,x) = g(x). \tag{4.1.3}$$

Recall the formula

$$f(t,x) = \int_{\mathbb{R}^d} \Pi_t(x-y)g(y)\,dy, \tag{4.1.4}$$

known from the theory of partial differential equations. A comparison with

$$\mathbb{E}^x[g(B_t)] = \int_{\mathbb{R}^d} \Pi_t(x-y)g(y)\,dy \tag{4.1.5}$$

shows that, as we have said above, the solution of (4.1.3) can be obtained by running a Brownian motion, and

$$f(t,x) = \mathbb{E}^x[g(B_t)] \tag{4.1.6}$$

is obtained. Using operator semigroup notation, (4.1.3) implies $f = e^{t(\Delta/2)}g$. In what follows, we ask whether a similar representation is possible when $-\frac{1}{2}\Delta$ is replaced by $-\frac{1}{2}\Delta + V$, with suitable V. Let now $f : \mathbb{R} \times \mathbb{R}^d \to \mathbb{R}$ be a solution of

$$\partial_t f(t,x) = \frac{1}{2}\Delta f(t,x) - V(x)f(t,x). \tag{4.1.7}$$

https://doi.org/10.1515/9783112231463-004

In semigroup notation, this problem translates then to studying $f = e^{-t(-\frac{1}{2}\Delta+V)}g$. In what follows, instead of equations in our considerations the basic input will be linear operators and one-parameter semigroups generated by them as above, although we will remember the deep relations with partial differential equations.

4.1.2 Linear operators and their spectra

In what follows we will be interested in various spectral properties of sums of self-adjoint operators. In the present section our main object will be Schrödinger operators

$$H = -\frac{1}{2}\Delta + V \tag{4.1.8}$$

defined on a suitable space as a sum of the Laplacian and the multiplication operator V viewed in a specific sense as a perturbation of the Laplacian. As we progress to take magnetic fields and spin into account, further operator terms will appear, and when we discuss relativistic quantum mechanics, the Laplacian will be replaced by its square root. In Volume 2 of the book, we then turn to quantum field theory, where further types of operators will be added to Schrödinger operators.

We start by briefly reviewing key concepts and facts of general linear operator theory used throughout this book, and next we will focus first on Laplacians and then on their perturbations.

Let $\mathcal{K}$ be a Hilbert space over the complex field $\mathbb{C}$ with inner product $(\cdot,\cdot)$ and norm $\|\cdot\|$. When inner products or norms of several distinct Hilbert spaces are compared, we indicate the specific space in the subscript whenever necessary; for L^p spaces, we use the standard notation $\|\cdot\|_p$. Let $\mathcal{H}$ be also a Hilbert space. The domain of a linear operator $A : \mathcal{K} \to \mathcal{H}$ will be denoted by $D(A)$, its range by $\operatorname{Ran}(A)$, and its kernel by $\operatorname{Ker}(A)$. Recall that for two linear operators A and B we have $A = B$ if and only if $D(A) = D(B)$ and $A\varphi = B\varphi$ for all $\varphi \in D(A)$, and $A \subset B$ if and only if $D(A) \subset D(B)$ and $A\varphi = B\varphi$ for all $\varphi \in D(A)$; in the latter case, B is said to be an *extension* of A, or from the opposite point of view, A is the *restriction* of B to $D(A)$. An operator A on $\mathcal{K}$ is *densely defined* if $D(A)$ is a dense subset of $\mathcal{K}$.

Definition 4.1 (Bounded/unbounded operators). Let $A : \mathcal{K} \to \mathcal{H}$ be a linear operator. Whenever there exists $c \geq 0$ such that $\|A\varphi\|_{\mathcal{H}} \leq c\|\varphi\|_{\mathcal{K}}$ for all $\varphi \in D(A)$, the operator A is called *bounded*, and otherwise it is called *unbounded*.

A linear operator A is called *strongly continuous at* $\varphi \in D(A)$ when there is a sequence $(\varphi_n)_{n\in\mathbb{N}} \subset D(A)$ such that $\varphi_n \to \varphi$ in strong convergence sense implies $A\varphi_n \to A\varphi$ also in strong convergence sense. If A is strongly continuous for all $\varphi \in D(A)$, A is called *strongly continuous*. A linear operator is bounded if and only if it is strongly continuous. The norm of a bounded operator A is defined as

$$\|A\| = \|A\|_{\mathcal{K},\mathcal{H}} = \sup_{\varphi\in D(A),\varphi\neq 0} \frac{\|A\varphi\|_{\mathcal{H}}}{\|\varphi\|_{\mathcal{K}}}. \tag{4.1.9}$$

A bounded operator is a *contraction* if $\|A\| \leq 1$. When $\mathcal{K}$ is finite-dimensional, any linear operator on it is bounded.

Next we give the notions of bounded operator topology we will often use.

Definition 4.2 (Topologies of bounded operators). Let $(A_n)_{n\in\mathbb{N}}$ be a sequence of bounded operators from $\mathcal{K}$ to $\mathcal{H}$ such that $D(A_n) = D$, $n \in \mathbb{N}$, and A another bounded operator with $D(A) = D$. Then $(A_n)_{n\in\mathbb{N}}$ is

(1) *weakly convergent* to A, denoted $A = \text{w-}\lim_{n\to\infty} A_n$, if $\lim_{n\to\infty}(\psi, A_n\varphi - A\varphi) = 0$, for all $\varphi \in D$ and for all $\psi \in \mathcal{H}$;

(2) *strongly convergent* to A, denoted $A = \text{s-}\lim_{n\to\infty} A_n$, if $\lim_{n\to\infty} \|A_n\varphi - A\varphi\|_{\mathcal{H}} = 0$, for all $\varphi \in D$;

(3) *uniformly convergent* to A if $\lim_{n\to\infty} \|A_n - A\| = 0$.

If A is a densely defined bounded operator with $\|A\varphi\|_{\mathcal{H}} \leq c\|\varphi\|_{\mathcal{K}}$, then it can be uniquely extended to an operator B such that $A \subset B$, $D(B) = \mathcal{K}$ and $\|B\varphi\|_{\mathcal{H}} \leq c\|\varphi\|_{\mathcal{K}}$ hold. In what follows, we assume that the domain of a bounded operator is the whole space. Recall that the *graph* of a linear operator $A : \mathcal{K} \to \mathcal{H}$ is the subset

$$G(A) = \Big\{(\varphi, A\varphi) \mid \varphi \in D(A)\Big\} \subset \mathcal{K} \oplus \mathcal{H}. \tag{4.1.10}$$

Definition 4.3 (Closed/closable operator). Let $A : \mathcal{K} \to \mathcal{H}$ be a linear operator.

(1) Operator A is *closed* if $G(A)$ is closed in $\mathcal{K}\oplus\mathcal{H}$, in other words, if the sequence $(\varphi_n)_{n\in\mathbb{N}}$ satisfies $A\varphi_n \to \phi$ and $\varphi_n \to \varphi$ as $n \to \infty$, then $\varphi \in D(A)$ and, moreover, $\phi = A\varphi$ holds.

(2) Operator A is *closable* if there exists a closed operator $B : \mathcal{K} \to \mathcal{H}$ such that $A \subset B$.

(3) The *closure* of A, denoted by $\bar{A}$, is the smallest closed extension of A, i. e., $\bar{A}$ is a closed extension of A and if B is another closed extension, then $\bar{A} \subset B$.

(4) A subspace $\mathcal{D} \subset D(A)$ is a *core* of A if $\overline{A\lceil_{\mathcal{D}}} = A$, in other words, if for every $\varphi \in D(A)$ there is a sequence $(\varphi_n)_{n\in\mathbb{N}} \subset \mathcal{D}$ such that $\varphi_n \to \varphi$ and $A\varphi_n \to A\varphi$ strongly as $n \to \infty$.

The bounded operators are closed. The next proposition ensures the existence of the closure of closable operators.

Proposition 4.4. *Let A be closable. Then the closure of A exists and $\overline{G(A)} = G(\bar{A})$ holds.*

A linear operator A is *injective* if and only if $A\varphi = 0$ implies $\varphi = 0$. In this case the *inverse* A^{-1} of A is defined by

$$D(A^{-1}) = \mathrm{Ran}(A), \quad A^{-1}\varphi = \phi \tag{4.1.11}$$

for $A\phi = \varphi$. From this definition, it follows that for $\varphi \in \operatorname{Ran}(A)$ and $\phi \in D(A)$,

$$AA^{-1}\varphi = \varphi, \quad A^{-1}A\phi = \phi. \tag{4.1.12}$$

Definition 4.5 (Invertibility). A linear operator A is called *invertible* if it is injective and has a bounded inverse on $\operatorname{Ran} A$.

Note that invertibility of an operator and the existence of an inverse are properties that do not necessarily coincide.

Definition 4.6 (Addition and multiplication of operators). The *sum* $A + B$ of two linear operators A and B is defined by

$$D(A + B) = D(A) \cap D(B), \quad (A + B)\varphi = A\varphi + B\varphi. \tag{4.1.13}$$

The *product* AB is defined by

$$D(AB) = \Big\{\varphi \in D(B) \mid B\varphi \in D(A)\Big\}, \quad (AB)\varphi = A(B\varphi). \tag{4.1.14}$$

Note that in general it may happen that $D(A) \cap D(B)$ is not a dense subset in $D(A)$ or $D(B)$.

Let $B(\mathcal{K}, \mathcal{H})$ denote the set of bounded operators from $\mathcal{K}$ to $\mathcal{H}$ with the whole space $\mathcal{K}$ as their domain. For $A, B \in B(\mathcal{K}, \mathcal{H})$, $A + B \in B(\mathcal{K}, \mathcal{H})$ and $\|A + B\| \le \|A\| + \|B\|$ hold, and for $B \in B(\mathcal{K}, \mathcal{H})$ and $A \in B(\mathcal{H}, \mathcal{L})$, $AB \in B(\mathcal{K}, \mathcal{L})$ and $\|AB\| \le \|A\|\|B\|$ hold.

In the following, a central issue will be the spectral properties of some linear operators.

Definition 4.7 (Resolvent and spectrum). Let $A : \mathcal{K} \to \mathcal{H}$ be a closed operator. The *resolvent set* of A is

$$\rho(A) = \Big\{\lambda \in \mathbb{C} \mid \operatorname{Ran}(\lambda - A) = \mathcal{H},\ \lambda - A \text{ is injective and } (\lambda - A)^{-1} \text{ is bounded}\Big\} \tag{4.1.15}$$

and its *spectrum* is defined as

$$\operatorname{Spec} A = \mathbb{C} \setminus \rho(A). \tag{4.1.16}$$

For $\lambda \in \rho(A)$, the bounded operator $R_A(\lambda) = (\lambda - A)^{-1}$ is the *resolvent* of A.

In the definition of the resolvent, A must be assumed to be closed. The reason is as follows. If $\lambda \in \rho(A)$, then $R_A(\lambda) = (\lambda - A)^{-1}$ is bounded and hence closed. Let $j : \mathcal{K} \oplus \mathcal{H} \to \mathcal{H} \oplus \mathcal{K}$ be defined by $j(\varphi, \phi) = (\phi, \varphi)$. This is an isomorphism between $\mathcal{K} \oplus \mathcal{H}$ and $\mathcal{H} \oplus \mathcal{K}$, and $G(B^{-1}) = jG(B)$ follows. Let B be closed. Then $G(B)$ is a closed set and hence $G(B^{-1})$ is also closed. We conclude that B^{-1} is closed. Thus we proved that in general if B is closed and injective, then B^{-1} is also closed. In particular, $\lambda - A$ is closed since such is $(\lambda - A)^{-1}$, from which it follows that A must be closed.

Furthermore, it can be also seen that if A is closable and injective, then $(\lambda - A)^{-1}$ is also closable and $\overline{(\lambda - A)^{-1}} = (\lambda - \bar{A})^{-1}$ follows. For a closable operator $A : \mathcal{K} \to \mathcal{H}$, its resolvent set is defined by

$$\rho(A) = \left\{\lambda \in \mathbb{C} \mid \operatorname{Ran}(\lambda - A) \text{ is dense}, \ \lambda - A \text{ is injective}, \ (\lambda - A)^{-1} \text{ is bounded}\right\}.$$

For a closed operator A, the conditions (1) $\operatorname{Ran}(\lambda - A)$ is dense, (2) $\lambda - A$ is injective, and (3) $(\lambda - A)^{-1}$ is bounded automatically imply that $\operatorname{Ran}(\lambda - A) = \mathcal{H}$, and we can consequently see that $\rho(A) = \rho(\bar{A})$.

The spectrum of A can be decomposed into three parts:

(1) The *point spectrum* $\operatorname{Spec}_{\mathrm{p}}(A)$ of A consists of the elements $\lambda \in \operatorname{Spec}(A)$ such that $\lambda - A$ is not injective; the elements of the point spectrum are called *eigenvalues*, i. e., for every such λ there exists $\varphi_\lambda \in D(A) \setminus \{0\}$, such that $A\varphi_\lambda = \lambda\varphi_\lambda$, and φ_λ is an *eigenvector* (or *eigenfunction* when the vector space is a set of functions), while the number $\dim \operatorname{Ker}(\lambda - A)$ is the *multiplicity* of eigenvalue λ.

(2) The *continuous spectrum* $\operatorname{Spec}_{\mathrm{c}}(A)$ of A consists of all $\lambda \in \operatorname{Spec} A$ for which $\lambda - A$ has an inverse with a dense domain but it is not a bounded operator.

(3) The *residual spectrum* $\operatorname{Spec}_{\mathrm{r}}(A)$ of A is given by $\lambda \in \operatorname{Spec}(A)$ such that $\lambda - A$ has an inverse, i. e., injective, but its domain is not dense.

Thus the complex field $\mathbb{C}$ is decomposed into the disjoint sets

$$\mathbb{C} = \rho(A) \cup \operatorname{Spec}_{\mathrm{p}}(A) \cup \operatorname{Spec}_{\mathrm{c}}(A) \cup \operatorname{Spec}_{\mathrm{r}}(A). \tag{4.1.17}$$

The resolvent set of a bounded operator A contains $\{z \in \mathbb{C} \mid |z| > \|A\|\}$ since

$$(\lambda - A)^{-1} = \frac{1}{\lambda} \sum_{n=0}^{\infty} \lambda^{-n} A^n$$

holds in the uniform topology and the right-hand side of the equality is bounded for $\lambda > \|A\|$. Thus the spectrum of bounded operators is a closed nonempty subset of $\{z \in \mathbb{C} \mid |z| \leq \|A\|\}$; however, in the case of unbounded operators both the spectrum and the resolvent set may be empty.

The operators we will consider below are self-adjoint, therefore we do not need to address spectral theory in its full generality.

Definition 4.8 (Adjoint operator). Let $A : \mathcal{K} \to \mathcal{H}$ be a densely defined linear operator. The *adjoint operator* A^* of A is a linear operator from $\mathcal{H}$ to $\mathcal{K}$ defined by

$$D(A^*) = \left\{\varphi \in \mathcal{H} \mid \exists \psi \in \mathcal{H} \text{ such that } (\varphi, A\phi) = (\psi, \phi), \ \forall \phi \in D(A)\right\}$$

and $A^*\varphi = \psi$.

Since A is densely defined, A^* is well defined. The adjoint operator A^* is always closed, and whenever A is closed, $A = (A^*)^*$ also holds. If A is closable, then $\bar{A} = (A^*)^*$ holds. It can be also checked that if $D(A^*)$ is dense, then A is closable. For a closable operator A, it follows that $\overline{(A^*)} = (\bar{A})^*$. If $A, B \in B(\mathcal{K}, \mathcal{K})$, then $A^*, B^* \in B(\mathcal{K}, \mathcal{K})$ and the properties $\|A^*\| = \|A\|$, $(A^*)^* = A$, $(aA + bB)^* = \bar{a}A^* + \bar{b}B^*$, and $(AB)^* = B^*A^*$ hold.

Definition 4.9 (Symmetric and self-adjoint operators). Let $A : \mathcal{K} \to \mathcal{K}$ be a densely defined linear operator. Then A is said to be a *symmetric operator* if $A \subset A^*$, a *self-adjoint operator* if $A^* = A$, and an *essentially self-adjoint operator* if $\bar{A}$ is self-adjoint.

Whenever $D(A^*)$ is dense, A is closable; moreover, $(A^*)^* = \bar{A}$ for a closable operator. If A is symmetric, $D(A^*)$ is dense by definition. Hence A is closable and $\bar{A}$ is a closed symmetric operator.

In general, infinitely many self-adjoint extensions of a given symmetric operator exist. Note that if $A \subset B$, then $B^* \subset A^*$. Let A be symmetric and B a self-adjoint extension of A. We have $A \subset B = B^* \subset A^*$. Thus $B = A^* \lceil_D$ with some domain D. Next let A be essentially self-adjoint and B a self-adjoint extension of A. Since $A \subset B$, it follows that $(A^*)^* \subset B$. Thus $B = B^* \subset ((A^*)^*)^* = (A^*)^*$ and therefore $B = (A^*)^* = \bar{A}$. We summarize this in the following proposition.

Proposition 4.10 (Self-adjoint extensions).
(1) *Let A be symmetric and B a self-adjoint extension of A. Then B is a restriction of A^*.*
(2) *Let A be essentially self-adjoint. Then $\bar{A}$ is the only self-adjoint extension of A, i. e., if B is self-adjoint and $A \subset B$, then $\bar{A} = B$.*
(3) *Let A be a symmetric operator and suppose that it has only one self-adjoint extension. Then A is essentially self-adjoint.*

A bounded operator A is symmetric if and only if it is self-adjoint. However, for an unbounded operator, these two properties are in general different; indeed, a symmetric operator need not be closed and a closed symmetric operator need not be self-adjoint. Equivalent conditions to self-adjointness and essential self-adjointness of a symmetric operator are as follows.

Proposition 4.11 (Equivalent conditions to self-adjointness). *Let A be a symmetric operator in $\mathcal{K}$. Then the following properties are equivalent:*
(1) *A is self-adjoint;*
(2) *A is closed and* $\mathrm{Ker}(A^* \pm i) = \{0\}$;
(3) $\mathrm{Ran}(A \pm i) = \mathcal{K}$;
(4) *A is closed and* $\overline{\mathrm{Ran}(A \pm i)} = \mathcal{K}$.

Proposition 4.12 (Equivalent conditions to essential self-adjointness). *Let A be a symmetric operator in $\mathcal{K}$. Then the following properties are equivalent:*
(1) *A is essentially self-adjoint;*
(2) $\mathrm{Ker}(A^* \pm i) = \{0\}$;

(3) $\overline{\mathrm{Ran}(A \pm i)} = \mathcal{K}$.

Self-adjointness removes some of the complications of the structure of the spectrum. The following facts are fundamental on the spectrum and resolvent of self-adjoint operators.

Proposition 4.13 (Spectral properties of self-adjoint operators). *Let A be a self-adjoint operator. Then*
(1) $\mathrm{Spec}(A) \subset \mathbb{R}$;
(2) *the residual spectrum* $\mathrm{Spec}_r(A)$ *is empty;*
(3) *the eigenvectors for distinct eigenvalues are orthogonal;*
(4) *for* $\lambda \in \mathrm{Spec}(A)$, *there exists a sequence* $(\varphi_n)_{n\in\mathbb{N}} \subset D(A)$ *such that* $\|\varphi_n\| = 1$ *and* $\lim_{n\to\infty} \|(\lambda - A)\varphi_n\| = 0$;
(5) *if there exists* $c > 0$ *such that* $\|(\lambda - A)\varphi\| \geq c\|\varphi\|$ *for all* $\varphi \in D(A)$, *then* $\lambda \in \rho(A)$ *and* $\{z \in \mathbb{C} \mid |z - \lambda| < c\} \subset \rho(A)$;
(6) $\|R_A(\lambda)\| \leq 1/|\Im\lambda|$ *holds.*

A self-adjoint operator A with the property that $(\varphi, A\varphi) \geq 0$, for all $\varphi \in D(A)$, is called a *positive* operator and denoted by $A \geq 0$. Positivity of A is equivalent to $\mathrm{Spec}(A) \subset [0, \infty)$.

Definition 4.14 (Unitary operator). A bounded operator $U : \mathcal{K} \to \mathcal{H}$ is called a *unitary operator* if it is an isometry, i. e., $\|U\varphi\|_{\mathcal{H}} = \|\varphi\|_{\mathcal{K}}$ for all $\varphi \in \mathcal{K}$, and $\mathrm{Ran}(U) = \mathcal{H}$.

The isometry condition can be equivalently formulated as $U^*U = 1$. Unitary operators can be used to define a *conjugate* operator to any given operator by keeping its spectrum unchanged. More precisely, if A is a closed operator and U is a unitary operator, then UAU^{-1} is well defined on $UD(A)$ and closed; moreover, $\mathrm{Spec}(UAU^{-1}) = \mathrm{Spec}(A)$. For instance, Fourier transform is a unitary operator on $L^2(\mathbb{R}^d)$ and this equality can be used to compute the spectrum of the Laplacian; see Example 4.21. Furthermore, if A is self-adjoint, then UAU^{-1} is also self-adjoint on $UD(A)$.

Definition 4.15. Let $(A_n)_{n\in\mathbb{N}}$ be a sequence of unbounded operators.
(1) $A_n \to A$ is said to converge in the sense of *strong resolvent convergence* if $R_{A_n}(\lambda)$ strongly converges to $R_A(\lambda)$, for all $\lambda \in \rho(A) \cap \bigcap_{n=1}^{\infty} \rho(A_n)$.
(2) $A_n \to A$ is said to converge in the sense of *uniform resolvent convergence* if $R_{A_n}(\lambda)$ uniformly converges to $R_A(\lambda)$, for all $\lambda \in \rho(A) \cap \bigcap_{n=1}^{\infty} \rho(A_n)$.

Convergence in strong and uniform resolvent sense are natural generalizations of strong and uniform convergence of bounded operators, respectively, to unbounded operators.

4.1.3 Spectral resolution

The *spectral theorem* establishes a fundamental relationship between a linear operator and a measure on its spectrum. Let $\mathscr{K}$ be a Hilbert space and the set of projections on $\mathscr{K}$ be denoted by $P(\mathscr{K})$. Consider the map $\mathcal{B}(\mathbb{R}) \ni B \mapsto E(B) \in P(\mathscr{K})$ with the properties

(1) $E(\emptyset) = 0$ and $E(\mathbb{R}) = 1$;

(2) if $B = \bigcup_{n=1}^{\infty} B_n$ with $B_m \cap B_n = \emptyset$, $m \neq n$, then

$$E(B) = \underset{n\to\infty}{\text{s-lim}} \sum_{k=1}^{n} E(B_k);$$

(3) $E(B_1 \cap B_2) = E(B_1)E(B_2)$.

A family of projections indexed by $\mathcal{B}(\mathbb{R})$ satisfying properties (1)–(3) above is called a *projection-valued measure*. For $\varphi \in \mathcal{K}$, the Borel measure $(\varphi, E(B)\varphi)$ on $\mathbb{R}$ can be defined, and, by the *polarization identity*, it gives rise to the complex measure $(\varphi, E(B)\psi)$, $\varphi, \psi \in \mathcal{K}$ through

$$(\varphi, E(B)\psi) = \frac{1}{4}\sum_{n=1}^{4} \frac{1}{i^n}((\varphi + i^n\psi), E(B)(\varphi + i^n\psi)). \tag{4.1.18}$$

The spectral theorem says that there is a one-to-one correspondence between self-adjoint operators and projection-valued measures. For a self-adjoint operator A and the corresponding $E_A(\lambda)$, we define $f(A)$ for a Borel measurable function f by

$$(\varphi, f(A)\psi)_{\mathcal{K}} = \int_{\mathbb{R}} f(\lambda) d(E_A(\lambda)\varphi, \psi), \tag{4.1.19}$$

with domain

$$D(f(A)) = \left\{\varphi \in \mathcal{K} \,\middle|\, \int_{\mathbb{R}} |f(\lambda)|^2 d\|E_A(\lambda)\varphi\|^2 < \infty\right\}.$$

Note that the support of the measure E_A coincides with Spec(A). Equality (4.1.19) is formally written as $f(A) = \int_{\mathbb{R}} f(\lambda) dE_A(\lambda)$ and is called the *spectral resolution* of $f(A)$.

The following algebraic relations hold.

Proposition 4.16. *Let A be a self-adjoint operator. Then for Borel measurable functions f and g,*

(1) $D(f(A) + g(A)) = D((f + g)(A)) \cap D(g(A)) = D((f + g)(A)) \cap D(f(A))$ *and* $f(A) + g(A) \subset (f + g)(A)$;

(2) $D(f(A)g(A)) = D((fg)(A)) \cap D(g(A))$ *and* $f(A)g(A) \subset (fg)(A)$.

We will use the terminology of also another classification of the spectrum.

Definition 4.17 (Discrete spectrum and essential spectrum). Let A be a self-adjoint operator. If $\lambda \in \operatorname{Spec}(A)$ is such that there is $\varepsilon > 0$ for which $\dim \operatorname{Ran} E_A(\lambda - \varepsilon, \lambda + \varepsilon) < \infty$, then λ is said to be an element of the *discrete spectrum* of A, denoted $\operatorname{Spec}_{\mathrm{d}}(A)$. If $\lambda \in \operatorname{Spec}(A)$ is such that $\dim \operatorname{Ran} E_A(\lambda - \varepsilon, \lambda + \varepsilon) = \infty$ for all $\varepsilon > 0$, then λ is in the *essential spectrum* of A, denoted $\operatorname{Spec}_{\mathrm{ess}}(A)$.

Note that the discrete and the essential spectra form a partition of the spectrum, i. e.,

$$\operatorname{Spec}(A) = \operatorname{Spec}_{\mathrm{ess}}(A) \sqcup \operatorname{Spec}_{\mathrm{d}}(A),$$

where this is a disjoint union. Moreover, while $\operatorname{Spec}_{\mathrm{ess}}(A)$ is a closed set, $\operatorname{Spec}_{\mathrm{d}}(A)$ need not be such. Furthermore, the following facts hold.

Proposition 4.18. *A point $\lambda \in \operatorname{Spec}_{\mathrm{d}}(A)$ if and only if both of the following properties hold:*
(1) *λ is an isolated point of $\operatorname{Spec}(A)$, i. e., there exists $\varepsilon > 0$ such that*

$$\operatorname{Spec}(A) \cap (\lambda - \varepsilon, \lambda + \varepsilon) = \{\lambda\};$$

(2) *λ is an eigenvalue of finite multiplicity, i. e., $\dim \{\varphi \in \mathcal{K} \mid A\varphi = \lambda\varphi\} < \infty$.*

An eigenvalue of arbitrary multiplicity which is not an isolated point in the spectrum is called an *embedded eigenvalue.*

There is a second possible breakup of the spectrum of a self-adjoint operator, which follows the canonical decomposition of a measure. Let A be a self-adjoint operator on a Hilbert space $\mathcal{H}$, and E_A be the spectral resolution as above. By definition, a vector $\psi \in \mathcal{H}_{\mathrm{ac}}$ if and only if the measure $(\psi, E_A(\cdot)\psi)$ is absolutely continuous with respect to Lebesgue measure on $\mathbb{R}$. Similarly, $\psi \in \mathcal{H}_{\mathrm{sing}}$ if and only if the measure $(\psi, E_A(\cdot)\psi)$ is singular with respect to Lebesgue measure. We then have $\mathcal{H} = \mathcal{H}_{\mathrm{ac}} \oplus \mathcal{H}_{\mathrm{sing}}$. Every eigenvector ψ belongs to $\mathcal{H}_{\mathrm{sing}}$, and by definition $\mathcal{H}_{\mathrm{p}}$ is the closed linear hull of the set $\{\psi \mid \psi \text{ is eigenvector}\}$ and $\mathcal{H}_{\mathrm{sc}} = \mathcal{H}_{\mathrm{sing}} \cap \mathcal{H}_{\mathrm{p}}^{\perp}$. With these notations,

$$\mathcal{H} = \mathcal{H}_{\mathrm{ac}} \oplus \mathcal{H}_{\mathrm{p}} \oplus \mathcal{H}_{\mathrm{sc}}$$

holds. Consider the restrictions $A_{\mathrm{ac}} = A\lceil_{\mathcal{H}_{\mathrm{ac}}}$, $A_{\mathrm{p}} = A\lceil_{\mathcal{H}_{\mathrm{p}}}$, and $A_{\mathrm{sc}} = A\lceil_{\mathcal{H}_{\mathrm{sc}}}$. Furthermore, consider $\operatorname{Spec}_{\mathrm{ac}}(A) = \operatorname{Spec}(A_{\mathrm{ac}})$ and $\operatorname{Spec}_{\mathrm{sc}}(A) = \operatorname{Spec}(A_{\mathrm{sc}})$.

Definition 4.19 (Absolutely continuous and singular continuous spectra).
The set $\operatorname{Spec}_{\mathrm{ac}}(A)$ is called the *absolutely continuous spectrum* of A, and $\operatorname{Spec}_{\mathrm{sc}}(A)$ the *singular continuous spectrum* of A.

The set $\operatorname{Spec}_{\mathrm{sing}}(A) = \operatorname{Spec}(A\lceil_{\mathcal{H}_{\mathrm{sing}}})$ is called the *singular spectrum* of A. Recall that $\operatorname{Spec}_{\mathrm{p}}(A) = \{\lambda \mid \lambda \text{ is an eigenvalue}\}$ is the point spectrum. Note that $\overline{\operatorname{Spec}_{\mathrm{p}}(A)} = \operatorname{Spec}(A_p)$. It is possible to show that

$$\mathrm{Spec}(A) = \mathrm{Spec}_{\mathrm{ac}}(A) \cup \overline{\mathrm{Spec}_{\mathrm{p}}(A)} \cup \mathrm{Spec}_{\mathrm{sc}}(A),$$

where the above sets need not be disjoint.

Example 4.20 (Multiplication operator). Let V be a Lebesgue measurable real-valued function, called *potential*, and define a multiplication operator by $(V\varphi)(x) = V(x)\varphi(x)$ for a. e. $x \in \mathbb{R}^d$, with domain $D(V) = \{\varphi \in L^2(\mathbb{R}^d) \mid Vf \in L^2(\mathbb{R}^d)\}$. Note that this operator is densely defined. Since V is real-valued, it is symmetric, and in fact it is a self-adjoint operator.

Example 4.21 (Laplace operator). Recall that the *weak partial derivative* $\partial_j\varphi$ of $\varphi \in L^1_{\mathrm{loc}}(\mathbb{R}^d)$ is a function $h \in L^1_{\mathrm{loc}}(\mathbb{R}^d)$ such that

$$\int_{\mathbb{R}^d} h(x)f(x)dx = -\int_{\mathbb{R}^d} \varphi(x)\partial_j f(x)dx, \quad f \in C_0^\infty(\mathbb{R}^d),$$

and is the jth component of the gradient operator in the weak sense. This notion reduces to the partial derivative (gradient) known from elementary calculus whenever $\varphi \in C^1(\mathbb{R}^d)$. The *Laplacian* Δ in $L^2(\mathbb{R}^d)$ is the symmetric operator defined by

$$\Delta : \varphi \mapsto \sum_{j=1}^{d} \partial_j^2\varphi \tag{4.1.20}$$

for $\varphi \in C_0^\infty(\mathbb{R}^d)$. It is seen that $T = \Delta\restriction_{C_0^\infty(\mathbb{R}^d)}$ is essentially self-adjoint. We denote the closure of T by the same symbol, i. e., $\Delta = \overline{T}$. The domain of Δ is the Sobolev space

$$D(-(1/2)\Delta) = H^2(\mathbb{R}^d) = \left\{f \in L^2(\mathbb{R}^d) \mid |k|^2\hat{f} \in L^2(\mathbb{R}^d)\right\},$$

where $\hat{f}$ and $\check{f}$ denote the Fourier transform and inverse Fourier transform of f, respectively. Moreover, $\widehat{-\Delta f}(k) = |k|^2\hat{f}(k)$ holds for every $f \in H^2(\mathbb{R}^d)$. Since the range of the multiplication operator $|k|^2$ is the positive semiaxis, it follows that $-\Delta$ has continuous spectrum and $\mathrm{Spec}(-\Delta) = \mathrm{Spec}_{\mathrm{ess}}(-\Delta) = \mathrm{Spec}_{\mathrm{ac}}(-\Delta) = [0, \infty)$; in particular, $-\Delta$ is a positive operator.

Example 4.22 (Fractional Laplacian). The operator $(-\Delta)^{\alpha/2}$ in $L^2(\mathbb{R}^d)$ with domain $H^\alpha(\mathbb{R}^d) = \{f \in L^2(\mathbb{R}^d) \mid |k|^\alpha\hat{f} \in L^2(\mathbb{R}^d)\}$, $0 < \alpha < 2$, defined by $\widehat{(-\Delta)^{\alpha/2}f}(k) = |k|^\alpha\hat{f}(k)$, is called a *fractional Laplacian* with exponent $\alpha/2$. It is essentially self-adjoint on $C_0^\infty(\mathbb{R}^d)$, and its spectrum is $\mathrm{Spec}((-\Delta)^{\alpha/2}) = \mathrm{Spec}_{\mathrm{ess}}((-\Delta)^{\alpha/2}) = \mathrm{Spec}_{\mathrm{ac}}((-\Delta)^{\alpha/2}) = [0, \infty)$. This is a case of a pseudodifferential operator. A related operator is a *fractional relativistic Laplacian* $(-\Delta + m^{2/\alpha})^{\alpha/2} - m$, with $m > 0$, see Example 4.267 below.

Finally, we discuss the so-called *min–max principle*, which is used to determine the eigenvalues of a self-adjoint operator bounded from below.

Proposition 4.23 (Min–max principle). *Let A be a self-adjoint operator in a Hilbert space $\mathscr{H}$, and suppose it is bounded from below. Let $n \in \mathbb{N}$ be arbitrary but fixed. Define*

$$\mu_n(A) = \sup_{\psi_1,\dots,\psi_{n-1} \in \mathscr{H}} \inf_{\substack{\psi \in D(A); \|\psi\|=1 \\ \psi \in [\psi_1,\dots,\psi_{n-1}]^\perp}} (\psi, A\psi), \tag{4.1.21}$$

where $[\psi_1,\dots,\psi_{n-1}]^\perp = \{\phi \in \mathscr{H} \mid (\psi_j,\phi) = 0, j = 1,\dots,n-1\}$. Exactly one of the following properties holds:

1. *there are n eigenvalues (counting multiplicity) below* $\inf \operatorname{Spec}_{\mathrm{ess}}(A)$ *and $\mu_n(A)$ is the nth eigenvalue, counting multiplicity;*
2. *$\mu_n(A) = \inf \operatorname{Spec}_{\mathrm{ess}}(A)$, $\mu_n(A) = \mu_{n+1}(A) = \mu_{n+2}(A) = \cdots$, and there are at most $n-1$ eigenvalues (counting multiplicity) below $\mu_n(A)$.*

The following useful tool is called the *Rayleigh–Ritz method*, applied in practice in order to get actual numerical estimates of eigenvalues.

Proposition 4.24 (Rayleigh–Ritz method). *Let A be a self-adjoint operator bounded from below in a Hilbert space $\mathscr{H}$. Let $D \subset D(A)$ be an n-dimensional subspace, and $P : \mathscr{H} \to D$ the projection to this subspace. We identify $A_D = PAP$ as a self-adjoint operator from D to itself, and let $e_1 \le \cdots \le e_n$ be the ordered eigenvalues of A_D. Let $\mu_j(A)$ be as given in* (4.1.21). *Then $\mu_j(A) \le e_j$ for $j = 1,\dots,n$. In particular, if A has eigenvalues $E_1,\dots,E_n$ at the bottom of its spectrum with $E_1 \le \cdots \le E_n$, then $E_j \le e_j$ for $j = 1,\dots,n$.*

An application of the min–max principle and the Rayleigh–Ritz method is in obtaining upper and lower bounds of the nth eigenvalue E_n of a self-adjoint operator A. By the min–max principle, we have the lower bound

$$E_n \ge \inf_{\substack{\psi \in D(A); \|\psi\|=1 \\ \psi \in [\psi_1,\dots,\psi_{n-1}]^\perp}} (\psi, A\psi),$$

and the Rayleigh–Ritz method gives the upper bound $E_n \le e_n$.

4.1.4 Compact operators and trace ideals

Many problems in classical mathematical physics can be reformulated in terms of integral equations. Consider the integral operator

$$(Kf)(x) = \int_{\mathbb{R}^d} K(x,y) f(y) dy$$

with integral kernel $K(x,y)$. If this kernel is in $L^2(\mathbb{R}^d \times \mathbb{R}^d)$, then K belongs to the following class of operators.

Definition 4.25 (Compact operator). A bounded operator $A : \mathcal{K} \to \mathcal{H}$ is a *compact operator* whenever A takes an arbitrary bounded set in $\mathcal{K}$ into a precompact set in $\mathcal{H}$. Equivalently, A is compact if and only if for every bounded sequence $(\varphi_n)_{n\in\mathbb{N}}$, $(A\varphi_n)_{n\in\mathbb{N}}$ has a subsequence convergent in $\mathcal{H}$.

Some important properties of compact operators are summarized below.

Proposition 4.26 (Riesz–Schauder theorem). *Let A be a compact operator.*
(1) *The set* $\mathrm{Spec}(A)$ *is discrete and has no limit points except possibly zero.*
(2) *Every $\lambda \in \mathrm{Spec}(A)$ is an eigenvalue of finite multiplicity.*

Proposition 4.27.
(1) *Let $C(\mathcal{K})$ be the set of compact operators from $\mathcal{K}$ to itself. Then $C(\mathcal{K})$ is an ideal of $B(\mathcal{K},\mathcal{K})$, i. e., $AB, BA \in C(\mathcal{K})$ for $A \in C(\mathcal{K})$ and $B \in B(\mathcal{K},\mathcal{K})$. Furthermore, $K(\mathcal{K}) = B(\mathcal{K},\mathcal{K})/C(\mathcal{K})$ is simple, i. e., $K(\mathcal{K})$ has no nontrivial ideal.*
(2) *The set $C(\mathcal{K})$ is closed in the uniform operator topology.*

The space $K(\mathcal{K})$ is called the *Calkin algebra*.

Example 4.28 (Integral operators). Let $k \in L^2(\mathbb{R}^d \times \mathbb{R}^d)$. Then the integral operator $A : f \mapsto \int_{\mathbb{R}^d} k(x,y)f(y)dy$ is a compact operator on $L^2(\mathbb{R}^d)$.

We present some useful inequalities and criteria for compact operators without proof.

Proposition 4.29 (Young's inequality). *Let $1 \le p,q,r \le \infty$ such that $\frac{1}{p} + \frac{1}{q} + \frac{1}{r} = 2$. Also, let $f \in L^p(\mathbb{R}^d)$, $g \in L^q(\mathbb{R}^d)$, and $h \in L^r(\mathbb{R}^d)$. Then there exists a constant $C(p,q,r,d)$ such that*

$$\left| \int_{\mathbb{R}^d\times\mathbb{R}^d} f(x)g(x-y)h(y)dxdy \right| \le C(p,q,r,d)\|f\|_p\|g\|_q\|h\|_r. \tag{4.1.22}$$

The next result is called *Hardy–Littlewood–Sobolev inequality* (or *weak Young inequality*). It relates with the situation above in which g is replaced by $|x|^{-\lambda} \notin L^p(\mathbb{R}^d)$, and is thus not covered by the Young inequality.

Proposition 4.30 (Hardy–Littlewood–Sobolev inequality). *Let $p,r > 1$ and $0 < \lambda < d$, with $\frac{1}{p} + \frac{\lambda}{d} + \frac{1}{r} = 2$. Let $f \in L^p(\mathbb{R}^d)$ and $h \in L^r(\mathbb{R}^d)$. Then there exists a constant $C(\lambda,p,r,d)$ such that*

$$\left| \int_{\mathbb{R}^d\times\mathbb{R}^d} f(x)|x-y|^{-\lambda}h(y)dxdy \right| \le C(\lambda,p,r,d)\|f\|_p\|h\|_r. \tag{4.1.23}$$

Example 4.31. Let $k(x,y) = V^{1/2}(x)|x-y|^{2-d}V^{1/2}(y)$ for $d \geq 2$ with $0 \leq V \in L^{d/2}(\mathbb{R}^d)$. Then $\int_{\mathbb{R}^d\times\mathbb{R}^d} k(x,y)^2 dxdy < \infty$ by the Hardy–Littlewood–Sobolev inequality, and the integral operator $A : f \mapsto \int_{\mathbb{R}^d} k(x,y)f(y)dy$ is compact. Further, below we will see that it is actually of Hilbert–Schmidt class, see Definition 4.33.

Next we introduce the *weak* L^p *space*

$$L^p_w(\mathbb{R}^d) = \Big\{f : \mathbb{R}^d \to \mathbb{C} \mid \sup_{a>0} a|\{x \in \mathbb{R}^d \mid |f(x)| > a\}|^{1/p} < \infty\Big\}. \tag{4.1.24}$$

Here $|E|$ denotes the Lebesgue measure of the measurable set $E \subset \mathbb{R}^d$. We note that $L^p(\mathbb{R}^d) \subset L^p_w(\mathbb{R}^d)$, since

$$\sup_{a>0} a^p|\{x \in \mathbb{R}^d \mid |f(x)| > a\}| \leq \int_{|f(x)|>a} |f(x)|^p dx \leq \|f\|_p^p.$$

The expression $\sup_{a>0} a|\{x \in \mathbb{R}^d \mid |f(x)| > a\}|^{1/p}$ does not define a norm on the space $L^p_w(\mathbb{R}^d)$, since the triangle inequality is not satisfied. For $p > 1$, however, there is an alternative expression which is equivalent to $\sup_{a>0} a|\{x \mid |f(x)| > a\}|^{1/p}$ and is a norm. It is given by

$$\|f\|_{p,w} = \sup_{A\subset\mathbb{R}^d, |A|<\infty} |A|^{-1/p^*} \int_A |f(x)|dx, \tag{4.1.25}$$

where $1/p + 1/p^* = 1$. Using this larger space leads to a weaker version of the Young inequality.

Proposition 4.32 (Weak Young inequality). *Let* $1 < p, q, r < \infty$ *with* $1/p + 1/q + 1/r = 2$ *and* $f \in L^p(\mathbb{R}^d)$, $h \in L^r(\mathbb{R}^d)$, *and* $g \in L^q_w(\mathbb{R}^d)$. *Then there exists a constant* $K(p,q,r,d)$ *such that*

$$\left| \int_{\mathbb{R}^d\times\mathbb{R}^d} f(x)g(x-y)h(y)dxdy \right| \leq K(p,q,r,d)\|f\|_p\|g\|_{q,w}\|h\|_r. \tag{4.1.26}$$

We note that $g_\lambda(x) = |x|^{-\lambda} \in L^{d/\lambda}_w(\mathbb{R}^d)$ and

$$\|g_\lambda\|_{d/\lambda,w} = \frac{d}{d-\lambda}\left(\frac{|S_{d-1}|}{d}\right)^{d/\lambda}, \tag{4.1.27}$$

where $|S_{d-1}|$ denotes the surface area of the $(d-1)$-dimensional unit sphere S_{d-1}. The Hardy–Littlewood–Sobolev inequality is a special case of the weak Young inequality.

Consider next the following class of operators, called *trace ideal* or *Schatten class*.

Definition 4.33 (Trace ideal/Schatten class). Let A be a compact operator on $\mathcal{K}$ and $\mu_1(A), \mu_2(A), \ldots$ the sequence of the eigenvalues of $|A|$ arranged in nonincreasing order with each eigenvalue occurring as often as its multiplicity, where $|A| = \sqrt{A^*A}$. Let $1 \leq p < \infty$; then $A \in \mathcal{I}_p$ if and only if

$$\|A\|_p = \left(\sum_{n=1}^{\infty} \mu_n(A)^p\right)^{1/p} < \infty.$$

In particular, $\mathcal{I}_1$ is called the *trace class*, and $\mathcal{I}_2$ the *Hilbert–Schmidt class.*

The trace ideal $\mathcal{I}_p$ is a normed space with norm $\|\cdot\|_p$ and it is complete. Since $\mu_1(A) = \|A\|$, we have $\|A\| \leq \|A\|_p$, and also $\mathcal{I}_p \subset \mathcal{I}_q$ whenever $1 \leq q \leq p$. For $A \in \mathcal{I}_1$, the *trace* of an operator A is defined by

$$\operatorname{Tr} A = \sum_{j=1}^{\infty} (\phi_j, A\phi_j),$$

for any complete orthonormal system $\{\phi_j\}_{j=1}^{\infty}$ of $\mathcal{K}$, and its value is independent of the choice of this system. Also, $|\operatorname{Tr}(A)| \leq \|A\|_1$ holds. Some basic properties of $\mathcal{I}_p$ are summarized next.

Proposition 4.34. *Let $1 \leq p < \infty$. Then $\mathcal{I}_p$ is a $*$-ideal of $B(\mathcal{K}, \mathcal{K})$, i. e.,*
(1) *$A \in \mathcal{I}_p$ implies that $A^* \in \mathcal{I}_p$ and $\|A\|_p = \|A^*\|_p$;*
(2) *$A \in \mathcal{I}_p$ and $B \in B(\mathcal{K}, \mathcal{K})$ imply that AB, $BA \in \mathcal{I}_p$, $\|AB\|_p \leq \|A\|_p\|B\|$, and $\|BA\|_p \leq \|B\|\|A\|_p$.*

We also define the *weak trace ideal* $\mathcal{I}_{p,w}$, which satisfies $\mathcal{I}_{p,w} \subset \mathcal{I}_q$ for all $p < q$.

Definition 4.35 (Weak trace ideal). Let A be a compact operator. Then $A \in \mathcal{I}_{p,w}$ for $1 \leq p < \infty$ whenever $\|A\|_{p,w}^* = \sup_{n\in\mathbb{N}} n^{1/p}\mu_n(A) < \infty$.

Note that $\|A\|_{p,w}^*$ is not a norm on $\mathcal{I}_{p,w}$. For $p \neq 1$, however, we have the *Calderón norm* defined by

$$\|A\|_{p,w} = \sup_{k\in\mathbb{N}} \left(k^{\frac{1}{p}} \frac{1}{k} \sum_{n=1}^{k} \mu_n(A)\right),$$

and

$$\|A\|_{p,w}^* \leq \|A\|_{p,w} \leq \frac{p}{p-1}\|A\|_{p,w}^*$$

holds. Hence it is seen that $\|\cdot\|_{p,w}^*$ and $\|\cdot\|_{p,w}$ are equivalent.

Proposition 4.36. *Let $A \in \mathcal{I}_{p,w}$ for $1 \leq p < \infty$. Then $A \in \mathcal{I}_q$ for $q > p$.*

Proof. Since there exists a constant C such that $\mu_n(A) \le Cn^{-1/p}$, it is immediate that $\|A\|_q \le \sum_{n=1}^{\infty} \mu_n(A)^q \le C^q \sum_{n=1}^{\infty} n^{-q/p} < \infty$. □

Proposition 4.37 (Hölder inequality for trace ideals). *Let $1/p = 1/q+1/r$, $p \ge 1$, and $A \in \mathscr{I}_q$, $B \in \mathscr{I}_r$. Then $AB \in \mathscr{I}_p$ and $\|AB\|_p \le \|A\|_q\|B\|_r$ holds. If $p > 1$, $A \in \mathscr{I}_{q,w}$, $B \in \mathscr{I}_{r,w}$, then $AB \in \mathscr{I}_{p,w}$ and $\|AB\|_{p,w} \le \frac{p}{p-1}\|A\|_{q,w}\|B\|_{r,w}$.*

From Proposition 4.37, we see that for $1 < p < \infty$ and $q = p/(p-1)$, $A \in \mathscr{I}_p$ and $B \in \mathscr{I}_q$ satisfy $AB \in \mathscr{I}_1$ and $|\mathrm{Tr}(AB)| \le \|AB\|_1 \le \|A\|_p\|B\|_q$. Thus the map $\mathscr{I}_p \ni A \mapsto \mathrm{Tr}(AB) \in \mathbb{C}$ defines the dual of $\mathscr{I}_p$. It is, however, known that any $\ell \in \mathscr{I}_p^*$ is of the form $\ell(\cdot) = \mathrm{Tr}\,(\cdot B)$ with some $B \in \mathscr{I}_q$. We state this in the following proposition.

Proposition 4.38 (Duality of $\mathscr{I}_p$). *Let $1 < p < \infty$ and $q = p/(p-1)$. Then $\mathscr{I}_p^* = \mathscr{I}_q$ in the sense that every $\ell \in \mathscr{I}_p^*$ has the form $\ell(A) = \mathrm{Tr}(AB)$ with some $B \in \mathscr{I}_q$ and $\|\ell\|_{\mathscr{I}_p^*} = \|B\|_q$.*

By this result we can identify $\mathscr{I}_p^*$ with $\mathscr{I}_q$ for all $1 < p < \infty$.

Next we consider trace ideals on the Hilbert space $L^2(\mathbb{R}^d)$. The most fundamental and useful examples of trace ideals on $L^2(\mathbb{R}^d)$ are obtained in the form $f(x)g(-i\nabla)$. When $f, g \in L^2(\mathbb{R}^d)$, $A = f(x)g(-i\nabla)$ is an integral operator with integral kernel $A(x,y) = (2\pi)^{-d/2}f(x)\check{g}(x-y)$, satisfying $A(\cdot,\cdot) \in L^2(\mathbb{R}^d \times \mathbb{R}^d)$ by Young's inequality. Hence $A \in \mathscr{I}_2$ on $L^2(\mathbb{R}^d)$. A more general fact is described by the following result.

Proposition 4.39. *Let $f \in L^p(\mathbb{R}^d)$ and $g \in L^p(\mathbb{R}^d)$ with $2 \le p < \infty$. Then it follows that $A = f(x)g(-i\nabla) \in \mathscr{I}_p$ on $L^2(\mathbb{R}^d)$ and $\|A\|_p \le (2\pi)^{-d/p}\|f\|_p\|g\|_p$.*

In the following we construct a compact operator by a combination of elements of $L^p(\mathbb{R}^d)$ and $L^p_w(\mathbb{R}^d)$.

Proposition 4.40. *Let $2 < p < \infty$, $g \in L^p(\mathbb{R}^d)$ and $u \in L^p_w(\mathbb{R}^d)$. Define the operator*

$$B_{u,g}h(k) = \frac{1}{(2\pi)^{d/2}} u(k) \int_{\mathbb{R}^d} e^{ik\cdot x} g(x)h(x)dx \tag{4.1.28}$$

on $L^2(\mathbb{R}^d)$, with $D(B_{u,g}) = C_0(\mathbb{R}^d)$. Then $B_{u,g}$ can be extended to a compact operator on $L^2(\mathbb{R}^d)$, and $B_{u,g} \in \mathscr{I}_{p,w}$; moreover, there exists a constant K_p such that

$$\|B_{u,g}\|^*_{p,w} \le K_p\|u\|_{p,w}\|g\|_p. \tag{4.1.29}$$

Let F be the Fourier transform on $L^2(\mathbb{R}^d)$. We denote $FB_{u,g}$ formally by $u(-i\nabla_x)g(x)$; see Proposition 4.73.

Corollary 4.41. *Let $2 < p < \infty$, $g \in L^p(\mathbb{R}^d)$ and $u \in L^p_w(\mathbb{R}^d)$. Define the operator $B_{u,g}$ on $L^2(\mathbb{R}^d)$ by (4.1.28). Then $B_{u,g} \in \mathscr{I}_q$ for all $q > p$.*

Proof. By Proposition 4.40, it follows that $B_{u,g} \in \mathscr{I}_{p,w}$. The corollary results then from Proposition 4.36. □

Example 4.42. Let $d \geq 3$ and $g \in L^d(\mathbb{R}^d)$. Since $u(k) = 1/|k| \in L^d_w(\mathbb{R}^d)$, by Proposition 4.40 and Corollary 4.41, we have $B_{u,g} \in \mathcal{I}_{d,w}$, and $B_{u,g} \in \mathcal{I}_p$ for $p > d$. Furthermore, $B^*_{u,g}B_{u,g} \in \mathcal{I}_{d/2,w}$ and, in particular, $B^*_{u,g}B_{u,g} \in \mathcal{I}_p$ for any $p > d/2$.

Finally, we discuss a convergence theorem. An $N \times N$ matrix $(a_{nm})_{1\leq n,m\leq N}$ (possibly $N = \infty$) is called *doubly substochastic* if $\sum_{n=1}^N |a_{nm}| < 1$ for $1 \leq m \leq N$, and $\sum_{m=1}^N |a_{nm}| < 1$ for $1 \leq n \leq N$. For example, if $\{f_n\}_{n=1}^N$, $\{g_m\}_{m=1}^N$ are orthonormal sets on a Hilbert space $\mathcal{K}$, the matrix $(a_{nm})_{1\leq n,m\leq N} = (|(f_n, g_m)_{\mathcal{K}}|^2)_{1\leq n,m\leq N}$ is doubly substochastic.

Proposition 4.43. *Let A and B be compact operators. Then there exists a doubly substochastic matrix $a = (a_{nm})_{1\leq n,m\leq\infty}$ such that*

$$\mu_n(A) - \mu_n(B) = \sum_{m=1}^{\infty} a_{nm}\mu_m(A - B),$$

for all $n \geq 1$, where μ_n is as given in Definition 4.33. In particular, $|\mu_n(A) - \mu_n(B)| \leq \|A - B\|$ for all $n \geq 1$, with uniform norm.

The next corollary can be immediately derived from Proposition 4.43.

Corollary 4.44 (Convergence of $\mu_n(A_m)$). *Suppose that $(A_n)_{n\in\mathbb{N}}$ and A are compact operators such that $\lim_{m\to\infty} \|A_m - A\| = 0$. Then $\mu_n(A_m) \to \mu_n(A)$ as $m \to \infty$, for all $n \geq 1$.*

Under an extra condition combined with $\lim_{m\to\infty} \|A_m - A\| = 0$, also $\mathcal{I}_p$-convergence is obtained.

Proposition 4.45 ($\mathcal{I}_p$-convergence). *Let $1 \leq p < \infty$.*

(1) *Suppose that $(A_m)_{m\in\mathbb{N}}$, A, and B are compact operators such that for all $n \geq 1$, $\mu_n(A_m) \leq \mu_n(B)$ and $B \in \mathcal{I}_p$. Suppose, moreover, that $A_m \to A$ in the operator norm as $m \to \infty$. Then $\lim_{m\to\infty} \|A_m - A\|_p = 0$.*
(2) *Suppose that $(A_m)_{m\in\mathbb{N}}, A \in \mathcal{I}_p$, and $A_m \to A$ and $A^*_m \to A^*$ strongly as $m \to \infty$ and $\|A_m\|_p \to \|A\|_p$. Then $\lim_{m\to\infty} \|A_m - A\|_p = 0$.*
(3) *Suppose that $(A_m)_{m\in\mathbb{N}}, A \in \mathcal{I}_p$, and $A_m \to A$, $|A_m| \to |A|$, and $|A^*_m| \to |A^*|$ weakly as $m \to \infty$. Then $\lim_{m\to\infty} \|A_m - A\|_p = 0$.*

The following result gives useful criteria guaranteeing that the spectrum of a self-adjoint operator is purely discrete.

Proposition 4.46. *Let A be a self-adjoint operator and bounded below. Then the following properties are equivalent:*

(1) *$(\lambda - A)^{-1}$ is a compact operator for some $\lambda \in \rho(A)$;*
(2) *$(\lambda - A)^{-1}$ is a compact operator for all $\lambda \in \rho(A)$;*
(3) *the set $\{\varphi \in D(A) \mid \|\varphi\| < 1, \|A\varphi\| \leq b\}$ is compact for all b;*
(4) *the set $\{\varphi \in D(|A|^{1/2}) \mid \|\varphi\| < 1, \||A|^{1/2}\varphi\| \leq b\}$ is compact for all b.*

Below we will consider perturbations of given operators and address the question whether and how the essential spectrum of a self-adjoint operator changes under a perturbation. Perturbations by operators satisfying the condition below leave the essential spectrum invariant.

Definition 4.47 (Relative compactness). Let A be self-adjoint. An operator B such that $D(A) \subset D(B)$ is called *relatively compact* with respect to A or A-compact if and only if $B(i + A)^{-1}$ is compact.

Then we have

Proposition 4.48. *Let A be self-adjoint and B be relatively compact with respect to A. Then $A + B$ with domain $D(A)$ is closed and* $\mathrm{Spec}_{\mathrm{ess}}(A + B) = \mathrm{Spec}_{\mathrm{ess}}(A)$.

4.1.5 Schrödinger operators

Now we turn to discussing the specific class of operators announced at the start of this chapter, and which will be a main object throughout this book.

Definition 4.49 (Schrödinger operator). Let V be a real-valued multiplication operator in $L^2(\mathbb{R}^d)$, called *potential*. The linear operator

$$H = -\frac{1}{2}\Delta + V \tag{4.1.30}$$

with the dense domain

$$D(H) = D\Big(-\frac{1}{2}\Delta\Big) \cap D(V) \tag{4.1.31}$$

is called a *Schrödinger operator*.

By the definition of Δ, it is readily seen that every $\varphi \in D(-(1/2)\Delta)$ is a bounded continuous function whenever $d \le 3$. Indeed, by the estimate

$$\|\hat{\varphi}\|_1 = \|(1 + |k|^2)^{-1}(1 + |k|^2)\hat{\varphi}\|_1 \le \|(1 + |k|^2)^{-1}\|_2\|(1 + |k|^2)\hat{\varphi}\|_2 < \infty,$$

the Fourier transform of φ lies in $L^1(\mathbb{R}^d)$. Thus φ is a continuous bounded function by the Riemann–Lebesgue lemma. Moreover, from the scaling $\hat{\varphi}_r(k) = r^d\hat{\varphi}(rk)$, we have $\|\hat{\varphi}\|_1 \le cr^{\frac{d-4}{2}}\||k|^2\hat{\varphi}\|_2 + cr^{d/2}\|\hat{\varphi}\|_2$, with a constant c independent of r. Since $d \le 3$, for arbitrary $a > 0$ there exists b such that

$$\|\varphi\|_\infty \le \|\hat{\varphi}\|_1 \le a\|\Delta\varphi\|_2 + b\|\varphi\|_2. \tag{4.1.32}$$

This bound is useful in estimating the relative bound of a given potential V with respect to the Laplacian.

We are interested in defining Schrödinger operators as self-adjoint operators. One reason is that then the spectrum is guaranteed to be real, matching with the uses of quantum mechanics. Secondly, since the solution of the Schrödinger equation (4.1.1) is given by $e^{-itH}\varphi_0$ for $\varphi(x,0) = \varphi_0(x)$, in order to make sure that e^{-itH} is a unitary map, the question has to be addressed whether the symmetric operator H is self-adjoint or essentially self-adjoint. From the point of view of quantum mechanics, distinct self-adjoint extensions lead to different time evolutions; however, since several self-adjoint extensions of H may exist, a particular one must be chosen to describe the evolution of the system. Since for any self-adjoint extension K of H it is true that $K \subset H^*$, an important step is to find a core of H as then $T = \overline{H\lceil_{\mathcal{D}}}$ is the unique self-adjoint extension of $H\lceil_{\mathcal{D}}$.

In general, it is difficult to show self-adjointness or essential self-adjointness of a Schrödinger operator. Here we review some basic criteria; however, as we are interested in defining and studying the properties of perturbations of the Laplacian, first we give a notion of perturbation.

Definition 4.50 (Relative boundedness). Given an operator A, an operator B is called *A-bounded* whenever $D(A) \subseteq D(B)$ and

$$\|Bf\| \leq a\|Af\| + b\|f\| \tag{4.1.33}$$

with some numbers $a, b \geq 0$, for all $f \in D(A)$. In this case a is called a *relative bound.* Moreover, if the relative bound a can be chosen arbitrarily small, B is said to be *infinitesimally small* with respect to A.

A classic result of self-adjoint perturbations of a given self-adjoint operator is offered by the following theorem.

Theorem 4.51 (Kato–Rellich theorem). *Suppose that A is a self-adjoint operator on $\mathcal{H}$, and B is a symmetric A-bounded operator with relative bound strictly less than* 1. *Then $A + B$ is self-adjoint on $D(A)$ and essentially self-adjoint on any core of A with*

$$\overline{(A+B)\lceil_D} = \overline{A\lceil_D} + \overline{B\lceil_D},$$

where D is any core of A. Furthermore, if A is bounded from below by M, then the self-adjoint operator $A + B$ is also bounded from below by $M - \max\left\{\frac{b}{1-a}, a|M| + b\right\}$, where a and b are given by (4.1.33).

Proof. It is sufficient to show that $\operatorname{Ran}(A + B \pm ix) = \mathcal{H}$ for some $x > 0$. Since A is self-adjoint, $\operatorname{Ran}(A \pm ix) = \mathcal{H}$. Since

$$A + B + ix = (B(A+ix)^{-1} + 1)(A + ix),$$

it is sufficient to show that $\operatorname{Ran}(B(A+ix)^{-1} + 1) = \mathcal{H}$. By the assumption, we have

$$\|B(A+ix)^{-1}f\| \leq a\|A(A+ix)^{-1}f\| + b\|(A+ix)^{-1}f\|.$$

From the identity $\|(A+ix)f\|^2 = \|Af\|^2 + x^2\|f\|^2$, it follows that $\|A(A+ix)^{-1}f\| \le \|f\|$ and $\|(A+ix)^{-1}f\| \le \|f\|/x$. Thus we have $\|B(A+ix)^{-1}f\| \le (a+b/x)\|f\|$ and $\|B(A+ix)^{-1}\| < 1$ for sufficiently large x. Hence $B(A+ix)^{-1}+1$ is invertible and $\mathrm{Ran}(A+B+ix) = \mathcal{H}$ follows for sufficiently large x. Similarly, we can show that $\mathrm{Ran}(A+B-ix) = \mathcal{H}$. We next show the second statement. Let D be a core of A. Setting $A' = A\lceil_D$ and $B' = B\lceil_D$, we obtain

$$\|B'f\| \le a\|A'f\| + b\|f\| \tag{4.1.34}$$

for $f \in D(A')$. For $f \in D(\overline{A'})$, there exists a sequence $(f_n)_{n=1}^{\infty} \subset D(A')$ such that $f_n \to f$ and $A'f_n \to \overline{A'}f$ as $n \to \infty$. By (4.1.34), we can also see that $B'f_n$ is a Cauchy sequence, hence $B'f_n \to \overline{B'}f$ as $n \to \infty$. Thus from the inequality $\|B'f_n\| \le a\|A'f_n\| + b\|f_n\|$ and a limiting argument it follows that

$$\|\overline{B'}f\| \le a\|\overline{A'}f\| + b\|f\| \tag{4.1.35}$$

for $f \in D(\overline{A'})$. Thus $\overline{A'}+\overline{B'}$ is self-adjoint on $D(\overline{A'})$. By the above limiting argument, we also see that $(A'+B')f_n \to \overline{A'}f + \overline{B'}f$ and $f_n \to f$, which implies that $\overline{A'+B'}f = \overline{A'}f + \overline{B'}f$ for $f \in D(\overline{A'})$. Hence $\overline{A'+B'} \supset \overline{A'}+\overline{B'}$ follows. Since $\overline{A'}+\overline{B'}$ is a self-adjoint extension of $A'+B'$, it is trivial to see that $\overline{A'}+\overline{B'} \supset \overline{A'+B'}$. Thus $\overline{A'+B'} = \overline{A'}+\overline{B'}$. We see that $\overline{A'+B'}$ is self-adjoint and hence $A+B$ is essentially self-adjoint on D. Finally, we estimate of $A+B$ from below. Let $-t < M$. We can show that $\|B(A+t)^{-1}\| < 1$ if $-t < M - \max\left\{\frac{b}{1-a}, a|M| + b\right\}$. Thus for such t, $\mathrm{Ran}(A+B+t) = \mathcal{H}$ and $(A+B+t)^{-1} = (A+t)^{-1}(B(A+t)^{-1}+1)^{-1}$ is bounded. This implies that $-t \notin \mathrm{Spec}(A+B)$. □

Applied to Schrödinger operators, the natural class of V for the Kato–Rellich theorem is $L^p(\mathbb{R}^d) + L^\infty(\mathbb{R}^d)$.

Example 4.52. Let $d = 3$ and $V = V_1 + V_2 \in L^2(\mathbb{R}^3) + L^\infty(\mathbb{R}^3)$. Since $\|V\varphi\|_2 \le \|V_1\|_2\|\varphi\|_\infty + \|V_2\|_\infty\|\varphi\|_2$ and $\|\varphi\|_\infty \le a\|\Delta\varphi\|_2 + b\|\varphi\|_2$, for arbitrary $a > 0$ and some $b > 0$ by (4.1.32), we have

$$\|V\varphi\|_2 \le a\|V_1\|_2\,\|\Delta\varphi\|_2 + (b + \|V_2\|_\infty)\|\varphi\|_2.$$

Thus by Theorem 4.51, the Schrödinger operator $-(1/2)\Delta + V$ is self-adjoint on $D(-(1/2)\Delta)$ and bounded from below. Moreover, it is essentially self-adjoint on any core of $-\Delta$. In particular, $-(1/2)\Delta - |x|^{-1}$ is self-adjoint on $D(-(1/2)\Delta)$ for $d = 3$.

Example 4.53. The previous example can be extended to arbitrary d-dimension. Let

$$V \in L^p(\mathbb{R}^d) + L^\infty(\mathbb{R}^d) \begin{cases} p = 2, & d \le 3 \\ p > d/2, & d \ge 4. \end{cases} \tag{4.1.36}$$

In the case $d \ge 4$, consider $V \in L^p(\mathbb{R}^d)$. Choose $q = 2p/(p-2)$ and $r = 2p/(p+2)$. From the Hölder and Hausdorff–Young inequalities, it follows that

$$\|Vf\|_2 \le \|V\|_p\|f\|_q \le C\|V\|_p\|\hat{f}\|_r.$$

Thus

$$\begin{aligned}\|\hat{f}\|_r &= \|(1+\varepsilon|k|^2)^{-1}(1+\varepsilon|k|^2)\hat{f}\|_r \le \|(1+\varepsilon|k|^2)^{-1}\|_p\|(1+\varepsilon|k|^2)\hat{f}\|_2 \\ &\le C'\varepsilon^{-d/(2p)}(\varepsilon\|\Delta f\|_2 + \|f\|_2).\end{aligned}$$

For a given $\delta > 0$ and a sufficiently small ε, we see that

$$\|Vf\|_2 \le \delta\|(-1/2)\Delta f\| + C_\delta\|f\|_2.$$

Thus V is infinitesimally small with respect to $-\Delta$. In particular, H is self-adjoint on $D(-(1/2)\Delta)$ and essentially self-adjoint on any core of $-\Delta$.

From the discussion in Example 4.53, we conclude the following result.

Proposition 4.54. *Let $V \in L^p(\mathbb{R}^d)$, with $p = 2$ for $d \le 3$, and $p > d/2$ for $d \ge 4$. Then there exists a constant C such that*

$$\|Vf\|_2 \le C\|V\|_p\|(-(1/2)\Delta + 1)f\| \tag{4.1.37}$$

for $f \in D(-(1/2)\Delta)$.

The Kato–Rellich theorem fails to be useful in investigating the self-adjointness of Schrödinger operators with a polynomial potential such as $V(x) = |x|^2$. In this case the following can be applied.

Theorem 4.55 (Kato's inequality). *Let $u \in L^1_{\mathrm{loc}}(\mathbb{R}^d)$ and suppose the distributional Laplacian is $\Delta u \in L^1_{\mathrm{loc}}(\mathbb{R}^d)$. Then the inequality*

$$\Delta|u| \ge \Re[(\operatorname{sgn} u)\Delta u] \tag{4.1.38}$$

holds in distributional sense.

Above we have the sign function $\operatorname{sgn} u = u(x)/|u(x)|$ whenever $u(x) \ne 0$, and $\operatorname{sgn} u = 0$ for $u(x) = 0$, defined as usual.

The natural class for Kato's inequality is $L^2_{\mathrm{loc}}(\mathbb{R}^d)$.

Corollary 4.56. *Let $V \in L^2_{\mathrm{loc}}(\mathbb{R}^d)$ be bounded below. Then $H = (-1/2)\Delta + V$ is essentially self-adjoint on $C_0^\infty(\mathbb{R}^d)$.*

Proof. By (3) of Proposition 4.11, it suffices to show that

$$\overline{\operatorname{Ran}(H\lceil_{C_0^\infty(\mathbb{R}^d)}+1)} = L^2(\mathbb{R}^d),$$

or equivalently, $(u, ((-1/2)\Delta + V + 1)f) = 0$ for $\forall f \in C_0^\infty(\mathbb{R}^d)$ implies that $u = 0$. Suppose that $((-1/2)\Delta + V + 1)u = 0$ in distributional sense. It follows that $(1/2)\Delta u = Vu + u$.

By Kato's inequality, $\Delta|u| \geq \Re([\operatorname{sgn} u]\Delta u) = 2(V+1)|u| \geq 0$. In particular, $\Delta|u| \geq 0$. Let $v_\delta(x) = v(x/\delta)\delta^{-d}$, where $v \geq 0$, $v \in C_0^\infty(\mathbb{R}^d)$ and $\int_{\mathbb{R}^d} v(x)dx = 1$. Consider $w_\delta = |u| * v_\delta$. We see that $w_\delta \in D(-(1/2)\Delta)$, $\Delta w^\delta = |u| * \Delta v_\delta = \Delta|u| * v_\delta \geq 0$, and $(w_\delta, \Delta w_\delta) \leq 0$ is trivial. Thus $w_\delta = 0$. Since $w_\delta \to |u|$ as $\delta \to 0$ in $L^2(\mathbb{R}^d)$, we have $|u| = 0$. Hence the statement follows. □

Corollary 4.56 can be extended to more general cases.

Corollary 4.57. *Let $V = V_1 + V_2$ be such that $V_1 \geq 0$, $V_1 \in L^2_{\mathrm{loc}}(\mathbb{R}^d)$, and V_2 is $-(1/2)\Delta$-bounded with a relative bound less than 1. Then $H = (-1/2)\Delta + V$ is essentially self-adjoint on $C_0^\infty(\mathbb{R}^d)$.*

Example 4.58. By Corollary 4.56, the operator $-(1/2)\Delta + P(x)$ with a polynomial $P(x)$ bounded from below is essentially self-adjoint on $C_0^\infty(\mathbb{R}^d)$ and bounded from below.

Example 4.59. By combining the Kato–Rellich theorem with Kato's inequality, it is seen that $H = (-1/2)\Delta + V$ with potential $V = V_1 + V_2$ such that $V_1 \in L^2_{\mathrm{loc}}(\mathbb{R}^d)$, $V_1 \geq 0$, and $(-(1/2)\Delta)^{1/2}$-bounded V_2, is self-adjoint on $D(-(1/2)\Delta + V_1)$ and essentially self-adjoint on $C_0^\infty(\mathbb{R}^d)$. To see this, first note that $-(1/2)\Delta + V_1$ is essentially self-adjoint by Kato's inequality. Secondly, V_2 is $(-(1/2)\Delta + V_1)$-bounded with infinitesimally small relative bound. Hence H is self-adjoint on $D(-(1/2)\Delta + V_1)$ by the Kato–Rellich theorem.

4.1.6 Schrödinger operators through quadratic forms

When the potential is too singular and $D(-\Delta) \cap D(V)$ is too small (in the extreme case, it may reduce to $\{0\}$), then a Schrödinger operator cannot be defined as an operator sum, however, it still can be defined in terms of a form sum. Recall that a *quadratic form* q on a Hilbert space $\mathcal{K}$ is a map $q : Q(q) \times Q(q) \to \mathbb{C}$ such that $q(f,g)$ is linear in g and antilinear in f for $f, g \in Q(q)$, where $Q(q)$ denotes the *form domain* of q. The form is *symmetric* if $\overline{q(f,g)} = q(g,f)$. If $q(f,f) \geq -M\|f\|^2$ for some M, then q is said to be *semibounded*. Note that a semibounded quadratic form is automatically symmetric if $\mathcal{K}$ is defined over the complex field $\mathbb{C}$. A semibounded quadratic form q is closed if $Q(q)$ is complete in the norm

$$\|f\|_{+1} = (q(f,f) + (M+1)\|f\|^2)^{1/2}$$

with M as above. Moreover, if q is closed and $D \subset Q(q)$ is dense in $Q(q)$ in the norm $\|\cdot\|_{+1}$, then D is called a *form core* for q.

By Riesz's theorem, there is a one-to-one correspondence between semibounded quadratic forms and self-adjoint operators bounded below. Let A be a self-adjoint operator bounded below. It is clear that A defines the semibounded quadratic form

$$q_A(f,g) = \int_{\mathbb{R}} \lambda d(E_A(\lambda)f,g)$$

with form domain $Q(q_A) = D(|A|^{1/2})$.

Proposition 4.60. *If q is a closed semibounded quadratic form, then there exists a unique self-adjoint operator A such that $q = q_A$.*

Using this fact, we can define the Schrödinger operator with a singular potential such as $V(x) = \delta(x)$. Also, the concept of relative boundedness of operators can be extended to forms as well.

Definition 4.61 (Relative form-boundedness). Let A be a positive self-adjoint operator and B a self-adjoint operator such that $Q(q_B) \supset Q(q_A)$ and

$$|q_B(f,f)| \leq aq_A(f,f) + b(f,f) \tag{4.1.39}$$

for some $a, b \geq 0$. Then B is said to be *relatively form-bounded* with respect to A or *A-form-bounded.*

If B is A-bounded, then B is relatively form-bounded with respect to A with the same relative bound as the A-bound. Next we present a form analogue of the Kato–Rellich theorem.

Theorem 4.62 (KLMN theorem). *Let A be a positive self-adjoint operator and suppose that β is a symmetric quadratic form on $Q(q_A)$ such that $|\beta(f,f)| \leq aq_A(f,f) + b(f,f)$ holds with some $a < 1$ and $b \geq 0$. Then there exists a unique self-adjoint operator C such that $Q(q_C) = Q(q_A)$ and $q_C(f,g) = q_A(f,g) + \beta(f,g)$ for $f,g \in Q(q_C)$. Furthermore, C is bounded from below by $-b$ and any domain of essential self-adjointness for A is a form core for C.*

Let now A and B be self-adjoint operators bounded from below such that $Q(q_A) \cap Q(q_B)$ is dense. The quadratic form $q(f,g) = q_A(f,g) + q_B(f,g)$ with $Q(q) = Q(q_A) \cap Q(q_B)$ defines a closed semibounded quadratic form. Its associated self-adjoint operator is then denoted by $C = A \dot{+} B$. If a nonnegative self-adjoint operator B is A-form-bounded with relative bound strictly smaller than 1, then the KLMN theorem implies that a unique self-adjoint operator associated with the quadratic form $q_A - q_B$ exists, and this is denoted by $A \dot{-} B$.

Example 4.63. Let $V(x) = |x|^{-\alpha}$, $0 \leq \alpha < 2$, and $d = 3$. In quantum mechanics, the relation $\||x|^{-1}f\| \leq 4\|\Delta f\|$ is known as the *uncertainty principle.* It can be seen that V is infinitesimally small form-bounded with respect to $-\Delta$, i. e., for every ε there exists $b_\varepsilon > 0$ such that $\|V^{1/2}f\| \leq \varepsilon\|(-\Delta)^{1/2}f\| + b_\varepsilon\|f\|$ for $f \in Q(-\Delta)$. Therefore there exists a self-adjoint operator K such that $(f, Kg) = (f, (-\Delta + V)g)$ for $f, g \in C_0^\infty(\mathbb{R}^3)$.

Example 4.64. Let $q(f,g) = (\nabla f, \nabla g) + \int_{\mathbb{R}^d} V(x)\bar{f}(x)g(x)dx$. The natural class of V such that $\inf\{q(f,f) \mid \|f\|_2 = 1, f \in H^1(\mathbb{R}^d)\} > -\infty$ is

$$V \in \begin{cases} L^{d/2}(\mathbb{R}^d) + L^\infty(\mathbb{R}^d), & d \geq 3, \\ L^p(\mathbb{R}^2) + L^\infty(\mathbb{R}^2),\ p > 1, & d = 2, \\ L^1(\mathbb{R}^1) + L^\infty(\mathbb{R}^1), & d = 1. \end{cases} \tag{4.1.40}$$

Note that for such V the fact that $f \in H^1(\mathbb{R}^d)$ implies by the Sobolev inequality that $\int_{\mathbb{R}^d} |V(x)||f(x)|^2 dx \leq \|V\|_{d/2}\|f\|^2_{2d/(d-2)} \leq C\|V\|_{d/2}\|\nabla f\|_2^2 < \infty$. Since q is semibounded on the form domain $H^1(\mathbb{R}^d)$, there exists a self-adjoint operator associated with q.

The relationship between the convergence of sequences of quadratic forms and associated self-adjoint operators is the following. Let q_1 and q_2 be symmetric quadratic forms bounded from below; $q_1 \geq q_2$ means that $Q(q_1) \subset Q(q_2)$ and $q_1(f,f) \geq q_2(f,f)$, for all $f \in Q(q_1)$. A sequence $(q_n)_{n\in\mathbb{N}}$ of symmetric quadratic forms bounded from below is nonincreasing (resp. nondecreasing) if $q_n \geq q_{n+1}$ (resp. $q_n \leq q_{n+1}$), for all n. It is well known that for any positive quadratic form q there exists a largest closable quadratic form q_r that is smaller than q.

Theorem 4.65 (Monotone convergence for forms).

(1) *(Case of nonincreasing sequence) Let $(q_n)_{n\in\mathbb{N}}$ be a nonincreasing sequence of densely defined, closed symmetric quadratic forms uniformly bounded from below, i. e., $q_n \geq \gamma$ with a constant γ, and let H_n be the self-adjoint operator associated with q_n. Consider the symmetric quadratic form q_∞ defined by*

$$q_\infty(f,f) = \lim_{n\to\infty} q_n(f,f), \quad Q(q_\infty) = \bigcup_{n=1}^\infty Q(q_n). \tag{4.1.41}$$

Let H be the self-adjoint operator associated with $\overline{(q_\infty)_r}$, the closure of $(q_\infty)_r$. Then $\text{s-lim}_{n\to\infty}(H_n - \zeta)^{-1} = (H - \zeta)^{-1}$ for $\Re\zeta < \gamma$.

(2) *(Case of nondecreasing sequence) Let $(q_n)_{n\in\mathbb{N}}$ be a nondecreasing sequence of densely defined, closed symmetric quadratic forms, and let H_n be the self-adjoint operator associated with q_n. Consider the symmetric quadratic form q_∞ defined by*

$$q_\infty(f,f) = \lim_{n\to\infty} q_n(f,f), \quad Q(q_\infty) = \left\{f \mid \lim_{n\to\infty} q_n(f,f) < \infty\right\}. \tag{4.1.42}$$

Then q_∞ is a closed symmetric quadratic form. Suppose, furthermore, that $Q(q_\infty)$ is dense, and let H be the self-adjoint operator associated with q_∞. Then we have $\text{s-lim}_{n\to\infty}(H_n - \zeta)^{-1} = (H - \zeta)^{-1}$, for $\operatorname{Im}\zeta \neq 0$.

Next we discuss a counterpart of relative compactness for forms.

Definition 4.66 (Relative form-compactness). Let q be a closed semibounded quadratic form on $\mathcal{K}$, and β a symmetric quadratic form such that $Q(\beta) \supset Q(q)$ with the property that there exists a compact operator C such that $\beta((A-\gamma)^{-1/2}f, (A-\gamma)^{-1/2}g) = (f, Cg)$. Here A is the unique self-adjoint operator such that $q = q_A$ and $\gamma \in \mathbb{R}$ is such that

$A - \gamma \geq 0$. Then β is said to be *relatively form-compact* with respect to A or *A-form-compact*. If $\beta = q_B$ with a self-adjoint operator B, then B is said to be relatively form-compact with respect to A or A-form-compact.

We have already seen in Proposition 4.48 that perturbations by relatively compact operators leave the essential spectrum invariant. A counterpart for relatively form-compact operators is the following result.

Proposition 4.67. *Let β be relatively form-compact with respect to a self-adjoint operator A bounded from below, and C be a self-adjoint operator such that $q_A + \beta = q_C$. Then* $\mathrm{Spec}_{\mathrm{ess}}(C) = \mathrm{Spec}(A)$.

Example 4.68. Let $\mathcal{K} = L^2(\mathbb{R}^d)$ and $A = (-1/2)\Delta$. Thus $q_A(f,g) = (A^{1/2}f, A^{1/2}g)$ with form domain $Q(q_A) = D(A^{1/2})$. Consider the perturbation by a real-valued nonpositive potential $V = -W$. Define the quadratic form $\beta(f,g) = -(W^{1/2}f, W^{1/2}g)$, and write $R = W^{1/2}(A+E)^{-1/2}$ for $E \geq 0$. Suppose that β is relatively form-compact with respect to A. Since

$$\beta((A+E)^{-1/2}f, (A+E)^{-1/2}g) = -(Rf, Rg),$$

we see that R^*R is compact, and thus $W^{1/2}(A+E)^{-1}W^{1/2} = RR^*$ is also compact. Conversely, if $W^{1/2}(A+E)^{-1}W^{1/2}$ is compact, then the operator R^*R is also compact, and the quadratic form β is relatively form-compact with respect to A. Thus we conclude that if $W^{1/2}(A+E)^{-1}W^{1/2}$ is compact, then $\mathrm{Spec}_{\mathrm{ess}}((-1/2)\Delta \dot{-} W) = \mathrm{Spec}_{\mathrm{ess}}((-1/2)\Delta) = [0,\infty)$.

4.1.7 Confining potentials and decaying potentials

A natural question is how the spectrum of the Laplacian changes under a perturbation by a potential V. It is not our aim to discuss general spectral theory in this book; instead, we discuss some situations which will be relevant in our considerations further below.

As a general feature, the structure of the spectrum of a Schrödinger operator $H = -\frac{1}{2}\Delta + V$ is strongly influenced by the asymptotic properties of V at infinity, and there is a split in the behavior according to the two main classes as defined below.

Definition 4.69 (Confining/decaying potentials). Let $V : \mathbb{R}^d \to \mathbb{R}$ be a potential. We say that V is a *confining potential* if $V(x) \to \infty$ as $|x| \to \infty$, and a *decaying potential* if $V(x) \to 0$ as $|x| \to \infty$.

We single out some basic examples often used in mathematical physics.

Example 4.70. Some typical cases of confining potentials are given by the following choices:

(1) *Harmonic and anharmonic oscillators.* Let $V(x) = |x|^{2n}$, $n \in \mathbb{N}$. The case $n = 1$ describes the potential of the harmonic oscillator, and $n \geq 2$ give anharmonic oscillators.
(2) *Double and multiple well potentials.* The potential $V(x) = |x|^4 - b|x|^2$, $b > 0$, is a symmetric double well potential. Multiple well potentials can be obtained by higher order polynomials.

For decaying potentials we single out the following specific cases:
(1) *Potential wells.* Let $V(x) = -v(x)$ with a compactly supported, nonnegative bounded Borel function $v \not\equiv 0$. Specifically, we can choose $V(x) = -a1_{B(0,1)}(bx)$, for $a, b > 0$.
(2) *Coulomb-type potentials,* $V(x) = -c|x|^{-\beta}$, with $c, \beta > 0$.
(3) *Yukawa-type potentials,* $V(x) = -c|x|^{-\beta}e^{-b|x|}$, with $b, c, \beta > 0$.
(4) *Pöschl–Teller potential,* $V(x) = -a/\cosh^2(b|x|)$, with $a, b > 0$.
(5) *Morse potential,* $V(x) = a\left((1 - e^{-b(|x|-r_0)})^2 - 1\right)$, with $a, b, r_0 > 0$.

Next we discuss some results which can be applied to these types of potentials to determine the structure of $\mathrm{Spec}(H)$.

Theorem 4.71 (Rellich's criterion). *Let F and G be functions on $\mathbb{R}^d$ such that $\lim_{|x|\to\infty} F(x) = \infty$ and $\lim_{|x|\to\infty} G(x) = \infty$. Then the set*

$$\left\{f \in L^2(\mathbb{R}^d) \,\middle|\, \int_{\mathbb{R}^d} |f(x)|^2 dx \leq 1, \int_{\mathbb{R}^d} F(x)|f(x)|^2 dx \leq 1, \int_{\mathbb{R}^d} G(p)|\hat{f}(p)|^2 dp \leq 1\right\}$$

is compact.

Using this criterion the following interesting result is obtained.

Theorem 4.72 (Confining potential). *Let $V \in L^1_{\mathrm{loc}}(\mathbb{R}^d)$ be bounded from below such that $V(x) \to \infty$ as $|x| \to \infty$. Then $H = (-1/2)\Delta \dot{+} V$ has compact resolvent. In particular, e^{-tH} is also compact for all $t > 0$ and H has a purely discrete spectrum, i. e.,* $\mathrm{Spec}(H) = \mathrm{Spec_d}(H)$.

Proof. By Proposition 4.46, it suffices to show that the set

$$S = \{f \in D(|H|^{1/2}) \mid \|f\| \leq 1, \||H|^{1/2}f\|^2 \leq b\}$$

is compact for all b. Since S is closed, we need only prove that S is contained in a compact set. Indeed,

$$S \subset \left\{f \in L^2(\mathbb{R}^d) \,\middle|\, \int_{\mathbb{R}^d} |f(x)|^2 dx \leq 1, \int_{\mathbb{R}^d} V(x)|f(x)|^2 dx \leq b, \int_{\mathbb{R}^d} \frac{1}{2}|p|^2|\hat{f}(p)|^2 dp \leq b\right\}$$

and the right-hand side is compact by Rellich's criterion. From the identity $e^{-tH} = e^{-tH}(H - z)(H - z)^{-1}$ and the boundedness of $e^{-tH}(H - z)$, it follows that e^{-tH} is also compact. □

From this theorem, we see that Schrödinger operators with confining potentials have a purely discrete spectrum. Next we consider Schrödinger operators with decaying potentials. The spectral properties of this class are markedly different from those with confining potentials. Let

$$L^{\infty,0}(\mathbb{R}^d) = \left\{ f \in L^{\infty}(\mathbb{R}^d) \mid f(x) \to 0 \text{ as } |x| \to \infty \right\}.$$

Define the operator $P(-i\nabla)$ by $P(-i\nabla)f = (2\pi)^{-d/2} \int_{\mathbb{R}^d} P(k)e^{ik\cdot x}\hat{f}(k)dk$.

Proposition 4.73 (Compactness). *Let $P, Q \in L^{\infty,0}(\mathbb{R}^d)$. Then $Q(x)P(-i\nabla)$ is compact.*

Proof. Let $1_\Lambda \in C_0^\infty(\mathbb{R}^d)$ be given by

$$1_\Lambda(x) = \begin{cases} 1, & |x| < \Lambda, \\ 0, & |x| > \Lambda + 1. \end{cases}$$

The product $1_\Lambda(x)1_\Lambda(-i\nabla)$ is an integral operator with kernel

$$k(x,y) = (2\pi)^{-d/2} 1_\Lambda(x)\hat{1}_\Lambda(y-x) \in L^2(\mathbb{R}^d \times \mathbb{R}^d).$$

Thus $1_\Lambda(x)1_\Lambda(-i\nabla)$ is compact by Example 4.28. Since $Q(x)1_\Lambda(x) \to Q(x)$ and $1_\Lambda(\xi)P(\xi) \to P(\xi)$ as $\Lambda \to \infty$ in $L^\infty(\mathbb{R}^d)$, we have

$$Q(x)P(-i\nabla) = \lim_{\Lambda\to\infty} Q(x)1_\Lambda(x)1_\Lambda(-i\nabla)P(-i\nabla)$$

uniformly. Since $Q(x)P(-i\nabla)$ is a uniform limit of a sequence of compact operators, it is compact by Proposition 4.26. □

Theorem 4.74 (Decaying potential). *Let V be such that*

$$V \in L^p(\mathbb{R}^d) + L^{\infty,0}(\mathbb{R}^d) \quad \begin{cases} p = 2, & d \le 3, \\ p > d/2, & d \ge 4. \end{cases}$$

Then $H = -\frac{1}{2}\Delta + V$ is essentially self-adjoint on $C_0^\infty(\mathbb{R}^d)$ and $\mathrm{Spec}_{\mathrm{ess}}(H) = \mathrm{Spec}_{\mathrm{ess}}(-\Delta) = [0,\infty)$.

Proof. Denote $H_0 = -\frac{1}{2}\Delta$. Essential self-adjointness follows by the Kato–Rellich theorem. By Proposition 4.73, $V(H_0+1)^{-1}$ is compact for $V \in L^{\infty,0}(\mathbb{R}^d)$. Thus it suffices to show that $V \in L^p(\mathbb{R}^d)$ is relatively compact with respect to H_0. Let $V \in L^p(\mathbb{R}^d)$. For each $\varepsilon > 0$, we decompose $V = V_\varepsilon + W_\varepsilon$, where $\|V_\varepsilon\|_p < \varepsilon$ and $W_\varepsilon \in L^{\infty,0}(\mathbb{R}^d)$. Indeed, let

$$1_\Lambda(x) = \begin{cases} 1, & |x| \le \Lambda, \\ 0, & |x| > \Lambda \end{cases} \quad \text{and} \quad g^{(N)} = \begin{cases} g, & |g| < N, \\ N, & g \ge N, \\ -N, & g \le -N. \end{cases}$$

Then V is decomposed as

$$V = \underbrace{(V1_\Lambda)^{(N)}}_{=V_\Lambda^N} + \underbrace{V - (V1_\Lambda)^{(N)}}_{=W_\Lambda^N}. \tag{4.1.43}$$

Thus $V_\Lambda^N \in L^{\infty,0}(\mathbb{R}^d)$, $W_\Lambda^N \in L^p(\mathbb{R}^d)$, and $\|W_\Lambda^N\|_p < \varepsilon$ for sufficiently large Λ and N. Furthermore, $V_\Lambda^N \to V$ in L^p as $\Lambda, N \to \infty$. We have by (4.1.37) that

$$\|(V(1+H_0)^{-1} - V_\Lambda^N(1+H_0)^{-1})f\| \le C\|V - V_\Lambda^N\|_p\|f\|_2.$$

Thus $V_\Lambda^N(1+H_0)^{-1} \to V(1+H_0)^{-1}$ in the uniform topology and $V_\Lambda^N(1+H_0)^{-1}$ is compact since $V_\Lambda^N \in L^{\infty,0}(\mathbb{R}^d)$. Hence $V(1+H_0)^{-1}$ is compact, and this completes the proof. □

Example 4.75. The following are two notable examples of Schrödinger operators with explicitly known spectra.

(1) *Harmonic oscillator.* Choose, for simplicity, $d = 1$, and let $\omega > 0$. The Schrödinger operator H with $V(x) = \frac{\omega^2}{2}x^2 - \frac{\omega}{2}$ describes the harmonic oscillator. In this case the spectrum is purely discrete. We have

$$\mathrm{Spec}(H) = \mathrm{Spec}_\mathrm{d}(H) = \left\{\omega n \mid n \in \mathbb{N} \cup \{0\}\right\}.$$

(2) *Hydrogen atom.* Choose $d = 3$, and let $\gamma > 0$. The Schrödinger operator H with $V(x) = -\frac{\gamma}{|x|}$ describes the hydrogen atom. In this case

$$\mathrm{Spec}_\mathrm{ess}(H) = [0, \infty), \quad \mathrm{Spec}_\mathrm{p}(H) = \mathrm{Spec}_\mathrm{d}(H) = \left\{-\frac{\gamma^2}{2n^2} \mid n \in \mathbb{N}\right\}.$$

4.1.8 Strongly continuous operator semigroups

We conclude this section with a summary of basic facts for operator semigroups. Recall from Definition 2.85 that a one-parameter family of bounded operators $\{S_t : t \ge 0\}$ on a Banach space $\mathcal{B}$ is said to be a *C_0-semigroup* whenever (1) $S_0 = 1$, (2) $S_sS_t = S_{s+t}$ for all s, t, and (3) S_t is strongly continuous in t, i. e., $\text{s-lim}_{t\to 0} S_t = S_0 = 1$. Every C_0-semigroup is generated by a uniquely associated operator.

Definition 4.76 (Generator). Let $\{S_t : t \ge 0\}$ be a C_0-semigroup on a Banach space $\mathcal{B}$. Its *generator* is defined by the linear operator

$$Tf = \operatorname*{s-lim}_{t\downarrow 0} \frac{1}{t}(S_t f - f)$$

with domain $D(T) = \left\{f \in \mathcal{B} \mid \text{s-lim}_{t\downarrow 0} \frac{1}{t}(S_t f - f) \text{ exists}\right\}$.

Note that T is densely defined, closed, and determines the semigroup uniquely by the expression $S_t = e^{-tT}$, for every $t \ge 0$.

Proposition 4.77. *Let $\{S_t : t \geq 0\}$ be a C_0-semigroup on Banach space $\mathcal{B}$ with generator T. The following properties hold for all $t \geq 0$:*

(1) $S_t D(T) \subset D(T)$ *and* $\frac{d}{dt} S_t \varphi = T S_t \varphi$;
(2) *there exist* $M \geq 1$ *and* $a > 0$ *such that* $\|S_t\| \leq M e^{at}$;
(3) $\int_0^t S_r \varphi dr \in D(T)$ *for all* $\varphi \in \mathcal{B}$;
(4) $S_t \varphi - \varphi = T \int_0^t S_r \varphi dr$; *moreover,* $S_t \varphi - \varphi = \int_0^t S_r T \varphi dr$ *if* $\varphi \in D(T)$;
(5) *if T is a bounded operator, then* $D(T) = \mathcal{B}$ *and* $\|S_t - S_0\| \leq \frac{M}{a}(e^{at} - 1)\|T\|$. *In particular, the map* $t \mapsto S_t$ *is uniformly continuous.*

With a given self-adjoint operator A, the semigroups e^{-itA} and e^{-tA} can be defined through the spectral resolution (4.1.19). In particular, $\{e^{-itA} : t \in \mathbb{R}\}$ is a strongly continuous one-parameter unitary group, and $\{e^{-tA} : t \geq 0\}$ is a strongly continuous one-parameter semigroup whenever A is bounded below.

Example 4.78 (Heat kernels). Some specific choices of interest of self-adjoint generators T in $L^2(\mathbb{R}^d)$ giving rise to C_0-semigroups are of the form $T = T(-i\nabla)$. The Fourier transform F on $L^2(\mathbb{R}^d)$ gives $e^{-tT} = F e^{-tT(k)} F^{-1}$, which is represented by an integral operator with an integral kernel $e^{-tT}(x,y) = (F e^{-tT(k)})(x-y)$, i. e.,

$$(e^{-tT} f)(x) = \int_{\mathbb{R}^d} e^{-tT}(x,y) f(y) dy.$$

(1) Operator $T = -\frac{1}{2}\Delta$ has the integral kernel

$$e^{(1/2)t\Delta}(x,y) = \frac{1}{(2\pi t)^{d/2}} \exp(-|x-y|^2/2t);$$

(2) Operator $T = (-\Delta)^{\alpha/2}$, $0 < \alpha < 2$, has the integral kernel

$$e^{-t(-\Delta)^{\alpha/2}}(x,y) = \frac{1}{(2\pi)^d} \int_{\mathbb{R}^d} e^{-t|k|^\alpha + ik\cdot(x-y)} dk;$$

moreover, for $\alpha = 1$, the explicit formula

$$e^{-t(-\Delta)^{1/2}}(x,y) = \Gamma\left(\frac{d+1}{2}\right) \frac{1}{\pi^{(d+1)/2}} \frac{t}{(t^2 + |x-y|^2)^{(d+1)/2}}$$

holds;
(3) Operator $T = (-\Delta + m^2)^{1/2} - m$, $m > 0$, has the integral kernel

$$e^{-t((-\Delta+m^2)^{1/2}-m)}(x,y) = \frac{1}{(2\pi)^d} \frac{t}{\sqrt{t^2 + |x-y|^2}} \int_{\mathbb{R}^d} e^{mt - \sqrt{(t^2+|x-y|^2)(|k|^2+m^2)}} dk.$$

The cases above are examples of Feller semigroups associated with Brownian motion, rotationally symmetric α-stable processes, and relativistic Cauchy processes, respectively.

Conversely, given a semigroup, here we present some basic results on their generators, of which we will often make use.

Proposition 4.79 (Stone's theorem). *Let $\{U_t : t \in \mathbb{R}\}$ be a strongly continuous one-parameter unitary group on a Hilbert space. Then there exists a unique self-adjoint operator A such that $U_t = e^{itA}$, $t \in \mathbb{R}$.*

Proposition 4.80 (Hille–Yosida theorem). *A linear operator T is the generator of a C_0-semigroup $\{S_t : t \geq 0\}$ on a Banach space if and only if*
(1) *T is a densely defined closed operator;*
(2) *there exists $a \in \mathbb{R}$ such that $(a, \infty) \subset \rho(T)$ and*

$$\|(a - T)^{-m}\| \leq \frac{M}{(\lambda - a)^m}, \quad \lambda > a, \quad m = 1, 2, \ldots,$$

where $\rho(T)$ denotes the resolvent set of T.

Proposition 4.81 (Semigroup version of Stone's theorem). *Let $\{S_t : t \geq 0\}$ be a symmetric C_0-semigroup on a Hilbert space. Then there exists a self-adjoint operator A bounded below such that $S_t = e^{-tA}$, $t \geq 0$.*

Proof. Define the operator A by

$$-Af = \operatorname*{s-lim}_{t\to 0} \frac{1}{t}(S_t - 1)f, \quad D(A) = \left\{ f \;\middle|\; \operatorname*{s-lim}_{t\to 0} \frac{1}{t}(S_t - 1)f \text{ exists} \right\}.$$

By the Hille–Yosida theorem and the fact that S_t is symmetric, A is a closed symmetric operator and for some $a \in \mathbb{R}$ the resolvent $\rho(A)$ contains $(-\infty, a)$. It is a standard fact that the spectrum of a closed symmetric operator is one of the following cases: (a) the closed upper half-plane, (b) the closed lower-half plane, (c) the entire plane, or (d) a subset of $\mathbb{R}$. Hence the spectrum of A must be as in (d), which implies that A is self-adjoint and bounded below. Since a given generator gives rise to a unique C_0-semigroup, we have $S_t = e^{-tA}$. □

In case $\{S_t : t \geq 0\}$ is not a semigroup while $\operatorname{s-lim}_{t\downarrow 0} t^{-1}(S_t - 1)$ does exist, the following result can be used.

Proposition 4.82. *Let A be a positive self-adjoint operator and $\varrho(t)$ a family of self-adjoint operators with $0 \leq \varrho(t) \leq 1$. Suppose that $S_t = t^{-1}(1 - \varrho(t))$ converges to A as $t \downarrow 0$ in the strong resolvent sense. Then*

$$\operatorname*{s-lim}_{n\to\infty} \varrho(t/n)^n = e^{-tA}.$$

Proof. By the strong resolvent convergence of $S_{t/n}$ to A, see Definition 4.15, we have that $\text{s-lim}_{n\to\infty} \exp(-tS_{t/n}) = e^{-tA}$. It suffices to show that

$$\operatorname*{s-lim}_{n\to\infty}(\exp(-tS_{t/n}) - \varrho(t/n)^n) = \operatorname*{s-lim}_{n\to\infty} G_n(tS_{t/n}) = 0,$$

where

$$G_n(x) = \begin{cases} e^{-x} - (1 - x/n)^n, & 0 \le x \le n, \\ e^{-x}, & x \ge n. \end{cases}$$

It is easily seen that $\lim_{n\to\infty} \|G_n\|_\infty = 0$ and the proposition follows. □

Next we consider semigroups generated by Schrödinger operators.

Definition 4.83 (Schrödinger semigroup). The one-parameter operator semigroup $\{e^{-tH} : t \ge 0\}$ generated by a Schrödinger operator H is a *Schrödinger semigroup.*

Due to the additive structure of Schrödinger operators, the following result will also be useful.

Proposition 4.84 (Trotter product formula). *Let A and B be positive self-adjoint operators such that $A + B$ is essentially self-adjoint on $D(A) \cap D(B)$. Then*

$$\operatorname*{s-lim}_{n\to\infty}(e^{-(t/n)A}e^{-(t/n)B})^n = e^{-t(\overline{A+B})}, \tag{4.1.44}$$

where $\overline{A+B}$ denotes the closure of $(A + B)\restriction_{D(A)\cap D(B)}$.

Proof. Let $\varrho(t) = e^{-(t/2)A}e^{-tB}e^{-(t/2)A}$. We have

$$(e^{-(t/n)A}e^{-(t/n)B})^n = e^{-(t/2n)A}\varrho(t/n)^{n-1}e^{-(t/2n)A}e^{-(t/n)B}.$$

Since e^{-tA} and e^{-tB} are uniformly bounded in $t \ge 0$ and strongly converge to the identity operator as $t \downarrow 0$, it suffices to show that

$$\operatorname*{s-lim}_{n\to\infty} \varrho(t/n)^{n-1} = e^{-tC},$$

where $C = \overline{A+B}$. We know that $t^{-1}(1 - \varrho(t))$ is a bounded self-adjoint operator and

$$\begin{aligned} t^{-1}(1 - \varrho(t))\phi &= e^{-(t/2)A}e^{-tB}t^{-1}(1 - e^{-(t/2)A})\phi + e^{-(t/2)A}t^{-1}(1 - e^{-tB})\phi \\ &\quad + t^{-1}(1 - e^{-(t/2)A})\phi \to C\phi \end{aligned}$$

as $t \to \infty$, for $\phi \in D(A)\cap D(B)$, which implies that $t^{-1}(1-\varrho(t)) \to C$ in the strong resolvent sense. Thus $\text{s-lim}_{n\to\infty} \varrho(t/n)^n = e^{-tC}$ follows by Proposition 4.82. Furthermore,

$$\operatorname*{s-lim}_{n\to\infty}(\varrho(t/n)^{n-1} - e^{-tC}) = \operatorname*{s-lim}_{n\to\infty} \varrho(t/n)^{n-1}(1 - \varrho(t/n)) + \operatorname*{s-lim}_{n\to\infty}(\varrho(t/n)^n - e^{-tC}) = 0.$$

Hence the proposition follows. □

The Trotter product formula can further be extended to more general cases.

Proposition 4.85. *Let A and B be positive self-adjoint operators in a Hilbert space* $\mathscr{H}$ *with form domains* $Q(q_A)$ *and* $Q(q_B)$. *Denote by* P_{AB} *the projection to* $\overline{Q(q_A) \cap Q(q_B)}$. *Then*

$$\operatorname*{s-lim}_{n\to\infty}(e^{-(t/n)A}e^{-(t/n)B})^n = e^{-t(A \dotplus B)}P_{AB}, \tag{4.1.45}$$

where $A \dotplus B$ *is the quadratic form sum of A and B, i. e., the self-adjoint operator in* $P_{AB}\mathscr{H}$ *associated with the densely defined closed quadratic form* $f \mapsto \|A^{1/2}f\|^2 + \|B^{1/2}f\|^2$.

4.2 Feynman–Kac formula for Schrödinger operators

4.2.1 Bounded smooth external potentials

Now we turn to discussing probabilistic representations of Schrödinger semigroups given by *Feynman–Kac formulae*, which are the main object of this book. We begin by assuming that $V \in C_0^\infty(\mathbb{R}^d)$, which is the simplest case. For such a choice of potential, the Schrödinger operator H is self-adjoint on $D(H) = H^2(\mathbb{R}^d)$ and bounded from below by the Kato–Rellich theorem. Therefore the semigroup $\{e^{-tH} : t \geq 0\}$ can be defined by the spectral resolution and the solution of (4.1.7) with initial condition g is given by $f(x,t) = (e^{-tH}g)(x)$. We will denote Brownian motion throughout by $(B_t)_{t\geq 0}$ as discussed in Chapter 2, and its augmented filtration $(\mathscr{F}_t^{\mathrm{BM}})_{t\geq 0}$ as given by Definition 2.109 by simply $(\mathscr{F}_t)_{t\geq 0}$, unless any confusion may arise.

Theorem 4.86 (Feynman–Kac formula for Schrödinger operator). *If* $V \in C_0^\infty(\mathbb{R}^d)$, *then for* $f, g \in L^2(\mathbb{R}^d)$,

$$(f, e^{-tH}g) = \int_{\mathbb{R}^d} \mathbb{E}^x[\overline{f(B_0)}e^{-\int_0^t V(B_s)\,ds}g(B_t)]dx \tag{4.2.1}$$

holds. In particular,

$$(e^{-tH}g)(x) = \mathbb{E}^x[e^{-\int_0^t V(B_s)\,ds}g(B_t)]. \tag{4.2.2}$$

Proof. Define the map K_t on $L^2(\mathbb{R}^d)$ by

$$(K_t f)(x) = \mathbb{E}^x[e^{-\int_0^t V(B_s)\,ds}f(B_t)].$$

Since $\|K_t f\|^2 \leq e^{-2t\inf_{x\in\mathbb{R}^d}V(x)}\|f\|^2$, the operator K_t is a bounded operator on $L^2(\mathbb{R}^d)$ for every $t \geq 0$. First, we show that $\{K_t : t \geq 0\}$ is a symmetric C_0-semigroup on $L^2(\mathbb{R}^d)$. Define $\tilde{B}_s = B_{t-s} - B_t$, $s < t$, for a fixed $t > 0$. By Proposition 2.100, we have $\tilde{B}_s \stackrel{\mathrm{d}}{=} B_s$. Thus

$$(f, K_t g) = \int_{\mathbb{R}^d} \mathbb{E}[\overline{f(x)} e^{-\int_0^t V(B_s+x)ds} g(B_t + x)]dx$$
$$= \mathbb{E}\left[\int_{\mathbb{R}^d} \overline{f(x)} e^{-\int_0^t V(\tilde{B}_s+x)ds} g(\tilde{B}_t + x)dx \right].$$

Changing the variable x to $y = \tilde{B}_t + x$, we obtain

$$(f, K_t g) = \mathbb{E}\left[\int_{\mathbb{R}^d} \overline{f(y - \tilde{B}_t)} e^{-\int_0^t V(\tilde{B}_s - \tilde{B}_t + y)ds} g(y)dy \right]$$
$$= \int_{\mathbb{R}^d} \mathbb{E}[\overline{f(y + B_t)} e^{-\int_0^t V(B_{t-s}+y)ds} g(y)]dy = (K_t f, g), \tag{4.2.3}$$

i. e., K_t is symmetric. Write now $Z_t = e^{-\int_0^t V(B_s)ds}$. The semigroup property follows directly from the Markov property of Brownian motion, i. e., for all $s, t \geq 0$,

$$(K_s K_t f)(x) = \mathbb{E}^x[Z_s \mathbb{E}^{B_s}(Z_t f(B_t))]$$
$$= \mathbb{E}^x[\mathbb{E}^x[Z_s e^{-\int_0^t V(B_{s+u})du} f(B_{s+t}) | \mathscr{F}_s]]$$
$$= \mathbb{E}^x[Z_{s+t} f(B_{s+t})] = K_{s+t} f(x).$$

Strong continuity is implied by

$$\|K_t f - f\| \leq \mathbb{E}[\|e^{-\int_0^t V(\cdot + B_s)ds} f(\cdot + B_t) - f\|] \to 0$$

as $t \downarrow 0$, by using the dominated convergence theorem. We now know that on both sides of (4.2.1) there is a C_0-semigroup. What remains to show is that their generators are equal, i. e., $\frac{1}{t}(K_t - 1)$ converges to $-H$ strongly in L^2 on $D(H)$. We will use Itô calculus for this. Suppose that $f \in C_0^\infty(\mathbb{R}^d)$ and put $Y_t = f(B_t)$; then the Itô formula gives $dZ_t = -VZ_t dt$ and $dY_t = \nabla f \cdot dB_t + \frac{1}{2}\Delta f dt$. By the product rule (2.4.44),

$$d(Z_t Y_t) = Z_t \left(-Vf(B_t) + \frac{1}{2}\Delta f(B_t) \right) dt + Z_t \nabla f(B_t) \cdot dB_t$$
$$= -Z_t (Hf)(B_t)dt + Z_t \nabla f(B_t) \cdot dB_t. \tag{4.2.4}$$

Taking expectation with respect to Wiener measure on both sides of the integrated form of (4.2.4) and using $\mathbb{E}^x[\int_0^t Z_s \nabla f \cdot dB_s] = 0$, we obtain

$$(K_t f)(x) = f(x) - \int_0^t \mathbb{E}^x[Z_s (Hf)(B_s)]ds = f(x) - \int_0^t K_s Hf(x)ds. \tag{4.2.5}$$

For $f \in D(H)$ and $\varepsilon > 0$, there exists $s_0 > 0$ with $\|K_s Hf - Hf\| \leq \varepsilon$, for all $s < s_0$. Thus by (4.2.5), with $t < s_0$,

$$\left\|\frac{1}{t}(K_t f - f) + Hf\right\| \leq \frac{1}{t}\int_0^t \|-K_s Hf + Hf\| ds \leq \varepsilon$$

is obtained. Since $C_0^\infty(\mathbb{R}^d)$ is a core of H by the Kato–Rellich theorem, for $f \in D(H)$ there exists a sequence $(f_n)_{n\in\mathbb{N}} \subset C_0^\infty(\mathbb{R}^d)$ such that $f_n \to f$ and $Hf_n \to Hf$ strongly as $n \to \infty$. Let $M_t = \frac{1}{t}(K_t - I)$. We have $\frac{1}{t}(K_t - I)f = \frac{1}{t}(K_t - I)f_n + M_t(f - f_n)$. It can be seen that

$$\begin{aligned}
\|M_t(f - f_n)\|^2 &= \frac{1}{t^2}\left\|\mathbb{E}[(e^{-\int_0^t V(B_s+x)ds} - 1)((f - f_n)(B_t + x))]\right\|^2 \\
&\leq \int_{\mathbb{R}^d} \mathbb{E}\left[\frac{1}{t^2}|e^{-\int_0^t V(B_s+x)ds} - 1|^2\right] \mathbb{E}[|(f - f_n)(B_t + x)|^2]dx \\
&\leq \delta^2 e^{2t\delta}\|f - f_n\|^2,
\end{aligned}$$

where $\delta = \|V\|_\infty$. Let $\eta > 0$, and suppose that $\|f - f_n\| \leq \eta$ and $\|Hf - Hf_n\| \leq \eta$ for all $n > N$ with some N. We conclude that for $n > N$,

$$\begin{aligned}
\left\|\frac{1}{t}(K_t - I)f + Hf\right\| &\leq \left\|\frac{1}{t}(K_t - I)f_n + Hf_n\right\| + \|M_t(f - f_n)\| + \|Hf - Hf_n\| \\
&\leq \left\|\frac{1}{t}(K_t - I)f_n + Hf_n\right\| + \delta e^{t\delta}\|f - f_n\| + \|Hf - Hf_n\| \leq \left\|\frac{1}{t}(K_t - I)f_n + Hf_n\right\| + (\delta e^{t\delta} + 1)\eta
\end{aligned}$$

uniformly in t. Thus we have $\lim_{t\to 0} \|\frac{1}{t}(K_t - I)f + Hf\| \leq (\delta + 1)\eta$ for every $\eta > 0$. This shows strong convergence of $\frac{1}{t}(K_t - 1)f$ to $-Hf$ for $f \in D(H)$ as $t \downarrow 0$. □

Definition 4.87 (Feynman-Kac semigroup). The symmetric C_0-semigroup defined by $L^2(\mathbb{R}^d) \ni f \mapsto K_t f = \mathbb{E}^x[e^{-\int_0^t V(B_s)ds} f(B_t)]$, $t \geq 0$, is called the *Feynman–Kac semigroup* for the given V.

It is seen that the solution of (4.1.7), like the solution of the heat equation, can be obtained by running a Brownian motion with an additional exponential weight applied. This gives a probabilistic representation of the solution or, equivalently, of the kernel of the semigroup e^{-tH}. Using this picture, we see that Brownian paths spending a long time in a region where V is large are exponentially penalized and thus their contribution to the expected value will be accordingly small. The dominant contribution will therefore come from paths that spend most of their time near the lowest values of V.

4.2.2 Derivation through the Trotter product formula

By using the Trotter product formula+ given in Proposition 4.84, we can give another proof of Theorem 4.86.

Alternate proof of Theorem 4.86. Write $H_0 = -(1/2)\Delta$, let $f_1, \ldots, f_{n-1} \in L^\infty(\mathbb{R}^d)$, $f_0, f_n \in L^2(\mathbb{R}^d)$, and write $e^{-(t_1-t_0)H_0} f_1 e^{-(t_2-t_1)H_0} f_2 \cdots e^{-(t_n-t_{n-1})H_0} f_n$ as

$$\prod_{j=1}^n e^{-(t_j-t_{j-1})H_0} f_j$$

for notational convenience. By the finite-dimensional distributions of Brownian motion, see (2.3.4)–(2.3.5), for $0 = t_0 \le t_1 \le \cdots \le t_n$ we have

$$\Big(\prod_{j=1}^n e^{-(t_j-t_{j-1})H_0} f_j \Big)(x) = \mathbb{E}^x \Big[\prod_{j=1}^n f_j(B_{t_j}) \Big] \tag{4.2.6}$$

and

$$\Big(\bar{f}_0, \prod_{j=1}^n e^{-(t_j-t_{j-1})H_0} f_j \Big) = \int_{\mathbb{R}^d} \mathbb{E}^x \Big[\prod_{j=0}^n f_j(B_{t_j}) \Big] dx. \tag{4.2.7}$$

The Trotter product formula combined with (4.2.7) gives

$$(f, e^{-tH} g) = \lim_{n\to\infty} (f, (e^{-(t/n)H_0} e^{-(t/n)V})^n g) = \lim_{n\to\infty} \int_{\mathbb{R}^d} \mathbb{E}^x [\overline{f(B_0)} g(B_t) e^{-(t/n)\sum_{j=1}^n V(B_{tj/n})}] dx.$$

Since V is continuous and B_s is almost surely continuous in s,

$$\frac{t}{n} \sum_{j=1}^n V(B_{tj/n}) \to \int_0^t V(B_s) ds$$

almost surely as $n \to \infty$. Thus we have

$$(f, e^{-tH} g) = \int_{\mathbb{R}^d} \mathbb{E}^x [\overline{f(B_0)} g(B_t) e^{-\int_0^t V(B_s) ds}] dx. \qquad \square$$

Having the Feynman–Kac formula for smooth bounded potentials at hand, we are now interested in extending it to wider classes of V. We present here a first extension covering potentials as singular as $V(x) = -1/|x|^3$ (see Section 4.3.3).

Theorem 4.88 (Feynman–Kac formula for Schrödinger operator with singular external potential). *Suppose that for $V = V_+ - V_-$ we have $V_+ \in L^1_{\mathrm{loc}}(\mathbb{R}^d \setminus S)$, where S is a closed set of measure zero, and V_- is relatively form-bounded with respect to $-(1/2)\Delta$ with a relative bound strictly less than 1. Define $H = -(1/2)\Delta \dotplus V_+ \dot{-} V_-$. Then* (4.2.1) *holds.*

Proof. Suppose that $V \in L^\infty$ and $V_n = \phi(x/n)(V * j_n)$, where $j_n = n^d \phi(xn)$ with $\phi \in C_0^\infty(\mathbb{R}^d)$ such that $0 \le \phi \le 1$, $\int_{\mathbb{R}^d} \phi(x) dx = 1$ and $\phi(0) = 1$. We see that $V_n(x) \to V(x)$

almost everywhere. There is a set $\mathcal{N} \subset \mathbb{R}^d$ of Lebesgue measure zero such that $V_n(x) \to V(x)$ for $x \notin \mathcal{N}$, and

$$\mathbb{E}^x[1_{\{B_s \in \mathcal{N}\}}] = \int_{\mathbb{R}^d} 1_{\mathcal{N}}(y)\Pi_t(y-x)dy = 0$$

for every $x \in \mathcal{N}$. We see that for every $x \in \mathcal{N}$,

$$0 = \int_0^t \mathbb{E}^x[1_{\{B_s \in \mathcal{N}\}}]ds = \mathbb{E}^x\left[\int_0^t 1_{\{B_s \in \mathcal{N}\}}\right]ds$$

by Fubini's theorem. Thus for every $x \in \mathcal{N}$, it can be seen that the Lebesgue measure of the set $\{t \in [0,\infty) \mid B_t(\omega) \in \mathcal{N}\}$ is zero for almost every $\omega \in \mathcal{X}$. Hence it follows that $\int_0^t V_n(B_s)ds \to \int_0^t V(B_s)ds$ almost surely under $\mathcal{W}^x$ for $x \notin \mathcal{N}$. Therefore, by the dominated convergence theorem,

$$\int_{\mathbb{R}^d} \mathbb{E}^x[\overline{f(B_0)}g(B_t)e^{-\int_0^t V_n(B_s)ds}]dx \to \int_{\mathbb{R}^d} \mathbb{E}^x[\overline{f(B_0)}g(B_t)e^{-\int_0^t V(B_s)ds}]dx$$

as $n \to \infty$. On the other hand, $e^{-t(-(1/2)\Delta+V_n)} \to e^{-t(-(1/2)\Delta+V)}$ strongly, since $-(1/2)\Delta + V_n$ converges to $-(1/2)\Delta + V$ on a common core. Thus (4.2.1) holds for $V \in L^\infty(\mathbb{R}^d)$. Now assume that $V_+ \in L^1_{\mathrm{loc}}(\mathbb{R}^d \setminus S)$ and V_- is relatively form-bounded with respect to $-(1/2)\Delta$. Let

$$V_{+n}(x) = \begin{cases} V_+(x), & V_+(x) < n, \\ n, & V_+(x) \geq n, \end{cases} \quad \text{and} \quad V_{-m}(x) = \begin{cases} V_-(x), & V_-(x) < m, \\ m, & V_-(x) \geq m. \end{cases}$$

Write $V_{n,m} = V_{+n} - V_{-m}$. We have

$$(f, e^{-t(-(1/2)\Delta+V_{n,m})}g) = \int_{\mathbb{R}^d} \mathbb{E}^x[\overline{f(B_0)}e^{-\int_0^t V_{n,m}(B_s)\,ds}g(B_t)]dx. \tag{4.2.8}$$

Denote $h = -(1/2)\Delta$, and define the closed quadratic forms

$$\begin{aligned} \mathfrak{q}_{n,m}(f,f) &= (h^{1/2}f, h^{1/2}f) + (V_{+n}^{1/2}f, V_{+n}^{1/2}f) - (V_{-m}^{1/2}f, V_{-m}^{1/2}f), \\ \mathfrak{q}_{n,\infty}(f,f) &= (h^{1/2}f, h^{1/2}f) + (V_{+n}^{1/2}f, V_{+n}^{1/2}f) - (V_-^{1/2}f, V_-^{1/2}f), \\ \mathfrak{q}_{\infty,\infty}(f,f) &= (h^{1/2}f, h^{1/2}f) + (V_+^{1/2}f, V_+^{1/2}f) - (V_-^{1/2}f, V_-^{1/2}f), \end{aligned}$$

whose form domains are respectively $Q(\mathfrak{q}_{n,m}) = Q(h)$, $Q(\mathfrak{q}_{n,\infty}) = Q(h)$, and $Q(\mathfrak{q}_{\infty,\infty}) = Q(h) \cap Q(V_+)$. Clearly, $\mathfrak{q}_{n,m} \geq \mathfrak{q}_{n,m+1} \geq \mathfrak{q}_{n,m+2} \geq \cdots \geq \mathfrak{q}_{n,\infty}$ and $\mathfrak{q}_{n,m} \to \mathfrak{q}_{n,\infty}$ in the sense of quadratic forms on $\bigcup_{m=1}^\infty Q(\mathfrak{q}_{n,m}) = Q(h)$. Since $\mathfrak{q}_{n,\infty}$ is closed on $Q(h)$, by Theorem 4.65

the associated positive self-adjoint operators satisfy $h + V_{+n} - V_{-m} \to h + V_{+n} \dot{-} V_{-}$ in the strong resolvent sense, which implies that for all $t \geq 0$,

$$\exp\left(-t\left(h + V_{+n} - V_{-m}\right)\right) \to \exp\left(-t\left(h + V_{+n} \dot{-} V_{-}\right)\right) \tag{4.2.9}$$

strongly as $m \to \infty$. Similarly, we have $\mathfrak{q}_{n,\infty} \leq \mathfrak{q}_{n+1,\infty} \leq \mathfrak{q}_{n+2,\infty} \leq \cdots \leq \mathfrak{q}_{\infty}$ and $\mathfrak{q}_{n,\infty} \to \mathfrak{q}_{\infty,\infty}$ in the form sense on $\{f \in \bigcap_{n=1}^{\infty} Q(\mathfrak{q}_{n,\infty}) \mid \sup_n \mathfrak{q}_{n,\infty}(f,f) < \infty\} = Q(h) \cap Q(V_+)$. Hence by Theorem 4.65 again, we obtain

$$\exp(-t(h \dot{+} V_{+n} \dot{-} V_{-})) \to \exp(-t(h \dot{+} V_{+} \dot{-} V_{-})), \quad t \geq 0, \tag{4.2.10}$$

in the strong sense as $n \to \infty$. By taking first $m \to \infty$ and then $n \to \infty$, it can be proven that both sides of (4.2.8) converge, i. e., the left-hand side of (4.2.8) converges by monotone convergence for forms by (4.2.9) and (4.2.10). On the other hand, let $f, g \in L^2(\mathbb{R}^d)$ and $f, g \geq 0$. We have

$$\lim_{n\to\infty} \lim_{m\to\infty} \int_{\mathbb{R}^d} \mathbb{E}^x[\overline{f(B_0)} e^{-\int_0^t V_{n,m}(B_s)ds} g(B_t)]dx = \int_{\mathbb{R}^d} \mathbb{E}^x[\overline{f(B_0)} e^{-\int_0^t V(B_s)ds} g(B_t)]dx,$$

which follows from the monotone convergence theorem for integrals for the limit $m \to \infty$ and the dominated convergence theorem for the limit $n \to \infty$, and hence it can be seen that the right-hand side of (4.2.8) converges to $\int_{\mathbb{R}^d} \mathbb{E}^x[\overline{f(B_0)} e^{-\int_0^t V(B_s)ds} g(B_t)]dx$ for any $f, g \in L^2(\mathbb{R}^d)$ by the decomposition $f = \Re f + i\Im f$, $\Re f = \Re f_+ - \Re f_-$, $\Im f = \Im f_+ - \Im f_-$, $g = \Re g + i\Im g$, $\Re g = \Re g_+ - \Re g_-$, and $\Im g = \Im g_+ - \Im g_-$. □

4.2.3 Kato-class potentials

The Feynman–Kac formula we have derived so far covers only a limited range of interesting models of mathematical physics. It is desirable to allow local singularities, like for instance in the case of the Coulomb potential $V(x) = -1/|x|$ in three dimensions. To include these and further cases of interest, we introduce here a large class of potentials and use it as a space of reference as far as we can. This space is Kato-class which on several counts is a natural function space, as will be made clear below. However, occasionally we will have to leave this space in order to draw stronger conclusions or because the conditions of Kato-class are too strong themselves for a particular statement we want to prove.

Our aim in this section is to introduce Kato-class and discuss its basic properties, then define Schrödinger operators with Kato-class potentials as self-adjoint operators, and finally derive a corresponding Feynman–Kac formula.

Definition 4.89 (Kato-class).

(1) A potential $V : \mathbb{R}^d \to \mathbb{R}$ is called a *Kato-class potential* whenever

$$\lim_{r\downarrow 0} \sup_{x\in\mathbb{R}^d} \int_{B_r(x)} |g(x-y)V(y)|\, dy = 0 \tag{4.2.11}$$

holds, where $B_r(x)$ is the closed ball of radius r centered at x, and

$$g(x) = \begin{cases} |x|, & d = 1, \\ -\log|x|, & d = 2, \\ |x|^{2-d}, & d \geq 3. \end{cases} \tag{4.2.12}$$

We denote this linear space by $\mathcal{K}(\mathbb{R}^d)$.

(2) Potential V belongs to *local Kato-class* if and only if $1_K V \in \mathcal{K}(\mathbb{R}^d)$ for any compact set $K \subset \mathbb{R}^d$. The set of local Kato-class potentials is denoted by $\mathcal{K}_{\mathrm{loc}}(\mathbb{R}^d)$.

(3) Potential V is *Kato-decomposable* whenever $V = V_+ - V_-$ with $V_- \in \mathcal{K}(\mathbb{R}^d)$ and $V_+ \in \mathcal{K}_{\mathrm{loc}}(\mathbb{R}^d)$.

For $d = 1$, we have an equivalent definition of Kato-class which we will use below.

Proposition 4.90. *Let $d = 1$. The potential V is of Kato-class if and only if*

$$\sup_{x\in\mathbb{R}} \int_{B_1(x)} |V(y)| dy < \infty. \tag{4.2.13}$$

Proof. Suppose (4.2.13) holds. We have

$$\sup_{x\in\mathbb{R}} \int_{B_r(x)} |x-y||V(y)| dy \leq r \sup_{x\in\mathbb{R}} \int_{B_1(x)} |V(y)| dy$$

for $0 < r < 1$, thus V is of Kato-class. Next suppose that V is of Kato-class. For every $\varepsilon > 0$, there exists $r > 0$ such that $\sup_{x\in\mathbb{R}} \int_{B_r(x)} |x-y||V(y)| dy < \varepsilon$. It follows that

$$\frac{r}{2} \sup_{x\in\mathbb{R}} \int_{r/2\leq|x-y|\leq r} |V(y)| dy \leq \sup_{x\in\mathbb{R}} \int_{r/2\leq|x-y|\leq r} |x-y||V(y)| dy \leq \varepsilon$$

and therefore

$$\sup_{x\in\mathbb{R}} \int_{r/2\leq|x-y|\leq r} |V(y)| dy \leq \frac{2\varepsilon}{r}.$$

Since $B_1(0)$ can be covered by a finite number N of rings, we have $B_1(0) \subset \bigcup_{j=1}^N A_j$ and $A_j = \{y \mid r/2 \leq |x_j - y| \leq r\}$ with suitable points x_j. This then gives

$$\sup_{x\in\mathbb{R}} \int_{B_1(x)} |V(y)|dy \le \sum_{j=1}^{N} \sup_{x\in\mathbb{R}} \int_{A_j+x} |V(y)|dy \le \frac{2N\varepsilon}{r} < \infty. \qquad \square$$

Before deriving a Feynman–Kac formula for such potentials, it is useful to discuss some basic properties of Kato-class. We begin by explaining the definition. Let Π_t be the probability transition kernel of a Brownian motion given by (2.3.3), and f a nonnegative function. By Fubini's theorem, we have

$$\mathbb{E}^x\left[\int_0^\infty f(B_t)dt\right] = \int_{\mathbb{R}^d} H(x,y)f(y)dy, \tag{4.2.14}$$

with the integral kernel $H(x,y) = \int_0^\infty \Pi_t(x,y)dt$. In the limit $t \to \infty$, $\Pi_t(x)$ behaves like $(2\pi t)^{-d/2}$, thus H diverges for $d = 1$ and 2. For $d \ge 3$, making the change of variable $|x-y|^2/(2t) \mapsto t$,

$$H(x,y) = \frac{\Gamma(d/2-1)}{2\pi^{d/2}} \frac{1}{|x-y|^{d-2}} \tag{4.2.15}$$

is obtained, where Γ is the Gamma function. For lower dimensions, we proceed by compensation with a suitable additive term. Let $\tilde{H}(x,y) = \int_0^\infty (\Pi_t(x,y) - h(t))\,dt$. For $d = 1$, we choose $h(t) = \Pi_t(0,0) = (2\pi t)^{-1/2}$, yielding

$$\tilde{H}(x,y) = \frac{1}{\sqrt{2\pi}} \int_0^\infty \frac{e^{-|x-y|^2/(2t)} - 1}{\sqrt{t}}\, dt = -|x-y|,$$

by making the same change of variable as above. For $d = 2$, write $h(t) = \Pi_t(0,v)$, with unit vector $v = \binom{1}{0}$. We have

$$\tilde{H}(x,y) = \frac{1}{2\pi} \int_0^\infty \frac{e^{-|x-y|^2/(2t)} - e^{-1/(2t)}}{t}\, dt = -\frac{1}{\pi} \log|x-y|.$$

This shows that g in Definition 4.89 is actually given by the potential kernels of the Laplacian.

The conditions in (4.2.12) limit the growth at infinity ($\mathcal{K}(\mathbb{R}^d)$) and the severity of local singularities ($L^1_{\text{loc}}(\mathbb{R}^d)$). These are illustrated by the following examples.

Example 4.91. Let $d \ge 2$. Kato-class allows singularities of the type $|x|^{-\alpha}$ with any $0 \le \alpha < 2$, including in particular Coulomb potential $V(x) = 1/|x|$. In contrast, the function $V(x) = |x|$ is not contained in $\mathcal{K}(\mathbb{R}^d)$.

Example 4.92. An application of the Hölder inequality gives an idea about the growth at infinity allowed. Let $d = 3$. Then

$$\left| \int_{B_r(x)} \frac{1}{|x-y|} |V(y)|\, dy \right| \leq \left(\int_{B_r(x)} \frac{1}{|x-y|^p}\, dy \right)^{1/p} \left(\int_{B_r(x)} |V(y)|^q\, dy \right)^{1/q},$$

with $1/p + 1/q = 1$. The first factor is independent of x and when $p < 3$, it tends to zero as $r \to 0$. Thus $V \in \mathcal{K}(\mathbb{R}^3)$ if $\sup_{x \in \mathbb{R}^3} \int_{B_r(x)} |V(y)|^q dy < \infty$ for some $r > 0$ and some $q > 3/2$. A direct calculation shows, for instance, that

$$V(x) = |x_1|^{\alpha} e^{-|x_1|^{\beta}(|x_2|^2+|x_3|^2)}, \quad x = (x_1, x_2, x_3)$$

is in $\mathcal{K}(\mathbb{R}^3)$ if $\beta > 3\alpha/2$. This shows that Kato-class is essentially an integrability condition; pointwise divergence is allowed as long as the given integrals can be controlled.

Example 4.93. Let $V \in L^p(\mathbb{R}^d) + L^\infty(\mathbb{R}^d)$ with $p > d/2$ and $1 \leq p < \infty$. Then $V \in \mathcal{K}(\mathbb{R}^d)$; see Lemma 4.105.

From the definition of $\mathcal{K}(\mathbb{R}^d)$, we furthermore obtain the properties below.

Proposition 4.94 (Properties of $\mathcal{K}(\mathbb{R}^d)$).
(1) *Let $V, W \in \mathcal{K}(\mathbb{R}^d)$ and f, g be real-valued continuous functions on $\mathbb{R}^d$. Then $fV + gW \in \mathcal{K}(\mathbb{R}^d)$.*
(2) $\mathcal{K}(\mathbb{R}^d) \subset L^1_{\mathrm{loc}}(\mathbb{R}^d)$.
(3) *If $V \in \mathcal{K}(\mathbb{R}^d)$, then $\sup_{x \in \mathbb{R}^d} \int_{B_1(x)} |V(y)|\, dy < \infty$.*

Proof. Claim (1) follows directly from the definition of $\mathcal{K}(\mathbb{R}^d)$. Consider statement (2). For $d = 1$, the statement in trivial by (4.2.13). Let $d \geq 2$, $K \subset \mathbb{R}^d$ be a compact set, and take $\varepsilon > 0$. There exists $r > 0$ such that $\sup_{x \in \mathbb{R}^d} \int_{B_r(x)} |g(x-y)V(y)|\, dy < \varepsilon$. Let $\{a_1, \dots, a_n\} \subset K$ be such that $\cup_{j=1}^n B_\varepsilon(a_j) \supset K$. We see that

$$\int_K |V(y)| dy \leq \sum_{j=1}^n \int_{B_\varepsilon(a_j)} |V(y)| dy = C \sum_{j=1}^n \int_{B_\varepsilon(a_j)} g(a_j - y)|V(y)| dx \leq nC\varepsilon < \infty$$

with a constant C independent of $\{a_1, \dots, a_n\}$. Hence (2) follows. To show statement (3), set $K = B_1(x)$ in the proof of (2). There exists a natural number n independent of x and $\{b_1, \dots, b_n\} \subset B_1(x)$ such that $\bigcup_{j=1}^n B_\varepsilon(b_j) \supset B_1(x)$. In the same way as for (2), we have $\int_{B_1(x)} |V(y)| dy \leq nC\varepsilon < \infty$. □

If V is Kato-decomposable, the positive part of V is in $L^1_{\mathrm{loc}}(\mathbb{R}^d)$. We also have the following property.

Lemma 4.95. *Suppose that $0 \leq V \in L^1_{\mathrm{loc}}(\mathbb{R}^d)$. Then $\mathcal{W}^x(\int_0^t V(B_s) ds = \infty) = 0$ for a. e. $x \in \mathbb{R}^d$. In particular, $\mathbb{E}^x[e^{-\int_0^t V(B_s) ds}] > 0$ for a. e. $x \in \mathbb{R}^d$.*

Proof. Let $N \in \mathbb{N}$ and

$$1_N(x) = \begin{cases} 1, & |x| < N, \\ 0, & |x| \geq N. \end{cases}$$

Since $1_N V \in L^1(\mathbb{R}^d)$, we have

$$\int_{\mathbb{R}^d} \mathbb{E}^x\left[\int_0^t 1_N V(B_s) ds\right] dx \leq t\|1_N V\|_{L^1}.$$

Hence for almost every fixed $x \in \mathbb{R}^d$, $\mathcal{W}^x(\int_0^t 1_N V(B_s)ds < \infty) = 1$, and then there exists $\mathcal{N}_N$, dependent on x, such that $\mathcal{W}^x(\mathcal{N}_N) = 0$ and $\int_0^t 1_N V(B_s(\omega))ds < \infty$, for every $\omega \in \mathcal{X} \setminus \mathcal{N}_N$. Let $\mathcal{N} = \bigcup_{N=1}^{\infty} \mathcal{N}_N$. Note that $\mathcal{N}$ depends on x. Thus $\int_0^t 1_N V(B_s(\omega))ds < \infty$, for every $N \in \mathbb{N}$ and $\omega \in \mathcal{X} \setminus \mathcal{N}$. Since $B_s(\omega)$ is continuous in s almost surely, for every $\omega \in \mathcal{X} \setminus \mathcal{N}$ there exists $N = N(\omega) \in \mathbb{N}$ such that $V(B_s(\omega)) = 1_N V(B_s(\omega))$, for all $0 \leq s \leq t$. Thus

$$\int_0^t V(B_s(\omega))ds = \int_0^t 1_N V(B_s(\omega))ds < \infty, \quad \omega \in \mathcal{X} \setminus \mathcal{N}. \qquad \square$$

Next we present a first equivalent characterization of Kato-class.

Proposition 4.96. *A nonnegative function V is in $\mathcal{K}(\mathbb{R}^d)$ if and only if*

$$\lim_{t \downarrow 0} \sup_{x \in \mathbb{R}^d} \mathbb{E}^x\left[\int_0^t V(B_s)\, ds\right] = 0. \tag{4.2.16}$$

Proof. As in (4.2.14), write for $t \in \mathbb{R}$,

$$\mathbb{E}^x\left[\int_0^t V(B_s)\, ds\right] = \int_0^t ds \int_{\mathbb{R}^d} \Pi_s(x-y)V(y)dy.$$

By the same change of variable as above,

$$\int_0^t \Pi_s(x-y)ds = \frac{1}{(2\pi)^{d/2}} \frac{1}{|x-y|^{d-2}} \int_{|x-y|^2/(2t)}^{\infty} e^{-u} u^{d/2-2}\, du = H(x,y;t)$$

readily follows. The key observation now is that $H(x,y;t)$ behaves like $g(x-y)$ in Definition 4.89 whenever $|x-y|^2/2t$ is small and vanishes rapidly when $|x-y|^2/2t$ goes to infinity. This is obvious in the cases $d \geq 3$ and is seen through integration by parts if

$d = 1$ and 2. We give the details only in the simplest case $d = 4$; the others proceed by the same arguments, though require further estimates.

A convenient feature of the case $d = 4$ is that we have an explicit formula for $H(x, y; t)$ and

$$\sup_{x\in\mathbb{R}^4} \mathbb{E}^x\left[\int_0^t V(B_s)\,ds\right] = \sup_{x\in\mathbb{R}^4} \int_{\mathbb{R}^4} \frac{1}{(2\pi|x-y|)^2} e^{-\frac{|x-y|^2}{2t}} V(y)dy \tag{4.2.17}$$

follows. We split up the domain of integration on the right-hand side above into $A_q = \{y \mid |x-y| < t^{1/q}\}$ and $B_q = \mathbb{R}^4 \setminus A_q$ for $q > 0$. Note that A_q shrinks to $\{x\}$ when $t \to 0$. If (4.2.16) holds, then the integral over A_2 on the right-hand side of (4.2.17) vanishes with $t \to 0$, uniformly in x. On A_2, $\exp(-|x-y|^2/2t)$ is bounded away from zero, hence $V \in \mathcal{K}(\mathbb{R}^4)$ follows. Conversely, assume $V \in \mathcal{K}(\mathbb{R}^4)$. Then by definition and the fact that the negative exponential is bounded from above, we obtain $\lim_{t\downarrow 0}\sup_{x\in\mathbb{R}^4}\int_{A_3}\frac{1}{(2\pi|x-y|)^2}e^{-\frac{|x-y|^2}{2t}}V(y)dy = 0$. On the other hand, on B_3 we have $\exp(-|x-y|^2/2t) \le \exp(-|x-y|/t^{1/6})$, thus the right-hand side of (4.2.17) tends to zero on this set by dominated convergence. □

We present a second equivalent characterization of Kato-class recalling the asymptotic properties of the resolvent of the Laplacian. For $d = 1$, the kernel of the resolvent operator $(-\Delta+\lambda)^{-1}$ is explicitly given by

$$(-\Delta+\lambda)^{-1}(x,y) = \frac{e^{-\sqrt{\lambda}|x-y|}}{\sqrt{\lambda}}. \tag{4.2.18}$$

In what follows, we assume $d \ge 2$. We have

$$(-\Delta+\lambda)^{-1}(x,y) = C_\lambda(x-y), \tag{4.2.19}$$

where

$$C_\lambda(x) = (2\pi)^{-d/2}\left(\frac{\sqrt{\lambda}}{|x|}\right)^{(d-2)/2} K_{(d-2)/2}(\sqrt{\lambda}|x|), \tag{4.2.20}$$

with $K_\nu(z)$ being the *modified Bessel function of the third kind* defined by

$$K_\nu(z) = \frac{1}{2}\left(\frac{z}{2}\right)^{-\nu}\int_0^\infty \exp\left(-\frac{z^2}{4t}-t\right)t^{\nu-1}dt. \tag{4.2.21}$$

We note that $K_\nu(z)$ also has the integral representation

$$K_\nu(z) = \frac{1}{2}\int_0^\infty \exp\left(-\frac{1}{2}\left(\frac{1}{t}+t\right)z\right)t^{\nu-1}dt.$$

We recall some further properties of K_ν. It is well defined on the domain

$$D = \{z \in \mathbb{C} \mid \Re z > 0, \Re(z^2) > 0\} = \{z \in \mathbb{C} \mid |\arg z| < \pi/4\}$$

for any $\nu \in \mathbb{C}$. The function $K_\nu : D \to \mathbb{C}$ is holomorphic and $K_\nu = K_{-\nu}$ follows. Furthermore, it can be shown that for $\Re\nu > -1/2$ and $z \in D$,

$$K_\nu(z) = \frac{\sqrt{\pi}}{\sqrt{2}\Gamma(\nu + \frac{1}{2})} \frac{e^{-z}}{\sqrt{z}} \int_0^\infty e^{-s} s^{\nu - \frac{1}{2}} \left(1 + \frac{s}{2z}\right)^{\nu - \frac{1}{2}} ds.$$

Inserting this identity into (4.2.20), for $d \geq 2$, $x \neq 0$, and $\lambda > 0$, we have

$$C_\lambda(x) = \frac{(\sqrt{\lambda})^{(d-3)/2}}{2^{(d+1)/2}\pi^{(d-1)/2}\Gamma(\frac{d-1}{2})} \frac{e^{-\sqrt{\lambda}|x|}}{|x|^{(d-1)/2}} \int_0^\infty e^{-s} s^{(d-3)/2} \left(1 + \frac{s}{2\sqrt{\lambda}|x|}\right)^{(d-3)/2} ds, \qquad (4.2.22)$$

where $z^a = \exp(a(\log|z| + i\arg z))$. In particular, for $d = 3$ we have

$$C_\lambda(x) = \frac{e^{-\sqrt{\lambda}|x|}}{4\pi|x|}.$$

Take $\lambda = 1$. The kernel of the resolvent operator $C_1(x)$, is called the *Bessel potential*. For $\lambda = 0$, we have $C_0(x)$, which is the *Riesz potential*, that will be discussed in Section 4.9. The following estimates hold for $C_\lambda(x)$ when $\lambda > 0$.

Lemma 4.97. *Let $\lambda > 0$. Then for every $x \in \mathbb{R}^d \setminus \{0\}$, we have*

$$|C_\lambda(x)| \leq \begin{cases} \frac{1}{2\sqrt{\pi}} \frac{1}{|\lambda|^{1/4}|x|^{1/2}} e^{-\sqrt{\lambda}|x|}, & d = 2, \\ \frac{3^{(d-3)/2}}{2^{d-1}\pi^{(d-1)/2}} \left(1 + \frac{\Gamma(d-2)}{\Gamma(\frac{d-1}{2})}\right) \frac{1}{|x|^{d-2}}, & d \geq 3,\ \sqrt{\lambda}|x| \leq 1, \\ \frac{3^{(d-3)/2}}{2^{d-1}\pi^{(d-1)/2}} \left(1 + \frac{\Gamma(d-2)}{\Gamma(\frac{d-1}{2})}\right) \frac{|\lambda|^{(d-3)/4}}{|x|^{(d-1)/2}} e^{-\sqrt{\lambda}|x|}, & d \geq 3,\ \sqrt{\lambda}|x| \geq 1. \end{cases}$$

Proof. Let $a > 0$. By direct calculation,

$$\int_0^\infty e^{-s} s^{(d-3)/2} \left(1 + \frac{s}{2a}\right)^{(d-3)/2} ds \leq \begin{cases} (3/2)^{(d-3)/2}(\Gamma(\frac{d-1}{2}) + \Gamma(d-2)a^{(3-d)/2}), & d \geq 3, \\ \sqrt{\pi}, & d = 2 \end{cases}$$

follows. Combining this with (4.2.22), we obtain the desired results. □

For $d = 2$ and $\sqrt{\lambda}|x| \leq 1$, there is a more explicit estimate.

Lemma 4.98. *Let $d = 2$, $\lambda > 0$, and $\sqrt{\lambda}|x| \leq 1$. Then for $x \in \mathbb{R}^2 \setminus \{0\}$, we have*

$$|C_\lambda(x)| \leq \frac{1}{2\pi} \log\left(\frac{2}{\sqrt{\lambda}|x|}\right).$$

Proof. We have directly

$$\frac{1}{4\pi}\int_0^\infty \frac{1}{t}e^{-\frac{\lambda|x|^2}{4t}-t}dt \le \int_0^{1/\lambda|x|^2} \frac{1}{t}e^{-\frac{1}{t}}dt + e^{-\lambda|x|^2}\int_{1/\lambda|x|^2}^\infty \frac{1}{t}e^{-\frac{4t}{\lambda|x|^2}}dt.$$

By the assumption $\sqrt{\lambda}|x| \le 1$, we furthermore have

$$\begin{aligned}
&\le \int_0^1 \frac{1}{t}e^{-\frac{1}{t}}dt + \int_1^{1/\lambda|x|^2} \frac{1}{t}e^{-\frac{1}{t}}dt + e^{-\lambda|x|^2}\int_1^\infty \frac{e^{-4t}}{t}dt \\
&\le \int_1^\infty \frac{1}{t}e^{-t}dt + \int_1^{1/\lambda|x|^2} \frac{dt}{t} + e^{-\lambda|x|^2}\int_1^\infty e^{-4t}dt \le 1 + \log\left(\frac{1}{\lambda|x|^2}\right) + \frac{1}{4}.
\end{aligned}$$

Hence

$$|C_\lambda(x)| \le \frac{1}{4\pi}\left(\frac{5}{4} + \log\left(\frac{1}{\lambda|x|^2}\right)\right) = \frac{1}{4\pi}\left(\log\left(\frac{e^{5/4}}{\lambda|x|^2}\right)\right) \le \frac{1}{2\pi}\left(\log\left(\frac{2}{\sqrt{\lambda}|x|}\right)\right),$$

and the lemma follows. □

As $|x-y| \to 0$, by Lemmas 4.97–4.98, we have

$$C_\lambda(x-y) \sim \begin{cases} |x-y|^{2-d}, & d \ge 3, \\ -\log|x-y|, & d = 2, \\ C, & d = 1; \end{cases}$$

moreover, for every $\delta > 0$,

$$\lim_{\lambda\to\infty} \sup_{|x-y|>\delta} e^{|x-y|}C_\lambda(x-y) = 0.$$

Note that the kernel of the resolvent operator $(-(1/2)\Delta + \lambda)^{-1}$ is $2C_{2\lambda}(x-y)$. We readily have $\sup_{x\in\mathbb{R}^d}\int_{\mathbb{R}^d} C_E(x-y)|V(y)|dy \to 0$ as $E \to \infty$ if and only if $V \in \mathcal{K}(\mathbb{R}^d)$, and by comparison with Definition 4.89 the proposition below follows. We see that $(-\Delta+E)^{-1}|V|(x) = \int_{\mathbb{R}^d} C_E(x-y)|V(y)|dy$, and we use

$$\|(-\Delta+E)^{-1}|V|\|_\infty = \sup_{x\in\mathbb{R}^d}\left|(-\Delta+E)^{-1}|V|(x)\right|. \tag{4.2.23}$$

Proposition 4.99. *We have $V \in \mathcal{K}(\mathbb{R}^d)$ if and only if*

$$\lim_{E\to\infty} \|(-\Delta+E)^{-1}|V|\|_\infty = 0. \tag{4.2.24}$$

Proof. For the sake of completeness, we present a proof of equivalence between (4.2.24) and (4.2.16). Take τ in a suitable neighborhood of 0 and denote $H_0 = -\frac{1}{2}\Delta$. Clearly, $\mathbb{E}^x[|V(B_s)|] = (e^{-sH_0}|V|)(x)$. Applying Laplace transform,

$$((H_0+E)^{-1}|V|)(x) = \sum_{n=0}^{\infty} \int_{nT}^{(n+1)T} e^{-sE}(e^{-sH_0}|V|)(x)ds.$$

A change of variable furthermore gives

$$((H_0+E)^{-1}|V|)(x) = \sum_{n=0}^{\infty} e^{-nTE} \int_0^T e^{-sE}(e^{-(s+nT)H_0}|V|)(x)ds.$$

By the semigroup property and using the integral kernel $e^{-tH_0}(x,y)$, we obtain

$$((H_0+E)^{-1}|V|)(x) = \sum_{n=0}^{\infty} e^{-nTE} \int_{\mathbb{R}^d} (e^{-nTH_0})(x,y)dy \int_0^T e^{-sE}(e^{-sH_0}|V|)(y)ds.$$

Take the supremum with respect y, which gives

$$((H_0+E)^{-1}|V|)(x) \le \left(\sup_{y\in\mathbb{R}^d} \int_0^T e^{-sE}(e^{-sH_0}|V|)(y)ds \right) \sum_{n=0}^{\infty} e^{-nTE} \int_{\mathbb{R}^d} (e^{-nTH_0})(x,y)dy.$$

Since $\int_{\mathbb{R}^d}(e^{-nTH_0})(x,y)dy = 1$, we have

$$\begin{aligned}\int_{\mathbb{R}^d} (e^{-nTH_0})(x,y)dy &\le \frac{1}{1-e^{-TE}} \sup_{y\in\mathbb{R}^d} \int_0^T e^{-sE}(e^{-sH_0}|V|)(y)ds \\ &\le \frac{1}{1-e^{-TE}} \|(H_0+E)^{-1}|V|\|_\infty.\end{aligned}$$

Taking the supremum, this yields

$$(1-e^{-TE})\|(H_0+E)^{-1}|V|\|_\infty \le \sup_{y\in\mathbb{R}^d} \int_0^T e^{-sE}(e^{-sH_0}|V|)(y)ds \le \|(H_0+E)^{-1}|V|\|_\infty.$$

Furthermore, trivially we have

$$e^{-TE} \int_0^T (e^{-sH_0}|V|)(y)ds \le \int_0^T e^{-sE}(e^{-sH_0}|V|)(y)ds \le \int_0^T (e^{-sH_0}|V|)(y)ds.$$

Thus $\sup_{y\in\mathbb{R}^d}\int_0^T(e^{-sH_0}|V|)(y)ds$ is bounded by $e^{TE}\|(H_0+E)^{-1}|V|\|_\infty$, for all $T \geq 0$, and hence

$$\lim_{T\downarrow 0}\sup_{x\in\mathbb{R}^d}\int_0^T(e^{-sH_0}|V|)(x)ds \leq \|(H_0+E)^{-1}|V|\|_\infty,$$

for all E. Therefore, if $\lim_{E\to\infty}\|(H_0+E)^{-1}|V|\|_\infty = 0$, then the right-hand side decreases with E, and optimizing over E implies (4.2.16). Conversely, if we assume (4.2.16), then by the above

$$\lim_{E\to\infty}\|(H_0+E)^{-1}|V|\|_\infty \leq \lim_{T\downarrow 0}\lim_{E\to\infty}\frac{1}{1-e^{-TE}}\sup_{x\in\mathbb{R}^d}\int_0^T(e^{-sH_0}|V|)(x)ds = 0,$$

i. e., (4.2.24) holds. □

The latter characterization is particularly interesting in view of the following property, which through the KLMN theorem implies that if V is Kato-class, then H is self-adjoint at least in the form sense.

Proposition 4.100. *If there exist $a, b > 0$ and $0 < \delta < 1$ such that for all $0 < \varepsilon < 1$,*

$$(f, |V|f) \leq \varepsilon\|\sqrt{-\Delta}f\|^2 + a\exp(b\varepsilon^{-\delta})\|f\|^2, \quad f \in D(\sqrt{-\Delta}), \tag{4.2.25}$$

then $V \in \mathcal{K}(\mathbb{R}^d)$. Conversely, if $V \in \mathcal{K}(\mathbb{R}^d)$, then $D(\sqrt{-\Delta}) \subset D(V^{1/2})$ and V is $-\Delta$-form-bounded with an infinitesimally small relative bound.

Proof. Suppose that (4.2.25) holds. Since

$$(-\Delta+E)^{-1}|V|(x) = \int_0^\infty e^{-tE}dt\int_{\mathbb{R}^d}\Pi_t(x-y)|V|(y)dy,$$

we estimate $\Pi_t(x-y)|V(y)|$ for a fixed x. Let $\phi(\cdot) = \sqrt{\Pi_t(x-\cdot)}$. By the assumption,

$$\begin{aligned}(\phi, |V|\phi) &\leq \varepsilon\|\sqrt{-\Delta}\phi\|^2 + a\exp(b\varepsilon^{-\delta})\|\phi\|^2\\ &= (\phi, -\Delta\phi) + a\exp(b\varepsilon^{-\delta}) \leq ct^{-2} + a\exp(b\varepsilon^{-\delta})\end{aligned}$$

with a constant c independent of x. Hence

$$\left|(-\Delta+E)^{-1}|V|(x)\right| \leq \left(\int_0^1 + \int_1^\infty\right)e^{-tE}(ct^{-2} + a\exp(b\varepsilon^{-\delta}))dt. \tag{4.2.26}$$

Taking $\varepsilon = (1+\log|t|)^{2/(1+\delta)}$, we have $a\exp(b\varepsilon^{-\delta}) \leq t^b ae^b$, thus (4.2.24) follows.

Conversely, for $V \in \mathcal{K}(\mathbb{R}^d)$ by duality it follows that $\|(-\Delta + E)^{-1}|V|\|_{\infty,\infty} = \||V|(-\Delta + E)^{-1}\|_{1,1}$, where $\|\cdot\|_{p,p}$ denotes bounded operator norm on L^p. Note that $\|(-\Delta+E)^{-1}|V|\|_{\infty,\infty} = \|(-\Delta + E)^{-1}|V|\|_{\infty}$. By applying the Stein interpolation theorem to the operator-valued function $F(z) = |V|^z(-\Delta + E)^{-1}|V|^{1-z}$ on $\{z \in \mathbb{C} \mid \Re z \in [0,1]\}$, it is seen that

$$\||V|^{1/2}(-\Delta + E)^{-1}|V|^{1/2}\|_{2,2} \leq \|(-\Delta + E)^{-1}|V|\|_{1,1} = \|(-\Delta + E)^{-1}|V|\|_{\infty,\infty}.$$

Hence

$$\||V|^{1/2}(-\Delta + E)^{-1/2}\|_{2,2}^2 \leq \|(-\Delta + E)^{-1}|V|\|_{\infty,\infty} \to 0$$

as $E \to \infty$. Since

$$\||V|^{1/2}f\| \leq \|(-\Delta + E)^{-1}|V|\|_{\infty,\infty}^{1/2}\|(-\Delta + E)^{1/2}f\|,$$

V is $-\Delta$-form-bounded with an infinitesimally small relative bound. □

Remark 4.101 (Stummel-class). Kato-class $\mathcal{K}(\mathbb{R}^d)$ defined above can be regarded as the quadratic form analogue of Stummel-class. A potential V is in *Stummel class* whenever

$$\lim_{r\to 0} \sup_{x\in\mathbb{R}^d} \int_{B_r(x)} h(x-y)|V(y)|^2\, dy = 0 \tag{4.2.27}$$

holds, where

$$h(x) = \begin{cases} 1 & \text{if } d \leq 3, \\ -\log|x| & \text{if } d = 4, \\ |x|^{4-d} & \text{if } d \geq 5. \end{cases} \tag{4.2.28}$$

The counterparts of Propositions 4.99–4.100 hold similarly. An equivalent condition to (4.101) is

$$\lim_{E\to\infty} \|(-\Delta + E)^{-2}|V|^2\|_{\infty} = 0.$$

Suppose that there are $a, b > 0$ and $0 < \delta < 1$ such that for all $\varepsilon < 1$ and all $f \in D(-\Delta)$,

$$\|Vf\|^2 \leq \varepsilon\| - \Delta f\|^2 + a\exp(b\varepsilon^{-\delta})\|f\|^2.$$

Then $V \in S(\mathbb{R}^d)$.

Remark 4.102. From the perspective of self-adjointness Kato-class is not a natural space, and certainly not the largest possible. A well-known example is the potential $V(x) = |x|^{-2}|\log|x||^{-\gamma}$, which is $-\Delta$-form-bounded with relative bound 0 exactly when $\gamma = 0$, while it is of Kato-class if and only if $\gamma > 1$.

Proposition 4.96 says that given $V \in \mathcal{K}(\mathbb{R}^d)$, one can make $\mathbb{E}^x[\int_0^t V(B_s)ds]$ arbitrarily small by taking t small, uniformly in x. We will now see that this implies exponential integrability of the random variable $\int_0^t V(B_s)\,ds$. The proof of this fact is based on Khasminskii's lemma, which follows directly from the Markov property of Brownian motion but is so useful that we state it as a separate result. First, we show a preliminary result.

Lemma 4.103. *Let $t \geq 0$. Let $f, g : \mathscr{X} \to \mathbb{R}$ be functions such that f is $\mathscr{F}_t^{\mathrm{BM}}$-measurable, $\mathbb{E}^x[|f|] < \infty$, and $\sup_{y\in\mathbb{R}} \mathbb{E}^y[|g|] < \infty$. Let $\theta_t : \mathscr{X} \to \mathscr{X}$ be a time shift given by $(\theta_t\omega)(\cdot) = \omega(t+\cdot)$ and define $\theta_t^* g(\omega) = g \circ \theta_t(\omega) = g(\omega(\cdot + t))$. Then*

$$|\mathbb{E}^x[f\theta_t^* g]| \leq \mathbb{E}^x[|f|] \sup_{y\in\mathbb{R}} \mathbb{E}^y[|g|].$$

Proof. Note that $|\mathbb{E}^x[g(B_{\cdot+t})|\mathscr{F}_t^{\mathrm{BM}}]| = |\mathbb{E}^{B_t}[g(B_\cdot)]| \leq \sup_{y\in\mathbb{R}} \mathbb{E}^y[|g|]$. Since f is $\mathscr{F}_t^{\mathrm{BM}}$-measurable,

$$|\mathbb{E}^x[f\theta_t^* g]| = |\mathbb{E}^x[f\mathbb{E}^x[\theta_t^* g|\mathscr{F}_t^{\mathrm{BM}}]]| \leq \mathbb{E}^x[|f| \sup_{y\in\mathbb{R}} \mathbb{E}^y[|g|]] = \sup_{y\in\mathbb{R}} \mathbb{E}^y[|g|]\mathbb{E}^x[|f|],$$

where we used the Markov property of Brownian motion. □

Lemma 4.104 (Khasminskii's lemma). *Let $V \geq 0$ be a measurable function on $\mathbb{R}^d$ with the property that for some $t > 0$ and $a < 1$,*

$$\sup_{x\in\mathbb{R}^d} \mathbb{E}^x\left[\int_0^t V(B_s)\,ds\right] = a. \tag{4.2.29}$$

Then

$$\sup_{x\in\mathbb{R}^d} \mathbb{E}^x[e^{\int_0^t V(B_s)\,ds}] \leq \frac{1}{1-a}. \tag{4.2.30}$$

Proof. By expanding the exponential in (4.2.30), it suffices to show that

$$a^n \geq \sup_{x\in\mathbb{R}^d} \mathbb{E}^x\left[\frac{1}{n!}\left[\int_0^t V(B_s)\,ds\right]^n\right] = \frac{1}{n!}\sup_{x\in\mathbb{R}^d} \mathbb{E}^x\left[\int_0^t ds_1 \cdots \int_0^t ds_n\, V(B_{s_1})\cdots V(B_{s_n})\right].$$

Since there are $n!$ ways of ordering $\{s_1, \ldots, s_n\}$, it suffices to prove

$$a^n \geq \sup_{x\in\mathbb{R}^d} \mathbb{E}^x\left[\int_{\Delta_n} V(B_{s_1})\cdots V(B_{s_n})ds_1\cdots ds_n\right], \tag{4.2.31}$$

where $\Delta_n = \{(s_1, \ldots, s_n) \in \mathbb{R}^n \mid 0 < s_1 < \cdots < s_n\}$. For fixed $s_1 < \cdots < s_{n-1}$, Lemma 4.103 gives

$$\mathbb{E}^x\left[V(B_{s_1})\cdots V(B_{s_{n-1}})\int_{s_{n-1}}^{t} V(B_{s_n})\,ds_n\right]$$
$$\leq \mathbb{E}^x\left[V(B_{s_1})\cdots V(B_{s_{n-1}})\right]\sup_{x\in\mathbb{R}^d}\mathbb{E}^x\left[\int_{s_{n-1}}^{t} V(B_{s_n})\,ds_n\right]$$
$$\leq \alpha\mathbb{E}^x\left[V(B_{s_1})\cdots V(B_{s_{n-1}})\right].$$

The result follows by induction and an application of Fubini's theorem. □

Now we are in a position to prove the exponential integrability of the integral over the potential. As will be seen below, this is essentially all we need to prove the Feynman–Kac formula for Kato-decomposable potentials. When $V_- \in \mathcal{K}(\mathbb{R}^d)$, it can be seen that the exponent $e^{\int_0^t V(B_s)ds}$ is integrable with respect to Wiener measure so that $\mathbb{E}^x[e^{\int_0^t V(B_s)ds}]$ is bounded for all x.

Lemma 4.105. *Let $0 \leq V \in \mathcal{K}(\mathbb{R}^d)$. Then there exist $\beta, \gamma > 0$ such that*

$$\sup_{x\in\mathbb{R}^d}\mathbb{E}^x[e^{\int_0^t V(B_s)ds}] < \gamma e^{\beta t}. \tag{4.2.32}$$

Furthermore, if $V \in L^p(\mathbb{R}^d)$ with $p > d/2$ and $1 \leq p < \infty$, then

$$\beta \leq c(p)^{1/\varepsilon}\Gamma(\varepsilon)^{1/\varepsilon}\|V\|_p^{1/\varepsilon}, \tag{4.2.33}$$

where $\varepsilon = 1 - \frac{d}{2p}$ and

$$c(p) = \begin{cases} (2\pi)^{-d/2}, & p = 1, \\ (2\pi)^{-d/2p}q^{d/(2q)}, & p > 1 \end{cases} \tag{4.2.34}$$

with $\frac{1}{p} + \frac{1}{q} = 1$. In particular, $L^p(\mathbb{R}^d) \subset \mathcal{K}(\mathbb{R}^d)$ for $p > d/2$ and $1 \leq p < \infty$.

Proof. There exists $t^* > 0$ such that $\alpha_t = \sup_{x\in\mathbb{R}^d}\mathbb{E}^x[\int_0^t V(B_s)ds] < 1$, for all $t \leq t^*$, and $\alpha_t \to 0$ as $t \to 0$. By Khasminskii's lemma, we have

$$\sup_{x\in\mathbb{R}^d}\mathbb{E}^x[e^{\int_0^t V(B_s)ds}] < \frac{1}{1-\alpha_t} \tag{4.2.35}$$

for all $t \leq t^*$. By means of Lemma 4.103, we obtain

$$\mathbb{E}^x[e^{\int_0^{2t^*} V(B_s)ds}] = \mathbb{E}^x[e^{\int_0^{t^*} V(B_s)ds}\mathbb{E}^{B_{t^*}}[e^{\int_0^{t^*} V(B_s)ds}]] \leq \left(\frac{1}{1-\alpha_{t^*}}\right)^2.$$

Repeating this procedure, we see that

$$\sup_{x\in\mathbb{R}^d}\mathbb{E}^x[e^{\int_0^t V(B_s)ds}] \leq \left(\frac{1}{1-\alpha_{t^*}}\right)^{[t/t^*]+1} \tag{4.2.36}$$

for all $t > 0$, where $[z] = \max\{w \in \mathbb{Z} \mid w \le z\}$. Choosing $\gamma = \frac{1}{1-a_{t^*}}$ and $\beta = \log\{(\frac{1}{1-a_{t^*}})^{1/t^*}\}$, we arrive at (4.2.32).

Next we prove (4.2.33). Suppose $V \in L^p(\mathbb{R}^d)$ with $p > d/2$ and $1 \le p < \infty$. Take $p > 1$. By the Hölder inequality, we have

$$\mathbb{E}^x[V(B_t)] \le \frac{1}{(2\pi t)^{d/2}} \left(\int_{\mathbb{R}^d} e^{-|x-y|^2 q/(2t)} \right)^{1/q} \|V\|_p.$$

By direct computation,

$$\frac{1}{(2\pi t)^{d/2}} \left(\int_{\mathbb{R}^d} e^{-q|x-y|^2/(2t)} dy \right)^{1/q} = \frac{q^{d/(2q)}}{(2\pi t)^{d/(2p)}};$$

in particular, we have

$$\|\mathbb{E}^x[V(B_t)]\|_\infty \le c(p) t^{-d/(2p)} \|V\|_p, \quad p > 1.$$

Recall the *Mittag-Leffler function* defined by

$$m_b(x) = \sum_{k=0}^{\infty} \frac{x^k}{\Gamma(1+kb)},$$

where Γ denotes the Gamma function, $x \in \mathbb{R}$, and $b > 0$. It is known $m_b(x)$ satisfies $\lim_{x\to\infty}(m_b(x) - \frac{1}{b} e^{x^{1/b}}) = 0$ and there exists $k_b > 0$ such that

$$m_b(x) \le k_b e^{x^{1/b}} \tag{4.2.37}$$

for all $x > 0$. Let $0 \le s_1 \le s_2 \le \cdots \le s_k$. By the Markov property of Brownian motion,

$$\begin{aligned}
&\mathbb{E}^x[V(B_{s_1}) \cdots V(B_{s_k})] \\
&= \mathbb{E}^x\left[V(B_{s_1}) \cdots V(B_{s_{k-1}}) \mathbb{E}^x[V(B_{s_k}) | \mathcal{F}^{\mathrm{BM}}_{s_{k-1}}]\right] \\
&= \mathbb{E}^x\left[V(B_{s_1}) \cdots V(B_{s_{k-1}}) \mathbb{E}^{B_{s_{k-1}}}[V(B_{s_k - s_{k-1}})]\right] \\
&\le \mathbb{E}^x[V(B_{s_1}) \cdots V(B_{s_{k-1}})] c(p) \|V\|_p (s_k - s_{k-1})^{-d/(2p)} \\
&\le \cdots \le c(p)^k \|V\|_p^k s_1^{-d/(2p)} (s_2 - s_1)^{-d/(2p)} \cdots (s_k - s_{k-1})^{-d/(2p)}.
\end{aligned}$$

Hence it follows that

$$\mathbb{E}^x\left[\frac{1}{k!}\left[\int_0^t V(B_s) ds\right]^k\right] = \int_0^t ds_1 \int_{s_1}^t ds_2 \cdots \int_{s_{k-1}}^t \mathbb{E}^x[V(B_{s_1}) \cdots V(B_{s_k})] ds_k$$

$$\le c(p)^k \|V\|_p^k \int_0^t ds_1 \cdots \int_{s_{k-1}}^t s_1^{-d/(2p)} (s_2 - s_1)^{-d/(2p)} \cdots (s_k - s_{k-1})^{-d/(2p)} ds_k$$
$$= \frac{(c(p)\|V\|_p t^\varepsilon \Gamma(\varepsilon))^k}{\Gamma(1 + k\varepsilon)},$$

where $\varepsilon = 1 - \frac{d}{2p} > 0$ and we used the formula

$$\int_0^1 z^{1+a}(1-z)^a dz = \frac{\Gamma(a)\Gamma(1+a)}{\Gamma(1+2a)},$$

giving the relationship between the Gamma and Beta functions.

By (4.2.37), it further follows that

$$\sup_{x\in\mathbb{R}^d} \mathbb{E}^x[e^{\int_0^t V(B_s)ds}] \le \sum_{k=0}^\infty \frac{(c(p)\|V\|_p t^\varepsilon \Gamma(\varepsilon))^k}{\Gamma(1+k\varepsilon)} = m_\varepsilon\left(c(p)\|V\|_p \Gamma(\varepsilon) t^\varepsilon\right) \le k_\varepsilon e^{c(p)^{1/\varepsilon}\|V\|_p^{1/\varepsilon}\Gamma(\varepsilon)^{1/\varepsilon} t}.$$

Hence (4.2.33) is proven. Let $p = 1$. Then $d = 1$, and it follows directly that

$$\mathbb{E}^x[V(B_t)] \le (2\pi t)^{-1/2}\|V\|_1 = c(1)t^{-1/2}\|V\|_1. \tag{4.2.38}$$

Thus (4.2.33) is obtained in a similar way as for $p > 1$. □

4.2.4 Feynman–Kac formula for Kato-decomposable potentials

When V is Kato-decomposable, the operator $-\frac{1}{2}\Delta + V$ is not always an easy object to define. We have seen in Sections 4.1.5–4.1.6 that much analytic machinery is needed to decide whether it is essentially self-adjoint and on which domain. The strategy we will adopt here is to define the Feynman–Kac semigroup as given in Definition 4.87 for Kato-decomposable potentials, and identify Schrödinger operators for such potentials as generators of this semigroup. Let

$$(K_t f)(x) = \mathbb{E}^x[e^{-\int_0^t V(B_s)\,ds} f(B_t)].$$

By Lemma 4.105, the map $K_t : L^\infty \to L^\infty$ is bounded and linear whenever V is Kato-decomposable. In fact, an even stronger property holds.

Definition 4.106. Let (M, dm) be a measurable space. An operator T is said to be L^p–L^q *bounded* if there exists $D \subset L^p(M)$ such that it is dense for all $1 \le p \le \infty$, and $T\lceil_D$: $L^p(M) \to L^q(M)$ is bounded for all $1 \le p \le q \le \infty$. In other words, $T_p = \overline{T\lceil_D}^{L^p}$ is bounded from $L^p(M)$ to $L^q(M)$ with domain $L^p(M)$.

Theorem 4.107 (L^p–L^q boundedness). *Let V be Kato-decomposable. Then for every $1 \leq p \leq q \leq \infty$, the semigroup $\{K_t : t \geq 0\}$ exists and its elements act from $L^p(\mathbb{R}^d)$ to $L^q(\mathbb{R}^d)$ as bounded operators.*

Proof. By making use of the Riesz–Thorin theorem quoted below in Lemma 4.108, it suffices to prove that K_t is bounded from $L^1(\mathbb{R}^d)$ to $L^1(\mathbb{R}^d)$, from $L^1(\mathbb{R}^d)$ to $L^\infty(\mathbb{R}^d)$, and from $L^\infty(\mathbb{R}^d)$ to $L^\infty(\mathbb{R}^d)$. Boundedness from $L^\infty(\mathbb{R}^d)$ to $L^\infty(\mathbb{R}^d)$ has already been shown in Lemma 4.105.

We prove boundedness from $L^1(\mathbb{R}^d)$ to $L^1(\mathbb{R}^d)$ by using duality. For $f \in L^1(\mathbb{R}^d)$ and $g \in L^\infty(\mathbb{R}^d)$, we have

$$\int_{\mathbb{R}^d} f(x) K_t g(x)\, dx = \int_{\mathbb{R}^d} \mathbb{E}^x[g(B_0) e^{-\int_0^t V(B_s)\,ds} f(B_t)]dx. \tag{4.2.39}$$

This follows from (4.2.3) in the proof of Theorem 4.86. By taking $g = 1$ and reading (4.2.39) from right to left, we find

$$\|K_t f\|_1 \leq \|f\|_1 \|K_t 1\|_\infty;$$

hence K_t is bounded from $L^1(\mathbb{R}^d)$ to $L^1(\mathbb{R}^d)$. Since K_t is bounded from $L^\infty(\mathbb{R}^d)$ to $L^\infty(\mathbb{R}^d)$ and from $L^1(\mathbb{R}^d)$ to $L^1(\mathbb{R}^d)$, the Riesz–Thorin theorem implies that K_t is bounded from $L^p(\mathbb{R}^d)$ to $L^p(\mathbb{R}^d)$ for all $p \geq 1$, and by the Markov property of Brownian motion, it is a semigroup on all of these spaces.

It remains to show that K_t is bounded from $L^1(\mathbb{R}^d)$ to $L^\infty(\mathbb{R}^d)$. Consider the diagram

$$L^1(\mathbb{R}^d) \xrightarrow{K_t} L^2(\mathbb{R}^d) \xrightarrow{K_t} L^\infty(\mathbb{R}^d). \tag{4.2.40}$$

Assume first that $f \in L^2(\mathbb{R}^d)$. By the Schwarz inequality,

$$\|K_t f\|_\infty^2 \leq \sup_{x \in \mathbb{R}^d} (\mathbb{E}^x[e^{-2\int_0^t V(B_s)\,ds}]\mathbb{E}^x[f(B_t)^2]).$$

The first factor on the right-hand side above is bounded by Khasminskii's lemma, while the second is bounded due to $\mathbb{E}^x[f(B_t)^2] = (2\pi t)^{-d/2} \int_{\mathbb{R}^d} f(y)^2 e^{-|x-y|^2/2t} dy$. Hence it follows that $K_t : L^2(\mathbb{R}^d) \to L^\infty(\mathbb{R}^d)$ and $\|K_t f\|_\infty \leq C\|f\|_2$. Now for $0 \leq f \in L^1(\mathbb{R}^d)$ choose $g \in L^2(\mathbb{R}^d)$, and by a similar reasoning as in (4.2.39), we find

$$\int_{\mathbb{R}^d} (K_t f)(x) g(x)\, dx = \int_{\mathbb{R}^d} f(x)(K_t g)(x)\, dx \leq \|K_t g\|_\infty \|f\|_1.$$

This can be extended to any $f \in L^1(\mathbb{R}^d)$ by separating $f = (\Re f)_+ - (\Re f)_- + i(\Im f)_+ - i(\Im f)_-$, where $f_+ = \max\{f(x), 0\}$, $f_- = -\min\{f(x), 0\}$, and

$$\left| \int_{\mathbb{R}^d} (K_t f)(x) g(x)\, dx \right| \le 4 \|K_t g\|_\infty \|f\|_1.$$

Hence $K_t f \in L^2(\mathbb{R}^d)$ for $f \in L^1(\mathbb{R}^d)$, and for $f \in L^1(\mathbb{R}^d)$ we have

$$\|K_t f\|_\infty = \|K_{t/2} K_{t/2} f\|_\infty \le C \|K_{t/2} f\|_2 \le \tilde{C} \|f\|_1.$$

Therefore K_t maps $L^1(\mathbb{R}^d)$ to $L^\infty(\mathbb{R}^d)$ and the proof is complete. □

Lemma 4.108 (Riesz–Thorin theorem). *Let $T : L^{p_j}(\mathbb{R}^d) \to L^{q_j}(\mathbb{R}^d)$ be a bounded operator, $j = 1, 2$, and $1 \le p_1, p_2, q_1, q_2 \le \infty$. Then T is bounded from $L^r(\mathbb{R}^d)$ to $L^s(\mathbb{R}^d)$ for every r, s such that*

$$\left\{ (r, s) \,\middle|\, \left(\frac{1}{r}, \frac{1}{s}\right) = \left(\frac{1-\theta}{p_1} + \frac{\theta}{p_2}, \frac{1-\theta}{q_1} + \frac{\theta}{q_2}\right),\ 0 < \theta < 1 \right\}.$$

By Theorem 4.107, we see that $K_t f$ is a function in $L^q(\mathbb{R}^d)$ for $f \in L^p(\mathbb{R}^d)$. Hence as a function in $L^q(\mathbb{R}^d)$, $K_t f(x)$ is defined for $x \in \mathbb{R}^d \setminus N_{t,f}$ with a null set $N_{t,f}$. Using the representation $(K_t f)(x) = \mathbb{E}^x[e^{-\int_0^t V(B_s)\, ds} f(B_t)]$, however, we can see in the theorem below that $x \mapsto K_t f(x)$ is continuous for all $x \in \mathbb{R}^d$ in this version. To see the continuity of $K_t f(x)$ in x, we make use of the smoothing effect of $e^{-t(-\Delta/2)}$. We define $e^{-t(-\Delta/2)} f(x)$ for every $x \in \mathbb{R}^d$ by

$$e^{-t(-\Delta/2)} f(x) = \int_{\mathbb{R}^d} \Pi_t(x - y) f(y)\, dy$$

for $f \in L^p(\mathbb{R}^d)$ with $1 \le p \le \infty$.

Proposition 4.109 (Smoothing effect). *Let $f \in L^p(\mathbb{R}^d)$ with some $1 \le p \le \infty$. Then $e^{-t(-\Delta/2)} f$ is continuous.*

Proof. Suppose $p = 1$. Since $e^{-t(-\Delta/2)} f(x) = \int_{\mathbb{R}^d} \Pi_t(x - y) f(y)\, dy$, the statement follows by the dominated convergence theorem. Let $p = \infty$ and define

$$f_n(x) = \begin{cases} f(x), & |x| \le n, \\ 0, & |x| > n. \end{cases}$$

Since $f_n \in L^1(\mathbb{R}^d)$ for each $n \in \mathbb{N}$, the function $e^{-t(-\Delta/2)} f_n(x)$ is continuous in x. Choose $n > |x|$. We then have

$$\begin{aligned} |e^{-t(-\Delta/2)} f_n(x) - e^{-t(-\Delta/2)} f(x)| &\le \int_{|y| \ge n} \Pi_t(x - y) |f(y)|\, dy \\ &\le \|f\|_\infty \int_{|z| \ge n + |x|} \Pi_t(z)\, dz \le \|f\|_\infty \int_{|z| \ge n} \Pi_t(z)\, dz. \end{aligned}$$

The right-hand side converges to zero as $n \to \infty$, uniformly in x. Hence the uniform limit $e^{-t(-\Delta/2)}f(x)$ is continuous in x. Finally, take $1 < p < \infty$ and let $g \in L^p(\mathbb{R}^d)$. Define

$$g_n(x) = \begin{cases} g(x), & |g(x)| \le n, \\ n, & |g(x)| > n. \end{cases}$$

We see that $e^{-t(-\Delta/2)}g_n(x)$ is continuous in x for each n. By the Hölder inequality, we have

$$\begin{aligned} |e^{-t(-\Delta/2)}g_n(x) - e^{-t(-\Delta/2)}g(x)| &\le \left(\int_{\mathbb{R}^d} \Pi_t(x-y)^q dy \right)^{1/q} \left(\int_{\mathbb{R}^d} |g_n(y) - g(y)|^p dy \right)^{1/p} \\ &\le \|\Pi_t\|_{L^q(\mathbb{R}^d)} \|g_n - g\|_{L^p(\mathbb{R}^d)} \end{aligned}$$

for $1/p + 1/q = 1$. The right-hand side converges to zero as $n \to \infty$, uniformly in x. Hence $e^{-t(-\Delta/2)}g(x)$ is continuous in x. □

Lemma 4.110. *Let V be a Kato-decomposable potential and $f \in L^\infty(\mathbb{R}^d)$. Then*

$$e^{-r(-\Delta/2)}K_{t-r}f(x) = \mathbb{E}^x[e^{-\int_r^t V(B_s)ds} f(B_t)], \quad t \ge r > 0. \tag{4.2.41}$$

In particular, (4.2.41) *is continuous in x for every $r > 0$.*

Proof. We have $e^{-r(-\Delta/2)}K_{t-r}f(x) = \mathbb{E}^x[\mathbb{E}^{B_r}[e^{-\int_0^{t-r} V(B_s)ds} f(B_{t-r})]]$. The Markov property implies the lemma. Furthermore, continuity follows by Proposition 4.109. □

Theorem 4.111. *Let V be Kato-decomposable and $f \in L^p(\mathbb{R}^d)$ for $1 \le p \le \infty$. Then $K_t f$ is a continuous function for every $t > 0$.*

Proof. First, we make a preliminary observation. Suppose that $V \in \mathcal{K}(\mathbb{R}^d)$. In the proof of Lemma 4.104, we have

$$\begin{aligned} \sup_{x\in\mathbb{R}^d} \mathbb{E}^x[|1 - e^{-\int_0^t V(B_s)\,ds}|] &\le \sum_{k=1}^\infty \frac{1}{k!} \sup_{x\in\mathbb{R}^d} \mathbb{E}^x\left[\left(\int_0^t |V(B_s)|\,ds\right)^k\right] \\ &\le \frac{\sup_{x\in\mathbb{R}^d} \mathbb{E}^x[\int_0^t |V(B_s)|ds]}{1 - \sup_{x\in\mathbb{R}^d} \mathbb{E}^x[\int_0^t |V(B_s)|ds]} \stackrel{t\to 0}{\longrightarrow} 0 \end{aligned} \tag{4.2.42}$$

by Proposition 4.96. Note that Khasminskii's lemma implies that for small enough T,

$$\sup_{0\le t\le T} \sup_{x\in\mathbb{R}^d} \mathbb{E}^x[e^{\int_0^t |V(B_s)|\,ds}] < \infty. \tag{4.2.43}$$

We can now proceed to prove the theorem. Since $K_t f = K_{t/2}K_{t/2}f$ and $K_{t/2}f \in L^\infty(\mathbb{R}^d)$ whenever $f \in L^p(\mathbb{R}^d)$, $1 \le p \le \infty$, it suffices to consider $f \in L^\infty(\mathbb{R}^d)$. The essential idea is to apply the smoothing property seen in Proposition 4.109. We have

$$K_tf(x) - K_tf(x_n) = K_tf(x) - e^{-r(-\Delta/2)}K_{t-r}f(x) + e^{-r(-\Delta/2)}K_{t-r}f(x)$$
$$- e^{-r(-\Delta/2)}K_{t-r}f(x_n) + e^{-r(-\Delta/2)}K_{t-r}f(x_n) - K_tf(x_n).$$

By the smoothing effect, we see that

$$e^{-r(-\Delta/2)}K_{t-r}f(x) - e^{-r(-\Delta/2)}K_{t-r}f(x_n) \to 0$$

as $n \to \infty$ for every $r > 0$. It is sufficient to show that $K_tf(x) - e^{-r(-\Delta/2)}K_{t-r}f(x) \to 0$ as $r \to 0$, uniformly in x. First, assume $V \in \mathcal{K}(\mathbb{R}^d)$. For every $r > 0$, let $g_r(x) = e^{-r(-\Delta/2)}K_{t-r}f(x)$. By Lemma 4.110, we have

$$\|g_r - K_tf\|_\infty = \sup_{x\in\mathbb{R}^d} \mathbb{E}^x[(1 - e^{-\int_0^T V(B_s)\,ds})e^{-\int_r^t V(B_s)\,ds}f(B_t)]$$
$$\leq \|f\|_\infty\Big(\sup_{x\in\mathbb{R}^d}\sup_{r\leq t}\mathbb{E}^x[e^{-\int_0^r V(B_s)\,ds}]\Big)\Big(\sup_{x\in\mathbb{R}^d}\mathbb{E}^x[1 - e^{-\int_0^r V(B_s)\,ds}]\Big) \overset{r\to 0}{\longrightarrow} 0.$$

In the last lines we have used Lemma 4.103 and (4.2.42). Thus K_tf is continuous.

Next let $V = V_+ - V_-$ be Kato-decomposable, and consider for $R > 0$ the function $V_R = V_{+R} - V_-$ with $V_{+R}(x) = V_+(x)$ if $|x| \leq R$ and $V_{+R}(x) = 0$ if $|x| > R$. We have $V_R \in \mathcal{K}(\mathbb{R}^d)$ by the definition of the Kato-decomposable property, and the above reasoning applies. Let $M \subset B_R(0)$ be compact, where $B_R(0)$ denotes the ball of radius R centered at the origin. We now claim that $\int_0^t V_R(B_s)\,ds = \int_0^t V(B_s)\,ds$ for each path $\omega \in \mathscr{X}$ starting in M and not leaving the ball up to time t. Hence

$$\sup_{x\in M}|\mathbb{E}^x[e^{-\int_0^t V(B_s)\,ds}f(B_t)] - \mathbb{E}^x[e^{-\int_0^t V_R(B_s)\,ds}f(B_t)]|$$
$$\leq \sup_{x\in M}\mathbb{E}^x[e^{\int_0^t V_-(B_s)\,ds}\left|e^{-\int_0^t V_+(B_s)ds} - 1\right||f(B_t)|1_{\{\sup_{0\leq s\leq t}|B_s|\geq R\}}]$$
$$\leq 2\|f\|_\infty \sup_{x\in M}\mathbb{E}^x[e^{\int_0^t V_-(B_s)\,ds}1_{\{\sup_{0\leq s\leq t}|B_s|\geq R\}}]$$
$$\leq 2\|f\|_\infty \sup_{x\in M}\left(\mathbb{E}^x[e^{-2\int_0^t V_-(B_s)\,ds}]\right)^{1/2}\left(\mathbb{E}^x[1_{\{\sup_{0\leq s\leq t}|B_s|\geq R\}}]\right)^{1/2}.$$

The first factor above is bounded since V_- is Kato-class, and the second goes to zero as $R \to \infty$ by Lévy's maximal inequality

$$\mathbb{E}^x[1_{\{\sup_{0\leq s\leq t}|B_s|\geq R\}}] \leq 2\mathbb{E}^x[1_{\{|B_t|>R\}}].$$

Hence K_tf is a locally uniform limit of continuous functions, and $K_tf(x)$ is continuous in x. □

The last part of the above argument is of some independent interest, and we present it separately.

Corollary 4.112. *Let V be Kato-decomposable and $f \in L^p(\mathbb{R}^d)$ with $1 \le p \le \infty$. Then $e^{-tH}f$ is the local uniform limit of the sequence $(e^{-tH_n}f)_{n\in\mathbb{N}}$ with $H_n = -\frac{1}{2}\Delta + V_n$ and V_n with compact support.*

One possible choice in the corollary above is $V_n = 1_{\{|x|\le n\}}V$. In particular, each V_n is in $\mathcal{K}(\mathbb{R}^d)$ as opposed to being only Kato-decomposable.

We are now ready to define H with Kato-class potential as a self-adjoint operator and give the Feynman–Kac formula for e^{-tH}.

Theorem 4.113 (Feynman–Kac formula for Schrödinger operator with Kato-decomposable potential). *Assume that V is Kato-decomposable. Then $\{K_t : t \ge 0\}$ defined by $(K_t f)(x) = \mathbb{E}^x[e^{-\int_0^t V(B_s)\,ds} f(B_t)]$ is a symmetric C_0-semigroup on $L^2(\mathbb{R}^d)$. Moreover, there exists a unique self-adjoint operator K bounded from below such that $K_t = e^{-tK}$, for every $t \ge 0$.*

Proof. We already know that $\{K_t : t \ge 0\}$ is a semigroup of bounded operators on $L^2(\mathbb{R}^d)$, and $(g, K_t f) = (K_t g, f)$ can be proven in the same way as (4.2.3). To see strong continuity, by denseness it suffices to prove $\|K_t f - f\| \to 0$ as $t \downarrow 0$ for all bounded L^2 functions f with compact support. Let $Q_t = e^{t\Delta/2}$. We see that $\|Q_t f - f\| \to 0$ as $t \downarrow 0$. We only need to show that

$$x \mapsto K_t f(x) - Q_t f(x) = \mathbb{E}^x[(e^{-\int_0^t V(B_s)\,ds} - 1) f(B_t)]$$

converges to 0 in $L^2(\mathbb{R}^d)$ as $t \downarrow 0$. Using the proof of Theorem 4.111 and Corollary 4.112, it follows that $K_t f - Q_t f \to 0$ as $t \downarrow 0$ pointwise in $\mathbb{R}^d$. Moreover,

$$(K_t f(x) - Q_t f(x))^2 \le \mathbb{E}^x[(e^{-\int_0^t V(B_s)\,ds} - 1)^2]\mathbb{E}^x[f(B_t)^2].$$

Here the first factor is bounded uniformly in t and x by Lemma 4.105, while the t-supremum over the second one is bounded and exponentially decaying at infinity, thus integrable. Thus $K_t f - Q_t f \to 0$ in $L^2(\mathbb{R}^d)$ by the dominated convergence theorem. Hence by the triangle inequality $\|K_t f - f\| \le \|K_t f - Q_t f\| + \|Q_t f - f\|$, the statement follows. Furthermore, by the Hille–Yosida theorem, there exists a unique self-adjoint operator K bounded below such that $K_t = e^{-tK}$, for every $t \ge 0$. □

Definition 4.114 (Schrödinger operator with Kato-decomposable potential). Let V be a Kato-decomposable potential. We define the *Schrödinger operator for the Kato-decomposable potential V* by K given in Theorem 4.113.

By Proposition 4.100, a Kato-decomposable potential is the sum of an $L^1_{\mathrm{loc}}(\mathbb{R}^d)$-potential and a $-\Delta$-form-bounded potential with infinitesimally small relative bound. This implies that K with a Kato-decomposable potential V coincides with the Schrödinger operator $-\frac{1}{2}\Delta + V$ defined by the quadratic form sum.

Remark 4.115. Suppose that $V = V_+ - V_-$ such that $V_+ \in L^1_{\mathrm{loc}}(\mathbb{R}^d)$ and $V_- \in \mathcal{K}(\mathbb{R}^d)$. We can also prove that $(K_t f)(x) = \mathbb{E}^x[e^{-\int_0^t V(B_s)\,ds} f(B_t)]$ is a symmetric C_0-semigroup on $L^2(\mathbb{R}^d)$, and for every $1 \le p \le q \le \infty$ the semigroup $\{K_t : t \ge 0\}$ exists and its elements act from $L^p(\mathbb{R}^d)$ to $L^q(\mathbb{R}^d)$ as bounded operators. The function $K_t f(x)$ is, however, not necessarily continuous in x.

4.3 Properties of Schrödinger operators and semigroups

4.3.1 Kernel of the Schrödinger semigroup

From the perspective of functional analysis and the theory of partial differential equations, it is a crucial point that for every $t > 0$ the operator e^{-tH} has an integral kernel given by the Feynman–Kac formula. This is addressed in the next theorem.

Theorem 4.116. *Let V be a Kato-decomposable potential. Then e^{-tK} is an integral operator in $L^p(\mathbb{R}^d)$ for every $1 \le p \le \infty$. Moreover, the integral kernel $\mathbb{R}^d \times \mathbb{R}^d \ni (x,y) \mapsto K_t(x,y)$ is jointly continuous in x and y, and is given by*

$$K_t(x,y) = \Pi_t(x-y)\mathbb{E}^{x,y}_{[0,t]}[e^{-\int_0^t V(B_s)\,ds}]. \tag{4.3.1}$$

Here $\mathcal{W}^{x,y}_{[0,t]}$ is the Brownian bridge measure defined in (2.3.20).

Proof. By Theorem 4.113, we have $(e^{-tK}f)(x) = \mathbb{E}^x[e^{-\int_0^t V(B_s)\,ds} f(B_t)]$, for every $f \in L^p(\mathbb{R}^d)$; and by Theorem 4.111, the above equality holds pointwise as the right-hand side is a continuous function. Thus

$$(e^{-tK}f)(x) = \int_{\mathbb{R}^d} \Pi_t(x-y) f(y) \mathbb{E}^{x,y}_{[0,t]}[e^{-\int_0^t V(B_s)\,ds}]\,dy,$$

which shows that (4.3.1) holds for all x and almost all y. Next we show continuity. Let $s = t/3$. We have

$$K_t(x,y) = \int_{\mathbb{R}^{2d}} K_s(x,v)K_s(v,w)K_s(w,y)\,dw\,dv = \int_{\mathbb{R}^{2d}} K_s(x,v)K_s(y,w)K_s(v,w)\,dw\,dv.$$

The product

$$K_s(x,v)K_s(y,w) = \Pi_s(x-v)\Pi_s(y-w)\mathbb{E}^{(x,v),(y,w)}_{[0,s]}[e^{-\int_0^s (V(B^{(1)}_r)+V(B^{(2)}_r))\,dr}]$$

is the kernel of the Schrödinger operator with Kato-class potential

$$\tilde{K} = -\frac{1}{2}\Delta_x - \frac{1}{2}\Delta_y + V(x) + V(y),$$

and $(x,y) \mapsto V(x)+V(y)$ is Kato-decomposable in $\mathbb{R}^{2d}$ since V is in $\mathbb{R}^d$, where $B_t^{(i)}$, $i = 1, 2$, denote two independent d-dimensional Brownian motions. On the other hand, the map $(v, w) \mapsto K_s(v, w)$ is bounded. One way to see this is to recall that e^{-sK} is bounded from $L^1(\mathbb{R}^d)$ to $L^\infty(\mathbb{R}^d)$ and $\sup_{(v,w)\in\mathbb{R}^{2d}} |K_s(v, w)| = \sup_{f\in L^1(\mathbb{R}^d),\|f\|_1=1} \|e^{-tK} f\|_\infty < \infty$. Hence

$$K_t(x,y) = (e^{-t\tilde{K}} K_s)(x,y)$$

with $K_s \in L^\infty(\mathbb{R}^d \times \mathbb{R}^d)$ and is thus jointly continuous in x and y by Theorem 4.111. □

The above proof has the virtue of simplicity; using more sophisticated arguments, it can be shown that $K_t(x,y)$ is even jointly continuous in x, y and t for $t > 0$.

4.3.2 Positivity improving and uniqueness of ground state

In this section we start our discussion of eigenfunctions of Schrödinger operators by using Feynman–Kac formulae. Our first goal is to discuss the uniqueness of the ground state, and then we turn briefly to their possible higher multiplicities, and furthermore to fundamental questions of existence and absence of ground states in the sections to follow.

Definition 4.117 (Ground state). Let H be a self-adjoint operator bounded from below. An eigenfunction φ_0 satisfying

$$H\varphi_0 = E(H)\varphi_0, \quad E(H) = E_0 = \inf \operatorname{Spec}(H) \tag{4.3.2}$$

is called a *ground state* of H, and the bottom of the spectrum E_0 is called the *ground state energy*. The nonnegative integer

$$m(H) = \dim \operatorname{Ker}(H - E(H)) \tag{4.3.3}$$

is called the multiplicity of the ground state. If $m(H) = 1$, the ground state is said to be *unique*.

A simple, but useful, fact is that the kernel $e^{-tH}(x,y)$ is strictly positive. This is readily implied by (4.3.1) whenever $e^{-\int_0^t V(B_s)ds}$ is strictly positive for almost every $\omega \in \mathscr{X}$. In particular, e^{-tH} is positivity improving in the sense of the following definition.

Definition 4.118 (Positivity preserving/improving operator). Let (M,μ) be a σ-finite measure space.

(1) A nonzero function $f \in L^2(M, d\mu)$ is called *positive* if $f \geq 0$ holds μ-almost everywhere. Moreover, f is called *strictly positive* if $f > 0$ holds μ-almost everywhere.
(2) A bounded operator A on $L^2(M, d\mu)$ is called a *positivity preserving operator* if $(f, Ag) \geq 0$, for all positive $f, g \in L^2(M, d\mu)$. A is called a *positivity improving operator* if $(f, Ag) > 0$, for all positive $f, g \in L^2(M, d\mu)$.

Example 4.119. Since $e^{t(\Delta/2)}f(x) = \int_{\mathbb{R}^d} \Pi_t(x-y)f(y)dy$, the operator $e^{t(\Delta/2)}$ is positivity improving. Moreover, $e^{-ia\cdot(-i\nabla)}$ is also a positivity preserving operator, as it produces the shift $e^{-ia(-i\nabla)}f(x) = f(x-a)$.

Positivity preserving operators of the form $e^{-tE(-i\nabla)}$ have a deep connection with Lévy processes. As seen in Example 3.29, the square root of the Laplacian generates a stable Lévy process. The Lévy–Khintchine formula (3.1.8) has the following important relationship with positivity preserving operators.

Proposition 4.120. *Let* $E : \mathbb{R}^d \to \mathbb{C}$.

(1) *Suppose that the real part of E is bounded from below. Then the following statements are equivalent:*
   (a) $e^{-tE(-i\nabla)}$ *is a positivity preserving semigroup;*
   (b) $e^{-tE(u)}$ *is a positive definite distribution, i. e., for all* $\phi \in C_0^\infty(\mathbb{R}^d)$ *we have* $\int_{\mathbb{R}^d}\int_{\mathbb{R}^d} e^{-tE(u)}\overline{\phi(x-u)}\phi(x)dudx \geq 0$;
   (c) $E(-u) = \overline{E(u)}$, *and it is conditionally negative definite, i. e.,*

$$\sum_{i,j=1}^{n} E(p_i - p_j)\bar{z}_i z_j \leq 0,$$

for every $n \in \mathbb{N}$ *and* $z = (z_1, \ldots, z_n) \in \mathbb{C}^n$ *with* $\sum_{i=1}^n z_i = 0$;
   (d) *there exists a triplet* (b, A, ν), *where* $b \in \mathbb{R}^d$, *A is a positive definite symmetric matrix, and* ν *is a Lévy measure such that*

$$-E(u) = a + iu\cdot b - \frac{1}{2}u\cdot Au + \int_{\mathbb{R}^d\setminus\{0\}} (e^{iu\cdot y} - 1 - iu\cdot y1_{\{|y|\leq 1\}})\nu(dy) \tag{4.3.4}$$

with some $a \in \mathbb{R}$.

(2) *Suppose that E is a spherically symmetric and polynomially bounded continuous function with* $E(0) = 0$, *with its real part bounded from below. Suppose, moreover, that* ΔE *in distributional sense is positive definite. Then E is conditionally negative definite.*

Write

$$E(u) = \sqrt{|u|^2 + m^2} - m, \quad m \geq 0;$$

then by a straightforward calculation

$$\Delta_u E(u) = (d-1)(|u|^2 + m^2)^{-1/2} + m^2(|u|^2 + m^2)^{-3/2}.$$

Since $(|u|^2+m^2)^{-\beta} = d_\beta \int_0^\infty t^{\beta-1}e^{-t(|u|^2+m^2)}dt$ and $e^{-t|u|^2}$ is positive definite, we conclude by Proposition 4.120 that $E(u)$ is conditionally negative definite, which implies that $-E(u)$ can be represented as in (4.3.4). Since the left-hand side is real and $E(u)/|u|^2 \to 0$ as $|u| \to 0$, its Lévy triplet is $(b, A, \nu) = (0, 0, \nu)$. Thus we have

$$-\sqrt{|u|^2+m^2}+m=\int_{|y|>0}(e^{iu\cdot y}-1-iu\cdot y1_{\{|y|<1\}})\nu(dy). \tag{4.3.5}$$

The following result gives the Feynman–Kac formula for $e^{-E(-i\nabla)}$.

Proposition 4.121. *Let* $(X_t)_{t\geq 0}$ *be a Lévy process on a probability space* $(\Omega,\mathscr{F},P)$ *with characteristic triplet* $(0,0,\nu)$ *in* (4.3.5). *Then*

$$(g,e^{-tE(-i\nabla)}f)=\int_{\mathbb{R}^d}\mathbb{E}_P[\overline{g(x)}f(x+X_t)]dx.$$

The Lévy measure ν in (4.3.5) can be determined exactly. The integral kernel of the operator $e^{-tE(-i\nabla)}$ is given by the Fourier transform $Fe^{-tE(u)}F^{-1}$ and thus

$$e^{-tE(-i\nabla)}(x,y)=2\left(\frac{m}{2\pi}\right)^{(d+1)/2}\frac{tK_{(d+1)/2}(m\sqrt{|x-y|^2+t^2})}{(t^2+|x-y|^2)^{(d+1)/4}}, \tag{4.3.6}$$

where K_ρ is the modified Bessel function of the second kind (4.2.21). Using that $K_\rho(z)\sim\frac{1}{2}\Gamma(\rho)(\frac{1}{2}z)^{-\rho}$ as $|z|\downarrow 0$, we obtain

$$\begin{aligned}(g,E(-i\nabla)f)&=\lim_{t\to 0}\frac{1}{t}[(g,f)-(g,e^{-tE(-i\nabla)}f)]\\&=\left(\frac{m}{2\pi}\right)^{(d+1)/2}\int_{\mathbb{R}^d}\int_{\mathbb{R}^d}\frac{\overline{(g(x)-g(y))}(f(x)-f(y))}{|x-y|^{(d+1)/2}}K_{(d+1)/2}(m\sqrt{|x-y|^2})dxdy.\end{aligned}$$

For $m=0$, the integral form of $(g,E(-i\nabla)f)$ is given by

$$(g,E(-i\nabla)f)=\frac{\Gamma((d+1)/2)}{2\pi^{(d+1)/2}}\int_{\mathbb{R}^d}\int_{\mathbb{R}^d}\frac{\overline{(g(x)-g(y))}(f(x)-f(y))}{|x-y|^{d+1}}dxdy+m(g,f).$$

Hence it is seen that the Lévy measure associated with $E(u)$ for $m=0$ is

$$\nu(dy)=\frac{1}{\pi^{(d+1)/2}}\Gamma\left(\frac{d+1}{2}\right)\frac{1}{|y|^{d+1}}dy$$

and for $m>0$,

$$\nu^m(dy)=2\left(\frac{m}{2\pi}\right)^{(d+1)/2}\frac{1}{|y|^{(d+1)/2}}K_{(d+1)/2}(m|y|)dy. \tag{4.3.7}$$

Remark 4.122. Let

$$\nu_\alpha(dy)=\frac{2^\alpha}{\pi^{d/2}}\frac{\Gamma(\frac{d+\alpha}{2})}{|\Gamma(-\frac{\alpha}{2})|}\frac{1}{|y|^{d+\alpha}}dy,\quad 0<\alpha<2,$$

where $\Gamma(-x) = -\frac{1}{x}\Gamma(1-x)$ for $x > 0$. The Lévy process $(X_t)_{t\geq 0}$ with triplet $(0, 0, \nu_\alpha)$ is a rotationally symmetric stable process with index α and its generator is $-(-\Delta)^{\alpha/2}$.

A useful application of the positivity properties of operators is in showing uniqueness of the ground state of self-adjoint operators.

Theorem 4.123 (Perron–Frobenius theorem). *Let (M, μ) be a σ-finite measure space and K be a self-adjoint operator bounded from below in $L^2(M, d\mu)$. Suppose that* $\inf \mathrm{Spec}(K)$ *is an eigenvalue and e^{-K} is positivity improving. Then* $\inf \mathrm{Spec}(K)$ *is nondegenerate and the eigenfunction corresponding to the eigenvalue* $\inf \mathrm{Spec}(K)$ *is strictly positive.*

Proof. Put $\lambda = \inf \mathrm{Spec}(K)$. We have $\|e^{-K}\| = e^{-\lambda}$. Let f be an eigenfunction corresponding to the eigenvalue λ. Since e^{-K} maps real-valued functions to real-valued functions, $\Re f$ is also an eigenfunction and therefore we may assume f to be real-valued from the outset. Since e^{-K} is positivity improving and linear, we find

$$(g, e^{-K} f) = (g, e^{-K} f^+ - e^{-K} f^-) \leq (g, e^{-K} f^+ + e^{-K} f^-) = (g, e^{-K} |f|)$$

for all $g \geq 0$ and all $f = f^+ - f^- \in L^2$ with $f^+ > 0$ and $f^- > 0$. Thus

$$e^{-\lambda}\|f\|^2 = (e^{-K} f, f) \leq (e^{-K}|f|, |f|) \leq \|e^{-K}\|\|f\|^2 = e^{-\lambda}\|f\|^2,$$

implying $(e^{-K} f, f) = (e^{-K}|f|, |f|)$. This further implies

$$(e^{-K} f^+, f^-) = -(e^{-K} f^-, f^+). \tag{4.3.8}$$

Since $f^+, f^-, e^{-K} f^+$, and $e^{-K} f^-$ are all nonnegative, both sides of (4.3.8) must vanish, and this can only happen if either f^+ or f^- vanishes almost everywhere as e^{-K} improves positivity. Hence f has a definite sign and, for instance, $f \geq 0$ can be assumed. However, since $f = e^{-K} e^{+\lambda} f$, it follows that f is in fact strictly positive.

The above reasoning applies to any eigenfunction corresponding to the eigenvalue λ. This shows that there cannot be two linearly independent such eigenfunctions since then they could be chosen orthogonal which is impossible as they both would have to be strictly positive. Hence f is unique. □

Theorem 4.124 (Positivity improving operators). *Suppose that $V_+ \in L^1_{\mathrm{loc}}(\mathbb{R}^d)$ and V_- is relatively form bounded with respect to $-\frac{1}{2}\Delta$ with a relative bound strictly smaller than one. Let $H = \frac{1}{2}\Delta \dotplus V_+ \dot{-} V_-$. Then $\{e^{-tH} : t \geq 0\}$ is positivity improving. In particular, for $H = -\frac{1}{2}\Delta + V$ with a Kato-decomposable V, the semigroup $\{e^{-tH} : t \geq 0\}$ is positivity improving.*

Proof. Since $\{e^{t(\Delta/2)} : t \geq 0\}$ is positivity improving, $\int_{\mathbb{R}^d} \mathbb{E}^x[f(B_0)g(B_t)]dx > 0$ for every non-negative f and g. The measure of

$$\mathscr{M} = \left\{(x, \omega) \in \mathbb{R}^d \times \mathscr{X} \mid f(x)g(B_t(\omega) + x) > 0\right\}$$

is positive. For $V_+ \in L^1_{\rm loc}(\mathbb{R}^d)$, it is seen in Lemma 4.95 that $e^{-\int_0^t V(B_s+x)ds} > 0$ for almost every $(x,\omega) \in \mathbb{R}^d \times \mathscr{X}$. We have

$$(f, e^{-tH}g) = \int_{\mathbb{R}^d} \mathbb{E}[f(x)g(B_t+x)e^{-\int_0^t V(B_s+x)ds}]dx$$
$$\geq \int_{\mathscr{M}} f(x)g(B_t+x)e^{-\int_0^t V(B_s+x)ds}dxd\mathcal{W} > 0,$$

and e^{-tH} is positivity improving for all $t \geq 0$. □

Theorem 4.125 (Uniqueness of ground state). *Suppose that $V_+ \in L^1_{\rm loc}(\mathbb{R}^d)$ and V_- is relatively form bounded with respect to $-\frac{1}{2}\Delta$ with a relative bound strictly smaller than one. Let $H = \frac{1}{2}\Delta \dot{+} V_+ \dot{-} V_-$. Then the ground state of H is unique whenever it exists. In particular, for $H = -\frac{1}{2}\Delta + V$ with a Kato-decomposable V, the ground state of H is unique whenever it exists.*

Proof. By Theorem 4.124, e^{-tH} is positivity improving. Uniqueness follows by the Perron–Frobenius theorem. □

4.3.3 Degenerate ground state and Klauder phenomenon

Now we turn to presenting an example where e^{-tH} is not positivity improving. To break positivity down, we construct a potential V such that

$$\int_{\mathbb{R}^d} \mathcal{W}^x\left(e^{-\int_0^t V(B_s)ds} = 0\right)dx > 0.$$

This requires the potential V to be singular in a suitable form. Let

$$\mathscr{M}_1 = \left\{x = (x_1,x_2,x_3) \in \mathbb{R}^3 \mid x_3 > 0\right\},$$
$$\mathscr{M}_2 = \left\{x = (x_1,x_2,x_3) \in \mathbb{R}^3 \mid x_3 < 0\right\},$$
$$\mathscr{M} = \mathscr{M}_1 \cup \mathscr{M}_2.$$

Define

$$H(\nu) = -\frac{1}{2}\Delta + \frac{1}{2}|x|^2 + \frac{\nu}{|x-\partial\mathscr{M}|^3}. \tag{4.3.9}$$

Here $\partial\mathscr{M}$ is the boundary of $\mathscr{M}$ and $|x-\partial\mathscr{M}|$ denotes the distance between x and $\partial\mathscr{M}$. Since the potential in (4.3.9) is singular, the self-adjointness of $H(\nu)$, $\nu \neq 0$, is unclear. To see this, we use the following general result.

Proposition 4.126. *Let D be an open set in $\mathbb{R}^d$ and $V \geq 0$ on D. Suppose that there is a uniform Lipschitz continuous function h on every compact subset of D such that*

(1) $\sum_{i=1}^{d}(\partial_i h)^2 \le e^{2h}$ *on* D;
(2) $\lim_{x\to\partial D} h(x) = \infty$ *(if* D *is not bounded,* ∞ *is regarded as a point of* ∂D*);*
(3) *there exists* $\delta > 0$ *such that* $(f, (-\Delta + V)f) \ge (1+\delta)(f, e^{2h}f)$ *for* $f \in C_0^\infty(D)$.

Then $-\Delta + V$ *is essentially self-adjoint on* $C_0^\infty(D)$.

Lemma 4.127. *Operator* H *is essentially self-adjoint on* $C_0^\infty(\mathscr{M})$*, for all* $\nu \in \mathbb{R}$.

Proof. Let

$$V_\varepsilon(x) = \begin{cases} \frac{\nu}{|x-\partial\mathscr{M}|^3} + \frac{1}{2}|x|^2, & |x-\partial\mathscr{M}| < \varepsilon, \\ \frac{\nu}{\varepsilon^3} + \frac{1}{2}|x|^2, & \text{otherwise.} \end{cases}$$

Since $V = V_\varepsilon + T$, where T is a bounded operator, it suffices to show that $-\Delta + 2V_\varepsilon$ is essentially self-adjoint on $C_0^\infty(\mathscr{M})$. Define $h = \log V_\varepsilon^{1/2}$; h can be directly shown to satisfy (2) and (3) in Proposition 4.126. We check (1), which is equivalent to having $0 \le 4V_\varepsilon^3 - \sum_{i=1}^3(\partial_i V_\varepsilon)^2$. For $|x-\partial\mathscr{M}| \ge \varepsilon$, it is immediate to see that $4V_\varepsilon^3 - \sum_{i=1}^3(\partial_i V_\varepsilon)^2 > 0$ for sufficiently small ε. We have for $|x-\partial\mathscr{M}| < \varepsilon$,

$$4V_\varepsilon^3 - \sum_{i=1}^3(\partial_i V_\varepsilon)^2 \ge 4\left(\frac{\nu}{|x-\partial\mathscr{M}|^3} + \frac{1}{2}|x|^2\right)^3 - \frac{1}{2}\left(\frac{9\nu^2}{|x-\partial\mathscr{M}|^8} + |x|^2\right).$$

Take a sufficiently small ε. The right-hand side is positive, and the lemma follows. □

Since $H(0) \le H(\nu)$, we have $\mu_n(H(0)) \le \mu_n(H(\nu))$, where

$$\mu_n(K) = \sup_{\phi_1,\dots,\phi_{n-1}} \inf_{\substack{\psi\in D(K),\|\psi\|=1 \\ \psi\in[\phi_1,\dots,\phi_{n-1}]^\perp}} (\psi, K\psi).$$

Using the fact that $\mu_n(H(0)) = n + \frac{1}{2}$, it is seen that $\mu_n(H(\nu)) \to \infty$ as $n \to \infty$. In particular, by the min–max principle, $H(\nu)$ has a purely discrete spectrum. Further, by Theorem 4.88, we have

$$(f, e^{-tH(\nu)}g) = \int_{\mathbb{R}^3} \mathbb{E}^x\left[\overline{f(x)}g(B_t)\exp\left(-\int_0^t \frac{\nu}{|B_s-\partial\mathscr{M}|^3} + |B_s|^2 ds\right)\right]dx. \tag{4.3.10}$$

Lemma 4.127 gives $H(\nu) = H_D(\nu)$, where $H_D(\nu)$ denotes $H(\nu)$ with Dirichlet boundary condition. We see that

$$(f, e^{-tH_D(\nu)}g) = \int_{M_1\cup M_2} \overline{f(x)}g(B_t)\exp\left(-\int_0^t \frac{\nu}{|B_s-\partial\mathscr{M}|^3} + |B_s|^2 ds\right)d\mathcal{W}^x dx, \tag{4.3.11}$$

with $M_j = \{(x,\omega) \in \mathbb{R}^3 \times \mathscr{X} \mid x + B_s \in \mathscr{M}_j, s \in [0,t]\}$ denoting the set of paths starting at x and confined to $\mathscr{M}_j$. Comparing (4.3.10) with (4.3.11), we see that

$$\int_0^t \frac{ds}{|B_s(\omega) - \partial\mathcal{M}|^3} = \infty$$

for $\omega \in \mathbb{R}^3 \times \mathscr{X} \setminus (M_1 \cup M_2)$. Thus such paths do not contribute to (4.3.10) at all. Hence $(f, e^{-tH(v)}g) = 0$ for f and g for which $\operatorname{supp} f \subset \mathcal{M}_1$ and $\operatorname{supp} g \subset \mathcal{M}_2$. Thus $e^{-tH(v)}$ is not positivity improving. Moreover, we have

$$\lim_{v\to 0}(f, e^{-tH(v)}g) = \int_{M_1\cup M_2} \overline{f(x)}g(B_t)d\mathcal{W}^x dx \neq (f, e^{-tH(0)}g).$$

This vestigial effect of singular potentials is called *Klauder phenomenon* and shows that once a singular potential is turned on, its effect continues to be felt after the potential is turned off.

We have seen that for $e^{-tH(v)}$ positivity improving breaks down. This opens the possibility to construct a nonunique ground state.

Corollary 4.128. *The ground state of $H(v)$ is two-fold degenerate.*

Proof. It can be seen that $H(v)$ can be reduced by $L^2(\mathcal{M}_j)$. With $H_j(v) = H(v)\lceil_{L^2(\mathcal{M}_j)}$ we have $H(v) = H_1(v) \oplus H_2(v)$. Let $E_j = \inf \operatorname{Spec}(H_j(v))$; then $E_1 = E_2$ by symmetry. Thus the corollary follows. □

By replacing the potential $v|x - \partial\mathcal{M}|^{-3}$ with $v|x - \partial\mathcal{M}|^{-2-\varepsilon}$, $\varepsilon > 0$, we can show the same results as in Lemma 4.127 and Corollary 4.128.

4.3.4 Existence and nonexistence of ground states

In this section we present criteria of existence and nonexistence of ground states of Schrödinger operators with Kato-decomposable potentials by using the Feynman–Kac formula. Also, we show a representation formula for ground states in case they exist.

Let V be a Kato-decomposable potential, consider

$$H = -\frac{1}{2}\Delta + V,$$

and denote $\inf \operatorname{Spec}(H) = E_0$.

Lemma 4.129. *Let $\rho \in L^2(\mathbb{R}^d)$ such that* $\operatorname{ess\,inf}_{x\in K} \rho(x) > 0$ *for every compact set $K \subset \mathbb{R}^d$, and $E(\cdot)$ be the spectral measure of H associated with ρ. Then* $\inf \operatorname{supp} E = E_0$.

Proof. Let $\mathcal{O} = \{f \in L^2(\mathbb{R}^d) \mid \operatorname{supp} f \subset \bigcup_{N>0} B_N\}$, where $B_N = \{x \in \mathbb{R}^d \mid |x| \leq N\}$. Since e^{-tH} is positivity improving, we have for $f \in \mathcal{O}$ that

$$(f, e^{-tH}f) \leq C^2(\rho, e^{-tH}\rho),$$

where $C = \frac{\operatorname{ess\,sup}_{x\in\mathbb{R}^d} |f(x)|}{\operatorname{ess\,inf}_{x\in\operatorname{supp}|f|} |\rho(x)|}$. Hence we have

$$\inf\operatorname{supp} E = -\lim_{t\to\infty}\frac{1}{t}\log(\rho, e^{-tH}\rho) \le -\lim_{t\to\infty}\frac{1}{t}\log\frac{(f, e^{-tH}f)}{C^2} = \inf\operatorname{supp} E_f,$$

where E_f is the spectral measure of H associated with f. Thus we obtain

$$\inf\operatorname{supp} E \le \inf\operatorname{supp} E_f, \quad f\in\mathcal{O}.$$

Since $E_0 = \inf_{f\in\mathcal{O}}\inf\operatorname{supp} E_f$, we obtain $\inf\operatorname{supp} E \ge E_0 \ge \inf\operatorname{supp} E$. □

For $\rho\in L^2(\mathbb{R}^d)$, define

$$\gamma_\rho(t) = \frac{(\rho, e^{-tH}\rho)^2}{\|e^{-tH}\rho\|^2}.$$

Formally, $\gamma_\rho(t)\to|(\rho,\varphi_0)|^2 > 0$ as $t\to\infty$ if ρ is a strictly positive function and a ground state φ_0 exists. Using this quantity, we obtain criteria on the existence and absence of a ground state.

Theorem 4.130 (Criterion for existence of ground state). *Let $\rho\in L^2(\mathbb{R}^d)$ with the property that $\operatorname{ess\,inf}_{x\in K}\rho(x) > 0$ for every compact set $K\subset\mathbb{R}^d$, and denote $\lim_{t\to\infty}\gamma_\rho(t) = a$. If $a > 0$, then H has a ground state.*

Proof. Since $a > 0$, for sufficiently large t it follows that $\sqrt{\gamma_\rho(t)} > b$ with some $b > 0$. Let $E(\cdot)$ be the spectral measure of H associated with ρ, and write $\delta = \inf\operatorname{supp} E$. Then by Lemma 4.129, it follows that $\delta = \inf\operatorname{Spec}(H) = E_0$. By the definition of δ, we see that $E([\delta,\delta+\varepsilon)) \ne 0$ for every $\varepsilon > 0$, and $E([\delta,\delta+\varepsilon))$ is strongly convergent to $E(\{\delta\})$ as $\varepsilon\downarrow 0$. We have

$$\sqrt{\gamma_\rho(t)} = \frac{\int_{[\delta,\delta+\varepsilon)} e^{-(\lambda-\delta)t}dE + \int_{[\delta+\varepsilon,\infty)} e^{-(\lambda-\delta)t}dE}{\left(\int_{[\delta,\infty)} e^{-2(\lambda-\delta)t}dE\right)^{1/2}}.$$

By the Schwarz inequality, we furthermore have

$$\frac{\int_{[\delta,\delta+\varepsilon)} e^{-(\lambda-\delta)t}dE}{\left(\int_{[\delta,\infty)} e^{-2(\lambda-\delta)t}dE\right)^{1/2}} \le E([\delta,\delta+\varepsilon))^{1/2}\frac{\left(\int_{[\delta,\delta+\varepsilon)} e^{-2(\lambda-\delta)t}dE\right)^{1/2}}{\left(\int_{[\delta,\infty)} e^{-2(\lambda-\delta)t}dE\right)^{1/2}} \le E([\delta,\delta+\varepsilon))^{1/2}.$$

Next consider the ratio on the left-hand side above. Since the denominator and the numerator can be estimated as

$$\int_{[\delta+\varepsilon,\infty)} e^{-(\lambda-\delta)t}dE \le e^{-\varepsilon t},$$

$$\int_{[\delta,\infty)} e^{-2(\lambda-\delta)t}dE \ge \int_{[\delta,\delta+\varepsilon/2)} e^{-2(\lambda-\delta)t}dE \ge e^{-\varepsilon t}E([\delta,\delta+\varepsilon/2)),$$

we see that

$$\frac{\int_{[\delta+\varepsilon,\infty)} e^{-(\lambda-\delta)t} dE}{\left(\int_{[\delta,\infty)} e^{-2(\lambda-\delta)t} dE\right)^{1/2}} \leq \frac{e^{-\varepsilon t}}{e^{-\varepsilon t/2} E([\delta, \delta+\varepsilon/2))^{1/2}} \leq \frac{e^{-\varepsilon t/2}}{E([\delta, \delta+\varepsilon/2))^{1/2}}.$$

It then follows that

$$\sqrt{\gamma_\rho(t)} \leq E([\delta, \delta+\varepsilon))^{1/2} + \frac{e^{-\varepsilon t/2}}{E([\delta, \delta+\varepsilon/2))^{1/2}}.$$

Taking $t \to \infty$ on both sides above, we obtain $b \leq E([\delta, \delta+\varepsilon))$ for every $\varepsilon > 0$, which tends to $b \leq E(\{\delta\})$ in the limit $\varepsilon \downarrow 0$. Hence $0 \neq E(\{\delta\}) = E(\{E_0\})$ and thus a ground state of H exists. □

Theorem 4.130 gives a sufficient condition for H to have a ground state. A similar argument can also be used to show the absence of ground states of H.

Theorem 4.131 (Criterion for absence of ground state). *Let $\rho \in L^2(\mathbb{R}^d)$ be a nonnegative function, not identically zero. If $a = \lim_{t\to\infty} \gamma_\rho(t) = 0$, then H has no ground state.*

Proof. We may suppose that $\inf \operatorname{Spec}(H) = 0$, so that $\lim_{t\to\infty} e^{-tH} = 1_{\{0\}}(H)$ in the strong sense. If 0 is an eigenvalue of H, i. e., H has a ground state φ_0, then $a = (\rho, \varphi_0) > 0$ since φ_0 is strictly positive by the fact that e^{-tH} is positivity improving. Thus $a = 0$ implies that H has no ground state. □

Now we assume that a ground state φ_0 of H exists. Since it is strictly positive, $(f, \varphi_0) \neq 0$ for $0 < f \in L^2(\mathbb{R}^d)$. From this, it follows that

$$\text{s-}\lim_{t\to\infty} \frac{e^{-tH} f}{\|e^{-tH} f\|} = \varphi_0.$$

This expression can be rewritten in terms of Feynman–Kac formula like

$$\varphi_0(x) = \text{s-}\lim_{t\to\infty} \frac{\mathbb{E}^x[e^{-\int_0^t V(B_s)ds} f(B_t)]}{\left(\int_{\mathbb{R}^d} |\mathbb{E}^x[e^{-\int_0^t V(B_s)ds} f(B_t)]|^2 dx\right)^{1/2}}. \tag{4.3.12}$$

Next we show that under an appropriate condition, formula (4.3.12) can be extended also to $f \notin L^2(\mathbb{R}^d)$ and we can choose, for instance, $f = 1$.

Theorem 4.132 (Ground state representation formula). *Suppose that a ground state φ_0 of H exists and assume that the function $\mathbb{R}^d \ni x \mapsto \mathbb{E}^x[e^{-\int_0^t V(B_s)ds}] \in L^2(\mathbb{R}^d)$ and*

$$\lim_{R\to\infty} \lim_{t\to\infty} \mathbb{E}^x[e^{-\int_0^t (V(B_s)-E_0)ds} 1_{|B_t|\geq R}] = 0 \tag{4.3.13}$$

strongly in $L^2(\mathbb{R}^d)$. Then

$$\varphi_0(x) = s\text{-}\lim_{t\to\infty} \frac{\mathbb{E}^x[e^{-\int_0^t V(B_s)ds}]}{\left(\int_{\mathbb{R}^d} |\mathbb{E}^x[e^{-\int_0^t V(B_s)ds}]|^2 dx\right)^{1/2}}. \tag{4.3.14}$$

Proof. Let 1_N be the indicator function on $[-N, N]^d$ and denote

$$a_{t,N}(x) = \mathbb{E}^x[e^{-\int_0^t V(B_s)ds} 1_N] = (e^{-tH_{\mathrm{p}}} 1_N)(x),$$
$$a_t(x) = \mathbb{E}^x[e^{-\int_0^t V(B_s)ds}].$$

Thus

$$\left\| \frac{a_t}{\|a_t\|} - \varphi_0 \right\| \le \left\| \frac{a_t}{\|a_t\|} - \frac{a_{t,N}}{\|a_{t,N}\|} \right\| + \left\| \frac{a_{t,N}}{\|a_{t,N}\|} - \varphi_0 \right\|. \tag{4.3.15}$$

Since $1_N \in L^2(\mathbb{R}^d)$ and $(\varphi_0, 1_N) \neq 0$, it is seen that

$$\lim_{t\to\infty} \left\| \frac{a_{t,N}}{\|a_{t,N}\|} - \varphi_0 \right\| = 0.$$

We also have

$$\left\| \frac{a_t}{\|a_t\|} - \frac{a_{t,N}}{\|a_{t,N}\|} \right\| \le 2 \frac{\|a_t - a_{t,N}\|}{\|a_{t,N}\|}.$$

Note that

$$\|a_{t,N}\|^2 = \int_{[E_{\mathrm{p}},\infty)} e^{-2\lambda t} d\|E(\lambda)1_N\|^2 \ge e^{-2E_0 t}\|E(\{E_0\})1_N\|^2 = e^{-2E_0 t}|(1_N, \varphi_0)|^2,$$

where $E(\cdot)$ denotes the spectral measure for H. By using (4.3.13), it follows that

$$\lim_{N\to\infty}\lim_{t\to\infty} \frac{\|a_t - a_{t,N}\|}{\|a_{t,N}\|} \le \lim_{N\to\infty}\lim_{t\to\infty} \frac{\left\|\mathbb{E}^x[e^{-\int_0^t (V(B_s)-E_0)} 1_{|B_t|\ge N}]\right\|}{(1_N, \varphi_0)} = 0.$$

By (4.3.15), we obtain for every $\varepsilon > 0$ that

$$\lim_{t\to\infty} \left\| \frac{a_t}{\|a_t\|} - \varphi_0 \right\| \le \lim_{t\to\infty} \left\| \frac{a_t}{\|a_t\|} - \frac{a_{t,N}}{\|a_{t,N}\|} \right\| < \varepsilon \tag{4.3.16}$$

for every $N > N_0$ with some N_0, as required. □

Remark 4.133. The representation formula in Theorem 4.132 gives a probabilistic interpretation to ground states. Recall the local time process $(L_t^x)_{t\ge 0}$ at position $x \in \mathbb{R}^d$ as defined in (3.5.10) for Brownian motion $(B_t)_{t\geq 0}$. (We note that when local time does not exist, this observation can be applied to the occupation measure.) Then we have

$$\int_0^t V(B_s)ds = \int_{\mathbb{R}^d} V(y)L_t^y dy.$$

This shows that one can actually think of V as being a test function for local time and

$$\varphi_0(x) = \text{s-}\lim_{t\to\infty} \frac{\mathbb{E}^x[e^{-\int_{\mathbb{R}^d} V(y)L_t^y dy}]}{\|\mathbb{E}^x[e^{-\int_{\mathbb{R}^d} V(y)L_t^y dy}]\|_2}$$

holds, i. e., V is a weight function giving the proportion of time paths will spend in specific locations in space under the "potential landscape" determined by V. A ground state results then as a long-time phenomenon, during which paths would "scan" the potential landscape, and it forms under the condition that this landscape allows paths to spend sufficient times in local regions. This aspect will be further discussed in the next section.

For illustration, we show how to use the above formula to determine the ground state of the harmonic oscillator explicitly. The following lemma is a special case of Lemmas 5.26–5.27 below, where also the proofs are presented.

Lemma 4.134. *For every $x, y \in \mathbb{R}^d$ and $t > 0$, we have*

$$\mathbb{E}^x\left[e^{-\frac{1}{2}\int_0^t |B_s|^2 ds}\right] = \left(\frac{1}{\cosh t}\right)^{d/2} \exp\left(-\frac{|x|^2}{2\coth t}\right)$$

and

$$\mathbb{E}^x[e^{-\frac{1}{2}\int_0^t |B_s|^2 ds}|B_t = y] = \left(\frac{t}{\sinh t}\right)^{d/2} \exp\left(-\frac{|y - xe^{-t}|^2}{1 - e^{-2t}} + \frac{|y|^2 - |x|^2}{2} + \frac{|y|^2}{2t}\right).$$

Example 4.135. Choose $V(x) = \frac{|x|^2}{2}$. Then (4.3.14) holds, giving $\varphi_0(x) = \frac{1}{\pi^{d/4}} e^{-\frac{|x|^2}{2}}$.

Proof. Let $H_{\text{osc}} = -\frac{1}{2}\Delta + \frac{1}{2}|x|^2$ corresponding to the d-dimensional harmonic oscillator. A standard calculation shows that

$$\inf \operatorname{Spec}(H_{\text{osc}}) = E_0 = \frac{d}{2} \quad \text{and} \quad \varphi_0(x) = \frac{1}{\pi^{d/4}} e^{-\frac{|x|^2}{2}}$$

satisfy

$$H_{\text{osc}}\varphi_0 = \frac{d}{2}\varphi_0.$$

Then by applying Lemma 4.134, we obtain

$$\frac{\mathbb{E}^x\left[e^{-\frac{1}{2}\int_0^t |B_s|^2 ds}\right]}{\left(\int_{\mathbb{R}^d}\left|\mathbb{E}^x\left[e^{-\frac{1}{2}\int_0^t |B_s|^2 ds}\right]\right|^2 dx\right)^{1/2}} = \frac{e^{-\frac{|x|^2}{2}\frac{1}{\coth t}}}{\left(\int_{\mathbb{R}^d} e^{-|x|^2\frac{1}{\coth t}} dx\right)^{1/2}} \to \frac{1}{\pi^{d/4}} e^{-\frac{|x|^2}{2}}$$

as $t \to \infty$. □

Remark 4.136. We note that while in Example 4.135 we could do without condition (4.3.13), it holds also for this case. Since $E_0 = d/2$, by making use of Lemma 4.134 again, we see that

$$e^{E_0 t}\mathbb{E}^x[e^{-\frac{1}{2}\int_0^t |B_s|^2 ds}1_{|B_t|>R}] = e^{dt/2}\int_{|y|>R} \Pi_t(y)\mathbb{E}^x[e^{-\frac{1}{2}\int_0^t |B_s|^2 ds}|B_t = y]dy$$

$$= \frac{1}{(2\pi)^{d/2}}\int_{|y|>R}\left(\frac{e^t}{\sinh t}\right)^{d/2}\exp\left(-\frac{|y - xe^{-t}|^2}{1 - e^{-2t}} + \frac{|y|^2 - |x|^2}{2}\right)dy.$$

Thus

$$\lim_{R\to\infty}\lim_{t\to\infty}\left\|e^{E_0 t}\mathbb{E}^x[e^{-\frac{1}{2}\int_0^t |B_s|^2 ds}1_{|B_t|>R}]\right\|^2 = \lim_{R\to\infty}\left|\int_{|y|>R} e^{-\frac{|y|^2}{2}}dy\right|^2 = 0$$

and (4.3.13) is satisfied.

4.3.5 Sojourn times and existence of bound states

We know from Theorem 4.72 that confining potentials have discrete spectra, and thus the related Schrödinger operators have a countable set of bound states. The existence of bound states for Schrödinger operators with decaying potentials is, however, not similarly guaranteed and this depends on further properties of the potential. In this section we will see that bound states exist also for decaying potentials as soon as they allow the related processes to spend sufficiently long times in given regions of space.

For the discussion, it will be sufficient to consider compactly supported potentials, and we start by the Schrödinger operator with a one-dimensional *potential well*

$$H_v = -\frac{1}{2}\Delta - v1_{\{|x|\le a\}}(x) \tag{4.3.17}$$

on $L^2(\mathbb{R})$, with the well having boundaries at $\pm a$, $a > 0$, and depth $v > 0$. The Feynman–Kac formula gives

$$(f, e^{-tH_v}g) = \int_{\mathbb{R}} \overline{f(x)}\mathbb{E}^x[e^{vU_t(a)}g(B_t)]dx \tag{4.3.18}$$

for $f, g \in L^2(\mathbb{R})$, where

$$U_t(a) = \int_0^t 1_{\{|B_s| \le a\}} ds$$

is the occupation time of a one-dimensional Brownian motion in the potential well (see also (3.5.8)). From (4.3.18), it follows that

$$(f, e^{-tH_v} g) = \int_{\mathbb{R}} \int_{\mathbb{R}} \overline{f(x)} g(y) \Pi_t(x-y) \mathbb{E}^{x,y}_{[0,t]} [e^{vU_t(a)}] dx dy, \tag{4.3.19}$$

where $\mathbb{E}^{x,y}_{[0,t]}$ denotes expectation with respect a Brownian bridge measure. Thus the integral kernel of e^{-tH_v} is given by

$$e^{-tH_v}(x,y) = \Pi_t(x-y) \mathbb{E}^{x,y}_{[0,t]} \left[e^{vU_t(a)} \right]. \tag{4.3.20}$$

We recall without proof the following strong ergodic property of one- and two-dimensional Brownian motion, known as the *Kallianpur–Robbins laws.*

Proposition 4.137 (Kallianpur–Robbins laws). *Let $W \in L^1(\mathbb{R}^d)$ and suppose that $\bar{W} = \int_{\mathbb{R}^d} W(x)dx > 0$. Then for every $u > 0$, we have*
$(d = 1)$

$$\lim_{t\to\infty} \mathcal{W}^x \left(\frac{\int_0^t W(B_s) ds}{\bar{W}\sqrt{t}} \le u \right) = \sqrt{\frac{2}{\pi}} \int_0^u e^{-z^2/2} dz; \tag{4.3.21}$$

$(d = 2)$

$$\lim_{t\to\infty} \mathcal{W}^x \left(\frac{\int_0^t W(B_s) ds}{\bar{W} \log t} \le \log u \right) = 1 - e^{-u}. \tag{4.3.22}$$

Here the limits do not depend on $x \in \mathbb{R}^d$.

Consider $d = 1$ and $W(x) = 1_{\{|x| \le a\}}(x)$. From the Kallianpur–Robbins laws, we obtain

$$\lim_{t\to\infty} \mathcal{W}^x \left(\frac{U_t(a)}{2a\sqrt{t}} \le u \right) = \sqrt{\frac{2}{\pi}} \int_0^u e^{-z^2/2} dz.$$

In particular, we have

$$\lim_{t\to\infty} \mathcal{W}^x \left(U_t(a) \ge 2au\sqrt{t} \right) \ge \sqrt{\frac{2}{\pi}} \int_u^\infty e^{-z^2/2} dz > 0. \tag{4.3.23}$$

Using (4.3.23), we can give a path integration proof of the following theorem.

Theorem 4.138. *Let $d = 1$. For every $v > 0$, there exists a bound state of H_v.*

Proof. Consider the set $G = \{|B_t| > bt^a\}$, where $b > 0$ and $a > 0$. Note that

$$\mathcal{W}^x(G) = \sqrt{\frac{2}{\pi t}} \int_{bt^a - x}^{\infty} e^{-y^2/2t} dy \le \sqrt{\frac{2}{\pi t}} \frac{e^{-\frac{(bt^a - x)^2}{2t}}}{bt^a - x} \xrightarrow{t\to\infty} 0, \tag{4.3.24}$$

whenever $a > \frac{1}{2}$. Hence we have with such a

$$\begin{aligned}
&\lim_{t\to\infty} \mathcal{W}^x(U_t(a) \ge 2au\sqrt{t}) \\
&\quad = \lim_{t\to\infty} \mathcal{W}^x(U_t(a) \ge 2au\sqrt{t}\,|G^c)\mathcal{W}^x(G^c) + \lim_{t\to\infty} \mathcal{W}^x(U_t(a) \ge 2au\sqrt{t}\,|G)\mathcal{W}^x(G) \\
&\quad = \lim_{t\to\infty} \mathcal{W}^x(U_t(a) \ge 2au\sqrt{t}\,|G^c)
\end{aligned} \tag{4.3.25}$$

due to (4.3.24). Consider the set $E = \{B_0 = x\} \cup \{U_t(a) \ge 2au\sqrt{t}\}$, and choose the functions $f(x) = 1_{\{|x|\le b\}}$ and $g(y) = 1_{\{|y|\le bt^a\}}$ with some $b > 0$. In particular, on E paths satisfy $e^{-2auv\sqrt{t} - \int_0^t V(B_s)ds} \ge 1$. Since $\frac{1}{2b}\int_{\mathbb{R}} f(x)dx = 1$ and the probabilities above do not depend on x, we furthermore have

$$\begin{aligned}
&\lim_{t\to\infty} \mathcal{W}^x\,(U_t(a) \ge 2au\sqrt{t}) \\
&\quad = \frac{1}{2b}\int_{\mathbb{R}} f(x) \lim_{t\to\infty} \int_{\mathbb{R}} g(y)\mathcal{W}^x\,(U_t(a) \ge 2au\sqrt{t}\,|\,B_t \in dy)\,dx \\
&\quad \le \frac{1}{2b}\int_{\mathbb{R}} \lim_{t\to\infty} \int_{\mathbb{R}} f(x)g(y)\mathbb{E}^{x,y}_{[0,t]}[1]dxdy \\
&\quad \le \frac{1}{2b}\int_{\mathbb{R}} \lim_{t\to\infty} e^{-2auv\sqrt{t}} \int_{\mathbb{R}} f(x)g(y)\mathbb{E}^{x,y}_{[0,t]}[e^{-\int_0^t V(B_s)ds}]dxdy \\
&\quad = \frac{1}{2b}\lim_{t\to\infty} \int_{\mathbb{R}} e^{-2auv\sqrt{t}} f(x)dx \int_{\mathbb{R}} e^{-tH_v}(x,y)g(y)dy,
\end{aligned} \tag{4.3.26}$$

where the last step is obtained by using (4.3.20).

Next suppose, to the contrary, that H_v is a nonnegative operator. Then e^{-tH_v} is contraction and hence $|(f, e^{-tH_v}g)| \le \|f\|\,\|g\|$ follows. We note that $\mathrm{Spec}_{\mathrm{ess}}(H_v) = [0,\infty)$ by Theorem 4.74. The dominated convergence theorem implies by the expression above and (4.3.23) that

$$\begin{aligned}
0 &< \lim_{t\to\infty} \mathcal{W}^x\,(U_t(a) \ge 2au\sqrt{t}) \le \frac{1}{2b}\lim_{t\to\infty} e^{-2auv\sqrt{t}}(f, e^{-tH_v}g) \\
&\le \frac{1}{2b}\lim_{t\to\infty} e^{-2auv\sqrt{t}}\|f\|\|g\| = \lim_{t\to\infty} e^{-2auv\sqrt{t}}t^{a/2} = 0
\end{aligned}$$

for every $v > 0$. This is a contradiction showing that for every $v > 0$ there is a negative discrete spectrum of H_v and a bound state exists. □

This argument can be extended to the two-dimensional case using the second statement in the Kallianpur–Robbins laws. Consider now the Schrödinger operator with a two-dimensional potential well

$$H_\nu = -\frac{1}{2}\Delta - \nu 1_{\{|x|\le a\}}, \tag{4.3.27}$$

where we assume $a, \nu > 0$ again.

Theorem 4.139. *Let $d = 2$. For every $\nu > 0$, there exists a bound state of H_ν.*

Proof. Since the potential is now rotationally symmetric, we use polar coordinates. The operator (4.3.27) is unitary equivalent to

$$\tilde{H}_\nu = \left(-\frac{1}{2}\frac{\partial^2}{\partial r^2} - \frac{1}{2r}\frac{\partial}{\partial r} - \nu 1_{\{r\le a\}}\right)\otimes 1 - \frac{1}{2r^2}\otimes\frac{\partial^2}{\partial\theta^2}$$

acting in $L^2(\mathbb{R}^2) \cong L^2([0,\infty)) \otimes L^2(S_1)$. Here $\frac{\partial^2}{\partial\theta^2}$ is the Laplace–Beltrami operator on $L^2(S_1)$. It is known that $\operatorname{Spec}(\frac{\partial^2}{\partial\theta^2}) = \{-l^2\}_{l=0}^\infty$, each $-l^2$ is a two-fold degenerate eigenvalue, and $e^{\pm i\theta l}$ are eigenfunctions associated with $-l^2$. We have

$$\tilde{H}_\nu = \bigoplus_{l=0}^{\infty}\left(H_{\mathrm{r},\nu} + \frac{l^2}{2r^2}\right)$$

under $L^2(\mathbb{R}^2) \cong \bigoplus_{l=0}^\infty L^2([0,\infty))$, where

$$H_{\mathrm{r},\nu} = -\frac{1}{2}\frac{\partial^2}{\partial r^2} - \frac{1}{2r}\frac{\partial}{\partial r} - \nu 1_{\{r\le a\}}.$$

Thus to prove the theorem, it is sufficient to show that the Hamiltonian $H_{\mathrm{r},\nu}$ has a bound state in $L^2([0,\infty))$. Note that

$$H^0_{\mathrm{r},\nu} = \frac{1}{2}\frac{\partial^2}{\partial r^2} + \frac{1}{2r}\frac{\partial}{\partial r}$$

is the generator of a two-dimensional Bessel process $(R_t)_{t\ge 0}$ on a probability space $(\Omega, \mathscr{F}, P^y)$, starting from $y \ge 0$. It follows that for $f, g \in L^2([0,\infty))$,

$$(f, e^{-tH_{\mathrm{r},\nu}}g) = \int_{[0,\infty)} \overline{f(y)}\mathbb{E}_P^y[e^{\nu U_t(a)}g(R_t)]dy.$$

From the two-dimensional case, it follows from the Kallianpur–Robbins laws that

$$\lim_{t\to\infty} P^y\left(\frac{U_t(a)}{\pi a^2 \log t} \le u\right) = 1 - e^{-u}, \quad u \ge 0,\ y \in \mathbb{R}^2,$$

where

$$U_t(a) = \int_0^t 1_{\{R_s \le a\}}(R_s)ds.$$

Consider the set $G = \{R_t > bt^\alpha\}$. Using the probability transition kernel of $(R_t)_{t\ge 0}$, i. e.,

$$\frac{r}{t}e^{-\frac{y^2+r^2}{2t}} I_0\left(\frac{yr}{t}\right),$$

where $I_0(x) = \sum_{k=0}^\infty \frac{(x/2)^{2k}}{k!\Gamma(k+1)}$ is the modified Bessel function of the first kind, we have for all $\alpha > \frac{1}{2}$,

$$\lim_{t\to\infty} P^y(R_t \ge bt^\alpha) = \lim_{t\to\infty} \frac{1}{t}\int_{bt^\alpha - y}^\infty re^{-\frac{y^2+r^2}{2t}} I_0\left(\frac{yr}{t}\right)dr = 0.$$

Hence by taking conditional expectation on G and $f(y) = 1_{\{y\le b\}}$, $g(r) = 1_{\{r\le bt^\alpha\}}$, we have similarly as in (4.3.26) that

$$\begin{aligned}
\lim_{t\to\infty} P^y\left(U_t(a) \ge \pi a^2 u \log t\right) &= \lim_{t\to\infty} P^y\left(U_t(a) \ge \pi a^2 u \log t \,|\, G^c\right) \\
&\le \frac{1}{b}\int_0^\infty \lim_{t\to\infty} \int_0^\infty f(y)g(r)e^{-\pi a^2 uv\log t} \int_{\mathscr{X}_D} e^{-\int_0^t V(R_s)ds} dP^{y,r}_{[0,t]} dydr \\
&= \frac{1}{b}\lim_{t\to\infty} \int_0^\infty t^{-\pi a^2 uv} f(y)dy \int_0^\infty e^{-tH_{\mathrm{r},v}}(y,r)g(r)dr.
\end{aligned}$$

Here $P^{y,r}_{[0,t]}$ denotes bridge measure. We take a and u such that $2\pi a^2 v > \alpha/u$. Assuming to the contrary that $H_{\mathrm{r},v}$ is nonnegative and hence a contraction, we obtain like in the one-dimensional case

$$\begin{aligned}
0 < e^{-u} &\le \lim_{t\to\infty} P^y\left(U_t(a) \ge \pi a^2 u \log t\right) \le \frac{1}{b}\lim_{t\to\infty} e^{-\pi a^2 uv \log t}(f, e^{-tH_{\mathrm{r},v}} g) \\
&\le \frac{1}{b}\lim_{t\to\infty} t^{-\pi a^2 uv}\|f\|\|g\| = \lim_{t\to\infty} t^{-\pi a^2 uv + \alpha/2} = 0.
\end{aligned}$$

This is a contradiction, implying the statement. □

Remark 4.140. We note that by a calculation one can show that

$$\mathbb{E}[U_t(a)] = \begin{cases} c_1\sqrt{t}, & d = 1, \\ c_2 \log t, & d = 2, \end{cases}$$

where $c_1 = c_1(a) > 0$ and $c_2 = c_2(a) > 0$, which explains the time-scales appearing in the Kallianpur–Robbins laws. Thus the above theorems show that the existence of a bound state can be understood to be a consequence of paths spending sufficiently long times in particular subsets of space and it gives bounds on how long these average total times must be. The proofs use the fact that, since in one and two dimensions Brownian motion is recurrent, specific ergodic properties hold on its additive functionals (given here by the Kallianpur–Robbins laws). On the other hand, three- or higher-dimensional Brownian motion is transient and such a result cannot hold. In this case indeed, we know that the potential well needs to be deep enough for a bound state to occur.

Next we discuss the relationship of sojourn times and bound states from a different perspective, and show that actually the eigenfunctions for a potential well can even be reconstructed by using such random times. Consider H_ν defined by (4.3.17). The spectrum of this operator is $\mathrm{Spec}(H_\nu) = \mathrm{Spec}_{\mathrm{ess}}(H_\nu) \cup \mathrm{Spec}_{\mathrm{d}}(H_\nu)$ with $\mathrm{Spec}_{\mathrm{ess}}(H_\nu) = [0,\infty)$. By general theory we know that all eigenvalues $E_0 < E_1 < E_2 < \cdots$ are simple. Since the potential is even, it follows that $\varphi(-x)$ is an eigenfunction whenever $\varphi(x)$ is too. Since all the eigenvalues are simple, we have $\varphi(-x) = \pm\varphi(x)$, i. e., they are either even or odd. Furthermore, the eigenfunction φ_n corresponding to eigenvalue E_n has exactly n zeroes. To determine the discrete component $\mathrm{Spec}_{\mathrm{d}}(H_\nu)$ of the spectrum, we need to solve the system of differential equations

$$\begin{cases} -\varphi''(x) - \nu\varphi(x) = E\varphi(x), & |x| \le a, \\ -\varphi''(x) = E\varphi(x), & |x| > a, \end{cases}$$

with boundary conditions given by the requirement that φ and φ' are continuous functions at $\pm a$. A straightforward calculation shows that the eigenvalues $E = -|E|$ corresponding to even and odd eigenfunctions satisfy respectively the transcendental equations

$$\tan\left(a\sqrt{2(\nu - |E|)}\right) = \sqrt{\frac{|E|}{\nu - |E|}} \quad \text{and} \quad \cot\left(a\sqrt{2(\nu - |E|)}\right) = -\sqrt{\frac{|E|}{\nu - |E|}}. \tag{4.3.28}$$

It can be verified by elementary analysis that these equations have a unique solution $E_n \in (-\nu, 0)$ for each $n \in \mathbb{N}$ and for every $a, \nu > 0$. Moreover, the total number of eigenvalues is $N + 1$, $N = 0, 1, 2, \ldots$, determined by the inequality

$$\frac{N\pi}{2} \le \sqrt{2\nu}a < \frac{(N+1)\pi}{2}.$$

The eigenfunctions are

$$\varphi_n(x) = \begin{cases} A_n e^{-\sqrt{2|E_n|}|x|} 1_{\{|x|>a\}} + B_n \cos(\sqrt{2(\nu - |E_n|)}\,x) 1_{\{|x|\le a\}}, & n = 2m, \\ A_n e^{-\sqrt{2|E_n|}|x|} 1_{\{|x|>a\}} + B_n \sin(\sqrt{2(\nu - |E_n|)}\,x) 1_{\{|x|\le a\}}, & n = 2m+1 \end{cases}$$

with $m = 0, 1, 2, \ldots$, and normalization constants A_n and B_n. In particular, for the ground state we have

$$\varphi_0(x) = A_0 e^{-\sqrt{2|E|}|x|} 1_{\{|x|>a\}} + B_0 \cos(\sqrt{2(\nu - |E|)}\,x) 1_{\{|x|\le a\}}, \tag{4.3.29}$$

with

$$A_0 = \sqrt{\frac{\sqrt{2|E|}}{1 + a\sqrt{2|E|}}} e^{a\sqrt{2|E|}} \cos(a\sqrt{2(v - |E|)}), \quad B_0 = \sqrt{\frac{\sqrt{2|E|}}{1 + a\sqrt{2|E|}}},$$

where for simplicity we write $E_0 = E$.

Now we show that the eigenfunctions can be determined by (4.3.18) in terms of occupation times, i. e.,

$$\varphi_n(x) = \mathbb{E}^x[e^{-|E_n|t + vU_t(a)} \varphi_n(B_t)], \quad t > 0, \ n \in \dot{\mathbb{N}}. \tag{4.3.30}$$

For simplicity, we only discuss the ground state in detail. Consider the first hitting times defined by $\tau_a = \inf\{t > 0 \mid B_t = a\}$ for a Brownian motion starting at zero, and $\tau_{a,b} = \tau_a \wedge \tau_b$ for the levels $a < 0 < b$. Recall from Lemma 2.60 that these are almost surely finite stopping times with respect to the natural filtration of Brownian motion. From (4.3.30), we have

$$\varphi_0(x) = \mathbb{E}[e^{-|E_0|t + vU_t^x(a)} \varphi_0(B_t + x)],$$

where we denote

$$U_t^x(a) = \int_0^t 1_{\{|B_s + x| \le a\}} ds = \int_0^t 1_{\{-a-x \le B_s \le a-x\}} ds.$$

Note that $U^x_{\tau_{-a-x,a-x}}(a) = \tau_{-a-x,a-x}$ whenever $|x| < a$ and is zero otherwise. Since the stopping times are almost surely finite, we furthermore have

$$\begin{aligned} \varphi_0(x) &= \mathbb{E}[e^{-|E_0|t\wedge\tau_{-a-x,a-x} + vU^x_{t\wedge\tau_{-a-x,a-x}}(a)} \varphi_0(B_{t\wedge\tau_{-a-x,a-x}} + x)] \\ &\to \mathbb{E}[e^{-|E_0|\tau_{-a-x,a-x} + vU^x_{\tau_{-a-x,a-x}}(a)} \varphi_0(B_{\tau_{-a-x,a-x}} + x)] \quad \text{as } t \to \infty. \end{aligned} \tag{4.3.31}$$

Recall the Laplace transforms

$$\mathbb{E}[e^{-u\tau_a}] = e^{-\sqrt{2u}|a|} \quad \text{and} \quad \mathbb{E}[e^{-u\tau_{a,b}}] = \frac{\cosh\left(\sqrt{2u}\,\frac{b+a}{2}\right)}{\cosh\left(\sqrt{2u}\,\frac{b-a}{2}\right)}, \quad u \ge 0. \tag{4.3.32}$$

Let $x > a$. By path continuity $\tau_{-a-x,a-x} = \tau_{a-x}$, and we have

$$\varphi_0(x) = \mathbb{E}[e^{-|E_0|\tau_{a-x}} \varphi_0(B_{\tau_{a-x}} + x)] = \varphi_0(a)\mathbb{E}[e^{-|E_0|\tau_{a-x}}] = \varphi_0(a)e^{-\sqrt{2|E_0|}(x-a)}.$$

We obtain similarly for $x < -a$ that $\tau_{-a-x,a-x} = \tau_{-a-x}$ and

$$\varphi_0(x) = \mathbb{E}[e^{-|E_0|\tau_{-a-x}} \varphi_0(B_{\tau_{-a-x}} + x)] = \varphi_0(-a)e^{-\sqrt{2|E_0|}(-x-a)} = \varphi_0(a)e^{\sqrt{2|E_0|}(x+a)}$$

using $\varphi_0(-a) = \varphi_0(a)$. When $-a < x < a$ we apply the two-barrier formula in (4.3.32) and obtain

$$
\begin{aligned}
\varphi_0(x) &= \mathbb{E}[e^{(\nu-|E_0|)\tau_{-a-x,a-x}} \varphi_0(B_{\tau_{-a-x,a-x}} + x)] \\
&= \mathbb{E}[e^{(\nu-|E_0|)\tau_{-a-x,a-x}} \varphi_0(B_{\tau_{-a-x,a-x}} + x) 1_{\{\tau_{-a-x} < \tau_{a-x}\}}] \\
&\quad + \mathbb{E}[e^{(\nu-|E_0|)\tau_{-a-x,a-x}} \varphi_0(B_{\tau_{-a-x,a-x}} + x) 1_{\{\tau_{-a-x} > \tau_{a-x}\}}] \\
&= \varphi_0(a) \frac{\cos(\sqrt{2(\nu - |E_0|)}x)}{\cos(\sqrt{2(\nu - |E_0|)}a)}.
\end{aligned}
$$

The constant $\varphi_0(a)$ can be determined by the normalization condition, which then yields the expression of the ground state given by (4.3.29). Note that for odd eigenfunctions $\varphi(-a) = -\varphi(a)$ and therefore the $\cos(\sqrt{2(\nu - |E_0|)}x)$ factors are replaced by $\sin(\sqrt{2(\nu - |E_0|)}x)$.

Remark 4.141. The argument can be extended to higher dimensions. Denote by $B_r(a)$ a ball of radius r centered at a, and consider the Schrödinger operator $H = -\frac{1}{2}\Delta - \nu 1_{B_r(a)}$. Define the stopping times

$$
T_{B_a} = \inf\{t > 0 : X_t \in \overline{B}_a\} \quad \text{and} \quad \tau_{B_a} = \inf\{t > 0 : X_t \notin B_a\}.
$$

Using the Ciesielski–Taylor formulae,

$$
\begin{aligned}
\mathbb{E}^x[e^{-u\tau_{B_r(a)}}] &= \left(\frac{r}{|x-a|}\right)^{\frac{d-2}{2}} \frac{I_{\frac{d-2}{2}}(|x|\sqrt{2u})}{I_{\frac{d-2}{2}}(r\sqrt{2u})}, \\
\mathbb{E}^x[e^{-uT_{B_r(a)}}] &= \left(\frac{r}{|x-a|}\right)^{\frac{d-2}{2}} \frac{K_{\frac{d-2}{2}}(|x|\sqrt{2u})}{K_{\frac{d-2}{2}}(r\sqrt{2u})},
\end{aligned}
$$

where $I_{\frac{d-2}{2}}$ and $K_{\frac{d-2}{2}}$ are the modified Bessel functions of the first and second kind, respectively, i. e.,

$$
\begin{aligned}
I_\nu(z) &= \frac{z^\nu 2^{1-\nu}}{\Gamma(\nu + \frac{1}{2})\sqrt{\pi}} \int_0^1 e^{-zy}(1 - y^2)^{\nu - \frac{1}{2}}\, dy, \\
K_\nu(z) &= \frac{z^\nu \sqrt{\pi}}{2^\nu \Gamma(\nu + \frac{1}{2})} \int_1^\infty e^{-zy}(y^2 - 1)^{\nu - \frac{1}{2}}\, dy.
\end{aligned}
$$

Then by a similar argument as above, we obtain

$$
\varphi_0(x) = A_0 \left(\frac{a}{|x|}\right)^{\frac{d-2}{2}} K_{\frac{d-2}{2}}(|x|\sqrt{2|E_0|}) 1_{\{|x|>a\}} + B_0 \left(\frac{a}{|x|}\right)^{\frac{d-2}{2}} I_{\frac{d-2}{2}}(|x|\sqrt{2(\nu - |E_0|)}) 1_{\{|x|\le a\}},
$$

where the constants can be determined from L^2-normalization as before.

4.3.6 The number of eigenfunctions with negative eigenvalues

In this section we prove the celebrated *Lieb–Thirring inequality*, giving an estimate on the number of eigenfunctions in the level sets of nonpositive eigenvalues of a Schrödinger operator. From the point of view of quantum mechanics, these eigenfunctions describe *bound states*. Let $\{E_j\}_{j=1}^N$ be nonpositive eigenvalues of a Schrödinger operator $-\frac{1}{2}\Delta - W$ with a negative potential $-W$. A heuristic motivation of Lieb–Thirring inequality is the following semiclassical approximation:

$$\begin{aligned}\sum_{j=1}^N |E_j|^\beta &\sim \frac{1}{(2\pi)^d} \int_{\mathbb{R}^d\times\mathbb{R}^d} \Big[\Big(\frac{1}{2}|p|^2 - W(x)\Big)_-\Big]^\beta dp dx \\ &= \frac{1}{(2\pi)^d}\frac{1}{2^\beta}\int_{\mathbb{R}^d} dx \int_{0\le r\le\sqrt{2W(x)}} (2W(x)-r^2)^\beta r^{d-1}|S_{d-1}| dr \\ &= \frac{1}{(2\pi)^d}\frac{2^{\beta+(d/2)}}{2^\beta}|S_{d-1}|\int_0^{\pi/2}\sin^{2\beta+1}\theta\cos^{d-1}\theta d\theta \int_{\mathbb{R}^d} W(x)^{\beta+(d/2)}dx.\end{aligned}$$

Inserting the $(d-1)$-dimensional surface area of the unit sphere $|S_{d-1}| = 2\pi^{d/2}/\Gamma(d/2)$ and using the formula $\int_0^{\pi/2}\sin^{2\beta+1}\theta\cos^{d-1}\theta d\theta = \frac{1}{2}\Gamma(\beta+1)\Gamma(d/2)/\Gamma(\beta+1+d/2)$, we have

$$\sum_{j=1}^N |E_j|^\beta \sim a_{d,\beta}^c \int_{\mathbb{R}^d} |W(x)|^{\beta+(d/2)}dx,$$

where

$$a_{d,\beta}^c = \frac{1}{(2\pi)^{d/2}}\frac{\Gamma(\beta+1)}{\Gamma(\beta+1+d/2)}.$$

Throughout this section we denote

$$H_0 = -\frac{1}{2}\Delta,$$

so that $H = H_0 + V$.

Let $E_T(\cdot)$ be the spectral resolution of the self-adjoint operator T and write $1_{(\cdot)}(T) = E_T(\cdot)L^2(\mathbb{R}^d)$. Define

$$\begin{aligned}N_E(V) &= \dim 1_{(-\infty,-E]}(H) = \#\{\text{eigenvalues of } H \le -E\}, \quad E>0,\\ N_0(V) &= \lim_{E\downarrow 0} N_E(V).\end{aligned}$$

Remark 4.142. In this section we discuss H such that V is relatively form-compact with respect to H_0. Then the essential spectrum of H is identical with that of H_0 and it follows that $\mathrm{Spec}_{\mathrm{ess}}(H) = [0,\infty)$ and $(-\infty,0)\cap\mathrm{Spec}(H)$ is purely discrete. When 0 is an accumulation point of the discrete spectrum, $N_0(V) = \infty = \dim 1_{(-\infty,0)}(H)$ follows. On the other hand, when 0 is not an accumulation point, we have $N_0(V) = \dim 1_{(-\infty,0)}(H)$. Thus in all cases for a relatively form-compact V it follows that

$$N_0(V) = \dim 1_{(-\infty,0)}(H) = \#\{\text{eigenvalues of } H < 0\}.$$

Note that $N_0(V)$ does not include a count of zero eigenvalues.

Before estimating $N_0(V)$, we briefly discuss the *Birman–Schwinger principle*. Recall the kernel of $-\Delta + m^2$ introduced in (4.2.20). Let $C_m(x,y)$ denote its integral kernel, which is the Green function defined by

$$(-\Delta + m^2)C_m(x,y) = \delta(x-y).$$

Alternatively, by Fourier transform

$$C_m(x,y) = C_m(x-y) = \frac{1}{(2\pi)^d}\int_{\mathbb{R}^d} \frac{e^{-ip\cdot(x-y)}}{|p|^2+m^2}dp.$$

As was seen in (4.2.20) this can be computed explicitly, i. e.,

$$C_m(x) = \frac{1}{(2\pi)^{d/2}}\left(\frac{m}{|x|}\right)^{(d-2)/2} K_{(d-2)/2}(m|x|), \tag{4.3.33}$$

where $K_\nu(x) > 0$ is the modified Bessel function of the third kind (4.2.21). The fact

$$C_m(x) = \begin{cases} \frac{e^{-m|x|}}{2m}, & d=1,\\ \frac{e^{-m|x|}}{4\pi|x|}, & d=3 \end{cases}$$

was shown in Lemmas 4.97–4.98. Many properties of $C_m(x,y)$ can be deduced from (4.3.33) and the known properties of Bessel functions. Here we give a summary of the properties we need for our purposes.

Proposition 4.143. *The Green function $C_m(x,y) = C_m(x-y)$ has the following properties:*
(1) $C_m(x,y) > 0$;
(2) *for $m|x-y| > 0$ it behaves like $C_m(x,y) \le \mathrm{const}\, m^{(d-3)/2}\frac{e^{-m|x-y|}}{|x-y|^{(d-1)/2}}$, for all $d \ge 3$;*
(3) *for $m|x-y| \sim 0$ it behaves like* $C_m(x,y) \sim \begin{cases} |x-y|^{-d+2}, & d\ge 3,\\ -\log(m|x-y|), & d=2;\end{cases}$
(4) $\lim_{m\to 0} C_m(x,y) = C_0(x,y)$, *for all $d \ge 3$;*
(5) $C_0(x,y) = \frac{\Gamma(d/2-1)}{4\pi^{d/2}}\frac{1}{|x-y|^{d-2}}$, *for all $d \ge 3$.*

Now we look at potentials V for which $|V|^{1/2}(-\Delta+m^2)^{-1}|V|^{1/2}$ is a compact operator, i. e., when V is relatively form-compact with respect to $-\Delta$.

Lemma 4.144 (Compactness of Birman–Schwinger kernel). *Let*

$$K = |V|^{1/2}(-\Delta + m^2)^{-1}|V|^{1/2}.$$

$(d = 1)$ *Let* $V \in L^1(\mathbb{R})$ *and* $m > 0$. *Then* $K \in \mathscr{I}_2$ *and* $\|K\|_2 \le C\|V\|_1$.
$(d = 2)$ *Let* $V \in L^p(\mathbb{R}^2)$ *for* $p > 1$ *and* $m > 0$. *Then* $K \in \mathscr{I}_p$ *and* $\|K\|_p \le C\|V\|_p$.
$(d \ge 3)$ *Let* $V \in L^{d/2}(\mathbb{R}^d)$ *and* $m \ge 0$. *Then* $K \in \mathscr{I}_2$ *and* $\|K\|_2 \le C\|V\|_{d/2}$.

Proof. For $d = 1$, it is trivial to see that

$$\int_{\mathbb{R}\times\mathbb{R}} |V(x)|C_m(x-y)^2|V(y)|dxdy \le C\|V\|_1^2.$$

Thus it is immediate that $K \in \mathscr{I}_2$ and $\|K\|_2^2 \le C\|V\|_1^2$. Consider $d = 2$. By Proposition 4.39, we see that $(-\Delta + m^2)^{-1/2}|V|^{1/2} \in \mathscr{I}_{2p}$ for all $p > 1$ since $(|k|^2 + m^2)^{-1/2} \in L^{2p}(\mathbb{R}^2)$ and $|V|^{1/2} \in L^{2p}(\mathbb{R}^2)$. Thus by Proposition 4.37, $K \in \mathscr{I}_p$ and $\|K\|_p \le C\|V^{1/2}\|_{2p}^2 = C\|V\|_p$. Finally, let $d \ge 3$. It suffices to discuss the case of $m = 0$ only, since

$$(f, |V|^{1/2}(-\Delta + m^2)^{-1}|V|^{1/2}f) \le (f, |V|^{1/2}(-\Delta)^{-1}|V|^{1/2}f).$$

By the Hardy–Littlewood–Sobolev inequality, we have

$$\int_{\mathbb{R}^d\times\mathbb{R}^d} |V(x)|C_0(x-y)^2|V(y)|dxdy \le C\|V\|_{d/2}^2.$$

This gives $K \in \mathscr{I}_2$ and $\|K\|_2 \le C\|V\|_{d/2}$. □

For $d \ge 3$, a similar estimate to Lemma 4.144 can be obtained due to Proposition 4.40, studying the map $L^p_w(\mathbb{R}^d) \times L^p(\mathbb{R}^d) \ni (u, g) \to B_{u,g} \in \mathscr{I}_{p,w}$.

Lemma 4.145. *Let* $d \ge 3$ *and* $V \in L^{d/2}(\mathbb{R}^d)$. *Then for* $p > d/2$,

$$|V|^{1/2}(-\Delta)^{-1}|V|^{1/2} \in \mathscr{I}_p \tag{4.3.34}$$

and

$$\||V|^{1/2}(-\Delta)^{-1}|V|^{1/2}\|_p \le D_p\|V\|_{d/2} \tag{4.3.35}$$

follow with suitable $D_p > 0$.

Proof. Note that $u(k) = 2|k|^{-1} \in L^d_w(\mathbb{R}^d)$. We have $B_{u,|V|^{1/2}} \in \mathscr{I}_{d,w}$, hence $(-\Delta)^{-1/2}|V|^{1/2} = FB_{u,|V|^{1/2}} \in \mathscr{I}_{d,w}$. It follows that $|V|^{1/2}(-\Delta)^{-1}|V|^{1/2} \in \mathscr{I}_{d/2,w}$ by Proposition 4.37, and then (4.3.34). Furthermore, by (4.1.29) we have $\|(-\Delta)^{-1/2}|V|^{1/2}\|_{d,w} \le C\||V|^{1/2}\|_d$ and by Proposition 4.37, we see again that $\||V|^{1/2}(-\Delta)^{-1}|V|^{1/2}\|_{d/2,w} \le C^2\||V|^{1/2}\|_d^2 = C^2\|V\|_{d/2}$. Using (4.36), we then arrive at (4.3.35). □

Let $V = V_+ - V_-$ be such that $V_- \in L^\infty(\mathbb{R}^d)$. Since

$$H_0 - V_- \leq H_0 + V$$

in the sense of quadratic forms, the number of negative eigenvalues of $H_0 + V$ is smaller than that of $H_0 - V_-$. Therefore, instead of $H_0 + V$ we may consider $H_0 - V_-$.

We assume in what follows that $V(x) = -W(x) \leq 0$ such that V is relatively form-compact with respect to H_0. Let

$$H(\alpha) = H_0 - \alpha W.$$

Since $-W$ is H_0-form-compact, the essential spectrum of $H(\alpha)$ stays unchanged, i. e.,

$$\mathrm{Spec}_{\mathrm{ess}}(H(\alpha)) = \mathrm{Spec}_{\mathrm{ess}}(H_0) = [0, \infty).$$

Furthermore, $H(\alpha)$ is bounded from below since $-W$ is also relatively form-bounded with an infinitesimally small relative bound. Notice that for $\lambda \geq 0$ we have $H(\alpha)\psi = -\lambda\psi$ exactly when $(H_0 + \lambda)\psi = \alpha W\psi$. From this, we get that $W\psi \in D((H_0 + \lambda)^{-1})$ and $\psi = \alpha(H_0 + \lambda)^{-1}W\psi$. Since $\psi \in D(W^{1/2})$, the identity $W^{-1/2}W^{1/2}\psi = \alpha(H_0 + \lambda)^{-1}W^{1/2}W^{1/2}\psi$ gives $W^{1/2}\psi = \alpha W^{1/2}(H_0 + \lambda)^{-1}W^{1/2}W^{1/2}\psi$. We then conclude that for $\lambda \geq 0$ we have $H(\alpha)\psi = -\lambda\psi$ if and only if

$$W^{1/2}(H_0 + \lambda)^{-1}W^{1/2}\varphi = \frac{1}{\alpha}\varphi,$$

where $\varphi = W^{1/2}\psi$. Put

$$K_\lambda = W^{1/2}(H_0 + \lambda)^{-1}W^{1/2}.$$

This is also called *Birman–Schwinger kernel*, for which

$$-\lambda \in \mathrm{Spec}(H(\alpha)) \iff \frac{1}{\alpha} \in \mathrm{Spec}(K_\lambda)$$

holds.

Lemma 4.146. *Let $e_j(\lambda)$ be the jth eigenvalue of K_λ, i. e., $\|K_\lambda\| = e_1(\lambda) \geq e_2(\lambda) \geq \cdots \geq 0$. Then the map $(0, \infty) \ni \lambda \mapsto e_j(\lambda)$ is continuous, nonincreasing, i. e., $e_j(\lambda + a) \leq e_j(\lambda)$ for any $a > 0$, and $\lim_{\lambda\to\infty} e_j(\lambda) = 0$.*

Proof. Let $\lambda \in (0, \infty)$. We show that $\lim_{\lambda'\to\lambda} \|K_\lambda - K_{\lambda'}\| = 0$, i. e., $\lambda \mapsto K_\lambda$ is uniformly continuous on $(0, \infty)$. It is direct to see that

$$\begin{aligned}\|K_\lambda - K_{\lambda'}\| &= |\lambda - \lambda'|\|W^{1/2}(H_0 + \lambda')^{-1}(H_0 + \lambda)^{-1}W^{1/2}\| \\ &\leq \frac{|\lambda - \lambda'|}{\sqrt{\lambda\lambda'}}\|W^{1/2}(H_0 + \lambda')^{-1/2}\|\|(H_0 + \lambda)^{-1/2}W^{1/2}\|.\end{aligned}$$

Note that

$$\|W^{1/2}(H_0+\lambda')^{-1/2}\| \le \|(H_0+\lambda')^{-1/2}(H_0+\lambda)^{1/2}\|\|(H_0+\lambda)^{-1/2}W^{1/2}\| \\ \le C\|(H_0+\lambda)^{-1/2}W^{1/2}\|,$$

where C is independent of λ'. Hence we have

$$\|K_\lambda - K_{\lambda'}\| \le C|\lambda-\lambda'|\frac{1}{\lambda}\|(H_0+\lambda)^{-1/2}W^{1/2}\|^2.$$

This yields that $\lambda \mapsto K_\lambda$ is uniformly continuous and for $\varepsilon > 0$ there exists $\delta > 0$ such that $|(\phi,(K_\lambda - K_{\lambda'})\phi)| \le \varepsilon\|\phi\|^2$ for $|\lambda-\lambda'| \le \delta$ and for arbitrary ϕ. We have

$$(\phi, K_{\lambda'}\phi) - \varepsilon\|\phi\|^2 \le (\phi, K_\lambda\phi) \le (\phi, K_{\lambda'}\phi) + \varepsilon\|\phi\|^2$$

for all ϕ. From this and the min–max principle, we see that

$$e_j(\lambda') - \varepsilon \le e_j(\lambda) \le e_j(\lambda') + \varepsilon$$

and thus $|e_j(\lambda)-e_j(\lambda')| \le \varepsilon$ for $|\lambda-\lambda'| < \delta$; so the continuity of $e_j(\cdot)$ follows. It is immediate to see that $(\phi, K_\lambda\phi) \ge (\phi, K_{\lambda'}\phi)$ for $\lambda \le \lambda'$. Consequently, $e_j(\cdot)$ is decreasing by the min–max principle again. Finally, let $e_j(\lambda) \in \operatorname{Spec}(K_\lambda)$ if and only if $-\lambda \in \operatorname{Spec}(H(1/e_j(\lambda)))$. Since $H(1/e_j(\lambda)) = H_0 \dot{-} 1/e_j(\lambda)W$ is bounded below by $-1/e_j(\lambda) \times c$ with some c by the KLMN theorem, $-\lambda \ge -1/e_j(\lambda) \times c$ and from this it follows that $c/\lambda \ge e_j(\lambda)$. Thus $\lim_{\lambda\to\infty} e_j(\lambda) = 0$. □

Proposition 4.147 (Birman–Schwinger principle). *Suppose that $V \le 0$ and it is H_0-form-compact. Then*

$$N_E(V) = \dim 1_{[1,\infty)}(K_E), \quad -E < 0, \tag{4.3.36}$$

$$N_0(V) \le \dim 1_{[1,\infty)}(K_0). \tag{4.3.37}$$

Proof. For $-E < 0$,

$$N_E(V) = \#\{e \in \operatorname{Spec}(H(1)) \mid e \le -E\} = \#\{\lambda \in [E,\infty) \mid \operatorname{Spec}(K_\lambda) \ni 1\} \\ = \#\{e \in \operatorname{Spec}(K_E) \mid e \ge 1\} = \dim 1_{[1,\infty)}(K_E). \tag{4.3.38}$$

The second equality of (4.3.38) can be derived from the fact that

$$-\lambda \in \operatorname{Spec}(H(1)) \iff 1 \in \operatorname{Spec}(K_\lambda), \tag{4.3.39}$$

and the third is derived from Lemma 4.146. The relation (4.3.36) follows from (4.3.38). To obtain (4.3.37) note that $N_0(E) = \lim_{E\downarrow 0} N_E(V)$ and $K_E \le K_0$. □

Next we extend the Birman–Schwinger principle to a version including zero eigenvalues. We proceed by more abstract arguments. Let $\mathcal{K}$ be a separable Hilbert space over $\mathbb{C}$ and T a self-adjoint operator in $\mathcal{K}$ such that $\inf \operatorname{Spec}(T) = 0$ and $0 \notin \operatorname{Spec}_\mathrm{p}(T)$. Let S be a self-adjoint operator such that $S \le 0$ and

$$L_E = |S|^{1/2}(T+E)^{-1}|S|^{1/2}$$

is compact for $E \ge 0$. The Birman–Schwinger principle says that

$$\dim 1_{(-\infty,0)}(T+S) \le \dim 1_{[1,\infty)}(L_0).$$

We want to extend this to

$$\dim 1_{(-\infty,0]}(T+S) \le \dim 1_{[1,\infty)}(L_0), \tag{4.3.40}$$

so that the left-hand side in (4.3.40) includes the number of nonpositive eigenvalues of $T+S$. In particular, the bound $\dim 1_{[1,\infty)}(L_0) < 1$ yields that $T+S$ has no nonpositive eigenvalues. To prove (4.3.40), we use the following lemma.

Lemma 4.148. *Let $a > 1$. Then* $\dim 1_{(-\infty,0]}(T+S) \le \dim 1_{(-\infty,0)}(T+aS)$.

Proof. We proceed by contradiction. Suppose that

$$\dim 1_{(-\infty,0)}(T+aS) < \dim 1_{(-\infty,0]}(T+S).$$

Since the left-hand side above is finite, we write $\dim 1_{(-\infty,0)}(T+aS) = n \ge 0$. Then $\dim 1_{(-\infty,0]}(T+S) \ge n+1$. Let $\mathcal{K} \supset \{\phi_1,\dots,\phi_{n+1}\}$ be an arbitrary orthonormal system of $1_{(-\infty,0]}(T+S)$, and $\mathscr{L}$ = linear hull of $\{\phi_1,\dots,\phi_{n+1}\}$ an $(n+1)$-dimensional subspace of $\mathcal{K}$. Consider a projection $P : \mathcal{K} \to \mathscr{L}$. The restriction

$$A = P(T+aS)P\lceil_{\mathscr{L}}$$

can be regarded as a self-adjoint operator from $\mathscr{L}$ to itself. Denote the eigenvalues of A by $\tilde{\mu}_j, j = 1,\dots,n+1$, ordered as $\tilde{\mu}_1 \le \tilde{\mu}_2 \le \dots \le \tilde{\mu}_{n+1}$. We can compare the eigenvalues of $T+aS$ and A. Define

$$\mu_j = \sup_{\psi_1,\dots,\psi_{j-1}\in\mathscr{K}} \inf_{\substack{\psi\in D(A);\|\psi\|=1\\ \psi\in[\psi_1,\dots,\psi_{j-1}]^\perp}} (\psi, A\psi).$$

Inequality $\mu_j \le \tilde{\mu}_j$ for $j = 1,\dots,n+1$ can be derived by the Rayleigh–Ritz technique in Proposition 4.24. We have $(f,(T+aS)f) = (f,(T+S)f) + (a-1)(f,Sf) \le 0$ for all $f \in \mathscr{L}$. Suppose that $0 = (g,(T+aS)g)$ for some $g \in \mathscr{L}$. Then $(g,(T+S)g) = (a-1)(g,Sg) = 0$ and $(g,Tg) = 0$ follow. However, $\operatorname{Spec}_\mathrm{p}(T) \not\ni 0$ implies that $g = 0$. Consequently, A is strictly negative and $\tilde{\mu}_j < 0$ for $j = 1,\dots,n+1$ follows. In particular, $\mu_{n+1} \le \tilde{\mu}_{n+1} < 0$. This yields

$\dim 1_{(-\infty,0)}(T + aS) \geq n + 1$, contradicting $\dim 1_{(-\infty,0)}(T + aS) = n$, and thus the lemma follows. □

Theorem 4.149. *Suppose that L_0 is compact. Then*

$$\dim 1_{(-\infty,0]}(T + S) \leq \dim 1_{[1,\infty)}(L_0).$$

Proof. By Lemma 4.148, we have

$$\dim 1_{(-\infty,0]}(T + S) \leq \lim_{a\uparrow 1} \dim 1_{(-\infty,0)}(T + aS). \tag{4.3.41}$$

Note that $\dim 1_{(-\infty,0)}(T+aS) = \lim_{E\uparrow 0} \dim 1_{(-\infty,E]}(T+aS)$, and by the Birman–Schwinger principle, we have $\dim 1_{T+aS}((-\infty, E]) \leq \dim 1_{aL_E}([1,\infty))$ for all $E < 0$. Since $L_E \leq L_0$, $\dim 1_{aL_E}([1,\infty)) \leq \dim 1_{[1,\infty)}(aL_0)$ follows. Using these results and (4.3.41), we have

$$\dim 1_{(-\infty,0]}(T + S) \leq \lim_{a\uparrow 1} \dim 1_{[1,\infty)}(aL_0) \leq \dim 1_{[1,\infty)}(L_0),$$

and the theorem follows. □

An immediate consequence leads to an extension of Proposition 4.147. Define

$$\bar{N}_E(V) = \dim 1_{(-\infty,-E]}(H) = \#\left\{\text{eigenvalues of } H \leq -E\right\}, \quad E \geq 0.$$

Note that $\bar{N}_E(V) = N_E(V)$ for $-E < 0$ and $\bar{N}_0(V) \geq N_0(V)$.

Corollary 4.150 (Extended Birman–Schwinger principle). *Suppose that $V \leq 0$ and it is H_0-form-compact. Then*

$$\bar{N}_E(V) = \dim 1_{[1,\infty)}(K_E), \quad -E < 0,$$
$$\bar{N}_0(V) \leq \dim 1_{[1,\infty)}(K_0).$$

Proof. We replace T, S, and L_0 by H_0, V, and K_0, respectively, in Theorem 4.149. □

Corollary 4.150 has the following immediate implication.

Corollary 4.151 (Absence of ground state). *Let $d \geq 3$. Suppose that $V \leq 0$ is H_0-form-compact and the operator norm of K_0 satisfies $\|K_0\| < 1$. Then H has no nonpositive eigenvalues. In particular, the bottom of the spectrum of H is not an eigenvalue, i. e., H has no ground state.*

Proof. By the extended Birman–Schwinger principle, $\bar{N}_0(V) \leq \dim 1_{[1,\infty)}(K_0) = 0$ follows, and $\mathrm{Spec}_{\mathrm{ess}}(H) = [0,\infty)$. □

Note that in general it is difficult to verify whether zero is an eigenvalue of H. We give a sufficient condition for H to have no zero eigenvalue in Proposition 4.168 below. However, as shown in Corollary 4.151, $\|K_0\| < 1$ implies that zero is no eigenvalue of H.

From now on, we will estimate $\bar{N}_0(V)$ from above. It can be seen directly that

$$W^{1/2}(H_0 + \alpha W + \lambda)^{-1}W^{1/2} = K_\lambda(1 + \alpha K_\lambda)^{-1}. \tag{4.3.42}$$

On the other hand, the integral kernel of

$$W^{1/2}(H_0 + \alpha W + \lambda)^{-1}W^{1/2} = W^{1/2}\int_0^\infty e^{-\lambda t}e^{-t(H_0+\alpha W)}W^{1/2}dt$$

can be expressed by using the conditional Wiener measure,

$$\begin{aligned}&(W^{1/2}(H_0 + \alpha W + \lambda)^{-1}W^{1/2})(x,y)\\&= W^{1/2}(x)W^{1/2}(y)\int_0^\infty e^{-\lambda t}\mathbb{E}^{x,y}_{[0,t]}\left[e^{-\alpha\int_0^t W(B_s)ds}\right]\Pi_t(x-y)dt.\end{aligned} \tag{4.3.43}$$

Let $g(y) = e^{-\alpha y}$ and $F(x) = x(1 + \alpha x)^{-1}$, i. e.,

$$F(x) = x\int_0^\infty e^{-y}g(xy)dy. \tag{4.3.44}$$

Together with (4.3.42), (4.3.43) can be reformulated as

$$F(K_\lambda)(x,y) = W^{1/2}(x)W^{1/2}(y)\int_0^\infty e^{-\lambda t}\mathbb{E}^{x,y}_{[0,t]}\left[g\left(\int_0^t W(B_s)ds\right)\right]\Pi_t(x-y)dt. \tag{4.3.45}$$

Moreover, when W is continuous, the right-hand side of (4.3.43) is also continuous in x and y. By a general result, $\operatorname{Tr} F(K_\lambda)$ can be evaluated by setting $x = y$. Hence

$$\operatorname{Tr} F(K_\lambda) = \int_0^\infty e^{-\lambda t}\Pi_t(0)dt\int_{\mathbb{R}^d} W(x)\mathbb{E}^{x,x}_{[0,t]}\left[g\left(\int_0^t W(B_s)ds\right)\right]dx. \tag{4.3.46}$$

Lemma 4.152. *Suppose that W is H_0-form-compact and continuous, and let $f(y) = yg(y) = ye^{-\alpha y}$ and $F(x) = x(1 + \alpha x)^{-1}$. Then it follows that*

$$\operatorname{Tr} F(K_\lambda) = \int_0^\infty \frac{dt}{t}e^{-\lambda t}\Pi_t(0)\int_{\mathbb{R}^d}\mathbb{E}^{x,x}_{[0,t]}\left[f\left(\int_0^t W(B_s)ds\right)\right]dx. \tag{4.3.47}$$

Proof. It suffices to show that

$$\frac{1}{t}\int_{\mathbb{R}^d}\mathbb{E}^{x,x}_{[0,t]}\left[f\left(\int_0^t W(B_s)ds\right)e^{-\alpha\int_0^t W(B_s)ds}\right]dx$$

$$= \int_{\mathbb{R}^d} W(x)\mathbb{E}^{x,x}_{[0,t]}\left[g\left(\int_0^t W(B_s)ds\right)e^{-\alpha\int_0^t W(B_s)ds}\right]dx. \tag{4.3.48}$$

Let $U_r = e^{-rH(-\alpha)}We^{-(t-r)H(-\alpha)}$. Note that U_r is compact, and its integral kernel can be computed as

$$\begin{aligned}(U_r f)(x) &= \mathbb{E}^x[e^{-\alpha\int_0^r W(B_s)ds}W(B_r)\mathbb{E}^{B_r}[e^{-\alpha\int_0^{t-r} W(B_s)ds}f(B_{t-r})]]\\ &= \mathbb{E}^x[e^{-\alpha\int_0^t W(B_s)ds}W(B_r)f(B_t)]\\ &= \int_{\mathbb{R}^d}\Pi_t(x-y)\mathbb{E}^{x,y}_{[0,t]}\left[e^{-\alpha\int_0^t W(B_s)ds}W(B_r)\right]f(y)dy.\end{aligned}$$

Hence the kernel is given by

$$U_r(x,y) = \Pi_t(x-y)\mathbb{E}^{x,y}_{[0,t]}\left[e^{-\alpha\int_0^t W(B_s)ds}W(B_r)\right]$$

and it is continuous in x and y. Thus it follows that

$$\operatorname{Tr} U_r = \int_{\mathbb{R}^d} U_r(x,x)dx = \Pi_t(0)\int_{\mathbb{R}^d}\mathbb{E}^{x,x}_{[0,t]}\left[e^{-\alpha\int_0^t W(B_s)ds}W(B_r)\right]dx. \tag{4.3.49}$$

From the identity $\operatorname{Tr} U_r = \operatorname{Tr} U_0$ it follows that $\int_0^t \frac{1}{t}\operatorname{Tr} U_r dr = \operatorname{Tr} U_0$. Inserting (4.3.49) into this yields thus (4.3.48). □

Furthermore, a more general formula holds.

Lemma 4.153. *Let* $0 \le W \in L^p(\mathbb{R}^d)$ *with*

$$p = \begin{cases} d/2, & d \ge 3,\\ > 1, & d = 2,\\ 1, & d = 1.\end{cases}$$

Fix $\lambda > 0$ *for* $d = 1,2$ *and* $\lambda \ge 0$ *for* $d \ge 3$. *Let* f *be a nonnegative lower semicontinuous function on* $[0,\infty)$ *with* $f(0) = 0$ *and let* f, g, *and* F *be related by*

$$F(x) = x\int_0^\infty e^{-y}g(xy)dy, \quad f(y) = g(y)y, \tag{4.3.50}$$

and $F(\|K_\lambda\|) < \infty$. *We have*

$$\operatorname{Tr} F(K_\lambda) = \int_0^\infty \frac{1}{t}e^{-\lambda t}\Pi_t(0)dt\int_{\mathbb{R}^d}\mathbb{E}^{x,x}_{[0,t]}\left[f\left(\int_0^t W(B_s)ds\right)\right]dx \tag{4.3.51}$$

$$= \int_0^\infty e^{-\lambda t}\Pi_t(0)dt \int_{\mathbb{R}^d} W(x)\mathbb{E}^{x,x}_{[0,t]}\left[g\left(\int_0^t W(B_s)ds\right)\right]dx, \tag{4.3.52}$$

where both sides above may be infinite.

We prove this lemma after Lemmas 4.154–4.157. In Lemmas 4.154–4.157, we assume that $d \geq 3$. In the cases $d = 1$ and 2, the proof is similar. We also write $K_\lambda(W)$ for K_λ when we emphasize the dependence of W, and $(a,b) \subset (0,\infty)$ in the proofs of Lemmas 4.154–4.157. Note that $W \in L^p(\mathbb{R}^d)$ implies that $(-1/2)\Delta + W$ is bounded from below; see Example 4.64. Furthermore, we see that $F(x) < \infty$ for every $0 \leq x \leq \|K_\lambda\|$.

Lemma 4.154. *Suppose that $\lambda > 0$, $W \in C_0^\infty(\mathbb{R}^d)$, and $g \in C_0((a,b))$. Let f, g, and F be related by* (4.3.50). *Then* (4.3.51) *and* (4.3.52) *hold.*

Proof. Denote the finite linear sum of functions of the form $g(y) = e^{-ay}$ by $\mathscr{C} =$ linear hull of $\{e^{-ay} \mid a \geq 0\}$, which by the Stone–Weierstrass theorem is dense in the space of continuous functions on $[0,\infty)$ vanishing at infinity. Hence there exists a sequence $(g_n)_{n\in\mathbb{N}} \subset \mathscr{C}$ such that $\|g_n - g\|_\infty \to 0$ as $n \to \infty$. We write $F(x) = F_g(x) = x\int_0^\infty e^{-y}g(xy)dy$ to emphasize the dependence of g. Since $F_g(x) = \int_0^\infty e^{-z/x}g(z)dz$, $g \mapsto F_g$ is linear and $F_g(x)$ is monotone increasing. Hence

$$F_{g_n}(K_\lambda)(x,y) = W^{1/2}(x)W^{1/2}(y)\int_0^\infty e^{-\lambda t}\mathbb{E}^{x,y}_{[0,t]}\left[g_n\left(\int_0^t W(B_s)ds\right)\right]\Pi_t(x-y)dt \tag{4.3.53}$$

for $g_n \in \mathscr{C}$ by (4.3.45). The integral operator defined by the integral kernel given by the left-hand side (resp. the right-hand side) of (4.3.53) is denoted by A_n (resp. B_n). We have for $\Phi, \Psi \in L^2(\mathbb{R}^d)$,

$$\begin{aligned}(\Phi, A_n\Psi)_{L^2(\mathbb{R}^d)} &= \int_0^\infty e^{-y}(\Phi, K_\lambda g_n(yK_\lambda)\Psi)_{L^2(\mathbb{R}^d)}dy \\ &\to \int_0^\infty e^{-y}(\Phi, K_\lambda g(yK_\lambda)\Psi)_{L^2(\mathbb{R}^d)}dy = (\Phi, F_g(K_\lambda)\Psi)_{L^2(\mathbb{R}^d)}\end{aligned}$$

as $n \to \infty$. On the other hand, from $\|g_n\|_\infty \leq \|g_n - g\|_\infty + \|g\|_\infty$ it follows that for all $n \geq 1$, $\|g_n\|_\infty < C$ with some C, and together with it we see that

$$\begin{aligned}(\Phi, B_n\Psi) &= \int_{\mathbb{R}^d\times\mathbb{R}^d} \overline{\Phi(x)}\Psi(y)W(x)^{1/2}W(y)^{1/2}\int_0^\infty e^{-\lambda t}\mathbb{E}^{x,y}_{[0,t]}\left[g_n\left(\int_0^t W(B_s)ds\right)\right]\Pi_t(x-y)dxdy \\ &\to \int_{\mathbb{R}^d\times\mathbb{R}^d} \overline{\Phi(x)}\Psi(y)W(x)^{1/2}W(y)^{1/2}\int_0^\infty e^{-\lambda t}\mathbb{E}^{x,y}_{[0,t]}\left[g\left(\int_0^t W(B_s)ds\right)\right]\Pi_t(x-y)dxdy\end{aligned}$$

as $n \to \infty$ by the dominated convergence theorem. We see that $F_g(K_\lambda)$ is an integral operator with integral kernel

$$F_g(K_\lambda)(x,y) = W^{1/2}(x)W^{1/2}(y) \int_0^\infty e^{-\lambda t}\mathbb{E}_{[0,t]}^{x,y}\left[g\left(\int_0^t W(B_s)ds\right)\right]\Pi_t(x-y)dt \qquad (4.3.54)$$

for $g \in C_0((a,b))$. It can be also checked that $F_g(K_\lambda) \in \mathscr{I}_1$ for $g \in C_0((a,b))$ by the fact $F_g(x) \le C_m x^m$ for any $m \ge 1$. To see this, let $h(x) = g(x)/x^m$. We have $\|h\|_\infty < \infty$, since $\operatorname{supp} g \subset (a,b)$ and $a > 0$. Note that

$$F_g(x) = x\int_0^\infty e^{-y}(xy)^m h(xy)dy \le \left(\|h\|_\infty \int_0^\infty e^{-y}y^m dy\right)x^{m+1}.$$

Thus for $m \ge 1$, $\operatorname{Tr} F_g(K_\lambda) \le C_{m+1}\operatorname{Tr} K_\lambda^{m+1} = C_{m+1}\|K_\lambda\|_{m+1}^{m+1} < \infty$, since $K_\lambda \in \mathscr{I}_2$. We have

$$F_g(K_\lambda) \in \mathscr{I}_1 \quad \text{for } g \in C_0((a,b)). \qquad (4.3.55)$$

Since the right-hand side of (4.3.54) is jointly continuous in x and y which can be seen by the expression in terms of Brownian bridge,

$$F_g(K_\lambda)(x,y)$$
$$= W^{1/2}(x)W^{1/2}(y)\int_0^\infty e^{-\lambda t}\mathbb{E}_{\mathcal{W}}^0\left[g\left(\int_0^t W\left(\left(1-\frac{s}{t}\right)x + \frac{s}{t}y - \frac{s}{t}B_t + B_s\right)ds\right)\right]\Pi_t(x-y)dt,$$

it follows that $\operatorname{Tr}(F(K_\lambda)) = \int_{\mathbb{R}^d} F(K_\lambda)(x,x)dx < \infty$ and (4.3.52). Equality (4.3.51) can be proven in the same way as (4.3.52). □

Lemma 4.155. *Suppose that $\lambda > 0$, $W \in L^\infty(\mathbb{R}^d)$ with compact support, and $g \in C_0((a,b))$. Let f, g, and F be related by* (4.3.50). *Then* (4.3.51) *and* (4.3.52) *hold.*

Proof. There exists a sequence $W_n \in C_0^\infty(\mathbb{R}^d)$ such that $W_n \uparrow W$ almost everywhere and $W_n \uparrow W$ in $L^{d/2}(\mathbb{R}^d)$ as $n \to \infty$. Since $K_\lambda(W) \in \mathscr{I}_2$, in the same way as (4.3.55) we can show that

$$F(K_\lambda(W)) \in \mathscr{I}_1. \qquad (4.3.56)$$

Since the operator norm $\|K_\lambda(W_n) - K_\lambda(W)\|$ is estimated as

$$\|K_\lambda(W_n) - K_\lambda(W)\| \le C\|W - W_n\|_{d/2}$$

by Lemma 4.144 with a constant $C > 0$, $K_\lambda(W_n) \to K_\lambda(W)$ as $n \to \infty$ in the operator norm, and then it also follows that

$$F(K_\lambda(W_n)) \to F(K_\lambda(W)) \tag{4.3.57}$$

as $n \to \infty$ in the operator norm. Since $W_n \le W$, we also see that

$$\mu_m(F(K_\lambda(W_n))) \le \mu_m(F(K_\lambda(W))) \tag{4.3.58}$$

for each m, where $\mu_m(A)$ denotes the mth eigenvalue of $\sqrt{A^*A}$ arranged in nonincreasing order with each eigenvalue occurring by their multiplicity; (4.3.56), (4.3.57), and (4.3.58) imply that

$$\operatorname{Tr} F(K_\lambda(W_n)) \to \operatorname{Tr} F(K_\lambda(W)) \tag{4.3.59}$$

as $n \to \infty$ by Proposition 4.45. On the other hand, there exist a compact set $S \subset \mathbb{R}^d$ and $A > 0$ such that $W_n(x) \le A1_S(x)$ for $x \in \mathbb{R}^d$. Let $g(x) = h(x)x^m$ be as in Lemma 4.154. Thus we have

$$e^{-\lambda t}W_n(x)g\left(\int_0^t W_n(B_s)ds\right) \le \|h\|_\infty e^{-\lambda t}W_n(x)\left(\int_0^t W_n(B_s)ds\right)^m \le \|h\|_\infty e^{-\lambda t}1_S(x)A^{m+1}t^m$$

and $e^{-\lambda t}1_S(x)A^{m+1}t^m \in L^1([0,\infty) \times \mathbb{R}^d, dt \otimes dx)$. Using this, we can show that

$$\begin{aligned}&\int_0^\infty e^{-\lambda t}\Pi_t(0)dt \int_{\mathbb{R}^d} W_n(x)\mathbb{E}^{x,x}_{[0,t]}\left[g\left(\int_0^t W_n(B_s)ds\right)\right]dx\\ &\quad\to \int_0^\infty e^{-\lambda t}\Pi_t(0)dt \int_{\mathbb{R}^d} W(x)\mathbb{E}^{x,x}_{[0,t]}\left[g\left(\int_0^t W(B_s)ds\right)\right]dx\end{aligned}$$

by the dominated convergence theorem, and hence (4.3.52) is proven. Equality (4.3.51) can be shown similarly. □

Lemma 4.156. *Suppose that $\lambda > 0$, $W \in L^{d/2}(\mathbb{R}^d)$, and $g \in C_0((a,b))$. Let f, g, and F be related by* (4.3.50). *Then* (4.3.51) *and* (4.3.52) *hold.*

Proof. Let $W_n \in L^\infty(\mathbb{R}^d)$ with compact support be such that $W_n \uparrow W$ almost everywhere and in $L^{d/2}(\mathbb{R}^d)$. In a similar manner to (4.3.55), we can see that $F(K_\lambda(W)) \in \mathscr{I}_1$. Since $\operatorname{Tr} F(K_\lambda(W_n))$ is monotone increasing,

$$\operatorname{Tr} F(K_\lambda(W_n)) \to \operatorname{Tr} F(K_\lambda(W)) \tag{4.3.60}$$

is shown in a similar way to (4.3.59). On the other hand, the identity

$$\operatorname{Tr} F(K_\lambda(W_n)) = \int_0^\infty e^{-\lambda t}\Pi_t(0)dt \int_{\mathbb{R}^d} W_n(x)\mathbb{E}^{x,x}_{[0,t]}\left[g\left(\int_0^t W_n(B_s)ds\right)\right]dx \tag{4.3.61}$$

implies that the right-hand side of (4.3.61) is monotone increasing with respect to n. Then the monotone convergence theorem yields

$$\begin{aligned}&\lim_{n\to\infty}\int_0^\infty e^{-\lambda t}\Pi_t(0)dt\int_{\mathbb{R}^d} W_n(x)\mathbb{E}^{x,x}_{[0,t]}\left[g\left(\int_0^t W_n(B_s)ds\right)\right]dx\\&=\int_0^\infty e^{-\lambda t}\Pi_t(0)dt\int_{\mathbb{R}^d} W(x)\mathbb{E}^{x,x}_{[0,t]}\left[g\left(\int_0^t W(B_s)ds\right)\right]dx<\infty.\end{aligned}$$

Together with (4.3.60), (4.3.52) follows, and (4.3.51) can be similarly proven. □

Lemma 4.157. *Suppose that* $\lambda > 0$, $W \in L^{d/2}(\mathbb{R}^d)$, *and* g *is a lower semicontinuous positive function. Let* f, g, *and* F *be related by* (4.3.50), *and* $F_g(\|K_\lambda\|) < \infty$. *Then* (4.3.51) *and* (4.3.52) *hold.*

Proof. Since any lower semicontinuous function is a monotone limit of continuous functions, there exist positive continuous functions g_n, $n \in \mathbb{N}$, such that $g_n(x) \uparrow g(x)$ for each x. Let $\chi_n \in C_0^\infty((1/n, n))$ be such that $0 \le \chi_n(x) \le 1$, $\chi_n(x) \uparrow 1$ for $x \in (0,\infty)$ as $n \to \infty$. Let $\bar{g}_n(x) = \chi_n(x)g_n(x)$. We have $\bar{g}_n(x) \uparrow g(x)$ and $\operatorname{supp}\bar{g}_n \subset (1/n, n)$. Hence $F_{\bar{g}_n}(K_\lambda) \in \mathscr{I}_1$ and

$$\operatorname{Tr} F_{\bar{g}_n}(K_\lambda)=\int_0^\infty e^{-\lambda t}\Pi_t(0)dt\int_{\mathbb{R}^d} W(x)\mathbb{E}^{x,x}_{[0,t]}\left[\bar{g}_n\left(\int_0^t W(B_s)ds\right)\right]dx. \tag{4.3.62}$$

Note that $\operatorname{Tr} F_{\bar{g}_n}(K_\lambda) \in \mathscr{I}_1$ is monotone increasing with respect to n, and condition $F_g(\|K_\lambda\|) < \infty$ implies that $F_g(K_\lambda)$ is compact and $F_{\bar{g}_n}(K_\lambda) \to F_g(K_\lambda)$ as $n \to \infty$ in the operator norm. We have $\mu_m(F_{\bar{g}_n}(K_\lambda)) \to \mu_m(F_g(K_\lambda))$ as $n \to \infty$ for each m by Proposition 4.44. It is sufficient to consider two cases.

(1) Case $\lim_{n\to\infty}\operatorname{Tr} F_{\bar{g}_n}(K_\lambda) < \infty$. By the monotone convergence theorem, $F_g(K_\lambda) \in \mathscr{I}_1$ and $e^{-\lambda t}W(x)g\left(\int_0^t W(B_s)ds\right) \in L^1([0,\infty)\times\mathbb{R}^d\times\mathscr{X}, dt\otimes dx\otimes\mathcal{W}^{x,x}_{[0,t]})$, and then

$$\operatorname{Tr} F_g(K_\lambda)=\int_0^\infty e^{-\lambda t}\Pi_t(0)dt\int_{\mathbb{R}^d} W(x)\mathbb{E}^{x,x}_{[0,t]}\left[g\left(\int_0^t W(B_s)ds\right)\right]dx \tag{4.3.63}$$

follows; so (4.3.52) is obtained.

(2) Case $\lim_{n\to\infty}\operatorname{Tr} F_{\bar{g}_n}(K_\lambda) = \infty$. Equality $\operatorname{Tr} F_g(K_\lambda) = \infty$ follows and the right-hand side of (4.3.63) is also infinity, showing (4.3.52). Equality (4.3.51) is similarly proven. □

Proof of Lemma 4.153. We see that $K_\delta = |K_\delta| = |K_\delta^*|$ and

$$(\Phi, K_\delta\Psi) = ((-\Delta)^{-1/2}W^{1/2}\Phi, (-\Delta)^{1/2}(-\Delta+\delta)^{-1}(-\Delta)^{1/2}(-\Delta)^{-1/2}W^{1/2}\Psi).$$

Both $(-\Delta)^{-1/2}W^{1/2}\Phi$ and $(-\Delta)^{-1/2}W^{1/2}\Psi$ are well defined since $d \geq 3$, and then we see that

$$\lim_{\delta\downarrow 0}(\Phi, K_\delta\Psi) = ((-\Delta)^{-1/2}W^{1/2}\Phi, (-\Delta)^{-1/2}W^{1/2}\Psi) = (\Phi, W^{1/2}(-\Delta)^{-1}W^{1/2}\Psi),$$

which implies that $K_\delta \to K_0$ weakly as $\delta \downarrow 0$. Furthermore, $K_\delta \in \mathscr{I}_2$ and $\|K_\delta\|_2$ can be expressed as

$$\|K_\delta\|_2^2 = \int_{\mathbb{R}^d\times\mathbb{R}^d} W(x)C_\delta(x-y)^2W(y)dxdy = \sum_{n=1}^{\infty}(W^{1/2}\phi_n, (-\Delta+\delta)^{-1}W^{1/2}\phi_n)^2$$

for any complete orthonormal system $\{\phi_n\}_{n\in\mathbb{N}}$ in $L^2(\mathbb{R}^d)$. From this, we see that the integral $\int_{\mathbb{R}^d\times\mathbb{R}^d} W(x)C_\delta(x-y)^2W(y)dxdy$ is monotone increasing as $\delta \downarrow 0$, and

$$\int_{\mathbb{R}^d\times\mathbb{R}^d} W(x)C_\delta(x-y)^2W(y)dxdy \leq \sum_{n=1}^{\infty}(W^{1/2}\phi_n, (-\Delta)^{-1}W^{1/2}\phi_n)^2 = \|K_0\|_2^2 < \infty.$$

We see that

$$\lim_{\delta\downarrow 0}\int_{\mathbb{R}^d\times\mathbb{R}^d} W(x)C_\delta(x-y)^2W(y)dxdy = \int_{\mathbb{R}^d\times\mathbb{R}^d} W(x)C_0(x-y)^2W(y)dxdy$$

by the monotone convergence theorem. This identically implies that

$$\lim_{\delta\downarrow 0}\|K_\delta\|_2 = \|K_0\|_2. \tag{4.3.64}$$

Hence $\|K_\delta - K_0\|_2 \to 0$ as $\delta \downarrow 0$ by Proposition 4.45. In particular, K_δ converges to K_0 in the operator norm, and then $F(K_\delta) \to F(K_0)$ in the operator norm. Hence we can conclude that $\mu_n(F(K_\delta)) \uparrow \mu_n(F(K_0))$ as $\delta \downarrow 0$ for each n.

(1) Case $\operatorname{Tr} F(K_\delta) < \infty$ and $\lim_{\delta\downarrow 0} \operatorname{Tr} F(K_\delta) < \infty$. By the monotone convergence theorem, we have $F(K_0) \in \mathscr{I}_1$ and $W(x)g\left(\int_0^t W(B_s)ds\right) \in L^1([0,\infty)\times\mathbb{R}^d\times\mathscr{X}, dt\otimes dx\otimes \mathcal{W}_{[0,t]}^{x,x})$, and then

$$\operatorname{Tr} F_g(K_0) = \int_0^\infty \Pi_t(0)dt \int_{\mathbb{R}^d} W(x)\mathbb{E}_{[0,t]}^{x,x}\left[g\left(\int_0^t W(B_s)ds\right)\right]dx. \tag{4.3.65}$$

(2) Case $\operatorname{Tr} F(K_\delta) < \infty$ and $\lim_{\delta\downarrow 0} \operatorname{Tr} F(K_\delta) = \infty$, or $\operatorname{Tr} F(K_\delta) = \infty$ for some $\delta > 0$. This time $\operatorname{Tr} F_g(K_0) = \infty$ follows and the right-hand side of (4.3.65) is also infinity, which shows (4.3.52). □

Corollary 4.158. *Let the conditions of Lemma* 4.153 *hold, and suppose that there exists an* ℓ *such that* $g(x) \leq \ell(x)x^m$ *and* $\int_0^\infty e^{-y}y^m\ell(xy)dy < \infty$ *for* $x \in [0,\infty)$ *with* $m = 2$ *for* $d \geq 3$, $m > 1$ *for* $d = 2$, *and* $m = 1$ *for* $d = 1$. *Then* $\operatorname{Tr} F(K_\lambda) < \infty$ *and* (4.3.51)–(4.3.52) *are satisfied.*

Proof. In Lemmas 4.154–4.157, $\int_0^\infty e^{-y} y^m \ell(xy)dy < \infty$ for arbitrary $x \in [0,\infty)$ implies that $F_g(K_\lambda) \in \mathscr{I}_1$ since $F_g(x) \le C_m x^m$. Then the corollary follows. □

We can now state the Lieb–Thirring inequality.

Theorem 4.159 (Lieb–Thirring inequality). *Let $V = V_+ - V_-$. Suppose that $d \ge 3$ and $V_- \in L^{d/2}(\mathbb{R}^d)$. Then there exists a constant a_d, independent of V, such that*

$$\bar{N}_0(V) \le a_d \int_{\mathbb{R}^d} |V_-(x)|^{d/2} dx.$$

Proof. Let F, f, and g be functions related by (4.3.50), and suppose that f is convex and $\int_0^\infty f(x)x^{-1}x^{-d/2}dx < \infty$. Hence $\operatorname{Tr} F(K_\lambda) < \infty$ and (4.3.51)–(4.3.52) are satisfied. Note that F is monotone increasing, thus $F(x)/F(1) \ge 1$, for $x \ge 1$. By Lemma 4.153, we have

$$\begin{aligned}
\bar{N}_0(V) &\le F(1)^{-1} \operatorname{Tr} F(K_0) \\
&= F(1)^{-1} \int_0^\infty \Pi_t(0)\frac{dt}{t} \int_{\mathbb{R}^d} \mathbb{E}^{x,x}_{[0,t]}\left[f\left(\int_0^t tW(B_s)\frac{ds}{t} \right) \right] dx \\
&\le F(1)^{-1} \int_0^\infty \Pi_t(0)\frac{dt}{t} \int_{\mathbb{R}^d} dx \int_0^t \frac{1}{t}\mathbb{E}^{x,x}_{[0,t]}\left[f\left(tW(B_s)\right)\right] ds,
\end{aligned}$$

where we applied Jensen's inequality to the convex function f. By using the definition of Brownian bridge

$$\mathbb{E}^{x,x}_{[0,t]}[h(B_s)] = \frac{1}{\Pi_t(0)} \int_{\mathbb{R}^d} h(y)\Pi_s(x-y)\Pi_{t-s}(y-x)dy,$$

the right-hand side above can be computed as

$$\begin{aligned}
&F(1)^{-1} \int_0^\infty \frac{dt}{t} \int_{\mathbb{R}^d} dx \int_0^t \frac{ds}{t} \int_{\mathbb{R}^d} f(tW(y))\Pi_s(x-y)\Pi_{t-s}(y-x)dy \\
&= F(1)^{-1} \int_0^\infty \Pi_t(0)\frac{dt}{t} \int_0^t \frac{ds}{t} \int_{\mathbb{R}^d} f(tW(y))dy \\
&= \frac{1}{(2\pi)^{d/2}F(1)} \int_0^\infty t^{-1-\frac{d}{2}} dt \int_{\mathbb{R}^d} f(tW(y))dy.
\end{aligned} \tag{4.3.66}$$

Changing the variable t to $s/W(y)$, we furthermore have

$$= \frac{1}{(2\pi)^{d/2}F(1)} \int_0^\infty f(s)s^{-1}s^{-d/2}ds \int_{\mathbb{R}^d} W(y)^{d/2}dy = a_d \int_{\mathbb{R}^d} W(y)^{d/2}dy,$$

where

$$a_d = \frac{1}{(2\pi)^{d/2}F(1)} \int_0^\infty f(s)s^{-1}s^{-d/2}ds = \frac{(2\pi)^{-d/2}\int_0^\infty f(s)s^{-1}s^{-d/2}ds}{\int_0^\infty f(s)s^{-1}e^{-s}ds}. \tag{4.3.67}$$

□

For $d = 1, 2$, we have the following corollary.

Corollary 4.160. *Let $V \in L^1(\mathbb{R})$ for $d = 1$, and $V \in L^p(\mathbb{R}^2)$ with $p > 1$ for $d = 2$. Suppose that V is negative and not identically zero. Then $H_0 + aV$ has a negative eigenvalue for every $a > 0$.*

Proof. Let $f = |V|^{1/2}\phi$ be such that $\phi \in L^2(\mathbb{R}^d)$ and $f \in L^2(\mathbb{R}^d) \cap L^1(\mathbb{R}^d)$. We have

$$(f, (H_0 + \lambda)^{-1}f) = \int_{\mathbb{R}^d} \frac{|\hat{f}(p)|^2}{|p|^2/2 + \lambda} dp \to \infty$$

as $\lambda \downarrow 0$ for $d = 1, 2$, since $\hat{f}$ is continuous and $\hat{f}(p) \neq 0$ for p near zero. Thus $(|V|^{1/2}\phi, (H_0 + \lambda)^{-1}|V|^{1/2}\phi)/\|\phi\|^2 \leq \|K_\lambda\| \to \infty$ as $\lambda \downarrow 0$, which means, together with the fact $-\lambda \in \mathrm{Spec}(H(\alpha)) \iff 1/\alpha \in \mathrm{Spec}(K_\lambda)$, that for every $\alpha > 0$ there exists $\lambda > 0$ such that $1/\alpha = \|K_\lambda\|$, since $\lambda \mapsto \|K_\lambda\|$ is continuous. Then $1/\alpha \in \mathrm{Spec}(K_\lambda)$, which implies that $-\lambda \in \mathrm{Spec}(H(\alpha))$. Hence for every $\alpha > 0$, there exists $-\lambda < 0$ such that $-\lambda \in \mathrm{Spec}(H(\alpha))$. □

Remark 4.161. For dimensions $d = 1$ and 2, the Schrödinger operator $H = H_0 + V$ with $V \leq 0$ (not identically zero) has at least one negative eigenvalue by Corollary 4.160. This may be compared with the case of $d \geq 3$ where, by the Lieb–Thirring inequality, there are no nonpositive eigenvalues if $\|V\|_{d/2}$ is sufficiently small.

The above ideas also allow studying other sums of negative eigenvalues of H such as the Riesz means

$$N^\beta(V) = \sum_{j=1}^{m} |E_j|^\beta, \quad m \leq \infty,$$

for $\beta > 0$. This is a byproduct of the estimate of $N_E(V)$ which can be used to define the counting measure $dN_E(V)$ on $[0, \infty)$. The map $[0, \infty) \ni E \mapsto N_E(V) \in \mathbb{N}$ is decreasing and we have

$$N^\beta(V) = -\int_0^\infty E^\beta dN_E(V).$$

Integration by parts gives

$$N^\beta(V) = -\int_0^\infty E^\beta \frac{dN_E(V)}{dE} dE = \beta \int_0^\infty E^{\beta-1} N_E(V) dE, \quad \beta > 0. \tag{4.3.68}$$

First, we make the following observation. Let $E > 0$. Since $(|k|^2 + E)^{-1/2} \in L^p(\mathbb{R}^d)$ for all $p > d$, Proposition 4.39 yields $(-\Delta + E)^{-1/2} V_-^{1/2} \in \mathscr{I}_{2\beta+d}$ for $V_- \in L^{\beta+(d/2)}(\mathbb{R}^d)$. Thus we have

$$V_-^{1/2}(-\Delta + E)^{-1} V_-^{1/2} \in \mathscr{I}_{\beta+(d/2)}.$$

In particular, $N_E(V) \le N_E(V_-) < \infty$ for every $E > 0$, and by the Birman–Schwinger principle, $N_E(V_-) \le \dim 1_{[1,\infty)}(V_-^{1/2}(-\Delta + E)^{-1} V_-^{1/2})$.

Theorem 4.162 (Lieb–Thirring inequality). *Suppose $d \ge 3$, $\beta > 0$, and let $V = V_+ - V_-$, with $V_- \in L^{\beta+(d/2)}(\mathbb{R}^d)$. Then*

$$N^\beta(V) \le a_{d,\beta} \int_0^\infty |V_-(x)|^{\beta+(d/2)} dx,$$

where $a_{d,\beta} = \beta a_d \int_0^1 z^{\beta-1}(1-z)^{d/2} dz$, with a_d in (4.3.67).

Proof. We denote the negative part of V by $-W$.

Step 1. First, we suppose that supp W is compact and let $E > 0$. We write the eigenvalue equation $Hf = -Ef$ as $(H + E)f = 0$. Since $V + E \ge -W + E \ge -(-W + E)_-$, we have

$$N_E(V) \le \bar{N}_0(-(-W + E)_-).$$

We also see that $W^{1/2} \in L^d(\mathbb{R}^d) \cap L^{d+2\beta}(\mathbb{R}^d)$. In particular, $(-W + E)_-^{1/2} \in L^d(\mathbb{R}^d)$ and it follows that $(-W+E)_-^{1/2}(-\Delta)^{-1}(-W+E)_-^{1/2}$ is compact. Consequently, in a similar manner to (4.3.66) in the proof of Theorem 4.159, we see that

$$N_E(V) \le \bar{N}_0(-(-W + E)_-) = F(1)^{-1} \int_0^\infty \frac{\Pi_t(0)}{t} dt \int_{\mathbb{R}^d} f(t(-(-W(y) + E)_-)) dy. \tag{4.3.69}$$

Here f is a nonnegative lower semicontinuous function on $[0, \infty)$ with $f(0) = 0$, and f, F are related by $F(x) = x \int_0^\infty e^{-y} g(xy) dy$ and $f(y) = g(y)y$. Combining (4.3.68) and (4.3.69), we obtain

$$N^\beta(V) = \beta \int_0^\infty E^{\beta-1} N_E(V) dE \le \beta \int_0^\infty E^{\beta-1} \bar{N}_0(-(-W(y) + E)_-) dE.$$

Inserting (4.3.69) into the right-hand side above, we have

$$N^\beta(V) \le \beta F(1)^{-1} \int_0^\infty E^{\beta-1} dE \int_0^\infty \frac{\Pi_t(0)}{t} dt \int_{\mathbb{R}^d} f(t(-(-W(y) + E)_-)) dy.$$

Since the integration domain with respect to E is actually $[0, W(y)]$, we have

$$N^\beta(V) \le \beta F(1)^{-1} \int_0^\infty \frac{\Pi_t(0)}{t} dt \int_{\mathbb{R}^d} dy \int_0^{W(y)} E^{\beta-1} f(t(W(y)-E)) dE$$
$$= \beta F(1)^{-1} \int_0^\infty \frac{\Pi_t(0)}{t} dt \int_{\mathbb{R}^d} dy \int_0^1 W(y)^\beta z^{\beta-1} f(tW(y)(1-z)) dz.$$

Changing the variable t to $s = t/(1-z)$ and using $\Pi_t(0) = (2\pi t)^{-d/2}$, we see that

$$N^\beta(V) \le \beta F(1)^{-1} \int_0^\infty \frac{\Pi_s(0)}{s} ds \int_{\mathbb{R}^d} W(y)^\beta f(sW(y)) dy \int_0^1 z^{\beta-1}(1-z)^{d/2} dz.$$

Thus as in the proof of Theorem 4.159, we obtain

$$N^\beta(V) \le a_{d,\beta} \int_{\mathbb{R}^d} W(y)^{\beta+(d/2)} dy.$$

This then shows the statement for W with compact support.

Step 2. Next assume that $W^{1/2} \in L^{d+2\beta}(\mathbb{R}^d)$, and let $W_n = W 1_{[-n,n]}$. We show that

$$N_E(-W_n) \to N_E(-W) \tag{4.3.70}$$

as $n \to \infty$ for every $E > 0$. Let $K_E(n) = W_n^{1/2}(-\Delta + E)^{-1} W_n^{1/2}$ and $K_E = W^{1/2}(-\Delta+E)^{-1}W^{1/2}$ for $E > 0$, and $\{\mu_j(K_E(n))\}_{j=1}^n$ (resp. $\{\mu_j(K_E)\}_{j=1}^n$) denote the eigenvalues of $K_E(n)$ (resp. K_E). From Proposition 4.39, it directly follows that

$$\|(-\Delta + E)^{-1/2} W_n^{1/2} - (-\Delta + E)^{-1/2} W^{1/2}\|_{d+2\beta}$$
$$\le (2\pi)^{-d/(d+2\beta)} \|W^{1/2} - W_n^{1/2}\|_{L^{d+2\beta}(\mathbb{R}^d)} \|(|k|^2 + E)^{-1/2}\|_{L^{d+2\beta}(\mathbb{R}^d)} \to 0$$

as $n \to \infty$. This implies that $\|(-\Delta + E)^{-1/2} W_n^{1/2} - (-\Delta + E)^{-1/2} W^{1/2}\| \to 0$ as $n \to \infty$, and hence $K_E(n) \to K_E$ uniformly as $n \to \infty$. Thus by Corollary 4.44, we see that $\mu_j(K_E(n)) \to \mu_j(K_E)$ as $n \to \infty$ for $j \ge 1$ and $E > 0$. Recall the identity $N_E(-W_n) = \dim 1_{[1,\infty)}(K_E(n))$. We conclude that $\dim 1_{[1,\infty)}(K_E(n)) \to \dim 1_{[1,\infty)}(K_E) < \infty$ as $n \to \infty$, and (4.3.70) follows. Finally, together with (4.3.70) and taking the limit $n \to \infty$ on both sides of

$$N^\beta(-W_n) = \beta \int_0^\infty E^{\beta-1} N_E(-W_n) dE \le a_{d,\beta} \int_{\mathbb{R}^d} |W_n(x)|^{\beta+(d/2)} dx,$$

we have

$$N^\beta(-W) \le a_{d,\beta} \int_{\mathbb{R}^d} |W(x)|^{\beta+(d/2)} dx.$$

□

4.3.7 Application to canonical commutation relations

The Lieb–Thirring inequality discussed in Theorem 4.162 can be applied to construct pairs of operators satisfying the *canonical commutation relations (CCR)*, which are a fundamental principle in quantum theory. In one-dimensional quantum mechanics, the momentum operator $P = -id/dx$ and the position operator Q given by multiplication by x satisfy the CCR in the form $[P, Q] = -i1$ on $D(PQ) \cap D(QP)$. There is the idea in the physics literature that time and energy are a similar complementary pair to position and momentum. To make mathematical sense of this, we aim to define a *time operator* as a symmetric operator satisfying the CCR with respect to the Hamiltonian of a quantum system. More precisely, we are interested in finding an operator T associated with a given self-adjoint operator S such that

$$[S, T] = -i1 \tag{4.3.71}$$

on $D(ST) \cap D(TS)$, and then call T a time operator associated with S. When the pair (S, T) satisfies CCR, it is known that either S or T is unbounded. Hence it may happen that $D(ST) \cap D(TS)$ is empty or not dense.

A *weak CCR* can be defined by replacing CCR (4.3.71) with a bilinear form

$$(S\phi, T\psi) - (T\phi, S\psi) = -i(\phi, \psi), \quad \phi, \psi \in D(S) \cap D(T). \tag{4.3.72}$$

A *weak time operator* T associated with S is then a symmetric operator satisfying (4.3.72).

A weak time operator can furthermore be extended to the so-called *ultraweak time operator*. A densely defined symmetric quadratic form $\mathfrak{t}(\cdot, \cdot) : L^2(\mathbb{R}^d) \times L^2(\mathbb{R}^d) \to \mathbb{C}$ is said to be an ultraweak time operator associated with a self-adjoint operator S whenever

$$\mathfrak{t}(S\phi, \psi) - \mathfrak{t}(\phi, S\psi) = -i(\phi, \psi) \tag{4.3.73}$$

for all $\psi, \phi \in \mathscr{D}$ with some domain $\mathscr{D}$, where $\mathscr{D}$ is called an *ultraweak CCR domain* for $(S, \mathfrak{t})$. Note that $\mathscr{D}$ is not necessarily dense. If T is a weak time operator, then T defines the ultraweak time operator $\hat{\mathfrak{t}} : L^2(\mathbb{R}^d) \times L^2(\mathbb{R}^d) \to \mathbb{C}$ by $\hat{\mathfrak{t}}(\phi, \psi) = (\phi, T\psi)$, with ultraweak CCR domain $D(S) \cap D(T)$.

The CCR are closely related to pairs of unitary groups. A pair of self-adjoint operators (A, B) is said to satisfy the *Weyl relation* whenever

$$e^{-isA} e^{-itB} = e^{ist} e^{-itB} e^{-isA}$$

holds for all $s, t \in \mathbb{R}$. The Weyl relation implies the CCR, and the pair (P, Q) satisfies the Weyl relation; in general, however, CCR does not imply the Weyl relation. If (A, B) satisfies the Weyl relation, B is called the *ultrastrong time operator* associated with A. It is known as the *von Neumann uniqueness theorem* that if a pair (A, B) satisfies the Weyl relation and there is no invariant domain with respect to e^{-isA} and e^{-itB}, then necessarily

$A \cong P$ and $B \cong Q$, where $\cong$ denotes unitary equivalence. When S is bounded from below, this theorem implies that there exists no symmetric operator T such that the pair (S,T) satisfies the Weyl relation, since $S \not\cong P$. Thus instead of the Weyl relation the so-called *weak Weyl relation* is used. A pair (A,B) is said to satisfy the weak Weyl relation whenever

(1) A is self-adjoint;
(2) B is symmetric;
(3) $e^{-itA}D(B) \subset D(B)$ and

$$Be^{-itA}\psi = e^{-itA}(B+t)\psi, \quad \forall\psi \in D(B),\ t \in \mathbb{R}.$$

Clearly, the Weyl relation implies the weak Weyl relation, and the weak Weyl relation implies the CCR. A symmetric operator T is a *strong time operator* associated with a self-adjoint operator S if and only if the pair (S,T) satisfies the weak Weyl relation.

Lemma 4.163. *Suppose that a strong time operator T associated with a self-adjoint operator S exists. Then the following hold:*
(1) *The closure $\bar{T}$ is also a strong time operator.*
(2) *If S is bounded below, then T has no self-adjoint extension.*
(3) $\mathrm{Spec}(S) = \mathrm{Spec}_{\mathrm{ac}}(S)$.

By this lemma, in what follows we may assume that the strong time operator is a closed symmetric operator, and there is no strong time operator if S has point spectrum.

Example 4.164 (Aharonov–Bohm operator). Let $P_j = -id/dx_j$ and Q_j be the multiplication by x_j for $j = 1,\dots,d$ in $L^2(\mathbb{R}^d)$. A strong time operator associated with P_j is Q_j. An important example includes the *Aharonov–Bohm operator* T_{AB}, which is a strong time operator associated with $\frac{1}{2}\sum_{j=1}^d P_j^2$ and defined by

$$T_{\mathrm{AB}} = \frac{1}{2}\sum_{j=1}^{d} \overline{(Q_jP_j^{-1} + P_j^{-1}Q_j)\lceil_{D_j}}, \tag{4.3.74}$$

with $D_j = \{\rho(P_j^2)D(Q_j) \mid \rho \in C_0^\infty(\mathbb{R}^d \setminus \{0\})\}$. We note that T_{AB} has no self-adjoint extensions.

In summary, the following hierarchy of relations can be seen:

$$\text{Weyl relation} \to \text{weak Weyl relation} \to \text{CCR} \to \text{weak CCR} \to \text{CCR}$$

in the quadratic form sense as in (4.3.73). Furthermore, this gives rise to several classes of time operators:

$$\text{ultrastrong time operator} \subset \text{strong time operator} \subset \text{time operator}$$
$$\subset \text{weak time operator} \subset \text{ultraweak time operator}.$$

Next we review time operators associated with self-adjoint operators S such that $\mathrm{Spec}(S) = \{E_j\}_{j=1}^{\infty}$ with $E_1 < E_2 < \cdots$ and $\lim_{n\to\infty} E_n = \infty$. In this case there exists no strong time operator by (3) of Lemma 4.163. Let $Se_{n\alpha} = E_n e_{n\alpha}$, $\alpha = 1, \ldots, M_n$, and $(e_{n\alpha}, e_{m\beta}) = \delta_{nm}\delta_{\alpha\beta}$, where M_n denotes the multiplicity of E_n. Let $\bar{e}_n = \frac{1}{\sqrt{M_n}} \sum_{\alpha=1}^{M_n} e_{n\alpha}$. Note that $(\bar{e}_n, \bar{e}_m) = \delta_{nm}$. Suppose that $\sum_{j=J}^{\infty} \frac{1}{E_j^2} < \infty$ for some $J \geq 1$. Then for a fixed n, we can see that

$$\sum_{m\neq n} \frac{1}{|E_n - E_m|^2} = \sum_{m\neq n} \frac{E_m^2}{|E_n - E_m|^2} \frac{1}{E_m^2} \leq M \sum_{m\neq n} \frac{1}{E_m^2} < \infty$$

with some constant M. It follows that

$$\sum_{m\neq n} \left| \frac{(\bar{e}_m, \phi)}{E_n - E_m} \right| \leq \left(\sum_{m\neq n} \frac{1}{|E_n - E_m|^2} \right)^{1/2} \left(\sum_{m\neq n} |(\bar{e}_m, \phi)|^2 \right)^{1/2} < \infty.$$

This means that $\sum_{m\neq n} \frac{(\bar{e}_m, \phi)}{E_n - E_m}$ is well defined for any n and we define T by

$$T\phi = i \sum_{n=1}^{\infty} \left(\sum_{m\neq n} \frac{(\bar{e}_m, \phi)}{E_n - E_m} \right) \bar{e}_n, \tag{4.3.75}$$

with

$$D(T) = \left\{ \phi \,\middle|\, \sum_{n=1}^{\infty} \left| \sum_{m\neq n} \frac{(\bar{e}_m, \phi)}{E_n - E_m} \right|^2 < \infty \right\}.$$

We can directly check the following lemma.

Lemma 4.165 (Time operator). *Suppose that* $\mathrm{Spec}(S) = \{E_j\}_{j=1}^{\infty}$*, where* E_j *is simple for all* $j \in \mathbb{N}$*, and assume* $\sum_{j=J}^{\infty} \frac{1}{E_j^2} < \infty$ *for some* $J \geq 1$*. Then* $[S, T] = -i\mathbb{1}$ *holds on* $\mathscr{E}$*, where* $\mathscr{E} =$ *linear hull of* $\{\bar{e}_n - \bar{e}_m \mid n, m \in \mathbb{N}\}$.

Note that $\mathscr{E}$ is not necessarily dense.

Now we construct ultraweak time operators associated with Schrödinger operators. Let $H = H_0 + V$ be a Schrödinger operator and assume that $\mathrm{Spec}(H) = \{E_j\}_{j=1}^{N} \cup [0, \infty)$ with $N = \infty$ and $\sum_{j=1}^{\infty} E_j^2 < \infty$. Let

$$L^2(\mathbb{R}^d) = \mathcal{H}_{\mathrm{ac}} \oplus \mathcal{H}_{\mathrm{p}} \oplus \mathcal{H}_{\mathrm{sc}}$$

be the decomposition of $L^2(\mathbb{R}^d)$ into the absolutely continuous part $\mathcal{H}_{\mathrm{ac}}$, point spectrum part $\mathcal{H}_{\mathrm{p}}$, and singular continuous part $\mathcal{H}_{\mathrm{sc}}$. Denote $H_{\mathrm{ac}} = H\lceil_{\mathcal{H}_{\mathrm{ac}}}$, $H_{\mathrm{sc}} = H\lceil_{\mathcal{H}_{\mathrm{sc}}}$, and $H_{\mathrm{p}} = H\lceil_{\mathcal{H}_{\mathrm{p}}}$. We have

$$H = H_{\mathrm{ac}} \oplus H_{\mathrm{p}} \oplus H_{\mathrm{sc}}.$$

We assume the following conditions on H.

Assumption 4.166. Let $H = H_0 + V$. We assume the following properties:
(1) $\mathrm{Spec}_{\mathrm{sc}}(H) = \emptyset$, i. e., $H = H_{\mathrm{ac}} \oplus H_{\mathrm{p}}$;
(2) $\mathrm{Spec}_{\mathrm{ac}}(H) = [0, \infty)$ and a strong time operator T_{ac} associated with H_{ac} exists;
(3) $\overline{\mathrm{Spec}_{\mathrm{p}}(H)} = \{0\} \cup \{E_j\}_{j=1}^{\infty}$, where E_j is simple and $E_j < 0$ for every $j \in \mathbb{N}$, $\{E_j\}_{j=1}^{\infty} = \mathrm{Spec}_{\mathrm{d}}(H)$ and $\sum_{j=1}^{\infty} E_j^2 < \infty$;
(4) $0 \notin \mathrm{Spec}_{\mathrm{p}}(H)$.

We note the following on Assumption 4.166. First, consider condition (2). We can construct a strong time operator associated with H_{ac} by a wave operator Ω playing an important role in the spectral scattering theory of Schrödinger operator. We present an outline below. Suppose that the wave operator $\Omega = \Omega^{-}(H, H_0) = \text{s-}\lim_{t\to+\infty} e^{itH} e^{-itH_0}$ exists. Then Ω satisfies the following properties: (1) $\Omega L^2(\mathbb{R}^d) \subset \mathcal{H}_{\mathrm{ac}}$, (2) $e^{-itH}\Omega = \Omega e^{-itH_0}$ for all $t \in \mathbb{R}$, (3) $\Omega^*\Omega = 1$, and (4) $\Omega\Omega^* =$ the projection onto $\mathcal{H}_{\mathrm{ac}}$. The strong time operator associated with H_{ac} can be constructed through wave operator Ω and the Aharonov–Bohm operator given in (4.3.74).

Lemma 4.167. *Let $T_{\mathrm{ac}} = \Omega T_{\mathrm{AB}} \Omega^*$ with $D(T_{\mathrm{ac}}) = \Omega D(T_{\mathrm{AB}})$. Then T_{ac} is the strong time operator associated with H_{ac}.*

Proof. Let $\phi' = \Omega\phi \in \Omega D(T_{\mathrm{AB}})$. Since $\Omega^*\Omega = 1$, $T_{\mathrm{ac}}\phi' = \Omega T_{\mathrm{AB}}\phi$ is well defined. It is seen that $e^{-itH} T_{\mathrm{ac}}\phi' = \Omega e^{-itH_0} T_{\mathrm{AB}}\phi = \Omega(T_{\mathrm{AB}} - t)e^{-itH_0}\phi$. Since $e^{-itH_0}\phi = \Omega^* e^{-itH}\Omega\phi$, we have $e^{-itH} T_{\mathrm{ac}}\phi' = (\Omega T_{\mathrm{AB}}\Omega^* - t\Omega\Omega^*)e^{-itH}\phi'$. Since $\Omega\Omega^*$ is the projection to $\mathcal{H}_{\mathrm{ac}}$, which is denoted by P_{ac}, and $\phi' = \Omega\phi \in \mathcal{H}_{\mathrm{ac}}$ and $\mathrm{Ran} T_{\mathrm{ac}} \subset \mathcal{H}_{\mathrm{ac}}$, we have $T_{\mathrm{ac}} e^{-itH_{\mathrm{ac}}}\phi' = e^{-itH_{\mathrm{ac}}}(T_{\mathrm{ac}} + t)\phi'$ and the proposition follows. □

Next consider (3) of Assumption 4.166. The condition $\sum_{j=1}^{\infty} E_j^2 < \infty$ is ensured by the Lieb–Thirring inequality

$$\sum_{j=1}^{\infty} E_j^2 \le \sum_{j=1}^{\infty} m_j E_j^2 \le a_{d,2} \int_{\mathbb{R}^d} |V_-(x)|^{2+\frac{d}{2}} dx,$$

where m_j denotes the multiplicity of E_j.

Finally, consider part (4) of Assumption 4.166. Although it is not trivial to check, the next statement gives a sufficient condition for the absence of zero eigenvalues of H.

Proposition 4.168. *We assume the following:* (1) $d \ge 3$, $V \in L^{d/2}_{\mathrm{loc}}(\mathbb{R}^d)$ *and* (2) *V can be decomposed into $V = V_1 + V_2$ such that for some $s \in [0, 1)$, some $R, C > 0$, and a positive differentiable function $h(r)$ defined on $[R, \infty)$ it follows that*
(a) *V_1 and V_2 are locally bounded on $S_R = \{x \in \mathbb{R}^d \mid |x| > R\}$, and V_1 is strictly negative on S_R;*
(b) *the condition*

$$\sup_{w \in S_{d-1}} \frac{d}{dr}(r^{s+1} V_1(rw)) \le -r^s h(r)$$

holds for $r > R$, where w is a coordinate on the $(d-1)$-dimensional unit sphere S_{d-1};

(c) *the limit*

$$\lim_{r\to\infty} \frac{r^{-1} + r\sup_{w\in S_{d-1}} |V_2(rw)|}{h(r)} = 0;$$

(d) *$dh(r)/dr \le Ch^2(r)$ on S_R;*

(e) *$\int_{S_R} h(|x|)^2|f(x)|^2dx < \infty$ and $\int_{S_R} |V_1(x)||f(x)|^2dx < \infty$ for any $f \in D(H)$.*

Then 0 is not an eigenvalue of H.

Example 4.169. We discuss two specific cases:

(1) Suppose that $V_1(x) = v(w)r^{-\mu} + R(x)$ and $V_2(x) = o(r^{-1-\mu/2})$, where $\mu \in [0,2)$, $\sup_{w\in S_{d-1}} v(w) < 0$, $R(x) = o(r^{-\mu})$, and $\sup_{w\in S_{d-1}} R(rw) = o(r^{-\mu-1})$. Then $V = V_1 + V_2$ satisfies assumptions stated above and $0 \notin \mathrm{Spec}_\mathrm{p}(H)$ follows.

(2) Let $a > 0$ be arbitrary. Putting $\mu = 0$ and $v(w) = -a < 0$ in (1), we can see that 0 is not an eigenvalue of $-\frac{1}{2}\Delta + V_2 - a$, i. e., $-\frac{1}{2}\Delta + V_2$ has no positive eigenvalues.

Let conditions (3)–(4) of Assumption 4.166 hold. Then

$$T_\mathrm{p}\phi = i\sum_{n=1}^{\infty}\left(\sum_{m\neq n}\frac{(\bar{e}_m,\phi)}{\frac{1}{E_n}-\frac{1}{E_m}}\right)\bar{e}_n$$

is a time operator associated with H_p^{-1} by Lemma 4.165. Since $H_\mathrm{p} = f(H_\mathrm{p}^{-1})$, where $f(x) = x^{-1}$, formally the time operator T associated with H_p is given by

$$T = \frac{1}{2}\left(T_\mathrm{p}\frac{1}{f'(H_\mathrm{p}^{-1})} + \frac{1}{f'(H_\mathrm{p}^{-1})}T_\mathrm{p}\right) = -\frac{1}{2}(T_\mathrm{p}H_\mathrm{p}^{-2} + H_\mathrm{p}^{-2}T_\mathrm{p}).$$

To verify this formula rigorously, instead of considering operators, we define the symmetric quadratic form $\mathfrak{t}_\mathrm{p} : D(T_\mathrm{p}) \times D(T_\mathrm{p}) \to \mathbb{C}$ on $\mathcal{H}_\mathrm{p}$ by

$$\mathfrak{t}_\mathrm{p}(\phi,\psi) = \begin{cases} -\frac{1}{2}\left((T_\mathrm{p}\phi, H_\mathrm{p}^{-2}\psi) + (H_\mathrm{p}^{-2}\phi, T_\mathrm{p}\psi)\right), \\ 0, \quad \text{otherwise.}\end{cases}$$

We formally write $\mathfrak{t}_\mathrm{p}(\phi,\psi) = (\phi, T\psi)$ and $T = -\frac{1}{2}(T_\mathrm{p}H_\mathrm{p}^{-2} + H_\mathrm{p}^{-2}T_\mathrm{p})$. Note that it is, however, not clear whether $D(H_\mathrm{p}^{-2}) \supset T_\mathrm{p}D(T_\mathrm{p})$. Hence we may not be able to define $\mathfrak{t}_\mathrm{p}$ as a symmetric operator. Now we define the densely defined symmetric quadratic form $\mathfrak{t}_H(\cdot,\cdot) : L^2(\mathbb{R}^d) \times L^2(\mathbb{R}^d) \to \mathbb{C}$ under the decomposition $L^2(\mathbb{R}^d) = \mathcal{H}_\mathrm{ac} \oplus \mathcal{H}_\mathrm{p}$ by

$$\mathfrak{t}_H(\phi_1 \oplus \phi_2, \psi_1 \oplus \psi_2) = (\phi_1, T_\mathrm{ac}\psi_1) + \mathfrak{t}_\mathrm{p}(\phi_2,\psi_2) \tag{4.3.76}$$

for $\phi_1, \psi_1 \in D(T_\mathrm{ac})$ and $\phi_2, \psi_2 \in D(T_\mathrm{p})$.

Theorem 4.170 (Ultraweak time operator). *Suppose Assumption* 4.166. *Then* $\mathfrak{t}_H$ *is an ultraweak time operator associated with* H *with ultraweak CCR domain* $D(T_{\mathrm{ac}}) \oplus H^{-1}\mathscr{E}$, *where* $H^{-1}\mathscr{E} =$ *linear hull of* $\{\frac{1}{E_n}\bar{e}_n - \frac{1}{E_m}\bar{e}_m \mid n, m \in \mathbb{N}\}$.

Note that $H^{-1}\mathscr{E}$ is not necessarily dense.

Proof. It is sufficient to show that $\mathfrak{t}_{\mathrm{p}}$ is an ultraweak time operator associated with H_{p} with ultraweak CCR domain $H_{\mathrm{p}}^{-1}\mathscr{E}$. We shall prove this. Let $\mathfrak{t} = -2\mathfrak{t}_{\mathrm{p}}$. Let $\phi' = H_{\mathrm{p}}^{-1}\phi$, $\psi' = H_{\mathrm{p}}^{-1}\psi \in H_{\mathrm{p}}^{-1}\mathscr{E}$. We see that

$$\mathfrak{t}(H_{\mathrm{p}}\phi', \psi') - \mathfrak{t}(\phi', H_{\mathrm{p}}\psi') = \mathfrak{t}(\phi, H_{\mathrm{p}}^{-1}\psi) - \mathfrak{t}(H_{\mathrm{p}}^{-1}\phi, \psi).$$

By the definition of $\mathfrak{t}$, we have

$$\begin{aligned}
&\mathfrak{t}(H_{\mathrm{p}}\phi', \psi') - \mathfrak{t}(\phi', H_{\mathrm{p}}\psi') \\
&= (T_{\mathrm{p}}\phi, H_{\mathrm{p}}^{-3}\psi) + (H_{\mathrm{p}}^{-2}\phi, T_{\mathrm{p}}H_{\mathrm{p}}^{-1}\psi) - (H_{\mathrm{p}}^{-3}\phi, T_{\mathrm{p}}\psi) - (T_{\mathrm{p}}H_{\mathrm{p}}^{-1}\phi, H_{\mathrm{p}}^{-2}\psi) \\
&= (H_{\mathrm{p}}^{-1}T_{\mathrm{p}}\phi, H_{\mathrm{p}}^{-2}\psi) - (H_{\mathrm{p}}^{-2}\phi, H_{\mathrm{p}}^{-1}T_{\mathrm{p}}\psi) + (H_{\mathrm{p}}^{-2}\phi, T_{\mathrm{p}}H_{\mathrm{p}}^{-1}\psi) - (T_{\mathrm{p}}H_{\mathrm{p}}^{-1}\phi, H_{\mathrm{p}}^{-2}\psi).
\end{aligned}$$

The first two terms of the rightmost side above can be computed by using $[H_{\mathrm{p}}^{-1}, T_{\mathrm{p}}] = -i\mathbb{1}$ on $\mathscr{E}$ as

$$\begin{aligned}
&(H_{\mathrm{p}}^{-1}T_{\mathrm{p}}\phi, H_{\mathrm{p}}^{-2}\psi) - (H_{\mathrm{p}}^{-2}\phi, H_{\mathrm{p}}^{-1}T_{\mathrm{p}}\psi) \\
&\quad = 2i(H_{\mathrm{p}}^{-1}\phi, H_{\mathrm{p}}^{-1}\psi) + (T_{\mathrm{p}}H_{\mathrm{p}}^{-1}\phi, H_{\mathrm{p}}^{-2}\psi) - (H_{\mathrm{p}}^{-2}\phi, T_{\mathrm{p}}H_{\mathrm{p}}^{-1}\psi).
\end{aligned}$$

Hence we conclude that $\mathfrak{t}(H_{\mathrm{p}}\phi', \psi') - \mathfrak{t}(\phi', H_{\mathrm{p}}\psi') = 2i(\phi', \psi')$ and the theorem follows. □

Now we give some examples of V such that H has an ultraweak time operator. It is sufficient to check Assumption 4.166. Suppose that V is of the form

$$V(x) = \frac{W(x)}{(|x|^2 + 1)^{1/2+\varepsilon}} \tag{4.3.77}$$

for some $\varepsilon > 0$, where $W : \mathbb{R}^d \to \mathbb{R}$ and $W(-\Delta + i)^{-1}$ is compact. If V is of the form (4.3.77), then V is called the *Agmon potential*. Agmon potentials form a linear space of $-\Delta$-bounded perturbations of relative bound zero. In particular, H is self-adjoint on $D(H_0)$. The perturbation by Agmon potential V leaves the essential spectrum of H_0 invariant, i. e., $\mathrm{Spec}_{\mathrm{ess}}(H) = \mathrm{Spec}_{\mathrm{ess}}(H_0) = [0, \infty)$. The following facts are known.

Lemma 4.171. *Let V be an Agmon potential. Then we have the following:*
(1) $\mathrm{Spec}_{\mathrm{sc}}(H) = \emptyset$;
(2) *the wave operator* $\Omega(H, H_0) = s\text{-}\lim_{t\to\infty} e^{-itH}e^{itH_0}$ *exists and is complete; in particular,* $\mathrm{Spec}_{\mathrm{ac}}(H) = [0, \infty)$;
(3) *the set of positive eigenvalues of H is a discrete subset in* $(0, \infty)$.

Example 4.172 (Agmon potential).

(1) It is known that any $G \in L^p(\mathbb{R}^d)$ for $d/2 < p < \infty$ and $p \geq 2$ is relatively compact with respect to $-\Delta$. Then $V(x) = \frac{G(x)}{(1+|x|^2)^{1/2+\varepsilon}}$, $\varepsilon > 0$, is an Agmon potential.

(2) $V(x) = \frac{U(x)}{(1+|x|^2)^{1/2+\varepsilon}}$, $\varepsilon > 0$, with $U \in L^\infty(\mathbb{R}^d)$ with $d \geq 3$ is an Agmon potential.

Theorem 4.173. *Let $d = 3$ and V be a nonpositive continuous spherically symmetric Agmon potential, i. e., $V(x) = W(r)$, where $r = |x|$. Suppose that*

(1) $V(x) \leq -\frac{a}{|x|^{2-\delta}}$ *for $|x| > R$ with some $R > 0$, $a > 0$ and $\delta > 0$;*

(2) $\int_b^\infty |W(r)|dr < \infty$ *for some $b > 0$ and $V \in L^2_{\mathrm{loc}}(\mathbb{R}^d \setminus \{0\})$;*

(3) $\int_{\mathbb{R}^3} |V(x)|^{7/2}dx < \infty$.

Then an ultraweak time operator associated with H exists.

Proof. By Proposition 4.171, $\mathrm{Spec}_{\mathrm{sc}}(H) = \emptyset$ and then $H = H_{\mathrm{ac}} \oplus H_{\mathrm{p}}$. Under (1) and (2), we can see that $\mathrm{Spec}(H) = \{E_j\}_{j=1}^\infty \cup [0,\infty)$, $\overline{\mathrm{Spec}_{\mathrm{p}}(H)} = \{0\} \cup \{E_j\}_{j=1}^\infty$, $0 \notin \mathrm{Spec}_{\mathrm{p}}(H)$, and $\mathrm{Spec}_{\mathrm{ac}}(H) = [0,\infty)$. Since the wave operator $\Omega(H, H_0)$ exists, $T_{\mathrm{ac}} = \Omega T_{\mathrm{AB}} \Omega^*$ is a strong time operator associated with H_{ac} by Proposition 4.167. Furthermore, (3) implies $\sum_{j=1}^\infty E_j^2 < \infty$ by the Lieb–Thirring inequality. Then the ultraweak time operator $\mathfrak{t}_{\mathrm{p}}$ associated with H_{p} exists with the ultraweak CCR domain $H_{\mathrm{p}}^{-1}\mathscr{E}$. We set

$$\mathfrak{t}(\phi, \psi) = (\phi_1, T_{\mathrm{ac}}\psi_1) + \mathfrak{t}_{\mathrm{p}}(\phi_2, \psi_2)$$

for $\phi = \phi_1 \oplus \phi_2$ and $\psi = \psi_1 \oplus \psi_2$ under the identification $L^2(\mathbb{R}^3) = \mathcal{H}_{\mathrm{ac}} \oplus \mathcal{H}_{\mathrm{p}}$. Hence symmetric quadratic form $\mathfrak{t}$ is the ultraweak time operator associated with H with ultraweak CCR domain $D(T_{\mathrm{ac}}) \cap H^{-1}\mathscr{E}$. □

Example 4.174. Let $d = 3$. Suppose that $U \in L^\infty(\mathbb{R}^3)$. Then $V(x) = \frac{U(x)}{(1+|x|^2)^{1/2+\varepsilon}}$ is an Agmon potential for all $\varepsilon > 0$. Suppose that U is negative, continuous, and spherically symmetric, as well as satisfies $U(x) = 1/|x|^a$ for $|x| > R$ with $0 < a < 1$ and some $R > 0$. For each a, we can choose $\varepsilon > 0$ such that $2\varepsilon + a < 1$. Hence V satisfies (1), (2), and (3) of Theorem 4.173. Hence an ultraweak time operator $\mathfrak{t}_H$ associated with H_{p} exists, and an ultraweak time operator associated with H exists.

Example 4.175. Let $d = 3$. The hydrogen atom Schrödinger operator is defined by $H_{\mathrm{hyd}} = H_0 - \frac{1}{|x|}$. It is known that $\mathrm{Spec}_{\mathrm{sc}}(H_{\mathrm{hyd}}) = \emptyset$, $\mathrm{Spec}_{\mathrm{p}}(H_{\mathrm{hyd}}) = \{-\frac{1}{2}j^{-2}\}_{j=1}^\infty$, and $\mathrm{Spec}_{\mathrm{ac}}(H_{\mathrm{hyd}}) = [0,\infty)$. Thus $\sum_{j=1}^\infty j^{-4} < \infty$ and $0 \notin \mathrm{Spec}_{\mathrm{p}}(H_{\mathrm{hyd}})$. Then an ultraweak time operator associated with $H_{\mathrm{hyd,p}}$ exists. Moreover, the modified wave operator $\Omega_D(H_{\mathrm{hyd}}, H_0)$ is defined by $\Omega_D(H_{\mathrm{hyd}}, H_0) = \mathrm{s} - \lim_{t\to\infty} e^{itH_{\mathrm{hyd}}} U_D(t)$ with some unitary operator $U_D(t)$. Operator $\Omega = \Omega_D(H, H_0)$ plays the roll of Ω in Proposition 4.167. Hence there exists a strong time operator associated with $H_{\mathrm{hyd,ac}}$. Together with them, we can conclude that an ultraweak time operator associated with H_{hyd} exists.

4.3.8 Exponential decay of eigenfunctions

A sufficient condition for the existence of a ground state of a Schrödinger operator is that $\liminf_{|x|\to\infty} V(x) = \infty$. However, ground states exist also for many other choices of V, such as Coulomb potential in three dimensions. In many cases these ground states decay exponentially at infinity, while for potentials growing at infinity they may decay even faster.

An intuitive derivation of the order of decay of the eigenfunctions is obtained as follows. Suppose the potential is of the form $V(x) \sim |x|^{2m}$ and the ground state of the form $\psi(x) \sim e^{-|x|^n}$. From the eigenvalue equation $-\frac{1}{2}\Delta\psi + |x|^{2m}\psi = E\psi$, we have

$$-\frac{1}{2}(n(n-1)|x|^{n-2} + n^2|x|^{2n-2})\psi + |x|^{2m}\psi \sim E\psi.$$

Comparing the leading order terms for large $|x|$ at the two sides, we expect that $n = m+1$. This can, in fact, be proven by using the Feynman–Kac formula. Suppose $\psi \in L^2(\mathbb{R}^d)$ with $H\psi = \lambda\psi$ for some $\lambda \in \mathbb{R}$. By adding a constant to V, we may assume $\lambda = 0$. Thus

$$\psi(x) = e^{-tH}\psi(x) = \mathbb{E}^x[e^{-\int_0^t V(B_s)\,ds}\psi(B_t)].$$

Therefore the value of $\psi(x)$ is obtained by running a Brownian motion from x under the potential V. This suggests that starting in regions where V is large, the effect will be quite small as most of the Brownian paths take a long time to leave a finite region. Formally,

$$|\psi(x)|^2 \leq \|\psi\|_\infty^2 \mathbb{E}^x[e^{-2\int_0^t V_+(B_s)\,ds}]\mathbb{E}^x[e^{-2\int_0^t V_-(B_s)\,ds}]$$

by the Schwarz inequality. By Khasminskii's lemma, the expectation involving V_- grows at most exponentially in t with a constant that can be estimated conveniently.

Now we present Carmona's argument in detail. This will work for any eigenfunction of H and not just the ground state; therefore we assume $\varphi \in L^2(\mathbb{R}^d)$ with

$$H\varphi = E\varphi.$$

For the present purposes, we introduce two further classes of potentials V.

Definition 4.176. The potential classes $\mathbb{V}^{\text{upper}}$ and $\mathbb{V}^{\text{lower}}$ are defined by the following properties:

($\mathbb{V}^{\text{upper}}$) $V \in \mathbb{V}^{\text{upper}}$ if and only if $V = W - U$ such that
  (1) $U \geq 0$ and $U \in L^p(\mathbb{R}^d)$ for $p > d/2$ and $1 \leq p < \infty$;
  (2) $W \in L^1_{\text{loc}}(\mathbb{R}^d)$ and $W_\infty = \inf_{x\in\mathbb{R}^d} W(x) > -\infty$.

($\mathbb{V}^{\text{lower}}$) $V \in \mathbb{V}^{\text{lower}}$ if and only if $V = W - U$ such that
  (1) $U \geq 0$ and $U \in L^p(\mathbb{R}^d)$ for $p > d/2$ and $1 \leq p < \infty$;
  (2) $W \geq 0$ and $W \in \mathcal{K}_{\text{loc}}(\mathbb{R}^d)$.

Let $V = W - U \in \mathbb{V}^{\mathrm{upper}}$. We have $W \in L^1_{\mathrm{loc}}(\mathbb{R}^d)$ and $0 \geq -U \in L^p(\mathbb{R}^d)$ implies $-U \in \mathcal{K}(\mathbb{R}^d)$. Hence $e^{-tH} : L^2(\mathbb{R}^d) \to L^\infty(\mathbb{R}^d)$ by Theorem 4.107 and Remark 4.115; in particular, $\varphi \in L^\infty(\mathbb{R}^d)$.

A fundamental estimate of which we will make use in showing the spatial decay of eigenfunctions is given in the following lemma.

Lemma 4.177 (Carmona's estimate). *Let $V \in \mathbb{V}^{\mathrm{upper}}$. Then for every $t, a > 0$ and every $0 < \alpha < 1/2$, there exist constants $D_1, D_2, D_3 > 0$ such that*

$$|\varphi(x)| \leq t^{-d/2} D_1 e^{D_2 \|U\|_p^m t} e^{Et} (D_3 e^{-\frac{\alpha}{4}\frac{a^2}{t}} e^{-tW_\infty} + e^{-tW_a(x)}) \|\varphi\|, \tag{4.3.78}$$

where $m = (1 - \frac{d}{2p})^{-1}$ and $W_a(x) = \inf\{W(y) \mid |x - y| < a\}$.

Proof. By the Schwarz inequality,

$$|\varphi(x)| \leq e^{tE} (\mathbb{E}^x[e^{-4\int_0^t W(B_s)ds}])^{1/4} (\mathbb{E}^x[e^{+4\int_0^t U(B_s)ds}])^{1/4} \mathbb{E}[|\varphi(x + B_t)|^2]^{1/2}. \tag{4.3.79}$$

Note that

$$\mathbb{E}[|\varphi(x + B_t)|^2] = \int_{\mathbb{R}^d} \Pi_t(y) |\varphi(x + y)|^2 dy \leq (2\pi t)^{-d/2} \|\varphi\|^2.$$

Let $A = \{\omega \in \mathscr{X} \mid \sup_{0 \leq s \leq t} |B_s(\omega)| > a\}$. It follows from Lévy's maximal inequality that

$$\mathbb{E}[1_A] \leq 2\mathbb{E}[1_{\{|B_t| \geq a\}}] = \frac{2}{(2\pi)^{d/2}} |S_{d-1}| \int_{a/\sqrt{t}}^\infty e^{-r^2/2} r^{d-1} dr \leq \xi_\alpha e^{-\alpha a^2/t}$$

with some ξ_α, for $0 < \alpha < 1/2$, where $|S_{d-1}|$ is the surface area of the $(d-1)$-dimensional unit sphere. The first factor in (4.3.79) is estimated as

$$\begin{aligned} \mathbb{E}^x[e^{-4\int_0^t W(B_s)ds}] &= \mathbb{E}^0[1_A e^{-4\int_0^t W(B_s + x)ds}] + \mathbb{E}^x[1_{A^c} e^{-4\int_0^t W(B_s)ds}] \\ &\leq e^{-4tW_\infty} \mathbb{E}^0[1_A] + e^{-4tW_a(x)} \leq \xi_\alpha e^{-\alpha a^2/t} e^{-4tW_\infty} + e^{-4tW_a(x)}. \end{aligned} \tag{4.3.80}$$

Next we estimate the second factor. Since U is in Kato-class, there exist constants $D_1, D_2 > 0$ such that

$$\mathbb{E}^x[e^{+4\int_0^t U(B_s)ds}] \leq D_1 e^{D_2 \|U\|_p^m t} \tag{4.3.81}$$

by Lemma 4.105. Setting $D_3 = \xi_\alpha^{1/4}$, we obtain the lemma by using the inequality $(a + b)^{1/4} \leq a^{1/4} + b^{1/4}$ for $a, b \geq 0$. □

Corollary 4.178 (Confining potential). *Let $V = W - U \in \mathbb{V}^{\text{upper}}$. Suppose that $W(x) \geq \gamma|x|^{2n}$ outside a compact set K, for some $n > 0$ and $\gamma > 0$. Take $0 < a < 1/2$. Then there exists a constant $C_1 > 0$ such that*

$$|\varphi(x)| \leq C_1\|\varphi\|e^{-\frac{ac}{16}|x|^{n+1}}, \tag{4.3.82}$$

where $c = \inf_{x\in\mathbb{R}^d\setminus K} W_{|x|/2}(x)/|x|^{2n}$.

Proof. Since $\sup_{x\in\mathbb{R}^d}|\varphi(x)| < \infty$, it suffices to show all the statements for sufficiently large $|x|$. Note that $W_{|x|/2}(x) \geq c|x|^{2n}$ for $x \in \mathbb{R}^d \setminus K$. We have the bounds

$$|x|W_{|x|/2}(x)^{1/2} \geq c|x|^{n+1} \quad \text{and} \quad |x|W_{|x|/2}(x)^{-1/2} \leq c|x|^{1-n}$$

for $x \in \mathbb{R}^d \setminus K$. Inserting $t = t(x) = |x|W_{|x|/2}(x)^{-1/2}$ and $a = a(x) = \frac{|x|}{2}$ into (4.3.78), we have

$$|\varphi(x)| \leq \|\varphi\|(c|x|^{n+1})^{-d/2}e^{-\frac{a}{16}c|x|^{n+1}}D_1e^{(D_2\|U\|_p^m+E_p)c|x|^{1-n}}(D_3e^{c|x|^{1-n}|W_\infty|} + e^{-(1-\frac{a}{16})c|x|^{n+1}})$$

for $x \in \mathbb{R}^d \setminus K$, where $m = (1-\frac{d}{2p})^{-1}$. Hence (4.3.82) follows. □

For $V = W - U \in \mathbb{V}^{\text{upper}}$, define $\Sigma = \liminf_{|x|\to\infty} V(x)$. Since $-U \in L^p(\mathbb{R}^d)$, $\liminf_{|x|\to\infty}(-U(x)) = 0$ and hence $\Sigma = \liminf_{|x|\to\infty} W(x)$. Moreover, $\Sigma \geq W_\infty$ holds.

Corollary 4.179 (Decay property: upper bound). *Let $V = W - U \in \mathbb{V}^{\text{upper}}$.*

(1) *(Decaying potential) Suppose that $\Sigma > E$, $\Sigma > W_\infty$, and let $0 < \beta < 1$. Then there exists a constant $C_2 > 0$ such that*

$$|\varphi(x)| \leq C_2\|\varphi\|\, e^{-\frac{\beta}{8\sqrt{2}}\frac{\Sigma-E}{\sqrt{\Sigma-W_\infty}}|x|}. \tag{4.3.83}$$

(2) *(Confining potential) Suppose that $\lim_{|x|\to\infty} W(x) = \infty$. Then there exist constants $C, \delta > 0$ such that*

$$|\varphi(x)| \leq C\|\varphi\|\, e^{-\delta|x|}. \tag{4.3.84}$$

Proof. Since $\sup_{x\in\mathbb{R}^d}|\varphi(x)| < \infty$, it is again sufficient to show the statements for large enough $|x|$.

Decaying case. Rewrite formula (4.3.78) as

$$|\varphi(x)| \leq t^{-d/2}D_1e^{D_2\|U\|_p^m t}(D_3e^{-\frac{a}{4}\frac{a^2}{t}}e^{-t(W_\infty-E)} + e^{-t(W_a(x)-E)})\|\varphi\|, \tag{4.3.85}$$

where $m = (1-\frac{d}{2p})^{-1}$. Upon flipping signs, with $\Sigma = \liminf_{|x|\to\infty}(-W_-(x))$ and $\Sigma > W_\infty$, it is possible to choose a decomposition $V = W - U \in \mathbb{V}^{\text{upper}}$ such that $D_2\|U\|_p^m \leq (\Sigma - E_p)/2$ since $\liminf_{|x|\to\infty}(-U(x)) = 0$. Inserting $t = t(x) = \varepsilon|x|$ and $a = a(x) = \frac{|x|}{2}$ into (4.3.85), we have

$$\begin{aligned}|\varphi(x)| &\le \frac{D_1 e^{D_2\|U\|_p^m \varepsilon|x|}}{(\varepsilon|x|)^{d/2}}(D_3 e^{-\frac{a}{16\varepsilon}|x|}e^{-\varepsilon|x|(W_\infty - E)} + e^{-\varepsilon|x|(W_{|x|/2}(x)-E)})\|\varphi\| \\ &\le \frac{D_1}{(\varepsilon|x|)^{d/2}}\left(D_3 e^{-(\frac{a}{16\varepsilon}+\varepsilon(W_\infty - E)-\frac{1}{2}\varepsilon(\Sigma-E))|x|} + e^{-\varepsilon((W_{|x|/2}(x)-E)-\frac{1}{2}(\Sigma-E))|x|}\right)\|\varphi\|.\end{aligned}$$

Choosing $\varepsilon = \sqrt{a/16}/\sqrt{\Sigma - W_\infty}$, the exponent in the first term above becomes

$$\frac{a}{16\varepsilon} + \varepsilon(W_\infty - E) - \frac{1}{2}\varepsilon(\Sigma - E) = \frac{1}{2}\varepsilon(\Sigma - E).$$

Moreover, we see that $\liminf_{|x|\to\infty} W_{|x|/2}(x) = \Sigma$, and we obtain

$$|\varphi(x)| \le C_2\|\varphi\|\, e^{-\frac{\varepsilon}{2}(\Sigma-E)|x|}$$

for sufficiently large $|x|$. Thus (4.3.83) follows.

Confining case. In this case for any $c > 0$, there exists large $N > 0$ such that $W_{|x|/2}(x) \ge c$, for all $|x| > N$. Inserting $t = t(x) = \varepsilon|x|$ and $a = a(x) = \frac{|x|}{2}$ into (4.3.78), we obtain

$$\begin{aligned}|\varphi(x)| &\le \frac{D_1 e^{D_2\|U\|_p^m \varepsilon|x|}}{(\varepsilon|x|)^{d/2}}(D_3 e^{-\frac{a}{16\varepsilon}|x|}e^{-\varepsilon|x|(W_\infty - E)} + e^{-\varepsilon|x|(W_{|x|/2}(x)-E)})\|\varphi\| \\ &\le \frac{D_1}{(\varepsilon|x|)^{d/2}}(D_3 e^{-(\frac{a}{16\varepsilon}-\varepsilon D_2\|U\|_p^m+\varepsilon(W_\infty - E))|x|} + e^{-\varepsilon(c-E-D_2\|U\|_p^m)|x|})\|\varphi\|\end{aligned}$$

for $|x| > N$. Choosing a sufficiently large c and a sufficiently small ε such that

$$\frac{a}{16\varepsilon} - \varepsilon D_2\|U\|_p^m + \varepsilon(W_\infty - E) > 0 \quad \text{and} \quad c - E - D_2\|U\|_p^m > 0,$$

we obtain $|\varphi(x)| \le C' e^{-\delta'|x|}$ for large enough $|x|$. Thus (4.3.84) follows. □

By using the Feynman–Kac formula, we can also establish a lower bound on positive eigenfunctions.

Lemma 4.180. *Let $V = W - U \in \mathbb{V}^{\text{lower}}$. Then the ground state φ is continuous and strictly positive.*

Proof. We have $\varphi(x) = e^{-t(H-E(H))}\varphi(x) = \mathbb{E}^x[e^{-\int_0^t (V(B_s)-E(H))ds}\varphi(B_t)]$, for all $t \ge 0$. Note that V is Kato-decomposable. Hence $\varphi(x)$ is continuous in x by Theorem 4.111. Since e^{-tH} is positivity improving, φ is strictly positive a. e. Let $N = \{x \in \mathbb{R}^d \mid \varphi(x) = 0\}$, which is a set of Lebesgue measure zero. Suppose that $x \in N$ and then $0 = \varphi(x) = \mathbb{E}^x[e^{-\int_0^t (V(B_s)-E(H))ds}\varphi(B_t)]$. Since $e^{-\int_0^t (V(B_s)-E(H))ds} > 0$ a. e., we have $\mathbb{E}^x[\varphi(B_t)] = 0$. This implies $0 = \int_{\mathbb{R}^d}\varphi(y)\Pi_t(x-y)dy$ and we see that $\varphi(y) = 0$ a. e. This contradicts $\varphi \ne 0$. Hence $N = \emptyset$ and the statement follows. □

Next we derive a preliminary estimate to the lower decay bound below. For $a, t \ge 0$ and $G \subset \mathbb{R}$, define

$$P(a, G, t) = \left\{ \sup_{0\le s\le t} |B_s| \le a, B_t \in G \right\}.$$

It is a known result that

$$\mathcal{W}(P(a, [c,d], t)) = \frac{1}{\sqrt{2\pi t}} \int_{[c,d]} \sum_{k=-\infty}^{\infty} (-1)^k \exp\left(-\frac{(u-2ka)^2}{2t}\right) du. \tag{4.3.86}$$

Note that $\sum_{k=-\infty}^{\infty} \exp\left(-(u-2ka)^2/2t\right) < \infty$ for every u, and

$$\int_{[c,d]} \sum_{k=-\infty}^{\infty} \exp\left(-\frac{(u-2ka)^2}{2t}\right) du < \infty.$$

Lemma 4.181. *Suppose that $a, \alpha, t > 0$ are such that $a > 2\alpha$ and $a^2/t > \beta$. Here β is the unique positive solution of*

$$F(\xi) = 1 - (e^{-21\xi/8} + e^{-5\xi/8} + e^{-165\xi/8}) = 0.$$

Then for every interval $[x-\alpha, x+\alpha] \subset [-a, a]$ it follows that

$$\mathcal{W}(P(a, [x-\alpha, x+\alpha], t)) \ge \frac{\alpha}{\sqrt{2\pi t}} F\left(\frac{a^2}{t}\right) e^{-\frac{a^2}{2t}}. \tag{4.3.87}$$

Proof. Let

$$g(x) = \mathcal{W}(P(a, [x-\alpha, x+\alpha], t) = \frac{1}{\sqrt{2\pi t}} \int_{x-\alpha}^{x+\alpha} \sum_{k=-\infty}^{\infty} (-1)^k \exp\left(-\frac{(u-2ka)^2}{2t}\right) du.$$

It can be seen that $g(x) = g(-x)$. Choose $x \in [0, a-\alpha]$ and let

$$f(x) = \mathcal{W}(P(a, [x, x+\alpha], t)) = \frac{1}{\sqrt{2\pi t}} \int_{x}^{x+\alpha} \sum_{k=-\infty}^{\infty} (-1)^k \exp\left(-\frac{(u-2ka)^2}{2t}\right) du.$$

We have $g(x) \ge f(x)$. Since $df(x)/dx \le 0$, f is monotone decreasing and we get

$$f(x) \ge f(a-\alpha) \ge \frac{1}{\sqrt{2\pi t}} \int_{a-\alpha}^{a} \sum_{k=-\infty}^{\infty} (-1)^k \exp\left(-\frac{(u-2ka)^2}{2t}\right) du.$$

Estimating the right-hand side, we obtain

$$\int_{a-\alpha}^{a} \sum_{k=-\infty}^{\infty} (-1)^k \exp\left(-\frac{(u-2ka)^2}{2t}\right) du \ge \alpha \left(e^{-\frac{a^2}{2t}} + I(a,t) - J(a,t) \right),$$

where

$$I(a,t) = \sum_{m\in\mathbb{Z}\setminus\{0\}} I(m) = \sum_{m\in\mathbb{Z}\setminus\{0\}} \exp\left(-\frac{|(2-8m)a|^2}{8t}\right),$$

$$J(a,t) = \sum_{m\in\mathbb{Z}} J(m) = \sum_{m\in\mathbb{Z}} \exp\left(-\frac{|(5-8m)a|^2}{8t}\right).$$

We see that $I(-m) - J(m+1) > 0$ for $m \geq 1$, and $I(m) - J(-m-1) > 0$ for $m \geq 1$. Hence we have

$$\int_{a-a}^{a} \sum_{k=-\infty}^{\infty} (-1)^k \exp\left(-\frac{(u-2ka)^2}{2t}\right) du \geq a\left(e^{-\frac{a^2}{2t}} - J(0) - J(1) - J(-1)\right).$$

Write $\xi = a^2/t$; so $e^{-\frac{a^2}{2t}} - J(0) - J(1) - J(-1) = e^{-\xi/2}F(\xi)$, and the result follows. □

Recall that eigenfunctions corresponding to $E(H)$ are strictly positive.

Lemma 4.182. *Let $V = W - U \in \mathbb{V}^{\text{lower}}$. Suppose that φ is a strictly positive eigenfunction of H at eigenvalue E. Also, let $x \in \mathbb{R}^d \setminus \{0\}$, and $a_1, \ldots, a_d$, $\alpha_1, \ldots, \alpha_d$, $b_1, \ldots, b_d$ and $t > 0$ satisfy, for $j = 1, \ldots, d$,*

$$a_j^2/t > \beta, \quad a_j > 2\alpha_j, \quad |[-a_j, a_j] \cap [-x_j - b_j, -x_j + b_j]| > 2\alpha_j, \tag{4.3.88}$$

where $|A|$ denotes the Lebesgue measure of A. Then

$$\varphi(x) \geq \varepsilon(b)e^{tE}e^{-t\tilde{W}_a(x)} \prod_{j=1}^{d} \frac{\alpha_j}{\sqrt{2\pi t}} F\left(\frac{a_j^2}{t}\right) e^{-\frac{a_j^2}{2t}}, \tag{4.3.89}$$

where

$$\tilde{W}_a(x) = \sup\left\{W(y) \mid |y_j - x_j| < a_j, j = 1, \ldots, d\right\}$$
$$\varepsilon(b) = \inf\left\{\varphi(y) \mid |y_j| \leq b_j, j = 1, \ldots, d\right\}.$$

Proof. Note that φ is continuous and $\varphi(x) > 0$ for all $x \in \mathbb{R}^d$. This can be shown as in Lemma 4.180, and implies that $0 < \varepsilon(b) < \infty$. Hence

$$\begin{aligned}\varphi(x) &= e^{tE}\mathbb{E}^x[\varphi(B_t)e^{-\int_0^t W(B_s)ds}e^{\int_0^t U(B_s)ds}] \geq e^{tE}\mathbb{E}^x[\varphi(B_t)e^{-\int_0^t W(B_s)ds}1_A] \\ &\geq \varepsilon(b)e^{tE}e^{-t\tilde{W}_a(x)}\mathbb{E}^x[1_A].\end{aligned} \tag{4.3.90}$$

Here

$$\begin{aligned}A &= \left\{|B_t^j + x_j| \leq b_j,\ \sup_{0\leq s\leq t}|B_s^j| \leq a_j,\ j = 1, \ldots, d\right\} \\ &= \left\{B_t^j \in [-b_j - x_j, b_j - x_j],\ \sup_{0\leq s\leq t}|B_s^j| \leq a_j,\ j = 1, \ldots, d\right\}.\end{aligned}$$

There exists k_j such that $[k_j - a_j, k_j + a_j] \subset [-a_j, a_j] \cap [-x_j - b_j, x_j + b_j]$, and hence $\bigcap_{j=1}^d P(a_j, [k_j - a_j, k_j + a_j], t) \subset A$. The independence of $P(a_j, [k_j - a_j, k_j + a_j], t), j = 1, \ldots, d$, and the estimate (4.3.88) yield

$$\mathbb{E}^x[1_A] \geq \prod_{j=1}^d \mathbb{E}^x[1_{P(a_j,[k_j-a_j,k_j+a_j],t)}] \geq \prod_{j=1}^d \frac{a_j}{\sqrt{2\pi t}} F\left(\frac{a_j^2}{t}\right) e^{-\frac{a_j^2}{2t}}. \tag{4.3.91}$$

Combining this with (4.3.90) and (4.3.91), we obtain (4.3.89). □

Theorem 4.183 (Lower bound on ground state). *Let $V = W - U \in \mathbb{V}^{\text{lower}}$. Suppose that $W(x) \leq \gamma |x|^{2m}$ outside a compact set with $\gamma > 0$ and $m > 1$, and $\varphi(x) \geq 0$. Then there exist constants $\delta, D > 0$ such that $\varphi(x) \geq De^{-\delta|x|^{m+1}}$.*

Proof. Since $\|\varphi\|_\infty < \infty$, it suffices to prove the corollary for large enough $|x|$. Set $t = |x|^{-(m-1)}$, $a_j = 1 + |x_j|$, $a_j = 1/2$, and $b_j = 1$ for $j = 1, \ldots, d$. It can be directly checked that with these choices (4.3.88) is satisfied. Note that $\tilde{W}_a(x) \leq \gamma 2^{4m} |x|^{2m}$. We see that $F(\xi) < 1$ and $\lim_{\xi\to\infty} F(\xi) = 1$. Hence $F(a_j^2/t) = F((1 + |x_j|)^2 |x|^{m-1}) \geq 1/2$ for sufficiently large $|x|$. Inserting into (4.3.89), we obtain

$$\begin{aligned}
\varphi(x) &\geq \varepsilon(b) \left(\frac{|x|^{(m-1)/2}}{4\sqrt{2\pi}}\right)^d e^{|x|^{-(m-1)}E} e^{-\gamma 2^{4m}|x|^{m+1}} e^{-\frac{1}{2}\sum_{j=1}^d (1+|x_j|)^2 |x|^{m-1}} \\
&\geq \frac{\varepsilon(b)}{2} \left(\frac{|x|^{(m-1)/2}}{4\sqrt{2\pi}}\right)^d e^{-(\gamma 2^{4m}+2)|x|^{m+1}}.
\end{aligned}$$

Here we assumed that $e^{|x|^{-(m-1)}E} > 1/2$ for sufficiently large $|x|$. We have

$$\frac{\varepsilon(b)}{2} \left(\frac{|x|^{(m-1)/2}}{4\sqrt{2\pi}}\right)^d > De^{-(\delta - \gamma 2^{4m} - 2)|x|^{m+1}},$$

with some D for sufficiently large $|x|$ and $\delta > \gamma 2^{4m} + 2$, and conclude that $\varphi(x) \geq De^{-\delta|x|^{m-1}}$. □

4.4 Feynman-Kac formula for Schrödinger operators with vector potentials

4.4.1 Feynman-Kac-Itô formula

The Schrödinger operator (4.1.30) gives the total energy of a quantum particle in a force field $-\nabla V$. In some cases of electrodynamics, there is an interaction with a magnetic field that has to be taken into account. This is done by adding a vector potential $a = (a_1, \ldots, a_d)$ to the Schrödinger operator. A formal definition of the *Schrödinger operator with vector potentials* is then given by

$$H(a) = \frac{1}{2}(-i\nabla - a)^2 + V. \tag{4.4.1}$$

Lemma 4.184. *Suppose that $a \in (L^{2p}(\mathbb{R}^d)+L^\infty(\mathbb{R}^d))^d$, $V \in L^p(\mathbb{R}^d)+L^\infty(\mathbb{R}^d)$, and $(\nabla\cdot a) \in L^p(\mathbb{R}^d) + L^\infty(\mathbb{R}^d)$, where $p = 2$ for $d \le 3$ and $p > d/2$ for $d \ge 4$. Then $H(a)$ is self-adjoint on $D(-(1/2)\Delta)$ and essentially self-adjoint on any core of $-\frac{1}{2}\Delta$.*

Proof. Under the assumption,

$$H(a) = -\frac{1}{2}\Delta + \frac{1}{2}(i\nabla\cdot a) + a\cdot(i\nabla) + \frac{1}{2}a\cdot a + V \tag{4.4.2}$$

follows. It is easy to see that $a\cdot a$, $\nabla\cdot a$, and V are infinitesimally small multiplication operators with respect to $-\frac{1}{2}\Delta$. Furthermore, $a\cdot\nabla$ is also infinitesimally small. Indeed, we have

$$\|a_\mu\partial_\mu f\| \le \|a_\mu\|_{2p}\|\partial_\mu f\|_q \le \|a_\mu\|_{2p}(\varepsilon\|\Delta f\|_2 + b_\varepsilon\|f\|)$$

for every $\varepsilon > 0$, where $q = 2p/(p-1)$. The second inequality follows from the Hausdorff–Young inequality as

$$\|\partial_\mu f\|_q \le \|k_\mu\hat{f}\|_{q'} \le \|(1+|k|)^{-\alpha}\|_{2p}\|(1+|k|)^\alpha k_\mu\hat{f}\|_2,$$

where $q' = 2p/(p+1)$ and $\alpha \in (\frac{d}{2p}, 1)$. By the Kato–Rellich theorem, $H(a)$ is self-adjoint on $D(-(1/2)\Delta)$, and essentially self-adjoint on any core of $-\frac{1}{2}\Delta$. □

In this section our goal is to construct a Feynman–Kac formula for $e^{-tH(a)}$. A drift transformation applied to Brownian motion gives

$$\mathbb{E}^x[f(B_t)e^{\int_0^t a(B_s)\cdot dB_s - \frac{1}{2}\int_0^t a(B_s)^2 ds}] = e^{-t(-\frac{1}{2}\Delta - a\cdot\nabla)}f(x).$$

Replacing a by $-ia$, formally,

$$\mathbb{E}^x[f(B_t)e^{-i\int_0^t a(B_s)\cdot dB_s + \frac{1}{2}\int_0^t a(B_s)^2 ds}] = e^{-t(-\frac{1}{2}\Delta + ia\cdot\nabla)}f(x)$$

is obtained. Let $W = V + \frac{1}{2}a\cdot a + \frac{1}{2}(i\nabla\cdot a)$. By (4.4.2), the "imaginary" drift transformation yields

$$\begin{aligned}
&\mathbb{E}^x\left[f(B_t)e^{-i(\int_0^t a(B_s)\cdot dB_s + \frac{1}{2}\int_0^t \nabla\cdot a(B_s)ds) - \int_0^t V(B_s ds}\right] \\
&= \mathbb{E}^x\left[f(B_t)e^{-i\int_0^t a(B_s)\cdot dB_s + \frac{1}{2}\int_0^t a(B_s)^2 ds - \int_0^t W(B_s)ds}\right] = (e^{-tH(a)}f)(x).
\end{aligned}$$

We make this rigorous in the next theorem.

Theorem 4.185 (Feynman–Kac formula for Schrödinger operator with vector potential, Feynman–Kac–Itô formula). *Suppose that $a \in (C_b^2(\mathbb{R}^d))^d$ and $V \in L^\infty(\mathbb{R}^d)$. Then for $f, g \in L^2(\mathbb{R}^d)$,*

$$(f, e^{-tH(a)}g) = \int_{\mathbb{R}^d} \mathbb{E}^x[\overline{f(B_0)}e^{-i\int_0^t a(B_s)\circ dB_s} e^{-\int_0^t V(B_s)ds} g(B_t)]dx. \tag{4.4.3}$$

In particular,

$$(e^{-tH(a)}f)(x) = \mathbb{E}^x[e^{-i\int_0^t a(B_s)\circ dB_s} e^{-\int_0^t V(B_s)ds} f(B_t)]. \tag{4.4.4}$$

Recall that the notation $\circ dB_s$ in the stochastic integral above stands for Stratonovich integral.

Proof. Write $Z_t = e^{X_t}$, $X_t = -i\int_0^t a(B_s)\circ dB_s - \int_0^t V(B_s)ds$, and $Y_t = f(B_t)$. Define

$$(Q_t f)(x) = \mathbb{E}^x[Z_t f(B_t)].$$

Note that $V(B_t)$ and $(\nabla\cdot a)(B_t)$ are $L^1_{\mathrm{loc}}(\mathbb{R}, dt)$ almost surely, and $a_j(B_t) \in M^2(0,t)$. Thus, since

$$dX_t = -ia\cdot dB_t - \frac{i}{2}(\nabla\cdot a)dt - Vdt,$$

we arrive at

$$dZ_t = Z_t dX_t + \frac{1}{2}Z_t(dX_t)^2 = Z_t\Big(-\frac{i}{2}(\nabla\cdot a) - \frac{1}{2}a\cdot a - V\Big)dt + Z_t(-ia)\cdot dB_t$$

by the Itô formula. Suppose that $f \in C_0^\infty(\mathbb{R}^d)$. By $dY_t = \frac{1}{2}\Delta f dt + \nabla f\cdot dB_t$ and the product formula,

$$\begin{aligned} d(Z_tY_t) &= Z_t\Big(-\frac{i}{2}(\nabla\cdot a)dt - \frac{1}{2}(a\cdot a)dt + (-ia)\cdot dB_t - Vdt\Big)Y_t \\ &\quad + Z_t\Big(\frac{1}{2}\Delta f dt + \nabla f\cdot dB_t\Big) + Z_t(-ia)\cdot\nabla f dt \\ &= -H(a)fZ_t dt + Z_t(\nabla f + (-ia)f)\cdot dB_t \end{aligned}$$

is obtained. This gives

$$(Q_t f)(x) - f(x) = -\int_0^t (Q_s H(a)f)(x)ds. \tag{4.4.5}$$

Now suppose that $f \in D(H(a))$. Since $C_0^\infty(\mathbb{R}^d)$ is a core of $H(a)$, there exists a sequence $(f_n)_{n\in\mathbb{N}} \subset C_0^\infty(\mathbb{R}^d)$ such that $f_n \to f$ and $H(a)f_n \to H(a)f$ as $n \to \infty$ strongly. Operator Q_t is bounded with $\|Q_t\| \le e^{t\delta}(2\pi t)^{-d/2}$ and $\delta = \sup_{x\in\mathbb{R}^d}|V(x)|$. We have $Q_t f_n - f_n \to Q_t f - f$

and $\int_0^t Q_s H(a) f_n ds \to \int_0^t Q_s H(a) f ds$ as $n \to \infty$ in $L^2(\mathbb{R}^d)$. Taking a subsequence n', for almost every $x \in \mathbb{R}^d$, $(Q_t f_{n'})(x) - f_{n'}(x) \to (Q_t f)(x) - f(x)$ and $\int_0^t (Q_s H(a) f_{n'})(x) ds \to \int_0^t (Q_s H(a) f)(x) ds$. Hence we have (4.4.5) for $f \in L^2(\mathbb{R}^d)$. In a similar manner as in Theorem 4.86, we see that

$$\operatorname*{s-lim}_{t\to 0} \frac{1}{t}(Q_t f - f) = -H(a) f.$$

Then it is sufficient to show that Q_t is a symmetric C_0-semigroup. We directly see that

$$\begin{aligned} Q_s Q_t f(x) &= \mathbb{E}^x[e^{-i\int_0^s a(B_r)\circ dB_r} \mathbb{E}^{B_s}[f(B_t) e^{-i\int_0^t a(B_s)\circ dB_s}]] \\ &= \mathbb{E}^x[e^{-i\int_0^s a(B_r)\circ dB_r} \mathbb{E}^x[f(B_{s+t}) e^{-i\int_s^{s+t} a(B_r)\circ dB_r} | \mathscr{F}_s]] \\ &= Q_{s+t} f(x). \end{aligned}$$

Recall that $(\mathscr{F}_t)_{t\geq 0}$ is the natural filtration of Brownian motion. Hence the semigroup property follows. Let $\tilde{B}_s = B_{t-s} - B_t$, $0 \leq s \leq t$. We have

$$\begin{aligned} (f, Q_t g) &= \int_{\mathbb{R}^d} \overline{f(x)} \mathbb{E}^x[e^{-i\int_0^t a(\tilde{B}_s)\circ d\tilde{B}_s} e^{-\int_0^t V(\tilde{B}_s) ds} g(\tilde{B}_t)] dx \\ &= \mathbb{E}^0\left[\int_{\mathbb{R}^d} \overline{f(x)} e^{-i\int_0^t a(x+\tilde{B}_s)\circ d\tilde{B}_s} e^{-\int_0^t V(x+\tilde{B}_s) ds} g(x+\tilde{B}_t)\right] dx. \end{aligned}$$

Let

$$I_j = \frac{1}{2}(a(x + \tilde{B}_{tj/n}) + a(x + \tilde{B}_{t(j-1)/n}))(\tilde{B}_{tj/n} - \tilde{B}_{t(j-1)/n}).$$

We have $\sum_{j=1}^n I_j \to \int_0^t a(x + \tilde{B}_s) \circ d\tilde{B}_s$ in $L^2(\mathscr{X})$ as $n \to \infty$, and a subsequence $(\sum_{j=1}^{n'} I_j)_{n'}$ converges to $\int_0^t a(x + \tilde{B}_s) \circ d\tilde{B}_s$ almost surely. We reset n' as n. Changing the variable x to $y - \tilde{B}_t$, we have

$$(f, Q_t g) = \lim_{n\to\infty} \mathbb{E}^0\left[\int_{\mathbb{R}^d} \overline{f(y - \tilde{B}_t)} e^{-i\sum_{j=1}^n \tilde{I}_j} e^{-\int_0^t V(y-\tilde{B}_t+\tilde{B}_s) ds} g(y)\right] dy,$$

where

$$\tilde{I}_j = \frac{1}{2}(a(y - \tilde{B}_t + \tilde{B}_{jt/n}) + a(y - \tilde{B}_t + \tilde{B}_{t(j-1)/n}))(\tilde{B}_{tj/n} - \tilde{B}_{t(j-1)/n}).$$

Recall that $a \in (C_b^2(\mathbb{R}^d))^d$. In the $L^2(\mathscr{X})$ sense it follows that

$$\lim_{n\to\infty} \sum_{j=1}^n \tilde{I}_j = -\int_0^t a(y + B_s) \circ dB_s.$$

This gives

$$(f, Q_t g) = \int_{\mathbb{R}^d} \overline{\mathbb{E}^x[f(B_t)e^{-i\int_0^t a(B_s)\circ dB_s} e^{-\int_0^t V(B_s)ds}]} g(x)dx = (Q_t f, g). \tag{4.4.6}$$

Hence Q_t is symmetric. The strong continuity of Q_t with respect to t is obtained in the same way as in the proof of Theorem 4.86. □

4.4.2 Alternative proof of the Feynman–Kac–Itô formula

As before, the Trotter product formula can once again be employed to prove the Feynman–Kac–Itô formula. Here we give this proof of Theorem 4.185, which will be useful in constructing the functional integral representation of the Pauli–Fierz model in Volume 2.

Proof. Suppose $a \in (C_b^2(\mathbb{R}^d))^d$ and $V = 0$. Write

$$K_s(x, y) = \Pi_s(x - y)e^{ih}, \tag{4.4.7}$$

where

$$h = h(x, y) = \frac{1}{2}(a(x) + a(y)) \cdot (x - y).$$

Define the family of integral operators $\varrho_s : L^2(\mathbb{R}^d) \to L^2(\mathbb{R}^d)$,

$$\varrho_s f(x) = \int_{\mathbb{R}^d} K_s(x, y) f(y) dy, \quad s \geq 0, \tag{4.4.8}$$

where $\varrho_0 f(x) = f(x)$. Note that ϱ_s is symmetric and $\|\varrho_s\| \leq 1$. Let $f, g \in C_0^\infty(\mathbb{R}^d)$. It is directly seen that

$$\frac{d}{ds}(g, \varrho_s f) = \int_{\mathbb{R}^d} \overline{g(x)} dx \int_{\mathbb{R}^d} \Pi_s(x - y) \left(\frac{1}{2}\Delta_y e^{ih} f(y)\right) dy.$$

Here we used that $\frac{d\Pi_s}{ds}(x - y) = \frac{1}{2}\Delta_y \Pi_s(x - y)$. Since

$$\begin{aligned}
\partial_j e^{ih} f &= (i\partial_j h \cdot f + \partial_j f)e^{ih}, \\
\partial_j^2 e^{ih} f &= \{\partial_j^2 f + 2i\partial_j h \cdot \partial_j f + (i\partial_j^2 h + (i\partial_j h)^2)f\}e^{ih}, \\
\partial_j h &= \frac{1}{2}\{\partial_j a(y) \cdot (x - y) - (a_j(x) + a_j(y))\}, \\
\partial_j^2 h &= \frac{1}{2}\partial_j^2 a(y) \cdot (x - y) - \partial_j a_j(y),
\end{aligned}$$

where $\partial_j = \partial_{y_j}, j = 1, \ldots, d$, we have

$$\lim_{s \to 0} \int_{\mathbb{R}^d} \Pi_s(x-y)(\Delta_y e^{ih} f(y))dy = -(-\Delta f - 2a \cdot (-i\nabla f) - (-i\nabla a - a \cdot a)f)$$
$$= -2H(a)f.$$

Thus $\lim_{s \to 0} \frac{d}{ds}(g, \varrho_s f) = (g, -H(a)f)$ and hence

$$\lim_{t \to 0}(g, t^{-1}(1-\varrho_t)f) = (g, H(a)f). \tag{4.4.9}$$

We show next that $\varrho_{t/2^n}^{2^n}$ converges to a symmetric semigroup. For $f, g \in L^2(\mathbb{R}^d)$,

$$(g, \varrho_{t/2^n}^{2^n} f) = \int_{\mathbb{R}^d} \overline{g(x)} e^{i\sum_{j=1}^{2^n} h(x_{j-1}, x_j)} f(x_n) \Big(\prod_{j=1}^{2^n} \Pi_{t/2^n}(x_j - x_{j-1}) \Big) \prod_{j=0}^{2^n} dx_j$$

with $x_0 = x$. Therefore,

$$\begin{aligned} &\lim_{n \to \infty}(g, \varrho_{t/2^n}^{2^n} f) \\ &= \lim_{n \to \infty} \int_{\mathbb{R}^d} \mathbb{E}^x \Big[\overline{g(B_0)} f(B_t) \exp\Big(-i \sum_{j=1}^{2^n} \frac{1}{2}(a(B_{tj/2^n}) + a(B_{t(j-1)/2^n}))(B_{tj/2^n} - B_{t(j-1)/2^n}) \Big) \Big] dx \\ &= \int_{\mathbb{R}^d} \mathbb{E}^x \Big[\overline{g(B_0)} f(B_t) \exp\Big(-i \int_0^t a(B_s) \circ dB_s \Big) \Big] dx \end{aligned} \tag{4.4.10}$$

by Corollary 2.134. Hence $|\lim_{n \to \infty}(g, \varrho_{t/2^n}^{2^n} f)| \leq \|g\| \, \|f\|$. By the Riesz theorem, there is a bounded operator S_t such that $\lim_{n \to \infty}(g, \varrho_{t/2^n}^{2^n} f) = (g, S_t f)$. It is symmetric since ϱ_s is symmetric. Equation (4.4.10) implies that

$$S_t f(x) = \mathbb{E}^x[f(B_s) e^{-i\int_0^t a(B_s) \circ dB_s}].$$

Hence $S_0 = 1$, s-$\lim_{t \to 0} S_t = 1$, and $S_s S_t = S_{s+t}$ follow. Thus S_t is a symmetric C_0-semigroup, so there exists a self-adjoint operator A such that $e^{-tA} = S_t$ by Proposition 4.81. It suffices to show that $A = H(a)$. This can be checked without using the Itô formula. We have

$$\begin{aligned} &\Big(\frac{1}{t}(e^{-tA} - 1)g, f \Big) = \lim_{n \to \infty} \Big(\frac{1}{t}(\varrho_{t/2^n}^{2^n} - 1)g, f \Big) \\ &= \lim_{n \to \infty} \sum_{j=0}^{2^n - 1} \frac{1}{2^n} \Big(\frac{2^n}{t}(\varrho_{t/2^n} - 1)g, \varrho_{t/2^n}^{2^n(j/2^n)} f \Big) = \lim_{n \to \infty} \int_0^1 \Big(\frac{2^n}{t}(\varrho_{t/2^n} - 1)g, \varrho_{t/2^n}^{[2^n s]} f \Big) ds, \end{aligned}$$

where the brackets denote the integer part in this formula. Since for $g \in D(-(1/2)\Delta)$,

$$\text{w-}\lim_{n\to\infty} \frac{2^n}{t}(\varrho_{t/2^n} - 1)g = -H(a)g,$$

the norm $\|(2^n/t)(\varrho_{t/2^n} - 1)g\|$ is uniformly bounded in n. Moreover, we see that $\text{s-lim}_{n\to\infty} \varrho_{t/2^n}^{[2^n s]} = e^{-stA}$, $s \in [0,1]$, by Corollary 4.187. Thus we conclude that

$$\left(\frac{1}{t}(e^{-tA} - 1)g, f\right) = \int_0^1 (-H(a)g, e^{-tsA}f)ds. \tag{4.4.11}$$

Letting $t \to 0$ on both sides above, we get $(g, Af) = (H(a)g, f)$, implying that $Ag = H(a)g$. Hence $A = H(a)$ and the proposition follows for $V = 0$.

Let now V be continuous and bounded. Note that for $f_0, f_n \in L^2(\mathbb{R}^d)$ and $f_j \in L^\infty(\mathbb{R}^d)$ for $j = 1, \ldots, n-1$, it follows that for $0 = t_0 < t_1 < \cdots < t_n$,

$$\left(f_0, \prod_{j=1}^n e^{-(t_j t_{j-1})H(a)} f_j\right) = \int_{\mathbb{R}^d} \mathbb{E}^x\left[\overline{f_0(B_0)}\left(\prod_{j=1}^n f_j(B_{t_j})\right) e^{-i\int_0^{t_n} a(B_s)\circ dB_s}\right] dx. \tag{4.4.12}$$

This can be proven by using the Markov property of Brownian motion as follows:

$$\begin{aligned}
&\left(f_0, \prod_{j=1}^n e^{-(t_j - t_{j-1})H(a)} f_j\right) \\
&= \int_{\mathbb{R}^d} \mathbb{E}^x\left[\overline{f_0(B_0)} f_1(B_{t_1}) e^{-i\int_0^{t_1} a(B_s)\circ dB_s} \mathbb{E}^{B_{t_1}}[f_1(B_{t_1}) e^{-i\int_0^{t_2 - t_1} a(B_s)\circ dB_s} G_2(B_{t_2 - t_1})]\right] dx \\
&= \int_{\mathbb{R}^d} \mathbb{E}^x[\overline{f_0(B_0)} f_1(B_{t_1}) e^{-i\int_0^{t_2} a(B_s)\circ dB_s} f_1(B_{t_1}) G_2(B_{t_2 - t_1})] dx,
\end{aligned}$$

where $G_j(x) = f_j(x)(\prod_{i=j+1}^n e^{-(t_i - t_{i-1})H(a)} f_i)(x)$. Repeating this procedure, we arrive at (4.4.12). From this and the Trotter product formula, we have

$$\begin{aligned}
(f, e^{-tH(a)}g) &= \lim_{n\to\infty} (f, (e^{-(t/n)H(a)} e^{-(t/n)V})^n g) \\
&= \lim_{n\to\infty} \int_{\mathbb{R}^d} \mathbb{E}^x[\overline{f(B_0)} e^{-i\int_0^t a(B_s)\circ dB_s} e^{-\sum_{j=1}^n V(B_{jt/n})}] dx \\
&= \int_{\mathbb{R}^d} \mathbb{E}^x[\overline{f(B_0)} e^{-i\int_0^t a(B_s)\circ dB_s} e^{-\int_0^t V(B_s) ds}] dx.
\end{aligned}$$

Finally, we can extend the result for smooth V in a similar way as done in Theorem 4.88. □

Now we prove the statements used in the proof above.

Lemma 4.186. *Let A_n be a positive and contracting self-adjoint operator such that* $\text{s-lim}_{n\to\infty} A_n = A$. *Then for all* $s \geq 0$ *we have that* $\text{s-lim}_{n\to\infty} A_n^s = A^s$.

Proof. Suppose that $0 < s \leq 1/2$. Since A_n is positive, we have

$$A_n^{2s} f = \frac{\sin(2\pi s)}{\pi} \int_0^\infty \lambda^{2s-1}(A_n + \lambda)^{-1} A_n f d\lambda.$$

From this follows that $A_n^{2s} \to A^{2s}$ weakly as $n \to \infty$, thus $A_n^s \to A^s$ strongly. For $0 \leq s \leq 1$, $\text{s-lim}_{n\to\infty} A_n^s = \text{s-lim}_{n\to\infty}(A_n^{s/2})^2 = A^s$, since A_n is a contraction. Repeating this procedure the result follows. □

Corollary 4.187. *We have* $\text{s-lim}_{n\to\infty} \varrho_{t/n}^{[ns]} = e^{-stA}$ *for* $s \geq 0$.

Proof. Write $A_n = \varrho_{t/n}^{2n}$. The operator A_n is positive, contracting, and self-adjoint such that $\text{s-lim}_{n\to\infty} A_n = e^{-2tA}$ holds. We have $A_n^{(ns+1)/n} \leq \varrho_{t/n}^{2[ns]} \leq A_n^s$ in the form sense. From this estimate, Lemma 4.186, and the fact that $\text{s-lim}_{n\to\infty} A_n^{1/n} = 1$, we obtain $(f, \varrho_{t/n}^{2[ns]} f) \to (f, e^{-2tsA} f)$ as $n \to \infty$. Hence $\text{s-lim}_{n\to\infty} \varrho_{t/n}^{[ns]} = e^{-tsA}$ follows. □

4.4.3 Extension to singular external and vector potentials

The Feynman–Kac–Itô formula derived in Sections 4.4.1 and 4.4.2 can further be extended to singular vector potentials $a = (a_1, \ldots, a_d)$. By using the form technique, we are able to define $H(a)$ as a self-adjoint operator. Let

$$D_\mu = -i\partial_\mu - a_\mu : L^2(\mathbb{R}^d) \to \mathscr{D}'(\mathbb{R}^d), \tag{4.4.13}$$

with Schwartz distribution space $\mathscr{D}'(\mathbb{R}^d)$ over $\mathbb{R}^d$. Define the quadratic form

$$q_a(f, g) = \sum_{\mu=1}^d (D_\mu f, D_\mu g) + (V^{1/2} f, V^{1/2} g) \tag{4.4.14}$$

with domain $Q(q_a) = \bigcap_{\mu=1}^d \{f \in L^2(\mathbb{R}^d) \mid D_\mu f \in L^2(\mathbb{R}^d)\} \cap D(V^{1/2})$.

Lemma 4.188. *Suppose that* $a \in (L^2_{\text{loc}}(\mathbb{R}^d))^d$ *and* $V \in L^1_{\text{loc}}(\mathbb{R}^d)$ *with* $V \geq 0$. *Then* q_a *is a closed symmetric form.*

Proof. Let $(f_n)_n \subset Q(q_a)$ be a Cauchy sequence with respect to the norm

$$\|\cdot\|_0 = (q_a(\cdot,\cdot) + \|\cdot\|^2)^{1/2};$$

then $V^{1/2} f_n$, $D_\mu f_n$ are Cauchy sequences in $L^2(\mathbb{R}^d)$. Thus there exist $g, f \in L^2(\mathbb{R}^d)$ such that $D_\mu f_n \to g$ and $V^{1/2} f_n \to V^{1/2} f$ as $n \to \infty$ in $L^2(\mathbb{R}^d)$. This gives $(f, D_\mu \phi) = (g, \phi)$, for all $\phi \in C_0^\infty(\mathbb{R}^d)$. Therefore $g = D_\mu f$ in $\mathscr{D}'(\mathbb{R}^d)$ distribution sense, and hence $Q(q_a)$ is complete with respect to $\|\cdot\|_0$. □

By Lemma 4.188, there exists a self-adjoint operator h such that

$$(f, hg) = q_a(f, g), \quad f \in Q(q_a), \, g \in D(h). \tag{4.4.15}$$

Definition 4.189 (Schrödinger operator with singular external and singular vector potential (form version)). Suppose that $a \in (L^2_{\mathrm{loc}}(\mathbb{R}^d))^d$ and $V \in L^1_{\mathrm{loc}}(\mathbb{R}^d)$. We denote the self-adjoint operator h in (4.4.15) by $H(a)$.

A sufficient condition for $C_0^\infty(\mathbb{R}^d)$ to be a core of $H(a)$ is the following.

Proposition 4.190.
(1) *Let $V \in L^1_{\mathrm{loc}}(\mathbb{R}^d)$ with $V \geq 0$. If $a \in (L^2_{\mathrm{loc}}(\mathbb{R}^d))^d$, then $C_0^\infty(\mathbb{R}^d)$ is a form core of $H(a)$.*
(2) *Let $V \in L^2_{\mathrm{loc}}(\mathbb{R}^d)$ with $V \geq 0$. If $a \in (L^4_{\mathrm{loc}}(\mathbb{R}^d))^d$ and $\nabla \cdot a \in L^2_{\mathrm{loc}}(\mathbb{R}^d)$, then $C_0^\infty(\mathbb{R}^d)$ is an operator core.*

We can also construct a Feynman–Kac–Itô formula for the form-defined $H(a)$.

Lemma 4.191. *Let $0 \leq V$ and $V \in L^1_{\mathrm{loc}}(\mathbb{R}^d)$. Suppose that $a^{(n)}, a \in (L^2_{\mathrm{loc}}(\mathbb{R}^d))^d$ and $a_\mu^{(n)} \to a_\mu$ in $L^2_{\mathrm{loc}}(\mathbb{R}^d)$ as $n \to \infty$. Then $H(a^{(n)})$ converges to $H(a)$ in the strong resolvent sense.*

Proof. Put $H_n = H(a^{(n)})$, $H = H(a)$, and $D_\mu^{(n)} = -i\partial_\mu - a_\mu^{(n)}$. Note that $C_0^\infty(\mathbb{R}^d)$ is a form core of $H(a)$ by Proposition 4.190. Write $u_n = (H_n + i)^{-1} f$ for $f \in L^2(\mathbb{R}^d)$. Thus $\|u_n\| \leq \|f\|$ and

$$\sum_{\mu=1}^d \|D_\mu^{(n)} u_n\|^2 + \|V^{1/2} u_n\|^2 \leq \|f\|^2.$$

Since $u_n, D_\mu^{(n)} u_n$, and $V^{1/2} u_n$ are bounded, there exist a subsequence $u_{n'}$ and some vectors $u, v, w_\mu \in L^2(\mathbb{R}^d)$ such that $u_{n'} \to u$, $V^{1/2} u_{n'} \to v$, and $D_\mu^{(n')} u_{n'} \to w_\mu$ weakly. Since, furthermore, $V^{1/2} u_{n'} \to V^{1/2} u$ and $D_\mu^{(n')} u_{n'} \to D_\mu u$ in $\mathscr{D}'(\mathbb{R}^d)$, we have $v = V^{1/2} u$ and $D_\mu u = w_\mu$, and hence $u \in Q(q_a)$. For $\phi \in C_0^\infty(\mathbb{R}^d)$,

$$q_a(u, \phi) = \lim_{n \to \infty} (u_n, H_n \phi) = (f - iu, \phi). \tag{4.4.16}$$

As $C_0^\infty(\mathbb{R}^d)$ is a form core of H, (4.4.16) implies that $u \in D(H)$ and $f = (H + i)u$. Hence $u = (H + i)^{-1} f$ follows. Therefore, $(H_n + i)^{-1} \to (H + i)^{-1}$ in the weak sense, and similarly, $(H_n - i)^{-1} \to (H - i)^{-1}$ weakly. Since

$$\|(H_n + i)^{-1} f\|^2 = \frac{i}{2}(f, (H_n + i)^{-1} f - (H_n - i)^{-1} f) \to \|(H + i)^{-1} f\|^2,$$

the strong resolvent convergence of H_n follows. □

Lemma 4.192. *Let $V \geq 0$ in $L^1_{\mathrm{loc}}(\mathbb{R}^d)$, $a \in (L^2_{\mathrm{loc}}(\mathbb{R}^d))^d$, and $\nabla \cdot a \in L^1_{\mathrm{loc}}(\mathbb{R}^d)$. Then $(f, e^{-tH(a)}g)$ has the same functional integral representation as in* (4.4.3).

Proof. First, suppose that $V \in L^\infty(\mathbb{R}^d)$. Note that since $a_\mu(B_s) \in \bigcup_{T>0} \mathbb{M}^2_T$, by the assumption that $a \in (L^2_{\mathrm{loc}}(\mathbb{R}^d))^d$, we have $\mathcal{W}^x(|\int_0^t a(B_s) \cdot dB_s| < \infty) = 1$; moreover, by Lemma 4.95, $\mathcal{W}^x(|\int_0^t \nabla \cdot a(B_s) ds| < \infty) = 1$ also holds. By using a mollifier, we can choose a sequence $a^{(n)} \in C_0^\infty(\mathbb{R}^d)$, $n = 1, 2, \ldots$, such that $a^{(n)}_\mu \to a_\mu$ in $L^2_{\mathrm{loc}}(\mathbb{R}^d)$ and $\nabla \cdot a^{(n)} \to \nabla \cdot a$ in $L^1_{\mathrm{loc}}(\mathbb{R}^d)$ as $n \to \infty$. Let $1_R = \chi(x_1/R) \cdots \chi(x_d/R)$, $R \in \mathbb{N}$, where $\chi \in C_0^\infty(\mathbb{R})$ is such that $0 \leq \chi \leq 1$, $\chi(x) = 1$ for $|x| < 1$ and $\chi(x) = 0$ for $|x| \geq 2$. Since $1_R a^{(n)} \to a^{(n)}$ as $R \to \infty$ in $(L^2_{\mathrm{loc}}(\mathbb{R}^d))^d$ and $a^{(n)} \to a$ as $n \to \infty$ in $(L^2_{\mathrm{loc}}(\mathbb{R}^d))^d$, it follows from Lemma 4.191 that $e^{-tH(1_R a^{(n)})} \to e^{-tH(1_R a)}$, as $n \to \infty$ and $e^{-tH(1_R a)} \to e^{-tH(a)}$ as $R \to \infty$ in the strong sense. Furthermore, (4.4.3) remains true for a replaced by $1_R a^{(n)} \in C_0^\infty(\mathbb{R}^d)$. Since $1_R a^{(n)} \in (C_0^\infty(\mathbb{R}^d))^d$ and $1_R a^{(n)} \to 1_R a$ in $(L^2_{\mathrm{loc}}(\mathbb{R}^d))^d$ as $n \to \infty$, it follows that

$$\int_0^t 1_R(B_s) a^{(n)}(B_s) \cdot dB_s \to \int_0^t 1_R(B_s) a(B_s) \cdot dB_s \tag{4.4.17}$$

and since $\nabla \cdot (1_R a^{(n)}) = (\nabla 1_R) \cdot a^{(n)} + 1_R(\nabla \cdot a^{(n)}) \to (\nabla 1_R) \cdot a + 1_R(\nabla \cdot a)$ in $L^1(\mathbb{R}^d)$, it follows that

$$\int_0^t \nabla \cdot (1_R(B_s) a^{(n)}(B_s)) ds \to \int_0^t \{(\nabla 1_R(B_s)) \cdot a(B_s) + 1_R(B_s)(\nabla \cdot a(B_s))\} \, ds \tag{4.4.18}$$

strongly in $L^1(\mathscr{X}, d\mathcal{W}^x)$. Thus there is a subsequence n' such that (4.4.17) and (4.4.18) hold almost surely with n replaced by n', and therefore (4.4.3) follows by a limiting argument for a replaced by $1_R a$. Let

$$\Omega_+(R) = \Big\{\omega \in \mathscr{X} \mid \max_{0 \leq s \leq t, 1 \leq \mu \leq d} B^\mu_s(\omega) \leq R\Big\},$$

$$\Omega_-(R) = \Big\{\omega \in \mathscr{X} \mid \min_{0 \leq s \leq t, 1 \leq \mu \leq d} B^\mu_s(\omega) \geq -R\Big\},$$

and

$$I(R) = \left| \int_0^t 1_R(B_s) a(B_s) \cdot dB_s - \int_0^t a(B_s) \cdot dB_s \right|.$$

Since

$$\mathcal{W}^x(\Omega_-(R)) = \mathcal{W}^x(\Omega_+(R)) = \prod_{\mu=1}^d \mathcal{W}^x(|B^\mu_t| \leq R) = \left(\frac{2}{\sqrt{2\pi t}} \int_0^R e^{-y^2/(2t)} dy \right)^d$$

and $I(R) = 0$ on $\Omega_+ \cap \Omega_-$, we have

$$\mathcal{W}^x(I(R) \geq \varepsilon) = \mathcal{W}^x(I(R) \geq \varepsilon, \Omega_+(R)^c \cup \Omega_-(R)^c) \leq 2\left(\frac{2}{\sqrt{2\pi t}}\int_R^\infty e^{-y^2/(2t)}dy\right)^d.$$

Hence $\lim_{R\to\infty}\mathcal{W}^x(I(R) \geq \varepsilon) = 0$. Thus there exists a subsequence R' such that $\int_0^t 1_{R'}(B_s)a(B_s)\cdot dB_s \to \int_0^t a(B_s)\cdot dB_s$ as $R' \to \infty$ almost surely. It can be seen in a similar way that with a subsequence (4.4.18) converges to $\int_0^t \nabla\cdot a(B_s)ds$ almost surely. Thus $\int_0^t 1_{R''}(B_s)a(B_s)\circ dB_s \to \int_0^t a(B_s)\circ dB_s$ almost surely. Since

$$|\overline{f(B_0)}g(B_t)e^{-\int_0^t V(B_s)ds}e^{-i\int_0^t 1_{R''}(B_s)a(B_s)\circ dB_s}| = |\overline{f(B_0)}g(B_t)|e^{-\int_0^t V(B_s)ds}$$

and the right-hand side is integrable, (4.4.3) holds for $a \in (L^2_{\mathrm{loc}}(\mathbb{R}^d))^d$ and $\nabla\cdot a \in L^1_{\mathrm{loc}}(\mathbb{R}^d)$ by the dominated convergence theorem. Finally, by a similar argument as in the proof of Theorem 4.88, we can extend (4.4.3) to $V \in L^1_{\mathrm{loc}}(\mathbb{R}^d)$. □

Define $H^0(a)$ by $H(a)$ with $V = 0$.

Lemma 4.193. *Let $a \in (L^2_{\mathrm{loc}}(\mathbb{R}^d))^d$, $\nabla\cdot a \in L^1_{\mathrm{loc}}(\mathbb{R}^d)$, and V be a real-valued multiplication operator.*

(1) *Suppose V is relatively form-bounded with respect to $-\frac{1}{2}\Delta$ with relative bound b. Then V is also relatively form-bounded with respect to $H^0(a)$ with a relative bound not larger than b.*
(2) *Suppose V is relatively bounded with respect to $-\frac{1}{2}\Delta$ with relative bound b. Then V is also relatively bounded with respect to $H^0(a)$ with a relative bound not larger than b.*

Proof. By virtue of Lemma 4.192, we have

$$|(f, e^{-tH^0(a)}g)| \leq (|f|, e^{t(1/2)\Delta}|g|). \tag{4.4.19}$$

Since $(H^0(a)+E)^{-1/2} = \frac{1}{\sqrt{\pi}}\int_0^\infty t^{-1/2}e^{-t(H^0(a)+E)}dt$, $E > 0$, (4.4.19) implies that

$$|(H^0(a)+E)^{-1/2}f|(x) \leq (-(1/2)\Delta + E)^{-1/2}|f|(x) \tag{4.4.20}$$

for almost every $x \in \mathbb{R}^d$. Hence we have

$$|V(x)|^{1/2}|(H^0(a)+E)^{-1/2}f|(x) \leq |V(x)|^{1/2}(-(1/2)\Delta + E)^{-1/2}|f|(x)$$

and

$$\frac{\| |V|^{1/2}(H^0(a)+E)^{-1/2}f\|}{\|f\|} \leq \frac{\| |V|^{1/2}(-(1/2)\Delta+E)^{-1/2}|f|\|}{\|f\|}. \tag{4.4.21}$$

Similarly, by using $(H^0(a)+E)^{-1} = \int_0^\infty e^{-t(H^0(a)+E)}dt$, $E > 0$, we have

$$\frac{\| |V|(H^0(a)+E)^{-1}f\|}{\|f\|} \le \frac{\| |V|(-(1/2)\Delta+E)^{-1}|f|\|}{\|f\|}. \tag{4.4.22}$$

Taking the limit $E \to \infty$, the right-hand sides of (4.4.21) and (4.4.22) converge to b. Hence (1) follows by (4.4.21) and (2) by (4.4.22). □

Let V be such that $V_+ \in L^1_{\text{loc}}(\mathbb{R}^d)$ and V_- is relatively form-bounded with respect to $-\frac{1}{2}\Delta$ with a relative bound strictly smaller than one. The KLMN theorem and Lemma 4.193 allow defining $H^0(a) \dotplus V_+ \dot{-} V_-$ as a self-adjoint operator.

Definition 4.194 (Schrödinger operator with singular external and singular vector potential). Let $a \in (L^2_{\text{loc}}(\mathbb{R}^d))^d$ and $\nabla \cdot a \in L^1_{\text{loc}}(\mathbb{R}^d)$. Let V be such that $V_+ \in L^1_{\text{loc}}(\mathbb{R}^d)$ and V_- is relatively form-bounded with respect to $-\frac{1}{2}\Delta$ with a relative bound strictly smaller than one. Then the Schrödinger operator with external potential V and vector potential a is defined by

$$H(a) = H^0(a) \dotplus V_+ \dot{-} V_-. \tag{4.4.23}$$

Theorem 4.195 (Feynman–Kac formula for Schrödinger operator with singular external potential and singular vector potential). *Let $a \in (L^2_{\text{loc}}(\mathbb{R}^d))^d$ and $\nabla \cdot a \in L^1_{\text{loc}}(\mathbb{R}^d)$. Suppose that $V_+ \in L^1_{\text{loc}}(\mathbb{R}^d)$ and V_- is relatively form-bounded with respect to $-\frac{1}{2}\Delta$ with a relative bound strictly smaller than one. Then the Feynman–Kac formula for $(f, e^{-tH(a)}g)$ is given by* (4.4.3).

Proof. The proof is a minor modification of the proof of Theorem 4.88. Let

$$V_{+n}(x) = \begin{cases} V_+(x), & V_+(x) < n, \\ n, & V_+(x) \ge n, \end{cases} \qquad V_{-m}(x) = \begin{cases} V_-(x), & V_-(x) < m, \\ m, & V_-(x) \ge m. \end{cases}$$

Write $V_{n,m} = V_{+n} - V_{-m}$. We have

$$(f, e^{-t(H^0(a)+V_{n,m})}g) = \int_{\mathbb{R}^d} \mathbb{E}^x[\overline{f(B_0)}e^{-i\int_0^t a(B_s)\circ dB_s} e^{-\int_0^t V_{n,m}(B_s)\,ds} g(B_t)]dx. \tag{4.4.24}$$

The only difference with Theorem 4.88 is that $e^{-i\int_0^t a(B_s)\circ dB_s}$ appears in the integrand. Write $h = H^0(a)$ and define the closed quadratic forms

$$\begin{aligned}
\mathfrak{q}_{n,m}(f,f) &= (h^{1/2}f, h^{1/2}f) + (V_{+n}^{1/2}f, V_{+n}^{1/2}f) - (V_{-m}^{1/2}f, V_{-m}^{1/2}f),\\
\mathfrak{q}_{n,\infty}(f,f) &= (h^{1/2}f, h^{1/2}f) + (V_{+n}^{1/2}f, V_{+n}^{1/2}f) - (V_-^{1/2}f, V_-^{1/2}f),\\
\mathfrak{q}_{\infty,\infty}(f,f) &= (h^{1/2}f, h^{1/2}f) + (V_+^{1/2}f, V_+^{1/2}f) - (V_-^{1/2}f, V_-^{1/2}f),
\end{aligned}$$

whose form domains are respectively $Q(\mathfrak{q}_{n,m}) = Q(h)$, $Q(\mathfrak{q}_{n,\infty}) = Q(h)$, and $Q(\mathfrak{q}_{\infty,\infty}) = Q(h) \cap Q(V_+)$. By the same arguments of monotone convergence of quadratic forms as in the proof of Theorem 4.88, we conclude that for all $t \ge 0$,

$$\exp\left(-t\left(h + V_{+n} - V_{-m}\right)\right) \to \exp\left(-t\left(h + V_{+n} \dot{-} V_{-}\right)\right) \tag{4.4.25}$$

as $m \to \infty$, and

$$\exp(-t(h \dot{+} V_{+n} \dot{-} V_{-})) \to \exp(-t(h \dot{+} V_{+} \dot{-} V_{-})) \tag{4.4.26}$$

as $n \to \infty$ strongly. By taking first $n \to \infty$ and then $m \to \infty$, it can be proven that both sides of (4.4.24) converge, i. e., the left-hand side of (4.4.24) converges by (4.4.25) and (4.4.26). On the other hand, we have the estimate

$$\begin{aligned}
&\int_{\mathbb{R}^d} \mathbb{E}^x[|\overline{f(B_0)}e^{-i\int_0^t a(B_s)\circ dB_s}e^{-\int_0^t V_{n,m}(B_s)\,ds}g(B_t)|]dx \\
&\leq \int_{\mathbb{R}^d} \mathbb{E}^x[|f(B_0)|e^{-\int_0^t V_{n,m}(B_s)\,ds}|g(B_t)|]dx \\
&\leq \int_{\mathbb{R}^d} \mathbb{E}^x[|f(B_0)|e^{-\int_0^t V_{n,\infty}(B_s)ds}|g(B_t)|]dx < \infty,
\end{aligned}$$

where the boundedness of the third term can be obtained from the proof Theorem 4.88. Moreover,

$$\left|\overline{f(B_0)}e^{-i\int_0^t a(B_s)\circ dB_s}e^{-\int_0^t V_{n,m}(B_s)\,ds}g(B_t)\right| \leq |f(B_0)|\,e^{-\int_0^t V_{n,\infty}(B_s)\,ds}\,|g(B_t)|,$$

and the dominated convergence theorem yields

$$\begin{aligned}
&\lim_{m\to\infty}\int_{\mathbb{R}^d} \mathbb{E}^x[\overline{f(B_0)}e^{-i\int_0^t a(B_s)\circ dB_s}e^{-\int_0^t V_{n,m}(B_s)\,ds}g(B_t)]dx \\
&= \int_{\mathbb{R}^d} \mathbb{E}^x[\overline{f(B_0)}e^{-i\int_0^t a(B_s)\circ dB_s}e^{-\int_0^t V_{n,\infty}(B_s)\,ds}g(B_t)]dx.
\end{aligned}$$

Furthermore, since

$$\begin{aligned}
&\int_{\mathbb{R}^d} \mathbb{E}^x[|\overline{f(B_0)}e^{-i\int_0^t a(B_s)\circ dB_s}e^{-\int_0^t V_{n,\infty}(B_s)\,ds}g(B_t)|]dx \\
&\leq \int_{\mathbb{R}^d} \mathbb{E}^x[|f(B_0)|e^{-\int_0^t V_{n,\infty}(B_s)\,ds}|g(B_t)|]dx < \infty
\end{aligned}$$

and

$$\left|\overline{f(B_0)}e^{-i\int_0^t a(B_s)\circ dB_s}e^{-\int_0^t V_{n',\infty}(B_s)\,ds}g(B_t)\right| \leq |f(B_0)|\,e^{-\int_0^t V_{n,\infty}(B_s)\,ds}\,|g(B_t)|$$

for $n \leq n'$, the dominated convergence theorem again yields

$$\lim_{n\to\infty}\int_{\mathbb{R}^d} \mathbb{E}^x[\overline{f(B_0)}e^{-i\int_0^t a(B_s)\circ dB_s}e^{-\int_0^t V_{n,\infty}(B_s)\,ds}g(B_t)]dx$$

$$= \int_{\mathbb{R}^d} \mathbb{E}^x[f(B_0)e^{-i\int_0^t a(B_s)\circ dB_s} e^{-\int_0^t V(B_s)\,ds} g(B_t)]dx.$$

Together with these, the right-hand side of (4.4.24) converges to

$$\int_{\mathbb{R}^d} \mathbb{E}^x[\overline{f(B_0)}e^{-i\int_0^t a(B_s)\circ dB_s} e^{-\int_0^t V(B_s)ds} g(B_t)]dx$$

for all $f, g \in L^2(\mathbb{R}^d)$ by taking first $m \to \infty$ and then $n \to \infty$. □

We used inequality (4.4.19) in the proof of Lemma 4.193. This can be extended to operators with singular external potentials.

Corollary 4.196 (Diamagnetic inequality). *Under the assumptions of Theorem* 4.195, *we have*

$$|(f, e^{-tH(a)}g)| \leq (|f|, e^{-tH(0)}|g|), \quad t \geq 0. \tag{4.4.27}$$

4.4.4 Kato-class potentials and L^p-L^q boundedness

We can also add Kato-class potentials to Schrödinger operators with vector potentials a. Let V be Kato-decomposable, $a \in (L^2_{\mathrm{loc}}(\mathbb{R}^d))^d$, and $\nabla \cdot a \in L^1_{\mathrm{loc}}(\mathbb{R}^d)$. Define

$$K_t f(x) = \mathbb{E}^x[e^{-i\int_0^t a(B_s)\circ dB_s} e^{-\int_0^t V(B_s)ds} f(B_t)]. \tag{4.4.28}$$

Since the absolute value of the first exponential is 1, the operator K_t is well defined and bounded on $L^2(\mathbb{R}^d)$, for every $t \geq 0$.

Theorem 4.197 (Feynman–Kac formula for Schrödinger operator with Kato-decomposable potential and vector potential). *Suppose that V is Kato-decomposable, $a \in (L^2_{\mathrm{loc}}(\mathbb{R}^d))^d$, and $\nabla \cdot a \in L^1_{\mathrm{loc}}(\mathbb{R}^d)$. Then $\{K_t : t \geq 0\}$ is a symmetric C_0-semigroup and there exists the unique self-adjoint operator $K(a)$ bounded from below such that*

$$K_t = e^{-tK(a)}, \quad t \geq 0.$$

Proof. The proof is similar to that of Theorem 4.185. □

Definition 4.198 (Schrödinger operator with Kato-decomposable potential and vector potential). Let $a \in (L^2_{\mathrm{loc}}(\mathbb{R}^d))^d$, $\nabla \cdot a \in L^1_{\mathrm{loc}}(\mathbb{R}^d)$, and V be Kato-decomposable. Then we define the *Schrödinger operator with Kato-decomposable potential V and vector potential a* by the self-adjoint operator $K(a)$ given in Theorem 4.197.

Proposition 4.199 (L^p–L^q boundedness and continuity of integral kernel). *Suppose that V is Kato-decomposable and $a \in (L^2_{\mathrm{loc}}(\mathbb{R}^d))^d$ with $\nabla \cdot a \in L^1_{\mathrm{loc}}(\mathbb{R}^d)$. Let $1 \leq p \leq q \leq \infty$. Then the following hold:*

(1) $e^{-tK(a)}$ *is* L^p–L^q *bounded, for every* $t \geq 0$;

(2) *if* $|a|^2 \in \mathcal{K}(\mathbb{R}^d)$ *and* $\nabla \cdot a \in \mathcal{K}(\mathbb{R}^d)$, *then* $K_t f(x)$ *is continuous in* x *for* $f \in L^p(\mathbb{R}^d)$ *with* $1 \leq p \leq \infty$;

(3) *if* $|a|^2 \in \mathcal{K}(\mathbb{R}^d)$ *and* $\nabla \cdot a \in \mathcal{K}(\mathbb{R}^d)$, *the integral kernel* $\mathbb{R}^d \times \mathbb{R}^d \ni (x,y) \mapsto K_t(x,y)$ *is jointly continuous in* x *and* y, *and is given by*

$$K_t(x,y) = \Pi_t(x-y)\mathbb{E}^{x,y}_{[0,t]}[e^{-\int_0^t V(B_s)\,ds}e^{-i\int_0^t a(B_s)\circ dB_s}]. \tag{4.4.29}$$

Here $\mathcal{W}^{x,y}_{[0,t]}$ *is the Brownian bridge measure defined in* (2.3.20).

Proof. By the diamagnetic inequality, we have $|e^{-tK(a)}f(x)| \leq e^{-tK(0)}|f|(x)$. In Theorem 4.107, we have shown that $e^{-tK(0)}$ is hypercontractive. Using the above estimate, it follows that $e^{-tK(a)}$ is bounded from $L^1(\mathbb{R}^d)$ to $L^1(\mathbb{R}^d)$, from $L^\infty(\mathbb{R}^d)$ to $L^\infty(\mathbb{R}^d)$, and from $L^1(\mathbb{R}^d)$ to $L^\infty(\mathbb{R}^d)$. This proves (1). Part (2) can be seen by a minor modification of the proof of Theorem 4.111. It suffices to consider $f \in L^\infty(\mathbb{R}^d)$. Let $x_n \to x$ as $n \to \infty$. We have

$$\begin{aligned}K_t f(x) - K_t f(x_n) = K_t f(x) &- e^{-r(-\Delta/2)}K_{t-r}f(x) + e^{-r(-\Delta/2)}K_{t-r}f(x)\\ &- e^{-r(-\Delta/2)}K_{t-r}f(x_n) + e^{-r(-\Delta/2)}K_{t-r}f(x_n) - K_t f(x_n)\end{aligned}$$

and

$$e^{-r(-\Delta/2)}K_{t-r}f(x) = \mathbb{E}^x[e^{-\int_r^t V(B_s)ds}e^{-i\int_r^t a(B_s)\circ dB_s}f(B_t)]. \tag{4.4.30}$$

By the smoothing effect, see Proposition 4.109,

$$e^{-r(-\Delta/2)}K_{t-r}f(x) - e^{-r(-\Delta/2)}K_{t-r}f(x_n) \to 0$$

as $n \to \infty$ for every $r > 0$. It is enough to show that $K_t f(x) - e^{-r(-\Delta/2)}K_{t-r}f(x) \to 0$ as $r \to 0$, uniformly in x. First, assume $V \in \mathcal{K}(\mathbb{R}^d)$. For every $r > 0$, let $g_r(x) = e^{-r(-\Delta/2)}K_{t-r}f(x)$. By (4.4.30) and the independence of

$$1 - e^{-\int_0^r V(B_s)\,ds}e^{-i\int_0^r a(B_s)\circ dB_s} \quad \text{and} \quad e^{-\int_r^t V(B_s)\,ds}e^{-i\int_r^t a(B_s)\circ dB_s}f(B_t),$$

we have

$$\begin{aligned}\|g_r - K_t f\|_\infty &\leq \sup_{x\in\mathbb{R}^d}\mathbb{E}^x\left[\left|1 - e^{-\int_0^r V(B_s)\,ds}e^{-i\int_0^r a(B_s)\circ dB_s}\right|\left|e^{-\int_r^t V(B_s)\,ds}e^{-i\int_r^t a(B_s)\circ dB_s}f(B_t)\right|\right]\\ &\leq A\sup_{x\in\mathbb{R}^d}\left(\mathbb{E}^x\left[\left|1 - e^{-i\int_0^r a(B_s)\circ dB_s}\right|\right] + \mathbb{E}^x\left[\left|1 - e^{-\int_0^r V(B_s)\,ds}\right|\right]\right).\end{aligned}$$

Here $A = \|f\|_\infty \sup_{x\in\mathbb{R}^d} \sup_{r\le t} \mathbb{E}^x[e^{-\int_r^t V(B_s)\,ds}]$. In the same way as in the proof of Theorem 4.111, we see that $\sup_{x\in\mathbb{R}^d} \mathbb{E}^x[1-e^{-\int_0^r V(B_s)\,ds}] \to 0$ as $r \to 0$. For the second term, we have

$$\mathbb{E}^x\left[\left|1-e^{-i\int_0^r a(B_s)\circ dB_s}\right|\right] \le \mathbb{E}^x\left[\left|\int_0^r a(B_s)dB_s\right|\right] + \mathbb{E}^x\left[\frac{1}{2}\left|\int_0^r \nabla\cdot a(B_s)ds\right|\right].$$

By Schwarz inequality and Itô-isometry, we have

$$\mathbb{E}^x\left[\left|\int_0^r a(B_s)dB_s\right|\right] \le \left(\mathbb{E}^x\left[\left|\int_0^r a(B_s)dB_s\right|^2\right]\right)^{1/2} = \left(\mathbb{E}^x\left[\int_0^r |a(B_s)|^2 ds\right]\right)^{1/2}.$$

Since $|a|^2 \in \mathcal{K}(\mathbb{R}^d)$, it follows that

$$\lim_{r\to 0}\sup_{x\in\mathbb{R}^d} \mathbb{E}^x\left[\left|\int_0^r a(B_s)dB_s\right|\right] = 0.$$

We also see that

$$\lim_{r\to 0}\sup_{x\in\mathbb{R}^d} \mathbb{E}^x\left[\left|\int_0^r \nabla\cdot a(B_s)ds\right|\right] \le \lim_{r\to 0}\sup_{x\in\mathbb{R}^d} \mathbb{E}^x\left[\int_0^r |\nabla\cdot a(B_s)|ds\right] = 0$$

since $\nabla\cdot a \in \mathcal{K}(\mathbb{R}^d)$. Thus we find that $K_t f(x)$ is continuous in x. In the case when V is Kato-decomposable, (2) can be proven in the same way as Theorem 4.111. Finally, we prove (3). The proof is similar to that of Theorem 4.116. We have

$$(e^{-tK}f)(x) = \int_{\mathbb{R}^d} \Pi_t(x-y)f(y)\mathbb{E}^{x,y}_{[0,t]}[e^{-\int_0^t V(B_s)\,ds}e^{-i\int_0^t a(B_s)\circ dB_s}]dy,$$

which shows that (4.4.29) holds for all x and almost all y. Next we show the continuity of integral kernel by a similar trick as in the proof of Theorem 4.116. Let $s = t/3$. We have

$$K_t(x,y) = \int_{\mathbb{R}^{2d}} K_s(x,v)K_s(v,w)K_s(w,y)\,dw\,dv = \int_{\mathbb{R}^{2d}} K_s(x,v)K_s(y,w)K_s(v,w)\,dw\,dv.$$

The product

$$\begin{aligned}&K_s(x,v)K_s(y,w)\\ &\quad = \Pi_s(x-v)\Pi_s(y-w)\mathbb{E}^{(x,v),(y,w)}_{[0,s]}[e^{-\int_0^s (V(B_r^{(1)})+V(B_r^{(2)}))\,dr}e^{-i\int_0^s a(B_r^{(1)})\circ dB_r^{(1)}-i\int_0^s a(B_r^{(2)})\circ dB_r^{(2)}}]\end{aligned}$$

is the kernel of the Schrödinger operator in $L^2(\mathbb{R}^{2d})$, namely

$$\tilde{K} = \frac{1}{2}(-i\nabla_x - a(x))^2 + \frac{1}{2}(-i\nabla_y - a(y))^2 + V(x) + V(y),$$

where $B_t^{(i)}$, $i = 1, 2$, denote two independent d-dimensional Brownian motions. We see that $V(x) + V(y)$ is Kato-decomposable, and $\nabla_x \cdot a(x) + \nabla_y \cdot a(y)$ and $|a(x)|^2 + |a(y)|^2$ are from Kato-class in $\mathbb{R}^{2d}$. On the other hand, the map $(v, w) \mapsto K_s(v, w)$ is bounded. Hence

$$K_t(x, y) = (e^{-t\tilde{K}} K_s)(x, y)$$

with $K_s \in L^\infty(\mathbb{R}^d \times \mathbb{R}^d)$ and is thus jointly continuous in x and y by statement (2). □

4.5 Feynman–Kac formula for unbounded semigroups and Stark effect

Throughout Sections 4.2–4.4, the Schrödinger operators H were required to be bounded from below, so that e^{-tH} is bounded for all $t \geq 0$. In what follows, we derive a Feynman–Kac formula for an unbounded semigroup e^{-tH} with interesting applications.

Definition 4.200 (Stark Hamiltonian). Let $E \in \mathbb{R}^3$. The operator

$$H(E) = -\frac{1}{2}\Delta + E \cdot x + V(x)$$

on $L^2(\mathbb{R}^3)$ is called a *Stark Hamiltonian.*

The Stark Hamiltonian describes the interaction between a charged quantum particle and an external electrostatic field E, leading to a shift and split-up of the particle's spectral lines. When $V(x) = -\frac{1}{|x|}$, the operator $H(0)$ has a point spectrum. In contrast, the spectrum of $H(E)$ is the whole of $\mathbb{R}$ when $E \neq 0$; in particular, it has no point spectrum at all.

Lemma 4.201 (Faris–Levine). *Let V and W be real-valued measurable functions on $\mathbb{R}^d$ such that $V(x) \geq -c|x|^2 - d$, with $c, d \geq 0$, and $V \in L^2_{\text{loc}}(\mathbb{R}^d)$. Suppose that*

(1) *there exists a dense subset D such that $D \subset D(-(1/2)\Delta) \cap D(V) \cap D(W)$, $x_j D \subset D$, and $\partial_j D \subset D$ so that $-\frac{1}{2}\Delta + V + W + 2c|x|^2$ is essentially self-adjoint on D;*
(2) *$-\frac{a}{2}\Delta + W$ is bounded from below on D for some $a < 1$.*

Then $-\frac{1}{2}\Delta + V + W$ is essentially self-adjoint on D.

Two immediate consequences of Lemma 4.201 are as follows.

Corollary 4.202. *Let V_1 and V_2 be such that*

(1) *$V_2 < -\frac{1}{2}\Delta$ with a relative bound less than 1;*
(2) *$V_1 \geq -c|x|^2 - d$ for some c and d, and $V_1 \in L^2_{\text{loc}}(\mathbb{R}^d)$.*

Then $-\frac{1}{2}\Delta + V_1 + V_2$ is essentially self-adjoint on $C_0^\infty(\mathbb{R}^d)$.

Proof. We apply Lemma 4.201. Let $V = -c|x|^2 - d$ and $W = V_1 + V_2 + c|x|^2 + d$. It trivially holds that $V \geq -c|x|^2 - d$ and $C_0^\infty \subset D(-(1/2)\Delta) \cap D(V) \cap D(W)$. The operator $-\frac{1}{2}\Delta + V + W + 2c|x|^2 = -\frac{1}{2}\Delta + (V_1 + 2c|x|^2) + V_2$ is essentially self-adjoint on $C_0^\infty(\mathbb{R}^d)$, since, by assumption (1) above, $V_1 + 2c|x|^2 + d > 0$ and $V_1 + 2c|x|^2 \in L^2_{\mathrm{loc}}(\mathbb{R}^d)$; see Corollary 4.57. Moreover, $-(a/2)\Delta + W$ is bounded from below. Hence by Lemma 4.201 the statement follows. □

The second corollary covers the Stark Hamiltonian.

Corollary 4.203. *Let $V < -\frac{1}{2}\Delta$ with a relative bound less than 1. Then the Stark Hamiltonian $H(E)$ is essentially self-adjoint on $C_0^\infty(\mathbb{R}^3)$.*

Lemma 4.204. *Consider the Coulomb potential $V(x) = -1/|x|$. Then for all $E \in \mathbb{R}^3$, $\inf \operatorname{Spec}(H(E)) = -\infty$. In particular, $e^{-tH(E)}$ is unbounded.*

Proof. Let $H_0(E) = -\frac{1}{2}\Delta + E \cdot x$. As $V < 0$, it suffices to show that $\operatorname{Spec}(H_0(E)) = \mathbb{R}$. Due to rotation symmetry, this can be reduced to considering the case $E = (\varepsilon, 0, 0)$. Let $x = (x_0, x_\perp) \in \mathbb{R} \times \mathbb{R}^2$ and $p = (p_0, p_\perp)$ be its conjugate momentum. The Hamiltonian becomes $H_0(E) = \frac{1}{2}(p_0^2 + p_\perp^2) + \varepsilon x_0$. Let $U_\parallel = e^{i(1/6\varepsilon)p_0^3}$ and $\mathrm{F}_\perp$ denote Fourier transform with respect to $x_\perp$. We see that $U = \mathrm{F}_\perp U_\parallel : L^2(\mathbb{R}_x \times \mathbb{R}^2_{k_\perp}) \to L^2(\mathbb{R}^3)$ maps $\mathscr{S}(\mathbb{R}^3)$ onto itself. Furthermore, we have

$$U^{-1} H_0(E) U = \frac{1}{2} k_\perp^2 + \varepsilon x_0 \tag{4.5.1}$$

on $\mathscr{S}(\mathbb{R}^3)$. Since $\mathscr{S}(\mathbb{R}^3)$ is a core of both $H_0(E)$ and the multiplication operator on the right-hand side of (4.5.1), $H_0(E)$ and $T = \frac{1}{2}k_\perp^2 + \varepsilon x_0$ are unitary equivalent as self-adjoint operators. Since $\operatorname{Spec}(T) = \mathbb{R}$, the lemma follows. □

Next we establish a Feynman–Kac formula for a class of unbounded semigroups including in particular the Stark Hamiltonian. Recall that $H = -\frac{1}{2}\Delta + V$. Our assumption on V is that for every $\varepsilon > 0$ there exists C_ε such that $V(x) \geq -\varepsilon|x|^2 - C_\varepsilon$. It is useful to rewrite the Feynman–Kac formula by using Brownian bridge starting from a at $t = 0$ and ending in b at $t = T$, i. e., the solution of the stochastic differential equation

$$dX_t = \frac{b - X_t}{T - t} dt + dB_t, \quad 0 \leq t < T, \quad X_0 = a. \tag{4.5.2}$$

This equation can be solved exactly to obtain

$$X_t = a\left(1 - \frac{t}{T}\right) + b\frac{t}{T} + (T - t)\int_0^t \frac{1}{T - s} dB_s, \quad t < T, \tag{4.5.3}$$

see Example 2.161. Note that $X_t \to b$ almost surely as $t \to T$. Hence, for continuous and bounded from below V,

$$(f, e^{-tH} g) = \int_{\mathbb{R}^d \times \mathbb{R}^d} \overline{f(x)} g(y) Q(x, y; V, t) \Pi_t(x - y) dx dy, \tag{4.5.4}$$

where by (4.5.3) we have

$$Q(x, y; V, t) = \mathbb{E}\left[\exp\left(-\int_0^t V\left(\left(1 - \frac{s}{t}\right)x + \frac{s}{t}y + \alpha_s\right)ds \right) \right], \tag{4.5.5}$$

and α_s is a Gaussian random variable defined by

$$\alpha_s = \begin{cases} (t-s)\int_0^s \frac{1}{t-r} dB_r, & s < t, \\ 0, & s = t. \end{cases}$$

Note that $\mathbb{E}[\alpha_s] = 0$ and $\mathbb{E}[\alpha_s \alpha_{s'}] = s(1 - \frac{s'}{t})$ for $s \le s' \le t$. In the remainder of this section, we prove the following theorem.

Theorem 4.205 (Feynman–Kac formula for unbounded semigroup). *Assume that V is continuous and for every $\varepsilon > 0$ there exists C_ε such that $V(x) \ge -\varepsilon |x|^2 - C_\varepsilon$. Let $f, g \in L^2(\mathbb{R}^d)$ have compact support. Then $f, g \in D(e^{-tH})$ and (4.5.4) holds.*

In order to prove this theorem, we first need the following lemma.

Lemma 4.206. *Let A_n and A be self-adjoint operators such that $A_n \to A$ in the strong resolvent sense. Also, let f be a continuous function and $\psi \in D(f(A_n))$ for all $n \in \mathbb{N}$. Then we have the following:*

(1) *If $\sup_{n \in \mathbb{N}} \|f(A_n)\psi\| < \infty$, then $\psi \in D(f(A))$.*
(2) *If $\sup_{n \in \mathbb{N}} \|f(A_n)^2 \psi\| < \infty$, then $f(A_n)\psi \to f(A)\psi$ in the strong sense as $n \to \infty$.*

Proof. (1) Let

$$f_m(x) = \begin{cases} m, & f(x) \ge m, \\ f(x), & |f(x)| \le m, \\ -m, & f(x) \le -m. \end{cases}$$

As f_m is bounded, $f_m(A_n) \to f_m(A)$ strongly as $n \to \infty$. Since

$$\|f_m(A)\psi\| = \lim_{n \to \infty} \|f_m(A_n)\psi\| \le \sup_{n \in \mathbb{N}} \|f_m(A_n)\| \le \sup_{n \in \mathbb{N}} \|f(A_n)\|,$$

we have $\sup_{m \in \mathbb{N}} \|f_m(A)\psi\| < \infty$. This implies $\psi \in D(f(A))$.

(2) Since $\|(f(A_n) - f_m(A_n))\psi\| \le m^{-1} \|f(A_n)^2 \psi\|$, it follows that $f_m(A_n)\psi \to f(A_n)\psi$, uniformly in n. Then $f(A_n)\psi \to f(A)\psi$ since $f_m(A_n)\psi \to f_m(A)\psi$. □

Proof of Theorem 4.205. Let $V_n = \max\{V(x), -n\}$ and $H_n = -\frac{1}{2}\Delta + V_n$. Since $C_0^\infty(\mathbb{R}^d)$ is a common core of H_n, it follows that $H_n \to H$ as $n \to \infty$ in the strong resolvent sense. For $f \in L^2(\mathbb{R}^d)$ with compact support, we prove

$$\sup_{n\in\mathbb{N}} \|e^{-tH_n} f\| < \infty. \tag{4.5.6}$$

Note that

$$e^{-\int_0^t V_n((1-\frac{s}{t})x+\frac{s}{t}y+\alpha_s)} \le e^{\int_0^t \varepsilon((1-\frac{s}{t})x+\frac{s}{t}y+\alpha_s)^2 + C_\varepsilon ds} \le e^{C_\varepsilon t + 2\varepsilon t(x^2+y^2) + 2\varepsilon t \int_0^t \alpha_s^2 \frac{ds}{t}}.$$

By Jensen's inequality,

$$e^{2\varepsilon t \int_0^t \alpha_s^2 \frac{ds}{t}} \le \int_0^t e^{2\varepsilon t \alpha_s^2} \frac{ds}{t},$$

and therefore $\mathbb{E}[e^{2\varepsilon t \int_0^t \alpha_s^2 \frac{ds}{t}}] \le \int_0^t \mathbb{E}[e^{2\varepsilon t \alpha_s^2}] \frac{ds}{t}$. A direct computation gives

$$\mathbb{E}[e^{2\varepsilon t \alpha_s^2}] = \frac{1}{\sqrt{2\pi\kappa_s}} \int_{\mathbb{R}} e^{2\varepsilon t y^2} e^{-y^2/2\kappa_s} dy = \frac{1}{\sqrt{\pi(1-4\varepsilon t\kappa_s)}},$$

where $\kappa_s = s(1-\frac{s}{t})$ is the covariance of α_s, and we assume $\varepsilon < 1$. Set $\varepsilon = \delta/t^2$ with some $\delta < 1$. Since $\kappa_s \le \kappa_{t/2} = t/4$, we have

$$Q(x, y; V, t) \le \frac{e^{C_\varepsilon t}}{\sqrt{\pi(1-\delta)}} e^{2\delta(x^2+y^2)/t}. \tag{4.5.7}$$

Hence

$$\|e^{-tH_n} f\|^2 \le \frac{e^{2C_\varepsilon t}}{\sqrt{\pi(1-\delta)}} \int_{\mathbb{R}^d\times\mathbb{R}^d} |f(x)f(y)|\Pi_{2t}(x-y) e^{\frac{\delta}{t}(x^2+y^2)} dxdy < \infty,$$

uniformly in n. Thus (4.5.6) follows. By Lemma 4.205 (1), we have $f \in D(e^{-tH})$ and $e^{-tH_n} f \to e^{-tH} f$ strongly as $n \to \infty$. Thus the left-hand side of (4.5.4) converges, and so does the right-hand side by (4.5.7) and the dominated convergence theorem. □

4.6 Feynman–Kac formula for relativistic Schrödinger operators

4.6.1 Relativistic Schrödinger operator

Relativity theory says that in the high-energy regime the kinetic energy of a body in motion should be proportional to the absolute value of its momentum p rather than its square. In fact, the energy of a classical relativistic particle of mass m is $\sqrt{|p|^2c^2 + m^2c^4}$, where c is the speed of light, and this justifies the definition of the *relativistic kinetic operator*

$$\sqrt{-\hbar^2 c^2 \Delta + m^2 c^4} - mc^2. \tag{4.6.1}$$

Since the constant mc^2 is subtracted, the bottom of its spectrum is zero. Using a basic convention in mathematical physics, from now on we choose units in which $c = 1 = \hbar$. In this section we consider Feynman–Kac formulae for relativistic Schrödinger operators of a more general form.

Definition 4.207 (Relativistic Schrödinger operator with vector potential). Suppose that $a \in (L^2_{\mathrm{loc}}(\mathbb{R}^d))^d$ and $V \in L^\infty(\mathbb{R}^d)$. Then the *relativistic Schrödinger operator with vector potential* a is defined by

$$H_{\mathrm{R}}(a) = (2H^0(a) + m^2)^{1/2} - m + V, \tag{4.6.2}$$

where $H^0(a)$ is the self-adjoint operator given in Definition 4.189 with $V = 0$.

Note that the square-root term in the definition above is given through the spectral projection of the self-adjoint operator $H^0(a)$. Denote by $H^0_{\mathrm{R}}(a)$ the operator $H_{\mathrm{R}}(a)$ with $V = 0$.

Proposition 4.208.

(1) *Suppose that* $a \in (L^4_{\mathrm{loc}}(\mathbb{R}^d))^d$ *and* $\nabla \cdot a \in L^2_{\mathrm{loc}}(\mathbb{R}^d)$. *Then* $C_0^\infty(\mathbb{R}^d)$ *is an operator core of* $H^0_{\mathrm{R}}(a)$.

(2) *Suppose that* $a \in (L^2_{\mathrm{loc}}(\mathbb{R}^d))^d$. *Then* $C_0^\infty(\mathbb{R}^d)$ *is a form core of* $H^0_{\mathrm{R}}(a)$.

Proof. For simplicity, in this proof we write $H^0_{\mathrm{R}}(a) = H_{\mathrm{R}}$ and $H^0(a) = H$.

(1) First, note that there exist nonnegative constants c_1 and c_2 such that

$$\|H_{\mathrm{R}} f\| \le c_1 \|Hf\| + c_2 \|f\| \tag{4.6.3}$$

for all $f \in D(H)$. Hence $C_0^\infty(\mathbb{R}^d)$ is contained in $D(H_{\mathrm{R}})$. Since H_{R} is a nonnegative self-adjoint operator, $H_{\mathrm{R}} + 1$ has a bounded inverse, and we use the fact that $C_0^\infty(\mathbb{R}^d)$ is a core of H_{R} if and only if $H_{\mathrm{R}} C_0^\infty(\mathbb{R}^d)$ is dense in $L^2(\mathbb{R}^d)$. Let $g \in L^2(\mathbb{R}^d)$ and suppose that $(g, (H_{\mathrm{R}} + 1)f) = 0$, for all $f \in C_0^\infty(\mathbb{R}^d)$. Then $C_0^\infty(\mathbb{R}^d) \ni f \mapsto (g, H_{\mathrm{R}} f) = -(g, f)$ defines a continuous functional which can be extended to $L^2(\mathbb{R}^d)$. Thus $g \in D(H_{\mathrm{R}})$ and $0 = ((H_{\mathrm{R}} + 1)g, f)$. Since $C_0^\infty(\mathbb{R}^d)$ is dense, we have $(H_{\mathrm{R}} + 1)g = 0$, and hence $g = 0$ since $H_{\mathrm{R}} + 1$ is one-to-one, proving the assertion.

(2) Note that $\|H_{\mathrm{R}}^{1/2} f\|^2 \le c_1 \|H^{1/2} f\|^2 + c_2 \|f\|^2$ for $f \in Q(h) = D(H^{1/2})$, and $C_0^\infty(\mathbb{R}^d)$ is contained in $Q(H_{\mathrm{R}}) = D(H_{\mathrm{R}}^{1/2})$. Since $H_{\mathrm{R}}^{1/2} + 1$ has also bounded inverse, it is seen by the same argument as above that $C_0^\infty(\mathbb{R}^d)$ is a core of $H_{\mathrm{R}}^{1/2}$ or a form core of H_{R}. □

A key element in deriving the Feynman–Kac formula is a use of the Lévy subordinator studied in Example 3.90. In the notation of the example, we choose $\delta = 1$ and $\gamma = m$ so that we have the subordinator $T_t = T_t(1, m)$ on a suitable probability space $(\mathcal{T}, \mathcal{B}_{\mathcal{T}}, P)$ such that

$$\mathbb{E}_P^0[e^{-uT_t}] = e^{-t(\sqrt{2u+m^2}-m)}. \tag{4.6.4}$$

For simplicity, we write $\mathbb{E}^{x,0}$ for $\mathbb{E}^{x,0}_{\mathcal{W}\times P}$.

Theorem 4.209 (Feynman–Kac formula for relativistic Schrödinger operator with vector potential). *Suppose that $V \in L^\infty(\mathbb{R}^d)$, $a \in (L^2_{\mathrm{loc}}(\mathbb{R}^d))^d$, and $\nabla \cdot a \in L^1_{\mathrm{loc}}(\mathbb{R}^d)$. Then*

$$(f, e^{-tH_{\mathrm{R}}(a)}g) = \int_{\mathbb{R}^d} \mathbb{E}^{x,0}[\overline{f(B_0)}g(B_{T_t})e^{-i\int_0^{T_t} a(B_s)\circ dB_s}e^{-\int_0^t V(B_{T_s})ds}]dx. \tag{4.6.5}$$

Proof. We divide the proof into four steps.

Step 1. Suppose $V = 0$. Then

$$(f, e^{-tH^0_{\mathrm{R}}(a)}g) = \int_{\mathbb{R}^d} \mathbb{E}^{x,0}[\overline{f(B_0)}g(B_{T_t})e^{-i\int_0^{T_t} a(B_s)\circ dB_s}]dx. \tag{4.6.6}$$

Denote by E_λ the spectral projection of the self-adjoint operator $H^0(a)$. We have

$$(f, e^{-tH^0_{\mathrm{R}}(a)}g) = \int_{[0,\infty)} e^{-t(\sqrt{2\lambda+m^2}-m)}d(f, E_\lambda g). \tag{4.6.7}$$

By (4.6.4), we obtain

$$(f, e^{-tH^0_{\mathrm{R}}(a)}g) = \int_0^\infty \mathbb{E}_P^0[e^{-\lambda T_t}]d(f, E_\lambda g) = \mathbb{E}_P^0[(f, e^{-T_t H^0(a)}g)].$$

Equality (4.6.6) follows from inserting the Feynman–Kac–Itô formula for $e^{-T_t H^0(a)}$, i. e.,

$$(f, e^{-T_t H^0_{\mathrm{R}}(a)}g) = \int_{\mathbb{R}^d} \mathbb{E}^{x,0}[\overline{f(B_0)}g(B_{T_t})e^{-i\int_0^{T_t} a(B_s)\circ dB_s}]dx, \tag{4.6.8}$$

discussed in Section 4.4.3.

Step 2. Let $0 = t_0 < t_1 < \cdots < t_n$ and $f_0, f_n \in L^2(\mathbb{R}^d)$ and $f_j \in L^\infty(\mathbb{R}^d)$ for $j = 1, \ldots, n-1$. We prove next that

$$\Big(f_0, \prod_{j=1}^n e^{-(t_j - t_{j-1})H^0_{\mathrm{R}}(a)}f_j\Big) = \int_{\mathbb{R}^d} \mathbb{E}^{x,0}\Big[\overline{f(B_0)}\Big(\prod_{j=1}^n f_j(B_{T_{t_j}})\Big)e^{-i\int_0^{T_t} a(B_s)\circ dB_s}\Big]dx. \tag{4.6.9}$$

We use the shorthand notation $G_j(x) = f_j(x)(\prod_{i=j+1}^{n} e^{-(t_i-t_{i-1})H_{\mathbb{R}}^0(a)} f_i)(x)$. The left-hand side of (4.6.9) can be written as

$$\int_{\mathbb{R}^d} \mathbb{E}^{x,0}[\overline{f(B_0)}e^{-i\int_0^{T_{t_1-t_0}} a(B_s)\circ dB_s} G_1(B_{T_{t_1-t_0}})]dx.$$

By the Markov property of $(B_t)_{t\geq 0}$, we have

$$\begin{aligned}
&\Big(f_0, \prod_{j=1}^{n} e^{-(t_j-t_{j-1})H_{\mathbb{R}}^0(a)} f_j\Big) \\
&= \int_{\mathbb{R}^d} \mathbb{E}^{x,0}\Big[\overline{f(B_0)}e^{-i\int_0^{T_{t_1}} a(B_s)\circ dB_s} \mathbb{E}_P^0 \mathbb{E}_{\mathcal{W}}^{B_{T_{t_1}}} [f_1(B_0)e^{-i\int_0^{T_{t_2-t_1}} a(B_s)\circ dB_s} G_2(B_{T_{t_2-t_1}})]\Big]dx \\
&= \int_{\mathbb{R}^d} \mathbb{E}^{x,0}\Big[\overline{f(B_0)}e^{-i\int_0^{T_{t_1}} a(B_s)\circ dB_s} \mathbb{E}_P^0 [f_1(B_{T_{t_1}})e^{-i\int_{T_{t_1}}^{T_{t_2-t_1}+T_{t_1}} a(B_s)\circ dB_s} G_2(B_{T_{t_1}+T_{t_2-t_1}})]\Big]dx.
\end{aligned}$$

The right-hand side above can be rewritten as

$$\int_{\mathbb{R}^d} \mathbb{E}^{x,0}\Big[\overline{f(B_0)}e^{-i\int_0^{T_{t_1}} a(B_s)\circ dB_s} f_1(B_{T_{t_1}})\mathbb{E}_P^{T_{t_1}} [e^{-i\int_0^{T_{t_2-t_1}} a(B_s)\circ dB_s} G_2(B_{T_{t_2-t_1}})]\Big]dx.$$

By the Markov property of $(T_t)_{t\geq 0}$ again, we can see that

$$\begin{aligned}
&= \int_{\mathbb{R}^d} \mathbb{E}^{x,0}\Big[\overline{f(B_0)}e^{-i\int_0^{T_{t_1}} a(B_s)\circ dB_s} f_1(B_{T_{t_1}})e^{-i\int_{T_{t_1}}^{T_{t_2}} a(B_s)\circ dB_s} G_2(B_{T_{t_2}})\Big]dx \\
&= \int_{\mathbb{R}^d} \mathbb{E}^{x,0}\Big[\overline{f(B_0)}e^{-i\int_0^{T_{t_2}} a(B_s)\circ dB_s} f_1(B_{T_{t_1}})G_2(B_{T_{t_2}})\Big]dx.
\end{aligned}$$

Continuing by the above procedure, (4.6.9) is obtained.

Step 3. Suppose now that $V \in L^\infty$ and it is continuous. We show (4.6.5) under this condition. By the Trotter product formula,

$$\begin{aligned}
(f, e^{-tH_{\mathbb{R}}(a)}g) &= \lim_{n\to\infty} (f, (e^{-(t/n)H_{\mathbb{R}}^0(a)}e^{-(t/n)V})^n g) \\
&= \lim_{n\to\infty} \int_{\mathbb{R}^d} \mathbb{E}^{x,0}[\overline{f(B_0)}g(B_{T_t})e^{-i\int_0^{T_t} a(B_s)\circ dB_s} e^{-\sum_{j=1}^{n}(t/n)V(B_{T_{tj/n}})}]dx \\
&= \text{r. h. s. (4.6.9)}.
\end{aligned}$$

Here we used the fact that for each path (ω, τ) the function $V(B_{T_s(\tau)}(\omega))$ is continuous in s except for at most finitely many points. Note that T_s almost surely has no accumulation points in any compact interval. Thus

$$\sum_{j=1}^{n} \frac{t}{n} V(B_{T_{tj/n}}) \to \int_0^t V(B_{T_s}) ds$$

as $n \to \infty$, for each path as a Riemann integral. This completes the proof of the claim.

Step 4. Finally, to complete the proof of the theorem, define V_n in the same manner as in the proof of Theorem 4.237. Note that V_n is bounded and continuous, and, moreover, $V_n(x) \to V(x)$ as $n \to \infty$ for $x \notin \mathscr{N}$, where $\mathscr{N}$ is a set of Lebesgue measure zero. Note that

$$\mathbb{E}^{x,0}[1_{\mathscr{N}}(B_{T_s})] = \mathbb{E}^0\left[1_{\{T_s>0\}} \int_{\mathbb{R}^d} 1_{\mathscr{N}}(x+y) k_s(y) dy\right] + 1_{\mathscr{N}}(x)\mathbb{E}^0[1_{\{T_s=0\}}] = 0$$

for $x \in \mathscr{N}$, where $k_s(x)$ is the distribution of the random variable B_{T_s} given by (4.3.6). Hence

$$0 = \int_0^t \mathbb{E}^{x,0}[1_{\mathscr{N}}(B_{T_s})] ds = \mathbb{E}^{x,0}\left[\int_0^t 1_{\mathscr{N}}(B_{T_s}) ds\right].$$

For a. e. (ω, τ), the Lebesgue measure of $\{t \in [0,\infty) \mid B_{T_t(\tau)}(\omega) \in \mathscr{N}\}$ is zero. Therefore $\int_0^t V_n(B_{T_s}) ds \to \int_0^t V(B_{T_s}) ds$ as $n \to \infty$ for a. e. (ω, τ). Moreover,

$$\int_{\mathbb{R}^d} \mathbb{E}^{x,0}[\overline{f(B_0)} g(B_{T_t}) e^{-i\int_0^{T_t} a(B_s)\circ dB_s} e^{-\int_0^t V_n(B_s) ds}] dx$$
$$\to \int_{\mathbb{R}^d} \mathbb{E}^{x,0}[\overline{f(B_0)} g(B_{T_t}) e^{-i\int_0^{T_t} a(B_s)\circ dB_s} e^{-\int_0^t V(B_s) ds}] dx$$

as $n \to \infty$. On the other hand, $e^{-t(H_{\mathrm{R}}^0(a)+V_n)} \to e^{-t(H_{\mathrm{R}}^0(a)+V)}$ strongly as $n \to \infty$, since $H_{\mathrm{R}}^0(a) + V_n$ converges to $H_{\mathrm{R}}^0(a) + V$ on the common domain $D(H_{\mathrm{R}}^0(a))$. □

The Feynman–Kac formula of $e^{-tH_{\mathrm{R}}(a)}$ can be extended to more singular external potentials. The key idea is similar to that applied in the case of $e^{-tH(a)}$ discussed in Theorem 4.88.

Lemma 4.210. *Let $a \in (L^2_{\mathrm{loc}}(\mathbb{R}^d))^d$, $\nabla \cdot a \in L^1_{\mathrm{loc}}(\mathbb{R}^d)$, and let V be a real-valued multiplication operator.*

(1) *Suppose V is relatively form-bounded with respect to $\sqrt{-\Delta + m^2} - m$ with relative bound b. Then V is also relatively form-bounded with respect to $H_{\mathrm{R}}^0(a)$ with a relative bound not larger than b.*
(2) *Suppose V is relatively bounded with respect to $\sqrt{-\Delta + m^2} - m$ with relative bound b. Then V is also relatively bounded with respect to $H_{\mathrm{R}}^0(a)$ with a relative bound not larger than b.*

Proof. By virtue of Theorem 4.209, we have

$$|(f, e^{-tH_{\mathrm{R}}^0(a)} g)| \le (|f|, e^{-t(\sqrt{-\Delta+m^2}-m)} |g|). \tag{4.6.10}$$

The proof is similar to that of Lemma 4.193. □

Let V be such that $V_+ \in L^1_{\mathrm{loc}}(\mathbb{R}^d)$ and V_- is relatively form-bounded with respect to $\sqrt{-\Delta + m^2} - m$ with a relative bound strictly smaller than one. The KLMN theorem and Lemma 4.210 can be used to define $H_{\mathrm{R}}^0(a) \dot{+} V_+ \dot{-} V_-$ as a self-adjoint operator.

Definition 4.211 (Relativistic Schrödinger operator with singular external potential and singular vector potential). Let $a \in (L^2_{\mathrm{loc}}(\mathbb{R}^d))^d$ and $\nabla \cdot a \in L^1_{\mathrm{loc}}(\mathbb{R}^d)$. Let V be such that $V_+ \in L^1_{\mathrm{loc}}(\mathbb{R}^d)$ and V_- is relatively form-bounded with respect to $\sqrt{-\Delta + m^2} - m$ with a relative bound strictly smaller than one. Then the relativistic Schrödinger operator with external potential V and vector potential a is defined by

$$H_{\mathrm{R}}(a) = H_{\mathrm{R}}^0(a) \dot{+} V_+ \dot{-} V_-. \tag{4.6.11}$$

Theorem 4.212 (Feynman–Kac formula for relativistic Schrödinger operator with singular external potential and singular vector potential). *Suppose $a \in (L^2_{\mathrm{loc}}(\mathbb{R}^d))^d$ and $\nabla \cdot a \in L^1_{\mathrm{loc}}(\mathbb{R}^d)$. Let V be such that $V_+ \in L^1_{\mathrm{loc}}(\mathbb{R}^d)$ and V_- is relatively form-bounded with respect to $\sqrt{-\Delta + m^2} - m$ with a relative bound strictly smaller than one. Then the Feynman–Kac representation of $(f, e^{-tH_{\mathrm{R}}(a)} g)$ is given by* (4.4.3).

Proof. The proof is similar to the limiting argument in Theorem 4.195. □

Corollary 4.213 (Diamagnetic inequality). *Under the same assumptions as in Theorem* 4.212, *we have*

$$|(f, e^{-tH_{\mathrm{R}}(a)} g)| \le (|f|, e^{-tH_{\mathrm{R}}(0)} |g|). \tag{4.6.12}$$

We can also show by using the Feynman–Kac formula that $e^{-tH_{\mathrm{R}}(0)}$, with $a = 0$, is positivity improving.

Theorem 4.214 (Positivity improving). *Suppose that $V_+ \in L^1_{\mathrm{loc}}(\mathbb{R}^d)$ and V_- is relatively form bounded with respect to $\sqrt{-\Delta + m^2} - m$ with a relative bound strictly smaller than one. Then $\{e^{-tH_{\mathrm{R}}(0)} : t \ge 0\}$ is positivity improving.*

Proof. Since $\{e^{-t(-\Delta/2)} : t \ge 0\}$ is positivity improving and $e^{-t(\sqrt{-\Delta+m^2}-m)} = \mathbb{E}[e^{-T_t(-\Delta/2)}]$, $\{e^{-t\sqrt{-\Delta+m^2}} : t \ge 0\}$ is also positivity improving, and we have $\int_{\mathbb{R}^d} \mathbb{E}^{x,0}[f(B_0) g(B_{T_t})] dx > 0$ for every nonnegative f, g. For $V_+ \in L^1_{\mathrm{loc}}(\mathbb{R}^d)$, in a similar way as in Lemma 4.95, it can be seen that $e^{-\int_0^t V(B_{T_s}+x) ds} > 0$ for almost every $(x, \omega, \varpi) \in \mathbb{R}^d \times \mathscr{X} \times \mathscr{T}$. The measure of

$$\mathscr{M} = \left\{(x, \omega, \varpi) \in \mathbb{R}^d \times \mathscr{X} \times \mathscr{T} \mid f(x) g(B_{T_t(\varpi)}(\omega) + x) > 0\right\}$$

is positive, and we have

$$(f, e^{-tH_{\mathrm{R}}(0)}g) \geq \int_{\mathscr{M}} f(x)g(B_{T_t}+x)e^{-\int_0^t V(B_{T_s}+x)ds}dxd\mathcal{W}dP > 0.$$

Thus $e^{-tH_{\mathrm{R}}(0)}$ is positivity improving for all $t \geq 0$. □

4.6.2 Relativistic Kato-class potentials

In Section 4.6.1 we constructed the Feynman–Kac formula of $e^{-tH_{\mathrm{R}}(a)}$. Similarly to $K(a)$, we can define the relativistic Schrödinger operator $K_{\mathrm{R}}(a)$ as the generator of a C_0-semigroup defined through Feynman–Kac formula with general external potentials. Let

$$p_t^R(x) = \frac{1}{(2\pi)^d}\int_{\mathbb{R}^d} e^{-ix\cdot\xi}e^{-t(\sqrt{|\xi|^2+m^2}-m)}d\xi$$

and define

$$g_R(x) = \int_0^\infty e^{-t}p_t^R(x)dt.$$

This $g_R(x)$ gives the kernel of $(\sqrt{-\Delta+m^2}-m+1)^{-1}$, i. e.,

$$(f, (\sqrt{-\Delta+m^2}-m+1)^{-1}h) = \int_{\mathbb{R}^d\times\mathbb{R}^d} \bar{f}(x)g_R(x-y)h(y)dxdy$$

holds. Formally, $g_R(x)$ can be represented as

$$g_R(x) = \frac{1}{(2\pi)^d}\int_{\mathbb{R}^d}\frac{e^{-ik\cdot x}}{\sqrt{|k|^2+m^2}-m+1}dk, \tag{4.6.13}$$

but it is not clear that the right-hand side of (4.6.13) is finite. In order to give a clear meaning to this, we need to define the right-hand side of (4.6.13) in the L^2 sense. We introduce three classes of external potentials given by

$$\mathcal{V}_1 = \left\{V:\mathbb{R}^d\to\mathbb{R}\,\middle|\,\lim_{t\downarrow 0}\sup_{x\in\mathbb{R}^d}\int_0^t \mathbb{E}^{x,0}_{\mathcal{W}\times P}[V(B_{T_s})]ds = 0\right\},$$

$$\mathcal{V}_2 = \left\{V:\mathbb{R}^d\to\mathbb{R}\,\middle|\,\lim_{\lambda\to\infty}\sup_{x\in\mathbb{R}^d}\left((\sqrt{-\Delta+m^2}-m+\lambda)^{-1}V\right)(x) = 0\right\},$$

$$\mathcal{V}_3 = \left\{ V : \mathbb{R}^d \to \mathbb{R} \,\middle|\, \lim_{\delta \downarrow 0} \sup_{x \in \mathbb{R}^d} \int_{|x-y|<\delta} g_R(x-y)V(y)dy = 0 \right\};$$

Also, we say that V is *uniformly locally integrable* whenever

$$\sup_{x \in \mathbb{R}^d} \int_{x+C_0} |V(y)|dy < \infty, \tag{4.6.14}$$

where C_0 denotes the d-dimensional cube centered at 0 with sides of unit length and $C_0 + x = \{x + y \mid y \in C_0\}$. In the nonrelativistic case, $V \in \mathcal{K}(\mathbb{R}^d)$ is shown to be also uniformly locally integrable, see Proposition 4.94.

Proposition 4.215. *Let V be a nonnegative function. Then we have the following:*
(1) $\mathcal{V}_2 \subset \mathcal{V}_1$ *and* $\mathcal{V}_2 \subset \mathcal{V}_3$*;*
(2) *let V be uniformly locally integrable; then* $\mathcal{V}_3 \subset \mathcal{V}_2$ *and* $\mathcal{V}_1 \subset \mathcal{V}_2$.

Proof. We give the proof in a more general case in Proposition 4.281 and Example 4.286 below. □

Next we define the relativistic Kato-class.

Definition 4.216 (Relativistic Kato-class).
(1) Potential V belongs to a *relativistic Kato-class* whenever it satisfies

$$\lim_{\delta \downarrow 0} \sup_{x \in \mathbb{R}^d} \int_{|x-y|<\delta} g_R(x-y)V(y)dy = 0. \tag{4.6.15}$$

We denote this linear space by $\mathcal{K}_R(\mathbb{R}^d)$.
(2) Potential $V = V_+ - V_-$ is *relativistic Kato-decomposable* whenever $V_+ 1_K \in \mathcal{K}_R(\mathbb{R}^d)$ for every compact set $K \subset \mathbb{R}^d$ and $V_- \in \mathcal{K}_R(\mathbb{R}^d)$.

Similarly to the nonrelativistic case in Propositions 4.96 and 4.99, we have equivalent descriptions for the relativistic Kato-class.

Proposition 4.217. *Suppose that V is uniformly locally integrable. Then the following are equivalent:*
(1) $\lim_{t \downarrow 0} \sup_{x \in \mathbb{R}^d} \int_0^t \mathbb{E}^{x,0}_{\mathcal{W}\times P}[V(B_{T_s})]ds = 0$;
(2) $\lim_{\lambda \to \infty} \sup_{x \in \mathbb{R}^d}((\sqrt{-\Delta + m^2} - m + \lambda)^{-1}V)(x) = 0$;
(3) *Potential V belongs to the relativistic Kato-class.*

Proof. This follows from Proposition 4.215. □

We can also show an exponential bound of relativistic Kato-class potentials by using Khasminskii's lemma and (4.6.15).

Lemma 4.218. *Suppose that V is uniformly locally integrable and from the relativistic Kato-class. Then for every $t \geq 0$,*

$$\sup_{x\in\mathbb{R}^d} \mathbb{E}^{x,0}\left[e^{\int_0^t V(B_{T_s})\,ds}\right] < \infty. \tag{4.6.16}$$

Proof. Since $\lim_{t\downarrow 0}\sup_{x\in\mathbb{R}^d}\int_0^t \mathbb{E}^{x,0}_{\mathcal{W}\times P}[V(B_{T_s})]ds = 0$, the lemma can be proven in a similar manner to Khasminskii's lemma. □

Define the operator $K_t^{\mathrm{R}} : L^2(\mathbb{R}^d) \to L^2(\mathbb{R}^d)$ by

$$(K_t^{\mathrm{R}} f)(x) = \mathbb{E}^{x,0}\left[e^{-i\int_0^{T_t} a(B_s)\circ dB_s} e^{-\int_0^t V(B_{T_s})ds} f(B_{T_t})\right]. \tag{4.6.17}$$

Theorem 4.219 (Feynman–Kac formula for relativistic Schrödinger operator with relativistic Kato-decomposable potential and vector potential). *Let $a \in (L^2_{\mathrm{loc}}(\mathbb{R}^d))^d$ and $\nabla\cdot a \in L^1_{\mathrm{loc}}(\mathbb{R}^d)$. Suppose that V is relativistic Kato-decomposable and uniformly locally integrable. Then the family of operators $\{K_t^{\mathrm{R}} : t \geq 0\}$ is a symmetric C_0-semigroup. In particular, there exists a self-adjoint operator $K_{\mathrm{R}}(a)$ such that*

$$K_t^{\mathrm{R}} = e^{-tK_{\mathrm{R}}(a)}, \quad t \geq 0. \tag{4.6.18}$$

Proof. Since subordinate Brownian motion is a Markov process, the proof of this result is similar to those of Theorems 4.197 and 4.185. By this and the Hille–Yosida theorem, we can define the corresponding operator $K_{\mathrm{R}}(a)$. □

Definition 4.220 (Relativistic Schrödinger operator with Kato-decomposable potential and vector potential). Let $a \in (L^2_{\mathrm{loc}}(\mathbb{R}^d))^d$ and $\nabla\cdot a \in L^1_{\mathrm{loc}}(\mathbb{R}^d)$. Suppose that V is relativistic Kato-decomposable and uniformly locally integrable. Then the *relativistic Schrödinger operator with relativistic Kato-decomposable potential V and vector potential a* is defined by $K_{\mathrm{R}}(a)$ in (4.6.18).

Similarly to the nonrelativistic case, we can show also the continuity of $K_t^{\mathrm{R}} f(x)$ in x. To see this, we similarly make use of the smoothing effect of $e^{-t(\sqrt{-\Delta+m^2}-m)}$. We define

$$e^{-t(\sqrt{-\Delta+m^2}-m)}f(x) = \int_{\mathbb{R}^d} p_t^X(x-y)f(y)dy = \mathbb{E}^{x,0}[f(B_{T_t})], \quad x\in\mathbb{R}^d,\ t\geq 0,$$

for $f \in L^p(\mathbb{R}^d)$ with $1 \leq p \leq \infty$, where $p_t^X(x)$ is given by (4.6.26)–(4.6.28).

Proposition 4.221 (Smoothing effect). *Let $f \in L^p(\mathbb{R}^d)$ with some $1 \leq p \leq \infty$. Then $e^{-t(\sqrt{-\Delta+m^2}-m)}f$ is continuous.*

Proof. The proof is similar to that in the nonrelativistic case in Proposition 4.109. Replacing Π_t by p_t^X, it is seen that $e^{-t(\sqrt{-\Delta+m^2}-m)}f(x)$ is continuous in x for $p = 1$ and $p = \infty$. Consider then case of $1 < p < \infty$. Let $g \in L^p(\mathbb{R}^d)$ and define

$$g_n(x) = \begin{cases} g(x), & |g(x)| \le n, \\ n, & |g(x)| > n. \end{cases}$$

We see that $e^{-t(\sqrt{-\Delta+m^2}-m)}g_n(x)$ is continuous in x for every $n \in \mathbb{N}$. By the Hölder inequality, we have

$$|e^{-t(\sqrt{-\Delta+m^2}-m)}g_n(x) - e^{-t(\sqrt{-\Delta+m^2}-m)}g(x)| \le \|p_t^X\|_{L^q(\mathbb{R}^d)}\|g_n - g\|_{L^p(\mathbb{R}^d)}$$

for $1/p + 1/q = 1$. Here we used that $p_t^X \in L^q(\mathbb{R}^d)$ for $1 \le p < \infty$. The right-hand side converges to zero as $n \to \infty$ uniformly in x, hence $x \mapsto e^{-t(\sqrt{-\Delta+m^2}-m)}g(x)$ is continuous. □

Lemma 4.222. *Suppose that V is a relativistic Kato-decomposable potential and let $f \in L^\infty(\mathbb{R}^d)$. Then*

$$e^{-r(\sqrt{-\Delta+m^2}-m)}K_{t-r}^{\mathrm{R}}f(x) = \mathbb{E}^{x,0}\left[e^{-\int_r^t V(B_{T_s})ds}e^{-i\int_{T_r}^{T_t} a(B_s)\circ dB_s}f(B_{T_t})\right], \quad t \ge r > 0. \qquad (4.6.19)$$

In particular, (4.6.19) *is continuous in x for every $r > 0$.*

Proof. We have

$$e^{-r(\sqrt{-\Delta+m^2}-m)}K_{t-r}^{\mathrm{R}}f(x) = \mathbb{E}^{x,0}\left[\mathbb{E}^{B_{T_r},0}\left[e^{-\int_0^{t-r} V(B_{T_s})ds}e^{-i\int_0^{T_{t-r}} a(B_s)\circ dB_s}f(B_{T_{t-r}})\right]\right].$$

For a fixed ϖ, consider

$$\mathbb{E}^{B_{T_r(\varpi)}}\left[e^{-\int_0^{t-r} V(B_{T_s(\varpi)})ds}e^{-i\int_0^{T_{t-r}(\varpi)} a(B_s)\circ dB_s}f(B_{T_{t-r}(\varpi)})\right].$$

For simplicity, write T_t for $T_t(\varpi)$. The Markov property of Brownian motion yields

$$\begin{aligned} &\mathbb{E}^{B_{T_r}}\left[e^{-\int_0^{t-r} V(B_{T_s})ds}e^{-i\int_0^{T_{t-r}} a(B_s)\circ dB_s}f(B_{T_{t-r}})\right] \\ &= \mathbb{E}^{x}\left[e^{-\int_0^{t-r} V(B_{T_s+T_r})ds}e^{-i\int_{T_r}^{T_{t-r}+T_r} a(B_s)\circ dB_s}f(B_{T_{t-r}+T_r})|\mathcal{F}_{T_t}\right], \end{aligned}$$

which implies that

$$\begin{aligned} &\mathbb{E}^{x}\left[\mathbb{E}^{B_{T_r}}\left[e^{-\int_0^{t-r} V(B_{T_s})ds}e^{-i\int_0^{T_{t-r}} a(B_s)\circ dB_s}f(B_{T_{t-r}})\right]\right] \\ &= \mathbb{E}^{x}\left[e^{-\int_0^{t-r} V(B_{T_s+T_r})ds}e^{-i\int_{T_r}^{T_{t-r}+T_r} a(B_s)\circ dB_s}f(B_{T_{t-r}+T_r})\right] \\ &= \mathbb{E}^{x}\left[e^{-\int_r^{t} V(B_{T_{s-r}+T_r})ds}e^{-i\int_{T_r}^{T_{t-r}+T_r} a(B_s)\circ dB_s}f(B_{T_{t-r}+T_r})\right]. \end{aligned}$$

For a fixed r, we notice that $T_t = T_t - T_r + T_r \doteq T_{t-r} + T_r$ for all $t \ge r$. Taking the expectation with respect to the subordinator, we have

$$\mathbb{E}^{x,0}\Big[\mathbb{E}^{B_{T_r},0}[e^{-\int_0^{t-r} V(B_{T_s})ds} e^{-i\int_0^{T_{t-r}} a(B_s)\circ dB_s} f(B_{T_{t-r}})]\Big]$$
$$= \mathbb{E}^{x,0}\Big[e^{-\int_r^t V(B_{T_s})ds} e^{-i\int_{T_r}^{T_t} a(B_s)\circ dB_s} f(B_{T_t})\Big],$$

which shows the lemma. □

Next we introduce the following condition. We say that a measurable function f on $\mathbb{R}^d$ is in $\mathcal{L}(\mathbb{R}^d)$ if

$$\lim_{r\downarrow 0}\sup_{x\in\mathbb{R}^d} \mathbb{E}^{x,0}\Bigg[\int_0^{T_r} |f(B_s)|ds\Bigg] = 0.$$

Proposition 4.223 (L^p–L^q boundedness and continuity of integral kernel). *Suppose that V is relativistic Kato-decomposable and $a \in (L^2_{\mathrm{loc}}(\mathbb{R}^d))^d$ with $\nabla\cdot a \in L^1_{\mathrm{loc}}(\mathbb{R}^d)$. Let $1 \le p \le q \le \infty$. Then the following hold:*
1. *$e^{-tK_{\mathrm{R}}(a)}$ is L^p–L^q bounded, for every $t \ge 0$;*
2. *If $|a|^2, \nabla\cdot a \in \mathcal{L}(\mathbb{R}^d)$ and V is uniformly locally integrable, then $K_t^{\mathrm{R}}f(x)$ is continuous in x for $f \in L^p(\mathbb{R}^d)$ with $1 \le p \le \infty$;*
3. *If $|a|^2, \nabla\cdot a \in \mathcal{L}(\mathbb{R}^d)$ and V is uniformly locally integrable, then the integral kernel $\mathbb{R}^d\times\mathbb{R}^d \ni (x,y) \mapsto K_t^{\mathrm{R}}(x,y)$ is jointly continuous in x and y, and is given by*

$$K_t^{\mathrm{R}}(x,y) = p_t^X(x-y)\mathbb{E}^{x,y}_{[0,t]}\Big[e^{-\int_0^t V(B_{T_s})\,ds} e^{-i\int_0^{T_t} a(B_s)\circ dB_s}\Big]. \tag{4.6.20}$$

   Here $P^{x,y}_{[0,t]}$ is the Lévy bridge measure for $(B_{T_t})_{t\ge0}$.
4. *Let $a = 0$ and V be uniformly locally integrable. Then $e^{-tK_{\mathrm{R}}(0)}$ is positivity improving.*

Proof. By the diamagnetic inequality, we have $|e^{-tK_{\mathrm{R}}(a)}f(x)| \le e^{-tK_{\mathrm{R}}(0)}|f|(x)$. Part (1) can be proven similarly to Theorem 4.107 with Brownian motion $(B_t)_{t\ge0}$ replaced by the subordinate Brownian motion $(B_{T_t})_{t\ge0}$. Claim (2) is also proven by a minor modification of the proof of Theorem 4.111. It suffices to consider $f \in L^\infty(\mathbb{R}^d)$. Let $x_n \to x$ as $n \to \infty$. We have

$$K_t^{\mathrm{R}}f(x) - K_t^{\mathrm{R}}f(x_n) = K_t^{\mathrm{R}}f(x) - e^{-r(\sqrt{-\Delta+m^2}-m)}K_{t-r}^{\mathrm{R}}f(x)$$
$$+ e^{-r(\sqrt{-\Delta+m^2}-m)}K_{t-r}^{\mathrm{R}}f(x) - e^{-r(\sqrt{-\Delta+m^2}-m)}K_{t-r}^{\mathrm{R}}f(x_n) + e^{-r(\sqrt{-\Delta+m^2}-m)}K_{t-r}^{\mathrm{R}}f(x_n) - K_t^{\mathrm{R}}f(x_n).$$

By the smoothing effect in Proposition 4.221, we see that

$$e^{-r(\sqrt{-\Delta+m^2}-m)}K_{t-r}^{\mathrm{R}}f(x) - e^{-r(\sqrt{-\Delta+m^2}-m)}K_{t-r}^{\mathrm{R}}f(x_n) \to 0$$

as $n \to \infty$, for every $r > 0$. It is sufficient to show that

$$K_t^{\mathrm{R}}f(x) - e^{-r(\sqrt{-\Delta+m^2}-m)}K_{t-r}^{\mathrm{R}}f(x) \to 0$$

as $r \to 0$, uniformly in x. Let $V \in \mathcal{K}_{\mathrm{R}}(\mathbb{R}^d)$ and let $g_r(x) = e^{-r(\sqrt{-\Delta+m^2}-m)}K^{\mathrm{R}}_{t-r}f(x)$. By the independence of

$$1 - e^{-\int_0^r V(B_{T_s(\varpi)})\,ds}e^{-i\int_0^{T_r(\varpi)} a(B_s)\circ dB_s} \quad \text{and}$$
$$e^{-\int_r^t V(B_{T_s(\varpi)})\,ds}e^{-i\int_{T_r(\varpi)}^{T_t(\varpi)} a(B_s)\circ dB_s}f(B_t)$$

for fixed ϖ and by Lemma 4.222, we have

$$\begin{aligned}\|g_r - K_t f\|_\infty &\le \sup_{x\in\mathbb{R}^d} \mathbb{E}^{x,0}\left[\left|1 - e^{-\int_0^r V(B_{T_s})\,ds}e^{-i\int_0^{T_r} a(B_s)\circ dB_s}\right|\left|e^{-\int_r^t V(B_{T_s})\,ds}e^{-i\int_{T_r}^{T_t} a(B_s)\circ dB_s}f(B_{T_t})\right|\right]\\ &\le B\sup_{x\in\mathbb{R}^d}\left(\mathbb{E}^{x,0}\left[\left|1 - e^{-i\int_0^{T_r} a(B_s)\circ dB_s}\right|\right] + \mathbb{E}^{x,0}\left[\left|1 - e^{-\int_0^r V(B_{T_s})\,ds}\right|\right]\right).\end{aligned}$$

Here $B = \|f\|_\infty \sup_{x\in\mathbb{R}^d}\sup_{r\le t}\mathbb{E}^{x,0}[e^{-\int_r^t V(B_{T_s})\,ds}]$. Since V is uniformly locally integrable,

$$\lim_{t\downarrow 0}\sup_{x\in\mathbb{R}^d}\int_0^t \mathbb{E}^{x,0}[V(B_{T_s})]ds = 0 \tag{4.6.21}$$

follows. We see then that $\sup_{x\in\mathbb{R}^d}\mathbb{E}^{x,0}[1 - e^{-\int_0^r V(B_{T_s})\,ds}] \to 0$ as $r \to 0$, in the same way as in the proof of Theorem 4.111. For the second term, we have

$$\mathbb{E}^{x,0}\left[\left|1 - e^{-i\int_0^{T_r} a(B_s)\circ dB_s}\right|\right] \le \mathbb{E}^{x,0}\left[\left|\int_0^{T_r} a(B_s)dB_s\right|\right] + \mathbb{E}^{x,0}\left[\frac{1}{2}\left|\int_0^{T_r}\nabla\cdot a(B_s)ds\right|\right].$$

For each $T_r = T_r(\varpi)$, by Schwarz inequality and Itô-isometry, we have

$$\mathbb{E}^x\left[\left|\int_0^{T_r} a(B_s)dB_s\right|\right] \le \left(\mathbb{E}^x\left[\left|\int_0^{T_r} a(B_s)dB_s\right|^2\right]\right)^{1/2} = \left(\mathbb{E}^x\left[\int_0^{T_r}|a(B_s)|^2 ds\right]\right)^{1/2}.$$

Since $|a|^2 \in \mathcal{L}(\mathbb{R}^d)$, it follows that

$$\lim_{r\to 0}\sup_{x\in\mathbb{R}^d}\mathbb{E}^{x,0}\left[\left|\int_0^{T_r} a(B_s)dB_s\right|\right] = 0.$$

We can also see that

$$\lim_{r\to 0}\sup_{x\in\mathbb{R}^d}\mathbb{E}^{x,0}\left[\left|\int_0^{T_r}\nabla\cdot a(B_s)ds\right|\right] \le \lim_{r\to 0}\sup_{x\in\mathbb{R}^d}\mathbb{E}^{x,0}\left[\int_0^{T_r}|\nabla\cdot a(B_s)|ds\right] = 0$$

since $\nabla \cdot a \in \mathcal{L}(\mathbb{R}^d)$. Thus $K_t^{\mathrm{R}} f(x)$ is continuous in x. Next, let $V = V_+ - V_-$ be relativistic Kato-decomposable, and consider for $R > 0$ the function $V_R = V_{+R} - V_-$ with $V_{+R}(x) = V_+(x)$ if $|x| \le R$ and $V_{+R}(x) = 0$ if $|x| > R$. We have $V_R \in \mathcal{K}_{\mathrm{R}}(\mathbb{R}^d)$. Let $M \subset B_R(0)$ be compact, where $B_R(0)$ denotes the ball of radius R centered at the origin. Hence

$$\begin{aligned}
&\sup_{x\in M} \left| \mathbb{E}^{x,0}\left[e^{-\int_0^t V(B_s)\,ds} e^{-i\int_0^{T_t} a(B_s)\circ dB_s} f(B_t)\right] - \mathbb{E}^{x,0}\left[e^{-\int_0^t V_R(B_s)\,ds} e^{-i\int_0^{T_t} a(B_s)\circ dB_s} f(B_t)\right]\right| \\
&\quad \le 2\|f\|_\infty \sup_{x\in M} \mathbb{E}^{x,0}\left[e^{\int_0^t V_-(B_{T_s})\,ds} 1_{\{\sup_{0\le s\le t} |B_{T_s}|\ge R\}}\right] \\
&\quad \le 2\|f\|_\infty \sup_{x\in M} \left(\mathbb{E}^{x,0}[e^{-2\int_0^t V_-(B_s)\,ds}]\right)^{1/2} \left(\mathbb{E}^{x,0}[1_{\{\sup_{0\le s\le t} |B_{T_s}|\ge R\}}]\right)^{1/2}.
\end{aligned}$$

The first factor above is bounded since V_- is of relativistic Kato-class, and the second goes to zero as $R \to \infty$ by Lévy's maximal inequality

$$\mathbb{E}^{x,0}[1_{\{\sup_{0\le s\le t} |B_{T_s}|\ge R\}}] = \mathbb{E}^0\mathbb{E}^x[1_{\{\sup_{0\le s\le t} |B_{T_s}|\ge R\}}] \le 2\mathbb{E}^0\mathbb{E}^x[1_{\{|B_{T_t}|>R\}}].$$

Hence $K_t^{\mathrm{R}} f(x)$ is continuous in x. Finally, to prove (3), we again proceed similarly as in the proof of Theorem 4.116. We have

$$(e^{-tK_{\mathrm{R}}} f)(x) = \int_{\mathbb{R}^d} p_t^X(x-y) f(y) \mathbb{E}_{[0,t]}^{x,y}[e^{-\int_0^t V(B_{T_s})\,ds} e^{-i\int_0^{T_t} a(B_s)\circ dB_s}]\,dy,$$

which shows that (4.4.29) holds for all x and almost all y. Next we show the continuity of integral kernel. Let $s = t/3$. We have

$$K_t^{\mathrm{R}}(x,y) = \int_{\mathbb{R}^{2d}} K_s^{\mathrm{R}}(x,v) K_s^{\mathrm{R}}(v,w) K_s^{\mathrm{R}}(w,y)\,dw\,dv = \int_{\mathbb{R}^{2d}} K_s^{\mathrm{R}}(x,v) K_s^{\mathrm{R}}(y,w) K_s^{\mathrm{R}}(v,w)\,dw\,dv.$$

The product

$$\begin{aligned}
&K_s^{\mathrm{R}}(x,v) K_s^{\mathrm{R}}(y,w) \\
&\quad = p_s^X(x-v) p_s^X(y-w) \mathbb{E}_{[0,s]}^{(x,v),(y,w)}\left[e^{-\int_0^s (V(B_{T_r}^{(1)}) + V(B_{T_r}^{(2)}))\,dr} e^{-i\int_0^{T_s} a(B_r^{(1)})\circ dB_r^{(1)} - i\int_0^{T_s} a(B_r^{(2)})\circ dB_r^{(2)}}\right]
\end{aligned}$$

is the kernel of the relativistic Schrödinger operator in $L^2(\mathbb{R}^{2d})$, namely

$$\tilde{K}_{\mathrm{R}} = \sqrt{(-i\nabla_x - a(x))^2 + m^2} - m + \sqrt{(-i\nabla_y - a(y))^2 + m^2} - m + V(x) + V(y),$$

where $B_t^{(i)}$, $i = 1, 2$, denote two independent d-dimensional Brownian motions. We see that $V(x) + V(y)$ is relativistic Kato-decomposable, and $\nabla_x \cdot a(x) + \nabla_y \cdot a(y)$ and $|a(x)|^2 + |a(y)|^2$ are in $\mathcal{L}(\mathbb{R}^{2d})$. On the other hand, the map $(v,w) \mapsto K_s^{\mathrm{R}}(v,w)$ is bounded. Hence $K_t^{\mathrm{R}}(x,y) = (e^{-t\tilde{K}_{\mathrm{R}}} K_s^{\mathrm{R}})(x,y)$ with $K_s^{\mathrm{R}} \in L^\infty(\mathbb{R}^{2d})$ and is jointly continuous in x and y by statement (2). Finally, to show (4), note that by (4.6.21) we have $e^{-\int_0^t V(B_{T_s})ds} > 0$ a. e. for sufficiently small t. Hence $(f, e^{-tH_R(0)}g) > 0$ for every nonnegative f and g for all $t > 0$ by the semigroup property. □

4.6.3 Decay of eigenfunctions

In this subsection we study the decay of eigenfunctions of relativistic Schrödinger operators. In Section 4.3.8 we have discussed the exponential decay of eigenfunctions of Schrödinger operators via Carmona's estimate. For relativistic Schrödinger operators, the study of the decay of eigenfunctions is more complicated than for classical Schrödinger operators, and our basic tool will be a martingale related to the Feynman–Kac semigroup.

For simplicity, we denote

$$X_t = B_{T_t}, \quad t \geq 0,$$

with the subordinator $(T_t)_{t\geq 0}$ used above. Recall that $(X_t)_{t\geq 0}$ is a Lévy process such that its generator is the self-adjoint operator $\sqrt{-\Delta + m^2} - m$. Let φ be an eigenfunction of the relativistic Schrödinger operator $H_R(0)$ such that

$$H_R(0)\varphi = E\varphi. \tag{4.6.22}$$

Also, as above, we have

$$e^{-tH_R(0)}f(x) = \mathbb{E}^0[f(X_t)e^{-\int_0^t V(X_s+x)ds}].$$

Consider the random process $(M_t(x))_{t\geq 0}$ with

$$M_t(x) = e^{Et}e^{-\int_0^t V(X_s+x)ds}\varphi(X_t + x), \quad t \geq 0. \tag{4.6.23}$$

Then it is seen that $\varphi(x)$ can be represented in terms of $(M_t(x))_{t\geq 0}$ as

$$\varphi(x) = (e^{-t(H_R(0)-E)}\varphi)(x) = \mathbb{E}^0[M_t(x)] = \mathbb{E}^x[M_t(0)] \tag{4.6.24}$$

for a. e. $x \in \mathbb{R}^d$ and for all $t \geq 0$.

Lemma 4.224 (Martingale properties). *The process $(M_t(x))_{t\geq 0}$ is a martingale with respect to the natural filtration of $(X_t)_{t\geq 0}$.*

Proof. Let $\mathscr{F}_s = \sigma(X_r, 0 \leq r \leq s)$. By the Markov property of $(X_t)_{t\geq 0}$, we readily see that for $t \geq s$ that

$$\mathbb{E}^0[M_t(x)|\mathscr{F}_s] = e^{Et}e^{-\int_0^s V(X_r+x)dr}\mathbb{E}^0\left[e^{-\int_s^t V(X_r)dr}\varphi(X_t + x)|\mathscr{F}_s\right]$$

$$= e^{Et}e^{-\int_0^s V(X_r+x)dr}\mathbb{E}^{X_s}\left[e^{-\int_0^{t-s}V(X_r)dr}\varphi(X_{t-s}+x)\right]$$
$$= e^{Es}e^{-\int_0^s V(X_r+x)dr}e^{(t-s)E}\mathbb{E}^{X_s}\left[e^{-\int_0^{t-s}V(X_r)dr}\varphi(X_{t-s}+x)\right]$$
$$= e^{Es}e^{-\int_0^s V(X_r+x)dr}\left(e^{-(t-s)(H_{\mathrm{R}}(0)-E)}\varphi\right)(X_s) = M_s. \qquad \square$$

Let τ be a stopping time with respect to the natural filtration $(\mathscr{F}_t)_{t\geq 0}$ of $(X_t)_{t\geq 0}$. Then $(M_{t\wedge\tau})_{t\geq 0}$ is also a martingale by Theorem 2.73, which implies by (4.6.24) that

$$\varphi(x) = \mathbb{E}^0[M_{t\wedge\tau}(x)]. \tag{4.6.25}$$

The distribution function p_t^X of X_t can be expressed in several ways. Let $t > 0$.

(1) It is given by

$$p_t^X(x) = \begin{cases} 2\left(\frac{m}{2\pi}\right)^2 \frac{t}{(t^2+|x|^2)^{(d+1)/4}} K_{(d+1)/2}(m\sqrt{|x|^2+t^2}), & m > 0, \\ \Gamma(\frac{d+1}{2})\pi^{-(d+1)/2}\frac{t}{(t^2+|x|^2)^{(d+1)/2}}, & m = 0, \end{cases} \tag{4.6.26}$$

where K_ν denotes the modified Bessel function of the second kind (4.2.21).

(2) We alternatively have the representation

$$p_t^X(x) = \frac{1}{(2\pi)^d}\frac{t}{\sqrt{|x|^2+t^2}}\int_{\mathbb{R}^d} e^{mt}e^{-\sqrt{(|x|^2+t^2)(|p|^2+m^2)}}dp. \tag{4.6.27}$$

(3) By subordination, explicitly using the distribution functions of Brownian motion $(B_t)_{t\geq 0}$ and of the subordinator $(T_t)_{t\geq 0}$, we also have

$$p_t^X(x) = e^{mt}\int_0^\infty e^{-m^2u}\Pi_{2u}(x)\theta_t(du), \tag{4.6.28}$$

where

$$\theta_t(du) = \pi^{-1/2}tu^{-3/2}e^{-t^2/(4u)}du.$$

Lemma 4.225. *If V is relativistic Kato-decomposable and uniformly locally integrable, then $\varphi \in L^\infty(\mathbb{R}^d)$.*

Proof. Since $\sup_{y\in\mathbb{R}^d} p_t^X(y) < \infty$ for $t > 0$ and V is relativistic Kato-decomposable, we have that

$$|\varphi(x)| \leq \left(\mathbb{E}^0[e^{-2\int_0^t V(X_r+x)dr}]\right)^{1/2}\left(\int_{\mathbb{R}^d}|\varphi(y+x)|^2 p_t^X(y)dy\right)^{1/2} \leq C\|\varphi\|$$

with a constant C. $\square$

Now we study the decay properties of the eigenfunction φ.

Theorem 4.226 (Confining potential: upper bound). *Let V be relativistic Kato-decomposable and uniformly locally integrable. Suppose that $\lim_{|x|\to\infty} V(x) = \infty$. Let φ be an eigenfunction corresponding to an eigenvalue E. Then for every positive constant $c > 0$, there exists a positive constant $D(c)$ such that*

$$|\varphi(x)| \le D(c)e^{-c|x|}.$$

Proof. For every $R > 0$, let $\tau_R = \inf\{t > 0 \mid |X_t| > R\}$. Then τ_R is a stopping time and we have

$$\varphi(x) = \mathbb{E}^0[e^{(t\wedge\tau_R)E}e^{-\int_0^{t\wedge\tau_R} V(X_s+x)ds}\varphi(X_{t\wedge\tau_R} + x)].$$

Hence

$$|\varphi(x)| \le \|\varphi\|_\infty \mathbb{E}^0[e^{-(W_R(x)-E)t\wedge\tau_R}],$$

where $W_R(x) = \inf_{|y-x|\le R} V(y)$. Note that $\lim_{|x|\to\infty} W_R(x) - E = \infty$ and $W_R(x) - E > 0$ for large $|x|$. Thus

$$\begin{aligned}|\varphi(x)| &\le \|\varphi\|_\infty \left(\mathbb{E}^0[1_{\{\tau_R<t\}}e^{-(W_R(x)-E)t\wedge\tau_R}] + \mathbb{E}^0[1_{\{\tau_R\ge t\}}e^{-(W_R(x)-E)t\wedge\tau_R}]\right)\\ &\le \|\varphi\|_\infty \left(e^{-t(W_R(x)-E)} + \mathbb{E}^0[1_{\{\tau_R<t\}}e^{-(W_R(x)-E)t\wedge\tau_R}]\right).\end{aligned}$$

We also see that

$$\mathbb{E}^0[1_{\{\tau_R<t\}}e^{-(W_R(x)-E)t\wedge\tau_R}] \le \mathbb{E}^0[1_{\{\tau_R<t\}}] = \mathbb{E}^0[1_{\{\sup_{s\in[0,t]}|X_s|\ge R\}}] \le \mathbb{E}^0[e^{a(\sup_{s\in[0,t]}|X_s|-R)}]$$

for every $a > 0$. The expectation $\mathbb{E}^0[e^{a\sup_{s\in[0,t]}|X_s|}]$ can be estimated for sufficiently small $a > 0$ by Theorem 3.78, leading to

$$|\varphi(x)| \le \|\varphi\|_\infty(e^{-(W_R(x)-E)t} + e^{-aR}De^{ct}),$$

where D is as in Theorem 3.78, $c = 1 + \int_{|z|>1}(e^{a|z|}-1)\nu(dz)$, and ν denotes the Lévy measure associated with $(X_t)_{t\ge0}$. Take $R = p_1|x|$ for $p_1 \in (0,1)$ and $t = p_2|x|$ for $p_2 > 0$ small enough such that $0 < ap_1 - cp_2$. We have

$$|\varphi(x)| \le \|\varphi\|_\infty(e^{-(W_{p_1|x|}(x)-E)p_2|x|} + e^{-(ap_1-cp_2)|x|}D),$$

and the theorem follows, since a is arbitrary and $\lim_{|x|\to\infty} W_{p_1|x|}(x) - E = \infty$ due to $p_1 < 1$. □

Next we consider the case when $\lim_{|x|\to\infty} V(x) = 0$.

Theorem 4.227 (Decaying potential: upper bound). *Let V be relativistic Kato-decomposable class and uniformly locally integrable. Suppose $\lim_{|x|\to\infty} V(x) = 0$. Let φ be an eigenfunction corresponding to an eigenvalue E.*

(1) *If $m > 0$, then for every $\varepsilon > 0$ there exists a constant $D_\varepsilon > 0$ such that*

$$|\varphi(x)| \leq D_\varepsilon e^{-m_\varepsilon |x|}, \tag{4.6.29}$$

where

$$m_\varepsilon = \begin{cases} m & \text{if } |E| > m, \\ \sqrt{2m|E| - E^2} - \varepsilon & \text{if } |E| \leq m. \end{cases}$$

(2) *If $m = 0$, then there exists a constant $c > 0$ such that*

$$|\varphi(x)| \leq \frac{c}{1 + |x|^{d+1}}. \tag{4.6.30}$$

Proof. Let $V = V_+ - V_-$ and $\tau_R(x) = \inf\{t > 0 \mid |X_t + x| \leq R\}$, which is a stopping time. For $\varepsilon > 0$, choose R large enough such that $\sup_{|y|>R} V_-(y) < \varepsilon$, and ε small enough such that $q = -(E + \varepsilon) > 0$. Thus

$$|\varphi(x)| = \left|\mathbb{E}^0\left[e^{(t\wedge\tau_R(x))E} e^{-\int_0^{t\wedge\tau_R(x)} V(X_s+x)ds} \varphi(X_{t\wedge\tau_R(x)} + x)\right]\right| \leq \|\varphi\|_\infty \mathbb{E}^0[e^{-q(t\wedge\tau_R(x))}].$$

Let μ_R^q be the q-capacitary measure of $\{y \in \mathbb{R}^d \mid |y| \leq R\}$. By (3.5.6), we have

$$\lim_{t\to\infty} \mathbb{E}^0[e^{-q(t\wedge\tau_R(x))}] = \mathbb{E}^0[e^{-q\tau_R(x)}] = \int_{|y|\leq R} u^q(x-y)\mu_R^q(dy), \tag{4.6.31}$$

where

$$u^q(x) = \int_0^\infty p_t^X(x) e^{-qt} dt.$$

Let $m > 0$. By (4.6.28), we have the upper bounds

$$u^q(x) \leq \begin{cases} \dfrac{Ce^{-\sqrt{2mq-q^2}|x|}}{|x|^{\frac{d}{2}-1}} & \text{if } q \leq m, \\[2ex] \dfrac{Ce^{-m|x|}}{|x|^{\frac{d}{2}-1}} & \text{otherwise}, \end{cases} \tag{4.6.32}$$

with a constant C, for all $|x| \geq 1$. For $|E| \leq m$, from (4.6.32) it follows that

$$\mathbb{E}^0[e^{-q\tau_R(x)}] \leq \begin{cases} C\int_{|y|\leq R} e^{-\sqrt{2mq-q^2}|x-y|}\mu_R^q(dy) \leq C' e^{-(\sqrt{2m|E|-E^2}+\varepsilon)|x|}, & |x| \geq R+1, \\ C_R, & |x| < R+1. \end{cases}$$

Here C, C', and C_R are positive constants. Hence (4.6.29) follows for the case $|E| \le m$. Similarly, we can show (4.6.29) for $|E| > m$ by (4.6.32).

Let $m = 0$. It is seen by (4.6.26) that

$$\frac{c_1}{|x|^{d+1}} \le u^q(x) \le \frac{c_2}{|x|^{d+1}} \tag{4.6.33}$$

for $|x| > 1$ with positive constants c_1 and c_2. Together with (4.6.31), this implies

$$\mathbb{E}^0[e^{-q\tau_R(x)}] \le \begin{cases} \int_{|y|\le R} \frac{c_2}{|x-y|^{d+1}} \mu_R^q(dy) \le \frac{c_2'}{|x|^{d+1}}, & |x| \ge R+1, \\ C_R, & |x| < R+1. \end{cases}$$

Here C_R is a constant. Thus (4.6.30) follows. □

Finally, we show a lower bound on the ground state of $H_R(0)$.

Theorem 4.228 (Lower bound). *Let φ_0 be the ground state of $H_R(0)$ and continuous. Suppose that V is continuous and $\lim_{|x|\to\infty} V(x) = 0$.*

(1) *Let $m = 0$. Then there exists a constant $c > 0$ such that*

$$\varphi_0(x) \ge \frac{c}{1+|x|^{d+1}}. \tag{4.6.34}$$

(2) *Let $m > 0$. Then there exists a constant $c > 0$ such that*

$$\varphi_0(x) \ge ce^{-m_\varepsilon |x|}. \tag{4.6.35}$$

Proof. Note that φ_0 is strictly positive. In a similar way as in the proof of Theorem 4.227, with $\tau_R(x) = \inf\{t > 0 \mid |X_t + x| \le R\}$ and $\varepsilon > 0$, we choose R large enough such that $\sup_{|y|>R} V_+(y) < \varepsilon$ and ε small enough such that $q = -(E-\varepsilon) > 0$. We see that

$$\begin{aligned} \varphi_0(x) &= \mathbb{E}^0\left[e^{(t\wedge\tau_R(x))E} e^{-\int_0^{t\wedge\tau_R(x)} V(X_s+x)ds} \varphi_0(X_{t\wedge\tau_R(x)}+x)\right] \\ &\ge \mathbb{E}^0\left[e^{(t\wedge\tau_R(x))E} e^{-\int_0^{t\wedge\tau_R(x)} V_+(X_s+x)ds} \varphi_0(X_{t\wedge\tau_R(x)}+x)\right] \\ &= \mathbb{E}^0\left[e^{\tau_R(x)E} e^{-\int_0^{\tau_R(x)} V_+(X_s+x)ds} \varphi_0(X_{\tau_R(x)}+x)\right] \\ &\ge \inf_{|y|\le R} \varphi_0(y)\mathbb{E}^0\left[e^{-q\tau_R(x)} 1_{\{\tau_R(x)<\infty\}}\right] \\ &= \inf_{|y|\le R} \varphi(y)\mathbb{E}^0\left[e^{-q\tau_R(x)}\right] \end{aligned}$$

with the usual convention $e^{-\infty} = 0$. Hence

$$\varphi_0(x) \ge C\mathbb{E}\left[e^{-q\tau_R(x)}\right] = C \int_{|y|\le R} u^q(x-y)\mu_R^q(dy) \tag{4.6.36}$$

with a suitable constant $C > 0$. By (4.6.27), we have

$$p_t^X(x) \geq \frac{1}{(2\pi)^d} \frac{t}{\sqrt{|x|^2 + t^2}} e^{-m|x|} \int_{\mathbb{R}^d} e^{mt} e^{-\sqrt{|x|^2+t^2}|p|} dp \geq c_d(|x|^2 + t^2)^{-(d+1)/2} t e^{-m|x|}$$

for a positive constant $c_d > 0$. From this, we have

$$u^q(x) = \int_0^\infty e^{-qt} p_t^X(x) dt \geq c_d e^{-m|x|} \int_0^\infty \frac{te^{-qt}}{(|x|^2 + t^2)^{(d+1)/2}} dt = c_n \phi(|x|) e^{-m|x|}. \tag{4.6.37}$$

Here $\phi(x)$ is comparable to $|x|^{-d}$ when $|x| \to \infty$. This lower bound improves for $q \leq m$. From (4.6.28), we have

$$u^q(x) \geq e_d \psi(|x|) e^{-\sqrt{2mq-q^2}|x|} \tag{4.6.38}$$

with a constant e_d and a function $\psi(x)$ which is comparable to $|x|^{-d}$ when $|x| \to \infty$. The lower bound for $m > 0$ follows from (4.6.38) and (4.6.37), and for $m = 0$ it follows from (4.6.33). □

Remark 4.229. Clearly, apart from asymptotic estimates of the eigenfunctions at infinity, there is also interest in their behavior at short to mid-range from the origin. In Section 4.9.9 below, we will present local estimates for a more general class of nonlocal Schrödinger operators, which can be simply particularized to the relativistic Schrödinger operators discussed in this section.

4.6.4 Nonrelativistic limit

To close this section, we consider a limit in which relativistic Schrödinger operators turn into classical Schrödinger operators studied previously. For the passage to this so-called nonrelativistic limit, we go back to (4.6.1) and reintroduce the speed of light c (while keeping $\hbar = 1$) as a parameter, and define

$$H_{\mathrm{R}}(0) = \sqrt{-c^2\Delta + m^2c^4} - mc^2 + V.$$

We write H_c for $H_{\mathrm{R}}(0)$ and, using the Feynman–Kac formula, next show that $H_c \to H_\infty$ as $c \to \infty$ in a specific sense, and the limit operator is the Schrödinger operator

$$H_\infty = -\frac{1}{2m}\Delta + V.$$

We call this a *nonrelativistic limit*.

For every $c > 0$, consider the subordinator $(T_t^c)_{t\geq 0}$ with parameter c such that

$$\mathbb{E}_P[e^{-uT_t^c}] = e^{-t(\sqrt{2c^2u+m^2c^4}-mc^2)}. \tag{4.6.39}$$

Then the related Feynman–Kac formulae become

$$(f, e^{-tH_c}g) = \int_{\mathbb{R}^d} \mathbb{E}_{\mathcal{W}}^x[\bar{f}(x)[\mathbb{E}_P[g(B_{T_t^c})e^{-\int_0^t V(B_{T_s^c})ds}]]dx,$$

$$(f, e^{-tH_\infty}g) = \int_{\mathbb{R}^d} \mathbb{E}_{\mathcal{W}}^x[\bar{f}(x)g(B_{t/m})e^{-\int_0^t V(B_{s/m})ds}]dx.$$

From comparison, one can expect that T_t^c converges to the constant (nonrandom) subordinator $\frac{t}{m}$ as $c \to \infty$, so that

$$\lim_{c\to\infty} \mathbb{E}_P[e^{-uT_t^c}] = e^{-tu/m}. \tag{4.6.40}$$

We show this convergence in the lemma below, extended to a class of functions of the subordinator.

Lemma 4.230. *If f is a bounded continuous function on $\mathbb{R}$, then*

$$\lim_{c\to\infty} \mathbb{E}_P[f(T_t^c)] = f(t/m).$$

Proof. Let $f \in \mathscr{S}(\mathbb{R})$. We have

$$\mathbb{E}_P[f(T_t^c)] = \frac{1}{\sqrt{2\pi}} \int_{\mathbb{R}} \check{f}(k)\mathbb{E}_P[e^{-ikT_t^c}]dk.$$

The distribution of T_t^c is given by

$$\rho_t^c(s) = \frac{ct}{\sqrt{2\pi}} e^{mc^2t} s^{-3/2} \exp\left(-\frac{1}{2}\left(\frac{c^2t^2}{s} + m^2c^2s\right)\right) 1_{[0,\infty)}(s).$$

We can compute the expectation $\mathbb{E}_P[e^{-ikT_t^c}]$ in terms of the Bessel function

$$K_\beta(z) = \frac{1}{2}\left(\frac{z}{2}\right)^{-\beta} \int_0^\infty \exp\left(-\frac{z^2}{4s} - s\right) s^{\beta-1} ds, \quad \beta \in \mathbb{C},$$

for $z \in \{z \in \mathbb{C} \mid \Re z > 0, \Re(z^2) > 0\} = \{z \in \mathbb{C} \mid |\arg z| < \pi/4\}$. We obtain

$$\mathbb{E}_P[e^{-ikT_t^c}] = \int_0^\infty \frac{ct}{\sqrt{2\pi}} e^{mc^2t} s^{-3/2} \exp\left(-\frac{1}{2}\left(\frac{c^2t^2}{s} + (2ki + m^2c^2)s\right)\right) ds.$$

Put $z = (2ki + m^2c^2)^{1/2}ct$. Now take the principal value of the square root so that

$$(2ki + m^2c^2)^{1/2} = (4k^2 + m^4c^4)^{1/4}e^{i\theta/2},$$

with $\tan\theta = 2k/m^2c^2$ and $|\theta| < \pi/2$. Using K_β, we have

$$\mathbb{E}_P[e^{-ikT_t^c}] = \sqrt{\frac{2}{\pi}}\sqrt{z}e^{mc^2t}K_{-1/2}(z).$$

Since $K_{-1/2}(z) = K_{1/2}(z) = \sqrt{\frac{\pi}{2}}\frac{e^{-z}}{\sqrt{z}}$, we see that

$$\mathbb{E}_P[e^{-ikT_t^c}] = e^{-(\sqrt{i2k+m^2c^2}-mc)ct}.$$

Furthermore, we have $c(\sqrt{2ki+m^2c^2} - mc) = c((4k^2+m^4c^4)^{1/4} - mc)e^{i\theta/2} + mc^2(e^{i\theta/2} - 1)$. The first term converges to zero as $c \to \infty$. For the second term, it can be seen that

$$mc^2(e^{i\theta/2} - 1) = mc^2(\cos\theta/2 - 1 + i\sin\theta/2).$$

Since $mc^2(\cos\theta/2 - 1) \sim mc^2(\theta/2)^2/2 \sim mc^2(k/m^2c^2)^2/2 \sim 0$ and $imc^2\sin\theta/2 \sim imc^2\theta/2 \sim ik/m$ as $c \to \infty$, we get

$$\lim_{c\to\infty}\mathbb{E}_P[e^{-ikT_t^c}] = e^{-ik(t/m)}$$

and

$$\lim_{c\to\infty}\mathbb{E}_P[f(T_t^c)] = \frac{1}{\sqrt{2\pi}}\int_{\mathbb{R}}\check{f}(k)e^{-ik(t/m)} = f(t/m).$$

When f is chosen to be a bounded continuous function, it can be uniformly approximated by functions in $\mathscr{S}(\mathbb{R})$, and the proof of the lemma is completed by a simple limiting argument. □

Remark 4.231. In Lemma 4.230, we actually show that $\rho_t^c(s) \to \delta(s - \frac{t}{m})$ as $c \to \infty$. This can also formally be seen through the following argument. We have the estimate $\frac{1}{2}\left(\frac{c^2t^2}{s} + m^2c^2s\right) \geq mtc^2$, in which equality holds only for $s = t/m$. Substituting $s = t/m$ in $\rho_t^c(s)$, we have $e^{mc^2t}\exp\left(-\frac{1}{2}\left(\frac{c^2t^2}{s} + m^2c^2s\right)\right) = 1$ and $\rho_t^c(t/m) = (t/m)^{-3/2}\frac{ct}{\sqrt{2\pi}} \to \infty$ as $c \to \infty$. On the other hand, $\rho_t^c(s) \to 0$ for $s \neq t/m$ as $c \to \infty$ due to having then $\frac{1}{2}\left(\frac{c^2t^2}{s} + m^2c^2s\right) > mtc^2$.

Next we derive the nonrelativistic limit of e^{-tH_c}. We put a strong restriction on the potential V just for the sake of simplicity.

Proposition 4.232. *If V is a bounded continuous function, then*

$$\text{s-}\lim_{c\to\infty} e^{-tH_c} = e^{-tH_\infty}. \tag{4.6.41}$$

Proof. We suppose that V is nonnegative without loss of generality. It is enough to show the weak limit

$$\lim_{c\to\infty}(f, e^{-tH_c}g) = (f, e^{-tH_\infty}g). \tag{4.6.42}$$

Since $H_c \geq 0$ for every $c > 0$, $\|e^{-tH_c}\| \leq 1$ for all $c > 0$. It is also sufficient to show (4.6.42) for arbitrary f, g in $\mathscr{S}(\mathbb{R})$ by a limiting argument. Note that by Lemma 4.230, it can be seen that $\mathbb{E}_\nu[|f(T_t^c) - f(t/m)|] \to 0$ as $c \to \infty$. It follows that

$$(f, e^{-tH_c}g) - (f, e^{-tH_\infty}g) = \int_{\mathbb{R}^d} \mathbb{E}^x_{\mathcal{W}}[\bar{f}(x)\mathbb{E}_P[(g(B_{T_t^c}) - g(B_{t/m}))e^{-\int_0^t V(B_{T_s^c})ds}]]dx$$
$$+ \int_{\mathbb{R}^d} \mathbb{E}^x_{\mathcal{W}}[\bar{f}(x)g(B_{t/m})\mathbb{E}_P[(e^{-\int_0^t V(B_{T_s^c})ds} - e^{-\int_0^t V(B_{s/m})ds})]]dx.$$

It is straightforward to see that

$$\lim_{c\to\infty}\left|\int_{\mathbb{R}^d} \mathbb{E}^x_{\mathcal{W}}[\bar{f}(x)\mathbb{E}_P[(g(B_{T_t^c}) - g(B_{t/m}))e^{-\int_0^t V(B_{T_s^c})ds}]]dx\right|$$
$$\leq e^{\delta t}\lim_{c\to\infty}\int_{\mathbb{R}^d} \mathbb{E}^x_{\mathcal{W}}[|\bar{f}(x)|\mathbb{E}_P[|g(B_{T_t^c}) - g(B_{t/m})|]]dx = 0,$$

where $\delta = \|V\|_\infty$. We also see that

$$\lim_{c\to\infty}\left|\int_{\mathbb{R}^d} \mathbb{E}^x_{\mathcal{W}}\left[\bar{f}(x)g(B_{t/m})\mathbb{E}_P\left[(e^{-\int_0^t V(B_{T_s^c})ds} - e^{-\int_0^t V(B_{s/m})ds})\right]\right]\right|$$
$$\leq \lim_{c\to\infty}\int_{\mathbb{R}^d} \mathbb{E}^x_{\mathcal{W}}\left[|\bar{f}(x)g(B_{t/m})|\int_0^t \mathbb{E}_P|V(B_{T_s^c}) - V(B_{s/m})|ds\right] = 0.$$

This yields the result. □

We note that Proposition 4.232 can also be shown by operator theory. Since $C_0^\infty(\mathbb{R}^d)$ is a core of H_c, $c > 0$, and H_∞, it follows that $\lim_{c\to\infty} H_c f = H_\infty f$ for all $f \in C_0^\infty(\mathbb{R}^d)$, implying $H_c \to H_\infty$ as $c \to \infty$ in the strong resolvent sense by the general result below.

Proposition 4.233. *Let $(A_n)_{n\in\mathbb{N}}$ and A be self-adjoint operators on a Hilbert space $\mathcal{H}$, and suppose that D is a common core to all A_n and A. If $A_n f \to Af$ as $n \to \infty$ for every $f \in D$, then $A_n \to A$ as $n \to \infty$ in strong resolvent sense.*

Proof. Let $z \in \mathbb{C}$ with $\Im z \neq 0$. Take $f \in D$ and $g = (A+z)f$. Then $(A_n+z)^{-1}g - (A+z)^{-1}g = (A_n + z)^{-1}(A - A_n)f$ converges to 0 as $n \to \infty$, since $Af - A_n f \to 0$ and $(A_n + z)^{-1}$ is uniformly bounded with $\|(A_n + z)^{-1}\| \leq 1/|\Im z|$. Since D is a core for A, the image $(A + z)D$ is dense, and we obtain that $(A_n + z)^{-1}g \to (A + z)^{-1}g$ as $n \to \infty$ for all $g \in \mathcal{H}$. □

4.7 Feynman–Kac formula for Schrödinger operators with spin

4.7.1 Schrödinger operators with spin $\frac{1}{2}$

In this section we choose $d = 3$ and consider the path integral representation of the Pauli operator, which includes spin $\frac{1}{2}$. The effect of spin will be described in terms of a ± 1-valued jump process, and the Schrödinger operator will be defined on $L^2(\mathbb{R}^3 \times \mathbb{Z}_2; \mathbb{C})$ instead of $L^2(\mathbb{R}^3; \mathbb{C}^2)$. This construction has the advantage of using scalar-valued functions rather than $\mathbb{C}^2$-valued functions, thus in the Feynman–Kac formula a scalar-valued integral kernel will occur.

Let $\sigma_1, \sigma_2, \sigma_3$ be the 2×2 *Pauli matrices* given by

$$\sigma_1 = \begin{pmatrix} 0 & 1 \\ 1 & 0 \end{pmatrix}, \quad \sigma_2 = \begin{pmatrix} 0 & -i \\ i & 0 \end{pmatrix}, \quad \sigma_3 = \begin{pmatrix} 1 & 0 \\ 0 & -1 \end{pmatrix}. \tag{4.7.1}$$

Each σ_μ is symmetric and traceless. They satisfy the anticommutation relations

$$\{\sigma_\mu, \sigma_\nu\} = 2\delta_{\mu\nu} \tag{4.7.2}$$

and thereby

$$\sigma_\mu \sigma_\nu = i \sum_{\lambda=1}^{3} \varepsilon^{\mu\nu\lambda} \sigma_\lambda, \tag{4.7.3}$$

where $\varepsilon^{\alpha\beta\gamma}$ denotes the Levi-Cività tensor given by

$$\varepsilon^{\alpha\beta\gamma} = \begin{cases} 1, & \alpha\beta\gamma \text{ is an even permutation of 123}, \\ -1, & \alpha\beta\gamma \text{ is an odd permutation of 123}, \\ 0, & \text{otherwise}. \end{cases}$$

Specifically, we have $\sigma_1\sigma_2 = i\sigma_3$, $\sigma_2\sigma_3 = i\sigma_1$, and $\sigma_3\sigma_1 = i\sigma_2$ and therefore, alternatively, $i\sigma_1, i\sigma_2, i\sigma_3$ can be seen as the infinitesimal generators of the Lie group $SU(2)$.

Definition 4.234 (Schrödinger operator with vector potential and spin 1/2). Consider the potentials $V \in L^\infty(\mathbb{R}^3)$ and $a \in (C_b^2(\mathbb{R}^3))^3$. The *Schrödinger operator with vector potential a and spin* $\frac{1}{2}$ is defined by

$$H_S(a) = \frac{1}{2}(\sigma \cdot (-i\nabla - a))^2 + V \tag{4.7.4}$$

acting on $L^2(\mathbb{R}^3; \mathbb{C}^2)$.

Expanding (4.7.4) by using the anticommutation relations (4.7.3), we obtain

$$H_S(a) = H(a) - \frac{1}{2}\sigma \cdot b, \tag{4.7.5}$$

where

$$b = (b_1, b_2, b_3) = \nabla \times a$$

is a magnetic field. Let $a \in (C_b^2(\mathbb{R}^3))^3$, $b \in (L^\infty(\mathbb{R}^3))^3$, and $V \in L^\infty(\mathbb{R}^3)$. Then $H_S(a)$ is self-adjoint on $D(-(1/2)\Delta)$ and bounded from below; moreover, it is essentially self-adjoint on any core of $-\frac{1}{2}\Delta$ as a consequence of the Kato–Rellich theorem.

Next we transform $H_S(a)$ into an operator on the set of $\mathbb{C}$-valued functions on a suitable space. Let $\mathbb{Z}_2$ be the set of the square roots of identity, i. e.,

$$\mathbb{Z}_2 = \{\theta_1, \theta_2\}, \tag{4.7.6}$$

where

$$\theta_\alpha = (-1)^\alpha = \begin{cases} -1, & \alpha = 1, \\ 1, & \alpha = 2. \end{cases} \tag{4.7.7}$$

By the identification

$$L^2(\mathbb{R}^3; \mathbb{C}^2) \ni \begin{pmatrix} f(x, +1) \\ f(x, -1) \end{pmatrix} \mapsto f(x, \theta) \in L^2(\mathbb{R}^3 \times \mathbb{Z}_2; \mathbb{C}) \tag{4.7.8}$$

and (4.7.5), $H_S(a)$ can be reduced to the self-adjoint operator $H_{\mathbb{Z}_2}(a)$ defined in the Hilbert space $L^2(\mathbb{R}^3 \times \mathbb{Z}_2)$ giving

$$H_{\mathbb{Z}_2}(a)f(x, \theta) = \left(H(a) - \frac{1}{2}\theta b_3(x)\right)f(x, \theta) - \frac{1}{2}(b_1(x) - i\theta b_2(x))f(x, -\theta), \tag{4.7.9}$$

where $x \in \mathbb{R}^3$ and $\theta \in \mathbb{Z}_2$. Thus $H_S(a)$ can be regarded as the operator $H_{\mathbb{Z}_2}(a)$ on the space of $\mathbb{C}$-valued functions with configuration space $\mathbb{R}^3 \times \mathbb{Z}_2$.

4.7.2 A jump process

In order to derive a Feynman–Kac formula for $e^{-tH_{\mathbb{Z}_2}(a)}$, in addition to Brownian motion we need a Lévy process accounting for the spin. We give this three-dimensional Lévy process $(Z_t)_{t\geq 0}$ on a probability space $(\mathcal{S}, \mathcal{B}_\mathcal{S}, P)$, with characteristics (b, A, ν) such that $\nu(\mathbb{R}^3 \setminus \{0\}) = 1$. For $I \in \mathcal{B}(\mathbb{R}^3)$, let

$$N(t, I) = |\{0 \leq s \leq t \mid \Delta Z_s \in I\}| \tag{4.7.10}$$

be the counting measure associated with $(Z_t)_{t\geq 0}$. Define the random process $(N_t)_{t\geq 0}$ on $(\mathcal{S}, \mathcal{B}_\mathcal{S}, P)$ by

$$N_t = N(t, \mathbb{R}^3 \setminus \{0\}). \tag{4.7.11}$$

Note that $(N_t)_{t\geq 0}$ is a Poisson process with intensity 1. Consider the measure

$$dN_t = \int_{\mathbb{R}^3\setminus\{0\}} N(dtdz) \tag{4.7.12}$$

on $\mathbb{R}^+$. The compensator of N_t is given by t and $\mathbb{E}_P[e^{-\alpha N_t}] = e^{t(e^{-\alpha}-1)}$. Since

$$\mathbb{E}_P\left[\int_0^{t+}\int_{\mathbb{R}^3\setminus\{0\}} f(s,\tau,z)N(dsdz)\right] = \mathbb{E}_P\left[\int_0^{t}\int_{\mathbb{R}^3\setminus\{0\}} f(s,\tau,z)ds\nu(dz)\right],$$

we have for $f = f(s,\tau)$, $(s,\tau) \in \mathbb{R}^+ \times \mathcal{S}$,

$$\mathbb{E}_P\left[\int_0^{t+} f(s,\tau)dN_s\right] = \mathbb{E}_P\left[\int_0^{t} f(s,\tau)ds\right], \tag{4.7.13}$$

independent of $z \in \mathbb{R}^3 \setminus \{0\}$. For each $\tau \in \mathcal{S}$, there exist $n = n(\tau) \in \mathbb{N}$ and points of discontinuity of $t \mapsto N_t$, $0 < s_1 = s_1(\tau), \ldots, s_n = s_n(\tau) \leq t$, such that

$$\int_0^{t+} f(s,N_s)dN_s = \sum_{j=1}^{n} f(s_j, N_{s_j}) = \sum_{j=1}^{n} f(s_j, j). \tag{4.7.14}$$

Since $\mathbb{E}_P[N_t] = t$ and $P(N_t = N) = t^N e^{-t}/N!$, the expectation of (4.7.14) reduces to a Lebesgue integral, i. e.,

$$\mathbb{E}_P\left[\int_0^{t+} f(s,N_s)dN_s\right] = \mathbb{E}_P\left[\int_0^{t} f(s,N_s)ds\right] = \int_0^t \sum_{n=0}^{\infty} f(s,n)\frac{s^n}{n!}e^{-s}ds.$$

Furthermore, Proposition 3.74 yields the following Itô-type formula.

Proposition 4.235. *Suppose that $h^i(t) = h^i(t,\tau) \in \mathbb{F}$, $i = 1,\ldots,d$, are independent of $z \in \mathbb{R}^d \setminus \{0\}$. Let $X^i = (X_t^i)_{t\geq 0}$ be the semimartingale given by*

$$dX_t^i = f^i \cdot dB_t + g^i dt + h^i dN_t, \quad i = 1,\ldots,d,$$

and $F \in C^2(\mathbb{R}^d)$, where $f^i = f^i(t,\omega)$ and $g^i = g^i(t,\omega)$. Then

$$\begin{aligned} F(X_t) = F(X_0) &+ \sum_{i=1}^{n}\int_0^t F_i(X_s)(f^i \cdot dB_s) + \sum_{i=1}^{n}\int_0^t F_i(X_s)g^i ds \\ &+ \frac{1}{2}\sum_{i,j=1}^{n}\int_0^t F_{ij}(X_s)(f^i \cdot f^j)ds + \int_0^{t+}(F(X_{s-} + h(s-)) - F(X_{s-}))dN_s. \end{aligned}$$

Here $F_i = \partial_i F$ and $F_{ij} = \partial_i\partial_j F$.

Proof. Note that $\int_0^{t+} h^i(s,\omega)dN_s = \int_0^{t+}\int_{\mathbb{R}^d\setminus\{0\}} h^i(s,\omega)N(dsdz)$ by the definition of dN_s. The result is obtained by a direct application of the Itô formula. □

Let $N + \alpha = (N_t + \alpha)_{t\geq 0}$ and $\mathbb{E}_P^\alpha[f(N)] = \mathbb{E}_P[f(N+\alpha)]$. It will be convenient to use the joint particle and spin process as a single $\mathbb{R}^3 \times \mathbb{Z}_2$-valued random process as given below.

Definition 4.236. Define the $\mathbb{R}^3\times\mathbb{Z}_2$-valued stochastic process $(q_t)_{t\geq 0}$ on the probability space $(\mathscr{X} \times \mathcal{S}, \mathcal{B}(\mathscr{X}) \times \mathcal{B}_{\mathcal{S}}, \mathcal{W}^x \times P)$ by

$$q_t = (B_t, \theta_{N_t}) \tag{4.7.15}$$

where $\theta_{N_t} = (-1)^{N_t}$.

For simplicity, we write $\mathbb{E}^{x,\alpha}_{\mathcal{W}\times P} = \mathbb{E}^{x,\alpha}$.

To determine the generator of the process $(q_t)_{t\geq 0}$, let σ_{F} be the fermionic harmonic oscillator defined by

$$\sigma_{\mathrm{F}} = \frac{1}{2}(\sigma_3 + i\sigma_2)(\sigma_3 - i\sigma_2). \tag{4.7.16}$$

Note that, in fact, $\sigma_{\mathrm{F}} = -\sigma_1 + 1_{2\times 2}$, where $1_{2\times 2}$ is the 2×2 identity matrix. A direct computation yields

$$(f, e^{-t(-(1/2)\Delta+\varepsilon\sigma_{\mathrm{F}})}g) = \sum_{\alpha=1,2}\int_{\mathbb{R}^3} \mathbb{E}^{x,\alpha}[\overline{f(q_0)}g(q_t)\varepsilon^{N_t}]dx. \tag{4.7.17}$$

Thus setting $\varepsilon = 1$, we see that the generator of q_t is $-(-\frac{1}{2}\Delta + \sigma_{\mathrm{F}})$ under the identification $L^2(\mathbb{R}^3 \times \mathbb{Z}_2) \cong L^2(\mathbb{R}^3; \mathbb{C}^2)$. In particular,

$$(f, e^{-t(-(1/2)\Delta-\sigma_1)}g) = e^t \sum_{\alpha=1,2}\int_{\mathbb{R}^3} \mathbb{E}^{x,\alpha}[\overline{f(q_0)}g(q_t)]dx.$$

4.7.3 Feynman–Kac formula for the jump process

In this section we derive the Feynman–Kac formula of $e^{-tH_{\mathbb{Z}_2}(a)}$ by making use of the random process $(q_t)_{t\geq 0}$ defined in (4.7.15).

Theorem 4.237 (Feynman–Kac formula for Schrödinger operator with vector potential and spin 1/2). *Let $a \in (C_{\mathrm{b}}^2(\mathbb{R}^3))^3$, $b \in (L^\infty(\mathbb{R}^3))^3$, and $V \in L^\infty(\mathbb{R}^3)$. If*

$$\int_0^t ds \int_{\mathbb{R}^3} \left|\log\frac{1}{2}\sqrt{b_1(y)^2 + b_2(y)^2}\right| \Pi_s(y-x)dy < \infty \tag{4.7.18}$$

for all $(x,t) \in \mathbb{R}^3 \times \mathbb{R}^+$*, then the Feynman–Kac-type formula*

$$(f, e^{-tH_{\mathbb{Z}_2}(a)}g) = e^t \sum_{\alpha=1,2} \int_{\mathbb{R}^3} \mathbb{E}^{x,\alpha}[\overline{f(q_0)}g(q_t)e^{Z_t}]dx \tag{4.7.19}$$

holds. Here

$$Z_t = -i\int_0^t a(B_s)\circ dB_s - \int_0^t V(B_s)ds + \int_0^t \frac{1}{2}\theta_{N_s}b_3(B_s)ds + \int_0^{t+} \log\left(\frac{1}{2}(b_1(B_s) - i\theta_{N_{s-}}b_2(B_s))\right) dN_s.$$

Before turning to the proof of Theorem 4.237, here is a heuristic argument. Since the diagonal part $-\frac{1}{2}\theta b_3(x)$ of $H_{\mathbb{Z}_2}(a)$ acts as an external potential up to the sign $\theta = \pm$, we formally have the integral $\int_0^t(-1/2)\theta_{N_s}b_3(B_s)ds$ in Z_t. This explains why the off-diagonal part $\int_0^t \log((1/2)(b_1(B_s) - i\theta_{N_s}b_2(B_s)))\, dN_s$ appears in Z_t. Let

$$W(x,-\theta) = \log\left(\frac{1}{2}(b_1(x) - i\theta b_2(x))\right) \tag{4.7.20}$$

and consider

$$(K_t^S f)(x,\sigma) = e^t\mathbb{E}^{x,\alpha}[f(B_t,\theta_{N_t})e^{\int_0^{t+} W(B_s,-\theta_{N_{s-}})dN_s}].$$

First assume, for simplicity, that W has no zeroes. The generator $-K^S$ of K_t^S can be computed by Itô's formula for semimartingales, yielding

$$d(e^{\int_0^{t+} W(B_s,-\theta_{N_s})dN_s}) = e^{\int_0^{t+} W(B_s,-\theta_{N_{s-}})dN_s}(e^{W(B_t,-\theta_{N_t})} - 1)dN_t. \tag{4.7.21}$$

On the other hand,

$$d(e^{-\int_0^t V(B_s)ds}) = e^{-\int_0^t V(B_s)ds}(-V(B_t))dt, \tag{4.7.22}$$

implying that $(e^{-t(-(1/2)\Delta+V)}f)(x) = \mathbb{E}^x[e^{-\int_0^t V(B_s)ds}f(B_t)]$. Comparing (4.7.21) and (4.7.22), it is clear that the Itô formula gives the differential for continuous processes and the difference for discontinuous ones. From (4.7.21), it follows that the generator K^S of K_t^S is

$$-K^S f(x,\theta) = \left(-\frac{1}{2}\Delta - e^{W(x,-\theta)} + 1\right)f(x,-\theta).$$

Thus $e^{-tK^S}f(x,\theta_\alpha) = e^t\mathbb{E}^{B_t,\alpha}[f(B_s,\theta_{N_t})e^{\int_0^t W(x,-\theta_{N_{s-}})dN_s}]$, giving rise to the special form of the off-diagonal part. We rewrite this term as

$$-\frac{1}{2}(b_1(x) - i\theta b_2(x)) = -e^{W(x,-\theta)} + 1 - 1.$$

Hence (4.7.20) follows.

Remark 4.238. We prove Proposition 4.237 by making use of the Itô formula. In order for the Itô formula to apply, however, the integrand in $\int_0^{t+} \cdots dN_s$ must be predictable with respect to the given filtration. Note that θ_{N_s} is right-continuous in s for each $(\omega, \tau) \in \mathscr{X} \times \mathcal{S}$, so we define $\theta_{N_{s-}} = \lim_{\varepsilon \uparrow 0} \theta_{N_{s-\varepsilon}}$. Thus $\theta_{N_{s-}}$ is left-continuous and $W(B_s, -\theta_{N_{s-}})$ is predictable, i. e., $W(B_s, -\theta_{N_{s-}})$ is $\sigma(B_r, N_r; 0 \le r \le s)$ measurable and left-continuous in s, for every $(\omega, \tau) \in \mathscr{X} \times \mathcal{S}$. This allows then an application of the Itô formula to $\int_0^{t+} W(B_s, -\theta_{N_{s-}}) dN_s$.

Proof of Theorem 4.237. Write $U(B_s, \theta_{N_s}) = -\frac{1}{2}\theta_{N_s} b_3(B_s) + V(B_s)$ for the diagonal part. The off-diagonal part $W(B_s, -\theta_{N_{s-}})$ is predictable and, in order to apply Itô's formula, we need to check that $|\int_0^{t+} W(B_s, -\theta_{N_{s-}}) dN_s|$ is bounded almost surely. Indeed,

$$\left| \mathbb{E}^{x,\alpha} \left[\int_0^{t+} W(B_s, -\theta_{N_{s-}}) dN_s \right] \right| \le \mathbb{E}^{x,\alpha} \left[\int_0^t \left| \log\left(\frac{1}{2} \sqrt{b_1(B_s)^2 + b_2(B_s)^2} \right) \right| dN_s \right]$$
$$= \int_0^t ds \int_{\mathbb{R}^3} \Pi_s(y - x) \left| \log\left(\frac{1}{2} \sqrt{b_1(y)^2 + b_2(y)^2} \right) \right| dy$$

is bounded by the assumption; hence $|\int_0^{t+} W(B_s, -\theta_{N_{s-}}) dN_s| < \infty$, almost surely.

Define $K_t^S : L^2(\mathbb{R}^3 \times \mathbb{Z}_2) \to L^2(\mathbb{R}^3 \times \mathbb{Z}_2)$ by

$$K_t^S g(x, \theta_\alpha) = \mathbb{E}^{x,\alpha}[e^{Z_t} g(B_t, \theta_{N_t})].$$

It can be seen that $\|K_t^S g\| \le V_M^{1/2} e^{M't} e^{(M-1)t/2} \|g\|$, where

$$M' = \frac{1}{2} \sup_{x \in \mathbb{R}^3} |b_3(x)|, \quad M = \frac{1}{4} \sup_{x \in \mathbb{R}^3} (b_1^2(x) + b_2^2(x)), \quad V_M = \sup_{x \in \mathbb{R}^3} \mathbb{E}^x[e^{-2\int_0^t V(B_s) ds}],$$

which are all finite. Thus K_t^S is bounded. Since Z_t is continuous at $t = 0$ for each $(\omega, \tau) \in \mathscr{X} \times \mathcal{S}$, dominated convergence yields

$$\|K_t^S g - g\| \le \sum_{\alpha=1,2} \int_{\mathbb{R}^3} \mathbb{E}^{x,\alpha}[|g(x, \theta) - g(B_t, \theta_{N_t}) e^{Z_t}|] dx \to 0$$

as $t \to 0$. Since $(B_t)_{t \ge 0}$ and $(N_t)_{t \ge 0}$ are independent Markov processes, it follows similarly as in the proof of Theorem 4.185 that

$$(K_s^S K_t^S g)(x, \theta_\alpha) = K_{s+t}^S g(x, \theta_\alpha),$$

i. e., $\{K_t^S : t \ge 0\}$ is a C_0-semigroup. Denote the generator of K_t^S by the closed operator $-h$. We will see below that $K_t^S = e^{-th} = e^{-t(H_{\mathbb{Z}_2}(a)+1)}$. By Proposition 4.235, it follows that

$$\theta_{N_t} - \theta_{N_0} = -2 \int_0^{t+} \theta_{N_{s-}} dN_s,$$

and

$$\begin{aligned} &g(B_t, \theta_{N_t}) - g(x, \theta_{N_0}) \\ &= \int_0^t \nabla g(B_s, \theta_{N_s}) \cdot dB_s + \frac{1}{2} \int_0^t \Delta g(B_s, \theta_{N_s}) ds + \int_0^{t+} (g(B_s, -\theta_{N_{s-}}) - g(B_s, \theta_{N_{s-}}))\, dN_s, \end{aligned}$$

as well as

$$\begin{aligned} e^{Z_t} - 1 = &\int_0^t e^{Z_s}(-ia(B_s)) \circ dB_s + \int_0^t e^{Z_s} \left(-V(B_s) + \frac{1}{2}(-ia(B_s))^2 - U(B_s, \theta_{N_s}) \right) ds \\ &+ \int_0^{t+} e^{Z_{s-}} (e^{W(B_s, -\theta_{N_{s-}})} - 1) dN_s. \end{aligned}$$

By the product rule, $d(e^{Z_t} g) = de^{Z_t} \cdot g + e^{Z_t} \cdot dg + de^{Z_t} \cdot dg$, and two identities above we have

$$\begin{aligned} &e^{Z_t} g(B_t, \theta_{N_t}) - g(x, \theta_{N_0}) \\ &= \int_0^t e^{Z_s} \left(\frac{1}{2}\Delta - ia(B_s) \cdot \nabla - \frac{1}{2} a(B_s)^2 - V(B_s) - U(B_s, \theta_{N_s}) \right) g(B_s, \theta_{N_s}) ds \\ &\quad + \int_0^t e^{Z_s} (\nabla g(B_s, \theta_{N_s}) - ia(B_s) \cdot g(B_s, \theta_{N_s})) \cdot dB_s \\ &\quad + \int_0^{t+} e^{Z_{s-}} (g(B_s, -\theta_{N_{s-}}) e^{W(B_s, -\theta_{N_{s-}})} - g(B_s, \theta_{N_{s-}})) dN_s. \end{aligned}$$

Take expectation on both sides above. The martingale part vanishes, and by (4.7.13) we obtain

$$\mathbb{E}^{x,\alpha}[e^{Z_t} g(B_t, \theta_{N_t}) - g(x, \theta_\alpha)] = \int_0^t \mathbb{E}^{x,\alpha}[G(s)] ds,$$

where

$$\begin{aligned} G(s) = &e^{Z_s} \left(\frac{1}{2}\Delta - ia(B_s) \cdot \nabla - \frac{i}{2} \nabla \cdot a(B_s) - \frac{1}{2} a(B_s)^2 - V(B_s) - U(B_s, \theta_{N_s}) \right) g(B_s, \theta_{N_s}) \\ &+ e^{Z_{s-}} \left(g(B_s, -\theta_{N_{s-}}) e^{W(B_s, -\theta_{N_{s-}})} - g(B_s, \theta_{N_{s-}}) \right), \end{aligned}$$

with $s > 0$, and

$$
\begin{aligned}
G(0) &= \left(\frac{1}{2}\Delta - ia(x)\cdot\nabla - \frac{i}{2}\nabla\cdot a(x) - \frac{1}{2}a(x)^2 - V(x) - U(x,\theta_\alpha) - 1\right)g(x,\theta_\alpha) \\
&\quad + e^{W(x,-\theta_\alpha)}g(x,-\theta_\alpha) \\
&= -(H_{\mathrm{S}}(a)+1)g(x,\theta_\alpha).
\end{aligned}
$$

At $s = 0$, $G(s)$ is continuous for each $(\omega,\tau) \in \mathscr{X} \times \mathcal{S}$, hence

$$
\begin{aligned}
\lim_{t\to 0}\frac{1}{t}(f,(K_t^S - 1)g) &= \lim_{t\to 0}\frac{1}{t}\int_0^t ds \sum_{\alpha=1,2}\int_{\mathbb{R}^3}\overline{f(x,\theta_\alpha)}\mathbb{E}^{x,\alpha}[G(s)]dx \\
&= (f, -(H_{\mathbb{Z}_2}(a)+1)g).
\end{aligned}
$$

Since $C_0^\infty(\mathbb{R}^3 \times \mathbb{Z}_2)$ is a core of $H_{\mathbb{Z}_2}(a)$, equality (4.7.19) follows. □

4.7.4 Extension to singular external potentials and singular vector potentials

Next we extend Theorem 4.237 to singular external potentials V and singular vector potentials a as in the spinless case, following a similar strategy as in Section 4.4.3.

Define the quadratic form

$$
q_a^{\mathrm{S}}(f,g) = \sum_{\mu=1}^{3}(\sigma_\mu D_\mu f, \sigma_\mu D_\mu g) + (V^{1/2}f, V^{1/2}g) \tag{4.7.23}
$$

with domain $Q(q_a^{\mathrm{S}}) = \bigcap_{\mu=1}^3 \{f \in L^2(\mathbb{R}^3) \mid D_\mu f \in L^2(\mathbb{R}^3)\} \cap D(V^{1/2})$.

Lemma 4.239. *Let $V \geq 0$ be from $L^1_{\mathrm{loc}}(\mathbb{R}^3)$ and $a \in (L^2_{\mathrm{loc}}(\mathbb{R}^3))^3$. Then q_a^{S} is a closed symmetric form.*

The proof is very similar to Lemma 4.188. From here it follows that there exists a self-adjoint operator h_S such that

$$
(f, h_S g) = q_a^S(f,g), \quad f \in Q(q_a^S),\ g \in D(h_S). \tag{4.7.24}
$$

A sufficient condition for $C_0^\infty(\mathbb{R}^3)$ to be a core of h_S can also be obtained in a similar way to the spinless case; see Proposition 4.190.

Proposition 4.240. *The following properties hold:*

(1) *Let $V \in L^1_{\mathrm{loc}}(\mathbb{R}^3)$ with $V \geq 0$. If $a \in (L^2_{\mathrm{loc}}(\mathbb{R}^3))^3$, then $C_0^\infty(\mathbb{R}^3)$ is a form core of h_S.*
(2) *Let $V \in L^2_{\mathrm{loc}}(\mathbb{R}^3)$ with $V \geq 0$. If $a \in (L^4_{\mathrm{loc}}(\mathbb{R}^3))^3$, $\nabla\cdot a \in L^2_{\mathrm{loc}}(\mathbb{R}^3)$, and $b \in (L^2_{\mathrm{loc}}(\mathbb{R}^3))^3$, then $C_0^\infty(\mathbb{R}^3)$ is an operator core of h_S.*

In case (2) above, the operator h_S can be realized as

$$h_S f = -\frac{1}{2}\Delta f - a\cdot(-i\nabla)f + \left(-\frac{1}{2}a\cdot a - (-i\nabla\cdot a) - \frac{1}{2}\sigma\cdot b\right)f + Vf. \tag{4.7.25}$$

Instead of h_S, we define the Schrödinger operator with spin by the sum of the kinetic term and spin. Assume that $a \in (L^2_{\mathrm{loc}}(\mathbb{R}^3))^3$, $0 \le V \in L^1_{\mathrm{loc}}(\mathbb{R}^3)$, and the magnetic field b satisfies $b \in (L^\infty(\mathbb{R}^3))^3$. Let $H^0(a)$ be given by Definition 4.189 with $V = 0$. It can be seen that $H^0(a) - \frac{1}{2}\sigma\cdot b$ is self-adjoint on $D(H^0(a))$. We regard b to be independent of a, and define

$$H_S(a,b) = H^0(a) \dotplus V - \frac{1}{2}\sigma\cdot b.$$

Let $H_{\mathbb{Z}_2}(a,b)$ be the unitary transformed operator to $L^2(\mathbb{R}^3\times\mathbb{Z}_2)$ of $H_S(a,b)$, and $H^0_{\mathbb{Z}_2}(a,b)$ denote $H_{\mathbb{Z}_2}(a,b)$ with $V = 0$. From now on, we discuss more general external potentials V and vector potentials a.

Consider general vector potentials possibly having zeroes. Note that (4.7.18) is a sufficient condition making sure that

$$\int_0^{t+} |W(B_s, -\theta_{N_{s-}})|dN_s < \infty, \quad \text{a.e. } (\omega,\tau) \in \mathscr{X}\times\mathcal{S}. \tag{4.7.26}$$

When, however, $b_1(x) - i\theta b_2(x)$ vanishes for some (x,θ), it is not clear that (4.7.26) holds. This case is relevant and Theorem 4.237 must be improved since we have to construct the path integral representation of $e^{-tH_{\mathbb{Z}_2}(a,b)}$ in which the off-diagonal part $b_1 - i\theta b_2$ of $H_{\mathbb{Z}_2}(a,b)$ has zeroes or a compact support. The off-diagonal part of $H_S(a,b)$, however, in general may have zeroes. For instance, a_μ for all $\mu = 1,2,3$ have a compact support, and so does the off-diagonal part of $b = \nabla\times a$. Therefore, in order to avoid that the off-diagonal part vanishes, we introduce

$$H^\varepsilon_{\mathbb{Z}_2}(a,b)f(x,\theta) = \left(H(a,b) - \frac{1}{2}\theta b_3(x)\right)f(x,\theta) - \Psi_\varepsilon\left(\frac{1}{2}(b_1(x) - i\theta b_2(x))\right)f(x,-\theta), \tag{4.7.27}$$

where

$$\Psi_\varepsilon(z) = z + \varepsilon\psi_\varepsilon(z) \tag{4.7.28}$$

for $z \in \mathbb{C}$ and $\varepsilon > 0$, with the indicator function

$$\psi_\varepsilon(z) = \begin{cases} 1, & |z| < \varepsilon/2, \\ 0, & |z| \ge \varepsilon/2. \end{cases} \tag{4.7.29}$$

Thus

$$\left|\Psi_\varepsilon\left(\frac{1}{2}(b_1 - i\theta b_2)\right)\right| > \frac{\varepsilon}{2} > 0.$$

Let $t > 0$ be fixed and consider the set $\mathscr{W} = \mathscr{W}_t$ defined by

$$\mathscr{W} = \left\{ \int_0^{t+} \log\left(\frac{1}{2}\sqrt{b_1(x+B_s)^2 + b_2(x+B_s)^2}\right) dN_s > -\infty \right\} \subset \mathbb{R}^3 \times \mathscr{X} \times \mathcal{S}. \tag{4.7.30}$$

Lemma 4.241. *Let $\mathscr{W}$ be given by* (4.7.30). *Then it follows that*

$$\lim_{\varepsilon\to 0} \left| e^{\int_0^{t+} \log(\Psi_\varepsilon(\frac{1}{2}(b_1(B_s+x)-i\theta_{s-}b_2(B_s+x))))dN_s} 1_{\mathscr{W}^c} \right| = 0.$$

Proof. Let $W(X) = \frac{1}{2}\sqrt{b_1(X)^2 + b_2(X)^2}$. We have

$$\left| e^{\int_0^{t+} \log(\Psi_\varepsilon(\frac{1}{2}(b_1(x+B_s)-i\theta_{N_s}b_2(x+B_s))))dN_s} \right| \le e^{\int_0^{t+} \log(W(x+B_s)+\varepsilon)dN_s}.$$

Note that for every $(x, \omega, \tau) \in \mathbb{R}^3 \times \mathscr{X} \times \mathcal{S}$ there exist a number $n = n(\tau)$ and random jump times $r_1(\tau), \dots, r_n(\tau)$ such that

$$\int_0^{t+} \log(W(x + B_s(\omega)) + \varepsilon)dN_s = \sum_{j=1}^{n} \log(W(x + B_{r_j(\tau)}(\omega)) + \varepsilon).$$

We see that $(x, \omega, \tau) \in \mathscr{W}^c$ if and only if there exists r such that $0 < r < t$, $s \mapsto N_s$ is discontinuous at $s = r$, and $b_1(x + B_r(\omega))^2 + b_2(x + B_r(\omega))^2 = 0$. Hence for $(x, \omega, \tau) \in \mathcal{W}^c$, there exists r_i such that

$$\int_0^{t+} \log(W(x + B_s) + \varepsilon)dN_s = \sum_{j\neq i}^{n} \log(W(x + B_{r_j}) + \varepsilon) + \log\varepsilon,$$

and

$$e^{\int_0^{t+} \log(W(x+B_s)+\varepsilon)dN_s} \le e^{\sum_{j\neq i}^{n} \log(W(x+B_{r_j})+\varepsilon)} e^{\log\varepsilon}.$$

Thus $\lim_{\varepsilon\to 0} |e^{\int_0^{t+} \log(W(x+B_s)+\varepsilon)dN_s} 1_{\mathscr{W}^c}| = 0$, and the lemma follows. □

Theorem 4.242 (Feynman–Kac formula for Schrödinger operator with singular vector potential and spin $\frac{1}{2}$). *Suppose that $V \in L^\infty(\mathbb{R}^3)$, $a \in (L^2_{\text{loc}}(\mathbb{R}^3))^3$, $\nabla \cdot a \in L^1_{\text{loc}}(\mathbb{R}^3)$, and $b \in (L^\infty(\mathbb{R}^3))^3$. Let $\mathscr{W}$ be given by* (4.7.30). *Then*

$$(f, e^{-tH_{\mathbb{Z}_2}(a,b)} g) = e^t \sum_{\alpha=1,2} \int_{\mathbb{R}^3} \mathbb{E}^{x,\alpha}[\overline{f(q_0)} g(q_t) e^{Z_t} 1_{\mathscr{W}}] dx. \tag{4.7.31}$$

Proof. We show this theorem through a similar limiting argument as in Lemma 4.192. Suppose first that $a^{(n)} \in (C_0^\infty(\mathbb{R}^3))^3$, $n \in \mathbb{N}$, are such that $a_\mu^{(n)} \to a_\mu$ in $L^2_{\mathrm{loc}}(\mathbb{R}^3)$ as $n \to \infty$. Take $1_R(x) = \chi(x_1/R)\chi(x_2/R)\chi(x_3/R)$, $R > 0$, where $\chi \in C_0^\infty(\mathbb{R})$ is such that $0 \le \chi \le 1$, $\chi(x) = 1$ for $|x| < 1$, and $\chi(x) = 0$ for $|x| \ge 2$. In the same way as in Lemma 4.191, we can show that

$$e^{-tH^\varepsilon_{\mathbb{Z}_2}(1_R a^{(n)},b)} \to e^{-tH^\varepsilon_{\mathbb{Z}_2}(1_R a,b)} \quad \text{as } n \to \infty, \tag{4.7.32}$$

$$e^{-tH^\varepsilon_{\mathbb{Z}_2}(1_R a,b)} \to e^{-tH^\varepsilon_{\mathbb{Z}_2}(a,b)} \quad \text{as } R \to \infty \tag{4.7.33}$$

in the strong sense. In the proof of Lemma 4.192, we have already shown that there exists a subsequence n' such that $\int_0^t 1_R a^{(n')}(B_s) \circ dB_s \to \int_0^t 1_R a(B_s) \circ dB_s$ almost surely as $n' \to \infty$, and R' such that $\int_0^t 1_{R'} a(B_s) \circ dB_s \to \int_0^t a(B_s) \circ dB_s$. We reset n' and R' by n and R, respectively. Let

$$\begin{aligned} Z_t^\varepsilon = {}& -i \int_0^t a(B_s) \circ dB_s - \int_0^t V(B_s)ds \\ & + \int_0^t \frac{1}{2}\theta_{N_s} b_3(B_s)ds + \int_0^{t+} \log\left(\Psi_\varepsilon\left(\frac{1}{2}(b_1(B_s) - i\theta_{N_s} b_2(B_s))\right)\right)dN_s \end{aligned}$$

and $Z_t^\varepsilon(n,R)$ be Z_t^ε with a replaced by $1_R a^{(n)}$. Hence we conclude that

$$\lim_{R\to\infty}\lim_{n\to\infty} \int_{\mathbb{R}^3} \mathbb{E}^{x,\alpha}[\overline{f(q_0)} g(q_t) e^{Z_t^\varepsilon(n,R)}]dx = \int_{\mathbb{R}^3} \mathbb{E}^{x,\alpha}[\overline{f(q_0)} g(q_t) e^{Z_t^\varepsilon}]dx.$$

Combining this with (4.7.32) and (4.7.33), we obtain

$$(f, e^{-tH^\varepsilon_{\mathbb{Z}_2}(a,b)} g) = e^t \sum_{\alpha=1,2} \int_{\mathbb{R}^3} \mathbb{E}^{x,\alpha}[\overline{f(q_0)} g(q_t) e^{Z_t^\varepsilon}]dx. \tag{4.7.34}$$

We also see that

$$\lim_{\varepsilon\to 0} \int_{\mathbb{R}^3} \mathbb{E}^{x,\alpha}[\overline{f(q_0)} g(q_t) e^{Z_t^\varepsilon}]dx = \int_{\mathbb{R}^3} \mathbb{E}^{x,\alpha}[\lim_{\varepsilon\to 0} \overline{f(q_0)} g(q_t) e^{Z_t^\varepsilon}]dx$$

and

$$\lim_{\varepsilon\to 0} \overline{f(q_0)} g(q_t) e^{Z_t^\varepsilon} = \lim_{\varepsilon\to 0} \overline{f(q_0)} g(q_t) e^{Z_t^\varepsilon} 1_{\mathscr{W}} + \lim_{\varepsilon\to 0} \overline{f(q_0)} g(q_t) e^{Z_t^\varepsilon} 1_{\mathscr{W}^c} = \overline{f(q_0)} g(q_t) e^{Z_t} 1_{\mathscr{W}}.$$

Since $e^{-tH^\varepsilon_{\mathbb{Z}_2}(a,b)}$ is strongly convergent to $e^{-tH_{\mathbb{Z}_2}(a,b)}$ as $\varepsilon \to 0$, (4.7.31) also follows. □

By Lemma 4.192, we have a diamagnetic inequality for $H_{\mathbb{Z}_2}(a,b)$.

Corollary 4.243 (Diamagnetic inequality). *Under the assumptions of Lemma* 4.240,

$$|(f, e^{-tH_{\mathbb{Z}_2}(a,b)}g)| \le (|f|, e^{-tH_{\mathbb{Z}_2}(0,b_0)}|g|), \tag{4.7.35}$$

where $b_0 = (\sqrt{b_1^2 + b_2^2}, 0, b_3)$.

Proof. Let $W(X) = \frac{1}{2}\sqrt{b_1(X)^2 + b_2(X)^2}$. The corollary follows from the fact that

$$|e^{\int_0^{t+} \log(\frac{1}{2}(b_1(B_s) - i\theta_{N_s} b_2(B_s))dN_s)}| \le e^{\int_0^{t+} \log W(B_s)dN_s}$$

and

$$|(f, e^{-tH_{\mathbb{Z}_2}(a,b)}g)| \le \sum_{\alpha=1,2} \int_{\mathbb{R}^3} \mathbb{E}^{x,\alpha}[|f(q_0)||g(q_t)|e^{\hat{Z}_t}]dx,$$

where the exponent $\hat{Z}_t$ is defined by

$$\hat{Z}_t = -\int_0^t V(B_s)ds + \int_0^t \frac{1}{2}\theta_{N_s} b_3(B_s)ds + \int_0^{t+} \log W(B_s)dN_s. \qquad \square$$

Theorem 4.244 (Feynman–Kac formula for Schrödinger operator with singular external potential, singular vector potential, and spin $\frac{1}{2}$). *Suppose that* $a \in (L^2_{\mathrm{loc}}(\mathbb{R}^3))^3$, $\nabla \cdot a \in L^1_{\mathrm{loc}}(\mathbb{R}^3)$, *and* b *satisfies* $b \in (L^\infty(\mathbb{R}^3))^3$.

(1) *Let* V *be a real-valued multiplication operator relatively bounded (resp. form-bounded) with respect to* $-\frac{1}{2}\Delta$ *with a relative bound* c. *Then* V *is also relatively bounded (resp. form-bounded) with respect to* $H^0_{\mathbb{Z}_2}(a, b)$ *with a relative bound not exceeding* c.

(2) *Let* V *be such that* $V_+ \in L^1_{\mathrm{loc}}(\mathbb{R}^3)$ *and* V_- *is relatively form-bounded with respect to* $-\frac{1}{2}\Delta$ *with a relative bound less than* 1. *Then*

$$H^0_{\mathbb{Z}_2}(a, b) \dotplus V_+ \dot{-} V_- \tag{4.7.36}$$

can be defined as a self-adjoint operator. This also will be denoted by $H_{\mathbb{Z}_2}(a, b)$.

(3) *Let* V *satisfy the same assumptions as in* (2). *Then the Feynman–Kac formula of* $(f, e^{-tH_{\mathbb{Z}_2}(a,b)}g)$ *is given by* (4.7.34) *and* (4.7.31).

Proof. By the diamagnetic inequality

$$\frac{\||W|^\alpha (H^0_{\mathbb{Z}_2}(a,b) + E)^{-\alpha} f\|}{\|f\|} \le \frac{\||W|^\alpha (H^0_{\mathbb{Z}_2}(0,b_0) + E)^{-\alpha} f\|}{\|f\|} \tag{4.7.37}$$

for $\alpha = 1$ and $\frac{1}{2}$, (1) follows by the same argument as in Lemma 4.193 and due to the boundedness of the magnetic field b_j. Statement (2) is implied by the KLMN theorem.

Thus by Theorem 4.237 and a minor modification of the proof of Theorem 4.195, also (3) follows. □

Definition 4.245 (Schrödinger operator with singular external potential, singular vector potential, and spin $\frac{1}{2}$). Assume that $a \in (L^2_{\mathrm{loc}}(\mathbb{R}^3))^3$ and that the magnetic field b satisfies $b \in (L^\infty(\mathbb{R}^3))^3$. Let V be such that $V_+ \in L^1_{\mathrm{loc}}(\mathbb{R}^3)$ and V_- is relatively form-bounded with respect to $-\frac{1}{2}\Delta$ with a relative bound less than 1. Then $H_{\mathbb{Z}_2}(a,b)$ is defined by (4.7.36).

4.7.5 Decay of eigenfunctions and martingale properties

In a similar way to the proof of the decay of eigenfunctions of Schrödinger operators, we can also show the decay of eigenfunctions of Schrödinger operators with spin. As before, we consider potentials in the classes $\mathbb{V}^{\mathrm{upper}}$ and $\mathbb{V}^{\mathrm{lower}}$, given in Definition 4.176. We note that, under the assumption that $b \in (L^\infty(\mathbb{R}^3))^3$, we have

$$b_3 \in L^\infty(\mathbb{R}^3), \quad \frac{1}{2}\sqrt{b_1^2 + b_2^2} \in L^\infty(\mathbb{R}^3). \tag{4.7.38}$$

Write

$$E^* = \frac{1}{2}\Big(\|b_3\|_\infty + \frac{1}{4}\|\sqrt{b_1^2 + b_2^2}\|_\infty^2 + 1\Big) \tag{4.7.39}$$

and consider

$$H_{\mathbb{Z}_2}(a,b)\varphi = E\varphi.$$

Let $V = W - U \in \mathbb{V}^{\mathrm{upper}}$. We have $W \in L^1_{\mathrm{loc}}(\mathbb{R}^d)$, and $0 \geq -U \in L^p(\mathbb{R}^d)$ implies that $-U \in \mathcal{K}(\mathbb{R}^d)$, thus V is Kato-decomposable.

Lemma 4.246. *Suppose that $b \in (L^\infty(\mathbb{R}^3))^3$. Let $V = W - U \in \mathbb{V}^{\mathrm{upper}}$. Then for every $t, a > 0$ and $0 < \alpha < 1/2$, there exist constants $D_1, D_2, D_3 > 0$ such that*

$$|\varphi(x)| \leq D_1\|\varphi\| e^{D_2\|U\|_p^m t} e^{(E+E^*)t}\big(D_3 e^{-\frac{\alpha}{4}\frac{a^2}{t}} e^{-tW_\infty} + e^{-tW_a(x)}\big), \tag{4.7.40}$$

where $m = (1 - \frac{d}{2p})^{-1}$ and $W_a(x) = \inf\{W(y) \mid |x-y| < a\}$.

Proof. Since

$$\varphi(x,\theta_\alpha) = e^{-t(H_{\mathbb{Z}_2}(a,b)-E)}\varphi(x,\theta_\alpha) = e^{(E+1)t}\mathbb{E}^{x,\alpha}[\varphi(B_t,\theta_{N_t})e^{Z_t}1_{\mathscr{W}}],$$

by the Schwarz inequality it follows that

$$\begin{aligned}|\varphi(x,\theta_\alpha)| &\leq e^{tE}e^t\big(\mathbb{E}^{x,\alpha}[|\varphi(B_t,\theta_{N_t})|^2]\big)^{1/2}\\ &\quad\times\big(\mathbb{E}^x[e^{-2\int_0^t V(B_s)ds}]\big)^{1/2}\big(\mathbb{E}^\alpha[e^{t\|b_3\|_\infty + 2N_t\log\|\frac{1}{2}\sqrt{b_1^2+b_2^2}\|_\infty}]\big)^{1/2}\end{aligned}$$

$$\leq e^{tE}e^{t}e^{t(E^*-1)}(\mathbb{E}^{x,\alpha}[|\varphi(B_t,\theta_{N_t})|^2])^{1/2}(\mathbb{E}^x[e^{-2\int_0^t V(B_s)ds}])^{1/2}.$$

Here we used $\mathbb{E}^\alpha[e^{N_tC}] = e^{t(e^C-1)}$. It can be seen that

$$\mathbb{E}^{x,\alpha}[|\varphi(B_t,\theta_{N_t})|^2] = \int_{\mathbb{R}^3} \Pi_t(y)\sum_{n=0}^\infty |\varphi(x+y,\theta_{\alpha+n})|^2\frac{t^n}{n!}e^{-t}dy$$
$$\leq \int_{\mathbb{R}^3} \Pi_t(y)(|\varphi(x+y,+1)|^2+|\varphi(x+y,-1)|^2)dy \leq \|\varphi\|^2.$$

Hence we obtain

$$|\varphi(x,\theta_\alpha)| \leq e^{t(E+E^*)}\left(\mathbb{E}^x[e^{-2\int_0^t V(B_s)ds}]\right)^{1/2}\|\varphi\|$$

and then

$$|\varphi(x)| \leq e^{t(E+E^*)}(\mathbb{E}^x[e^{-4\int_0^t W(B_s)ds}])^{1/4}(\mathbb{E}^x[e^{+4\int_0^t U(B_s)ds}])^{1/4}\mathbb{E}[|\varphi(x+B_t)|^2]^{1/2}.$$

In the same way as with the estimate (4.3.79), with E replaced by $E+E^*$, the lemma follows. □

The above Carmona-type estimate yields the required decay properties; the proof is similar to those of Corollaries 4.178–4.179.

Corollary 4.247 (Decay property). *Suppose* (4.7.38). *Let* $V = W - U \in \mathcal{V}^{\text{upper}}$.

(1) *Suppose that* $W(x) \geq \gamma|x|^{2n}$ *outside a compact set* K, *for some* $n > 0$ *and* $\gamma > 0$. *Take* $0 < a < 1/2$. *Then there exists a constant* $C_1 > 0$ *such that*

$$|\varphi(x,\theta_\alpha)| \leq C_1\|\varphi\|\exp\left(-\frac{ac}{16}|x|^{n+1}\right), \tag{4.7.41}$$

where $c = \inf_{x\in\mathbb{R}^d\setminus K} W_{|x|/2}(x)/|x|^{2n}$.

(2) *Let* $\Sigma = \liminf_{|x|\to\infty} V(x)$. *Suppose that* $\Sigma > E+E^*$, $\Sigma > W_\infty$, *and let* $0 < \beta < 1$. *Then there exists a constant* $C_2 > 0$ *such that*

$$|\varphi(x,\theta_\alpha)| \leq C_2\|\varphi\|\exp\left(-\frac{\beta}{8\sqrt{2}}\frac{\Sigma-E-E^*}{\sqrt{\Sigma-W_\infty}}|x|\right). \tag{4.7.42}$$

(3) *Suppose that* $\lim_{|x|\to\infty} W(x) = \infty$. *Then there exist constants* $C, \delta > 0$ *such that*

$$|\varphi(x,\theta_\alpha)| \leq C\|\varphi\|\exp(-\delta|x|). \tag{4.7.43}$$

Finally, we show a martingale property of the process associated with Schrödinger operators with spin, as a counterpart of Lemma 4.224. By using this martingale together with a suitable choice of stopping times, one can then estimate the decay of eigenfunctions of $H_{\mathbb{Z}_2}(a,b)$ by an alternative approach than above. For simplicity, we assume throughout that

$$\mathbb{E}_P^x\left[\int_0^t |\log\frac{1}{2}\sqrt{b_1(B_s)^2+b_2(B_s)^2}|ds\right]<\infty,\quad \text{a.e. } x\in\mathbb{R}^3,$$

i. e., the measure of $\mathscr{W}^c$ is zero. We recall that the exponent of the Feynman–Kac formula of $e^{-tH_{\mathbb{Z}_2}(a,b)}$ is given by $Z_t = Z_t^V + Z_t^{\mathrm{A}} + Z_t^{\mathrm{S}}$ with

$$Z_t^V = -\int_0^t V(B_s)ds,\quad Z_t^{\mathrm{A}} = -i\int_0^t a(B_s)\circ dB_s,$$

$$Z_t^{\mathrm{S}} = -\int_0^t U_d(B_s,\theta_{N_s})ds + \int_0^{t+}\log(-U_o(B_s,-\theta_{N_{s-}}))\,dN_s.$$

We write $U_d(x,Y) = -\frac{1}{2}Yb_3(x)$ and $U_o(x,-Y) = -\frac{1}{2}(b_1(x)-iYb_2(x))$ for notational simplicity. Let φ be an eigenfunction such that $H_{\mathbb{Z}_2}(a,b)\varphi = E\varphi$. Also, let $Z_t(x,\alpha) = Z_t^V(x) + Z_t^{\mathrm{A}}(x)+Z_t^{\mathrm{S}}(x,\alpha)$ be given by Z_t with B_s and N_s replaced by B_s+x and $N_s+\alpha$, respectively, i. e.,

$$Z_t^V(x) = -\int_0^t V(B_s+x)ds,\quad Z_t^{\mathrm{A}}(x) = -i\int_0^t a(B_s+x)\circ dB_s,$$

$$Z_t^{\mathrm{S}}(x,\alpha) = -\int_0^t U_d(B_s+x,\theta_{N_s+\alpha})ds + \int_0^{t+}\log(-U_o(B_s+x,-\theta_{N_{s-}+\alpha}))\,dN_s.$$

Define the random process $(M_t(x,\alpha))_{t\geq 0}$ by

$$M_t(x,\alpha) = e^{t(E+1)}e^{Z_t(x,\alpha)}\varphi(B_t+x,\theta_{N_t+\alpha}),\quad t\geq 0,$$

and the filtration

$$\mathscr{M}_t = \sigma(q_r, 0\leq r\leq t),\quad t\geq 0.$$

Note that $e^{-t(H_{\mathbb{Z}_2}(a,b)-E)}\Psi_{\mathrm{p}} = \varphi$ for all $t\geq 0$, and thus

$$\mathbb{E}^{x,\alpha}[M_t(0,0)] = \mathbb{E}^{0,0}[M_t(x,\alpha)] = \varphi(x,\theta_\alpha). \tag{4.7.44}$$

Theorem 4.248 (Martingale property). *The random process $(M_t(x,\alpha))_{t\geq 0}$ is a martingale with respect to $(\mathscr{M}_t)_{t\geq 0}$, i. e., $\mathbb{E}^{0,0}[M_t(x,\alpha)|\mathscr{M}_s] = M_s(x,\alpha)$ for $t\geq s$.*

Proof. We prove the case when $(x, \alpha) = (0, 0)$ for simplicity; the proof for $(x, \alpha) \neq (0, 0)$ is similar. Let $Z_t([u, v])$ be defined by Z_t with the integration domain $[0, t]$ replaced by $[u, v]$, and write $M_t = M_t(0, 0)$. We see that

$$\mathbb{E}^{0,0}[M_t|\mathcal{M}_s] = e^{t(E+1)}e^{Z_t([0,s])}\mathbb{E}^{0,0}\left[e^{Z_t([s,t])}\varphi(q_t)|\mathcal{M}_s\right].$$

By the Markov property of the $\mathbb{R}^d \times \mathbb{Z}_2$-valued random process $(B_t, N_t)_{t\geq 0}$ we have

$$\mathbb{E}^{0,0}\left[e^{Z_t([s,t])}\varphi(q_t)|\mathcal{M}_s\right] = \mathbb{E}^{B_s,N_s}\left[e^{-\int_0^{t-s}V(B_r)dr}e^{-i\int_0^{t-s}a(B_r)\circ dB_r}e^{\int_0^{t-s}U_d(B_r,\theta_r)dr}e^{K}\varphi(q_{t-s})\right]. \tag{4.7.45}$$

The off-diagonal part K in (4.7.45) is

$$\begin{aligned} K &= \sum_{\substack{s<u\leq t \\ N_{(u-s)+}\neq N_{(u-s)-}}} \log(-U_o(B_{u-s}, -\theta_{N_{(u-s)-}})) \\ &= \sum_{\substack{0<r\leq t-s \\ N_{r+}\neq N_{r-}}} \log(-U_o(B_r, -\theta_{N_{r-}})) = \int_0^{(t-s)+} \log(-U_o(B_r, -\theta_{N_{r-}}))dN_r. \end{aligned}$$

Hence we conclude that

$$\mathbb{E}^{0,0}[e^{Z_t([s,t])}\varphi(q_t)|\mathcal{M}_s] = \mathbb{E}^{B_s,N_s}[e^{Z_t([0,t-s])}\varphi(q_{t-s})],$$

which implies that

$$\mathbb{E}^{0,0}[M_t|\mathcal{M}_s] = e^{s(E+1)}e^{Z_t([0,s])}\mathbb{E}^{B_s,N_s}[M_{t-s}] = M_s.$$

□

4.8 Feynman–Kac formula for relativistic Schrödinger operators with spin

4.8.1 Relativistic Schrödinger operator with spin $\frac{1}{2}$

By the methods developed above, it is possible to derive a Feynman–Kac formula also for relativistic Schrödinger operators with spin $\frac{1}{2}$. This operator is formally given by

$$\sqrt{(-i\nabla - a)^2 - \sigma \cdot b + m^2} - m + V. \tag{4.8.1}$$

As seen above, the Feynman–Kac formula for the Schrödinger operator $-\frac{1}{2}\Delta + V$ features Brownian motion $(B_t)_{t\geq 0}$ on $(\mathcal{X}, \mathcal{B}(\mathcal{X}), \mathcal{W})$, and for the Schrödinger operator $\frac{1}{2}(-i\nabla - a)^2 - \frac{1}{2}\sigma \cdot b$ with spin, in addition to Brownian motion $(B_t)_{t\geq 0}$ also a Poisson process $(N_t)_{t\geq 0}$ on $(\mathcal{S}, \mathcal{B}_{\mathcal{S}}, P_N)$ appears in the expressions. Furthermore, for the relativistic Schrödinger operator $\sqrt{-\Delta + m^2} - m + V$, both Brownian motion $(B_t)_{t\geq 0}$ and a subordinator $(T_t)_{t\geq 0}$

on $(\mathcal{T}, \mathcal{B}_{\mathcal{T}}, P_T)$ are used. As it can be expected, in order to construct a Feynman–Kac formula for (4.8.1), we will need all three random processes $(B_t)_{t\geq 0}$, $(N_t)_{t\geq 0}$, and $(T_t)_{t\geq 0}$ on the product probability space

$$\left(\mathscr{X} \times \mathcal{S} \times \mathcal{T}, \mathcal{B}(\mathscr{X}) \times \mathcal{B}_{\mathcal{S}} \times \mathcal{B}_{\mathcal{T}}, \mathcal{W} \times P_N \times P_T\right).$$

We have seen that $C_0^\infty(\mathbb{R}^3)$ is a form core of $(\sigma\cdot(-i\nabla - a))^2$ under suitable conditions on a in Proposition 4.240. A counterpart to this is the following, which can be proven similarly to Proposition 4.208. Denote

$$H_{\mathrm{RS}}^0(a) = (2H_{\mathrm{S}}^0(a,b) + m^2)^{1/2} - m. \tag{4.8.2}$$

Proposition 4.249. *Let $b \in (L^\infty(\mathbb{R}^3))^3$. Then we have the following:*
(1) *If $a \in (L^2_{\mathrm{loc}}(\mathbb{R}^3))^3$, then $C_0^\infty(\mathbb{R}^3)$ is a form core of $H_{\mathrm{RS}}^0(a)$.*
(2) *If $a \in (L^4_{\mathrm{loc}}(\mathbb{R}^3))^3$ and $\nabla \cdot a \in L^2_{\mathrm{loc}}(\mathbb{R}^3)$, then $C_0^\infty(\mathbb{R}^3)$ is an operator core of $H_{\mathrm{RS}}^0(a)$.*

In particular, from Proposition 4.249, the relation $Q(H_{\mathrm{RS}}^0(a)) \supset C_0^\infty(\mathbb{R}^3)$ follows whenever $a \in (L^2_{\mathrm{loc}}(\mathbb{R}^3))^3$. Thus $Q(V) \cap Q(H_{\mathrm{RS}}^0(a))$ is dense whenever $V \in L^1_{\mathrm{loc}}(\mathbb{R}^3)$, and then the form sum of $H_{\mathrm{RS}}^0(a)$ and V can be densely defined.

Definition 4.250 (Relativistic Schrödinger operator with vector potential and spin $\frac{1}{2}$). Let $V \in L^\infty(\mathbb{R}^3)$, $a \in (L^2_{\mathrm{loc}}(\mathbb{R}^3))^3$, and $b \in (L^\infty(\mathbb{R}^3))^3$. Then the *relativistic Schrödinger operator with vector potential a and spin* 1/2 is defined by

$$H_{\mathrm{RS}}(a,b) = (2H_{\mathrm{S}}^0(a,b) + m^2)^{1/2} - m + V \tag{4.8.3}$$

acting in $L^2(\mathbb{R}^3; \mathbb{C}^2)$, where $m \geq 0$ is assumed to be sufficiently large so that $2H_{\mathrm{S}}^0(a,b) + m^2 \geq 0$ holds.

As done for $H_{\mathrm{S}}(a,b)$ before, we also transform $H_{\mathrm{RS}}(a,b)$ by a unitary operator to the self-adjoint operator of the form

$$H_{\mathrm{R}\mathbb{Z}_2}(a,b) = (2H_{\mathbb{Z}_2}^0(a,b) + m^2)^{1/2} - m + V \tag{4.8.4}$$

on $L^2(\mathbb{R}^3 \times \mathbb{Z}_2)$. In what follows, we consider $H_{\mathrm{R}\mathbb{Z}_2}(a,b)$ instead of $H_{\mathrm{RS}}(a,b)$.

To derive the Feynman–Kac formula for $e^{-tH_{\mathrm{R}\mathbb{Z}_2}(a,b)}$, we use the random process

$$(\tilde{q}_t)_{t\geq 0} = (B_{T_t}, \theta_{T_t})_{t\geq 0} \tag{4.8.5}$$

on $\mathscr{X} \times \mathcal{S} \times \mathcal{T}$ under the probability measure $\mathcal{W}^x \times P_N \times P_T$, where the subordinator $(T_t)_{t\geq 0}$ satisfies $\mathbb{E}_{P_T}[e^{-uT_t}] = e^{-t(\sqrt{2u+m^2}-m)}$. We write $\mathbb{E}^{x,\alpha,0} = \mathbb{E}^{x,\alpha,0}_{\mathcal{W}\times P_N \times P_T}$ for simplicity, and denote by $\rho(\cdot, t)$ the distribution of T_t on $\mathbb{R}$, given explicitly by

$$\rho(r,t) = \frac{t}{(2\pi r^3)^{1/2}} \exp\left(-\frac{1}{2}\left(\frac{t^2}{r} + m^2 r\right) + mt\right) 1_{[0,\infty)}(r).$$

Theorem 4.251 (Feynman–Kac formula for relativistic Schrödinger operator with vector potential and spin 1/2). *Let $V \in L^\infty(\mathbb{R}^3)$, $a \in (L^2_{\mathrm{loc}}(\mathbb{R}^3))^3$, and $\nabla \cdot a \in L^1_{\mathrm{loc}}(\mathbb{R}^3)$. Suppose that $b \in (L^\infty(\mathbb{R}^3))^3$ and*

$$\int_0^\infty \rho(r,t)dr \int_0^r ds \int_{\mathbb{R}^3} dy \left|\log\left(\frac{1}{2}\sqrt{b_1(y)^2 + b_2(y)^2}\right)\right| \Pi_s(y-x) < \infty \tag{4.8.6}$$

for all $(x,t) \in \mathbb{R}^3 \times \mathbb{R}^+$. Then

$$(f, e^{-tH_{\mathrm{R}\mathbb{Z}_2}(a,b)} g) = \sum_{\alpha=1,2} \int_{\mathbb{R}^3} \mathbb{E}^{x,\alpha,0}[e^{T_t}\overline{f(\tilde{q}_0)} g(\tilde{q}_t) e^{\tilde{Z}_t}]dx, \tag{4.8.7}$$

where

$$\tilde{Z}_t = -i\int_0^{T_t} a(B_s) \circ dB_s - \int_0^t V(B_{T_s})ds + \int_0^{T_t} \frac{1}{2}\theta_{N_s} b_3(B_s)ds + \int_0^{T_t+} \log\left(\frac{1}{2}(b_1(B_s) - i\theta_{N_{s-}} b_2(B_s))\right) dN_s.$$

Proof. The proof is a combination of the formulae already established for $H_{\mathrm{R}}(a) = \sqrt{(-i\nabla - a)^2 + m^2} - m + V$ and $H_{\mathbb{Z}_2}(a,b) \cong \frac{1}{2}(-i\nabla - a)^2 - \frac{1}{2}\sigma \cdot b + V$. We only give an outline. Note that

$$\begin{aligned}
&\mathbb{E}^{x,\alpha,0}\left[\left|\int_0^{T_t+} \log\left(\frac{1}{2}(b_1(B_s) - i\theta_{N_{s-}} b_2(B_s))\right) dN_s\right|\right] \\
&\quad \le \int_0^\infty \rho(r,t)dr \int_0^r ds \int_{\mathbb{R}^3} \Pi_s(y-x) \left|\log\left(\frac{1}{2}\sqrt{b_1(y)^2 + b_2(y)^2}\right)\right| dy,
\end{aligned}$$

which is bounded by (4.8.6). Thus $\int_0^{T_t+} \log(\frac{1}{2}(b_1(B_s) - i\theta_{N_{s-}} b_2(B_s)))dN_s$ is almost surely finite for every (x,α).

First, consider the case of $V = 0$. The Feynman–Kac formula for

$$(f, e^{-tH^0_{\mathrm{R}\mathbb{Z}_2}(a,b)} g) = \mathbb{E}_{P_T}[(f, e^{-T_t H^0_{\mathbb{Z}_2}(a,b)} g)]$$

is given by Theorem 4.242, namely, the time t in the Feynman–Kac formula of $(f, e^{-tH^0_{\mathbb{Z}_2}(a,b)} g)$ is replaced by a random time given by the subordinator T_t, and it is integrated with respect to P_T. Next, for a continuous bounded V, using the Trotter product formula

$$\lim_{n\to\infty} (e^{-(t/n)H^0_{\mathrm{R}\mathbb{Z}_2}(a,b)} e^{-(t/n)V})^n = e^{-tH_{\mathrm{R}\mathbb{Z}_2}(a,b)},$$

we arrive at (4.8.7). A limiting argument on V gives then (4.8.6). □

Remark 4.252. Recall that $\rho(r,t)$ is the distribution of T_t. Clearly,

$$\int_0^\infty \rho(r,t)dr \int_0^r ds \int_{\mathbb{R}^3} dy \left|\log\left(\frac{1}{2}\sqrt{b_1(y)^2+b_2(y)^2}\right)\right| \Pi_s(y-x)$$
$$\leq \int_0^\infty ds \int_{\mathbb{R}^3} \left|\log\left(\frac{1}{2}\sqrt{b_1(y)^2+b_2(y)^2}\right)\right| \Pi_s(y-x)dy. \tag{4.8.8}$$

Hence if the right-hand side of (4.8.8) is finite, then condition (4.8.6) is satisfied.

Condition (4.8.6) requires that the measure of $\{y \in \mathbb{R}^3 \mid b_1^2(y)+b_2^2(y)=0\}$ is zero. We will remove this condition to consider the Hamiltonian with b_1 and b_2 which have zeroes. To avoid zeroes of the off-diagonal part of the spin, we introduce the cutoff function Ψ_ε as defined in (4.7.28). Let $t>0$ be fixed and consider the set $\mathscr{R}=\mathscr{R}_t$ defined by

$$\mathscr{R} = \left\{ \int_0^{T_t+} \log\left(\frac{1}{2}\sqrt{b_1(x+B_s)^2+b_2(x+B_s)^2}\right) dN_s > -\infty \right\} \subset \mathbb{R}^3 \times \mathscr{X} \times \mathcal{S} \times \mathcal{T}. \tag{4.8.9}$$

Lemma 4.253. *Let $\mathscr{R}$ be given by* (4.8.9). *It follows that*

$$\lim_{\varepsilon\to 0}\left| e^{\int_0^{T_t+} \log(\Psi_\varepsilon(\frac{1}{2}(b_1(B_s+x)-i\theta_{s-}b_2(B_s+x))))dN_s} 1_{\mathscr{R}^c}\right| = 0.$$

Proof. The proof is a minor modification of that of Lemma 4.241. Let

$$W(X) = \frac{1}{2}\sqrt{b_1(X)^2+b_2(X)^2}.$$

We have

$$\left| e^{\int_0^{T_t+} \log\left(\Psi_\varepsilon(\frac{1}{2}(b_1(x+B_s)-i\theta_{N_s} b_2(x+B_s)))\right)dN_s}\right| \leq e^{\int_0^{T_t+} \log(W(x+B_s)+\varepsilon)dN_s}.$$

We note that for every $(x,\omega,\tau,\varpi) \in \mathbb{R}^3 \times \mathscr{X} \times \mathcal{S} \times \mathcal{T}$ there exist a number $n=n(\tau,\varpi)$ and random jump times $r_1(\tau),\dots,r_n(\tau)$ such that

$$\int_0^{T_t(\varpi)+} \log(W(x+B_s(\omega))+\varepsilon)dN_s = \sum_{j=1}^n \log(W(x+B_{r_j(\tau)}(\omega))+\varepsilon).$$

We see that $(x,\omega,\tau,\varpi) \in \mathscr{R}^c$ if and only if there exists r such that for $0<r<T_t(\varpi)$ the map $s \mapsto N_s$ is discontinuous at $s=r$, and $b_1(x+B_r(\omega))^2+b_2(x+B_r(\omega))^2=0$. Hence for $(x,\omega,\tau,\varpi) \in \mathscr{R}^c$, there exists r_i such that

$$\int_0^{T_t+} \log(W(x+B_s)+\varepsilon)dN_s = \sum_{j\neq i} \log(W(x+B_{r_j})+\varepsilon) + \log\varepsilon,$$

and $e^{\int_0^{T_t+}\log(W(x+B_s)+\varepsilon)dN_s} \le e^{\sum_{j\neq i}^n \log(W(x+B_{r_j})+\varepsilon)} e^{\log\varepsilon}$. Thus it follows that

$$\lim_{\varepsilon\to 0}\left| e^{\int_0^{T_t+}\log(W(x+B_s)+\varepsilon)dN_s} 1_{\mathscr{R}^c}\right| = 0. \qquad \square$$

Theorem 4.254 (Feynman–Kac formula for relativistic Schrödinger operator with singular vector potential and spin 1/2). *Let $V \in L^\infty(\mathbb{R}^3)$. Suppose that $a \in (L^2_{\mathrm{loc}}(\mathbb{R}^3))^3$, $\nabla\cdot a \in L^1_{\mathrm{loc}}(\mathbb{R}^3)$ and $b \in (L^\infty(\mathbb{R}^3))^3$. Then*

$$(f, e^{-tH_{\mathrm{R}\mathbb{Z}_2}(a,b)}g) = \sum_{\alpha=1,2}\int_{\mathbb{R}^3} \mathbb{E}^{x,\alpha,0}[e^{T_t}\overline{f(\tilde{q}_0)}g(\tilde{q}_t)e^{\tilde{Z}_t}1_{\mathscr{R}}]dx, \tag{4.8.10}$$

where $\mathscr{R}$ is given by (4.8.9).

Proof. The proof is a simple modification of Theorem 4.242, i. e., q_t, Z_t, and $\tilde{Z}_t^\varepsilon$ in the proof of Theorem 4.242 are replaced by $\tilde{q}_t$, $\tilde{Z}_t$, and

$$\begin{aligned}\tilde{Z}_t^\varepsilon = &-i\int_0^{T_t} a(B_s)\circ dB_s - \int_0^t V(B_{T_s})ds \\ &+ \int_0^{T_t}\frac{1}{2}\theta_{N_s}b_3(B_s)ds + \int_0^{T_t+}\log\left(\Psi_\varepsilon\left(\frac{1}{2}(b_1(B_s) - i\theta_{N_s}b_2(B_s))\right)\right)dN_s,\end{aligned}$$

respectively. $\square$

From the Feynman–Kac formula, we can derive a diamagnetic inequality for the Hamiltonian with spin $H_{\mathrm{R}\mathbb{Z}_2}(a,b)$. However, this expression is not as simple as that for $H_{\mathbb{Z}_2}(a,b)$ derived in Corollary 4.243, where we obtained the diamagnetic inequality

$$|(f, e^{-tH^0_{\mathbb{Z}_2}(a,b)}g)| \le (|f|, e^{-tH^0_{\mathbb{Z}_2}(0,b_0)}|g|)$$

with $b_0 = (\sqrt{b_1^2+b_2^2}, 0, b_3)$. Note that $H^0_{\mathbb{Z}_2}(0,b_0)$ can be negative since

$$\inf\mathrm{Spec}(H^0_{\mathbb{Z}_2}(0,b_0)) \le \inf\mathrm{Spec}(H^0_{\mathbb{Z}_2}(a,b)). \tag{4.8.11}$$

The difficulty for the relativistic case is that the formula

$$(f, e^{-t\sqrt{2H+m^2}-m}g) = \mathbb{E}_P[(f, e^{-T_tH}g)], \quad m \ge 0,$$

holds only for positive self-adjoint operators H. To derive the diamagnetic inequality for $H_{\mathrm{R}\mathbb{Z}_2}(a,b)$, we define

$$\tilde{H}_{\mathrm{R}\mathbb{Z}_2}(a,b) = \sqrt{2(H^0_{\mathbb{Z}_2}(a,b)-E)+m^2} - m + V, \tag{4.8.12}$$

where $E = \inf \operatorname{Spec}\left(H^0_{\mathbb{Z}_2}(0, b_0)\right)$, and $H^0_{\mathbb{Z}_2}(a, b) - E \geq 0$ follows from (4.8.11). Thus

$$\tilde{H}_{\mathrm{R}\mathbb{Z}_2}(0, b_0) = \sqrt{2(H_{\mathbb{Z}_2}(0, b_0) - E) + m^2} - m + V \tag{4.8.13}$$

is well defined. From the Feynman–Kac formula in Theorem 4.254, we immediately obtain

Corollary 4.255 (Diamagnetic inequality). *Under the assumptions of Theorem* 4.254,

$$|(f, e^{-t\tilde{H}_{\mathrm{R}\mathbb{Z}_2}(a,b)} g)| \leq (|f|, e^{-t\tilde{H}_{\mathrm{R}\mathbb{Z}_2}(0,b_0)} |g|) \tag{4.8.14}$$

holds.

Using this inequality, we can now give a similar result to Theorem 4.244.

Theorem 4.256 (Feynman–Kac formula for relativistic Schrödinger operator with singular external potential, singular vector potential, and spin 1/2). *Suppose that* $a \in (L^2_{\mathrm{loc}}(\mathbb{R}^3))^3$, $\nabla \cdot a \in L^1_{\mathrm{loc}}(\mathbb{R}^3)$, *and* $b \in (L^\infty(\mathbb{R}^3))^3$.

(1) *Let* V *be a relatively bounded (resp. form-bounded) potential with respect to* $\sqrt{-\Delta + m^2}$ *with a relative bound* c. *Then* V *is also relatively bounded (resp. form-bounded) with respect to* $H^0_{\mathrm{R}\mathbb{Z}^2}(a, b)$ *with a relative bound not exceeding* c.

(2) *Let* $V = V_+ - V_-$ *be such that* $V_+ \in L^1_{\mathrm{loc}}(\mathbb{R}^3)$ *and* V_- *is relatively form-bounded with respect to* $\sqrt{-\Delta + m^2}$ *with a relative bound less than* 1. *Then*

$$H^0_{\mathrm{R}\mathbb{Z}_2}(a, b) \dotplus V_+ \dot{-} V_- \tag{4.8.15}$$

can be defined as a self-adjoint operator. This is also denoted by $H_{\mathrm{R}\mathbb{Z}_2}(a, b)$.

(3) *Let* V *satisfy the same assumptions as in* (2). *Then the Feynman–Kac formula of* $(f, e^{-tH_{\mathrm{R}\mathbb{Z}_2}(a,b)} g)$ *is given by* (4.8.10).

Proof. Denote $S = -\frac{1}{2}\sigma \cdot b_0$ and $E = \inf \operatorname{Spec}(H^0_{\mathbb{Z}_2}(0, b_0))$. Note that

$$\|\sqrt{-\Delta + m^2} f\|^2 = \|\sqrt{2H^0_{\mathbb{Z}_2}(0, b_0) - 2E + m^2} f\|^2 + (f, -2(S - E)f).$$

Since $|(f, 2Sf)| \leq \kappa \|f\|^2$ with a constant κ, we have

$$\|\sqrt{-\Delta + m^2} f\|^2 \leq \|\sqrt{2H^0_{\mathbb{Z}_2}(0, b_0) - 2E + m^2} f\|^2 + (|2E| + \kappa)\|f\|^2.$$

Together with $\|Vf\| \leq c\|\sqrt{-\Delta + m^2} f\| + \varepsilon \|f\|^2$ with a constant ε, we have

$$\|Vf\| \leq c\|\sqrt{2H^0_{\mathbb{Z}_2}(0, b_0) - 2E + m^2} f\| + C\|f\| \tag{4.8.16}$$

with a constant C. By (4.8.16) and the diamagnetic inequality in Corollary 4.255, V is also relatively bounded with respect to $\sqrt{2H^0_{\mathbb{Z}_2}(a, b) - 2E + m^2}$ with the same relative bound c. Moreover, since the difference

$$T = \sqrt{2H^0_{\mathbb{Z}_2}(a,b) - 2E + m^2} - \sqrt{2H^0_{\mathbb{Z}_2}(a,b) + m^2}$$

is a bounded operator with $\|T\| \le \sqrt{|2E|}$, V is also relatively bounded with respect to $\sqrt{2H^0_{\mathbb{Z}_2}(a,b) + m^2}$ with relative bound c. Thus (1) follows. The form version is similarly proven. The proofs of (2) and (3) are the same as those of (2) and (3) of Theorem 4.244. □

Essential self-adjointness of $\sqrt{(\sigma \cdot (-i\nabla - a))^2 + m^2} - m + V$ on $\mathbb{C}^2 \otimes C_0^\infty(\mathbb{R}^3)$ can also be obtained from Proposition 4.240 and (1) of Theorem 4.256.

Theorem 4.257 (Essential self-adjointness of relativistic Schrödinger operator with spin). *Let V be relatively bounded with respect to $\sqrt{-\Delta + m^2}$ with a relative bound less than one. Suppose that $a \in (L^4_{\mathrm{loc}}(\mathbb{R}^3))^3$, $\nabla \cdot a \in L^2_{\mathrm{loc}}(\mathbb{R}^3)$, and $\nabla \times a = b$ satisfies $b \in (L^\infty(\mathbb{R}^3))^3$. Then $\sqrt{(\sigma \cdot (-i\nabla - a))^2 + m^2} - m + V$ is essentially self-adjoint on $\mathbb{C}^2 \otimes C_0^\infty(\mathbb{R}^3)$.*

Proof. By Proposition 4.249, we can see that $\sqrt{(\sigma \cdot (-i\nabla - a))^2 + m^2}$ is essentially self-adjoint on $\mathbb{C}^2 \otimes C_0^\infty(\mathbb{R}^3)$. Note that $\sqrt{(\sigma \cdot (-i\nabla - a))^2 + m^2} - m = H^0_{\mathrm{RS}}(a,b)$, since the operator $(\sigma \cdot (-i\nabla - a))^2$ can be expanded and equals $H^0_{\mathrm{S}}(a,b)$. Claim (1) of Theorem 4.256 yields that V is relatively bounded with respect to $\sqrt{(\sigma \cdot (-i\nabla - a))^2 + m^2}$ with a relative bound less than one. Hence the theorem follows by the Kato–Rellich theorem. □

This leads us to the following definition

Definition 4.258 (Relativistic Schrödinger operator with singular external potential, singular vector potential, and spin 1/2). Suppose that $a \in (L^2_{\mathrm{loc}}(\mathbb{R}^3))^3$ and the magnetic field b satisfies $b \in (L^\infty(\mathbb{R}^3))^3$. Let $V = V_+ - V_-$ be such that $V_+ \in L^1_{\mathrm{loc}}(\mathbb{R}^3)$ and V_- is relatively form-bounded with respect to $\sqrt{-\Delta + m^2}$ with a relative bound less than 1. Then $H_{\mathrm{R}\mathbb{Z}_2}(a,b)$ is defined by (4.8.15).

4.8.2 Martingale properties

In this section we consider martingale properties for the processes associated with $H_{\mathrm{R}\mathbb{Z}_2}$, similarly to relativistic Schrödinger operators and Schrödinger operators with spin discussed before. Let φ be an eigenfunction of $H_{\mathrm{R}\mathbb{Z}_2}(a,b)$ such that

$$H_{\mathrm{R}\mathbb{Z}_2}(a,b)\varphi = E\varphi$$

with eigenvalue $E \in \mathbb{R}$. Consider the stochastic process $(Y_t)_{t\ge 0}$ defined by

$$Y_t = e^{tE} e^{T_t} e^{\tilde{Z}_t} \varphi(\tilde{q}_t), \quad t \ge 0,$$

on $(\mathscr{X} \times \mathcal{S} \times \mathcal{T}, \mathcal{B}(\mathscr{X}) \times \mathcal{B}_{\mathcal{S}} \times \mathcal{B}_{\mathcal{T}}, \mathcal{W} \times P_N \times P_{\mathcal{T}})$. Furthermore, for $(x,\alpha) \in \mathbb{R}^3 \times \{1,2\}$, we define

$$Y_t(x,\alpha) = e^{tE} e^{T_t} e^{\tilde{Z}_t(x,\alpha)} \varphi(\tilde{q}_t(x,\alpha)), \quad t \geq 0,$$

where $\tilde{q}_t(x,\alpha) = (B_{T_t} + x, \theta_{\alpha+N_{T_t}})$ and $\tilde{Z}_t(x,\alpha) = \tilde{Z}_t^V(x) + \tilde{Z}_t^A(x) + \tilde{Z}_t^S(x,\alpha)$ is given by

$$\tilde{Z}_t^V(x) = -\int_0^t V(B_{T_s} + x)ds, \quad \tilde{Z}_t^A(x) = -i\int_0^{T_t} a(B_s + x) \circ dB_s,$$

$$\tilde{Z}_t^S(x,\alpha) = -\int_0^{T_t} U_d(B_s + x, \theta_{\alpha+N_s})ds + \int_0^{T_t+} \log\left(-U_o(B_s + x, -\theta_{\alpha+N_{s-}})\right) dN_s.$$

Here the functions U_o and U_d mapping $\mathbb{R}^3 \times \mathbb{Z}_2$ to $\mathbb{R}$ are given by $U_d(x,Y) = -\frac{1}{2}Yb_3(x)$ and $U_o(x,-Y) = -\frac{1}{2}(b_1(x) - iYb_2(x))$. We have

$$\mathbb{E}^{x,\alpha,0}[Y_t] = \mathbb{E}^{0,0,0}[Y_t(x,\alpha)] = \varphi(x,\theta_\alpha).$$

We introduce a filtration under which $(Y_t)_{t\geq 0}$ is a martingale. Define $Y_t(\varpi)$ and $Y_t(x,\alpha,\varpi)$ for every $\varpi \in \mathcal{T}$ by Y_t and $Y_t(x,\alpha)$, respectively, with the subordinator T_t replaced by the random number $T_t(\varpi) \geq 0$. For each $\varpi \in \mathcal{T}$, let

$$\mathcal{G}_t^{(1)}(\varpi) = \sigma\big((B_r, N_r), 0 \leq r \leq T_t(\varpi)\big) \subset \mathcal{B}(\mathscr{X}) \times \mathcal{B}_{\mathcal{S}}$$

and define

$$\mathcal{G}_t^{(1)} = \left\{E \in \mathcal{B}(\mathscr{X}) \times \mathcal{B}_{\mathcal{S}} \times \mathcal{B}_{\mathcal{T}} \mid E_\varpi^1 \in \mathcal{G}_t^{(1)}(\varpi)\ \forall \varpi \in \mathcal{T}\right\},$$

where

$$E_\varpi^1 = \left\{(\omega,\eta) \in \mathscr{X} \times \mathcal{B}_{\mathcal{S}} \mid (\omega,\eta,\varpi) \in E\right\}.$$

We also define

$$\mathcal{G}_t^{(2)} = \left\{E \in \mathcal{B}(\mathscr{X}) \times \mathcal{B}_{\mathcal{S}} \times \mathcal{B}_{\mathcal{T}} \mid E_{\omega,\eta}^2 \in \sigma(T_r : 0 \leq r \leq t)\ \forall (\omega,\eta) \in \mathscr{X} \times \mathcal{S}\right\},$$

where $E_{\omega,\eta}^2$ denotes the section of E at $(\omega,\eta) \in \mathscr{X} \times \mathcal{S}$, i. e.,

$$E_{\omega,\eta}^2 = \left\{\varpi \in \mathcal{T} \mid (\omega,\eta,\varpi) \in E\right\}.$$

We see that $\mathcal{G}_t^{(1)}$ and $\mathcal{G}_t^{(2)}$ are sub-σ-fields of $\mathcal{B}(\mathscr{X}) \times \mathcal{B}_{\mathcal{S}} \times \mathcal{B}_{\mathcal{T}}$. Define the filtration $(\mathcal{G}_t)_{t\geq 0}$ by

$$\mathcal{G}_t = \mathcal{G}_t^{(1)} \cap \mathcal{G}_t^{(2)}, \quad t \geq 0.$$

The conditional expectation $\mathbb{E}^{0,0,0}[Y_t(x,\alpha)|\mathcal{G}_t^{(1)}] = \mathbb{E}^{0,0,0}[Y_t(x,\alpha)|\mathcal{G}_t^{(1)}](\cdot,\cdot,\cdot)$ is a random process on $\mathscr{X} \times \mathcal{S} \times \mathcal{T}$ and $\mathbb{E}^{0,0}[Y_t(x,\alpha,\varpi)|\mathcal{G}_t^{(1)}(\varpi)](\cdot,\cdot)$ on $\mathscr{X} \times \mathcal{S}$. The relationship is given in the following lemma.

Lemma 4.259. *It follows that*

$$\mathbb{E}^{0,0,0}[Y_t(x,\alpha)|\mathcal{G}_t^{(1)}](\cdot,\cdot,\varpi) = \mathbb{E}^{0,0}[Y_t(x,\alpha,\varpi)|\mathcal{G}_t^{(1)}(\varpi)](\cdot,\cdot) \tag{4.8.17}$$

for a. e. $\varpi \in \mathcal{T}$.

Proof. Let $A \in \mathcal{G}_t^{(1)}$ with section $A_\varpi \in \mathcal{G}_t^{(1)}(\varpi)$, and $C \in \mathcal{B}_\mathcal{T}$. We have

$$\begin{aligned}\mathbb{E}^{0,0,0}[1_C 1_A Y_t(x,\alpha)] &= \int_\mathcal{T} 1_C \mathbb{E}^{0,0}[1_{A_\varpi} Y_t(x,\alpha,\varpi)]dP_\mathcal{T}(\varpi)\\ &= \int_\mathcal{T} 1_C \mathbb{E}^{0,0}\left[1_{A_\varpi}(\cdot,\cdot)\mathbb{E}^{0,0}\left[Y_t(x,\alpha,\varpi)|\mathcal{G}_t^{(1)}(\varpi)\right](\cdot,\cdot)\right]dP_\mathcal{T}(\varpi).\end{aligned}$$

On the other hand, we have

$$\begin{aligned}\mathbb{E}^{0,0,0}[1_C 1_A Y_t(x,\alpha)] &= \mathbb{E}^{0,0,0}\left[1_C 1_A \mathbb{E}^{0,0,0}[Y_t(x,\alpha)|\mathcal{G}_t^{(1)}]\right]\\ &= \int_\mathcal{T} 1_C \mathbb{E}^{0,0}\left[1_{A_\varpi}(\cdot,\cdot)\mathbb{E}^{0,0,0}\left[Y_t(x,\alpha)|\mathcal{G}_t^{(1)}\right](\cdot,\cdot,\varpi)\right]dP_\mathcal{T}(\varpi).\end{aligned}$$

Since $C \in \mathcal{B}_\mathcal{T}$ is arbitrary, a comparison of the two sides above gives

$$\mathbb{E}^{0,0}\left[1_{A_\varpi}(\cdot,\cdot)\mathbb{E}^{0,0}\left[Y_t(x,\alpha,\varpi)|\mathcal{G}_t^{(1)}(\varpi)\right](\cdot,\cdot)\right] = \mathbb{E}^{0,0}\left[1_{A_\varpi}(\cdot,\cdot)\mathbb{E}^{0,0,0}\left[Y_t(x,\alpha)|\mathcal{G}_t^{(1)}\right](\cdot,\cdot,\varpi)\right]. \tag{4.8.18}$$

Let $A = A' \times \mathcal{T}$ be a rectangle. We have $A_\varpi = A' \in \mathcal{B}(\mathscr{X}) \times \mathcal{B}_\mathcal{S}$ for all $\varpi \in \mathcal{T}$. Substituting A into (4.8.18), we have

$$\mathbb{E}^{0,0}\left[1_{A'}(\cdot,\cdot)\mathbb{E}^{0,0}\left[Y_t(x,\alpha,\varpi)|\mathcal{G}_t^{(1)}(\varpi)\right](\cdot,\cdot)\right] = \mathbb{E}^{0,0}\left[1_{A'}(\cdot,\cdot)\mathbb{E}^{0,0,0}\left[Y_t(x,\alpha)|\mathcal{G}_t^{(1)}\right](\cdot,\cdot,\varpi)\right].$$

Since $A' \in \mathcal{B}(\mathscr{X}) \times \mathcal{B}_\mathcal{S}$ is arbitrary, (4.8.17) follows. □

Lemma 4.260 (Martingale properties). *The random process* $(Y_t(x,\alpha))_{t\ge 0}$ *is a martingale with respect to* $(\mathcal{G}_t)_{t\ge 0}$, *i. e.,* $\mathbb{E}^{0,0,0}[Y_t(x,\alpha)|\mathcal{G}_s] = Y_s(x,\alpha)$ *for* $t \ge s$.

Proof. We show the case when $(x,\alpha) = (0,0)$ to keep the notation simple; the proof for $(x,\alpha) \neq (0,0)$ is similar. Note that

$$\mathbb{E}^{0,0,0}[Y_t|\mathcal{G}_s] = \mathbb{E}^{0,0,0}[Y_t|\mathcal{G}_s^{(1)} \cap \mathcal{G}_s^{(2)}] = \mathbb{E}^{0,0,0}[\mathbb{E}^{0,0,0}[Y_t|\mathcal{G}_s^{(1)}]|\mathcal{G}_s^{(2)}].$$

We first compute $\mathbb{E}^{0,0}[Y_t(\varpi)|\mathcal{G}_s^{(1)}(\varpi)]$. Write

$$\tilde{Z}_t([u,v]) = -\int_u^v V(B_{T_r})dr - i\int_{T_u}^{T_v} a(B_r)\circ dB_r - \int_{T_u}^{T_v} U_d(B_r,\theta_{N_r})dr + \int_{T_u}^{T_v+} \log(-U_o(B_r,-\theta_{N_{r-}}))dN_r$$

and, for every $\varpi \in \mathcal{T}$,

$$\tilde{Z}_t([u,v],\varpi) = -\int_u^v V(B_{T_r(\varpi)})dr - i\int_{T_u(\varpi)}^{T_v(\varpi)} a(B_r)\circ dB_r$$
$$-\int_{T_u(\varpi)}^{T_v(\varpi)} U_d(B_r,\theta_{N_r})dr + \int_{T_u(\varpi)}^{T_v(\varpi)+} \log(-U_o(B_r,-\theta_{N_{r-}}))dN_r$$

and $\tilde{q}_t(\varpi) = (B_{T_t(\varpi)}, \theta_{N_{T_t(\varpi)}})$, $t \geq 0$. Since $T_t(\varpi)$ is nonrandom, we see in a similar way to the nonrelativistic case that

$$\begin{aligned}
&\mathbb{E}^{0,0}[Y_t(\varpi)|\mathcal{G}_s^{(1)}(\varpi)]\\
&= e^{tE}e^{T_t(\varpi)}e^{\tilde{Z}_t([0,s],\varpi)}\mathbb{E}^{B_{T_s(\varpi)},N_{T_s(\varpi)}}\Big[e^{-\int_s^t V(B_{T_r(\varpi)-T_s(\varpi)})dr}e^{-i\int_{T_s(\varpi)}^{T_t(\varpi)} a(B_{r-T_s(\varpi)})\circ dB_r}\\
&\times e^{-\int_{T_s(\varpi)}^{T_t(\varpi)} U_d\left(B_{r-T_s(\varpi)},\theta_{N_{r-T_s(\varpi)}}\right)dr}e^{\int_{T_s(\varpi)}^{T_t(\varpi)+}\log\left(-U_o\left(B_{r-T_s(\varpi)},-\theta_{N_{(r-T_s(\varpi))-}}\right)\right)dN_r}\\
&\times \varphi\left(B_{T_t(\varpi)-T_s(\varpi)},\theta_{N_{T_t(\varpi)-T_s(\varpi)}}\right)\Big],
\end{aligned}$$

where we used the Markov property of $((B_t,N_t))_{t\geq 0}$. Hence by Lemma 4.259, we have

$$\mathbb{E}^{0,0,0}[Y_t|\mathcal{G}_s^{(1)}] = e^{tE_P}e^{T_s}e^{\tilde{Z}_t([0,s])}Z_{t,s},$$

where

$$\begin{aligned}
Z_{t,s} = e^{T_t-T_s}\mathbb{E}^{B_{T_s},N_{T_s}}\Big[&e^{-\int_s^t V(B_{T_r-T_s})dr}e^{-i\int_{T_s}^{T_t} a(B_{r-T_s})\circ dB_r}e^{-\int_{T_s}^{T_t} U_d(B_{r-T_s},\theta_{N_{r-T_s}})dr}\\
&\times e^{\int_{T_s}^{T_t+}\log(-U_o(B_{r-T_s},-\theta_{N_{(r-T_s)-}}))dN_r}\varphi(B_{T_t-T_s},\theta_{N_{T_t-T_s}})\Big].
\end{aligned}$$

Here $Z_{t,s}$ is given by

$$\begin{aligned}
e^{u-v}\mathbb{E}^{B_u,N_v}\Big[&e^{-\int_s^t V(B_{T_r-u})dr}e^{-i\int_v^u a(B_{r-v})\circ dB_r}e^{-\int_v^u U_d(B_{r-v},\theta_{N_{r-v}})dr}\\
&\times e^{\int_v^{u+}\log(-U_o(B_{r-v},-\theta_{N_{(r-v)-}}))dN_r}\varphi(B_{u-v},\theta_{N_{u-v}})\Big]
\end{aligned}$$

evaluated at $u = T_t$ and $v = T_s$. We note that

$$\mathbb{E}^{0,0,0}[f|\mathcal{G}_s^{(2)}](\omega_1,\omega_2,\cdot) = \mathbb{E}_P^0[f(\omega_1,\omega_2,\cdot)|\mathcal{N}_s](\cdot), \tag{4.8.19}$$

where $\mathcal{N}_s = \sigma(T_r, 0 \le r \le s)$. Since $e^{tE}e^{T_s}e^{\tilde{Z}_t([0,s])}$ is measurable with respect to $\mathcal{G}_s^{(2)}$, by (4.8.19) we consider the conditional expectation of $Z_{t,s}$, giving

$$\mathbb{E}^{0,0,0}\left[Z_{t,s}\,\middle|\,\mathcal{G}_s^{(2)}\right] = \mathbb{E}_P^0\Big[e^{T_t-T_s}\mathbb{E}^{B_{T_s},N_{T_s}}\Big[e^{-\int_s^t V(B_{T_r-T_s})dr}e^{-i\int_0^{T_t-T_s} a(B_r)\circ dB_r}$$
$$\times e^{-\int_0^{T_t-T_s} U_d(B_r,\theta_{N_r})dr}e^{\int_0^{(T_t-T_s)+}\log(-U_o(B_r,-\theta_{N_{r-}}))dN_r}\varphi(B_{T_t-T_s},\theta_{N_{T_t-T_s}})\Big]\Big|\mathcal{N}_s\Big].$$

By the Markov property of $(T_t)_{t\ge0}$, we have

$$= \mathbb{E}_P^{T_s}\Big[e^{T_{t-s}}\mathbb{E}^{B_{T_0},N_0}\Big[e^{-\int_s^t V(B_{T_{r-s}})dr}e^{-i\int_0^{T_{t-s}} a(B_r)\circ dB_r}$$
$$\times e^{-\int_0^{T_{t-s}} U_d(B_r,\theta_{N_r})dr}e^{\int_0^{T_{t-s}+}\log(-U_o(B_r,-\theta_{N_{r-}}))dN_r}\varphi(B_{T_{t-s}},\theta_{N_{T_{t-s}}})\Big]\Big].$$

Since $\mathbb{E}_P^u[f(T_\cdot)] = \mathbb{E}_P^0[f(T_\cdot + u)]$, we see that

$$= \mathbb{E}_P^0\Big[e^{T_{t-s}-T_0}\mathbb{E}^{B_{T_0+u},N_{T_0+u}}\Big[e^{-\int_s^t V(B_{T_{r-s}-T_0})dr}e^{-i\int_0^{T_{t-s}-T_0} a(B_r)\circ dB_r}$$
$$\times e^{-\int_0^{T_{t-s}-T_0} U_d(B_r,\theta_{N_r})dr}e^{\int_0^{(T_{t-s}-T_0)+}\log(-U_o(B_r,-\theta_{N_{r-}}))dN_r}\varphi(B_{T_{t-s}-T_0},\theta_{N_{T_{t-s}-T_0}})\Big]\Big]\Big\lceil_{u=T_s}$$
$$= \mathbb{E}^{B_{T_s},N_{T_s},0}\Big[e^{T_{t-s}}e^{-\int_0^{t-s} V(B_{T_r})dr}e^{-i\int_0^{T_{t-s}} a(B_r)\circ dB_r}$$
$$\times e^{-\int_0^{T_{t-s}} U_d(B_r,\theta_{N_r})dr}e^{\int_0^{T_{t-s}+}\log(-U_o(B_r,-\theta_{N_{r-}}))dN_r}\varphi(\tilde{q}_{t-s})\Big]$$
$$= \left(e^{-(t-s)H_{\mathrm{R}\mathbb{Z}_2}(a,b)}\varphi\right)(\tilde{q}_s).$$

Hence we conclude that

$$\mathbb{E}^{0,0,0}[Y_t|\mathcal{G}_s] = e^{sE}e^{T_s}e^{\tilde{Z}_t([0,s])}(e^{-(t-s)(H_{\mathrm{R}\mathbb{Z}_2}(a,b)-E)}\varphi)(\tilde{q}_s) = Y_s,$$

and the result follows. □

4.8.3 Decay of eigenfunctions

In this section we estimate the decay of eigenfunctions of $H_{\mathrm{R}\mathbb{Z}_2}$ by using the martingale properties established in Lemma 4.260. We will use the following assumptions.

Assumption 4.261. Let the following properties hold:
(1) $b \in (L^\infty(\mathbb{R}^3))^3$;
(2) $m^* < m^2/2$, where

$$m^* = \|b_3\|_\infty + \frac{1}{4}\|\sqrt{b_1^2 + b_2^2}\|_\infty^2 + 1;$$

(3) V is uniformly locally integrable and belongs to the relativistic Kato-class, i. e.,

$$\lim_{t\downarrow 0}\sup_{x\in\mathbb{R}^3}\mathbb{E}^{x,0}\left[\int_0^t V(B_{T_r})dr\right]=0.$$

In (4.7.39), we define $E^* = m^*/2$.

Lemma 4.262. *If Assumption* 4.261 *holds, then* $\varphi\in L^\infty(\mathbb{R}^d)$ *and*

$$|\varphi(x,\theta_\alpha)|^2\le C\|\varphi\|^2\mathbb{E}^{0,0}\left[e^{2(t\wedge\tau)E}e^{-2\int_0^{t\wedge\tau}V(B_{T_r}+x)dr}\right]\mathbb{E}^{0,0}\left[e^{m^*T_{t\wedge\tau}}\right],\tag{4.8.20}$$

for every stopping time τ *with respect to* $(\mathcal{G}_s)_{s\ge 0}$ *and* $t\ge 0$*. Here* $C>0$ *is a constant.*

Proof. Let $W(X)=\frac{1}{2}\sqrt{b_1(X)^2+b_2(X)^2}$. Recall that $Y_t=e^{tE}e^{T_t}e^{\tilde{Z}_t}\varphi(\tilde{q}_t)$, and note that

$$\varphi(x,\theta_\alpha)=\mathbb{E}^{x,\alpha,0}[Y_t]$$

for every t. Schwarz inequality yields

$$\begin{aligned}|\varphi(x,\theta_\alpha)|^2&\le e^{2tE}\mathbb{E}^{x,\alpha,0}\left[e^{2T_t}e^{-2\int_0^t V(B_{T_r})dr}e^{\int_0^{T_t}|b_3(B_r)|dr}e^{2\int_0^{T_t+}\log W(B_r)dN_r}\right]\mathbb{E}^{x,\alpha,0}\left[|\varphi(\tilde{q}_t)|^2\right]\\&\le e^{2tE}\mathbb{E}^{0,0}[e^{-2\int_0^t V(B_{T_r}+x)dr}e^{T_tm^*}]\mathbb{E}^{x,\alpha,0}[|\varphi(\tilde{q}_t)|^2].\end{aligned}\tag{4.8.21}$$

Note that

$$\begin{aligned}\mathbb{E}^{x,\alpha,0}[|\varphi(\tilde{q}_t)|^2]&=\int_0^\infty\rho(s,t)ds\int_{\mathbb{R}^3}\Pi_s(y)dy\sum_{n=0}^\infty|\varphi(x+y,\theta_{\alpha+n})|^2\frac{s^n}{n!}e^{-s}\\&\le\int_{\mathbb{R}^d}(|\varphi(x+y,1)|^2+|\varphi(x+y,-1)|^2)dy\int_0^\infty\rho(s,t)\Pi_s(y)ds\\&\le C\|\varphi\|^2\end{aligned}$$

with a constant C, and where $\rho(\cdot,t)$ is the distribution of T_t. Furthermore, take q and p such that $\frac{m^2}{2m^*}>q>1$ and $\frac{1}{p}+\frac{1}{q}=1$. By the Hölder inequality, we get

$$\mathbb{E}^{x,0}[e^{-2\int_0^t V(B_{T_r})dr}e^{T_tm^*}]\le\left(\mathbb{E}^{x,0}[e^{-2p\int_0^t V(B_{T_r})dr}]\right)^{1/p}\left(\mathbb{E}^{x,0}[e^{qT_tm^*}]\right)^{1/q}.$$

The first term on the right-hand side above satisfies

$$\sup_{x\in\mathbb{R}^d}\left(\mathbb{E}^{x,0}\left[e^{-2p\int_0^t V(B_{T_r})dr}\right]\right)^{1/p}<\infty,$$

since V is of relativistic Kato-class, and

$$\mathbb{E}^{x,0}[e^{qT_t m^*}] = \mathbb{E}^0_\nu[e^{qT_t m^*}] = e^{+t(m-\sqrt{m^2-2qm^*})} < \infty.$$

Hence $\varphi \in L^\infty(\mathbb{R}^d)$. Note that by the martingale property of $Y_t(x,\alpha)$,

$$\varphi(x,\theta_\alpha) = \mathbb{E}^{0,0,0}[Y_{t\wedge\tau}(x,\alpha)]$$

for every stopping time τ and $t \geq 0$. The estimate (4.8.20) follows from (4.8.21), with t replaced by $t \wedge \tau$. □

Choosing a suitable stopping time, we can estimate the decay of eigenfunctions of $H_{\mathrm{R}\mathbb{Z}_2}$.

Lemma 4.263. *Let* $\tau_R = \inf\{t \mid |B_{T_t}| > R\}$. *Then* τ_R *is a stopping time with respect to the filtration* $(\mathcal{G}_t)_{t\geq 0}$.

Proof. It suffices to show that $\{\tau_R \leq t\} \in \mathcal{G}_t$. Note that

$$\left\{\tau_R \leq t\right\} = \bigcup_{\varpi\in\mathcal{T}} (A(\varpi),\varpi),$$

where $A(\varpi) = \{\omega \in \mathscr{X} \mid \sup_{0\leq s\leq t} |B_{T_s(\varpi)}(\omega)| > R\} \times \mathcal{S} \in \mathcal{G}^{(1)}_t(\varpi)$. Thus it follows that $\{\tau_R \leq t\} \in \mathcal{G}^{(1)}_t$. Moreover,

$$\left\{\tau_R \leq t\right\} = \bigcup_{(\omega,\eta)\in\mathscr{X}\times\mathcal{S}} (\omega,\eta,B(\omega)),$$

where $B(\omega) = \{\varpi \in \mathcal{T} \mid \sup_{0\leq s\leq t} |B_{T_s(\varpi)}(\omega)| > R\}$. Therefore $\{\tau_R \leq t\} \in \mathcal{G}^{(2)}_t$ and hence $\{\tau_R \leq t\} \in \mathcal{G}_t$. □

Theorem 4.264 (Confining potentials). *Suppose that* $\lim_{|x|\to\infty} V(x) = \infty$ *and let Assumption* 4.261 *hold. Then for every* $a > 0$, *there exists* $b > 0$ *such that*

$$|\varphi(x,\theta_\alpha)| \leq be^{-a|x|}.$$

Proof. We have by Lemma 4.262, for $t \geq 0$,

$$|\varphi(x,\theta_\alpha)| \leq \sqrt{C}\|\varphi\|\left(\mathbb{E}^{0,0}\left[e^{2E(t\wedge\tau_R)}e^{-2\int_0^{t\wedge\tau_R} V(B_{T_r}+x)dr}\right]\right)^{1/2}\left(\mathbb{E}^{0,0}\left[e^{m^* T_{t\wedge\tau_R}}\right]\right)^{1/2}.$$

Let $W(x) = W_R(x) = \inf\{V(y) \mid |x-y| < R\}$, and note that $\lim_{|x|\to\infty} W(x) - E = \infty$. In particular, we may assume that $W(x) - E > 0$. This gives

$$\left(\mathbb{E}^{0,0}\left[e^{2E(t\wedge\tau_R)}e^{-2\int_0^{t\wedge\tau_R} V(B_{T_r}+x)dr}\right]\right)^{1/2}$$

$$\leq \left(\mathbb{E}^{0,0}\left[1_{\{\tau_R<t\}}e^{-2(W(x)-E)(t\wedge\tau_R)}\right]\right)^{1/2} + \left(\mathbb{E}^{0,0}\left[1_{\{\tau_R\geq t\}}e^{-2(W(x)-E)(t\wedge\tau_R)}\right]\right)^{1/2}$$
$$\leq \left(\mathbb{E}^{0,0}\left[1_{\{\tau_R<t\}}\right]\right)^{1/2} + e^{-(W(x)-E)t}.$$

We see that

$$\mathbb{E}^{0,0}\left[1_{\{\tau_R<t\}}\right] = \mathbb{E}^{0,0}\left[1_{\{\sup_{0\leq s\leq t}|B_{T_s}|-R\geq 0\}}\right] \leq \mathbb{E}^{0,0}\left[e^{\beta(\sup_{0\leq s\leq t}|B_{T_s}|-R)}\right]$$

for any $\beta \geq 0$. It can be shown that $(\mathbb{E}^{0,0}[e^{\beta\sup_{0\leq s\leq t}|B_{T_s}|}])^{1/2} \leq C_1e^{C_2t}$ for sufficiently small β by Theorem 3.78. Hence

$$\left(\mathbb{E}^{0,0}[1_{\{\tau_R<t\}}]\right)^{1/2} \leq e^{-\beta R/2}C_1e^{C_2t},$$

and

$$\left(\mathbb{E}^{0,0}\left[e^{2E(t\wedge\tau_R)}e^{-2\int_0^{t\wedge\tau_R}V(B_{T_r}+x)dr}\right]\right)^{1/2} \leq e^{-(W(x)-E)t} + e^{-\beta R/2}C_1e^{C_2t}. \tag{4.8.22}$$

We also see that

$$\mathbb{E}^{0,0}\left[e^{m^*T_{t\wedge\tau_R}}\right] \leq \mathbb{E}^{0,0}\left[1_{\{t<\tau_R\}}e^{m^*T_{t\wedge\tau_R}}\right] + \mathbb{E}^{0,0}\left[1_{\{t\geq\tau_R\}}e^{m^*T_{t\wedge\tau_R}}\right]$$
$$\leq \mathbb{E}^{0,0}\left[e^{m^*T_t}\right] + \mathbb{E}^{0,0}\left[1_{\{t\geq\tau_R\}}e^{m^*T_{\tau_R}}\right] \leq 2\mathbb{E}^0_{P_T}\left[e^{m^*T_t}\right],$$

where we used $T_{\tau_R} \leq T_t$ for $\tau_R \leq t$. Thus we have

$$\left(\mathbb{E}^{0,0}\left[e^{m^*T_{t\wedge\tau_R}}\right]\right)^{1/2} \leq \sqrt{2}e^{(m-\sqrt{m^2-2m^*})t/2} \tag{4.8.23}$$

by (4.3.82). Hence by (4.8.22) and (4.8.23), we have

$$|\varphi(x,\theta_\alpha)| \leq \sqrt{2}\|\varphi\|\left(e^{-(W(x)-E)t} + e^{-\beta R/2}C_1e^{C_2t}\right)e^{(m-\sqrt{m^2-2m^*})t/2}. \tag{4.8.24}$$

Note that by inserting $R = p|x|$ with any $0 < p < 1$, $W(x) - E = W_{p|x|}(x) - E \to \infty$ as $|x| \to \infty$. Thus substituting $t = \delta|x|$ for sufficiently small $\delta > 0$ and $R = p|x|$ with some $0 < p < 1$ in (4.8.24), the theorem follows. □

Next we consider the case of potentials decaying to zero as $|x| \to \infty$.

Theorem 4.265 (Decaying potentials). *Suppose that* $\lim_{|x|\to\infty} V(x) = 0$ *and let Assumption* 4.261 *hold. Also, assume that*

$$m - \sqrt{m^2 - 2m^*} < -2E. \tag{4.8.25}$$

Then there exist $a, b > 0$ *such that*

$$|\varphi(x,\theta_\alpha)| \leq be^{-a|x|}. \tag{4.8.26}$$

Proof. Define $\tau_R = \tau_R(x) = \inf\{t \geq 0 \mid |B_{T_t} + x| \leq R\}$. τ_R is a stopping time, which can be seen in the same way as in Lemma 4.263. Thus we have for all $t \geq 0$

$$\begin{aligned}|\varphi(x,\theta_a)| &\leq \|\varphi\| \left(\mathbb{E}^{0,0}\left[e^{2E(t\wedge\tau_R)}e^{-2\int_0^{t\wedge\tau_R} V(B_{T_r}+x)dr}\right]\right)^{1/2} \left(\mathbb{E}^{0,0}\left[e^{m^* T_{t\wedge\tau_R}}\right]\right)^{1/2} \\ &= \|\varphi\| \left(\mathbb{E}^{x,0}\left[e^{2E(t\wedge\tau_R(0))}e^{-2\int_0^{t\wedge\tau_R(0)} V(B_{T_r})dr}\right]\right)^{1/2} \left(\mathbb{E}^{x,0}\left[e^{m^* T_{t\wedge\tau_R(0)}}\right]\right)^{1/2}.\end{aligned}$$

We rewrite $\tau_R(0)$ to τ_R. Let $\varepsilon > 0$ be arbitrary. For a sufficiently large R, it follows that $\sup_{|x|>R}|V(x)| < \varepsilon$, and we see that $|\int_0^{t\wedge\tau_R} V(B_{T_r}+x)dr| \leq (t\wedge\tau_R)\varepsilon$. This gives

$$|\varphi(x,\theta_a)| \leq \left(\mathbb{E}^{x,0}[e^{2(E+\varepsilon)(t\wedge\tau_R)}]\right)^{1/2} \left(\mathbb{E}^{x,0}\left[e^{m^* T_{t\wedge\tau_R}}\right]\right)^{1/2} \|\varphi\|.$$

Thus in a similar way as in the proof of Theorem 4.227, we can see that

$$\begin{aligned}\mathbb{E}^{x,0}[e^{2(E+\varepsilon)(t\wedge\tau_R)}] &= \mathbb{E}^{x,0}[1_{\{t\leq\tau_R\}}e^{2(E+\varepsilon)t}] + \mathbb{E}^{x,0}[1_{\{t>\tau_R\}}e^{2(E+\varepsilon)\tau_R}] \\ &\leq e^{2(E+\varepsilon)t} + C_1 e^{-m_\varepsilon|x|}, \quad t \geq 0,\end{aligned}$$

where

$$m_\varepsilon = \begin{cases} m & \text{if } 2|E| > m, \\ 2\sqrt{m|E| - |E|^2} & \text{if } 2|E| \leq m. \end{cases}$$

Also, note that

$$\mathbb{E}^{x,0}[e^{m^* T_{t\wedge\tau_R}}] \leq 2e^{(m-\sqrt{m^2-2m^*})t}.$$

Therefore,

$$|\varphi(x,\theta_a)| \leq (e^{(E+\varepsilon)t} + C_1 e^{-m_\varepsilon|x|/2})\sqrt{2}e^{(m-\sqrt{m^2-2m^*})t/2}.$$

Upon inserting $t = \delta|x|$ with a sufficiently small δ, the theorem follows from (4.8.25). □

4.9 Feynman–Kac formula for nonlocal Schrödinger operators

4.9.1 Nonlocal Schrödinger operators

Schrödinger operators with other kinetic terms than $(-\Delta)^{1/2}$, which were discussed in Section 4.6, are interesting for further applications in mathematical physics and the theory of stochastic processes. In this section we consider more general cases which we call *nonlocal Schrödinger operators*. Recall that an operator A is called *local* whenever $\operatorname{supp} Af \subset \operatorname{supp} f$ for all $f \in D(A)$, and nonlocal otherwise. For definiteness, we will consider only a specific class of nonlocal Schrödinger operators. Recall from (3.6.16) that $\mathcal{B}_0$ is the set of Bernstein functions Ψ such that $\lim_{u\downarrow 0}\Psi(u) = 0$.

Definition 4.266 (Nonlocal Schrödinger operator). Let $\Psi \in \mathcal{B}_0$ and $V \in L^\infty(\mathbb{R}^d)$. We call the operator

$$H^\Psi = \Psi\left(-\frac{1}{2}\Delta\right) + V \tag{4.9.1}$$

a nonlocal Schrödinger operator with Bernstein function Ψ *of the Laplacian.*

Note that $\Psi\left(-\frac{1}{2}\Delta\right)$ is defined through the spectral projection of the Laplacian. Furthermore, since V is bounded, H^Ψ is self-adjoint on $D(\Psi(-\frac{1}{2}\Delta))$.

Example 4.267. We give two families of examples of Bernstein functions of the Laplacian of fundamental interest. Consider the Bernstein function $\Psi_{0,\alpha}(u) = (2u)^{\alpha/2}$ and the operator $\Psi_{0,\alpha}(-\frac{1}{2}\Delta) = (-\Delta)^{\alpha/2}$. This is the *fractional Laplacian* of order $\frac{\alpha}{2} \in (0,1)$, for $f \in C_0^\infty(\mathbb{R}^d)$ defined by

$$-L_{0,\alpha}f(x) = ((-\Delta)^{\alpha/2}f)(x) = \lim_{\varepsilon\downarrow 0} \int\limits_{|y-x|>\varepsilon} (f(y) - f(x))\nu_{0,\alpha}(y-x)dy, \tag{4.9.2}$$

$$\nu_{0,\alpha}(x) = \frac{2^\alpha\Gamma(\frac{d+\alpha}{2})}{\pi^{d/2}|\Gamma(-\frac{\alpha}{2})|}\frac{1}{|x|^{d+\alpha}}, \quad x \in \mathbb{R}^d \setminus \{0\}. \tag{4.9.3}$$

We note that this definition is equivalent on various function spaces to the definition in terms of Fourier transform given in Example 4.22. The operator $L_{0,\alpha}$ is the infinitesimal generator of a rotationally symmetric α-stable process, whose Lévy measure is $\nu_{0,\alpha}(dx) = \nu_{0,\alpha}(x)dx$. Also, consider the Bernstein function $\Psi_{m,\alpha}(u) = (2u+m^{2/\alpha})^{\alpha/2}$ and the operator $\Psi_{m,\alpha}(-\frac{1}{2}\Delta) = (-\Delta + m^{2/\alpha})^{\alpha/2} - m$. This is the *fractional relativistic Laplacian* of order $\frac{\alpha}{2} \in (0,1)$ and rest mass $m \geq 0$, for $f \in C_0^\infty(\mathbb{R}^d)$ defined by

$$-L_{m,\alpha}f(x) = \left((-\Delta + m^{2/\alpha})^{\alpha/2} - m\right)f(x) = \lim_{\varepsilon\downarrow 0} \int\limits_{|y-x|>\varepsilon} (f(y) - f(x))\nu_{m,\alpha}(y-x)dy, \tag{4.9.4}$$

$$\nu_{m,\alpha}(x) = \frac{K_{(d+\alpha)/2}(m^{1/\alpha}|x|)}{|x|^{(d+\alpha)/2}}, \quad x \in \mathbb{R}^d \setminus \{0\}. \tag{4.9.5}$$

The operator $L_{m,\alpha}$ is the infinitesimal generator of a rotationally symmetric fractional relativistic α-stable process whose Lévy measure is $\nu_{m,\alpha}(dx) = \nu_{m,\alpha}(x)dx$. Here

$$K_\rho(z) = \frac{1}{2}\left(\frac{z}{2}\right)^\rho \int\limits_0^\infty t^{-\rho-1}e^{-t-\frac{z^2}{4t}}dt, \quad z > 0,\ \rho > -\frac{1}{2},$$

is the modified Bessel function of the second kind, already used above. Since

$$K_\rho(z) \sim \sqrt{\frac{\pi}{2z}}\,e^{-z} \quad \text{as } z \to \infty \quad \text{and} \quad K_\rho(z) \sim \frac{\Gamma(\rho)}{2}\left(\frac{2}{z}\right)^\rho \quad \text{as } z \downarrow 0,$$

it follows that $v_{m,\alpha}$ and $v_{0,\alpha}$ are comparable for small enough $|x|$, whereas $v_{m,\alpha}$ decays exponentially as $|x| \to \infty$, while $v_{0,\alpha}$ stays polynomial. A consequence of the second asymptotic formula above is that $v_{m,\alpha} \to v_{0,\alpha}$ as $m \downarrow 0$, thus the operator $L_{0,\alpha}$ is the massless limit of $L_{m,\alpha}$ and they coincide when $m = 0$. Furthermore, it is possible to show that on a Zygmund–Hölder-type space

$$L_{m,\alpha}f(x) = L_{0,\alpha}f(x) + \frac{1}{2}\int_{\mathbb{R}^d} (f(x+h) - 2f(x) + f(x-h))\sigma_{m,\alpha}(|h|)dh,$$

where

$$\sigma_{m,\alpha}(r) = \frac{\alpha 2^{1-\frac{d-\alpha}{2}}}{\Gamma\left(1-\frac{\alpha}{2}\right)\pi^{d/2}}\left(\frac{2^{\frac{d+\alpha}{2}-1}\Gamma\left(\frac{d+\alpha}{2}\right)}{r^{d+\alpha}} - \frac{m^{\frac{d+\alpha}{2\alpha}}K_{\frac{d+\alpha}{2}}\left(m^{1/\alpha}r\right)}{r^{\frac{d+\alpha}{2}}}\right)$$

$$= \frac{\alpha 2^{1-\frac{d-\alpha}{2}}}{\Gamma\left(1-\frac{\alpha}{2}\right)\pi^{d/2}}\frac{1}{r^{d+\alpha}}\int_0^{m^{1/\alpha}r} w^{\frac{d+\alpha}{2}}K_{\frac{d+\alpha}{2}-1}(w)dw$$

is a finite measure such that $\int_{\mathbb{R}^d}\sigma_{m,\alpha}(|x|)dx = m$. Therefore $L_{m,\alpha} - L_{0,\alpha}$ is a bounded operator for all $m > 0$.

Let $(T_t^\Psi)_{t\geq 0}$ be the subordinator defined on a probability space $(\mathcal{T}, \mathcal{B}_{\mathcal{T}}, P)$ and uniquely associated with Ψ, see Proposition 3.99. Recall subordinate Brownian motion

$$(X_t^\Psi)_{t\geq 0} = (B_{T_t^\Psi})_{t\geq 0}, \tag{4.9.6}$$

as given in Definition 3.91. We have the following probabilistic representation of nonlocal Schrödinger operators with Bernstein functions of the Laplacian.

Theorem 4.268 (Feynman–Kac formula for nonlocal Schrödinger operator). *Suppose that $V \in L^\infty(\mathbb{R}^d)$. Then we have for $f, g \in L^2(\mathbb{R}^d)$,*

$$(f, e^{-tH^\Psi}g) = \int_{\mathbb{R}^d} \mathbb{E}^{x,0}_{\mathcal{W}\times P}\left[\overline{f(X_0^\Psi)}g(X_t^\Psi)e^{-\int_0^t V(X_s^\Psi)ds}\right]dx. \tag{4.9.7}$$

Proof. Replacing the subordinator $(T_t)_{t\geq 0}$ in Theorem 4.209 with $(T_t^\Psi)_{t\geq 0}$, the proof is similar. □

4.9.2 Vector potentials

In this section we consider nonlocal Schrödinger operators with vector potential a and their Feynman–Kac formulae. We assume throughout that $a \in (L^2_{\mathrm{loc}}(\mathbb{R}^d))^d$. By Definition 4.189, we denote the self-adjoint operator h in (4.4.15) by $h(a)$.

Definition 4.269 (Nonlocal Schrödinger operator with vector potential). Let $\Psi \in \mathcal{B}_0$ and assume $a \in (L^2_{\mathrm{loc}}(\mathbb{R}^d))^d$. Whenever V is a real-valued bounded potential, we define the *nonlocal Schrödinger operator with vector potential* a by

$$H^\Psi(a) = \Psi(h(a)) + V. \tag{4.9.8}$$

Lemma 4.270. *Take* $\Psi \in \mathcal{B}_0$.
(1) *If* $a \in (L^4_{\mathrm{loc}}(\mathbb{R}^d))^d$ *and* $\nabla \cdot a \in L^2_{\mathrm{loc}}(\mathbb{R}^d)$, *then* $C_0^\infty(\mathbb{R}^d)$ *is a core of* $\Psi(h(a))$.
(2) *If* $a \in (L^2_{\mathrm{loc}}(\mathbb{R}^d))^d$, *then* $C_0^\infty(\mathbb{R}^d)$ *is a form core of* $\Psi(h(a))$.

Proof. (1) There exist nonnegative constants c_1 and c_2 such that $\Psi(u) \le c_1 u + c_2$ for all $u \ge 0$. This gives the bound

$$\|\Psi(h(a))f\| \le c_1\|h(a)f\| + c_2\|f\| \tag{4.9.9}$$

for all $f \in D(h(a))$. Hence it can be proven that $C_0^\infty(\mathbb{R}^d)$ is contained in $D(\Psi(h(a)))$ and $(\Psi(h(a)) + 1)C_0^\infty(\mathbb{R}^d)$ is dense. Thus (1) follows.

(2) Note that $\|\Psi(h(a))^{1/2}f\|^2 \le c_1\|h(a)^{1/2}f\|^2 + c_2\|f\|^2$ for $f \in Q(h(a)) = D(h(a)^{1/2})$, and $C_0^\infty(\mathbb{R}^d)$ is contained in $Q(\Psi(h(a))) = D(\Psi(h(a))^{1/2})$. Due to the fact that $\Psi(h(a))^{1/2} + 1$ has also a bounded inverse, it is seen by the same argument as above that $C_0^\infty(\mathbb{R}^d)$ is a core of $\Psi(h(a))^{1/2}$ or a form core of $\Psi(h(a))$. □

Let $a \in (L^2_{\mathrm{loc}}(\mathbb{R}^d))^d$ and $\nabla \cdot a \in L^1_{\mathrm{loc}}(\mathbb{R}^d)$. We have

$$(f, e^{-th(a)}g) = \int_{\mathbb{R}^d} \mathbb{E}^x \left[\overline{f(B_0)}g(B_t)e^{-i\int_0^t a(B_s)\circ dB_s}\right] dx. \tag{4.9.10}$$

We turn to constructing a functional integral representation for nonlocal Schrödinger operators defined by (4.9.8).

Theorem 4.271 (Feynman–Kac formula for nonlocal Schrödinger operator with singular vector potential). *Suppose that* $\Psi \in \mathcal{B}_0$ *and* $V \in L^\infty(\mathbb{R}^d)$. *Let* $a \in (L^2_{\mathrm{loc}}(\mathbb{R}^d))^d$ *and* $\nabla \cdot a \in L^1_{\mathrm{loc}}(\mathbb{R}^d)$. *Then we have*

$$(f, e^{-tH^\Psi(a)}g) = \int_{\mathbb{R}^d} \mathbb{E}^{x,0}_{\mathcal{W}\times P} \left[\overline{f(B_0)}g(B_{T_t^\Psi})e^{-i\int_0^{T_t^\Psi} a(B_s)\circ dB_s} e^{-\int_0^t V(B_{T_s^\Psi})ds}\right] dx. \tag{4.9.11}$$

Proof. The proof is similar to that of Theorem 4.209, applying T_t^Ψ instead of T_t. □

Corollary 4.272 (Diamagnetic inequality). *Let* $\Psi \in \mathcal{B}_0$ *and* $V \in L^\infty(\mathbb{R}^d)$, *and suppose that* $a \in (L^2_{\mathrm{loc}}(\mathbb{R}^d))^d$ *and* $\nabla \cdot a \in L^1_{\mathrm{loc}}(\mathbb{R}^d)$. *Then*

$$|(f, e^{-tH^\Psi(a)}g)| \le (|f|, e^{-tH^\Psi(0)}|g|) \tag{4.9.12}$$

and the energy comparison inequality $\inf \mathrm{Spec}(H^\Psi(0)) \le \inf \mathrm{Spec}(H^\Psi(a))$ *holds.*

Proof. By Theorem 4.271, we have

$$|(f, e^{-tH^{\Psi}(a)}g)| \le \int_{\mathbb{R}^d} \mathbb{E}^{x,0}_{\mathcal{W}\times P}\left[|f(B_0)||g(B_{T_t^\Psi})|e^{-\int_0^t V(B_{T_s^\Psi})ds}\right] dx.$$

The right-hand side above coincides with that of (4.9.12), and the second statement follows directly from (4.9.12). □

By making use of the functional integral representation, we can now also consider more singular external potentials.

Lemma 4.273. *Let $a \in (L^2_{\mathrm{loc}}(\mathbb{R}^d))^d$ and $\nabla \cdot a \in L^1_{\mathrm{loc}}(\mathbb{R}^d)$.*

(1) *Suppose that V is relatively form-bounded with respect to $\Psi(-\frac{1}{2}\Delta)$ with relative bound b. Then V is also relatively form-bounded with respect to $\Psi(h(a))$ with a relative bound not larger than b.*
(2) *Suppose that V is relatively bounded with respect to $\Psi(-\frac{1}{2}\Delta)$ with relative bound b. Then V is also relatively bounded with respect to $\Psi(h(a))$ with a relative bound not larger than b.*

Proof. By virtue of Corollary 4.272, we have $|(f, e^{-t\Psi(h(a))}g)| \le (|f|, e^{-t\Psi(-\frac{1}{2}\Delta)}|g|)$ and the remainder of the proof is similar to that of Theorem 4.193. □

Theorem 4.274 (Self-adjointness for $H^{\Psi}(a)$).

(1) *Suppose that $a \in (L^2_{\mathrm{loc}}(\mathbb{R}^d))^d$ and $\nabla\cdot a \in L^1_{\mathrm{loc}}(\mathbb{R}^d)$, and let V be relatively bounded with respect to $\Psi(-\frac{1}{2}\Delta)$ with relative bound strictly smaller than one. Then $H^{\Psi}(a)$ is self-adjoint on $D(\Psi(h(a)))$ and bounded from below. Moreover, it is essentially self-adjoint on any core of $\Psi(h(a))$.*
(2) *Suppose, furthermore, that $a \in (L^4_{\mathrm{loc}}(\mathbb{R}^d))^d$ and $\nabla \cdot a \in L^2_{\mathrm{loc}}(\mathbb{R}^d)$. Then $C_0^\infty(\mathbb{R}^d)$ is an operator core of $H^{\Psi}(a)$.*

Proof. By (2) of Lemma 4.273, V is also relatively bounded with respect to $\Psi(h(a))$ with a relative bound strictly smaller than one. Then (1) follows from the Kato–Rellich theorem. Statement (2) follows from Lemma 4.270. □

Lemma 4.273 also allows $\Psi(h(a)) + V$ be defined in form sense. Let $V = V_+ - V_-$. Lemma 4.273 implies that whenever V_- is form-bounded with respect to $\Psi(-\frac{1}{2}\Delta)$ with a relative bound strictly smaller than one, it is also form-bounded with respect to $\Psi(h(a))$ with a relative bound strictly smaller than one. Moreover, assume that $V_+ \in L^1_{\mathrm{loc}}(\mathbb{R}^d)$. We see that for $a \in (L^2_{\mathrm{loc}}(\mathbb{R}^d))^d$, $Q(\Psi(h(a))) \cap Q(V_+) \supset C_0^\infty(\mathbb{R}^d)$ by Lemma 4.270. In particular, $Q(\Psi(h(a))) \cap Q(V_+)$ is dense. Define the quadratic form

$$\mathfrak{q}(f,f) = (\Psi(h(a))^{1/2}f, \Psi(h(a))^{1/2}f) + (V_+^{1/2}f, V_+^{1/2}f) - (V_-^{1/2}f, V_-^{1/2}f) \tag{4.9.13}$$

on $Q(\Psi(h(a))) \cap Q(V_+)$.

Definition 4.275 (Nonlocal Schrödinger operator with singular external potential and singular vector potential). Let $a \in (L^2_{\mathrm{loc}}(\mathbb{R}^d))^d$, $\nabla \cdot a \in L^1_{\mathrm{loc}}(\mathbb{R}^d)$, and $V = V_+ - V_-$ be such that $V_+ \in L^1_{\mathrm{loc}}(\mathbb{R}^d)$ and V_- is form-bounded with respect to $\Psi(-\frac{1}{2}\Delta)$ with a relative bound strictly smaller than 1. We denote the self-adjoint operator associated with the quadratic form q in (4.9.13) by $\Psi(h(a)) \dotplus V_+ \dot{-} V_-$.

Since we need the conditions $a \in (L^2_{\mathrm{loc}}(\mathbb{R}^d))^d$ and $\nabla\cdot a \in L^1_{\mathrm{loc}}(\mathbb{R}^d)$ to show the relative form-boundedness of V_- with respect to $\Psi(h(a))$, $a \in (L^2_{\mathrm{loc}}(\mathbb{R}^d))^d$, and $\nabla \cdot a \in L^1_{\mathrm{loc}}(\mathbb{R}^d)$, they are assumed in Definition 4.275. Now we extend Theorem 4.271 to potentials expressed in terms of form sums.

Theorem 4.276 (Feynman–Kac formula for nonlocal Schrödinger operator with singular external potential and singular vector potential). *Suppose that $a \in (L^2_{\mathrm{loc}}(\mathbb{R}^d))^d$ and $\nabla \cdot a \in L^1_{\mathrm{loc}}(\mathbb{R}^d)$. Let $V = V_+ - V_-$ be such that $V_+ \in L^1_{\mathrm{loc}}(\mathbb{R}^d)$ and V_- is infinitesimally small with respect to $\Psi(-\frac{1}{2}\Delta)$ in the form sense. Then the functional integral representation given by Theorem 4.271 also holds for $\Psi(h(a)) \dotplus V_+ \dot{-} V_-$.*

Proof. The proof is similar to that of Theorem 4.195. □

4.9.3 Ψ-Kato-class potentials

Next we consider an appropriate extension of Kato-class to subordinate Brownian motion. Since our input is an operator and not directly a process, we need the following conditions.

Assumption 4.277. Let $\Psi \in \mathcal{B}_0$ be such that

$$\int_{\mathbb{R}^d} e^{-t\Psi(|\xi|^2/2)} d\xi < \infty, \quad t > 0. \tag{4.9.14}$$

Let $\Psi \in \mathcal{B}_0$ and $(b, \lambda) \in \mathbb{R}_+ \times \mathcal{L}$ be its corresponding nonnegative drift coefficient and Lévy measure, i. e., $\Psi(u) = bu + \int_0^\infty (1 - e^{-uy})\, \lambda(dy)$. It is clear that if $b > 0$, then (4.9.14) is satisfied. If $b = 0$ but $\int_0^1 \lambda(dy) < \infty$, (4.9.14) is not satisfied, since $\sup_{u \ge 0} \Psi(u) < \infty$. Thus Ψ obeying (4.9.14) at least satisfies $\int_0^1 \lambda(dy) = \infty$ when $b = 0$. In this case we have

$$\Psi\left(\frac{|u|^2}{2}\right) \ge \int_0^1 (1 - e^{-|u|^2 y/2})\lambda(dy) \ge (1 - e^{-1}) \int_0^1 \left(\frac{|u|^2 y}{2} \wedge 1\right) \lambda(dy) \ge (1 - e^{-1}) \int_{2/|u|^2}^1 \lambda(dy).$$

Thus for $b = 0$ and $\int_0^1 \lambda(dy) = \infty$, assuming that there exists ρ such that $\int_{2/|u|^2}^1 \lambda(dy) \ge \rho(u)$ and $\int_{\mathbb{R}^d} e^{-t\rho(\xi)} d\xi < \infty$, Assumption 4.277 holds. Under this assumption, we define

$$p_t^\Psi(x) = \frac{1}{(2\pi)^d} \int_{\mathbb{R}^d} e^{-ix\cdot\xi} e^{-t\Psi(|\xi|^2/2)} d\xi \tag{4.9.15}$$

and

$$G_\lambda^\Psi(x) = \int_0^\infty e^{-\lambda t} p_t^\Psi(x) dt. \tag{4.9.16}$$

The function $p_t^\Psi(x)$ denotes the *probability transition kernel* of $(X_t^\Psi)_{t\geq 0}$ and the function G_λ^Ψ is the integral kernel of the resolvent $(\Psi(-\frac{1}{2}\Delta) + \lambda)^{-1}$ with $\lambda > 0$, i. e.,

$$\left(f, \left(\Psi(-\frac{1}{2}\Delta) + \lambda\right)^{-1} g\right) = \int_{\mathbb{R}^d \times \mathbb{R}^d} \overline{f(x)} g(y) G_\lambda^\Psi(x - y) dx dy.$$

Clearly, $p_t^\Psi(x)$ and $G_\lambda^\Psi(x)$ are spherically symmetric. For $f \in C_0^\infty(\mathbb{R}^d)$, it follows that

$$\mathbb{E}_{\mathcal{W}\times P}^{0,0}[f(X_t)] = \int_{\mathbb{R}^d} f(x) p_t^\Psi(x) dx. \tag{4.9.17}$$

Hence for nonnegative $f \in C_0^\infty(\mathbb{R}^d)$, the right-hand side of (4.9.17) is nonnegative and thus $p_t^\Psi(x) \geq 0$ for almost every $x \in \mathbb{R}^d$. By a limiting argument with $f \to 1$, we can also see that $p_t^\Psi \in L^1(\mathbb{R}^d)$ and $\|p_t^\Psi\|_{L^1(\mathbb{R}^d)} = 1$ by (4.9.17). We, moreover, compute G_λ^Ψ as

$$G_\lambda^\Psi(x) = \frac{1}{(2\pi)^d} \frac{1}{|x|^{(d-1)/2}} \int_0^\infty \frac{r^{(d-1)/2}}{\lambda + \Psi(r^2/2)} \sqrt{r|x|} J_{(d-2)/2}(r|x|) dr$$

with the Bessel function of the first kind

$$J_\nu(s) = \frac{(s/2)^\nu}{\sqrt{\pi}\Gamma(\nu + \frac{1}{2})} \int_0^\pi e^{is\cos\theta} (\sin\theta)^{2\nu} d\theta = \sum_{n=0}^\infty \frac{(-1)^n}{n!\Gamma(n + \nu + 1)} \left(\frac{s}{2}\right)^{2n+\nu}.$$

Note that $\sup_{u\geq 0} \sqrt{u} J_\nu(u) < \infty$. Let

$$\|f\|_{l^1(L^\infty)} = \sum_{z\in\mathbb{Z}^d} \sup_{x\in C_z} |f(x)|,$$

where C_z denotes the unit cube centered at $z \in \mathbb{Z}^d$.

We further introduce a second assumption on the probability transition kernel p_t^Ψ.

Assumption 4.278. The probability transition kernel p_t^Ψ satisfies, for every $\delta > 0$,

$$\sup_{t>0} \|1_{\{|x|>\delta\}} p_t^\Psi\|_{l^1(L^\infty)} < \infty. \tag{4.9.18}$$

Let f be a real-valued function on $\mathbb{R}^d$. When $r \mapsto f(rx)$ is nonincreasing on $[0, \infty)$ for all $x \in \mathbb{R}^d$, we say that f is *radially nonincreasing*. In $d = 1$, for a radially nonincreasing L^1

function f, it can be seen by the definition of $l^1(L^\infty)$ that there exists a constant $C_\delta = C_\delta(f)$ such that for each $\delta > 0$,

$$\|1_{\{|x|>\delta\}}f\|_{l^1(L^\infty)} \leq C_\delta \|f\|_{L^1}. \tag{4.9.19}$$

In the general case $d \geq 2$, it can be also seen that (4.9.19) holds for all radially nonincreasing f. In particular, Assumption 4.278 is satisfied whenever p_t^Ψ is radially nonincreasing, since $\|p_t^\Psi\|_{L^1} = 1$.

Example 4.279 ($\alpha/2$-stable subordinator). For $\Psi(u) = u^{\alpha/2}$, $0 < \alpha < 2$, it is clear that Assumption 4.277 is satisfied. It is also known that the distribution density of $B_{T_t^\Psi}$ (which in this case is a symmetric α-stable process) is radially nonincreasing. This is obtained by a unimodality argument of spherically symmetric distribution functions.

Example 4.280 (Relativistic Cauchy subordinator). Let $\Psi(u) = \sqrt{2u + m^2} - m$, $m \geq 0$. It is clear that Assumption 4.277 is satisfied. The distribution function p_t^Ψ of $B_{T_t^\Psi}$ is expressed as (4.6.27). Then p_t^Ψ is indeed radially nonincreasing.

We introduce three classes of potentials, given by

$$\mathcal{V}_1 = \left\{ V : \mathbb{R}^d \to \mathbb{R} \,\middle|\, \limsup_{t\downarrow 0} \sup_{x\in\mathbb{R}^d} \int_0^t \mathbb{E}^{x,0}_{\mathcal{W}\times\nu}[V(X_s^\Psi)]ds = 0 \right\}, \tag{4.9.20}$$

$$\mathcal{V}_2 = \left\{ V : \mathbb{R}^d \to \mathbb{R} \,\middle|\, \lim_{\lambda\to\infty} \sup_{x\in\mathbb{R}^d} \left| \left((\Psi(-\tfrac{1}{2}\Delta) + \lambda)^{-1} V \right)(x) \right| = 0 \right\}, \tag{4.9.21}$$

$$\mathcal{V}_3 = \left\{ V : \mathbb{R}^d \to \mathbb{R} \,\middle|\, \lim_{\delta\downarrow 0} \sup_{x\in\mathbb{R}^d} \int_{|x-y|<\delta} G_1^\Psi(x-y)V(y)dy = 0 \right\}. \tag{4.9.22}$$

Proposition 4.281. *Let Assumptions* 4.277–4.278 *hold and V be nonnegative. Then we have the following:*

(1) $\mathcal{V}_2 \subset \mathcal{V}_1$;

(2) *if V is uniformly locally integrable, then $\mathcal{V}_3 \subset \mathcal{V}_2$ and $\mathcal{V}_1 \subset \mathcal{V}_2$;*

(3) *if $\lim_{|x|\to 0} G_1^\Psi(x)/G_\lambda^\Psi(x) \leq c$ with a constant c independent of λ, then $\mathcal{V}_2 \subset \mathcal{V}_3$.*

Proof. In the sequence of steps below, we will assume that an arbitrary $\varepsilon > 0$ is given.

(1) Let $V \in \mathcal{V}_2$ and $\lambda > 0$. We have

$$\int_0^t \mathbb{E}^{x,0}_{\mathcal{W}\times P}[V(X_s^\Psi)]ds \leq \int_0^t e^{\lambda(t-s)}\mathbb{E}^{x,0}_{\mathcal{W}\times P}[V(X_s^\Psi)]ds \leq e^{\lambda t}\int_0^\infty e^{-\lambda s}\mathbb{E}^{x,0}_{\mathcal{W}\times P}[V(X_s^\Psi)]ds.$$

Also,

$$\sup_{x\in\mathbb{R}^d}\int_0^\infty e^{-\lambda s}\mathbb{E}^{x,0}_{\mathcal{W}\times P}[V(X_s^\Psi)]ds \le \sup_{x\in\mathbb{R}^d}\left((\Psi(-\frac{1}{2}\Delta)+\lambda)^{-1}V\right)(x) < \frac{\varepsilon}{2}$$

for sufficiently large λ. Let $e^{\lambda t} < 2$ with small enough t. Then $\int_0^t \mathbb{E}^{x,0}_{\mathcal{W}\times P}[V(X_s^\Psi)]ds \le \varepsilon$ for sufficiently small t, and $V \in \mathcal{V}_1$ follows.

(2) Let $a(V) = \sup_{x\in\mathbb{R}^d}\int_{x+C_0} V(y)dy$. We have

$$\begin{aligned}(p_s^\Psi * V)(x) &= \sum_{z\in\mathbb{Z}^d}\int_{C_z} p_s^\Psi(y)V(x-y)dy \\ &\le \sum_{z\in\mathbb{Z}^d}\left(\sup_{x\in C_z} p_s^\Psi(x)\right)\int_{C_z} V(x-y)dy \le a(V)\|p_s^\Psi\|_{l^1(L^\infty)}.\end{aligned}$$

Since $|p_s^\Psi(x)| \le \int_{\mathbb{R}^d} e^{-s\Psi(\xi^2/2)}d\xi < \infty$ by (4.9.14), we obtain $\|1_{|y|\le\delta}p_s^\Psi\|_{l^1(L^\infty)} \le C\|p_s^\Psi\|_\infty < \infty$ with $C = C(\delta)$. We furthermore have

$$\|p_s^\Psi\|_{l^1(L^\infty)} \le \|1_{|y|>\delta}p_s^\Psi\|_{l^1(L^\infty)} + \|1_{|y|\le\delta}p_s^\Psi\|_{l^1(L^\infty)} \le \|1_{|y|>\delta}p_s^\Psi\|_{l^1(L^\infty)} + C\|p_s^\Psi\|_\infty < \infty$$

by (4.9.18). In particular, we get

$$\sup_{x\in\mathbb{R}^d}\mathbb{E}^{x,0}_{\mathcal{W}\times P}[V(X_s^\Psi)] \le \sup_{x\in\mathbb{R}^d}(p_s^\Psi * V)(x) \le a(V)\|p_s^\Psi\|_{l^1(L^\infty)}. \tag{4.9.23}$$

Let $V \in \mathcal{V}_1$ and $\sup_{x\in\mathbb{R}^d}\int_0^t \mathbb{E}^{x,0}_{\mathcal{W}\times P}[V(X_s^\Psi)]ds \le \varepsilon/2$ for sufficiently small t. We see that by (4.9.23),

$$\begin{aligned}\sup_{x\in\mathbb{R}^d}\left((\Psi(-\frac{1}{2}\Delta)+\lambda)^{-1}V\right)(x) &= \sup_{x\in\mathbb{R}^d}\int_0^\infty e^{-\lambda s}\mathbb{E}^{x,0}_{\mathcal{W}\times P}[V(X_s^\Psi)]ds \\ &\le \sup_{x\in\mathbb{R}^d}\int_0^t e^{-\lambda s}\mathbb{E}^{x,0}_{\mathcal{W}\times P}[V(X_s^\Psi)]ds + \sup_{x\in\mathbb{R}^d}\int_t^\infty e^{-\lambda s}\mathbb{E}^{x,0}_{\mathcal{W}\times P}[V(X_s^\Psi)]ds \\ &\le \frac{\varepsilon}{2} + a(V)\sup_{s\ge t}\|p_s^\Psi\|_{l^1(L^\infty)}\int_t^\infty e^{-\lambda s}ds \le \varepsilon\end{aligned}$$

for sufficiently large λ. Thus $V \in \mathcal{V}_2$ follows.

Next let $V \in \mathcal{V}_3$. We split the integral as

$$\int_{\mathbb{R}^d} G_1^\Psi(x-y)V(y)dy = \int_{|x-y|\le\delta} G_1^\Psi(x-y)V(y)dy + \int_{|x-y|>\delta} G_1^\Psi(x-y)V(y)dy.$$

Since $\lambda \to G_\lambda^\Psi(z)$ is a nonincreasing function, it follows that

$$\frac{\varepsilon}{2} > \sup_{x\in\mathbb{R}^d} \int_{|x-y|<\delta} G_1^\Psi(x-y)V(y)dy \geq \sup_{x\in\mathbb{R}^d} \int_{|x-y|<\delta} G_\lambda^\Psi(x-y)V(y)dy$$

for every $\lambda > 1$ and sufficiently small $\delta > 0$. On the other hand, we have by (4.9.23) that

$$\int_{|x-y|>\delta} G_\lambda^\Psi(x-y)V(y)dy = \int_0^\infty e^{-\lambda s}ds \int_{|y|>\delta} p_s^\Psi(y)V(x-y)dy$$
$$\leq a(V)\left(\sup_{s>0} \|1_{|y|>\delta}p_s^\Psi\|_{l^1(L^\infty)}\right)\int_0^\infty e^{-\lambda s}ds < \frac{\varepsilon}{2}$$

for sufficiently large λ. Thus we obtain $\sup_{x\in\mathbb{R}^d}\int_{\mathbb{R}^d} G_\lambda^\Psi(x-y)V(y)dy \leq \varepsilon$ for large enough λ and $V \in \mathcal{V}_2$ follows.

(3) Let $V \in \mathcal{V}_2$. Choose λ such that $\sup_{x\in\mathbb{R}^d}((\Psi(-\frac{1}{2}\Delta)+\lambda)^{-1}V)(x) < \varepsilon$. Note that $|z| < \delta(\lambda)$ with sufficiently small $\delta(\lambda)$ implies that $G_1^\Psi(z) \leq cG_\lambda^\Psi(z)$. Thus we see that

$$\sup_{x\in\mathbb{R}^d} \int_{|x-y|<\delta(\lambda)} G_1^\Psi(x-y)V(y)dy \leq \sup_{x\in\mathbb{R}^d} c \int_{|x-y|<\delta(\lambda)} G_\lambda^\Psi(x-y)V(y)dy$$
$$\leq c \sup_{x\in\mathbb{R}^d}\left((\Psi(-\frac{1}{2}\Delta)+\lambda)^{-1}V\right)(x) \leq \varepsilon.$$

Hence $V \in \mathcal{V}_3$ follows. □

We give an example satisfying the condition in (3) above.

Proposition 4.282. *Suppose that $G_{\lambda_0}^\Psi(x) \to \infty$ as $|x| \to 0$ for some λ_0. Then for every $\lambda > 0$,*

$$\lim_{|x|\to 0}\frac{G_1^\Psi(x)}{G_\lambda^\Psi(x)} = 1. \tag{4.9.24}$$

Proof. Since $\sup_{s\geq t} p_s^\Psi(0) \leq \int_{\mathbb{R}^d} e^{-t\Psi(\xi^2/2)}d\xi < \infty$ for all $t > 0$, we have

$$\left|\int_t^\infty e^{-\lambda s}p_s^\Psi(x)ds\right| \leq \int_t^\infty e^{-\lambda s}p_s^\Psi(0)ds \leq \left(\sup_{s\geq t} p_s^\Psi(0)\right)\int_t^\infty e^{-\lambda s}ds < \infty.$$

Now $G_{\lambda_0}^\Psi(x) \to \infty$ implies that $\int_0^t e^{-\lambda_0 s}p_s^\Psi(x)ds \to \infty$ as $|x| \to 0$ for every $t > 0$. Hence $G_\lambda^\Psi(x) \to \infty$ as $|x| \to 0$ for all $\lambda > 0$. Moreover, from

$$\lim_{|x|\to 0}\frac{G_1^\Psi(x)}{G_\lambda^\Psi(x)}=\lim_{|x|\to 0}\frac{\int_0^t e^{-s}p_s^\Psi(x)ds+\int_t^\infty e^{-s}p_s^\Psi(x)ds}{\int_0^t e^{-\lambda s}p_s^\Psi(x)ds+\int_t^\infty e^{-\lambda s}p_s^\Psi(x)ds}$$

it follows that

$$\lim_{|x|\to 0}\frac{G_1^\Psi(x)}{G_\lambda^\Psi(x)}=\lim_{|x|\to 0}\frac{\int_0^t e^{-s}p_s^\Psi(x)ds}{\int_0^t e^{-\lambda s}p_s^\Psi(x)ds}.$$

Let $\lambda\geq 1$. We have $\lim_{|x|\to 0}G_1^\Psi(x)/G_\lambda^\Psi(x)\geq 1$. On the other hand,

$$\lim_{|x|\to 0}\frac{\int_0^t e^{-s}p_s^\Psi(x)ds}{\int_0^t e^{-\lambda s}p_s^\Psi(x)ds}\leq e^{(\lambda-1)t}.$$

Hence

$$1\leq\lim_{|x|\to 0}\frac{G_1^\Psi(x)}{G_\lambda^\Psi(x)}\leq e^{(\lambda-1)t}$$

and t is arbitrary. This means that (4.9.24) is satisfied. The case $\lambda<1$ can be discussed similarly. □

Now we are able to define Kato-class potentials with respect to the operator $\Psi(-\frac{1}{2}\Delta)$.

Definition 4.283 (Ψ-Kato-class and Ψ-Kato-decomposable potential). Let Assumptions 4.277–4.278 hold. A potential V is of *Ψ-Kato-class* if

$$\lim_{\delta\downarrow 0}\sup_{x\in\mathbb{R}^d}\int_{|x-y|<\delta}G_1^\Psi(x-y)V(y)dy=0.$$

Let $V=V_+-V_-$ in terms of its positive and negative parts. We say that a potential V is *Ψ-Kato-decomposable* whenever V_- is of Ψ-Kato-class and V_+1_K is of Ψ-Kato-class for every compact set $K\subset\mathbb{R}^d$.

We denote Kato-class by $\mathcal{K}^\Psi(\mathbb{R}^d)$. In Proposition 4.281, we give equivalent conditions with Ψ-Kato-class potential under some conditions. We summarize it next.

Proposition 4.284. *Suppose that $V\in\mathcal{K}^\Psi(\mathbb{R}^d)$ and is uniformly locally integrable. Then V satisfies*

$$\lim_{t\downarrow 0}\sup_{x\in\mathbb{R}^d}\int_0^t\mathbb{E}^{x,0}_{\mathcal{W}\times P}[V(X_s^\Psi)]ds=0,\tag{4.9.25}$$

$$\lim_{\lambda\to\infty}\sup_{x\in\mathbb{R}^d}\left((\Psi(-\frac{1}{2}\Delta)+\lambda)^{-1}V\right)(x)=0.\tag{4.9.26}$$

In addition, suppose that $\lim_{|x|\to 0} G_1^\Psi(x)/G_\lambda^\Psi(x) \le c$ *with constant c independent of* λ. *Then* (4.9.25), (4.9.26), *and* $V \in \mathcal{K}^\Psi(\mathbb{R}^d)$ *are equivalent.*

Example 4.285. Let $d = 3$. Since $J_{1/2}(x) = (2/\pi)^{1/2} x^{-1/2} \sin x$, we have

$$G_\lambda^\Psi(x) = \frac{1}{2\pi^2 |x|} \int_0^\infty \frac{r \sin r}{|x|^2 \left(\lambda + \Psi\left(\frac{r^2}{2|x|^2}\right)\right)} dr.$$

Example 4.286 (Relativistic Kato-class). Consider the relativistic Kato-class obtained as a Ψ-Kato class with $\Psi(u) = \sqrt{2u + m^2} - m$. We verify that $\lim_{|x|\to 0} \frac{G_1^\Psi(x)}{G_\lambda^\Psi(x)} = 1$. By (4.6.27), we have the representation

$$p_t^\Psi(x) = \frac{1}{(2\pi)^d} \frac{t}{\sqrt{|x|^2 + t^2}} \int_{\mathbb{R}^d} e^{mt} e^{-\sqrt{(|x|^2+t^2)(p^2+m^2)}} dp.$$

This gives

$$p_t^\Psi(x) \ge \frac{1}{(2\pi)^d} \frac{t}{\sqrt{|x|^2 + t^2}} \int_{\mathbb{R}^d} e^{-m|x|} e^{-\sqrt{|x|^2+t^2}|p|} dp \ge C \frac{t e^{-m|x|}}{(\sqrt{|x|^2 + t^2})^{d+1}}$$

with constant $C > 0$. We have

$$G_\lambda^\Psi(x) \ge C \int_0^\infty \frac{t e^{-\lambda t} e^{-m|x|}}{(\sqrt{|x|^2 + t^2})^{d+1}} dt = \frac{C e^{-m|x|}}{|x|^d} \int_0^\infty \frac{u e^{-\lambda u|x|}}{(1 + u^2)^{(d+1)/2}} du.$$

Hence it follows that $\lim_{|x|\to 0} G_\lambda^\Psi(x) = \infty$ for every $\lambda > 0$, and Proposition 4.282 yields that $\lim_{|x|\to 0} G_1^\Psi(x)/G_\lambda^\Psi(x) = 1$. This implies that $\mathcal{V}_2 \subset \mathcal{V}_3$, where $\mathcal{V}_2$ is given by (4.9.21) and $\mathcal{V}_3$ by (4.9.22).

Next we show a Feynman–Kac formula for nonlocal Schrödinger operators with Bernstein functions of the Laplacian and Ψ-Kato-decomposable potentials.

Lemma 4.287. *Let V be a uniformly locally integrable and positive* Ψ*-Kato-class potential with* $\Psi \in \mathcal{B}_0$, *and let Assumptions* 4.277–4.278 *hold. Then for every* $t \ge 0$,

$$\sup_{x\in\mathbb{R}^d} \mathbb{E}_{\mathcal{W}\times P}^{x,0} \left[e^{\int_0^t V(X_s^\Psi) ds} \right] < \infty. \tag{4.9.27}$$

Proof. Since V satisfies $\lim_{t\downarrow 0} \sup_{x\in\mathbb{R}^d} \int_0^t \mathbb{E}_{\mathcal{W}\times P}^{x,0}[V(X_s^\Psi)] ds = 0$, the lemma is proven in a similar way as Lemma 4.105 by an application of Khasminskii's lemma. □

The following expression then results in a similar way to Theorem 4.113.

Theorem 4.288 (Feynman–Kac formula for nonlocal Schrödinger operator with Ψ-Kato-decomposable potential). *Let $\Psi \in \mathcal{B}_0$ and Assumptions 4.277–4.278 hold. Let V be a uniformly locally integrable and Ψ-Kato-decomposable potential. Consider*

$$K_t^\Psi f(x) = \mathbb{E}^{x,0}_{\mathcal{W}\times P}\left[e^{-\int_0^t V(X_s^\Psi)ds} f(X_t^\Psi)\right] \tag{4.9.28}$$

for $f \in L^2(\mathbb{R}^d)$. Then $\{K_t^\Psi : t \geq 0\}$ is a symmetric C_0-semigroup on $L^2(\mathbb{R}^d)$. Moreover, there exists a unique self-adjoint operator K^Ψ bounded from below such that

$$K_t^\Psi = e^{-tK^\Psi}, \quad t \geq 0.$$

Definition 4.289 (Nonlocal Schrödinger operator with Ψ-Kato-decomposable potential). Let $\Psi \in \mathcal{B}_0$ and V be a Ψ-Kato-decomposable potential. We define the *nonlocal Schrödinger operator with Ψ-Kato-decomposable potential V* by the self-adjoint operator K^Ψ given in Theorem 4.288.

The next result says that we can define a Feynman–Kac semigroup for Ψ-Kato-decomposable potentials and singular external potentials.

Theorem 4.290 (Feynman–Kac formula for nonlocal Schrödinger operator with Ψ-Kato-decomposable potential and vector potential). *Choose $\Psi \in \mathcal{B}_0$ and let Assumptions 4.277 and 4.278 hold. Also, let V be Ψ-Kato-class and uniformly locally integrable potential, $a \in (L^2_{\mathrm{loc}}(\mathbb{R}^d))^d$, and $\nabla \cdot a \in L^1_{\mathrm{loc}}(\mathbb{R}^d)$. Consider*

$$K_t^{\Psi,a} f(x) = \mathbb{E}^{x,0}_{\mathcal{W}\times P}\left[e^{-i\int_0^{T_t^\Psi} a(B_s)\circ dB_s} e^{-\int_0^t V(B_{T_s^\Psi})ds} f(B_{T_t^\Psi})\right].$$

Then $\{K_t^{\Psi,a} : t \geq 0\}$ is a symmetric C_0-semigroup. In particular, there exists a self-adjoint operator $K^\Psi(a)$ bounded from below such that

$$K_t^{\Psi,a} = e^{-tK^\Psi(a)}, \quad t \geq 0.$$

Proof. The proof is a modification of the proofs of Theorems 4.197 and 4.185. □

Definition 4.291 (Nonlocal Schrödinger operator with Ψ-Kato-decomposable potential and vector potential). Let Assumptions 4.277–4.278 hold, and V be a Ψ-Kato-class and uniformly locally integrable potential. Suppose that $a \in (L^2_{\mathrm{loc}}(\mathbb{R}^d))^d$ and $\nabla \cdot a \in L^1_{\mathrm{loc}}(\mathbb{R}^d)$. Then the *nonlocal Schrödinger operator with Ψ-Kato-decomposable potential V and vector potential a* is defined by $K^\Psi(a)$ given in Theorem 4.290.

We can show also the continuity of $K_t^\Psi f(x)$ in x. To see this, we similarly make use of the smoothing effect of e^{-tK^Ψ}. Define

$$e^{-tK^\Psi} f(x) = \int_{\mathbb{R}^d} p_t^\Psi(x-y)f(y)dy = \mathbb{E}^{x,0}[f(B_{T_t})], \quad x \in \mathbb{R}^d,\ t \geq 0,$$

for $f \in L^p(\mathbb{R}^d)$ with $1 \leq p \leq \infty$.

Proposition 4.292 (Smoothing effect). *Let Assumptions* 4.277–4.278 *hold.*

(1) *Let* $f \in L^p(\mathbb{R}^d)$ *with* $p = 1$ *or* $p = \infty$. *Then* $e^{-tK^\Psi} f$ *is continuous.*
(2) *Let* $f \in L^p(\mathbb{R}^d)$ *with some* $1 < p < \infty$. *Suppose that* $p_t^\Psi \in L^q(\mathbb{R}^d)$ *for* $1/p + 1/q = 1$. *Then* $e^{-tK^\Psi} f$ *is continuous.*

Proof. The proof is similar to that of the relativistic case in Proposition 4.221. Note that $p_t^\Psi \in L^\infty(\mathbb{R}^d) \cap L^1(\mathbb{R}^d)$. Replacing p_t^X by p_t^Ψ, it is seen that $e^{-tK^\Psi} f(x)$ is continuous in x for $p = 1$ and $p = \infty$. Consider then case of $1 < p < \infty$. Let $g \in L^p(\mathbb{R}^d)$ and define $g_n(x) = \begin{cases} g(x), & |g(x)| \le n, \\ n, & |g(x)| > n. \end{cases}$ We see that $e^{-tK^\Psi} g_n(x)$ is continuous in x for every $n \in \mathbb{N}$. By the Hölder inequality, we have $|e^{-tK^\Psi} g_n(x) - e^{-tK^\Psi} g(x)| \le \|p_t^\Psi\|_{L^q(\mathbb{R}^d)} \|g_n - g\|_{L^p(\mathbb{R}^d)}$. The right-hand side converges to zero as $n \to \infty$ uniformly in x, hence $x \mapsto e^{-tK^\Psi} g(x)$ is continuous. □

Lemma 4.293. *Let Assumptions* 4.277–4.278 *hold. Suppose that* V *is a* Ψ*-Kato-decomposable potential and let* $f \in L^\infty(\mathbb{R}^d)$. *Then*

$$e^{-rK^\Psi} K_{t-r}^\Psi f(x) = \mathbb{E}^{x,0}\left[e^{-\int_r^t V(B_{T_s^\Psi})ds} e^{-i\int_{T_r^\Psi}^{T_t^\Psi} a(B_s)\circ dB_s} f(B_{T_t^\Psi})\right], \quad t \ge r > 0. \tag{4.9.29}$$

In particular, (4.9.29) *is continuous in* x *for every* $r > 0$.

Proof. The proof is similar to that of Lemma 4.222. □

We introduce the following condition. We say that a measurable function f on $\mathbb{R}^d$ is in $\mathcal{L}^\Psi(\mathbb{R}^d)$ if

$$\lim_{r\downarrow 0} \sup_{x\in\mathbb{R}^d} \mathbb{E}^{x,0}\left[\int_0^{T_r^\Psi} |f(B_s)|ds\right] = 0.$$

Proposition 4.294 (L^p-L^q boundedness and continuity of integral kernel). *Let Assumptions* 4.277–4.278 *hold. Suppose that* V *is* Ψ*-Kato-decomposable and* $a \in (L^2_{\mathrm{loc}}(\mathbb{R}^d))^d$ *with* $\nabla \cdot a \in L^1_{\mathrm{loc}}(\mathbb{R}^d)$. *Let* $1 \le p \le q \le \infty$. *Then the following hold:*

(1) $e^{-tK^\Psi(a)}$ *is* L^p–L^q *bounded, for every* $t \ge 0$.
(2) *If* $|a|^2, \nabla \cdot a \in \mathcal{L}(\mathbb{R}^d)$ *and* $p_t^\Psi \in L^q(\mathbb{R}^d)$ *for* $1/p + 1/q = 1$, *then* $K_t^\Psi f(x)$ *is continuous in* x *for* $f \in L^p(\mathbb{R}^d)$ *with* $1 \le p \le \infty$.
(3) *If* $|a|^2, \nabla \cdot a \in \mathcal{L}(\mathbb{R}^d)$, *then the integral kernel* $\mathbb{R}^d \times \mathbb{R}^d \ni (x,y) \mapsto K_t^\Psi(x,y)$ *is jointly continuous in* x *and* y, *and is given by*

$$K_t^\Psi(x,y) = p_t^\Psi(x-y)\mathbb{E}_{[0,t]}^{x,y}[e^{-\int_0^t V(B_{T_s^\Psi})\,ds} e^{-i\int_0^{T_t^\Psi} a(B_s)\circ dB_s}]. \tag{4.9.30}$$

Here $P_{[0,t]}^{x,y}$ *is the Lévy bridge measure for* $(B_{T_t^\Psi})_{t\ge 0}$.

(4) *If* $a = 0$, *then* $e^{-tK^\Psi(0)}$ *is positivity improving.*

Proof. Part (1) can be proven similarly to Theorem 4.107 with Brownian motion $(B_t)_{t\geq 0}$ replaced by the subordinate Brownian motion $(B_{T_t^\Psi})_{t\geq 0}$. By the Riesz–Thorin theorem, it suffices to show that $e^{-tK^\Psi(a)}$ is bounded as an operator of (1) $L^\infty(\mathbb{R}^d) \to L^\infty(\mathbb{R}^d)$, (2) $L^1(\mathbb{R}^d) \to L^1(\mathbb{R}^d)$, and (3) $L^1(\mathbb{R}^d) \to L^\infty(\mathbb{R}^d)$. By the diamagnetic inequality, we have $|e^{-tK_\mathrm{R}(a)}f(x)| \leq e^{-tK_\mathrm{R}(0)}|f|(x)$. Hence we can prove (1)–(3) for $e^{-tK^\Psi(0)}$ in a similar way as Theorem 4.107. Claims (2)–(4) also follow by a minor modification of the proof of Theorem 4.111. □

4.9.4 Fractional Kato-class potentials

In this section we consider a special class of Ψ-Kato class, which is related to stable processes and fractional Schrödinger operators. Furthermore, it includes an extension of classical potential theory.

Definition 4.295 (Fractional Kato-class potential). Let $\alpha \in (0,2)$ and $\Psi(u) = u^{\alpha/2}$. We say that V is in *fractional Kato-class* $\mathcal{K}^\alpha = \mathcal{K}^\alpha(\mathbb{R}^d)$ in the sense of Definition 4.283. We also say that $V = V_+ - V_-$ is *fractional Kato-decomposable* when $V_+ \in \mathcal{K}^\alpha$ and $V_- \in L^1_{\mathrm{loc}}(\mathbb{R}^d)$. When $1_K V \in \mathcal{K}^\alpha$ for every compact subset $K \subset \mathbb{R}^d$, we write $V \in \mathcal{K}^\alpha_{\mathrm{loc}}$.

Let $\Psi(u) = u^{\alpha/2}$. Then $(X_t^\Psi)_{t\geq 0}$ is a rotationally symmetric α-stable process generated by $-(-\Delta)^{\alpha/2}$, and we write G^α_λ for G^Ψ_λ. We compute the kernel function G^α_0 for every dimension d. Formally, it is given by

$$G^\alpha_0(x) = \frac{1}{(2\pi)^d}\int_{\mathbb{R}^d} \frac{e^{-ik\cdot x}}{|k|^\alpha} dk.$$

We want to vary α from 0 to 2, but now we consider G^α_0 also for $\alpha \in (0,d]$ with $d \geq 1$. To compute the Fourier transform $h_{\alpha,d}(x)$ of $|k|^{-\alpha}$, we use the equality

$$\int_{\mathbb{R}^d} \frac{\overline{\hat{f}(k)}}{|k|^\alpha} dk = \int_{\mathbb{R}^d} \bar{f}(x) h_{\alpha,d}(x) dx, \quad f \in \mathscr{S}(\mathbb{R}^d). \tag{4.9.31}$$

Note that the integrand at the left-hand side above has a singularity at $k = 0$ which is integrable whenever $\alpha < d$. In case when $\alpha = d$, we need to choose $f \in \mathscr{S}(\mathbb{R}^d)$ such that at least $\hat{f}(0) = 0$ is satisfied. It can be seen that the domain of the self-adjoint operator $|k|^{-\alpha}$ is

$$D(|k|^{-\alpha}) = \left\{ f \in L^2(\mathbb{R}^d) \,\middle|\, \int_{\mathbb{R}^d} \left||k|^{-\alpha}\hat{f}(k)\right|^2 dk < \infty \right\}.$$

Thus $f \in \mathscr{S}(\mathbb{R}^d)$ with $\hat{f}(0) \neq 0$ does not belong to $D(|k|^{-\alpha})$.

Let $0 < \alpha < d$. For the computation of the Fourier transform, it is useful to apply the equality

$$\int_0^\infty e^{-t|k|^2} t^\beta dt = \frac{\Gamma(1+\beta)}{|k|^{2\beta+2}}, \quad k \neq 0,$$

for $\beta > -1$. Let $f \in \mathscr{S}(\mathbb{R}^d)$. Putting $\beta = (\alpha-2)/2$, we have

$$\int_{\mathbb{R}^d} \frac{\overline{\hat{f}(k)}}{|k|^\alpha} dk = \frac{1}{\Gamma(\alpha/2)} \int_0^\infty t^{(\alpha-2)/2} \left(\int_{\mathbb{R}^d} \overline{\hat{f}(k)} e^{-t|k|^2} dk \right) dt. \tag{4.9.32}$$

The requirement $0 < \alpha < d$ yields $(\alpha-2)/2 > -1$ and thus the left-hand side of (4.9.32) is finite. Note also that

$$\int_{\mathbb{R}^d} \overline{\hat{f}(k)} e^{-t|k|^2} dk = \int_{\mathbb{R}^d} \bar{f}(x) \frac{e^{-|x|^2/4t}}{(2\pi t)^{d/2}} dx.$$

Hence

$$\int_{\mathbb{R}^d} \frac{\overline{\hat{f}(k)}}{|k|^\alpha} dk = \frac{1}{\Gamma(\alpha/2)} \int_{\mathbb{R}^d} \bar{f}(x) \left(\int_0^\infty t^{(\alpha-2)/2} \frac{e^{-|x|^2/4t}}{(2\pi t)^{d/2}} dt \right) dx = \frac{c_{d-\alpha}}{c_\alpha} \int_{\mathbb{R}^d} \bar{f}(x) \frac{1}{|x|^{d-\alpha}} dx. \tag{4.9.33}$$

Here $c_\kappa = \Gamma(\kappa/2)2^{\kappa/2}$. We conclude that

$$c_\alpha \widehat{|k|^{-\alpha}} = c_{d-\alpha}|x|^{-(d-\alpha)}$$

in the sense of (4.9.33). Since $\widehat{fg} = (2\pi)^{-d/2}\hat{f} * \hat{g}$ and $\widehat{(-\Delta)^{-\alpha/2}f} = |k|^{-\alpha}\hat{f}$, we have

$$(f, (-\Delta)^{-\alpha/2} g) = \mathcal{A}(d,\alpha) \int_{\mathbb{R}^d \times \mathbb{R}^d} \bar{f}(x) \frac{1}{|x-y|^{d-\alpha}} g(y) dx dy \tag{4.9.34}$$

for $f, g \in \mathscr{S}(\mathbb{R}^d)$, with

$$\mathcal{A}(d,\alpha) = \frac{1}{2\pi^{d/2}} \frac{c_{d-\alpha}}{c_\alpha} = \frac{\Gamma((d-\alpha)/2)}{\Gamma(\alpha/2)\pi^{d/2}2^\alpha}.$$

Equality (4.9.34) can further be extended to $f \in L^2(\mathbb{R}^d)$ and $g \in L^p(\mathbb{R}^d)$ with $p = \frac{2d}{d+2\alpha}$, and

$$(-\Delta)^{-\alpha/2} g(x) = \mathcal{A}(d,\alpha) \int_{\mathbb{R}^d} \frac{1}{|x-y|^{d-\alpha}} g(y) dy \in L^2(\mathbb{R}^d)$$

by a use of the Hardy–Littlewood–Sobolev inequality. We obtain

$$G_0^\alpha(x) = \mathcal{A}(d,\alpha) \frac{1}{|x|^{d-\alpha}}, \quad 0 < \alpha < d.$$

Let $\alpha = d$. The map $\alpha \mapsto \mathcal{A}(d,\alpha)$ has a singularity at $\alpha = d$, and for $f \in \mathscr{S}(\mathbb{R}^d)$, the integral $\int_{\mathbb{R}^d} \frac{\overline{\hat{f}(k)}}{|k|^d} dk$ is not finite in general. Define

$$\mathring{\mathscr{S}}(\mathbb{R}^d) = \left\{ f \in \mathscr{S}(\mathbb{R}^d) \,\middle|\, \int_{\mathbb{R}^d} \frac{\overline{\hat{f}(k)}}{|k|^d} dk < \infty \right\}.$$

In particular, $f \in \mathring{\mathscr{S}}(\mathbb{R}^d)$ implies that $\hat{f}(0) = 0$, i. e., $\int_{\mathbb{R}^d} f(x)dx = 0$. Let $\varepsilon > 0$ and $f \in \mathring{\mathscr{S}}(\mathbb{R}^d)$. By the results above, we have

$$\int_{\mathbb{R}^d} \overline{\hat{f}(k)} \frac{1}{|k|^{d-\varepsilon}} dk = \frac{c_\varepsilon}{c_{d-\varepsilon}} \int_{\mathbb{R}^d} \bar{f}(x) \frac{1}{|x|^\varepsilon} dx.$$

Consider the right-hand side above. As $\varepsilon \to 0$,

$$\varepsilon \frac{c_\varepsilon}{c_{d-\varepsilon}} = \frac{2^\varepsilon \Gamma(1+\varepsilon/2)}{2^{d/2-1}\Gamma(d/2-\varepsilon/2)} \to \frac{1}{2^{d/2-1}\Gamma(d/2)}.$$

Here we used the identity $\varepsilon\Gamma(\varepsilon/2) = 2\Gamma(1+\varepsilon/2)$. Hence by subtraction of this term

$$\lim_{\varepsilon \downarrow 0} \left(\frac{c_\varepsilon}{c_{d-\varepsilon}} |x|^{-\varepsilon} - \frac{1}{2^{d/2-1}\Gamma(d/2)} \frac{1}{\varepsilon} \right) = \lim_{\varepsilon \downarrow 0} \left(\varepsilon \frac{c_\varepsilon}{c_{d-\varepsilon}} |x|^{-\varepsilon} - \frac{1}{2^{d/2-1}\Gamma(d/2)} \right) \frac{1}{\varepsilon}.$$

The right-hand side above can be computed as

$$= \frac{d}{d\varepsilon} \left(\frac{2^\varepsilon \Gamma(1+\varepsilon/2)}{2^{d/2-1}\Gamma(d/2-\varepsilon/2)} |x|^{-\varepsilon} \right) \Big\lceil_{\varepsilon=0} = D(d) + \frac{1}{\Gamma(d/2)2^{d/2-1}} \log |x|^{-1},$$

where $D(d) = \frac{(\log 2 + \Gamma'(d/2)/2)\Gamma(d/2) + \Gamma'(d/2)/2}{2^{d/2-1}\Gamma(d/2)^2}$ and $\Gamma'(x) = \int_0^\infty e^{-u} u^x \log u du$ is the derivative of the Gamma function. By using this fact and $\int_{\mathbb{R}^d} f(x)dx = 0$, we see that

$$\begin{aligned} \lim_{\varepsilon \downarrow 0} \frac{c_\varepsilon}{c_{d-\varepsilon}} \int_{\mathbb{R}^d} \overline{\hat{f}(x)} |x|^{-\varepsilon} dx &= \lim_{\varepsilon \downarrow 0} \int_{\mathbb{R}^d} \bar{f}(x) \left(\frac{c_\varepsilon}{c_{d-\varepsilon}} |x|^{-\varepsilon} - \frac{1}{2^{d/2-1}\Gamma(d/2)} \frac{1}{\varepsilon} \right) dx \\ &= \int_{\mathbb{R}^d} \bar{f}(x) \left(D(d) + \frac{1}{\Gamma(d/2)2^{d/2-1}} \log \frac{1}{|x|} \right) dx = \frac{1}{\Gamma(d/2)2^{d/2-1}} \int_{\mathbb{R}^d} \bar{f}(x) \log \frac{1}{|x|} dx. \end{aligned}$$

Thus

$$\int_{\mathbb{R}^d} \overline{\hat{f}(k)} \frac{1}{|k|^d} dk = \frac{1}{\Gamma(d/2)2^{d/2-1}} \int_{\mathbb{R}^d} \bar{f}(x) \log \frac{1}{|x|} dx \tag{4.9.35}$$

for $f \in \mathring{\mathscr{S}}(\mathbb{R}^d)$, and we obtain

$$\widehat{|k|^{-d}} = -\frac{1}{\Gamma(d/2)2^{d/2-1}} \log |x|$$

in the sense of (4.9.35). This gives

$$G_0^d(x) = -\frac{1}{\Gamma(d/2)\pi^{d/2}2^{d-1}} \log |x|.$$

We summarize these results in the following proposition.

Proposition 4.296 (Riesz potential). *It follows that*

$$G_0^\alpha(x) = \begin{cases} \mathcal{A}(d,\alpha)|x|^{\alpha-d}, & 0 < \alpha < d, \\ \mathcal{A}(d,d)\log |x|^{-1}, & \alpha = d, \end{cases} \tag{4.9.36}$$

in the sense that

$$\int_{\mathbb{R}^d} \overline{\hat{f}(k)}|k|^{-\alpha}\hat{g}(k)dk = \int_{\mathbb{R}^d\times\mathbb{R}^d} \bar{f}(x)G_0^\alpha(x-y)g(y)dxdy,$$

whenever $f, g \in \mathscr{S}(\mathbb{R}^d)$ for $0 < \alpha < d$, and for $f, g \in \mathring{\mathscr{S}}(\mathbb{R}^d)$ for $\alpha = d$. Here

$$\mathcal{A}(d,\alpha) = \begin{cases} \frac{\Gamma((d-\alpha)/2)}{\Gamma(\alpha/2)\pi^{d/2}2^\alpha}, & 0 < \alpha < d, \\ \frac{1}{\Gamma(d/2)\pi^{d/2}2^{d-1}}, & \alpha = d. \end{cases}$$

The expression given by (4.9.36) is called the *Riesz potential.*

Example 4.297 (Newton potential for $d \geq 2$). The limit case $\alpha = 2$ yields the integral kernel of $(-\Delta)^{-1}$, and $G_0^2(x)$ is called the *Newton potential,* given by

$$G_0^2(x) = \begin{cases} \frac{1}{2\pi}\log |x|^{-1}, & d = 2, \\ \frac{1}{A_d(d-2)}|x|^{2-d}, & d \geq 3, \end{cases}$$

where A_d denotes the $(d-1)$-dimensional area of the unit sphere in $\mathbb{R}^d$, i. e., $A_d = 2\pi^{d/2}/\Gamma(d/2)$. For $d = 1$, we can define the Newton potential in the distributional sense by

$$G_0^2(x) = -\frac{1}{2}|x|, \quad d = 1.$$

It is direct to see that

$$-\frac{d^2}{dx^2}\left(-\frac{1}{2}|x|\right) = \delta(x) \quad \text{and} \quad -\frac{d^2}{dx^2}\left(-\frac{1}{2}|x| * f\right) = f(x)$$

for every $f \in \mathscr{S}(\mathbb{R})$ in the distributional sense.

Now we give a criterion for the fractional Kato-class.

Proposition 4.298. *Let V be a nonnegative function and $\alpha \in (0,2)$.*
(1) *If $0 < \alpha \le d$ and*

$$\limsup_{\delta \downarrow 0} \sup_{x \in \mathbb{R}^d} \int_{|x-y|<\delta} G_0^\alpha(x-y)V(y)dy = 0,$$

then $V \in \mathcal{K}^\alpha$.
(2) *If $1 = d < \alpha$ and*

$$\limsup_{\delta \downarrow 0} \sup_{x \in \mathbb{R}^d} \int_{|x-y|<\delta} V(y)dy = 0,$$

then $V \in \mathcal{K}^\alpha$.

Proof. Let $0 < \alpha \le d$. Since

$$\int_{|x-y|<\delta} G_1^\alpha(x-y)V(y)dy \le \int_{|x-y|<\delta} G_0^\alpha(x-y)V(y)dy,$$

part (1) follows. Let $1 = d < \alpha$. Then $|G_1^\alpha(x)| \le \frac{1}{2\pi}\int_{\mathbb{R}} \frac{1}{|k|^\alpha+1}dk < \infty$. Thus $G_1^\alpha(x)$ is bounded and this proves part (2). □

Some examples and counterexamples of the fractional Kato-class are as follows.

Example 4.299. Let $\alpha \in (0,2)$.
(1) Let $V_+ \in L^\infty_{\mathrm{loc}}(\mathbb{R}^d)$ and $V_- \in L^\infty(\mathbb{R}^d)$. Then $V \in \mathcal{K}^\alpha_{\mathrm{loc}}$.
(2) It follows that $\mathcal{K}^\alpha_{\mathrm{loc}} \subset L^1_{\mathrm{loc}}(\mathbb{R}^d)$.
(3) Let $\alpha > d = 1$. Then $|y|^{-\beta}$ with $\beta < 1$ belongs to $\mathcal{K}^\alpha$. This follows from

$$\limsup_{\delta \downarrow 0} \sup_{x \in \mathbb{R}} \int_{|x-y|<\delta} |y|^{-\beta} dy = 0$$

for $\beta < 1$.
(4) Let $\alpha < d$. Then $|y|^{-\beta}$ with $\beta < \alpha$ belongs to $\mathcal{K}^\alpha$. We have

$$\int_{|x-y|<\delta} \frac{1}{|x-y|^{d-\alpha}} \frac{1}{|y|^\beta} dy \le \left(\int_{|x-y|<\delta} \frac{1}{|x-y|^{p(d-\alpha)}} dy \right)^{1/p} \left(\int_{|x-y|<\delta} \frac{1}{|y|^{q\beta}} dy \right)^{1/q}$$

for $1/p + 1/q = 1$. We estimate

$$\int_{|x-y|<\delta} \frac{1}{|x-y|^{p(d-\alpha)}} dy = c \int_0^\delta \frac{1}{r^{p(d-\alpha)-d+1}} dr < \infty$$

for $p(d-\alpha) - d + 1 < 1$, and

$$\int_{|x-y|<\delta} \frac{1}{|y|^{q\beta}} dy \le \int_{|y|<\delta} \frac{1}{|y|^{q\beta}} dy = c \int_0^\delta \frac{1}{r^{q\beta-d+1}} dr < \infty$$

for $q\beta - d + 1 < 1$. For $\beta < \alpha$, there exist p and q such that $d/\alpha < q < d/\beta$, $1/p + 1/q = 1$, $p(d-\alpha) - d + 1 < 1$ and $q\beta - d + 1 < 1$. Thus we have

$$\lim_{\delta \downarrow 0} \sup_{x \in \mathbb{R}^d} \int_{|x-y|<\delta} \frac{1}{|x-y|^{d-\alpha}} \frac{1}{|y|^{\beta}} dy = 0,$$

implying the statement.

(5) Let $R_i \in \mathbb{R}^d$, $\beta_i > 0$, and $\varepsilon_i \in \{-1, 1\}$ for $1 \le i \le k$. Then the potential $\sum_{i=1}^k \varepsilon_i |x - R_i|^{-\beta_i}$ belongs to $\mathcal{K}^\alpha$ whenever each $\beta_i < \alpha$ for $\alpha < d$, and $\beta_i < 1$ for $\alpha \ge d = 1$.

(6) Let $d = 3$. In light of (3)–(5) above, the Coulomb potential $V(x) = -\frac{C}{|x|}$ belongs to Kato-class $\mathcal{K}^\alpha$ for $1 < \alpha < 2$ only. As in (4) above, we have

$$\lim_{\delta \downarrow 0} \sup_{x \in \mathbb{R}^d} \int_{|x-y|<\delta} \frac{1}{|x-y|^{3-\alpha}} \frac{1}{|y|} dy = 0$$

for $1 < \alpha < 2$.

4.9.5 Generalized spin

We may generalize the relativistic Schrödinger operator $H_{\mathrm{RS}}(\alpha)$ with spin $\frac{1}{2}$ on the space $L^2(\mathbb{R}^3 \times \mathbb{Z}_2)$ to higher-order spins by considering a similar operator on $L^2(\mathbb{R}^d \times \mathbb{Z}_p)$ for $d \ge 1$ and $p \ge 2$. Denote by $\mathbb{Z}_p$ the cyclic group of the pth roots of identity by $\mathbb{Z}_p = \{\theta_1^{(p)}, \dots, \theta_p^{(p)}\}$, where

$$\theta_\beta = \theta_\alpha^{(p)} = \exp\left(2\pi i \frac{\alpha}{p}\right), \quad \alpha \in \mathbb{N}.$$

Consider the finite-dimensional vector space $\ell^2(\mathbb{Z}_p)$. We define a spin operator with its diagonal part U and off-diagonal part U_β, $\beta = 1, \dots, p-1$, separately.

Definition 4.300 (Generalized spin operator). Let the following two functions be given:

(1) (Diagonal part) Let $U_p : \mathbb{R}^d \times \mathbb{Z}_p \to \mathbb{R}$ be such that $\max_{\theta \in \mathbb{Z}_p} |U_p(x, \theta)|$ is a multiplication operator, relatively bounded with respect to $-\frac{1}{2}\Delta$.

(2) (Off-diagonal part) Let $W_\beta : \mathbb{R}^d \times \mathbb{Z}_p \to \mathbb{C}, 1 \le \beta \le p-1$, be such that $\max_{\theta \in \mathbb{Z}_p} |W_\beta(x, \theta)|$ is a multiplication operator, relatively bounded with respect to $-\frac{1}{2}\Delta$. Moreover, let $U_\beta : \mathbb{R}^d \times \mathbb{Z}_p \to \mathbb{C}$ be defined as

$$U_\beta(x, \theta_\alpha) = \frac{1}{2}\left(W_\beta(x, \theta_{\alpha+\beta}) + \overline{W_{p-\beta}(x, \theta_\alpha)}\right), \quad \alpha = 1, \dots, p, \ \beta = 1, \dots, p-1. \quad (4.9.37)$$

We call the map $M_{\mathbb{Z}_p} : f(x,\theta_\alpha) \mapsto \sum_{\beta=1}^{p} U_\beta(x,\theta_\alpha) f(x,\theta_{\alpha+\beta})$ a *generalized spin operator* on $L^2(\mathbb{R}^d \times \mathbb{Z}_p)$.

Below we will use the notation

$$u_\beta(x) = \max_{\theta\in\mathbb{Z}_p} |U_\beta(x,\theta)|, \quad 1 \le \beta \le p. \tag{4.9.38}$$

Clearly, $u_\beta(x)$ is a multiplication operator relatively bounded with respect to $-\frac{1}{2}\Delta$, i. e., there exist $c_\beta > 0$ and $b_\beta \ge 0$ such that

$$\|u_\beta f\| \le c_\beta \| -\frac{1}{2}\Delta f\| + b_\beta \|f\|, \quad \beta = 1,\dots,p, \tag{4.9.39}$$

for all $f \in D(-\frac{1}{2}\Delta)$. These definitions of U_β cover, in particular, the $\mathbb{Z}_2$ case of the Schrödinger operator with spin $\frac{1}{2}$.

Example 4.301 (Spin 1/2). Let $d = 3$ and $p = 2$. Define $W_1(x,\theta) = -\frac{1}{2}(b_1(x) + i\theta b_2(x))$ for $\theta \in \mathbb{Z}_2$. Then $\theta_1 = -1$, $\theta_2 = 1$, and by (4.9.37) we see that

$$U_1(x,\theta) = \frac{1}{2}(W_1(x,\theta\theta_1) + \overline{W_1(x,\theta)}), \quad \theta \in \mathbb{Z}_2.$$

It is straightforward to see that $W_1(x,\theta\theta_1) = -\frac{1}{2}(b_1(x) - i\theta b_2(x)) = \overline{W_1(x,\theta)}$; hence the off-diagonal part is $U_1(x,\theta) = -\frac{1}{2}(b_1(x) - i\theta b_2(x))$, while the diagonal part is given by $U_2(x,\theta) = -\frac{1}{2}\theta b_3(x)$.

Example 4.302. Let $p \ge 2$, and $W_\beta(\theta) = W(\theta) = -\frac{1}{2}(b_1 + i\theta b_2)$ for $1 \le \beta \le p-1$. Then

$$U_\beta(\theta_\alpha) = \frac{1}{2}\left(W_\beta(\theta_{\alpha+\beta}) + \overline{W_{p-\beta}(\theta_\alpha)}\right) = -\frac{1}{2}\left(b_1 + i\frac{\theta_{\alpha+\beta} - \theta_{p-\alpha}}{2} b_2\right). \tag{4.9.40}$$

This gives one possible generalization of the case of spin $\frac{1}{2}$ in Example 4.301.

Definition 4.303 (Schrödinger operator with vector potential and generalized spin). We define the *Schrödinger operator with vector potential $a \in (L^2_{\mathrm{loc}}(\mathbb{R}^d))^d$ and generalized spin $M_{\mathbb{Z}_p}$* by $h_{\mathbb{Z}_p}(a) = 1 \otimes h(a) + M_{\mathbb{Z}_p}$. Here we made the identification $L^2(\mathbb{R}^d \times \mathbb{Z}_p) \cong \ell^2(\mathbb{Z}_p) \otimes L^2(\mathbb{R}^d)$.

Formally, $h_{\mathbb{Z}_p}(a)$ is written as

$$(h_{\mathbb{Z}_p}(a)f)(x,\theta_\alpha) = \left(\frac{1}{2}(-i\nabla - a(x))^2 + U_p(x,\theta_\alpha)\right) f(x,\theta_\alpha) + \sum_{\beta=1}^{p-1} U_\beta(x,\theta_\alpha) f(x,\theta_{\alpha+\beta}).$$

We will use the following conditions on the generalized spins.

Assumption 4.304. Let u_β be defined by (4.9.38) and c_β by (4.9.39). We consider the following conditions: (U1) $\sum_{\beta=1}^{p} c_\beta < 1$, (U2) $u_\beta \in L^\infty(\mathbb{R}^d)$ for $\beta = 1, \dots, p$.

Condition (U1) of Assumption 4.304 is a sufficient condition for self-adjointness of $h_{\mathbb{Z}_p}(a)$. Condition (U2) is a stronger assumption than (U1), and we will use it to construct a functional integral representation of $e^{-th_{\mathbb{Z}_p}(a)}$.

Theorem 4.305 (Self-adjointness). *Let $a \in (L^2_{\text{loc}}(\mathbb{R}^d))^d$, $\nabla \cdot a \in L^1_{\text{loc}}(\mathbb{R}^d)$, and take* (U1) *of Assumption 4.304. Then $h_{\mathbb{Z}_p}(a)$ is self-adjoint on $\ell^2(\mathbb{Z}_p) \otimes D(h(a))$ and bounded from below. Moreover, it is essentially self-adjoint on any core of $1 \otimes h(a)$. In particular, $\ell^2(\mathbb{Z}_p) \otimes C_0^\infty(\mathbb{R}^d)$ is an operator core of $h_{\mathbb{Z}_p}(a)$.*

Proof. It follows that $(g(x,\cdot), M_{\mathbb{Z}_p} f(x,\cdot))_{\ell^2(\mathbb{Z}_p)} = (M_{\mathbb{Z}_p} g(x,\cdot), f(x,\cdot))_{\ell^2(\mathbb{Z}_p)}$ and $M_{\mathbb{Z}_p}$ is symmetric. We have $\|M_{\mathbb{Z}_p} f\| \le \sum_{\beta=1}^{p} c_\beta \|(1 \otimes h(a)) f\| + b\|f\|$, where c_β is given by (4.9.39) and b is a constant. Thus the claim follows from the Kato–Rellich theorem. □

Definition 4.306 (Nonlocal Schrödinger operator with generalized spin). Suppose that $a \in (L^2_{\text{loc}}(\mathbb{R}^d))^d$, $\nabla \cdot a \in L^1_{\text{loc}}(\mathbb{R}^d)$, and (U1) of Assumption 4.304 holds. Let $\Psi \in \mathcal{B}_0$ and put

$$\overline{h_{\mathbb{Z}_p}(a)} = \begin{cases} h_{\mathbb{Z}_p}(a) & \text{if } \inf \operatorname{Spec}(h_{\mathbb{Z}_p}(a)) \ge 0, \\ h_{\mathbb{Z}_p}(a) - \inf \operatorname{Spec}(h_{\mathbb{Z}_p}(a)) & \text{if } \inf \operatorname{Spec}(h_{\mathbb{Z}_p}(a)) < 0. \end{cases} \tag{4.9.41}$$

We define the *nonlocal Schrödinger operator with vector potential a and spin $\mathbb{Z}_p$* by

$$H^\Psi_{\mathbb{Z}_p}(a) = \Psi\left(\overline{h_{\mathbb{Z}_p}(a)}\right) + V.$$

Suppose that $a \in (L^2_{\text{loc}}(\mathbb{R}^d))^d$, $\nabla \cdot a \in L^1_{\text{loc}}(\mathbb{R}^d)$, and (U1) of Assumption 4.304 holds. If $\Psi \in \mathcal{B}_0$, then $\ell^2(\mathbb{Z}_p) \otimes C_0^\infty(\mathbb{R}^d)$ is an operator core of $\Psi(h_{\mathbb{Z}_p}(a))$. Since $h_{\mathbb{Z}_p}(a)$ is essentially self-adjoint on $\ell^2(\mathbb{Z}_p) \otimes C_0^\infty(\mathbb{R}^d)$, this can be proven in the same way as Lemma 4.270.

Next we give a functional integral representation of $e^{-tH^\Psi_{\mathbb{Z}_p}(a)}$ by means of a joint process combining Brownian motion, a Poisson process, and a subordinator. Let $(N_t^\beta)_{t \ge 0}$, $\beta = 1, \dots, p-1$, be $(p-1)$ independent Poisson processes with unit intensity on a probability space $(\mathcal{S}, \mathcal{B}_\mathcal{S}, P_N)$, i. e., $P_N(N_t^\beta = n) = e^{-t} t^n / n!$. Define the random process $(N_t)_{t \ge 0}$ by

$$N_t = \sum_{\beta=1}^{p-1} \beta N_t^\beta. \tag{4.9.42}$$

Let $\mathscr{F}_t^N = \sigma(N_s, s \le t)$, $t \ge 0$, be the natural filtration. Since $(N_t)_{t\ge 0}$ is a Lévy process, it is Markov with respect to $(\mathscr{F}_t^N)_{t\ge 0}$. We write $\mathbb{E}_{P_N}[f(N_t+\alpha)]$ as $\mathbb{E}_{P_N}^{\alpha}[f(N_t)]$. Also, $\mathbb{E}_{P_N}^{\alpha}[1_{N_0=\alpha}] = 1$. Define $\int_v^{w+} g(N_{s-})dN_s^\beta = \sum_{v\le r\le w, N_{r+}^\beta \ne N_{r-}^\beta} g(N_{r-})$. It can be seen that

$$\mathbb{E}_{P_N}\left[\int_v^{w+} g(N_{s-})dN_s^\beta\right] = \mathbb{E}_{P_N}\left[\int_v^{w} g(N_s)ds\right]. \tag{4.9.43}$$

Lemma 4.307. *Suppose that $a \in (L^2_{\mathrm{loc}}(\mathbb{R}^d))^d$, $\nabla \cdot a \in L^1_{\mathrm{loc}}(\mathbb{R}^d)$, (U1) of Assumption 4.304 holds, and*

$$\int_0^t ds \int_{\mathbb{R}^d} \Pi_s(x-y)|\log u_\beta(y)|dy < \infty, \quad \beta = 1,\dots,p-1. \tag{4.9.44}$$

Then

$$(f, e^{-th_{\mathbb{Z}_p}(a)}g) = e^{(p-1)t}\sum_{\alpha=1}^{p}\int_{\mathbb{R}^d} \mathbb{E}^{x,\alpha}_{\mathcal{W}\times P_N}\left[\overline{f(B_0,\theta_{N_0})}g(B_t,\theta_{N_t})e^{\mathscr{S}}\right]dx, \tag{4.9.45}$$

where $\mathscr{S} = \mathscr{S}_a + \mathscr{S}_{\mathrm{spin}}$ and

$$\mathscr{S}_a = -i\int_0^t a(B_s)\circ dB_s, \quad \mathscr{S}_{\mathrm{spin}} = -\int_0^t U_p(B_s,\theta_{N_s})ds + \sum_{\beta=1}^{p-1}\int_0^{t+}\log(-U_\beta(B_s,\theta_{N_{s-}}))dN_s^\beta.$$

Here we take the principal branch of $\log z$ for $z \in \mathbb{C}$.

Proof. First, assume that both the diagonal part $U_p(x,\theta_\alpha)$ and the off-diagonal part $U_\beta(x,\theta_\alpha)$ are continuous in x and $a \in (C_0^\infty(\mathbb{R}^d))^d$. Since it follows from (4.9.44) and (4.9.43) that

$$\mathbb{E}^{x,\alpha}_{\mathcal{W}\times P_N}\left[\int_0^{t+}|\log(-U_\beta(B_s,\theta_{N_{s-}}))|dN_s^\beta\right] \le \int_0^t ds\int_{\mathbb{R}^d}\Pi_s(x-y)|\log u_\beta(y)|dy < \infty,$$

we note that $\int_0^{t+}|\log(-U_\beta(B_s,\theta_{N_{s-}}))|dN_s^\beta < \infty$ almost surely. We have for $c > 0$,

$$\left|\mathbb{E}^{x,\alpha}_{\mathcal{W}\times P_N}[e^{c\mathscr{S}_{\mathrm{spin}}}]\right| \le \exp\left(t\left(c\|u_p\|_\infty + \sum_{\beta=1}^{p-1}(\|u_\beta\|_\infty^c - 1)\right)\right). \tag{4.9.46}$$

Denote

$$Z_{[v,w]} = -i\int_v^w a(B_s)\circ dB_s - \int_v^w U_p(B_s,\theta_{N_s})ds + \sum_{\beta=1}^{p-1}\int_v^{w+}\log(-U_\beta(B_s,\theta_{N_{s-}}))dN_s^\beta$$

and let

$$(\mathrm{P}_t g)(x,\theta_a) = \mathbb{E}^{x,a}_{\mathcal{W}\times P_N}\left[e^{Z_{[0,t]}} g(B_t,\theta_{N_t})\right].$$

Let $g \in \ell^2(\mathbb{Z}_p)\otimes C_0^\infty(\mathbb{R}^d)$. By using the Schwarz inequality and setting $c = 2$ in (4.9.46), we have the estimate $\|\mathrm{P}_t g\|^2 \le \exp\left(t\left(2\|u_p\|_\infty + \sum_{\beta=1}^{p-1}(\|u_\beta\|_\infty^2 - 1)\right)\right)\|g\|^2$. Thus P_t is bounded. We show now that $\{\mathrm{P}_t : t \ge 0\}$ is a symmetric C_0-semigroup with generator $-(h_{\mathbb{Z}_p}(a) + p - 1)$, i. e., (1) $\mathrm{P}_0 = I$, (2) $\mathrm{P}_s\mathrm{P}_t = \mathrm{P}_{s+t}$, (3) $\mathrm{P}_t g$ is continuous in t, and (4) $\lim_{t\to 0}\frac{1}{t}(\mathrm{P}_t g - g) = -(h_{\mathbb{Z}_p}(a) + (p-1))g$ strongly. First, (1) is trivial. To check (2), note that $(\mathrm{P}_t\mathrm{P}_s g)(x,\theta_a) = \mathbb{E}^{x,a}_{\mathcal{W}\times P_N}\left[e^{Z_{[0,t]}}\mathbb{E}^{B_t,N_t}_{\mathcal{W}\times P_N}\left[e^{Z_{[0,s]}} g(B_s,\theta_{N_s})\right]\right]$. By the Markov property of $(B_t)_{t\ge 0}$ and that of $(N_t)_{t\ge 0}$, we see that $(\mathrm{P}_t\mathrm{P}_s g) = (\mathrm{P}_{s+t} g)(x,\theta_a)$. This proves (2). Next we obtain the generator of P_t. The product formula gives

$$\begin{aligned}
d\left(e^{Z_{[0,t]}} g\right)(B_t,\theta_{N_t}) = \int_0^t & e^{Z_{[0,s]}}\left\{\frac{1}{2}\Delta g(B_s,\theta_{N_s}) - ia(B_s)\cdot\nabla g(B_s,\theta_{N_s})\right. \\
&+ \left.\left(\frac{1}{2}(-i\nabla\cdot a)(B_s) - \frac{1}{2}a(B_s)^2 - U_p(B_s,\theta_{N_s})\right) g(B_s,\theta_{N_s})\right\} ds \\
&+ \int_0^t e^{Z_{[0,s]}}\left(\nabla g(B_s,\theta_{N_s}) - ia(B_s) g(B_s,\theta_{N_s})\right)\cdot dB_s \\
&+ \sum_{\beta=1}^{p-1}\int_0^{t+} e^{Z_{[0,s-]}}\left(g(B_s,\theta_{N_{s-}+\beta}) e^{\log(-U_\beta(B_s,\theta_{N_{s-}}))} - g(B_s,\theta_{N_{s-}})\right) dN_s^\beta .
\end{aligned}$$

Here we used that $dN_t = \sum_{\beta=1}^{p-1}\int_0^{t+}\beta dN_s^\beta$ and $d\theta_{N_t} = \sum_{\beta=1}^{p-1}(\theta_{N_t+\beta} - \theta_{N_t})$. Take expectations on both sides above. We have

$$\frac{1}{t}(f,(\mathrm{P}_t - 1)g) = \frac{1}{t}\int_0^t ds \int_{\mathbb{R}^d}\overline{f(x)}\mathbb{E}^{x,a}_{\mathcal{W}\times P_N}[G(s)]\,dx,$$

where $G(0) = -(h_{\mathbb{Z}_p}(a) + (p-1))g(x,\theta_a)$ and

$$\begin{aligned}
G(s) = {} & e^{Z_{[0,s]}}\left(\frac{1}{2}\Delta - ia(B_s)\cdot\nabla + \frac{1}{2}(-i\nabla\cdot a)(B_s) - \frac{1}{2}a(B_s)^2 - U_p(B_s,\theta_{N_s})\right) g(B_s,\theta_{N_s}) \\
&+ \sum_{\beta=1}^{p-1} e^{Z_{[0,s]}}\left(g(B_s,\theta_{N_s+\beta}) e^{\log(-U_\beta(B_s,\theta_{N_s}))} - g(B_s,\theta_{N_s})\right)
\end{aligned}$$

for $s \ne 0$. Note that $U_\beta(x,\theta)$ and $a_{P_N}(x)$ are continuous in x. Therefore $G(s)$ is continuous at $s = 0$ for each $(\omega,\varpi) \in \mathscr{X}\times S$, and $\mathbb{E}^{x,a}_{\mathcal{W}\times P_N}[G(s)]$ is continuous at $s = 0$ by the dominated convergence theorem. Thus $\lim_{t\to 0}\frac{1}{t}(f,(\mathrm{P}_t - 1)g) = (f, -(h_{\mathbb{Z}_p}(a) + (p-1))g)$ follows. Finally, the strong continuity (3) follows from (2) and (4), and hence we have

$e^{t(p-1)}\mathrm{P}_t g = e^{-th_{\mathbb{Z}_p}(a)}g$. By a simple limiting argument, this identity can be extended to $a \in (L^2_{\mathrm{loc}}(\mathbb{R}^d))^d$ satisfying $\nabla \cdot a \in L^1_{\mathrm{loc}}(\mathbb{R}^d)$, and U_β given in Definition 4.300. □

Now we can state and prove the functional integral representation of $e^{-th_{\mathbb{Z}_p}(a)}$.

Theorem 4.308 (Feynman–Kac formula for nonlocal Schrödinger operator with vector potential and generalized spin). *Let $\Psi \in \mathcal{B}_0$ and $V \in L^\infty(\mathbb{R}^d)$. Let* (U1) *of Assumption* 4.304 *hold, $a \in (L^2_{\mathrm{loc}}(\mathbb{R}^d))^d$, $\nabla \cdot a \in L^1_{\mathrm{loc}}(\mathbb{R}^d)$ and*

$$\int_{\mathbb{R}} \rho(r,t)dr \int_0^r ds \int_{\mathbb{R}^d} \Pi_s(x-y)|\log u_\beta(y)|dy < \infty, \quad \beta = 1,\dots,p-1, \tag{4.9.47}$$

where $\rho(r,t)$ is the distribution of T_t^Ψ on $\mathbb{R}$ and u_β is given in (4.9.38). *Then*

$$(f, e^{-tH^\Psi_{\mathbb{Z}_p}(a)}g) = \sum_{\alpha=1}^{p} \int_{\mathbb{R}^d} \mathbb{E}^{x,\alpha,0}_{\mathcal{W}\times P_N \times P_T}\left[e^{(p-1)T_t^\Psi}\overline{f(B_0,\theta_{N_0})}g(B_{T_t^\Psi},\theta_{N_{T_t^\Psi}})e^{\mathcal{S}^\Psi}\right]dx, \tag{4.9.48}$$

where $\mathcal{S}^\Psi = \mathcal{S}^\Psi_V + \mathcal{S}^\Psi_a + \mathcal{S}^\Psi_{\mathrm{spin}}$ with

$$\mathcal{S}^\Psi_V = -\int_0^t V(B_{T_s^\Psi})ds, \quad \mathcal{S}^\Psi_a = -i\int_0^{T_t^\Psi} a(B_s)\circ dB_s,$$

$$\mathcal{S}^\Psi_{\mathrm{spin}} = \begin{cases} -\int_0^{T_t^\Psi}(U_p(B_s,\theta_{N_s}) - \xi)\,ds + \sum_{\beta=1}^{p-1}\int_0^{T_t^\Psi+}\log(-U_\beta(B_s,\theta_{N_{s-}}))dN_s^\beta & \text{if } \xi < 0, \\ -\int_0^{T_t^\Psi}(U_p(B_s,\theta_{N_s}))\,ds + \sum_{\beta=1}^{p-1}\int_0^{T_t^\Psi+}\log(-U_\beta(B_s,\theta_{N_{s-}}))dN_s^\beta & \text{if } \xi \geq 0 \end{cases}$$

and $\xi = \inf \mathrm{Spec}(h_{\mathbb{Z}_p}(a))$.

Proof. Since from (4.9.47) it follows that

$$\mathbb{E}^{x,\alpha,0}_{\mathcal{W}\times P_N\times P_T}\left[\int_0^{T_t^\Psi+}|\log(-U_\beta(B_s,\theta_{N_{s-}}))|dN_s^\beta\right] \leq \int_{\mathbb{R}}\rho(r,t)dr\int_0^r ds\int_{\mathbb{R}^d}\Pi_s(x-y)|\log u_\beta(y)|dy$$

is finite, we note that $\int_0^{T_t^\Psi+}|\log(-U_\beta(B_s,\theta_{N_{s-}}))|dN_s^\beta < \infty$ almost surely. Using Lemma 4.307, we obtain similarly like in the proof of (4.9.7) that

$$\left(f, e^{-t\Psi(\overline{h_{\mathbb{Z}_p}(a)})}g\right) = \sum_{\alpha=1}^{p}\int_{\mathbb{R}^d}\mathbb{E}^{x,\alpha,0}_{\mathcal{W}\times P_N\times P_T}\left[e^{(p-1)T_t^\Psi}\overline{f(B_0,\theta_{N_0})}g(B_{T_t^\Psi},\theta_{N_{T_t^\Psi}})e^{\mathcal{S}^\Psi_a+\mathcal{S}^\Psi_{\mathrm{spin}}}\right]dx.$$

Note that when V is continuous, by the Trotter product formula it follows that

$$\left(f, e^{-tH^\Psi_{\mathbb{Z}_p}(a)}g\right) = \lim_{n\to\infty}\left(f, \left(e^{-(t/n)\Psi(h_{\mathbb{Z}_p}(a))}e^{-(t/n)V}\right)^n g\right) = (4.9.48)$$

Hence the theorem holds for continuous V. This can be extended to $V \in L^\infty(\mathbb{R}^d)$ in the same way as in Theorem 4.209. □

Now let $h^0_{\mathbb{Z}_p}$ be defined by $h_{\mathbb{Z}_p}(a)$ with a and U_β, $\beta = 1, \dots, p-1$, replaced by 0 and $|U_\beta|$, respectively. Write $\xi_0 = \inf \operatorname{Spec}(h^0_{\mathbb{Z}_p})$ and let

$$\overline{h^0_{\mathbb{Z}_p}} = \begin{cases} h^0_{\mathbb{Z}_p} & \text{if } \xi_0 \geq 0, \\ h^0_{\mathbb{Z}_p} - \xi_0 & \text{if } \xi_0 < 0. \end{cases}$$

An immediate corollary of Theorem 4.308 is as follows.

Corollary 4.309 (Diamagnetic inequality). *Under the assumptions of Theorem 4.308, we have $h_{\mathbb{Z}_p}(a) - \xi_0 \geq 0$. Moreover,*

(1) *if $\xi_0 \geq 0$, then* $\left|\left(f, e^{-t(\Psi(h_{\mathbb{Z}_p}(a))+V)} g\right)\right| \leq \left(|f|, e^{-t\left(\Psi(\overline{h^0_{\mathbb{Z}_p}})+V\right)} |g|\right)$;

(2) *if $\xi_0 < 0$, then* $\left|\left(f, e^{-t(\Psi(h_{\mathbb{Z}_p}(a)-\xi_0)+V)} g\right)\right| \leq \left(|f|, e^{-t\left(\Psi(\overline{h^0_{\mathbb{Z}_p}})+V\right)} |g|\right)$.

We have also the following consequences.

Theorem 4.310 (Self-adjointness). *Let* (U2) *of Assumption* 4.304 *be satisfied, and suppose that V is relatively bounded with respect to $\Psi(\overline{h^0_{\mathbb{Z}_p}})$ with a relative bound strictly less than 1. Moreover, assume* (4.9.44).

(1) *Let $a \in (L^2_{\mathrm{loc}}(\mathbb{R}^d))^d$ and $\nabla \cdot a \in L^1_{\mathrm{loc}}(\mathbb{R}^d)$. Then $H^\Psi_{\mathbb{Z}_p}(a)$ is self-adjoint on the domain $D(\Psi(\overline{h_{\mathbb{Z}_p}(a)}))$ and essentially self-adjoint on any core of $\Psi(\overline{h_{\mathbb{Z}_p}(a)})$. In particular, for $a \in (L^4_{\mathrm{loc}}(\mathbb{R}^d))^d$ and $\nabla \cdot a \in L^2_{\mathrm{loc}}(\mathbb{R}^d)$, $H^\Psi_{\mathbb{Z}_p}(a)$ is essentially self-adjoint on $C_0^\infty(\mathbb{R}^d)$.*

(2) *Let $a \in (L^4_{\mathrm{loc}}(\mathbb{R}^d))^d$ and $\nabla \cdot a \in L^2_{\mathrm{loc}}(\mathbb{R}^d)$. Then the functional integral representation of $e^{-tH^\Psi_{\mathbb{Z}_p}(a)}$ is given by* (4.9.48).

Proof. Applying Corollary 4.309, we see that if V is relatively bounded with respect to $\Psi(\overline{h^0_{\mathbb{Z}_p}})$ with a relative bound b, then V is relatively bounded with respect to $\Psi(\overline{h_{\mathbb{Z}_p}(a)})$ with a relative bound not larger than b. Hence (1) is trivial; (2) is proven in a similar way to the approximation argument in Step 4 of Theorem 4.209. □

4.9.6 Recurrence properties and existence of bound states

We revisit the problem discussed in Section 4.3.5 on the relationship between the existence of a bound state of Schrödinger operators with potential wells and sojourn times of one- and two-dimensional Brownian motion. Here we give a more general version to nonlocal Schrödinger operators. The discussion also relates to the ideas in Corollary 4.160.

Theorem 4.311. *Let $\Psi \in \mathcal{B}_0$ such that $\lim_{u\to\infty}\Psi(u) = \infty$. The following properties are equivalent:*

(1) *The subordinate Brownian motion $(X_t)_{t\geq 0} = (B_{T_t^\Psi})_{t\geq 0}$ is recurrent.*

(2) *The nonlocal Schrödinger operator $H_\nu^\Psi = \Psi(-\frac{1}{2}\Delta) - \nu 1_{\{|x|\leq a\}}$ has a bound state for all $\nu > 0$.*

Proof. Let $V(x) = -\nu 1_{\{|x|\leq a\}}(x)$. Consider first the implication (1) ⇒ (2). Let $f \in L^2(\mathbb{R}^d)$ with continuous Fourier transform $\hat{f}$ near the origin, and $E < 0$. We have

$$\left(f, |V|^{1/2}\left(\Psi(-\frac{1}{2}\Delta) + E\right)^{-1} |V|^{1/2} f\right) = \int_{\mathbb{R}^d} \frac{\nu|\hat{f}(y)|^2}{E + \Psi(|y|^2/2)} \prod_{j=1}^d \left(\frac{\sin a y_j}{y_j}\right)^2 dy_j$$

$$\geq \int_{|y|\leq\pi/2a} \frac{\nu|\hat{f}(y)|^2}{E + \Psi(|y|^2/2)} \prod_{j=1}^d \left(\frac{\sin a y_j}{y_j}\right)^2 dy_j \geq c \int_{|y|\leq\pi/2a} \frac{|\hat{f}(0)|^2}{E + \Psi(|y|^2/2)} dy$$

with a constant $c > 0$. Note that the operator $|V|^{1/2}\left(\Psi(-\frac{1}{2}\Delta) + E\right)^{-1}|V|^{1/2}$ is compact and thus the operator norm

$$\lambda(E) = \left\| |V|^{1/2}\left(\Psi(-\frac{1}{2}\Delta) + E\right)^{-1} |V|^{1/2} \right\| \tag{4.9.49}$$

is an eigenvalue. Since the right-hand side is larger than $c|\hat{f}(0)|^2 \int_{|y|\leq 1} \frac{1}{E+\Psi(|y|)} dy$ by monotonicity of Ψ, using the Chung–Fuchs criterion we have by recurrence

$$\liminf_{E\downarrow 0}\left(f, |V|^{1/2}\left(\Psi(-\frac{1}{2}\Delta) + E\right)^{-1} |V|^{1/2} f\right) = \infty,$$

which implies that $\lambda(E) \to \infty$ and in particular $\lambda(E) > 1$ for large E. Hence a bound state of H_ν^Ψ exists by the Birman–Schwinger principle.

For the converse implication (2) ⇒ (1), assume $\int_{|y|\leq 1} \frac{1}{\Psi(|y|)} dy < \infty$, which implies

$$\int_{\mathbb{R}^d} \frac{1}{\Psi(|y|^2/2)} \prod_{j=1}^d \left(\frac{\sin a y_j}{y_j}\right)^2 dy_j < \infty.$$

On the other hand, for every $E > 0$

$$\left(f, |V|^{1/2}\left(\Psi(-\frac{1}{2}\Delta) + E\right)^{-1} |V|^{1/2} f\right) \leq \int_{\mathbb{R}^d} \frac{\|\hat{f}\|_\infty^2}{\Psi(|y|^2/2)} \prod_{j=1}^d \left(\frac{\sin a y_j}{y_j}\right)^2 dy_j.$$

We can choose for all $\varepsilon > 0$ an $f > 0$, $\|f\| = 1$ such that $\|\hat{f}\|_\infty < \varepsilon$. Choosing ε so small that the right-hand side above is less than one, we obtain

$$\limsup_{E\downarrow 0}\left(f, |V|^{1/2}\left(\Psi(-\frac{1}{2}\Delta)+E\right)^{-1}|V|^{1/2}f\right) < 1.$$

By the Birman–Schwinger principle, this means that H_V^Ψ has no bound state. □

4.9.7 The number of eigenfunctions with negative eigenvalues

In this section we extend the Lieb–Thirring inequality discussed in Section 4.3.6 to nonlocal Schrödinger operators. Throughout this section we assume that

$$d \geq 3.$$

From now on, similarly to Section 4.3.6, we consider the number

$$\bar{N}_E(V) = \dim 1_{(-\infty,-E]}(H^\Psi) = \#\left\{\text{eigenvalues of } H^\Psi \leq -E\right\}, \quad E \geq 0.$$

Let $V = V_+ - V_-$ be such that $V_\pm \geq 0$ and $V_- \in L^\infty(\mathbb{R}^d)$. Since

$$H_-^\Psi = \Psi\left(-\frac{1}{2}\Delta\right) - V_- \leq H^\Psi,$$

the number of negative eigenvalues of H^Ψ is smaller than that of H_-^Ψ. Therefore instead of H^Ψ we may consider H_-^Ψ. Let $W = |V| = V_- \geq 0$ and set

$$K_\gamma^\Psi = W^{1/2}\left(\Psi\left(-\frac{1}{2}\Delta\right)+\gamma\right)^{-1}W^{1/2}, \quad \gamma \geq 0. \tag{4.9.50}$$

Using $F(x) = x(1+\lambda x)^{-1}$ and $g(x) = e^{-\lambda x}$ with $F(x) = x\int_0^\infty e^{-y}g(xy)dy$ as in Section 4.3.6, a direct computation gives

$$F(K_\gamma^\Psi) = K_\gamma^\Psi(1+\lambda K_\gamma^\Psi)^{-1} = W^{1/2}\left(\Psi\left(-\frac{1}{2}\Delta\right)+\lambda W+\gamma\right)^{-1}W^{1/2}. \tag{4.9.51}$$

Similar to the Schrödinger operator, we also see the extended Birman–Schwinger principle for nonlocal Schrödinger operators.

Proposition 4.312 (Extended Birman–Schwinger principle). *Suppose that $V \leq 0$ and it is relatively form-compact with respect to $-\Delta$. Then*

$$\bar{N}_E(V) = \dim 1_{[1,\infty)}(K_E^\Psi), \quad -E < 0, \tag{4.9.52}$$

$$\bar{N}_0(V) \leq \dim 1_{[1,\infty)}(K_0^\Psi). \tag{4.9.53}$$

Since F is monotone increasing, it follows that

$$\bar{N}_E(V) \le F(1)^{-1} \sum_{\substack{\mu \in \mathrm{Spec}(K_y^\Psi) \\ \mu \ge 1}} F(\mu). \tag{4.9.54}$$

Using this inequality, we will estimate the trace of $F(K_y^\Psi)$ in the same fashion as for classical Schrödinger operators. For simplicity, we make the following assumptions.

Assumption 4.313. The following hold:
(1) V is a continuous and nonpositive function;
(2) there exists $\lambda^* > 0$ such that $\|(\Psi(-\frac{1}{2}\Delta) + \lambda)^{-1/2}|V|^{1/2}\| < 1$ for all $\lambda \ge \lambda^*$;
(3) there exists $p \ge 1$ such that $|V|^{1/2}(\Psi(-\frac{1}{2}\Delta) + \lambda)^{-1}|V|^{1/2} \in \mathscr{I}_p$ for all $\lambda \ge 0$.

Part (2) of Assumption 4.313 implies that V is relatively form-bounded with respect to $\Psi(-\frac{1}{2}\Delta)$ with relative bound strictly smaller than 1. By Proposition 4.73, the operator $(\Psi(-\frac{1}{2}\Delta)+\lambda)^{-1/2}|V|^{1/2}$ is compact for $V \in L^{\infty,0}$, since Ψ is increasing. Part (3) ensures that the Birman–Schwinger principle holds and is used in the proof of Lemma 4.316 below.

Remark 4.314. Let $\Psi(u) = u$. Then $H^\Psi = -(1/2)\Delta + V$ becomes a classical Schrödinger operator. We also suppose that $V \in L^{d/2}(\mathbb{R}^d)$. Thus

$$|V|^{1/2}\left(\Psi\left(-\frac{1}{2}\Delta\right) + \lambda\right)^{-1} |V|^{1/2} \in \mathscr{I}_p$$

for all $\lambda \ge 0$ for $p > d/2$ and (3) of Assumption 4.313 is satisfied.

Recall that $(T_t^\Psi)_{t\ge 0}$ denotes the subordinator on a probability space $(\mathcal{T}, \mathcal{B}_\mathcal{T}, P)$ such that $\mathbb{E}_P[e^{uT_t^\Psi}] = e^{-t\Psi(u)}$.

Lemma 4.315. *Let Assumption 4.313 hold. Then $F(K_y^\Psi)$ is an integral operator with integral kernel $F(K_y^\Psi)(x,y)$, and the map $\mathbb{R}^d \times \mathbb{R}^d \ni (x,y) \mapsto F(K_y^\Psi)(x,y) \in \mathbb{R}$ is continuous.*

Proof. By the Feynman–Kac formula and using $X_s^\Psi = B_{T_s^\Psi}$, we see that for $f, g \in L^2(\mathbb{R}^d)$,

$$(f, e^{-t(\Psi(-\frac{1}{2}\Delta)+W)}g) = \int_{\mathbb{R}^d\times\mathbb{R}^d} \mathbb{E}_P^0\left[\Pi_{T_t^\Psi}(x-y)\mathbb{E}^{x,y}_{[0,T_t^\Psi]}[e^{-\int_0^t W(B_{T_s^\Psi})ds}]\right]\overline{f(x)}g(y)dxdy. \tag{4.9.55}$$

Note that the measure $\mathcal{W}^{x,y}_{[0,T_t^\Psi]}$ is defined by the Brownian bridge measure $\mathcal{W}^{x,y}_{[0,t]}$ evaluated at the random time $t = T_t^\Psi$. For every $T_t^\Psi(\omega)$ with $\omega \in \mathcal{T}$, we also define the Brownian bridge $(Z_t^{x,y})_{t\ge 0}$ starting from x at $t = 0$ and ending in y at $t = T_t^\Psi(\omega)$ by

$$Z_t^{x,y} = \left(1 - \frac{t}{T_t^\Psi}\right)x + \frac{t}{T_t^\Psi}y - \frac{t}{T_t^\Psi}B_{T_t^\Psi} + B_t.$$

Thus (4.9.55) is equal to

$$(f, e^{-t(\Psi(-\frac{1}{2}\Delta)+W)}g) = \int_{\mathbb{R}^d\times\mathbb{R}^d} \mathbb{E}_P^0\left[\Pi_{T_t^\Psi}(x-y)\mathbb{E}_{\mathcal{W}}^0[e^{-\int_0^t W(Z_s^{x,y})ds}]\right]\overline{f(x)}g(y)dxdy. \quad (4.9.56)$$

Hence the integral kernel of $e^{-t(\Psi(-\frac{1}{2}\Delta)+W)}$ is given by

$$e^{-t(\Psi(-\frac{1}{2}\Delta)+W)}(x,y) = \mathbb{E}_P^0\left[\Pi_{T_t^\Psi}(x-y)\mathbb{E}_{\mathcal{W}}^0[e^{-\int_0^t W(Z_s^{x,y})ds}]\right]$$

and this implies that the map

$$(x,y) \mapsto e^{-t(\Psi(-\frac{1}{2}\Delta)+W)}(x,y) \quad (4.9.57)$$

is jointly continuous. Applying Laplace transform, we obtain for all $f \in L^2(\mathbb{R}^d)$,

$$F(K_\gamma^\Psi)f(x) = W(x)^{1/2}\left(\int_0^\infty e^{-\gamma t}e^{-t(\Psi(-\frac{1}{2}\Delta)+\lambda W)}W^{1/2}f\right)(x)dt. \quad (4.9.58)$$

We have

$$F(K_\gamma^\Psi)f(x) = \int_{\mathbb{R}^d\times\mathbb{R}^d} F(K_\gamma^\Psi)(x,y)f(y)dxdy,$$

where

$$F(K_\gamma^\Psi)(x,y) = W(x)^{1/2}W(y)^{1/2}\int_0^\infty e^{-\gamma t}e^{-t(\Psi(-\frac{1}{2}\Delta)+\lambda W)}(x,y)dt.$$

The continuity of $(x,y) \mapsto F(K_\gamma^\Psi)(x,y)$ follows from (4.9.57). □

From the Feynman–Kac formula given in Theorem 4.268, we obtain the kernel

$$e^{-t(\Psi(-\frac{1}{2}\Delta)+\lambda W)}(x,y) = \mathbb{E}_{\mathcal{W}\times P}^{0,0}\left[e^{-\lambda\int_0^t W(X_s^\Psi+x)ds}\middle| X_t^\Psi + x = y\right]p_t^\Psi(x-y), \quad (4.9.59)$$

where p_t^Ψ is the distribution of X_t^Ψ given by (4.9.15), and it follows that

$$\begin{aligned}&F(K_\gamma^\Psi)(x,y)\\ &= W(x)^{1/2}W(y)^{1/2}\int_0^\infty e^{-\gamma t}\mathbb{E}_{\mathcal{W}\times P}^{0,0}\left[g\left(\int_0^t W(X_s^\Psi+x)ds\right)\middle| X_t^\Psi + x = y\right]p_t^\Psi(x-y)dt.\end{aligned} \quad (4.9.60)$$

By Lemma 4.315 and setting $x = y$ in (4.9.60), we have

$$\operatorname{Tr} F(K_\gamma^\Psi) = \int_{\mathbb{R}^d} F(K_\gamma^\Psi)(x,x)dx$$
$$= \int_{\mathbb{R}^d} W(x)dx \int_0^\infty e^{-\gamma t}\mathbb{E}^{0,0}_{\mathcal{W}\times P}\left[g\left(\int_0^t W(X_s^\Psi + x)ds\right)\Big|X_t^\Psi = 0\right] p_t^\Psi(0)dt. \quad (4.9.61)$$

For the rest, we can proceed as in the Schrödinger operators' case. The equality

$$\operatorname{Tr} F(K_\gamma^\Psi) = \int_{\mathbb{R}^d} dx \int_0^\infty \frac{e^{-\gamma t}}{t}\mathbb{E}^{0,0}_{\mathcal{W}\times P}\left[f\left(\int_0^t W(X_s^\Psi + x)ds\right)\Big|X_t^\Psi = 0\right] p_t^\Psi(0)dt \quad (4.9.62)$$

can be obtained in the same way as in Lemma 4.152. Varying F and f while keeping their relationship unchanged, we obtain the following lemma.

Lemma 4.316. *Let Assumption* 4.313 *hold and fix* $\gamma \geq 0$. *Let* f *be a nonnegative lower semicontinuous function on* $[0,\infty)$ *with* $f(0) = 0$, *and* f, g, *and* F *be related by*

$$F(x) = x\int_0^\infty e^{-y} g(xy)dy, \quad f(y) = g(y)y, \quad (4.9.63)$$

with $F(\|K_\gamma^\Psi\|) < \infty$. *Then*

$$\operatorname{Tr} F(K_\gamma^\Psi) = \int_0^\infty \frac{e^{-\gamma t}}{t} p_t^\Psi(0)dt \int_{\mathbb{R}^d} \mathbb{E}^{0,0}_{\mathcal{W}\times P}\left[f\left(\int_0^t W(X_s^\Psi + x)ds\right)\Big|X_t^\Psi = 0\right]dx$$
$$= \int_0^\infty e^{-\gamma t} p_t^\Psi(0)dt \int_{\mathbb{R}^d} W(x)\mathbb{E}^{0,0}_{\mathcal{W}\times P}\left[g\left(\int_0^t W(X_s^\Psi + x)ds\right)\Big|X_t^\Psi = 0\right]dx,$$

where both sides above are possibly infinite.

Proof. The proof is a minor modification of the proof of Lemma 4.153. □

Now we arrive at the Lieb–Thirring inequality for H^Ψ.

Theorem 4.317 (Lieb–Thirring inequality for H^Ψ). *Let Assumption* 4.313 *hold, and* f *be a nonnegative lower semicontinuous and convex function on* $[0,\infty)$, *with* $f(0) = 0$. *Then*

$$\bar{N}_0(V) \leq \frac{1}{\int_0^\infty e^{-s}s^{-1}f(s)ds}\int_0^\infty \frac{f(s)}{s}ds \int_{\mathbb{R}^d} p^\Psi_{s/|V(x)|}(0)1_{\{|V(x)|>0\}}dx, \quad (4.9.64)$$

where

$$p^\Psi_{s/|V(x)|}(0) = \frac{1}{(2\pi)^d}\int_{\mathbb{R}^d} e^{-s\Psi(|\xi|^2/2)/|V(x)|}d\xi.$$

We note that the right-hand side of (4.9.64) may not be finite; this depends on the choice of f.

Proof. Let F be a function as in (4.9.63). Since F is a monotone increasing function, we have

$$\bar{N}_0(V) \leq F(1)^{-1}\operatorname{Tr}(F(K_0^\Psi))$$
$$= F(1)^{-1}\int_0^\infty p_t^\Psi(0)\frac{dt}{t}\int_{\mathbb{R}^d}\mathbb{E}^{0,0}_{\mathcal{W}\times P}\left[f\left(\int_0^t t|V(X_s^\Psi + x)|\frac{ds}{t}\right)\Big|X_t^\Psi = 0\right]dx.$$

By Jensen's inequality,

$$\bar{N}_0(V) \leq F(1)^{-1}\int_0^\infty p_t^\Psi(0)\frac{dt}{t}\int_{\mathbb{R}^d}\mathbb{E}^{0,0}_{\mathcal{W}\times P}\left[\int_0^t f\left(t|V(X_s^\Psi + x)|\right)\frac{ds}{t}\Big|X_t^\Psi = 0\right]dx.$$

Using $\int_0^t \frac{ds}{t} = 1$ and swapping dx and $d\mathcal{W}\times dP$, we obtain

$$\bar{N}_0(V) \leq F(1)^{-1}\int_0^\infty p_t^\Psi(0)\frac{dt}{t}\int_{\mathbb{R}^d} f(t|V(x)|)dx.$$

When $V(x) = 0$, also $f(tV(x)) = 0$. This implies that the right-hand side above equals

$$F(1)^{-1}\int_0^\infty p_t^\Psi(0)\frac{dt}{t}\int_{\mathbb{R}^d} f(t|V(x)|)1_{\{|V(x)|>0\}}dx.$$

Changing the variable from $t|V(x)|$ to s and integrating with respect to s, the bound (4.9.64) follows from $F(1) = \int_0^\infty e^{-s}s^{-1}f(s)ds$. □

Next we discuss how the Lieb–Thirring inequality (4.9.64) in fact depends on the Bernstein function Ψ. To make this expression more explicit, we note that the diagonal part of the probability transition kernel has the representation

$$p_t^\Psi(0) = \frac{1}{(2\pi)^d}\int_0^\infty e^{-r}\left(\int_{\mathbb{R}^d} 1_{\{\sqrt{\Psi(\xi^2/2)}\leq\sqrt{r/t}\}}d\xi\right)dr. \tag{4.9.65}$$

Denote by $B_r^\Psi(x)$ a ball of radius r centered at x in the topology of the metric given by

$$d^\Psi(\xi,\eta) = \sqrt{\Psi(|\eta-\xi|^2/2)}, \tag{4.9.66}$$

i. e., $B_r^\Psi(x) = \{y \in \mathbb{R}^d \mid d^\Psi(x,y) \le r\}$. Note that $d^\Psi(\xi,\eta) = 0$ if and only if $\xi = \eta$, since Ψ is concave and a C^∞-function. The integral $\int_{\mathbb{R}^d} 1_{\{\sqrt{\Psi(\xi^2/2)} \le \sqrt{r/t}\}} d\xi$ is the volume of $B^\Psi_{\sqrt{r/t}}(0)$ in this metric. If d^Ψ satisfies the condition

$$\int_{\mathbb{R}^d} 1_{B^\Psi_{2r}(x)} dy \le c \int_{\mathbb{R}^d} 1_{B^\Psi_r(x)} dy, \quad x \in \mathbb{R}^d,\ r > 0, \tag{4.9.67}$$

with a constant $c > 0$ independent of x and r, then d^Ψ is said to have the *volume doubling property*. When d^Ψ has this property, then it furthermore follows that

$$c_1 \int_{\mathbb{R}^d} 1_{\{\sqrt{\Psi(\xi^2/2)} \le \sqrt{r/t}\}} d\xi \le p_t^\Psi(0) \le c_2 \int_{\mathbb{R}^d} 1_{\{\sqrt{\Psi(\xi^2/2)} \le \sqrt{r/t}\}} d\xi \tag{4.9.68}$$

with some constants c_1 and c_2. We use the following lemma without proof.

Lemma 4.318 (Volume doubling). *A necessary and sufficient condition for $\Psi \in \mathcal{B}_0$ to give rise to a volume doubling d^Ψ is*

$$\liminf_{u\to 0} \frac{\Psi(Cu)}{\Psi(u)} > 1, \quad \liminf_{u\to\infty} \frac{\Psi(Cu)}{\Psi(u)} > 1$$

for some $C > 1$.

In particular, this lemma implies that Ψ increases at infinity as a (possibly fractional) power.

Corollary 4.319. *Suppose that $\Psi \in \mathcal{B}_0$ is strictly monotone increasing. Then under the assumptions of Theorem 4.317, we have*

$$\bar{N}_0(V) \le A \int_0^\infty \frac{f(s)}{s} ds \int_{\mathbb{R}^d} dx \int_0^\infty \left(\Psi^{-1}\left(\frac{r|V(x)|}{s}\right)\right)^{d/2} e^{-r} dr, \tag{4.9.69}$$

where

$$A = \frac{1}{(2\pi)^{d/2} \frac{d}{2} \Gamma\left(\frac{d}{2}\right)} \frac{1}{\int_0^\infty e^{-s} s^{-1} f(s) ds}.$$

Furthermore, if d^Ψ has the volume doubling property, then

$$\bar{N}_0(V) \le c_2 A \int_0^\infty \frac{f(s)}{s} ds \int_{\mathbb{R}^d} \left(\Psi^{-1}\left(\frac{|V(x)|}{s}\right)\right)^{d/2} dx, \tag{4.9.70}$$

where c_2 is given in (4.9.68).

Proof. Since under the assumption the function $\Psi \in \mathcal{B}_0$ is invertible and its inverse is increasing, the proof is straightforward by using $\operatorname{Ker}\Psi = \{0\}$, (4.9.65), (4.9.68), and

$$\int_{\mathbb{R}^d} 1_{\{\sqrt{\Psi(\xi^2/2)} \le \sqrt{x}\}} d\xi = \int_{\mathbb{R}^d} 1_{\{\xi^2 \le 2\Psi^{-1}(x)\}} d\xi = \frac{\pi^{d/2}}{\frac{d}{2}\Gamma\left(\frac{d}{2}\right)} \left(2\Psi^{-1}(x)\right)^{d/2}. \qquad \square$$

In case when $\Psi \in \mathcal{B}_0$ has a scaling property, we can derive a more explicit formula.

Corollary 4.320. *Suppose that $\Psi \in \mathcal{B}_0$ is strictly monotone increasing and the assumptions of Theorem 4.317 hold. In addition, assume that there exists $\gamma > 0$ such that $\Psi(au) = a^\gamma \Psi(u)$ for all $a, u \ge 0$, and $\int_0^\infty f(s)s^{-1-\frac{d}{2\gamma}} ds < \infty$. Then*

$$\bar{N}_0(V) \le B \int_{\mathbb{R}^d} \left(\Psi^{-1}(|V(x)|)\right)^{d/2} dx, \tag{4.9.71}$$

where

$$B = \frac{\Gamma(\frac{d}{2\gamma}+1)}{(2\pi)^{d/2}\frac{d}{2}\Gamma\left(\frac{d}{2}\right)} \frac{\int_0^\infty s^{-1-\frac{d}{2\gamma}} f(s)ds}{\int_0^\infty e^{-s}s^{-1}f(s)ds}.$$

Proof. The inverse function Ψ^{-1} has the scaling property $\Psi^{-1}(av) = a^{1/\gamma}\Psi^{-1}(v)$. Thus the corollary follows. $\square$

Instead of the scaling property suppose now that there exists $\lambda > 0$ such that $\Psi(u) \ge Cu^\lambda$ with a constant $C > 0$. This inequality holds for at least large enough u if d^Ψ has the volume doubling property. We have a similar formula to Corollary 4.320.

Corollary 4.321. *Suppose that $\Psi \in \mathcal{B}_0$ is strictly monotone increasing and the assumptions of Theorem 4.317 hold. If $\Psi(u) \ge Cu^\lambda$, then*

$$\bar{N}_0(V) \le B' \int_{\mathbb{R}^d} |V(x)|^{d/2\lambda} dx, \tag{4.9.72}$$

where

$$B' = \frac{C^{-d/2\lambda}}{(2\pi)^{d/2}\frac{d}{2}\Gamma\left(\frac{d}{2}\right)} \frac{\int_0^\infty f(s)s^{-1-\frac{d}{2\lambda}} ds}{\int_0^\infty e^{-s}s^{-1}f(s)ds}.$$

Proof. The bound $\Psi(u) \ge Cu^\lambda$ gives $\Psi^{-1}(u) \le C^{-1/\lambda}u^{1/\lambda}$, and the corollary follows. $\square$

In some special cases of Bernstein functions Ψ, we can derive more explicit forms of the Lieb–Thirring inequality. Recall the Sobolev space

$$H^s(\mathbb{R}^d) = \{f \in L^2(\mathbb{R}^d) | \hat{f}(k)(1+|k|^2)^s \in L^2(\mathbb{R}^d)\} \tag{4.9.73}$$

for $s \geq 0$. In what follows we discuss some key specific cases of the Lieb–Thirring inequality, where we assume for simplicity that V is nonpositive and continuous.

Example 4.322 (Fractional Schrödinger operators). Let $\Psi(u) = (2u)^{\alpha/2}$ and $H^\Psi = (-\Delta)^{\alpha/2} + V$, $0 < \alpha \leq 2$. For $f, g \in H^{\alpha/2}(\mathbb{R}^d)$, we define the quadratic form

$$Q(f,g) = ((-\Delta)^{\alpha/4} f, (-\Delta)^{\alpha/4} g) - (|V|^{1/2} f, |V|^{1/2} g). \tag{4.9.74}$$

Let $V \in L^{d/\alpha}(\mathbb{R}^d)$ for $\alpha \in (0,2)$. From the Sobolev inequality

$$\|f\|_q \leq C\|(-\Delta)^{\alpha/2} f\|_p \tag{4.9.75}$$

for $q = pd/(d - \alpha p)$ and $d > \alpha p$ with some constant C, it follows that

$$\|(-\Delta)^{\alpha/4} f\|_2^2 \geq \frac{1}{C}\|f\|^2_{\frac{2d}{d-\alpha}} \geq \frac{1}{C}(|V|^{1/2} f, |V|^{1/2} f)\|V\|^{-1}_{d/\alpha}. \tag{4.9.76}$$

The estimate gives $Q(f,f) \geq 0$ when $\|V\|_{d/\alpha} < 1/C$. There is a bounded function $\lambda(x)$ such that $h = V - \lambda$ satisfies $\|h\|_{d/\alpha} < 1/C$. Thus $V = h + \lambda$ and $\lambda \in L^\infty(\mathbb{R}^d)$, thus V is form-bounded with respect to $(-\Delta)^{\alpha/2}$ with a relative bound strictly smaller than 1. In particular, $\inf_{f \in H^{\alpha/2}(\mathbb{R}^d)} Q(f,f) > -\infty$. Furthermore, $(-\Delta)^{\alpha/4}|V|^{1/2} \in \mathscr{I}_{2d/\alpha,w}$ by Proposition 4.40, and $|V|^{1/2}(-\Delta)^{\alpha/2}|V|^{1/2} \in \mathscr{I}_{d/\alpha,w}$ follows from Proposition 4.37. Hence

$$|V|^{1/2}(-\Delta)^{\alpha/2}|V|^{1/2} \in \mathscr{I}_p, \quad p > \frac{d}{\alpha}. \tag{4.9.77}$$

Thus Assumption 4.313 is verified. It follows that

$$p_t^\Psi(0) = \frac{1}{(2\pi)^d} \int_{\mathbb{R}^d} e^{-t|\xi|^\alpha} d\xi = \frac{C_1(\alpha,d)}{t^{d/\alpha}}, \tag{4.9.78}$$

where

$$C_1(\alpha,d) = \frac{2\pi^{d/2}}{\alpha(2\pi)^d} \frac{\Gamma(d/\alpha)}{\Gamma(d/2)}.$$

With the constant

$$L^{(1)}_{\alpha,d} = C_1(\alpha,d) \frac{\int_0^\infty s^{-1-d/\alpha} f(s) ds}{\int_0^\infty e^{-s} s^{-1} f(s) ds},$$

we have by Theorem 4.317 that

$$\bar{N}_0(V) \leq L^{(1)}_{\alpha,d} \int_{\mathbb{R}^d} |V(x)|^{d/\alpha} dx. \tag{4.9.79}$$

Example 4.323 (Fractional relativistic Schrödinger operators). Let $0 < \alpha < 2$ and $m > 0$. Choose $\Psi(u) = (2u + m^{2/\alpha})^{\alpha/2} - m$ and $H^\Psi = (-\Delta + m^{2/\alpha})^{\alpha/2} - m + V$. Using (4.9.76), we can also derive

$$\|(-\Delta + m^{2/\alpha})^{\alpha/4} f\|_2^2 \geq \|(-\Delta)^{\alpha/4} f\|_2^2 \geq \frac{1}{C}(|V|^{1/2} f, |V|^{1/2} f)\|V\|_{d/\alpha}^{-1}. \tag{4.9.80}$$

Hence $V \in L^{d/\alpha}(\mathbb{R}^d)$ is relatively form-bounded with respect to $(-\Delta + m^{2/\alpha})^{\alpha/2} - m$ with relative bound strictly smaller than 1. Suppose that $V \in L^{d/\alpha}(\mathbb{R}^d) \cap L^{d/2}(\mathbb{R}^d)$. We have

$$|V|^{1/2}(-\Delta + m^{2/\alpha})^{-\alpha/2}|V|^{1/2} \in \mathscr{I}_p, \quad p > d/\alpha$$

in a similar way to (4.9.77). Thus Assumption 4.313 is satisfied. For $m > 0$, we have

$$p_t^\Psi(0) = \frac{1}{(2\pi)^d} \int_{\mathbb{R}^d} e^{-t((|\xi|^2 + m^{2/\alpha})^{\alpha/2} - m)} d\xi$$

and the bound

$$p_t^\Psi(0) \leq \frac{|S_{d-1}|}{(2\pi)^d}\left(\int_{r \leq m^{1/\alpha}} e^{-\frac{\alpha r^2}{2(2m^{2/\alpha})^{1-\alpha/2}} t} r^{d-1} dr + \int_{r > m^{1/\alpha}} e^{-\frac{\alpha r^\alpha}{2^{2-\alpha/2}} t} r^{d-1} dr \right) \leq \frac{C_2(\alpha, d)}{t^{d/2}} + \frac{C_3(\alpha, d)}{t^{d/\alpha}},$$

with

$$C_2(\alpha, d) = \frac{1}{2}\left(\frac{2}{\alpha}(2m^{2/\alpha})^{1-\alpha/2}\right)^{d/2} \Gamma\left(\frac{d}{2}\right) \text{ and } C_3(\alpha, d) = \frac{1}{\alpha}\left(\frac{2^{2-\alpha/2}}{\alpha}\right)^{d/\alpha} \Gamma\left(\frac{d}{\alpha}\right).$$

In case $m = 0$, see (4.9.78). With

$$L_{\alpha,d}^{(2)} = C_2(\alpha, d) \frac{\int_0^\infty s^{-1-d/2} f(s) ds}{\int_0^\infty e^{-s} s^{-1} f(s) ds} \quad \text{and} \quad L_{\alpha,d}^{(3)} = C_3(\alpha, d) \frac{\int_0^\infty s^{-1-d/\alpha} f(s) ds}{\int_0^\infty e^{-s} s^{-1} f(s) ds}$$

which are independent of V, we have

$$\bar{N}_0(V) \leq L_{\alpha,d}^{(2)} \int_{\mathbb{R}^d} |V(x)|^{d/2} dx + L_{\alpha,d}^{(3)} \int_{\mathbb{R}^d} |V(x)|^{d/\alpha} dx. \tag{4.9.81}$$

Example 4.324 (Relativistic Schrödinger operators). Let $\Psi(u) = \sqrt{2u + m^2} - m$ and $H^\Psi = (-\Delta + m^2)^{1/2} - m + V$ for $m > 0$. This is the case of $\alpha = 1$ in Example 4.323. Hence $V \in L^{d/2}(\mathbb{R}^d)$ is relatively form-bounded with respect to $(-\Delta + m^2)^{1/2} - m$ with relative bound strictly smaller than 1. Let $V \in L^{d/2}(\mathbb{R}^d) \cap L^d(\mathbb{R}^d)$. Then we have

$$\bar{N}_0(V) \leq L_{1,d}^{(2)} \int_{\mathbb{R}^d} |V(x)|^{d/2} dx + L_{1,d}^{(3)} \int_{\mathbb{R}^d} |V(x)|^d dx.$$

Example 4.325 (Sums of different stable generators). Let $\Psi(u) = (2u)^{\alpha/2} + (2u)^{\beta/2}$, $0 < \alpha, \beta < 2$, and $H^\Psi = (-\Delta)^{\alpha/2} + (-\Delta)^{\beta/2} + V$. Relative boundedness of V follows similarly as above, whenever $V \in L^{d/\alpha}(\mathbb{R}^d) \cap L^{d/\beta}(\mathbb{R}^d)$. Let $\alpha \geq \beta$. We have

$$\begin{aligned} &|V|^{1/2}\left((-\Delta)^{\alpha/2} + (-\Delta)^{\beta/2}\right)^{-1}|V|^{1/2} \\ &= |V|^{1/2}\left((-\Delta)^{(\alpha-\beta)/2} + 1\right)^{-1}(-\Delta)^{-\beta/2}|V|^{1/2} \leq |V|^{1/2}(-\Delta)^{-\beta/2}|V|^{1/2}. \end{aligned}$$

In a similar way to (4.9.77), it is obtained that

$$|V|^{1/2}\left((-\Delta)^{\alpha/2} + (-\Delta)^{\beta/2}\right)^{-1}|V|^{1/2} \in \mathscr{I}_p, \quad p > \min\{d/\alpha, d/\beta\}.$$

Thus Assumption 4.313 is satisfied. By Theorem 4.317 and using

$$p_t^\Psi(0) \leq c\left(t^{-\frac{d}{\alpha}} \wedge t^{-\frac{d}{\beta}}\right), \quad t > 0 \tag{4.9.82}$$

with some constant $c > 0$, we obtain

$$\bar{N}_0(V) \leq L_\alpha \int_{\mathbb{R}^d} |V(x)|^{d/\alpha} dx + L_\beta \int_{\mathbb{R}^d} |V(x)|^{d/\beta} dx,$$

with

$$L_\gamma = \frac{c}{(2\pi)^{d/2}\frac{d}{2}\Gamma\left(\frac{d}{2}\right)} \frac{\int_0^\infty s^{-1-d/\gamma} f(s) ds}{\int_0^\infty e^{-s} s^{-1} f(s) ds}, \quad \gamma = \alpha, \beta.$$

Example 4.326 (Jump-diffusion operators). Let $\Psi(u) = u + bu^{\alpha/2}$, $0 < \alpha < 2$, and $b \in (0, 1]$. Then $H^\Psi = -\Delta + b(-\Delta)^{\alpha/2} + V$. By (4.9.76) whenever $V \in L^{d/2}(\mathbb{R}^d) \cap L^{d/\alpha}(\mathbb{R}^d)$, V is relatively form-bounded with respect to $-\Delta + b(-\Delta)^{\alpha/2}$ with relative bound strictly smaller than 1. We have

$$\begin{aligned} &|V|^{1/2}(-\Delta + b(-\Delta)^{\alpha/2})^{-1}|V|^{1/2} \\ &= |V|^{1/2}\left((-\Delta)^{1-\alpha/2} + b\right)^{-1}(-\Delta)^{-\alpha/2}|V|^{1/2} \leq |V|^{1/2}(-\Delta)^{-\alpha/2}|V|^{1/2}. \end{aligned}$$

Then

$$|V|^{1/2}(-\Delta + b(-\Delta)^{\alpha/2})^{-1}|V|^{1/2} \in \mathscr{I}_p, \quad p > d/\alpha$$

and Assumption 4.313 is satisfied. Using the estimate

$$p_t^\Psi(x-y) \leq \left(t^{-d/2} \wedge (bt)^{-d/\alpha}\right) \wedge \left(t^{-d/2} e^{-|x-y|^2/ct} + (bt)^{-d/\alpha} \wedge \frac{bt}{|x-y|^{d+\alpha}}\right), \tag{4.9.83}$$

with some $c > 0$, we have

$$p_t^\Psi(0) \le t^{-d/2} \wedge (bt)^{-d/\alpha}$$

and by Theorem 4.317 obtain

$$\bar{N}_0(V) \le D \int_{\mathbb{R}^d} |V(x)|^{d/2} dx + D_\alpha \int_{\mathbb{R}^d} |V(x)|^{d/\alpha} dx,$$

where

$$D = \frac{1}{(2\pi)^{d/2} \frac{d}{2} \Gamma\left(\frac{d}{2}\right)} \frac{\int_0^\infty s^{-1-d/2} f(s) ds}{\int_0^\infty e^{-s} s^{-1} f(s) ds} \quad \text{and} \quad D_\alpha = \frac{b^{-d/\alpha}}{(2\pi)^{d/2} \frac{d}{2} \Gamma\left(\frac{d}{2}\right)} \frac{\int_0^\infty s^{-1-d/\alpha} f(s) ds}{\int_0^\infty e^{-s} s^{-1} f(s) ds}.$$

4.9.8 Decay of eigenfunctions

In this section we briefly summarize results on the spatial decay at infinity of bound states for nonlocal Schrödinger operators. These have been obtained via Feynman–Kac type representations, however, by an extensive use of techniques of probabilistic potential theory which are beyond the scope of this book, therefore we refer the reader to the literature for details of proof.

In our framework here we use directly a random process as a basic input, however, we also discuss the related operators. Let $(X_t)_{t\ge0}$ be a Lévy process with values in $\mathbb{R}^d$, $d \ge 1$, with probability measure P^x of the process starting from $x \in \mathbb{R}^d$, and denote by $\mathbb{E}^x$ expectation with respect to this measure. Recall that $(X_t)_{t\ge0}$ is a Markov process with respect to its natural filtration satisfying the strong Markov property and having càdlàg paths. Denote by $p_t(x)$ its transition probability density, whenever it exists, let $D \subset \mathbb{R}^d$ be an open bounded set, and consider $\tau_D = \inf\{t \ge 0 \mid X_t \notin D\}$, i. e., the first exit time of $(X_t)_{t\ge0}$ from D. Then the transition densities $p_D(t,x,y)$ of the process killed on exiting D also exist and are given by the Dynkin–Hunt formula

$$p_D(t,x,y) = p_t(y-x) - \mathbb{E}^x\left[p_{t-\tau_D}(y - X_{\tau_D}) 1_{\{\tau_D<t\}}\right], \quad x,y \in D,\ t > 0.$$

The Green function of the process $(X_t)_{t\ge0}$ on D is thus $G_D(x,y) = \int_0^\infty p_D(t,x,y)dt$, for $x,y \in D$.

We will use the following class of processes. Below we adopt the notation $f \asymp g$ to mean that there exist $C_1, C_2 > 0$ such that $C_1 f \le g \le C_2 f$, and $C(X)$ for a constant dependent by a process $(X_t)_{t\ge0}$.

Definition 4.327 (Jump-paring Lévy processes). We call a Lévy process $(X_t)_{t\ge0}$ a symmetric *jump-paring Lévy process* if it satisfies the following properties:

(1) Its Lévy triplet is $(0, A, \nu)$ such that the Lévy measure $\nu(dx) = \nu(x)dx$ satisfies
   (i) $\nu(E) = \nu(-E)$, for every Borel set $E \subset \mathbb{R}^d \setminus \{0\}$;
   (ii) $\nu(\mathbb{R}^d \setminus \{0\}) = \infty$.

(2) There exists a nonincreasing function $g : (0, \infty) \to (0, \infty)$ such that
   (i) $\nu(x) \asymp g(|x|)$, for all $x \in \mathbb{R}^d \setminus \{0\}$;
   (ii) there is a constant $C_1 = C_1(X) > 0$ such that $g(r) \le C_1 g(r+1)$, for all $r \ge 1$;
   (iii) there is a constant $C_2 = C_2(X) > 0$ such that

$$\int\limits_{\substack{|x-y|>1 \\ |y|>1}} g(|x-y|)g(|y|)dy \le C_2 g(|x|), \quad |x| \ge 1. \tag{4.9.84}$$

(3) There exists $t_b = t_b(X) > 0$ such that $\sup_{x \in \mathbb{R}^d} p_{t_b}(x) < \infty$.
(4) For all $0 < p < q < r < \infty$, we have

$$\sup_{x \in B(0,p)} \sup_{y \in B(0,q)^c} G_{B(0,r)}(x,y) < \infty.$$

Condition 1 (i) above says that the process is symmetric, i. e., $-X_t \stackrel{d}{=} X_t$, for all $t \ge 0$, and 1 (ii) implies that P^X is absolutely continuous with respect to Lebesgue measure and thus $P^X(X_t \in dy)$ has a density, which we denote by $p_t(y-x)$. While the other conditions are technical and self-explanatory, condition 2 (iii) has a structural importance. It provides a control of the convolutions of ν with respect to large jumps, saying that double (and by iteration, any multiple) large jumps of the process are stochastically dominated by single large jumps. Heuristically this may be interpreted as a basic preference of the process to get to a location far away from the point of departure through a single large jump rather than via a sequence of large jumps, for which reason we refer to this as the *jump-paring property*.

Condition 1 (i) in Definition 4.327 implies that the Lévy process $(X_t)_{t\ge0}$ has characteristic function

$$\mathbb{E}^0[e^{i\xi\cdot X_t}] = e^{-t\psi(\xi)}, \quad \xi \in \mathbb{R}^d,\ t > 0,$$

whose exponent given by the Lévy–Khintchine formula simplifies to

$$\psi(\xi) = \frac{1}{2}A\xi\cdot\xi + \int\limits_{\mathbb{R}^d} (1-\cos(\xi\cdot z))\nu(dz), \tag{4.9.85}$$

where A is the diffusion matrix and $\nu(dz) = \nu(z)dz$ is the Lévy measure with intensity ν. Thus the generator L of the process is determined by

$$\widehat{Lf}(\xi) = -\psi(\xi)\hat{f}(\xi), \quad \xi \in \mathbb{R}^d,\ f \in D(L), \tag{4.9.86}$$

with domain $D(L) = \left\{ f \in L^2(\mathbb{R}^d) \mid \psi\hat{f} \in L^2(\mathbb{R}^d) \right\}$. The operator L is negative definite and self-adjoint with core $C_0^\infty(\mathbb{R}^d)$, and we have

$$Lf(x) = \int\limits_{\mathbb{R}^d\setminus\{0\}} \left(f(x+z) - f(x) - 1_{\{|z|\le1\}}(z)z\cdot\nabla f(x)\right)\nu(z)dz, \quad x \in \mathbb{R}^d,\ f \in C_0^\infty(\mathbb{R}^d).$$

The jump-paring class has a nontrivial overlap with subordinate Brownian motions in the sense that neither class contains the other. Next we give some examples and counterexamples to the jump-paring class.

Example 4.328. The following Lévy processes satisfy Definition 4.327:

(1) Specific cases of subordinate Brownian motions with characteristic exponents ψ such that $e^{-t_b\psi} \in L^1(\mathbb{R}^d)$ for some $t_b > 0$. Examples include:
  (i) *Rotationally symmetric α-stable process.* Let $\psi(\xi) = |\xi|^\alpha$, $\alpha \in (0,2)$. In this case $g(r) = r^{-d-\alpha}$ and $L = -(-\Delta)^{\alpha/2}$.
  (ii) *Mixtures of independent rotationally symmetric stable processes.* Let $\psi(\xi) = a|\xi|^\alpha + b|\xi|^\beta$, with $0 < \beta < \alpha < 2$ and $a, b > 0$. We have $g(r) = aC(\alpha)r^{-d-\alpha} + bC(\beta)r^{-d-\beta}$ and $L = -a(-\Delta)^{\alpha/2} - b(-\Delta)^{\beta/2}$.
  (iii) *Jump-diffusion process.* Let $\psi(\xi) = a|\xi|^\alpha + b|\xi|^2$, $0 < \alpha < 2$, $a, b > 0$, i. e., the process is a mixture of a rotationally symmetric α-stable process and an independent Brownian motion. In this case $g(r) = r^{-d-\alpha}$ and $L = -a(-\Delta)^{\alpha/2} + b\Delta$.
  (iv) *Rotationally symmetric geometric α-stable process.* Let $\psi(\xi) = \log(1 + |\xi|^{\alpha/2})$, $0 < \alpha < 2$. In this case, $g(r) = r^{-d}(1+r)^{-\alpha}$ and $L = \log(1 - \Delta^{\alpha/2})$. Notice that ψ is now a slowly varying function at infinity. In contrast to the previous examples, in this case there is $t_b(\alpha) > 0$ for which the transition probability densities are unbounded for $0 < t \le t_b(\alpha)$, though they are bounded for $t > t_b(\alpha)$.
  (v) *Relativistic rotationally symmetric α-stable process.* Let $\psi(\xi) = (|\xi|^2 + m^{2/\alpha})^{\alpha/2} - m$, $\alpha \in (0,2)$, $m > 0$. In this case, $g(r) = e^{-m^{1/\alpha}r}r^{-d-\alpha}(1 + r^{(d+\alpha-1)/2})$ and $L = -(-\Delta + m^{2/\alpha})^{\alpha/2} + m$.

(2) *Symmetric Lévy processes with subexponentially localized Lévy measures.* Let $(X_t)_{t\ge0}$ be a symmetric Lévy process with intensity $g(r) = e^{-ar^\beta}r^{-d-\delta}(1+r)^{d+\delta-\gamma}$, where $a > 0$, $\beta \in (0,1]$, $\delta \in [0,2)$ and $\gamma > (d+1)/2$. This class includes a large family of (exponentially) tempered symmetric stable processes ($a > 0$, $\beta = 1$, $\gamma = \delta + d$, $\delta \in (0,2)$) and the rotationally symmetric relativistic stable processes above.

In contrast, the following two examples do not belong to the jump-paring class:

(3) *Rotationally symmetric geometric 2-stable (gamma variance) process.* Let $\psi(\xi) = \log(1 + |\xi|^2)$. In this case, $g(r) = r^{-d}e^{-r}(1+r)^{(d-1)/2}$ and $L = -\log(1 - \Delta)$. In this example the Lévy measure does not satisfy the jump-paring condition 2 (iii).

(4) *Iterated symmetric geometric α-stable process.* Let $\psi(\xi) = \log(1 + \log^\alpha(1 + |\xi|^\alpha))$, $0 < \alpha < 2$. It can be checked directly that the transition densities are unbounded for every $t > 0$ and condition (3) fails.

Next we give the class of potentials which will be used in this context.

Definition 4.329 (X-Kato class). Let $(X_t)_{t\ge0}$ be a Lévy process. We say that the potential $V : \mathbb{R}^d \to \mathbb{R}$ belongs to *X-Kato class* $\mathcal{K}^X$ if it satisfies

$$\lim_{t\downarrow 0} \sup_{x\in\mathbb{R}^d} \mathbb{E}^x\left[\int_0^t |V(X_s)|ds\right] = 0.$$

Also, we say that V is an *X-Kato decomposable potential*, denoted $V \in \mathcal{K}_{\pm}^X$, whenever

$$V = V_+ - V_-, \quad \text{with } V_- \in \mathcal{K}^X \quad \text{and} \quad V_+ \in \mathcal{K}_{\mathrm{loc}}^X,$$

where V_+, V_- denote the positive and negative parts of V, respectively, and where $V_+ \in \mathcal{K}_{\mathrm{loc}}^X$ means that $V_+ 1_K \in \mathcal{K}^X$ for all compact sets $K \subset \mathbb{R}^d$.

It is straightforward to see that $L^\infty_{\mathrm{loc}}(\mathbb{R}^d) \subset \mathcal{K}_{\mathrm{loc}}^X$. Moreover, by stochastic continuity of $(X_t)_{t\geq 0}$ also $\mathcal{K}_{\mathrm{loc}}^X \subset L^1_{\mathrm{loc}}(\mathbb{R}^d)$, and thus an X-Kato decomposable potential is always locally absolutely integrable. Note that the condition defining X-Kato class allows local singularities of V.

Define

$$T_t f(x) = \mathbb{E}^x\left[e^{-\int_0^t V(X_s)ds} f(X_t)\right], \quad f \in L^2(\mathbb{R}^d),\ t > 0. \tag{4.9.87}$$

Using the Markov property and stochastic continuity of the process, it can be shown that $\{T_t : t \geq 0\}$ is a strongly continuous semigroup of symmetric operators on $L^2(\mathbb{R}^d)$, which we call the *Feynman–Kac semigroup* associated with the process $(X_t)_{t\geq 0}$ and potential V. We summarize basic properties of the Feynman–Kac semigroup, which we only state for the class defined in Definition 4.327, though they are valid more generally.

Proposition 4.330. *Let $(X_t)_{t\geq 0}$ be a symmetric jump-paring Lévy process and $V \in \mathcal{K}_{\pm}^X$. Then the following properties hold:*

(1) *For all $t > 0$, every T_t is a bounded operator on every $L^p(\mathbb{R}^d)$ space, $1 \leq p \leq \infty$. The operators $T_t : L^p(\mathbb{R}^d) \to L^p(\mathbb{R}^d)$ for $1 \leq p \leq \infty$, $t > 0$, and $T_t : L^p(\mathbb{R}^d) \to L^\infty(\mathbb{R}^d)$ for $1 < p \leq \infty$, $t \geq t_b$, and $T_t : L^1(\mathbb{R}^d) \to L^\infty(\mathbb{R}^d)$ for $t \geq 2t_b$ are bounded, with some $t_b > 0$.*

(2) *For all $t \geq 2t_b$, T_t has a bounded measurable kernel $u(t,x,y)$ symmetric in x and y, i. e., $T_t f(x) = \int_{\mathbb{R}^d} u(t,x,y)f(y)dy$, for all $f \in L^p(\mathbb{R}^d)$ and $1 \leq p \leq \infty$.*

(3) *For all $t > 0$ and $f \in L^\infty(\mathbb{R}^d)$, $T_t f$ is a bounded continuous function.*

(4) *For all $t \geq 2t_b$, the operators T_t are positivity improving, i. e., $T_t f(x) > 0$ for all $x \in \mathbb{R}^d$ and $f \in L^2(\mathbb{R}^d)$ such that $f \geq 0$ and $f \neq 0$ a. e.*

Furthermore, by the Hille–Yosida theorem there exists a self-adjoint operator H bounded from below such that $e^{-tH} = T_t$. We call the operator

$$H = -L + V$$

defined in the form sense and with dense domain in $L^2(\mathbb{R}^d)$ a *nonlocal Schrödinger operator* with kinetic term $-L$, where L is the infinitesimal generator of the process $(X_t)_{t\geq 0}$, and potential V.

Next we assume that H has a nonempty discrete spectrum, and consider the spatial decay of the related eigenfunctions. When V is a confining potential, then T_t are compact operators for all $t > 0$, and thus $\operatorname{Spec}(H) = \operatorname{Spec}_d(H)$. If V is a decaying potential then T_t are not compact operators, and we need to assume that $\operatorname{Spec}_d(H) \neq \emptyset$. Let φ_n be an eigenfunction at eigenvalue λ_n, $n = 0, 1, 2, \ldots$, solving the equation

$$H\varphi_n = \lambda_n \varphi_n.$$

As before, we denote by φ_0 the unique and strictly positive ground state corresponding to $\lambda_0 = \inf \operatorname{Spec}(H)$. Note that since $\varphi_n = e^{\lambda_n t_b} T_{t_b} \varphi_n$, by (1) and (3) of Proposition 4.330, we have that $T_t(L^p(\mathbb{R}^d)) \subset L^\infty(\mathbb{R}^d)$ and $T_t(L^\infty(\mathbb{R}^d)) \subset C_b(\mathbb{R}^d)$ for every $t > 2t_b$ and $p \geq 1$, it follows that each φ_n is a bounded continuous function and thus it makes sense to study their pointwise decay at infinity.

For an open set $D \subset \mathbb{R}^d$ and an X-Kato decomposable potential V, we will use the function

$$\Lambda_D^V(x) = \mathbb{E}^x \left[\int_0^{\tau_D} e^{-\int_0^t V(X_s) ds} dt \right], \quad x \in D,$$

where τ_D is the first exit time of $(X_t)_{t\geq 0}$ from D. Intuitively, $\Lambda_D^V(x)$ is the mean sojourn time in D under the potential of the process $(X_t)_{t\geq 0}$ starting from an interior point x before its first exit from D. Also, like before, we denote a ball of radius r centered at z by $B_r(z)$. In the statements below, we will not keep track of the constants, thus the same C may mean a different constant from line to line.

We then have the following results. First, we discuss the case of confining potentials.

Theorem 4.331 (Case of confining potentials). *If $(X_t)_{t\geq 0}$ is a jump-paring Lévy process with Lévy intensity ν and $V \in \mathcal{K}_\pm^X$ is a confining potential, then for every $\delta \geq 0$ for which $\lambda_0 + \delta > 0$, there exist constants $C_n = C_n(X, V, \delta)$ and $R_n = R_n(X, V, \delta) > 0$ such that*

$$|\varphi_n(x)| \leq C_n \|\varphi_n\|_\infty \Lambda_{B_1(x)}^{V+\delta}(x)\, \nu(x), \quad |x| \geq R_n,\ n = 0, 1, 2, \ldots \tag{4.9.88}$$

Moreover, for the ground state we have

$$\varphi_0(x) \asymp \Lambda_{B_1(x)}^{V+\delta}(x)\, \nu(x), \quad |x| \geq R_0. \tag{4.9.89}$$

The above result can be interpreted as saying that the decay of the ground state at far out points is driven by how soon on average paths leave small neighborhoods. Since for confining potentials $C_1 \frac{1}{\sup_D V} \leq \Lambda_D^V \leq C_2 \frac{1}{\inf_D V}$, with appropriate constants C_1, C_2, for sufficiently temperately growing potentials we can get more explicit bounds.

Corollary 4.332. *If further to the assumptions of Theorem 4.331 also $\sup_{y\in B_1(x)} V(y) \leq C \inf_{y \in B_1(x)} V(y)$ holds with some $C \geq 1$ for large enough $|x|$, then we have*

$$|\varphi_n(x)| \leq C_n' \frac{\nu(x)}{V(x)}, \quad n = 0, 1, 2, \ldots,$$

and

$$\varphi_0(x) \asymp \frac{\nu(x)}{V(x)},$$

whenever $|x|$ is large enough.

The above expression shows a neat separation of the contributions of the jump intensity and of the potential into the decay, and points to a mechanism of balance between the killing effect of the potential and jump activity.

Another consequence is the following surprising property.

Corollary 4.333. *If the assumptions of Theorem* 4.331 *hold, then there exists a constant $C_n = C_n(X, V)$ such that*

$$|\varphi_n(x)| \leq C_n \, \|\varphi_n\|_\infty \, \varphi_0(x), \quad x \in \mathbb{R}^d, \ n \in \mathbb{N}.$$

This ground state domination holds in the case of classical Schrödinger operators $H = -\frac{1}{2}\Delta + V$ if the semigroup $\{e^{-tH} : t \geq 0\}$ is intrinsically ultracontractive, while it is straightforward to see that it fails for the harmonic oscillator, when intrinsic ultracontractivity does not hold (see Definition 5.3 and Proposition 5.5 in Section 5.1.1 below). However, for nonlocal Schrödinger operators related to jump-paring Lévy processes and X-Kato decomposable potentials, it holds also in lack of intrinsic ultracontractivity.

Next we turn to decaying potentials. Now both $V(x)$ and $\nu(x)$ tend to zero as $|x| \to \infty$ so that far out the process under V behaves increasingly similar to the unperturbed process. Therefore an expression as in Corollary 4.332 resulting from the killing effect can no longer hold, and the contribution of the potential becomes more subtle.

Theorem 4.334 (Case of decaying potentials: lower bound). *Let $(X_t)_{t\geq 0}$ be a jump-paring Lévy process with Lévy intensity ν, $V \in \mathcal{K}_\pm^X$ a decaying potential, and φ an eigenfunction of H at eigenvalue $\lambda < 0$. Then the following hold:*

(1) *If property* 2 (ii) *in Definition* 4.327 *holds, then there exist $C = C(\lambda) > 0$ and $R = R(\lambda) > 0$ such that*

$$\varphi(x) \geq C\nu(x), \quad |x| \geq R_0.$$

(2) *If either property 2* (ii) *in Definition* 4.327 *holds but* 2 (iii) *does not hold, or neither hold, then*

$$\limsup_{|x|\to\infty} \frac{\varphi(x)}{\nu(x)} = \infty.$$

In the event of the eigenvalue being sufficiently low-lying below zero (i. e., the bottom of the essential spectrum), the following result holds.

Theorem 4.335 (Case of decaying potentials: upper bounds). *Let $(X_t)_{t\geq 0}$ be a jump-paring Lévy process with Lévy intensity ν, $V \in \mathcal{K}_{\pm}^X$ a decaying potential, and φ an eigenfunction of H at eigenvalue $\lambda < 0$. If there exists $\lambda^* = \lambda^*(X) > 0$ such that φ is an eigenfunction for an eigenvalue $\lambda \in (-\infty, -\lambda^*)$, then there exist $C = C(X, \lambda)$ and $R = R(X, \lambda)$ such that*

$$|\varphi(x)| \leq C \, \|\varphi\|_\infty \, \nu(x), \quad |x| \geq R. \tag{4.9.90}$$

For arbitrary eigenvalues, when thus they can be any near to zero, the situation is more complicated and depends more closely on the nature of the jumps. If ν is heavy-tailed, we have the following.

Theorem 4.336 (Upper bound for decaying potentials and heavy-tailed jump kernel). *Let $(X_t)_{t\geq 0}$ be a jump-paring Lévy process with Lévy intensity ν, $V \in \mathcal{K}_{\pm}^X$ a decaying potential, and φ an eigenfunction of H at eigenvalue $\lambda < 0$. Assume, furthermore, that there exists $c > 0$ such that for all $r \geq 1$ we have $\nu(x - y) \leq c\nu(x)$, whenever $|y| \leq r$ and $|x| \geq 2r$. Then* (4.9.90) *holds with suitable constants C and R.*

For lighter-tailed ν, the extra condition in Theorem 4.336 breaks down, and we need a more detailed control in terms of suitable parameter functions relating to some intrinsic jump preferences of the processes involved. Define

$$K_1^X(s) = \sup_{|x|\geq s} \frac{\int_{|x-y|>s,\, |y|>s} \nu(x-y)\nu(y)dy}{\nu(x)}, \quad s \geq 1. \tag{4.9.91}$$

This is a nonincreasing function and gives the optimal constant C in the bound

$$\int\limits_{\substack{|x-y|>s \\ |y|>s}} \nu(x-y)\nu(y)dy \leq C\nu(x), \quad |x| \geq s,$$

for any fixed $s \geq 1$. Thus $1/K_1^X(s)$ is the rate of preference of single jumps of size at least s over double jumps of size at least s each, i. e., when $K_1^X(s)$ decreases with $s \to \infty$, this preference improves. Next let $0 < s_1 < s_2 < s_3 \leq \infty$ and define

$$K_2^X(s_1, s_2, s_3) = \inf\left\{C \geq 1 \mid \nu(x-y) \leq C\,\nu(x),\ |y| \leq s_1,\ s_2 \leq |x| < s_3\right\}. \tag{4.9.92}$$

This function is nondecreasing in $s_1 \in (0, s_2)$, for every fixed $0 < s_2 < s_3$. Whenever $s_3 = \infty$ and $s_2 \gg s_1 \gg 1$, the ratio $1/K_2^X(s_1, s_2, s_3)$ measures the rate of preference of a direct large jump from $x \in B_{s_2}(0)^c$ to zero over a jump from a point z situated in the s_1-neighborhood of x to zero. In our applications, the second case corresponds to the situation when the process moving from x to zero first fluctuates inside $B_{s_1}(x)$ and then makes one large jump to the origin. Thus K_2^X is the inverse rate of preference of the scenario "no small steps but direct large jump" over the scenario "first small steps, then large jump." Then we have the following result covering this case.

Theorem 4.337 (Upper bound for decaying potentials and light-tailed jump kernel). *Let $(X_t)_{t\geq 0}$ be a jump-paring Lévy process with Lévy intensity ν and transition probability density $p_t(x)$, $V \in \mathcal{K}_{\pm}^X$ a decaying potential, and φ an eigenfunction of H at eigenvalue $\lambda < 0$. Denote by $\Phi(r) = \sup_{|\xi|\leq r} \psi(\xi)$, $r > 0$, the symmetrized Lévy exponent of $(X_t)_{t\geq 0}$. Furthermore, assume the following conditions:*
(1) $\int_{\mathbb{R}^d} |x|^2 \nu(dx) < \infty$;
(2) *there exists* $\kappa_1 \geq 2$ *such that* $\lim_{s\to\infty} K_1^X(\kappa_1 s) K_2^X(s, \kappa_1 s, \infty) = 0$;
(3) *there exists* $\kappa_2 < \infty$ *such that for all* $s_1 \geq 1$ *we have* $\limsup_{s\to\infty} K_2^X(s_1, s, \infty) \leq \kappa_2$;
(4) *there exists* $C_1 > 0$ *such that* $\sup_{|x|\geq r} p_t(x) \leq C_1 \frac{t\Phi(1/r)}{r^d}$, $t > 0$, $r \geq 1$;
(5) *there exist* $C_2 \geq 1$ *and* $R > 0$ *such that* $p_t(x) \leq C_2 p_t(y)$, *for every* $t > 0$ *and* $|x| \geq |y| \geq R$ *satisfying* $|x - y| \leq 1$.

Then (4.9.90) *holds with R and a suitable constant C.*

To understand what condition (2) in Theorem 4.337 means, first notice that by the definition of K_1^X it follows that

$$\int_{\substack{|x-z|>s_2 \\ |z|>s_2}} \nu(x-z)\nu(z)dz \leq K_1^X(s_2)\nu(x), \quad \text{for all } x \text{ such that } |x| > s_2,$$

and using the definition of K_2^X we have

$$\nu(x) \leq K_2^X(s_1, s_2, \infty)\nu(y), \quad \text{for all } y \text{ such that } |y| > s_2,\ |x-y| < s_1.$$

Thus $1/(K_1^X(s_2)K_2^X(s_1, s_2, \infty))$ measures the rate of preference of a single large jump from y to the origin ("direct single large jump" scenario) over two large jumps from $x \in B_{s_1}(y)$ to the origin ("first small steps, then two large jumps" scenario). This means that condition (2) corresponds to the situation when the first scenario outdoes the second at a rate which improves when the length of the long jumps increases appropriately with the scale of the smaller fluctuations. Secondly, condition (3) says that the "no small steps but direct large jump" scenario above outdoes the "first small steps, then large jump" scenario at a rate $1/\kappa_2$, no matter how large the smaller fluctuations are (i. e., there is always a suitably larger jump for which the former is preferred to occur). Conditions (4)–(5) are just technical and are met by a large selection of processes.

Remark 4.338. Note that for decaying potentials $\Lambda_D^V \asymp C(D)$, where $C(D)$ is a constant which in (4.9.90) appears amalgamated in a constant prefactor. Theorems 4.335–4.337 show that, under appropriate extra conditions, the decay continues to be driven by the same mechanism as for confining potentials. However, as Theorem 4.334 shows, when the jump-paring condition does not hold, the decay of bound states is necessarily slower than the fall-off rate of the jump kernel.

The decay rates for specific jump-paring Lévy processes and confining potentials are straightforward to derive from the above results. The most detailed results are known from explicit solutions of the eigenvalue problem, which will be summarized in Section 4.9.10 below. Here we discuss just one example for confining potentials, and a generic example for decaying potentials.

Example 4.339 (Jump-diffusions). Consider the operator $L = -(-\Delta)^{\alpha/2} + a\Delta$ generating the jump-diffusion process $(Z_t)_{t\geq 0}$, $Z_t = X_t + B_t$, where $(X_t)_{t\geq 0}$ is a rotationally symmetric α-stable process and $(B_t)_{t\geq 0}$ is an independent Brownian motion with diffusion matrix $A = aI_d$ (where I_d is the identity matrix in $\mathbb{R}^d$). For a confining potential $V \in \mathcal{K}_{\pm}^X$, the ground state of $H = -L + V$ decays as predicted by Theorem 4.331 or Corollary 4.332, i. e., the decay rate is determined by the fractional Laplacian and the jump component of the process. These results provide the leading order of decay only and it can be expected that a more refined asymptotics will contain a further term of smaller order coming from the Brownian component of the process.

Example 4.340. Consider a decaying potential $V \in \mathcal{K}_{\pm}^X$, and a symmetric jump-paring Lévy process $(X_t)_{t\geq 0}$ with generator L and profile function of the Lévy intensity of the form

$$g(r) = e^{-cr^{\beta}} r^{-\delta}, \tag{4.9.93}$$

with suitable parameters $c > 0$, $\beta, \delta \geq 0$. Assume that a ground state φ_0 of the nonlocal Schrödinger operator $H = -L + V$ exists. The following ground state decay behaviors occur:

(1) *Polynomially decaying Lévy intensities.* If $c = 0$ and $\delta > d$, then $\varphi_0(x) \asymp |x|^{-\delta}$ for all $\lambda_0 < 0$.

(2) *Stretched-exponentially decaying Lévy intensities.* If $c > 0$, $\beta \in (0,1)$ and $\delta \geq 0$, then $\varphi_0(x) \asymp e^{-c|x|^{\beta}} |x|^{-\delta}$ for all $\lambda_0 < 0$.

(3) *Exponentially decaying Lévy intensities.* If $c > 0$, $\beta = 1$, and $\delta > \frac{d+1}{2}$, and there exists λ_X^* such that $|\lambda_0| > \lambda_X^*$, then we have $\varphi_0(x) \asymp e^{-c|x|}|x|^{-\delta}$. When the eigenvalue λ_0 is arbitrary, then for large enough $|x|$ for every $\varepsilon > 0$ there is a $C_\varepsilon > 0$ such that

$$\varphi_0(x) \geq C_\varepsilon \max\left\{e^{-\bar{c}\sqrt{|\lambda_0|+\varepsilon}|x|}, e^{-c|x|}|x|^{-\delta}\right\}.$$

Moreover, if $\delta \leq \frac{d+1}{2}$, then $\limsup_{|x|\to\infty} \frac{\varphi_0(x)}{V(x)} = \infty$.

(4) *Superexponentially decaying Lévy intensities.* If $c > 0$, $\beta > 1$, and $\delta \geq 0$, then for every $\varepsilon > 0$ there is a $C_\varepsilon > 0$ such that

$$\varphi_0(x) \geq C_\varepsilon e^{-\bar{c}\sqrt{\min\{|\lambda_0|+\varepsilon, 1\}}\,|x|(\log|x|)^{(\beta-1)/\beta}}$$

for large enough $|x|$ and any $\lambda_0 < 0$.

Remark 4.341. It is seen that the above results improve on the decay estimates of eigenfunctions for the relativistic Schrödinger operator presented in Section 4.6.3.

Remark 4.342 (Regime change in the decay rates). The example above shows that there is a split in the decay behavior as the tails of the Lévy intensity are progressively made lighter from polynomial and subexponential through exponential to superexponential decays of ν. It is seen that if $|\lambda_0|$ is small with respect to the edge of the essential spectrum of the generator of the free process and ν decreases exponentially, then the decay rate of φ_0 is slower than ν, with essential contribution of λ_0. However, when $|\lambda_0|$ is large enough, the fall-off is again dominated by ν as long as the basic jump-paring condition holds. In particular, this includes the exponential case with $\delta > \frac{d+1}{2}$. On the other hand, when $\delta \leq \frac{d+1}{2}$ the jump-paring condition fails, the decay of φ_0 is slower than the decay of ν, no matter how large $|\lambda_0|$ is.

4.9.9 Local estimates of eigenfunctions

In this subsection we present qualitative and quantitative properties of ground states in the bulk. Since a detailed presentation would take much space, some of the proofs will need to be omitted, and we refer the reader to the literature discussed in Chapter 6 below.

Our main goal is to present some estimates on the ground states at short to mid-range from the origin for fractional relativistic Schrödinger operators, while extensions to more general nonlocal Schrödinger operators are possible, which, however, we will not present here. Taken together with the previous subsection, the description of the spatial behavior of ground states will then be reasonably complete. Before coming to this, we discuss some basic qualitative properties of ground states and other eigenfunctions, which will be essential information to the local spatial estimates.

First, we consider fractional relativistic Schrödinger operators with compactly supported potentials. Using the notations in Example 4.267, we take the fractional relativistic Laplacian $-L_{m,\alpha} = (-\Delta + m^{2/\alpha})^{\alpha/2} - m$, $\alpha \in (0, 2)$, $m \geq 0$, and define

$$H_{m,\alpha} = -L_{m,\alpha} - v1_{\mathcal{K}},$$

where $v > 0$, and $\mathcal{K} \subset \mathbb{R}^d$ is a bounded set with nonempty interior, which we call a potential well, like for classical Schrödinger operators above. We use the notation $\mathcal{B}_a$ for a ball of radius a centered at the origin; when $\mathcal{K} = \mathcal{B}_a$ with some $a > 0$, we call it a spherical potential well.

Throughout this subsection we assume that the operator $H_{m,\alpha}$ has a ground state φ_0 at eigenvalue $\lambda_0 < 0$, which as it has been seen before is unique and has a strictly positive version, which we will use without further notice. Also, we use the generic notation $(X_t)_{t\geq 0}$ of the Lévy processes generated by $L_{m,\alpha}$, which is a rotationally symmetric α-stable process if $m = 0$ and a rotationally symmetric relativistic α-stable process if $m > 0$. We denote expectation with respect to the path measure of either process starting at $x \in \mathbb{R}^d$ by $\mathbb{E}^x$. Then the Feynman-Kac representation

$$\varphi_0(x) = e^{\lambda_0 t} e^{-tH_{m,a}} \varphi_0(x) = e^{\lambda_0 t} \mathbb{E}^x [e^{-\int_0^t V(X_s)ds} \varphi_0(X_t)] = e^{\lambda_0 t} \mathbb{E}^x [e^{vU_t^{\mathcal{K}}(X)} \varphi_0(X_t)] \quad (4.9.94)$$

holds for all $x \in \mathbb{R}^d$ and $t \geq 0$, where $U_t^{\mathcal{K}}(X) = \int_0^t 1_{\mathcal{K}}(X_s)ds$ is the occupation measure up to time t of the set $\mathcal{K}$ by $(X_t)_{t\geq 0}$.

We begin by showing some invariance properties of the ground states of $H_{m,a}$ for potential wells with given symmetries.

Theorem 4.343. *Let $\mathcal{K} = \mathcal{B}_a$ with a given $a > 0$ and consider the ground state φ_0 of $H_{m,a}$. Then φ_0 is rotationally symmetric.*

Proof. First, observe that if another function $\widetilde{\varphi}_0$ existed satisfying (4.9.94), $\|\widetilde{\varphi}_0\|_2 = 1$, and $\widetilde{\varphi}_0 > 0$, then by the uniqueness of the ground state we would have $\widetilde{\varphi}_0 \equiv \varphi_0$ almost surely.

Fix a rotation $\mathcal{R} \in \mathrm{SO}(d)$ and consider $\widetilde{\varphi}_0(x) = \varphi_0(\mathcal{R}x)$. Clearly, since $\mathcal{R}$ is an isometry, it is immediate that $\|\widetilde{\varphi}_0\|_2 = 1$, $\widetilde{\varphi}_0 > 0$, and

$$\widetilde{\varphi}_0(x) = \mathbb{E}[e^{-\int_0^t (V(X_s+\mathcal{R}x)-\lambda_0)ds} \varphi_0(X_t + \mathcal{R}x)].$$

By rotational invariance of $(X_t)_{t\geq 0}$, we may furthermore write

$$\widetilde{\varphi}_0(x) = \mathbb{E}[e^{-\int_0^t (V(\mathcal{R}X_s+\mathcal{R}x)-\lambda_0)ds} \varphi_0(\mathcal{R}X_t + \mathcal{R}x)] = \mathbb{E}[e^{-\int_0^t (V(X_s+x)-\lambda_0)ds} \widetilde{\varphi}_0(X_t + x)],$$

where we used the fact that also V is rotationally invariant and $\mathcal{K} = \mathcal{B}_a$. Then by the observation above, $\widetilde{\varphi}_0 \equiv \varphi_0$ almost surely. Since $\mathcal{R} \in \mathrm{SO}(d)$ is arbitrary, the claim follows. □

We can also prove the reflection invariance of φ_0 when $\mathcal{K}$ has a mirror symmetry.

Theorem 4.344. *Let $\mathcal{K}$ be reflection symmetric with respect to a hyperplane $\mathcal{H} \ni 0$ and let $\mathcal{S} : \mathbb{R}^d \to \mathbb{R}^d$, $x \mapsto \mathcal{S}x$, be the reflection of x with respect to $\mathcal{H}$. Consider the ground state φ_0 of $H_{m,a}$. Then $\varphi_0(\mathcal{S}x) = \varphi_0(x)$, for all $x \in \mathbb{R}^d$.*

Proof. The proof is similar to Theorem 4.343. Consider $\widetilde{\varphi}_0(x) = \varphi_0(\mathcal{S}x)$. By the isometry property of $\mathcal{S}$, we have $\|\widetilde{\varphi}_0\|_2 = 1$, $\widetilde{\varphi}_0 > 0$, and

$$\widetilde{\varphi}_0(x) = \mathbb{E}[e^{-\int_0^t (V(X_s+\mathcal{S}x)-\lambda_0)ds} \varphi_0(X_t + \mathcal{S}x)].$$

Since $(X_t)_{t\geq 0}$ is isotropic, we get

$$\widetilde{\varphi}_0(x) = \mathbb{E}[e^{-\int_0^t (V(\mathcal{S}X_s+\mathcal{S}x)-\lambda_0)ds} \varphi_0(\mathcal{S}X_t + \mathcal{S}x)] = \mathbb{E}[e^{-\int_0^t (V(X_s+x)-\lambda_0)ds} \widetilde{\varphi}_0(X_t + x)],$$

where we used that if $x \in \mathcal{K}$, then also $\mathcal{S}x \in \mathcal{K}$. Like before, we obtain $\varphi_0(\mathcal{S}x) = \widetilde{\varphi}_0(x) = \varphi_0(x)$ for all $x \in \mathbb{R}^d$. □

We note that Theorems 4.343 and 4.344 hold respectively for any rotationally or reflection symmetric potential V once a ground state exists and is unique.

Next we show that for a spherical potential well the ground state of $H_{m,\alpha}$ is a radially nondecreasing function.

Theorem 4.345. *Let φ_0 be the ground state of $H_{m,\alpha}$. Then there exists a nonincreasing function $\rho : [0,\infty) \to \mathbb{R}$ such that $\varphi_0(x) = \rho(|x|)$ for every $x \in \mathbb{R}^d$.*

Proof. We only give a sketch of the proof. The idea is to construct a moving planes-type argument to estimate the effect of $L_{m,\alpha}$ on a reflection-antisymmetric function at its minimum, dependent on its distance in the antisymmetric direction from particular level sets. First, we can do this for the $m = 0$ case and then transfer it on to $m > 0$ by the decomposition discussed in Example 4.267.

The first step of the argument is crucially based on the following estimates. Denote $x = (x_1, x')$, where x_1 is its first component. For $\mu \in \mathbb{R}$, consider $x^\mu = (2\mu - x_1, x')$ and write $\mathcal{U}_\mu = \{x \in \mathbb{R}^d : x_1 < \mu\}$. We call w a μ-antisymmetric function whenever $w(x^\mu) = -w(x)$ for every $x \in \mathcal{U}_\mu$.

First, let $m = 0$, take a μ-antisymmetric function $w : \mathbb{R}^d \to \mathbb{R}$, and suppose that there exists $x \in \mathcal{U}_\lambda = \{x \in \mathbb{R}^d : x_1 < \lambda\}$ such that $w(x) = \inf_{x \in \mathcal{U}_\mu} w(y) < 0$, where $L_{0,\alpha} w(x)$ is defined via the integral representation (4.9.2). Denote $\delta = \lambda - x_1$. Then it can be shown that there exists a constant $C = C(d,\alpha) > 0$ such that

$$L_{0,\alpha} w(x) \le C\left(\delta^{-\alpha} w(x) - \delta \int_{\mathcal{U}_\lambda} \frac{(w(y) - w(x))(\lambda - y_1)}{|x - y^\mu|^{d+\alpha+2}} dy\right). \tag{4.9.95}$$

Next let $m > 0$ and take the operator $L_{m,\alpha}$ defined via the integral representation as in (4.9.4). Then extending (4.9.95) to the massive case gives

$$L_{m,\alpha} w(x) \le C_1\left((\delta^{-\alpha} - m) w(x) - \delta \int_{\mathcal{U}_\lambda} \frac{(w(y) - w(x))(\lambda - y_1) K_{(d+\alpha+2)/2}(m^{1/\alpha}|x - y^\mu|)}{|x - y^\mu|^{\frac{d+\alpha+2}{2}}} dy\right). \tag{4.9.96}$$

In the second step, we then apply a maximum principle and show monotonicity. □

Finally, we give a result on the location of maximizers of eigenfunctions in the potential well. Specifically, we show how deep inside the support of eigenfunction φ corresponding to an eigenvalue λ can a maximizer x^* be for a potential well. We denote by $\operatorname{Int} \mathcal{K}$ the interior of $\mathcal{K}$. Recall that the inradius $\operatorname{inrad} A$ of a set A is the radius of the largest ball contained in A.

Theorem 4.346. *Let $V = -v 1_{\mathcal{K}}$ with a bounded convex set $\mathcal{K}$, and φ be an eigenfunction of $H_{m,\alpha}$ corresponding to eigenvalue $\lambda = -|\lambda| < 0$. Then there exist two constants $\kappa_1 = \kappa_1(m, \alpha, \operatorname{inrad} \mathcal{K})$, $\kappa_2 = \kappa_2(m, \alpha, \operatorname{inrad} \mathcal{K}) > 0$ such that if*

$$\frac{\nu - |\lambda|}{|\lambda|} \le \kappa_1, \tag{4.9.97}$$

then $x^* \in \operatorname{Int} \mathcal{K}$ *and*

$$\operatorname{dist}(x^*, \partial\mathcal{K}) \ge \frac{1}{\sqrt{\Psi_{m,a}^{-1}(|\lambda|/\kappa_2)}}, \tag{4.9.98}$$

where $\Psi_{m,a}$ *are given in Example* 4.267.

Remark 4.347.

(1) We note that for a potential well $V = -\nu 1_{\mathcal{K}}$, $\nu > 0$, the bound

$$\nu - |\lambda| < \lambda_1^{\mathcal{K}}, \tag{4.9.99}$$

holds, where $\lambda_1^{\mathcal{K}}$ is the principal eigenvalue of $\Psi_{m,a}(-\Delta)$ in $\mathcal{K}$ with Dirichlet exterior condition on $\mathcal{K}^c$. Indeed, from the Feynman-Kac formula, we get that

$$\begin{aligned} \varphi(x) &\ge \mathbb{E}^x\left[e^{\int_0^t (\nu 1_{\mathcal{K}}(X_s)+\lambda)ds}\varphi(X_t)1_{\{t<\tau_{\mathcal{K}}\}}\right] \\ &\ge e^{t(\nu-|\lambda|)} \min_{y\in\mathcal{K}} \varphi(y)\, \mathbb{P}^x(t < \tau_{\mathcal{K}}), \quad x \in \mathcal{K}. \end{aligned}$$

By taking logarithms on both sides and dividing by $t > 0$, we get

$$\nu - |\lambda| \le -\limsup_{t\to\infty} \frac{1}{t} \log \mathbb{P}^x(t < \tau_{\mathcal{K}}) \le \lambda_1^{\mathcal{K}}.$$

Since the eigenvalue λ is simple, we obtain a strong inequality. Thus the numerator at the left-hand side of (4.9.97) is always bounded by $\lambda_1^{\mathcal{K}}$, and so for $|\lambda|$ large enough (4.9.97) holds.

(2) Also, notice that the result in Theorem 4.346 continues to hold for more general potentials V supported on $\mathcal{K}$ and $\lambda < 0$. In this situation, (4.9.97) will be replaced by

$$\frac{-\min_{x\in\mathcal{K}} V(x) - |\lambda|}{|\lambda|} \le \varrho_1.$$

Now we turn to local estimates of ground states for spherical potential wells. Our approach will be a combination of optional stopping, monotonicity, and estimates of first exit and entrance times. For arbitrary $x \in \mathbb{R}^d$, consider the random process $(M_t^x)_{t\ge 0}$,

$$M_t^x = e^{\lambda_0 t}\mathbb{E}^x[e^{\nu U_t^{\mathcal{K}}(X)}\varphi_0(X_t)]. \tag{4.9.100}$$

Since this is a martingale, with suitable counterparts of the stopping times used in (4.3.31) for Brownian motion, we can derive a representation which can be used to obtain sharp estimates in the bulk. Define the first exit time

$$\tau_a = \inf\left\{t > 0 : X_t \in \mathcal{B}_a^c\right\}$$

from the ball of the process when starting from the inside, and the first entrance time

$$T_a = \inf\left\{t > 0 : X_t \in \mathcal{B}_a\right\}$$

when starting from outside of the well. In contrast with Brownian motion, now the supports of X_{T_a} and X_{τ_a} spread over the full sets $\mathcal{B}_a$ and $\mathcal{B}_a^c$, respectively, and obtaining φ_0 in an explicit form becomes very difficult. Nevertheless, since $|X_{T_a}| \le a$ and $|X_{\tau_a}| \ge a$ almost surely, using that in a spherical potential well φ_0 is radially nonincreasing, we can derive good approximations.

Theorem 4.348. *Let φ_0 be the ground state of $H_{m,a}$ with $V = -v1_{\mathcal{B}_a}$. Then the estimates*

$$\frac{\varphi_0(x)}{\varphi_0(a)} \asymp \begin{cases} \mathbb{E}^x[e^{(v-|\lambda_0|)\tau_a}] & \text{if } |x| \le |a|, \\ \mathbb{E}^x[e^{-|\lambda_0|T_a}] & \text{if } |x| \ge |a| \end{cases} \tag{4.9.101}$$

hold, where the comparability constants depend on d, m, a, $|a|$, v, and λ_0.

In order to prove Theorem 4.348, we need some auxiliary results. First, we show that the martingale property leads to the following Feynman–Kac-type formula for the stopped process. Let $(M_t)_{t\ge0}^x$ be the process given in (4.9.100), and denote by $\mathbb{P}$ the measure of $(X_t)_{t\ge0}$.

Lemma 4.349. *The random process $(M_t^x)_{t\ge0}$ is a martingale with respect to the natural filtration $(\mathcal{F}_t^X)_{t\ge0}$ of $(X_t)_{t\ge0}$. Moreover, if τ is a $\mathbb{P}$-almost surely finite stopping time with respect to $(\mathcal{F}_t^X)_{t\ge0}$, then $\varphi_0(x) = M_\tau^x$.*

Proof. The martingale property follows as in Lemma 4.224 above. Since φ_0 is strictly positive, clearly M_t^x is almost surely nonnegative. Thus by the Feynman–Kac formula, $\mathbb{E}[(M_t^x)^+] = \mathbb{E}[M_t^x] = \varphi_0(x) \le \|\varphi_0\|_\infty$. The martingale convergence theorem implies that $(M_t^x)_{t\ge0}$ has a final element M_∞^x with $\mathbb{E}[|M_\infty^x|] < \infty$, and the optional stopping theorem then gives

$$\varphi_0(x) = \mathbb{E}[M_0^x] = \mathbb{E}[M_\tau^x] = \mathbb{E}^x[e^{-\int_0^\tau (V(X_s)-\lambda_0)ds}\varphi_0(X_\tau)]. \qquad \square$$

We state the next two lemmas without proof and refer to the literature. Recall $v_{m,a}$, $m \ge 0$, from Example 4.267.

Lemma 4.350. *For every $R > 0$, there exists a constant $C = C_{d,m,a,R} > 1$ such that*

$$\int_{\mathcal{B}_{CR}^c} v_{m,a}(|x-y|)dy \le \frac{1}{2}\int_{\mathcal{B}_R^c} v_{m,a}(|x-y|)dy.$$

Moreover, if $m = 0$, then $C_{d,0,a,R}$ does not depend on R.

The following lemma actually results from a combination of Lemma 4.350 and the Ikeda–Watanabe formula often used in the potential theory of Lévy jump processes.

Lemma 4.351. *For every $R > 0$, there exists a constant $\widetilde{C} = C_{d,m,a,R} > 0$ such that*

$$\mathbb{E}^x[g(\tau_R)1_{\{R \le |X_{\tau_R}| \le \widetilde{C}R\}}] \ge \frac{1}{2}\mathbb{E}^x[g(\tau_R)]$$

for every function $g \ge 0$ and $x \in \mathcal{B}_R$.

Now we are in the position to prove the local estimates in terms of the moment generating function of the exit time from a ball and the Laplace transform of the entrance time for a ball, respectively.

Proof of Theorem 4.348. Notice that by Theorem 4.343 the ground state φ_0 is rotationally symmetric, and by Theorem 4.345 it is nonincreasing.

Step 1. First, consider $|x| \le a$. Since τ_a is an almost surely finite stopping time, $X_{\tau_a} \in \mathcal{B}_a^c$, and $\varphi_0(X_{\tau_a}) \le \varphi_0(a)$ by monotonicity, Proposition 4.349 gives

$$\varphi_0(x) = \mathbb{E}^x[e^{(v-|\lambda_0|)\tau_a}\varphi_0(X_{\tau_a})] \le \varphi_0(a)\mathbb{E}^x[e^{(v-|\lambda_0|)\tau_a}]. \tag{4.9.102}$$

Let $C_1 = C_1(d, m, a, a) > 1$ given by Lemma 4.351. Then using that $|X_{\tau_a}| \le C_1|a|$, we get

$$\begin{aligned}\varphi_0(x) &\ge \mathbb{E}^x[e^{(v-|\lambda_0|)\tau_a}\varphi_0(X_{\tau_a})1_{\{|a| \le |X_{\tau_a}| \le C_1|a|\}}] \\ &\ge \varphi_0(C_1 a)\mathbb{E}^x[e^{(v-|\lambda_0|)\tau_a}1_{\{|a| \le |X_{\tau_a}| \le C_1|a|\}}].\end{aligned}$$

Define the almost surely finite stopping time $T_N = T_a \wedge N$ for every $N \in \mathbb{N}$. By Proposition 4.349, we obtain

$$\varphi_0(C_1 a) = \mathbb{E}^{C_1 a}[e^{-|\lambda_0|T_N}\varphi_0(X_{T_N})] \le \varphi_0(0)\mathbb{E}^{C_1 a}[e^{-|\lambda_0|T_N}].$$

Taking $N \to \infty$, dominated convergence gives

$$0 < \varphi_0(C_1 a) \le \varphi_0(0)\mathbb{E}^{C_1 a}[e^{-|\lambda_0|T_a}],$$

hence $\mathbb{P}^{C_1 a}(T_a = \infty) < 1$. Denote this probability by $C_2 = C_2(d, m, a, a)$. Thus there exists $C_3 = C_3(d, m, a, a) > 0$ such that $\mathbb{P}^{C_1 a}(T_a > C_3) < C_2$. By Proposition 4.349, we then get

$$\begin{aligned}\varphi_0(C_1 a) &= \mathbb{E}^{C_1 a}[e^{-|\lambda_0|T_N}\varphi_0(X_{T_N})] \\ &\ge \mathbb{E}^{C_1 a}[e^{-|\lambda_0|T_a}\varphi_0(X_{T_N})] \ge \mathbb{E}^{C_1 a}[e^{-|\lambda_0|T_a}\varphi_0(X_{T_N})1_{\{T_a \le C_3\}}].\end{aligned}$$

Taking $N \to \infty$ and using monotonicity, we obtain

$$\varphi_0(C_1 a) \ge \mathbb{E}^{C_1 a}[e^{-|\lambda_0|T_a}\varphi_0(X_{T_a})1_{\{T_a \le C_3\}}] \ge (1 - C_2)e^{-C_3|\lambda_0|}\varphi_0(a). \tag{4.9.103}$$

Since by Lemma 4.351 we have

$$\mathbb{E}^x[e^{(\nu-|\lambda_0|)\tau_a}1_{\{a\le|X_{\tau_a}|\le C_1 a\}}] \ge \frac{1}{2}\mathbb{E}^x[e^{(\nu-|\lambda_0|)\tau_a}], \tag{4.9.104}$$

a combination with (4.9.103) gives

$$\varphi_0(x) \ge \frac{C_4}{2}\varphi_0(a)\mathbb{E}^x[e^{(\nu-|\lambda_0|)\tau_a}],$$

with $C_4 = C_4(d, m, \alpha, a, |\lambda_0|) = (1 - C_2)e^{-C_3|\lambda_0|}$. Thus in the end

$$\varphi_0(x) \asymp \varphi_0(a)\mathbb{E}^x[e^{(\nu-|\lambda_0|)\tau_a}], \quad |x| \le a,$$

follows, where the proportionality constants depend on $d, m, \alpha, a, |\lambda_0|$.

Step 2. Next consider $|x| > a$, and let T_a and T_N be as in the previous step. Proposition 4.349 gives

$$\varphi_0(x) = \mathbb{E}^x[e^{-|\lambda_0|T_N}\varphi_0(X_{T_N})] \ge \mathbb{E}^x[e^{-|\lambda_0|T_a}\varphi_0(X_{T_N})] \ge \mathbb{E}^x[e^{-|\lambda_0|T_a}\varphi_0(X_{T_N})1_{\{T_a<\infty\}}],$$

since $T_N \le T_a$. Taking $N \to \infty$, we obtain

$$\varphi_0(x) \ge \mathbb{E}^x[e^{-|\lambda_0|T_a}\varphi_0(X_{T_a})1_{\{T_a<\infty\}}] \ge \varphi_0(a)\mathbb{E}^x[e^{-|\lambda_0|T_a}1_{\{T_a<\infty\}}] = \varphi_0(a)\mathbb{E}^x[e^{-|\lambda_0|T_a}]. \tag{4.9.105}$$

Moreover, by dominated convergence, the limit $N \to \infty$ gives

$$\varphi_0(x) \le \varphi_0(0)\mathbb{E}^x[e^{-|\lambda_0|T_N}] \to \varphi_0(0)\mathbb{E}^x[e^{-|\lambda_0|T_a}].$$

By a combination of Step 1, (4.9.109) below, and (4.9.99), it follows that there exists a constant $C_5 = C_5(d, m, \alpha, a, |\lambda_0|)$ such that

$$\varphi_0(0) \le C_5\varphi_0(a)\left(1 + \frac{\nu - |\lambda_0|}{\lambda_a - \nu + |\lambda_0|}\right) = C_6 = C_6(d, m, \alpha, a, \nu, |\lambda_0|)\varphi_0(a)$$

so that

$$\varphi_0(x) \le C_6\varphi_0(a)\mathbb{E}^x[e^{-|\lambda_0|T_a}]. \tag{4.9.106}$$

This then leads to $\varphi_0(x) \asymp \varphi_0(a)\mathbb{E}^x[e^{-|\lambda_0|T_a}]$ for all $|x| \ge a$, where the proportionality constants depend on d, m, α, a, ν, and $|\lambda_0|$.

Denote

$$\gamma = \frac{\nu - |\lambda_0|}{\lambda_a - \nu + |\lambda_0|} = \frac{1}{\frac{\lambda_a}{\nu-|\lambda_0|} - 1}, \tag{4.9.107}$$

which measures the ratio between the energy gap separating the ground state eigenvalue from the bottom value of the potential and the energy needed to climb the potential well. By (4.9.99), we have that $\gamma > 0$.

Corollary 4.352. *Under the conditions of Theorem* 4.348, *it follows, in particular, that*

$$\frac{\varphi_0(x)}{\varphi_0(a)} \asymp \begin{cases} 1 + \gamma \left(\frac{|a|-|x|}{|a|}\right)^{\alpha/2} & \text{if } |x| \leq |a|, \\ v_{m,\alpha}(|x|) & \text{if } |x| \geq |a|, \end{cases} \tag{4.9.108}$$

where γ is given by (4.9.107) *and the comparability constants depend on d, m, α, $|a|$, v, and* $|\lambda_0|$.

Proof. Let $|x| \leq a$. Also, let $R > 0$ and take $x \in \mathcal{B}_R$. Then for every $0 \leq \lambda < \lambda_R$, it is possible to show that

$$\mathbb{E}^x[e^{\lambda\tau_R} - 1] \asymp \frac{\lambda}{\lambda_R - \lambda}\left(\frac{R - |x|}{R}\right)^{\alpha/2}, \tag{4.9.109}$$

where the proportionality constant depends on d, m, α, and R. Moreover, $\mathbb{E}^x[e^{\lambda\tau_R}] = \infty$ for all $\lambda \geq \lambda_R$. Using Theorem 4.348 and (4.9.109), the result is straightforward.

Next let $|x| \geq a$. If $m = 0$, it can be shown that for some $R = R_{d,0,\alpha,a}$ there is a constant $C = C(d, 0, \alpha)$ such that

$$\varphi_0(x) \geq C v_{0,\alpha}(|x|), \quad |x| \geq R.$$

If $m > 0$, it can be shown that, similarly, for some $R = R_{d,m,\alpha,a}$,

$$\varphi_0(x) \geq C|x|^{-\frac{d+\alpha+1}{2}} e^{-m^{1/\alpha}|x|}, \quad |x| \geq R,$$

with a constant $C = C(d, m, \alpha, a)$. On the other hand, due to the behavior at infinity of the modified Bessel function of the second kind,

$$v_{m,\alpha}(x) \sim |x|^{-\frac{d+\alpha+1}{2}} e^{-m^{1/\alpha}|x|}, \quad \text{as } |x| \to \infty,$$

so with a suitable constant $\tilde{C} = \tilde{C}(d, m, \alpha)$ it follows that

$$\varphi_0(x) \geq \tilde{C} v_{m,\alpha}(|x|), \quad |x| \geq R.$$

Hence by (4.9.106), we get $\mathbb{E}^x[e^{-|\lambda_0|T_a}] \geq K\, v_{m,\alpha}(|x|)$ for all $|x| \geq R$, with a constant $K > 0$. Thus we finally obtain

$$\mathbb{E}^x[e^{-|\lambda_0|T_a}] \asymp v_{m,\alpha}(|x|), \quad |x| \geq a,$$

where the proportionality constants depend on d, a, m, α, v, and $|\lambda_0|$. □

Remark 4.353. From the proofs, it actually follows that

$$C_1\varphi_0(a)e^{-C_2|\lambda_0|}\mathbb{E}^x[e^{(\nu-|\lambda_0|)\tau_a}] \le \varphi_0(x) \le C_3\varphi_0(a)\mathbb{E}^x[e^{(\nu-|\lambda_0|)\tau_a}]$$

and

$$C_1\varphi_0(a)e^{-C_2|\lambda_0|}\Big(1+\gamma\Big(\frac{|a|-|x|}{|a|}\Big)^{\alpha/2}\Big) \le \varphi_0(x) \le C_3\varphi_0(a)\Big(1+\gamma\Big(\frac{|a|-|x|}{|a|}\Big)^{\alpha/2}\Big)$$

also hold for every $|x| \le |a|$, with positive constants $C_1 = C_1(d,m,\alpha,a)$, $C_2 = C_2(d,m,\alpha,a)$, and $C_3 = C_3(d,m,\alpha,a)$.

In the immediate neighborhood of the potential well, we can obtain a more refined result.

Proposition 4.354. *There exist constants* $\varepsilon = \varepsilon_{d,m,\alpha,a,\nu}$ *and* $C = C_{d,m,\alpha,a,\nu} > 0$ *such that*

$$\Big|\frac{\varphi(x)}{\varphi(a)} - 1\Big| \le C\big||x| - a\big|^{\alpha/2}, \quad x \in \mathcal{B}_{|a|+\varepsilon} \setminus \mathcal{B}_{|a|-\varepsilon}.$$

Proof. The case $x \in \partial\mathcal{B}_a$ is obvious. Let $x \in \mathcal{B}_a$. By (4.9.102), we have

$$\frac{\varphi(x)}{\varphi(a)} - 1 \le \mathbb{E}^x[e^{(\nu-|\lambda_0|)\tau_a} - 1] \le C(a-|x|)^{\alpha/2},$$

where we used (4.9.109). Let now $x \in \mathcal{B}_a^c$. By (4.9.105), we then have

$$1 - \frac{\varphi(x)}{\varphi(a)} \le \mathbb{E}^x[1 - e^{-|\lambda_0|T_a}].$$

It is possible to show that there exists $R^* = R^*(d,m,\alpha,|a|,\lambda) > |a|$ such that $\mathbb{E}^x[e^{-\lambda T_a}] \ge \frac{1}{2}$, for all $|a| \le |x| \le R^*$. Choosing this R^* and $\varepsilon = (R^* - |a|) \wedge |a|$, the result follows. □

Making use of the fact that $\|\varphi_0\|_2 = 1$, we can derive a two-sided bound on the value of $\varphi_0(a)$. We skip the proof.

Proposition 4.355. *Denote* $\mathscr{I} = \int_1^\infty r^{d-1}v_{m,\alpha}^2(|a|r)\,dr$ *and let* $B(x,y)$ *be the Beta-function. Then*

$$\varphi_0(a) \asymp \frac{1}{\sqrt{|a|^d d\omega_d\left(\frac{1}{d} + 2\gamma B\left(d, 1+\frac{\alpha}{2}\right) + \gamma^2 B(d,1+\alpha) + \mathscr{I}\right)}},$$

where γ *is given by* (4.9.107) *and* ω_d *is the volume of the d-dimensional unit ball.*

From an asymptotic analysis of the profile functions in Corollary 4.352, the difference of the leading terms indicates that, although the ground state φ_0 is continuous everywhere, around the boundary of the potential well it has different smoothness properties. We state the regularity properties of the ground state arbitrarily close to the boundary next. First, consider the operators without the potential.

Lemma 4.356. *Consider the operator $L_{m,\alpha}$ and the following cases:*
(1) $\alpha \in (0,1)$ *and* $f \in C^{\alpha+\delta}_{\mathrm{loc}}(\mathbb{R}^d) \cap L^\infty(\mathbb{R}^d)$ *for some* $\delta \in (0, 1-\alpha)$;
(2) $\alpha \in [1,2)$ *and* $f \in C^{1,\alpha+\delta-1}_{\mathrm{loc}}(\mathbb{R}^d) \cap L^\infty(\mathbb{R}^d)$ *for some* $\delta \in (0, 2-\alpha)$.

In either case above, the function $\mathbb{R}^d \ni x \mapsto L_{m,\alpha} f(x)$ *is continuous.*

Next consider the operator with the potential added.

Theorem 4.357. *Let φ_0 be the ground state of $H_{m,\alpha}$. The following hold:*
(1) *If $\alpha \in (0,1)$, then $\varphi_0 \notin C^{\alpha+\delta}_{\mathrm{loc}}(\mathbb{R}^d)$ for every $\delta \in (0, 1-\alpha)$. Moreover, we have that $\varphi_0 \notin C^{\alpha+\delta}_{\mathrm{loc}}(\mathcal{B}_{|a|+\varepsilon} \setminus \overline{\mathcal{B}}_{|a|-\varepsilon})$ for every $\varepsilon \in (0, |a|)$ and $\delta \in (0, 1-\alpha)$.*
(2) *If $\alpha \in [1,2)$, then $\varphi_0 \notin C^{1,\alpha+\delta-1}_{\mathrm{loc}}(\mathbb{R}^d)$ for every $\delta \in (0, 2-\alpha)$. Moreover, we have that $\varphi_0 \notin C^{1,\alpha+\delta-1}_{\mathrm{loc}}(\mathcal{B}_{|a|+\varepsilon} \setminus \overline{\mathcal{B}}_{|a|-\varepsilon})$ for every $\varepsilon \in (0, |a|)$ and $\delta \in (0, 2-\alpha)$.*

Part (1) of the above theorem implies, in particular, that if $\alpha \in (0,1)$, then $\varphi_0 \notin C^1$ on $\partial\mathcal{B}_a$, and part (2) implies that if $\alpha \in [1,2)$, then $\varphi_0 \notin C^2$ on $\partial\mathcal{B}_a$.

We can make use of the local estimates to study the L^p-size of ground states. Since $\|\varphi_0\|_2 = 1$, the expression $\varphi_0^2(x)dx$ defines a probability measure on $\mathbb{R}^d$. Let $p > 0$ and define the p-moment

$$\mathcal{M}_p(\varphi_0) = \left(\int_{\mathbb{R}^d} |x|^p \varphi_0^2(x) dx \right)^{1/p}.$$

Let $m \geq 0$, $\alpha \in (0,2)$, and write

$$p^*(m,\alpha) = \begin{cases} d + 2\alpha & \text{if } m = 0, \\ \infty & \text{if } m > 0. \end{cases}$$

Lemma 4.358. *The following cases occur:*
(1) *If $0 < p < p^*(m,\alpha)$, then $\mathcal{M}_p(\varphi_0) < \infty$.*
(2) *If $p \geq p^*(m,\alpha)$, then $\mathcal{M}_p(\varphi_0) = \infty$.*

Proof. Combining $\nu_{0,\alpha}(r) = C_{d,\alpha} r^{-d-\alpha}$ and $\nu_{m,\alpha}(r) \asymp r^{-(d+\alpha+1)/2} e^{-m^{1/\alpha} r}$ as $r \to \infty$ if $m > 0$ with Corollary 4.352 implies the result directly. □

Proposition 4.359. *Let $0 < p < p^*(m,\alpha)$. There exist constants $C_1 = C_1(d, m, \alpha, a, p)$, and $C_2 = C_2(d, m, \alpha, a) > 0$ such that*

$$\mathcal{M}_p(\varphi_0) \geq C_1 \gamma^{2/p} e^{-\frac{2}{p} C_2 |\lambda_0|} \varphi_0^{2/p}(a). \tag{4.9.110}$$

Proof. Remark 4.353 implies

$$\begin{aligned}\varphi_0^2(x) &\geq C_3^2\varphi_0^2(a)\left(1+2\gamma\left(\frac{a-|x|}{a}\right)^{\alpha/2}+\gamma^2\left(\frac{a-|x|}{a}\right)^{\alpha}\right)e^{-2C_2|\lambda_0|}\\ &\geq C_3^2\varphi_0^2(a)\gamma^2\left(\frac{a-|x|}{a}\right)^{\alpha}e^{-2C_2|\lambda_0|},\quad |x|\leq a,\end{aligned}$$

where the last step follows by the fact that $\frac{a-|x|}{a}\leq 1$. Hence

$$\begin{aligned}\int_{\mathbb{R}^d}|x|^p\varphi_0^2(x)dx &\geq \int_{\mathcal{B}_a}|x|^p\varphi_0^2(x)dx\\ &\geq C_3^2\varphi_0^2(a)\gamma^2e^{-2C_2|\lambda_0|}\int_{\mathcal{B}_a}|x|^p\left(\frac{a-|x|}{a}\right)^{\alpha}dx.\end{aligned}$$

Choosing $C_1^p = C_3^2\int_{\mathcal{B}_a}|x|^p\left(\frac{a-|x|}{a}\right)^{\alpha}dx$ completes the proof. □

Proposition 4.360. *Let $0<p<p^*(m,\alpha)$ and $v>\lambda_a+\delta$ for some $\delta>0$. Then there exists a constant $C=C(d,m,\alpha,\delta,a,p)>0$ such that*

$$\mathcal{M}_p(\varphi_0)\leq C\,\gamma^{2/p}\varphi_0^{2/p}(a).$$

Proof. As in Theorem 4.348, notice that for $|x|\geq|a|$ we have, by Proposition 4.345,

$$\varphi_0(x)\leq\varphi_0(0)\mathbb{E}^x[e^{-|\lambda_0|T_a}]. \tag{4.9.111}$$

Moreover, by Remark 4.353,

$$\varphi_0(0)\leq C^{(1)}_{d,m,\alpha,a}\gamma\varphi_0(a). \tag{4.9.112}$$

On the other hand, from $v-|\lambda_0|<\lambda_a$ we get $|\lambda_0|>v-\lambda_a>\delta$ and then

$$\mathbb{E}^x[e^{-|\lambda_0|T_a}]\leq\mathbb{E}^x[e^{-\delta T_a}]\leq C_2 v_{m,\alpha}(|x|),\quad |x|\geq|a|, \tag{4.9.113}$$

by using (4.9.109). Combining (4.9.111)–(4.9.113), we get

$$\varphi_0(x)\leq C_3\gamma\varphi_0(a)v_{m,\alpha}(|x|),\quad |x|\geq|a|, \tag{4.9.114}$$

where $C_3=C_1C_2$. For $|x|\leq|a|$, we have directly by Remark 4.353 that

$$\varphi_0(x)\leq C_1\varphi_0(a)\left(1+\gamma\left(\frac{|a|-|x|}{|a|}\right)^{\alpha/2}\right)\leq C_1\gamma\varphi_0(a), \tag{4.9.115}$$

where again we used that $\frac{|a|-|x|}{|a|}\leq 1$. Hence by (4.9.114)–(4.9.115), we get

$$\int_{\mathbb{R}^d} |x|^p \varphi_0^2(x)dx = \int_{\mathcal{B}_a} |x|^p \varphi_0^2(x)dx + \int_{\mathcal{B}_a^c} |x|^p \varphi_0^2(x)dx \leq C^p \gamma^2 \varphi_0^2(a),$$

where $C^p = \max\left\{C_1^2 \int_{\mathcal{B}_a} |x|^p dx,\ C_3^2 \int_{\mathcal{B}_a^c} |x|^p v_{m,\alpha}^2(x)dx\right\}$. □

We can further estimate the contribution to the moments of the probability mass separately inside and outside the potential well. For every $p \geq 1$, define

$$\mathcal{M}_p^{\text{in}}(\varphi_0) = \mathcal{M}_p(\varphi_0 1_{\mathcal{B}_a}) = \Big(\int_{\mathcal{B}_a} |x|^p \varphi_0^2(x)dx\Big)^{1/p},$$

$$\mathcal{M}_p^{\text{out}}(\varphi_0) = \mathcal{M}_p(\varphi_0 1_{\mathcal{B}_a^c}) = \Big(\int_{\mathcal{B}_a^c} |x|^p \varphi_0^2(x)dx\Big)^{1/p}.$$

Since φ_0 is continuous, it is clear that $\mathcal{M}_p^{\text{in}}(\varphi_0) < \infty$ for all $p \geq 1$, $m \geq 0$ and $\alpha \in (0, 2)$. We can provide a simple two-sided bound to $\mathcal{M}_p^{\text{in}}(\varphi_0)$.

Proposition 4.361. *Let $p > 0$. Then there exist positive constants $c_1 = c_1(d, m, \alpha, |a|, p)$, $c_2 = c_2(d, m, \alpha, |a|)$, and $c_3 = c_3(d, m, \alpha, |a|)$ such that*

$$c_1 e^{-2c_2|\lambda_0|/p} \gamma^{2/p} \varphi_0^{2/p}(a) \leq \mathcal{M}_p^{\text{in}}(\varphi_0) \leq c_3 \gamma^{2/p} \varphi_0^{2/p}(a).$$

Proof. The lower bound follows by a similar argument as in Proposition 4.359. For the upper bound, observe that (4.9.115) continues to hold and then

$$\int_{\mathcal{B}_a} |x|^p \varphi_0^2(x)dx \leq c_4^2 \frac{d}{d+p} \omega_d |a|^{d+p} \gamma^2 \varphi_0^2(a).$$

By choosing $c_3^p = c_4^2 \frac{d}{d+p} \omega_d |a|^{d+p}$, we get the statement. □

Remark 4.362. As in Theorem 4.348, we can prove that there exist constants $k_1 = k_1(d, m, \alpha, |a|) > 0$ and $k_2 = k_2(d, m, \alpha, |a|) > 1$ such that

$$\varphi_0(x) \geq k_1 \gamma \varphi_0(a) \left(\frac{|a| - |x|}{|a|}\right)^{\alpha/2} \mathbb{E}^{k_2 a}[e^{-|\lambda_0| T_a}],$$

and thus

$$\mathcal{M}_p^{\text{in}}(\varphi_0) \geq C_{d,m,\alpha,a,p} \gamma^{2/p} \varphi_0^{2/p}(a) \left(\mathbb{E}^{k_2 a}[e^{-|\lambda_0| T_a}]\right)^{2/p}.$$

Furthermore, by Lemma 4.358, it follows that $\mathcal{M}_p^{\text{out}}(\varphi_0) < \infty$ if and only if $p < p^*(m, \alpha)$. We give an upper bound for $\mathcal{M}_p^{\text{out}}(\varphi_0)$.

Proposition 4.363. *Let $0 < p < p^*(m, \alpha)$ and $v > \lambda_a + \delta$ for some $\delta > 0$. There exists a constant $C = C_{d,m,\alpha,\delta,|a|,p}$ such that*

$$\mathcal{M}_p^{\mathrm{out}}(\varphi_0) \le C\gamma^{2/p}\varphi_0(a)^{2/p}.$$

Moreover, $C \to 0$ as $\delta \to \infty$.

Proof. By (4.9.111) and (4.9.112), we have for $|x| \ge a$,

$$\varphi_0(x) \le C_1\gamma\varphi_0(a)\mathbb{E}^x[e^{-|\lambda_0|T_a}] \le C_1\gamma\varphi_0(a)\mathbb{E}^x[e^{-\delta T_a}],$$

and thus

$$\int_{\mathcal{B}_a^c} |x|^p\varphi_0^2(x)dx \le C_1\gamma^2\varphi_0^2(a)\int_{\mathcal{B}_a^c} |x|^p\left(\mathbb{E}^x[e^{-\delta T_a}]\right)^2 dx,$$

where, due to a combination of (4.9.109) and $p < p^*(m,\alpha)$, the integral at the right-hand side above is bounded above by $\int_{\mathcal{B}_a^c} |x|^p v_{m,\alpha}^2(|x|)dx < \infty$. The upper bound follows by choosing $C^p = C_1^2\int_{\mathcal{B}_a^c} |x|^p\left(\mathbb{E}^x[e^{-\delta T_a}]\right)^2 dx$. The second statement follows by the dominated convergence theorem. □

Remark 4.364. We note that by Propositions 4.361–4.363 and Remarks 4.362–4.364, we also have

$$\frac{\mathcal{M}_p^{\mathrm{out}}(\varphi_0)}{\mathcal{M}_p^{\mathrm{in}}(\varphi_0)} \le K_1\frac{\left(\int_{\mathcal{B}_a^c} |x|^p\left(\mathbb{E}^x[e^{-|\lambda_0|T_a}]\right)^2 dx\right)^{1/p}}{\left(\mathbb{E}^{C_2a}[e^{-|\lambda_0|T_a}]\right)^{2/p}} \le K_2\frac{\left(\int_{\mathcal{B}_a^c} |x|^p v_{m,\alpha}^2(|x|)dx\right)^{1/p}}{v_{m,\alpha}^{2/p}(C_2a)},$$

where $K_1 = K_1(d,m,\alpha,|a|,p) > 0$, $K_2 = K_2(d,m,\alpha,|a|,p,|\lambda_0|) > 0$, and C_2 is defined in Remark 4.362.

We complete this line of study by discussing the R-dependence of the probability distribution function

$$F(R) = \int_{\mathcal{B}_R} \varphi_0^2(x)dx, \quad R \ge 0. \tag{4.9.116}$$

Below we present a two-sided bound for F without proof by using the local estimates of the ground state obtained above. We will use the standard notation $B(u;x,y) = \int_0^u r^{x-1}(1-r)^{y-1}dr$, $0 \le u \le 1$, for the incomplete Beta-function, and $B(x,y)$ for the Beta-function, which equals the incomplete Beta-function for $u = 1$. Also, we write

$$\bar{\gamma} = \gamma - 1 = \frac{\nu - |\lambda_0|}{\lambda_a - \nu + |\lambda_0|}.$$

Theorem 4.365.

(1) *Let $R \le |a|$. There exist positive constants $C_1 = C_1(d,m,\alpha,|a|)$, $C_2 = C_2(d,m,\alpha,|a|)$, and $C_3 = C_3(d,m,\alpha,|a|)$ such that*

$$K_1\left(\frac{1}{d}\left(\frac{R}{|a|}\right)^d + 2\bar{\gamma}B\Big(\frac{R}{|a|};d,1+\frac{\alpha}{2}\Big) + \bar{\gamma}^2 B\Big(\frac{R}{|a|};d,1+\alpha\Big)\right)$$
$$\leq F(R) \leq K_2\left(\frac{1}{d}\left(\frac{R}{|a|}\right)^d + 2\bar{\gamma}B\Big(\frac{R}{|a|};d,1+\frac{\alpha}{2}\Big) + \bar{\gamma}^2 B\Big(\frac{R}{|a|};d,1+\alpha\Big)\right),$$

where $K_1 = (C_1\varphi_0(a))^2 e^{-2C_2|\lambda_0|} d\omega_d |a|^d$ *and* $K_2 = (C_3\varphi_0(a))^2 d\omega_d|a|^d$.

(2) *Let* $R \geq |a|$. *There exists a constant* $C_4 = C_4(d, m, \alpha, |a|, |\lambda_0|) > 0$ *such that*

$$F(R) \geq C_4\varphi_0^2(a) d\omega_d |a|^d \left(\frac{1}{d} + 2\bar{\gamma}B\left(d, 1+\frac{\alpha}{2}\right) + \bar{\gamma}^2 B\left(d, 1+\alpha\right) + \mathscr{I}(R)\right),$$

where $\mathscr{I}(R) = \int_1^{R/|a|} r^{d-1} v_{m,\alpha}^2(|a|r)\,dr$.

(3) *Let* $R \geq |a|$. *Suppose that* $v > \lambda_a$ *and write* $\delta = v - \lambda_a$. *Then there exists a constant* $C_5 = C_5(d, m, \alpha, \delta, |a|) > 0$ *such that*

$$F(R) \leq C_5\varphi_0^2(a)\left(\frac{1}{d} + 2\bar{\gamma}B\left(d, 1+\frac{\alpha}{2}\right) + \bar{\gamma}^2 B\left(d, 1+\alpha\right) + \gamma^2 \mathscr{I}(R)\right).$$

The above results show a concentration property of the ground state. Part (1) of Theorem 4.365 indicates that inside the potential well, for fixed R the paths select from three different independent sampling strategies in building up the stationary distribution φ_0^2. One corresponds to the uniform distribution $(\frac{R}{|a|})^d$, and the other two are Beta-distributions $\mathrm{B}(d, \beta_1)$, $\mathrm{B}(d, \beta_2)$, contributing with different weights. Their probability distribution functions are, respectively,

$$I_{R/|a|}(d,\beta_1) = \frac{B(\frac{R}{a}; d, \beta_1)}{B(d,\beta_1)}, \; \beta_1 = 1 + \frac{\alpha}{2} \quad \text{and} \quad I_{R/|a|}(d,\beta_2) = \frac{B(\frac{R}{a}; d, \beta_2)}{B(d,\beta_2)}, \; \beta_2 = 1+\alpha.$$

The contribution of $\mathrm{B}(d, \beta_1)$ with weight proportional to $w_1 = 2\bar{\gamma}B(d, \beta_1)$ corresponds to relatively few large jumps, while the contribution of $\mathrm{B}(d, \beta_2)$ with weight proportional to $w_2 = \bar{\gamma}^2 B(d, \beta_2)$ corresponds to relatively many small jumps. A similar observation can be made about the mix of random jump strategies outside the potential well.

The approach above to derive local estimates on the ground state of $H_{m,\alpha}$ with potential wells can be extended to potentials supported everywhere on $\mathbb{R}^d$. We illustrate this by a few statements.

Let $V(x) = -v(|x|)$ with a continuous nonincreasing function $v : \mathbb{R}^+ \to \mathbb{R}^+$ such that $\lim_{r\to\infty} v(r) = 0$. We continue to assume that $H_{m,\alpha}$ has a ground state $\varphi_0 > 0$ with eigenvalue $\lambda_0 < 0$. By extension of Theorem 4.345, the function φ_0 is radial and nonincreasing, thus we can write $\varphi_0(x) = \varrho_0(|x|)$ with a suitable nonincreasing function $\varrho_0 : \mathbb{R}^+ \to \mathbb{R}^+$.

Theorem 4.366. *Let* φ_0 *be the ground state of* $H_{m,\alpha}$ *with* $V(x) = -v(|x|)$, $v : \mathbb{R}^+ \to \mathbb{R}^+$ *nonincreasing and continuous. Choose any* $\beta > 0$ *such that the level set*

$$\mathcal{K}_\beta = \left\{x \in \mathbb{R}^d : V(x) < -\beta\right\} \neq \emptyset.$$

Then there exists a constant $C = C(d, m, \alpha, \beta, |\lambda_0|)$ *such that*

$$C\varphi_0(x_\beta)\mathbb{E}^x[e^{(\beta-|\lambda_0|)\tau_{r_\beta}}] \le \varphi_0(x) \le \varphi_0(x_\beta)\mathbb{E}^x[e^{(v(0)-|\lambda_0|)\tau_{r_\beta}}], \quad x \in \mathcal{K}_\beta,$$

where $\tau_{r_\beta} = \inf\{t > 0 : X_t \in \mathcal{K}_\beta^c\}$, $x_\beta \in \partial\mathcal{K}_\beta$ *is arbitrary and* $r_\beta = |x_\beta|$.

Proof. Let $x \in \mathcal{K}_\beta$. Since v is nonincreasing and continuous, $\mathcal{K}_\beta$ is an open ball centered at the origin, thus we can write it as $\mathcal{K}_\beta = \mathcal{B}_{r_\beta}$ with an appropriate $r_\beta > 0$. Take the stopping time $\tau_{r_\beta} = \inf\{t > 0 : X_t \in \mathcal{B}_{r_\beta}^c\}$. Since φ_0 is radial, it does not vary on $\partial\mathcal{B}_{r_\tau}$. Let $x_\beta \in \partial\mathcal{B}_{r_\beta}$ be arbitrary. By Proposition 4.349, we have

$$\varphi_0(x) = \mathbb{E}^x[e^{\int_0^{\tau_{r_\beta}} v(|X_s|)ds-|\lambda_0|\tau_{r_\beta}}\varphi_0(X_{\tau_{r_\beta}})] \le \varphi_0(x_\beta)\mathbb{E}^x[e^{(v(0)-|\lambda_0|)\tau_{r_\beta}}].$$

Take $C_1 = C_1(d, m, \alpha, r_\beta) > 1$ as defined in Lemma 4.351 and notice that

$$\begin{aligned}\varphi_0(x) &\ge \mathbb{E}^x[e^{\int_0^{\tau_{r_\beta}} v(|X_s|)ds-|\lambda_0|\tau_{r_\beta}}\varphi_0(X_{\tau_{r_\beta}})1_{\{r_\beta\le X_{\tau_{r_\beta}}\le C_1 r_\beta\}}] \\ &\ge \varphi_0(C_1x_\beta)\mathbb{E}^x[e^{(\beta-|\lambda_0|)\tau_{r_\beta}}1_{\{r_\beta\le X_{\tau_{r_\beta}}\le C_1 r_\beta\}}]. \end{aligned} \tag{4.9.117}$$

Also, by the definition of C_1,

$$\mathbb{E}^x[e^{(\beta-|\lambda_0|)\tau_{r_\beta}}1_{\{r_\beta\le X_{\tau_{r_\beta}}\le C_1 r_\beta\}}] \ge \frac{1}{2}\mathbb{E}^x[e^{(\beta-|\lambda_0|)\tau_{r_\beta}}]. \tag{4.9.118}$$

As in Theorem 4.348, we then obtain $\varphi_0(C_1x_\beta) \ge \mathbb{E}^{C_1x_\beta}[e^{-|\lambda_0|T_{r_\beta}}\varphi_0(X_{T_{r_\beta}})]$, where $T_{r_\beta} = \inf\{t > 0 : X_t \in \mathcal{K}_\beta\}$ and we used the fact that $v(|x|) \ge 0$ for all $x \in \mathbb{R}^d$. Hence by monotonicity,

$$\varphi_0(C_1x_\beta) \ge \varphi_0(x_\beta)\mathbb{E}^{C_2x_\beta}[e^{-|\lambda_0|T_{r_\beta}}] \ge C_2\varphi_0(x_\beta), \tag{4.9.119}$$

where $C_2 = C_2(d, m, \alpha, r_\beta, |\lambda_0|) = C_3 v_{m,\alpha}(C_1r_\beta)$ and $C_3 = C_3(d, m, \alpha, r_\beta, |\lambda_0|)$. A combination of (4.9.117)–(4.9.119) yields the result. □

Remark 4.367. Similarly to Step 1 in Theorem 4.348, we have that $\beta - |\lambda_0| < \lambda_{r_\beta}$ whenever $\mathcal{K}_\beta \neq \emptyset$. In particular, due to $\lim_{\beta\to v(0)} \lambda_{r_\beta} = \infty$, there is a constant $\beta_0 > 0$ such that $v(0) - |\lambda_0| < \lambda_{r_\beta}$ for every $\beta \in (\beta_0, v(0))$.

By a similar argument as in the proof of Corollary 4.352, we obtain more explicit bounds.

Corollary 4.368. *Under the conditions of Theorem 4.366, let* $\mathcal{K}_\beta \neq \emptyset$, *and furthermore assume that* $|\lambda_0| < \beta$ *and* $v(0) - |\lambda_0| < \lambda_{\mathcal{K}_\beta}$. *Then there exists a constant* $C = C(d, m, \alpha, \beta, |\lambda_0|)$ *such that*

$$C\varphi_0(x_\beta)\left(1+\frac{\beta-|\lambda_0|}{\lambda_{\mathcal{K}_\beta}-\beta+|\lambda_0|}\left(\frac{r_\beta-|x|}{r_\beta}\right)^{\alpha/2}\right)$$
$$\le \varphi_0(x)\le \varphi_0(x_\beta)\left(1+\frac{v(0)-|\lambda_0|}{\lambda_{\mathcal{K}_\beta}-v(0)+|\lambda_0|}\left(\frac{r_\beta-|x|}{r_\beta}\right)^{\alpha/2}\right),$$

for every $x\in\mathcal{K}_\beta$, *where* $\tau_{r_\beta}=\inf\{t>0: X_t\in\mathcal{K}_\beta^c\}$, $x_\beta\in\partial\mathcal{K}_\beta$ *is arbitrary, and* $r_\beta=|x_\beta|$.

Theorem 4.369. *Let* φ_0 *be the ground state of* $H_{m,\alpha}$ *with* $V(x)=-v(|x|)$, $v:\mathbb{R}^+\to\mathbb{R}^+$ *nonincreasing and continuous. Let* $\beta_1\le|\lambda_0|$ *and* $\beta_2\in(\beta_0,v(0))$, *where* β_0 *is defined as in Remark* 4.367, *such that* $\beta_1\le\beta_2$. *Define* $\mathcal{K}_{\beta_i}=\{x\in\mathbb{R}^d, V(x)<-\beta_i\}$, $i=1,2$. *Then*

$$\varphi_0(x_{\beta_1})\mathbb{E}^x[e^{-|\lambda_0|T_{r_{\beta_1}}}]\le\varphi_0(x)\le C_{d,m,\alpha,\beta_2,|\lambda_0|}\varphi_0(x_{\beta_2})\mathbb{E}^x[e^{(\beta_1-|\lambda_0|)T_{r_{\beta_1}}}],\quad x\in\mathcal{K}_{\beta_1},$$

where $x_{\beta_i}\in\partial\mathcal{K}_{\beta_i}$ *and* $r_{\beta_i}=|x_{\beta_i}|$, $i=1,2$.

Proof. By a similar argument as in Theorem 4.366, there exist r_{β_i} such that $\mathcal{K}_{\beta_i}=\mathcal{B}_{r_{\beta_i}}$, $i=1,2$. Moreover, $\mathcal{K}_{\beta_1}^c\subseteq\mathcal{K}_{\beta_2}^c$ since v is nonincreasing. Let $x\in\mathcal{K}_{\beta_1}^c$ and observe that, as in Theorem 4.348,

$$\varphi_0(x_{\beta_1})\mathbb{E}^x[e^{-|\lambda_0|T_{r_{\beta_1}}}]\le\varphi_0(x)\le\varphi_0(0)\mathbb{E}^x[e^{(\beta_1-|\lambda_0|)T_{r_{\beta_1}}}],$$

where $x_{\beta_1}\in\partial\mathcal{B}_{r_{\beta_1}}$. Using that $0\in\mathcal{K}_{\beta_2}$, by Corollary 4.368, we get

$$\varphi_0(0)\le\varphi_0(x_{\beta_2})\left(1+\frac{v(0)-|\lambda_0|}{\lambda_{r_{\beta_2}}-v(0)+|\lambda_0|}\right)=C_{d,m,\alpha,\beta_2,|\lambda_0|}\varphi_0(x_{\beta_2}),\quad x_{\beta_2}\in\partial\mathcal{B}_{r_{\beta_2}}.\qquad\square$$

As in Corollary 4.352 again, by using the asymptotics of the Laplace transform of the hitting times we get the following.

Corollary 4.370. *Let* φ_0 *be the ground state of* $H_{m,\alpha}$ *with* $V(x)=-v(|x|)$, $v:\mathbb{R}^+\to\mathbb{R}^+$ *nonincreasing and continuous. Choose* $\beta_1\le|\lambda_0|$ *and* $\beta_2\in(\beta_0,v(0))$, *where* β_0 *is defined in Remark* 4.367, *such that* $\beta_1\le\beta_2$. *Consider the level sets* $\mathcal{K}_{\beta_i}=\{x\in\mathbb{R}^d, V(x)<-\beta_i\}$, $i=1,2$. *Then*

$$C^{(1)}_{d,m,\alpha,\beta_1,|\lambda_0|}\varphi_0(x_{\beta_2})\nu_{m,\alpha}(|x|)\le\varphi_0(x)\le C^{(2)}_{d,m,\alpha,\beta_2,\beta_1,|\lambda_0|}\varphi_0(x_{\beta_2})\nu_{m,\alpha}(|x|),$$

where $x_{\beta_i}\in\partial\mathcal{K}_{\beta_i}$ *and* $r_{\beta_i}=|x_{\beta_i}|$, $i=1,2$.

4.9.10 Massless relativistic harmonic oscillator

In this section our aim is to discuss the specific one-dimensional nonlocal Schrödinger operator

$$H_{\text{rosc}} = \sqrt{-\Delta} + x^2. \tag{4.9.120}$$

We will derive its spectrum and various properties of its eigenfunctions. Note that $\sqrt{-\Delta}$ is essentially self-adjoint on $C_0^\infty(\mathbb{R})$ and $\operatorname{Spec}(\sqrt{-\Delta}) = \operatorname{Spec}_{\text{ess}}(\sqrt{-\Delta}) = [0,\infty)$. Since the potential is positive and growing at infinity, we know that $\operatorname{Spec}(H) \subset [0,\infty)$ is discrete.

Consider the eigenvalue equation

$$\sqrt{-\Delta}\varphi_n + x^2\varphi_n = E_n\varphi_n, \quad \varphi_n \in H^1(\mathbb{R}),\ n \in \mathbb{N}. \tag{4.9.121}$$

In this section we use another numbering and denote the ground state by φ_1. Theorem 4.226 directly implies that $x^2\varphi_n \in L^1(\mathbb{R})$ for $n \in \mathbb{N}$. This also means that $\varphi_n \in L^1(\mathbb{R})$ for every $n \in \mathbb{N}$. The Fourier transform of φ_n is denoted by $\hat{\varphi}_n$, and the eigenvalue equation (4.9.121) transforms into

$$-\hat{\varphi}_n''(y) + (|y| - E_n)\hat{\varphi}_n(y) = 0, \quad y \in \mathbb{R}. \tag{4.9.122}$$

We have

$$|\hat{\varphi}_n''(y)| \le \frac{1}{\sqrt{2\pi}}\|x^2\varphi_n\|_{L^1(\mathbb{R})}, \quad y \in \mathbb{R}. \tag{4.9.123}$$

Furthermore, by multiplying and dividing, and then using the Schwarz inequality,

$$\|\hat{\varphi}_n\|_{L^1(\mathbb{R})} \le \left(\int_{\mathbb{R}} \frac{1}{1+y^2}dy\right)^{1/2}\left(\int_{\mathbb{R}}(1+y^2)|\hat{\varphi}_n(y)|^2dy\right)^{1/2} = \sqrt{\pi}\|\varphi_n\|_{H^1(\mathbb{R})}.$$

Hence it follows that $\hat{\varphi}_n \in C^2(\mathbb{R}) \cap L^1(\mathbb{R})$ for all $n \in \mathbb{N}$. Note, moreover, that if $\hat{\varphi}_n(y)$ is a solution of (4.9.122), then so is $\hat{\varphi}_n(-y)$. Thus it suffices to consider (4.9.122) only for $y > 0$, and construct odd and even solutions on the whole of $\mathbb{R}$. We then know that the odd and even solutions satisfy, respectively,

$$\hat{\varphi}_n(0+) = 0,\ n = 2,4,6,\ldots, \quad \text{and} \quad \hat{\varphi}_n'(0+) = 0,\ n = 1,3,5,\ldots \tag{4.9.124}$$

Our goal in what follows is to study the spectrum and the eigenfunctions of H_{rosc} given by (4.9.121) through the L^1 solutions of (4.9.122) satisfying the boundary conditions (4.9.124). Let $y > 0$. Setting $\phi_n(y) = \hat{\varphi}_n(y - E_n)$ in (4.9.122) and making the linear transformation $y - E_n \mapsto y$, the eigenvalue equation reduces to the *Airy differential equation*

$$\phi_n''(y) - y\phi_n(y) = 0, \tag{4.9.125}$$

whose solutions are expressed in terms of the *Airy functions* as

$$\phi_n(y) = c_{1,n}\operatorname{Ai}(y) + c_{2,n}\operatorname{Bi}(y).$$

Here

$$\mathrm{Ai}(x) = \frac{1}{\pi}\int_0^\infty \cos\left(\frac{t^3}{3} + xt\right) dt,$$

$$\mathrm{Bi}(x) = \frac{1}{\pi}\int_0^\infty \left[\sin\left(\frac{t^3}{3} + xt\right) + \exp\left(-\frac{t^3}{3} + xt\right)\right] dt.$$

Since $\hat{\varphi}_n \in L^1(\mathbb{R})$ and the function Bi tends to infinity as $x \to \infty$, we have $c_{2,n} = 0$ for all n. Without loss of generality, we take $c_{1,n} = 1$ for all n, and we normalize these functions in a next step. For even solutions on the full real line, we thus obtain

$$\hat{\varphi}_n(y) = \mathrm{Ai}(|y| - E_n), \quad y \in \mathbb{R}. \tag{4.9.126}$$

The boundary conditions (4.9.124) for this case give $\mathrm{Ai}'(-E_n) = 0$. For the odd solutions, we have

$$\hat{\varphi}_n(y) = \mathrm{sgn}(y)\mathrm{Ai}(|y| - E_n), \quad y \in \mathbb{R}, \tag{4.9.127}$$

and the boundary conditions imply $\mathrm{Ai}(-E_n) = 0$. By the Parseval identity and the fact that

$$(y\mathrm{Ai}(y)^2 - (\mathrm{Ai}'(y))^2)' = \mathrm{Ai}(y)^2,$$

which is a straightforward consequence of the Airy equation (4.9.125), we obtain

$$\|\varphi_n\|^2 = \|\hat{\varphi}_n\|^2 = 2\int_0^\infty \mathrm{Ai}(y - E_n)^2 dy = 2\left((\mathrm{Ai}'(-E_n))^2 + E_n \mathrm{Ai}(-E_n)^2\right).$$

We have thus proved the following result.

Theorem 4.371. *The spectrum of H_{rosc} is purely discrete given by $\mathrm{Spec}(H_{\mathrm{rosc}}) = \{E_k\}_{k=0}^\infty$ with*

$$E_{2k-1} = -a_k' \quad \text{and} \quad E_{2k} = -a_k,$$

for $k \in \mathbb{N}$, where a_k and a_k' denote the zeroes of the functions Ai and Ai′ in decreasing order. They are all simple, the eigenfunctions $\varphi_{2k-1}(x)$ are even and $\varphi_{2k}(x)$ are odd. Furthermore, the Fourier transforms of the normalized eigenfunctions $\hat{\varphi}_n$ are given by

$$\hat{\varphi}_n(y) = \begin{cases} \frac{1}{\sqrt{2E_n}\mathrm{Ai}(-E_n)}\mathrm{Ai}(|y| - E_n), & n = 1, 3, 5, \ldots, \\ \frac{1}{\sqrt{2}\mathrm{Ai}'(-E_n)}\mathrm{sgn}(y)\mathrm{Ai}(|y| - E_n), & n = 2, 4, 6, \ldots \end{cases}$$

We present some further results on the eigenvalues and eigenfunctions without proof, and refer the interested reader to the literature. To have a more explicit idea of the dependence of the eigenvalues on the index k, we make use of the asymptotic expansions and estimates for the zeroes of the Airy function and its derivative.

Corollary 4.372 (Asymptotic behavior of eigenvalues). *We have as $k \to \infty$*

$$E_{2k-1} \sim g\left(\frac{3\pi}{8}(4k-1)\right) \quad \text{and} \quad E_{2k} \sim f\left(\frac{3\pi}{8}(4k-3)\right),$$

where

$$g(t) = t^{2/3}\left(1 - \frac{7}{48}t^{-2} + \frac{35}{288}t^{-4} - \frac{181223}{207360}t^{-6} + \frac{18683371}{1244160}t^{-8} - \cdots\right),$$
$$f(t) = t^{2/3}\left(1 + \frac{5}{48}t^{-2} - \frac{5}{36}t^{-4} + \frac{77125}{82944}t^{-6} - \frac{108056875}{6967296}t^{-8} + \cdots\right).$$

Moreover, for $k \in \mathbb{N}$,

$$E_{2k-1} \leq \left(\frac{3\pi}{8}(4k-1)\right)^{2/3} \leq E_{2k} \leq \left(\frac{3\pi}{8}(4k-1)\right)^{2/3}\left(1 + \frac{3}{2}\arctan\left(\frac{5}{18\pi(4k-1)}\right)\right).$$

A numerical calculation for the first few eigenvalues gives:

$$E_1 \cong 1.01879297164747, \quad E_2 \cong 2.33810741045976,$$
$$E_3 \cong 3.24819758217983, \quad E_4 \cong 4.08794944413097,$$
$$E_5 \cong 4.82009921117874, \quad E_6 \cong 5.52055982809555.$$

In particular, these expressions give the following spectral gap estimate.

Corollary 4.373 (Spectral gap). *It follows that $E_2 - E_1 \geq \left(\frac{3\pi}{8}\right)^{2/3}(3^{2/3} - 1)$.*

Another consequence is a result on the asymptotic behavior of the heat trace at zero and a Weyl-type formula on the asymptotic distribution of eigenvalues. Let $N(E) = |\{n \in \mathbb{N} \mid E_n \leq E\}|$ be the counting measure of the number of eigenvalues E_n of H_{rosc} up to level $E > 0$.

Corollary 4.374 (Weyl-type law). *It follows that*

$$\lim_{t\to 0+} t^{3/2}\sum_{n=1}^{\infty} e^{-E_n t} = \frac{1}{\sqrt{\pi}} \quad \text{and} \quad \lim_{E\to\infty}\frac{N(E)}{E^{3/2}} = \frac{4}{3\pi}.$$

Next we summarize some further information obtained on the eigenfunctions.

Corollary 4.375 (Asymptotic behavior of eigenfunctions). *For every $k \in \mathbb{N}$ and $N = 2, 3, \ldots$, we have as $|x| \to \infty$ the asymptotic expansions*

$$\varphi_{2k-1}(x) = \sqrt{\frac{2}{-a'_k}}\left(\frac{p_3(a'_k)}{x^4} - \frac{p_5(a'_k)}{x^6} + \cdots + (-1)^N\frac{p_{2N-1}(a'_k)}{x^{2N}}\right) + O\left(\frac{1}{x^{2N+2}}\right),$$
$$\varphi_{2k}(x) = \sqrt{2}\left(\frac{q_4(a_k)}{x^5} - \frac{q_6(a_k)}{x^7} + \cdots + (-1)^N\frac{q_{2N}(a_k)}{x^{2N+1}}\right) + O\left(\frac{1}{x^{2N+3}}\right),$$

where p_n and q_n are polynomials of degree n defined by the recursive relations

$$p_{n+1}(x) = p_n'(x) + xq_n(x), \quad q_{n+1}(x) = p_n(x) + q_n'(x)$$

with $p_0(x) \equiv 1$ and $q_0(x) \equiv 0$.

Furthermore, we have the following corollary.

Corollary 4.376 (Properties of eigenfunctions).
(1) *The eigenfunctions φ_n are real analytic functions on $\mathbb{R}$ with Maclaurin expansions*

$$\varphi_{2k-1}(x) = \sqrt{\frac{2}{-a_k'}}\frac{1}{\operatorname{Ai}(a_k')}\sum_{m=0}^{\infty}\frac{w_{2m}(a_k')(-1)^m}{(2m)!}x^{2m},$$

$$\varphi_{2k}(x) = \frac{\sqrt{2}}{\operatorname{Ai}'(a_k)}\sum_{m=0}^{\infty}\frac{w_{2m+1}(a_k)(-1)^m}{(2m+1)!}x^{2m+1},$$

where $k \in \mathbb{N}$ and $w_n(x) = \int_0^\infty \operatorname{Ai}(u+x)u^n du$.
(2) *The eigenfunctions φ_n have a finite number of zeroes.*
(3) *The eigenfunctions φ_n are uniformly bounded.*
(4) *The ground state φ_1 is decreasing on $(0,\infty)$, and there exist $x_1 > x_0 > 0$ such that φ_1 is concave on $[-x_0, x_0]$ and is convex on $(-\infty, -x_1]$ and $[x_1, \infty)$.*

Finally, we have the following heat kernel estimates.

Corollary 4.377 (Properties of semigroup).
(1) (Long times) *There exists a constant $c > 1$ such that*

$$\frac{1}{c}\frac{e^{a_1' t}}{(1+x^4)(1+y^4)} \le e^{-tH_{\text{rosc}}}(x,y) \le c\frac{e^{a_1' t}}{(1+x^4)(1+y^4)},$$

for every $t > 1$ and $x, y \in \mathbb{R}$.
(2) (Short times) *For every $t \in (0,1]$ and $x, y \in \mathbb{R}$, we have*

$$e^{-tH_{\text{rosc}}}(x,y) \approx \frac{t}{t^2 + |x-y|^2} \quad \textit{whenever } t < \frac{1}{1+(|x| \vee |y|)^2}.$$

Remark 4.378 (Massless quartic relativistic oscillator). Unlike in the classical case, in the relativistic case the eigenvalue equation

$$\left(-\frac{d^2}{dx^2}\right)^{1/2}\varphi_n + x^4\varphi_n = \lambda_n\varphi_n, \quad \varphi_n \in H^1(\mathbb{R}),\ n \in \mathbb{N},$$

can also be explicitly solved, describing the *massless relativistic quartic anharmonic oscillator*. The main observation is that this can be transformed to a study of the *fourth-order Airy equation*

$$\frac{d^4}{dy^4}\phi_n(y) + y\phi_n(y) = 0,$$

and the solutions can be obtained in terms of the *fourth-order Airy function*

$$\mathrm{Ai}_4(y) = \frac{1}{\pi}\int_0^\infty \cos\left(\frac{z^5}{5} + yz\right) dz$$

and its rotations. In this case a subtle spectral shift phenomenon occurs; for the details, we refer the reader to the literature (see discussion in Chapter 6 below).

4.9.11 Embedded eigenvalues

In this short section we illustrate that Feynman–Kac representations can be used also for a study of embedded eigenvalues and the related eigenfunctions. Like in Section 4.9.8 above, since proofs need further techniques than discussed in this book, we need to refer the reader to the literature. Here we summarize some results for nonocal Schrödinger operators, first by giving some examples of potentials generating embedded eigenvalues, and then by reviewing decay results showing the power of the probabilistic representation in understanding the mechanisms governing these delicate phenomena.

As we have seen before, the classical and nonlocal Schrödinger operators considered above have the property that when the potential is confining, their spectra are discrete. For decaying potentials, however, the spectrum of such an operator has a component of essential spectrum equal to $[0, \infty)$, and dependent on further properties of the potential it may or may not have a discrete component. Whenever a discrete nonempty spectrum exists, it consists of at most a countable set of negative eigenvalues, with possible accumulation point at zero. In particular, for such potentials the spectrum can be rather complicated and many interesting spectral phenomena may occur. One feature is the possibility of existence of embedded eigenvalues, i. e., eigenvalues which are not isolated (see the discussion in and following Definition 4.17). The related eigenfunctions correspond to unstable physical states, which during their lifetime have some localization properties comparable to bound states. Such states have also been created experimentally.

The classic example producing such an eigenvalue is due to von Neumann and Wigner, who discovered that the potential and function given by

$$V_{\mathrm{NW}}(x) = \frac{-32\sin|x|\left(g(|x|)^3\cos|x| - 3g(|x|)^2\sin^3|x| + g(|x|)\cos|x| + \sin^3|x|\right)}{(1+g(|x|)^2)^2}, \tag{4.9.128}$$

$$\varphi_{\mathrm{NW}}(x) = \frac{\sin|x|}{|x|(1+g(|x|)^2)}, \tag{4.9.129}$$

with $g(|x|) = 2|x| - \sin 2|x|$, $x \in \mathbb{R}^3$, satisfy the eigenvalue equation

$$-\frac{1}{2}\Delta\varphi_{\mathrm{NW}} + V_{\mathrm{NW}}\varphi_{\mathrm{NW}} = \varphi_{\mathrm{NW}}$$

in a pointwise sense. The given potential decays like

$$V_{\mathrm{NW}}(x) = -\frac{8\sin 2|x|}{|x|} + O\left(\frac{1}{|x|^2}\right), \quad \text{as } |x| \to \infty,$$

thus in this case an embedded eigenvalue $\lambda = +1$ occurs.

The mechanism behind the appearance of these embedded states is a carefully crafted interplay between the oscillations and the slow $O(1/|x|)$ decay of the potentials, which physically allow a specially controlled rate of transmission and reflection of the particle through this potential landscape. This observation inspired to construct further Neumann–Wigner type potentials and led to a line of research of its own. The number of eigenvalues is generally not necessarily limited to a small set, and it is possible to construct examples where a dense set of embedded eigenvalues exists. However, Kato showed that, roughly speaking, if $V(x) = o(1/|x|)$, then no embedded eigenvalues occur for classical Schrödinger operators, indicating that the margin between existence/nonexistence of embedded eigenvalues is actually quite narrow.

Embedded eigenvalues may also exist for nonlocal Schrödinger operators. Consider the relativistic Schrödinger operator on $L^2(\mathbb{R})$ given by

$$H = (-\Delta + m^2)^{1/2} - m + V \tag{4.9.130}$$

and assume $m > 0$. Denote $p = -i\frac{d}{dx}$ and define the functions

$$g(x) = 2x - \sin(2x), \quad h(x) = \frac{1}{1 + g(x)^2},$$
$$f(x) = \left(\sqrt{(p+1)^2 + m^2} + \sqrt{(p-1)^2 + m^2}\right) h(x),$$

and

$$\varphi(x) = f(x)\sin x, \tag{4.9.131}$$
$$V(x) = \lambda - \frac{1}{\varphi(x)}\left(\sqrt{p^2 + m^2} - m\right)\varphi(x), \tag{4.9.132}$$

where $\lambda = \sqrt{1 + m^2} - m > 0$.

Theorem 4.379. *Let H be given by (4.9.130), V by (4.9.132), and φ by (4.9.131). If $m \geq 146$, then V is a real-valued smooth potential with the property that $V(x) = O(1/|x|)$, and λ, φ satisfy the eigenvalue equation $H\varphi = \lambda\varphi$, $\varphi \in D(H)$. Furthermore, there exists a constant $C > 0$ such that*

$$\frac{2}{1 + x^2} \leq |\varphi(x)| \leq \frac{C}{1 + x^2}, \quad x \in \mathbb{R}.$$

We note that the restriction $m \geq 146$ is inessential, and by a scaling a similar result applies for all $m > 0$. Indeed, with $a > 0$ and $(U_a g)(x) = a^{1/2} g(ax)$, the operator H is unitary equivalent to $H_a = U_a H U_a^{-1} = \frac{1}{a}(\sqrt{p^2 + (am)^2} - am + aV(ax))$. Then by Theorem 4.379 we can construct a smooth decaying potential such that H_a has a positive eigenvalue for any a with $am > 146$.

We can use this basic result to derive a result also in three dimensions.

Corollary 4.380. *Let $m \geq 146$, write $W(x) = V(|x|)$, $x \in \mathbb{R}^3$, and consider the operator $H_r = \sqrt{-\Delta + m^2} - m + W(x)$ on $L^2(\mathbb{R}^3)$. Then*

$$\zeta(x) = \frac{\varphi(|x|)}{\sqrt{4\pi}|x|}$$

is in $D(H_r) = D(\sqrt{-\Delta})$ and satisfies the eigenvalue equation $H_r\zeta = \lambda\zeta$ with the same eigenvalue $\lambda = \sqrt{1 + m^2} - m$.

Next we note that in the nonrelativistic limit the potentials, the eigenvalues and eigenfunctions constructed in the previous section converge to the expressions obtained by von Neumann and Wigner. To obtain this, similarly to Section 4.6.4 we restore the speed of light $c > 0$ as a parameter in the operator. Let

$$\begin{aligned}
f_c(x) &= \frac{1}{2mc}\left(\sqrt{(p+1)^2 + m^2c^2} + \sqrt{(p-1)^2 + m^2c^2}\right) h(x),\\
\varphi_c(x) &= f_c(x)\sin x,\\
\lambda_c &= c\left(\sqrt{1 + m^2c^2} - mc\right),\\
V_c(x) &= \lambda_c - c\frac{\left(\sqrt{p^2 + m^2c^2} - mc\right)\varphi_c(x)}{\varphi_c(x)}.
\end{aligned}$$

Then we define the relativistic Hamiltonian with c by

$$H_c = \sqrt{c^2p^2 + m^2c^4} - mc^2 + V_c(x).$$

By Theorem 4.379, the eigenvalue equation $H_c\varphi_c = \lambda_c\varphi_c$ holds for all $c > \frac{146}{m}$.

Theorem 4.381. *For every fixed $m > 0$, we have the following nonrelativistic limit:*

$$\begin{aligned}
&\lim_{c\to\infty} \varphi_c(x) = \sin(x)h(x) = \varphi_\infty(x), \quad \textit{uniformly in } C^2(\mathbb{R}),\\
&\lim_{c\to\infty} \lambda_c = \frac{1}{2m},\\
&\lim_{c\to\infty} V_c(x) = \frac{1}{2m}\left(1 - \frac{p^2\varphi_\infty(x)}{\varphi_\infty(x)}\right), \quad x \in \mathbb{R}\setminus\pi\mathbb{N}.
\end{aligned}$$

In the three-dimensional case, we retrieve the expressions (4.9.128)–(4.9.129). With a similar notation as above, we obtain

Corollary 4.382. *For every fixed $m > 0$, we have $\lim_{c\to\infty} \lambda_c = \frac{1}{2m}$, $\lim_{c\to\infty} \zeta_c(x) = \varphi_{\mathrm{NW}}(x)$, uniformly in $C^2(\mathbb{R}^3)$, and $\lim_{c\to\infty} W_c(x) = V_{\mathrm{NW}}(x)$, for all $x \in \mathbb{R}^3$.*

Finally, we consider the delicate situation of eigenvalues at the continuum edge. In Corollary 4.151 and Proposition 4.168 we have seen some conditions ruling out zero eigenvalues for classical Schrödinger operators, and discussed that in general it is difficult to determine whether zero is an eigenvalue or not, which may be important information for specific applications. First, we show by an example that zero eigenvalues can occur for nonlocal Schrödinger operators as well and then we will study some intriguing behaviors of the related eigenfunctions. We note that there are many more examples of such potentials.

Let $\kappa > 0$, $\alpha \in (0,2)$, and P be a solid harmonic polynomial, homogeneous of degree $l \geq 0$, i. e., satisfying $P(cx) = c^l P(x)$ for all $c > 0$, and $\Delta P = 0$. Recall Gauss regularized hypergeometric function

$$ {}_2\mathbf{F}_1\left(\begin{matrix} a & b \\ & c \end{matrix} \,\middle|\, z\right) = \frac{1}{\Gamma(c+1)} \sum_{k=0}^{\infty} \frac{(a)_k (b)_k}{k!} z^k, \quad z \in \mathbb{C}, $$

for $a, b, c \in \mathbb{C}$ such that c is not a negative integer, with the Pochhammer symbol $(p)_k = p(p+1)\cdots(p+k-1)$.

Theorem 4.383. *Let P be a solid harmonic polynomial of degree $l \in \dot{\mathbb{N}}$, and write $\delta = d+2l$ for $d \in \mathbb{N}$. Also, let $\theta > 0$, and $0 < \kappa \leq \frac{\delta+\alpha}{2}$ with $\alpha \in (0,2)$. Consider*

$$ \begin{aligned} V_{\kappa,\alpha}(x) &= -\frac{2^\alpha \sqrt{\theta^\alpha}}{\Gamma(\kappa)} \Gamma\left(\frac{\delta+\alpha}{2}\right) \Gamma\left(\frac{\alpha}{2}+\kappa\right) (1+\theta|x|^2)^\kappa \, {}_2\mathbf{F}_1\left(\begin{matrix} \frac{\delta+\alpha}{2} & \frac{\alpha}{2}+\kappa \\ & \frac{\delta}{2} \end{matrix} \,\middle|\, -\theta|x|^2\right), \\ \varphi_\kappa(x) &= \frac{P(x)}{(1+\theta|x|^2)^\kappa} \end{aligned} \tag{4.9.133} $$

for $x \in \mathbb{R}^d$. Then the following hold:

(1) *The equality*

$$ H_\kappa^{(\alpha)} \varphi_\kappa = (-\Delta)^{\alpha/2} \varphi_\kappa + V_{\alpha,\kappa}(x) \varphi_\kappa = 0 $$

holds in the distributional sense. For $\kappa \in (0, \frac{\delta}{4}]$, the functions φ_κ are zero-resonances and for $\kappa \in (\frac{\delta}{4}, \frac{\delta+\alpha}{2})$ they are zero-eigenfunctions of $H_\kappa^{(\alpha)}$.

(2) *We have $\lim_{|x|\to\infty} V_{\kappa,\alpha}(x) = 0$ if and only if $0 < \kappa < \frac{\delta+\alpha}{2}$.*

(3) *(Asymptotic behavior at infinity) As $|x| \to \infty$, we have*

$$ |V_{\kappa,\alpha}(x)| = \begin{cases} O\left(|x|^{-\alpha}\right) & \text{if } \kappa \in (0, \frac{\delta}{2}) \setminus \{\frac{\delta-\alpha}{2}\}, \\ O\left(|x|^{-2\alpha}\right) & \text{if } \kappa = \frac{\delta-\alpha}{2}, \\ O\left(|x|^{-\alpha} \log|x|\right) & \text{if } \kappa = \frac{\delta}{2}, \\ O\left(|x|^{2\kappa-\delta-\alpha}\right) & \text{if } \kappa \in (\frac{\delta}{2}, \frac{\delta+\alpha}{2}). \end{cases} $$

Furthermore,

(i) $\lim_{|x|\to\infty} |x|^\alpha V_{\kappa,\alpha}(x) = -2^\alpha\sqrt{\theta^\alpha}\,\frac{\Gamma(\frac{\delta}{2}-\kappa)\Gamma(\frac{\alpha}{2}+\kappa)}{\Gamma(\kappa)\Gamma(\frac{\delta-\alpha}{2}-\kappa)}$, *for* $\kappa \in (0,\frac{\delta}{2})\setminus\{\frac{\delta-\alpha}{2}\}$;

(ii) $\lim_{|x|\to\infty} |x|^{2\alpha} V_{\kappa,\alpha}(x) = -2^\alpha\sqrt{\theta^\alpha}\,\frac{\Gamma(\frac{\delta+\alpha}{2})}{\Gamma(\frac{\delta-\alpha}{2})}$, *for* $\kappa = \frac{\delta-\alpha}{2}$;

(iii) $\lim_{|x|\to\infty} \frac{|x|^\alpha}{\log(1+|x|^2)} V_{\kappa,\alpha}(x) = -2^\alpha\sqrt{\theta^\alpha}\,\frac{\Gamma(\frac{\delta+\alpha}{2})}{\Gamma(-\frac{\alpha}{2})\Gamma(\frac{\delta}{2})}$, *for* $\kappa = \frac{\delta}{2}$;

(iv) $\lim_{|x|\to\infty} |x|^{2\kappa-\delta-\alpha} V_{\kappa,\alpha}(x) = -2^\alpha\sqrt{\theta^\alpha}\,\frac{\Gamma(\kappa-\frac{\delta}{2})\Gamma(\frac{\delta+\alpha}{2})}{\Gamma(\kappa)\Gamma(-\frac{\alpha}{2})}$, *for* $\kappa \in (\frac{\delta}{2},\frac{\delta+\alpha}{2})$.

(4) *For a sufficiently large* $|x|$, *we have* $V_{\kappa,\alpha}(x) > 0$ *if* $\kappa \in \left(\frac{\delta-\alpha}{2},\frac{\delta+\alpha}{2}\right)$ *and* $V_{\kappa,\alpha}(x) < 0$ *if* $\kappa \in \left(\frac{\delta}{4},\frac{\delta-\alpha}{2}\right]$.

For the remaining part of this section, we discuss the decay properties of zero-energy eigenfunctions and resonances by using Feynman–Kac representations. From the results in Sections 4.3.8, 4.6.3, and 4.9.8, we have seen that the decay rates for an eigenfunction φ crucially depend on the gap between the eigenvalue $\lambda < 0$ corresponding to φ and the continuum edge $\Sigma = 0$. Now we are interested in decay properties in the conditions when this gap is zero.

We will use the same notations and concepts as in Section 4.9.8, and only indicate here the differences. First, we give the definition of the class of processes that we will use, which slightly differs from the jump-paring class given by Definition 4.327. Consider the operator L with symbol ψ as defined by (4.9.85)–(4.9.86). We will use the symmetrization $\Phi(r) = \sup_{|\xi|\le r}\psi(\xi)$, $r > 0$, of the symbol ψ as above. We consider the following class of Lévy processes and related operators L.

Definition 4.384. Let $(X_t)_{t\ge0}$ be a Lévy process with Lévy–Khintchine exponent ψ as in (4.9.85) and Lévy triplet $(0,A,\nu)$, satisfying the following conditions:

(1) There exist a nonincreasing function $g:(0,\infty)\to(0,\infty)$ and constants $C_1, C_2 > 0$ such that

$$\nu(x) \asymp C_1 g(|x|), \quad x \in \mathbb{R}^d\setminus\{0\}, \quad \text{and} \quad g(|x|) \le C_2 g(2|x|), \quad |x| \ge 1.$$

(2) There exists $t_b > 0$ such that $\sup_{x\in\mathbb{R}^d} p(t_b,x) = p(t_b,0) < \infty$.

(3) There exists a constant $C_3 = C_3(X)$ such that

$$\sup_{x,y:\,|x-y|\ge s/8} G_{B(0,s)}(x,y) \le C_3\frac{\Phi(1/s)}{s^d}, \quad s \ge 1.$$

Note that the difference from Definition 4.327 is that condition (1) requires a doubling property of the profile function of ν instead of the more general jump-paring property. These conditions are satisfied by isotropic and anisotropic stable processes, layered stable processes, etc., and the related operators include the fractional Laplacian and generators of Lévy processes comparable with the fractional Laplacian.

Choosing X-Kato decomposable potentials from $\mathcal{K}_{\pm}^X$ as given in Definition 4.329, we define the Feynman–Kac semigroup as in (4.9.87). Also, define

$$Hf = -\lim_{t\downarrow 0} \frac{1}{t}(T_t f - f), \quad f \in L^p(\mathbb{R}^d),$$

such that the strong limit exists in $L^p(\mathbb{R}^d)$. We denote the set of all such functions by $D_{L^p}H$ and call it the L^p-domain on H. Then H is a closed unbounded operator such that $D_{L^p}H$ is dense in $L^p(\mathbb{R}^d)$. For $p = 2$, this operator can be identified with the self-adjoint operator $H = -L + V$ defined in the form sense.

Let $V \in \mathcal{K}_{\pm}^X$ be a decaying potential. We consider solutions $\varphi \in L^p(\mathbb{R}^d)$, $p \geq 1$, $\varphi \not\equiv 0$, of the equation

$$H\varphi = 0. \tag{4.9.134}$$

In the L^2-framework,

$$\Sigma = \inf \operatorname{Spec}_{\mathrm{ess}}(H) = \inf \operatorname{Spec}_{\mathrm{ess}}(-L) = 0$$

is the edge of the essential spectrum of H and (4.9.134) can be understood as the eigenvalue equation at Σ. Whenever the solution φ to (4.9.134) is such that $\varphi \in L^2(\mathbb{R}^d)$, we call it a *zero-energy eigenfunction* (or *zero-energy bound state*) and then 0 is an eigenvalue. Otherwise, we call both (by a slight abuse of language) a *zero-resonance.*

When the eigenvalue is zero, the decay no longer can benefit from the relative position of the eigenvalue from the continuum edge as in the situations discussed in Section 4.9.8. As it turns out, now it depends on vestigial effects of the potential like its sign at infinity. Thus we split the discussion according to this. First, we consider potentials V positive at infinity in the sense that

$$V(x) \to 0 \text{ as } |x| \to \infty, \text{ and there exists } r_0 > 0 \text{ such that } V(x) \geq 0 \text{ for } |x| \geq r_0. \tag{4.9.135}$$

We will use the notation

$$V_*(x) = \inf_{r_0 \leq |y| \leq \frac{3}{2}|x|} V(y), \quad |x| \geq r_0.$$

Notice that $V_*(x)$ is a radial and nonincreasing function such that $V_*(x) \geq 0$, $|x| \geq r_0$.

Theorem 4.385 (Upper bound). *Let $(X_t)_{t\geq 0}$ be a Lévy process with Lévy–Khintchine exponent ψ as in* (4.9.85) *and Lévy intensity ν, such that the assumptions in Definition* 4.384 *hold. Let $V \in \mathcal{K}_{\pm}^X$ such that* (4.9.135) *holds with a suitable $r_0 > 0$, and φ be a solution of* (4.9.134). *Then we have the following:*

(1) *If*

$$\lim_{|x|\to\infty} \frac{\Phi\left(\frac{1}{|x|}\right)}{V_*(x)} = 0 \quad \text{and} \quad \int_{|x|>2r_0} \frac{\nu(x)}{V_*(x)} dx < \infty,$$

then there exists $C > 0$ *and* $R \geq 2r_0$ *such that*

$$|\varphi(x)| \leq C \, \|\varphi\|_\infty \frac{v(x)}{V_*(x)}, \qquad |x| > R.$$

In particular, $\varphi \in L^1(\mathbb{R}^d)$.

(2) *If*

$$\lim_{|x|\to\infty} \frac{\Phi\left(\frac{1}{|x|}\right)}{V_*(x)} = 0 \quad \text{and} \quad \int\limits_{|x|>2r_0} \frac{v(x)}{V_*(x)} dx = \infty,$$

and

$$\sup_{|x|\geq 2r_0} \left[\frac{v(x)}{V_*(x)} \exp\left(\eta_* \int\limits_{2r_0\leq|z|\leq|x|} \frac{v(z)}{V_*(z)} dz \right) \right] < \infty,$$

with a constant $\eta_* > 0$, *then there exist* $C > 0$ *and* $R \geq 2r_0$ *such that*

$$|\varphi(x)| \leq C \, \|\varphi\|_\infty \frac{v(x)}{V^*(x)} \exp\left(\eta_* \int\limits_{r_0\leq|z|\leq|x|} \frac{v(z)}{V_*(z)} dz \right), \qquad |x| > R.$$

(3) *If* $\liminf_{|x|\to\infty} \frac{\Phi(\frac{1}{|x|})}{V_*(x)} > 0$, $\varphi \geq 0$, *and* $\varphi \in L^p(\mathbb{R}^d)$ *for some* $p \in (1, \infty)$, *then there exists* $C > 0$ *such that*

$$\varphi(x) \leq C \, \|\varphi\|_p \left(\frac{\left(\int_{|z|>|x|} v(z)^{\frac{p}{p-1}} dz\right)^{\frac{p-1}{p}}}{\Phi(\frac{1}{|x|})} + \frac{1}{|x|^{d/p}} \right), \qquad |x| > 2r_0.$$

The next result deals with the case when φ has no definite sign, but it is anti-symmetric with respect to a given $(d-1)$-dimensional hyperplane $\mathfrak{p} \subset \mathbb{R}^d$ with $0 \in \mathfrak{p}$, and has a definite sign on both of the corresponding half-spaces. By rotating the co-ordinate system if necessary, we choose $\mathfrak{p} = \{x \in \mathbb{R}^d \mid x_1 = 0\}$ and assume that

$$\begin{aligned} \varphi(-x_1, x_2, \ldots, x_d) &= -\varphi(x_1, x_2, \ldots, x_d), \; x = (x_1, \ldots, x_d) \in \mathbb{R}^d, \\ &\text{and} \quad \varphi(x_1, \ldots, x_d) \geq 0 \text{ whenever } x_1 > 0. \end{aligned} \tag{4.9.136}$$

Theorem 4.386. *Let the assumptions of Theorem* 4.385 *hold, and furthermore suppose that there exist* $C_0 > 0$ *and* $R_0 > 0$ *such that*

$$|v(z_1) - v(z_2)| \leq C_0 \frac{v(z_2)}{|z_2|} |z_1 - z_2|, \quad |z_1| \geq |z_2| \geq R_0. \tag{4.9.137}$$

Let φ *be a solution of* (4.9.134) *such that* (4.9.136) *is satisfied. Then the following hold:*

(1) *If* $\lim_{|x|\to\infty} \frac{\Phi(\frac{1}{|x|})}{V_*(x)} = 0$, $\int_{|x|>2r_0} \frac{v(x)}{V_*(x)} dx < \infty$, $\varphi \in L^1(\mathbb{R}^d)$, *and there exists a constant* $C_1 > 0$ *such that*

$$\frac{v(y)}{V_*(y)} \le C_1 \frac{v(x)}{V_*(x)}, \quad |y| \ge \frac{|x|}{2} \ge 2r_0, \tag{4.9.138}$$

then there exists $C > 0$ *and* $R \ge 2r_0 \vee 4R_0$ *such that*

$$|\varphi(x)| \le C(\|\varphi\|_\infty \vee \|\varphi\|_1) \frac{v(x)}{V_*(x)} \left(\frac{\Phi(1/|x|)}{V_*(x)} \vee \left(\frac{1}{|x|} \int_{2r_0<|z|<\frac{|x|}{2}} |z_1| \frac{v(z)}{V_*(z)} dz \right) \right),$$

for $|x_1| > R$.

(2) *If* $\liminf_{|x|\to\infty} \frac{\Phi(\frac{1}{|x|})}{V_*(x)} > 0$ *and* $\varphi \in L^p(\mathbb{R}^d)$, $p > 1$, *then there exists* $C > 0$ *such that*

$$|\varphi(x)| \le C\,\|\varphi\|_p \left(\frac{\left(\int_{|z|>|x|} v(z)^{\frac{p}{p-1}} dz \right)^{\frac{p-1}{p}}}{\Phi(\frac{1}{|x|})} + \frac{1}{|x|^{d/p}} \right), \quad |x_1| > 2r_0.$$

Now we give a lower estimate on the eigenfunctions. We use the notation

$$V^*(x) = \sup_{|y|\ge\frac{|x|}{2}} V(y), \quad |x| \ge 2r_0.$$

Theorem 4.387 (Lower bound). *Let the assumptions of Theorem* 4.385 *hold and* φ *be a positive solution of* (4.9.134). *Then we have the following:*

(1) *If* $\lim_{|x|\to\infty} \frac{\Phi(\frac{1}{|x|})}{V^*(x)} = 0$ *and* $\int_{|x|>2r_0} \frac{v(x)}{V^*(x)} dx < \infty$, *then there exist* $C > 0$ *and* $R \ge 2r_0$ *such that*

$$\varphi(x) \ge C \frac{v(x)}{V^*(x)}, \quad |x| > R.$$

(2) *If* $\lim_{|x|\to\infty} \frac{\Phi(\frac{1}{|x|})}{V^*(x)} = 0$ *and* $\int_{|x|>2r_0} \frac{v(x)}{V^*(x)} dx = \infty$, *then there exist* $\eta^*, C > 0$ *and* $R \ge 2r_0$ *such that*

$$\varphi(x) \ge C \frac{v(x)}{V^*(x)} \exp\left(\eta^* \int_{2r_0\le|y|\le|x|} \frac{v(y)}{V^*(y)} dy \right), \quad |x| > R.$$

(3) *If* $\liminf_{|x|\to\infty} \frac{\Phi(\frac{1}{|x|})}{V^*(x)} > 0$ *and* $\int_{|x|>2r_0} \frac{v(x)}{\Psi(\frac{1}{|x|})} dx < \infty$, *then there exist* $C > 0$ *and* $R \ge 2r_0$ *such that*

$$\varphi(x) \ge C \frac{v(x)}{\Phi\left(\frac{1}{|x|}\right)}, \quad |x| > R.$$

(4) *If* $\liminf_{|x|\to\infty} \frac{\Phi(\frac{1}{|x|})}{V^*(x)} > 0$ *and* $\int_{|x|>2r_0} \frac{\nu(x)}{\Psi(\frac{1}{|x|})} dx = \infty$, *then there exist* $\eta^*, C > 0$ *and* $R \geq 2r_0$ *such that*

$$\varphi(x) \geq C \frac{\nu(x)}{\Phi\left(\frac{1}{|x|}\right)} \exp\left(\eta^* \int\limits_{2r_0 \leq |y| \leq |x|} \frac{\nu(y)}{\Phi\left(\frac{1}{|y|}\right)} dy \right), \quad |x| > R.$$

Next we consider potentials negative at infinity in the sense that

$$\text{there exists } r_0 > 0 \text{ and } C > 0 \text{ such that } 0 \leq -V(x) \leq C\Phi(1/|x|), |x| \geq r_0. \tag{4.9.139}$$

Note that indeed $V(x) \to 0$ as $|x| \to \infty$, and the condition above also covers potentials with compact support such as potential wells. We present a counterpart of Theorem 4.386 (3) in the case when the potential is negative at infinity and the negative nodal domain of φ is a subset of a given half-space. By rotating the coordinate system if necessary, without loss of generality we assume that

$$\text{there exists } l \in \mathbb{R} \text{ such that } \operatorname{supp} \varphi_- \subset \left\{ y \in \mathbb{R}^d \mid y_1 < l \right\}. \tag{4.9.140}$$

Theorem 4.388. *Let* $(X_t)_{t\geq 0}$ *be a Lévy process with Lévy–Khintchine exponent* ψ *as in* (4.9.85) *and Lévy intensity* ν *such that the assumptions in Definition* 4.384 *hold. Consider* $V \in L^\infty(\mathbb{R}^d)$ *satisfying* (4.9.139). *Moreover, suppose that for every* $\varepsilon \in (0,1)$ *there exists* $M \geq 1$ *such that* $\Phi(r) \leq \varepsilon\Phi(Mr)$, *with* $r \in (0,1]$. *If* φ *is a solution of* (4.9.134) *such that* $\varphi \in L^p(\mathbb{R}^d)$, *for some* $p > 1$, *and* (4.9.140) *holds, then for every* $\varepsilon \in (0,1)$ *there exist* $C > 0$ *and* $R > 3r_0$ *such that*

$$\varphi(x) \leq C\left(\|\varphi\|_p \vee \|\varphi\|_\infty\right) \left(\frac{\left(\int_{|y|>|x|} \nu(y)^{\frac{p}{p-(1+\varepsilon)}} dy\right)^{\frac{p-(1+\varepsilon)}{p}}}{\Phi(1/|x|)} + \frac{1}{|x|^{\frac{d}{p}}} \right)^{1-\varepsilon}, \quad x_1 \geq R.$$

Moreover, if $\varphi(l + x_1, x_2, \ldots, x_d) = -\varphi(l - x_1, x_2, \ldots, x_d)$, $x \in \mathbb{R}^d$, *with* l *given by* (4.9.140), *then there exists* $\widetilde{R} > 3r_0 \vee |l|$ *such that the same upper bound is true for* $\varphi(x)$ *replaced by* $|\varphi(x)|$, *whenever* $|x_1| \geq \widetilde{R}$.

To highlight these results, we particularize to fractional Schrödinger operators. First, consider potentials that are positive at infinity.

Theorem 4.389. *Let* $L = -(-\Delta)^{\alpha/2}$, $0 < \alpha < 2$, *and* $V \in \mathcal{K}_\pm^X$ *for which there exists* $r_0 > 0$ *such that* $V(x) > 0$ *and* $V(x) \asymp |x|^{-\beta}$, *for* $|x| \geq r_0$, *with some* $\beta > 0$. *Suppose that there exists a positive function* $\varphi \in C_b(\mathbb{R}^d)$ *which is a solution of* (4.9.134). *Then the following hold:*

(1) *If $\beta < \alpha$, then there exist constants $C_1, C_2 > 0$ such that*

$$\frac{C_1}{(1+|x|)^{d+\alpha-\beta}} \le \varphi(x) \le \frac{C_2}{(1+|x|)^{d+\alpha-\beta}}, \quad x \in \mathbb{R}^d.$$

In particular, $\varphi \in L^p(\mathbb{R}^d)$, for every $p \ge 1$.

(2) *If $\beta \ge \alpha$, then there exist $\rho \in (0,1)$ and a constant $C_3 > 0$ such that*

$$\varphi(x) \ge \frac{C_3}{(1+|x|)^{d-\rho}}, \quad x \in \mathbb{R}^d.$$

In particular, $\varphi \notin L^p(\mathbb{R}^d)$, for every $p \in [1, \frac{d}{d-\rho}]$. On the other hand, if $\varphi \in L^p(\mathbb{R}^d)$ for some $p > 1$, then there exists $C_4 > 0$ such that

$$\varphi(x) \le \frac{C_4}{(1+|x|)^{d/p}}, \quad x \in \mathbb{R}^d.$$

In particular, if $\varphi \in L^p(\mathbb{R}^d)$ with $p = \frac{d}{d-\rho-\varepsilon}$ for some $\varepsilon \in (0,1)$, then there exists $C_4 > 0$ such that

$$\varphi(x) \le \frac{C_4}{(1+|x|)^{d-\rho-\varepsilon}}, \quad x \in \mathbb{R}^d.$$

Notice that under the assumptions of Theorem 4.389 it follows that $\varphi \in L^1(\mathbb{R}^d)$ if and only if $\alpha > \beta$. It can already be seen from the above theorem that there is a transition in the localization properties of φ when $\alpha > \beta$ changes to $\alpha \le \beta$. For a closer understanding of this transition around $\alpha \approx \beta$, we consider a more refined class of potentials.

Theorem 4.390. *Let $L = -(-\Delta)^{\alpha/2}$, $0 < \alpha < 2$, and $V \in \mathcal{K}_\pm^X$ for which there exists $r_0 > 0$ such that $V(x) > 0$ and $V(x) \asymp |x|^{-\alpha}(\log|x|)^\delta$, for $|x| \ge r_0$, with some $\delta > 0$. Suppose that there exists a positive function $\varphi \in C_b(\mathbb{R}^d)$ which is a solution of (4.9.134). Then the following hold:*

(1) *If $\delta > 1$, then there exist constants $C_1, C_2 > 0$ such that*

$$\frac{C_1}{(1+|x|)^d(\log(1+|x|))^\delta} \le \varphi(x) \le \frac{C_2}{(1+|x|)^d(\log(1+|x|))^\delta}, \quad x \in \mathbb{R}^d.$$

In particular, $\varphi \in L^p(\mathbb{R}^d)$, for every $p \ge 1$.

(2) *If $\delta = 1$, then there exist $0 < \rho_1 \le 1 \le \rho_2$ and constants $C_4, C_5 > 0$ such that*

$$\frac{C_1}{(1+|x|)^d(\log(1+|x|))^{1-\rho_1}} \le \varphi(x) \le \frac{C_2}{(1+|x|)^d(\log(1+|x|))^{1-\rho_2}}, \quad x \in \mathbb{R}^d.$$

In particular, $\varphi \in L^p(\mathbb{R}^d)$ for every $p > 1$, but $\varphi \notin L^1(\mathbb{R}^d)$.

(3) *If $\delta \in (0,1)$, then there exist $0 < \rho_1 \leq 1 \leq \rho_2$ and constants $C_6, C_7 > 0$ such that*

$$C_6 \frac{e^{\frac{\rho_1}{1-\delta}(\log|x|)^{1-\delta}}}{(1+|x|)^d(\log(1+|x|))^\delta} \leq \varphi(x) \leq C_7 \frac{e^{\frac{\rho_2}{1-\delta}(\log|x|)^{1-\delta}}}{(1+|x|)^d(\log(1+|x|))^\delta}, \quad x \in \mathbb{R}^d.$$

In particular, $\varphi \in L^p(\mathbb{R}^d)$, for every $p > 1$, but $\varphi \notin L^1(\mathbb{R}^d)$.

(4) *If $\delta \leq 0$, then exactly the same bounds and L^p-properties hold as in* (2) *of Theorem* 4.389.

Remark 4.391. From the results above, it is seen that the possible localization properties of the positive zero-energy eigenfunctions or zero-resonances for decaying potentials positive at infinity splits naturally into disjoint regimes representing the following three different scenarios. For simplicity, assume that V is a potential that is positive at infinity and temperately varying so that $V(x) \asymp V^*(x) \asymp V_*(x)$ for large enough $|x|$. Let $r_0 > 0$ be large enough, and define

$$h(r) = \int_{r_0<|x|<r} \frac{\nu(x)}{V(x)} dx \quad \text{and} \quad \widetilde{h}(r) = \int_{r_0<|x|<r} \frac{\nu(x)}{\Phi(1/|x|)} dx, \quad r > r_0.$$

The functions h and $\widetilde{h}$ are bounded on (r_0, ∞) if and only if the ratios ν/V and $\nu/\Phi(1/|x|)$, respectively, are integrable at infinity. The following situations occur:

Scenario (1). If $\lim_{|x|\to\infty} \frac{\Phi(1/|x|)}{V(x)} = 0$ and the ratio $\frac{\nu(x)}{V(x)}$ is integrable at infinity, then

$$\varphi(x) \asymp \frac{\nu(x)}{V(x)},$$

for large enough $|x|$. In particular, $\varphi \in L^1(\mathbb{R}^d)$. Clearly, in this case h is bounded.

Scenario (2). If $\lim_{|x|\to\infty} \frac{\Phi(1/|x|)}{V(x)} = 0$ and the integrability of $\frac{\nu(x)}{V(x)}$ at infinity breaks down, then $h(r) \to \infty$ as $r \to \infty$. Hence (1) is no longer true and h contributes into the behavior of φ giving

$$C_1 \frac{\nu(x)}{V(x)} e^{\rho_1 h(|x|)} \leq \varphi(x) \leq C_2 \frac{\nu(x)}{V(x)} e^{\rho_2 h(|x|)},$$

for large enough x, with some $0 < \rho_1 \leq 1 \leq \rho_2$ and $C_1, C_2 > 0$. We see that Scenario (1) differs from (2) by the unboundedness of h.

Scenario (3). If $\liminf_{|x|\to\infty} \frac{\Phi(1/|x|)}{V(x)} > 0$, then the fall-off rate of φ at infinity rapidly decreases so that

$$\varphi(x) \geq C_1 \frac{\nu(x)}{\Phi(1/|x|)} e^{\rho \widetilde{h}(|x|)},$$

for large x, with some $\rho, C_1 > 0$. Clearly, in case $\widetilde{h}$ is bounded and does not contribute to the lower rate.

By a direct inspection, one can see that these results compare well with the exact results in Theorem 4.383 (3). In particular, it can be checked that in the range $\frac{d}{2} < \kappa < \frac{d+\alpha}{2}$ Scenario 1 prevails, for $\kappa = \frac{d}{2}$ we have Scenario 2, and for $0 \vee \frac{d-\alpha}{2} < \kappa < \frac{d}{2}$ we have Scenario 3.

Finally, we present a result for the case of fractional Schrödinger operators with potentials that are negative at infinity.

Theorem 4.392. *Let $L = -(-\Delta)^{\alpha/2}$, $0 < \alpha < 2$, and $V \in \mathcal{K}_{\pm}^X$ for which there exists $r_0 > 0$ such that $V(x) \le 0$ and $|V(x)| \le C|x|^{-\alpha}$, for $|x| \ge r_0$, with some $C > 0$. Suppose that there exists a function $\varphi \in L^p(\mathbb{R}^d)$, for some $p > 1$, which is a solution to (4.9.134) with the property that there exists $i \in \{1, 2, \ldots, d\}$ such that $\varphi(x_1, \ldots, -x_i, \ldots, x_d) = -\varphi(x_1, \ldots, x_i, \ldots, x_d)$ and $\operatorname{supp} \varphi_- \subset \{x \in \mathbb{R}^d \mid x_i \le 0\}$. Then for every $q \in (0, \frac{d}{p})$, there exist $C = C(q)$ and $R > 0$ such that*

$$|\varphi(x)| \le C\left(\|\varphi\|_p \vee \|\varphi\|_\infty\right) \frac{1}{|x|^q}, \quad |x_i| \ge R.$$

Remark 4.393. From the above, it is seen that the decay of ground states at zero eigenvalue depends essentially on two factors. On the one hand, the sign of the potential at infinity makes a qualitative difference. From the decay results above, one can appreciate that a positive tail of the potential has a soft bouncing effect tending to contain paths in compact regions, while a negative potential allows the paths to spread more out to infinity. This difference makes the analysis of potentials negative at infinity much more difficult than of potentials positive at infinity.

On the other hand, the decay depends on mean sojourn times spent in some regions by the paths. Since (by assuming $V(x) \asymp V^*(x) \asymp V_*(x)$, for simplicity)

$$\mathbb{E}^0[\tau_{B_{|x|}(0)}] \asymp \frac{1}{\Phi(1/|x|)} \quad \text{and} \quad \Lambda_{B_{|x|/2}(x)}(x) = \mathbb{E}^x\left[\int_0^{\tau_{B_{|x|/2}(x)}} e^{-\int_0^t V(X_s)ds} dt\right] \asymp \frac{1}{V(x)},$$

we see that now the decay is governed by the quantity

$$\frac{\Phi\left(\frac{1}{|x|}\right)}{V(x)} \asymp \frac{\Lambda_{B_{|x|/2}(x)}(x)}{\mathbb{E}^0[\tau_{B_{|x|}(0)}]},$$

so that the decays relate to *global* survival times (in large balls of radii proportional to $|x|$ itself), rather than *local* survival times as in (4.9.89), which makes the occurrence of zero eigenvalues a very small and delicate effect. Like in the case of decaying potentials with strictly negative eigenvalues, see Remark 4.342, here we see further qualitative transitions between decay regimes. When $\Lambda_{B_{|x|/2}(x)}(x) = o(\mathbb{E}^0[\tau_{B_{|x|}(0)}])$ as in Scenarios (1)–(2) above, the potential has a relatively pronounced effect, making the paths have a slight but decisive preference for neighborhoods around the origin than for far out neighborhoods, even if these regions are now very large. This is especially reflected in the decay by V entering explicitly in Scenario (1). When, however, $\Lambda_{B_{|x|/2}(x)}(x) = O(\mathbb{E}^0[\tau_{B_{|x|}(0)}])$ as in Scenario (3), the effect of the potential is extremely weak also in relative terms, and the two lifetimes (one under the potential, the other free of the potential) evolve on the same scale, getting very near to, though still differing from, the situation of free fluctuations. This probabilistic picture also gives a hint about the very "birth" of bound states at zero energy level.

5 Gibbs measures associated with Feynman–Kac semigroups

5.1 Ground state transform and related processes

5.1.1 Ground state transform and the intrinsic semigroup

As we have seen in Chapter 4, the Feynman–Kac representations of the semigroups generated by standard Schrödinger operators $-\frac{1}{2}\Delta + V$ and nonlocal Schrödinger operators $\Psi(-\frac{1}{2}\Delta) + V$ with Bernstein functions $\Psi \in \mathcal{B}_0$ of the Laplacian are formally the same but with the essential difference of featuring different processes. This allows us to unify the discussion so that we obtain:

(1) standard Schrödinger operators and Brownian motion for $\Psi(u) = u$;
(2) fractional Schrödinger operators and rotationally symmetric α-stable processes for $\Psi(u) = (2u)^{\alpha/2}$, $0 < \alpha < 2$;
(3) relativistic Schrödinger operators and rotationally symmetric relativistic α-stable processes for $\Psi(u) = (2u + m^{2/\alpha})^{\alpha/2} - m$, $0 < \alpha < 2$, $m > 0$;
(4) other nonlocal Schrödinger operators and subordinate Brownian motions with further possible choices of Ψ.

In this section therefore H will be any of these Schrödinger operators, $\{K_t : t \geq 0\}$ the related Feynman–Kac semigroup, and $(X_t)_{t\geq 0}$ the Lévy process (Brownian motion or a jump process) generated by $-\Psi(-\frac{1}{2}\Delta)$, unless Ψ is further specified. Also, by Ψ-Kato decomposable potentials we mean any of the cases as given by Definitions 4.89, 4.283, or 4.295.

The Feynman–Kac semigroup $\{K_t : t \geq 0\}$ has the peculiarity that in general $K_t 1_{\mathbb{R}^d} \neq 1_{\mathbb{R}^d}$. Now we introduce a semigroup related to the Feynman–Kac semigroup which, however, does keep unit mass and thus becomes a Markov semigroup. This is called the *intrinsic Feynman–Kac semigroup.*

We assume that V is Ψ-Kato-decomposable, and the corresponding Schrödinger operator is defined through the Feynman–Kac formula, as in Definition 4.114 or (4.9.28). We suppose, moreover, that H has a normalized ground state $\varphi_0 > 0$, i. e.,

$$H\varphi_0 = E_0\varphi_0, \quad E_0 = \inf \mathrm{Spec}(H), \quad \|\varphi_0\|_2 = 1. \tag{5.1.1}$$

We write $\bar{H} = H - E_0$ so that $\bar{H}\varphi_0 = 0$.

Definition 5.1 (Ground state transform). Let H be a standard or nonlocal Schrödinger operator with Kato-decomposable potential V, and $\varphi_0 > 0$ be its normalized unique ground state. Consider the probability measure

$$d\mathrm{N}_0 = \varphi_0^2(x)dx \tag{5.1.2}$$

https://doi.org/10.1515/9783112231463-005

on $\mathbb{R}^d$, and the Hilbert space $L^2(\mathbb{R}^d, d\mathsf{N}_0)$. The *ground state transform* is the unitary map

$$\mathscr{U} : L^2(\mathbb{R}^d, d\mathsf{N}_0) \to L^2(\mathbb{R}^d, dx), \quad f \mapsto \varphi_0 f.$$

In fact, the ground state transform is a case of Doob's h-transform familiar from the theory of random processes. From now on, we use the simpler notation $L^2(d\mathsf{N}_0) = L^2(\mathbb{R}^d, d\mathsf{N}_0)$.

Define the operator

$$L = \mathscr{U}^{-1}\bar{H}\mathscr{U} \tag{5.1.3}$$

with domain

$$D(L) = \{f \in L^2(d\mathsf{N}_0) \mid \mathscr{U}f \in D(H)\}.$$

Also, let

$$\widetilde{K}_t(x,y) = \frac{e^{E_0 t} K_t(x,y)}{\varphi_0(x)\varphi_0(y)}, \tag{5.1.4}$$

where $K_t(x,y)$ denotes the integral kernel of K_t.

Definition 5.2 (Intrinsic Feynman–Kac semigroup). The one-parameter semigroup $\{\widetilde{K}_t : t \geq 0\}$ acting on $L^2(d\mathsf{N}_0)$ with integral kernel $\widetilde{K}_t(x,y)$, i. e.,

$$\widetilde{K}_t f(x) = \int_{\mathbb{R}^d} \widetilde{K}_t(x,y) f(y) d\mathsf{N}_0(y), \tag{5.1.5}$$

is called the *intrinsic Feynman–Kac semigroup* for the potential V.

The generator of the intrinsic semigroup is the operator L given by (5.1.3). Note that even though the L^p-norms of the operators K_t can be larger than 1, the operators $\widetilde{K}_t$ are always contractions. Moreover, the intrinsic Feynman–Kac semigroup is more natural than $\{K_t : t \geq 0\}$ since $\widetilde{K}_t 1_{\mathbb{R}^d} = 1_{\mathbb{R}^d}$, for every $t > 0$.

In view of Examples 4.72 and 4.74, it can be expected that there are important differences between Schrödinger (and therefore Feynman–Kac) semigroups depending on the behavior of V at infinity. The following is a strong regularizing property.

Definition 5.3 (Ultracontractivity properties). The semigroup $\{e^{-tH} : t \geq 0\}$ is called
(1) *ultracontractive* if e^{-tH} is contractive from $L^2(\mathbb{R}^d)$ to $L^\infty(\mathbb{R}^d)$ for all $t > 0$;
(2) *intrinsically ultracontractive* (IUC) if e^{-tL} is ultracontractive on $L^2(d\mathsf{N}_0)$, $t > 0$;
(3) *asymptotically intrinsically ultracontractive* (AIUC) if there exists $t_0 > 0$ such that e^{-tL} is ultracontractive on $L^2(d\mathsf{N}_0)$, for all $t > t_0$.

Theorem 5.4 (Intrinsic ultracontractivity). *Let H be a Schrödinger operator with a Kato-decomposable potential V and ground state φ_0. Consider its Feynman–Kac semigroup*

$\{K_t : t \geq 0\}$. *Then* $\{e^{-tH} : t \geq 0\}$ *is IUC (resp. AIUC) if and only if for every* $t > 0$ *(resp. if for every* $t > t_0$*, for some* $t_0 > 0$*) there is a constant* $C_{V,t} > 0$ *such that*

$$K_t(x,y) \leq C_{V,t}\, \varphi_0(x)\varphi_0(y), \quad x,y \in \mathbb{R}^d, \tag{5.1.6}$$

or equivalently,

$$\widetilde{K}_t(x,y) \leq C_{V,t}, \quad x,y \in \mathbb{R}^d. \tag{5.1.7}$$

A consequence of intrinsic ultracontractivity is that a similar lower bound on the kernel also holds, i. e., for every $t > 0$ there is a constant $C^{(1)}_{V,t} > 0$ such that

$$K_t(x,y) \geq C^{(1)}_{V,t}\varphi_0(x)\varphi_0(y), \quad x,y \in \mathbb{R}^d. \tag{5.1.8}$$

An immediate consequence of this is that if the semigroup is intrinsically ultracontractive, then $\varphi_0 \in L^1(\mathbb{R}^d)$.

A classic result for the Feynman–Kac semigroup generated by Schrödinger operators $H = -\frac{1}{2}\Delta + V$ identifying the borderline case for IUC in a specific growth-class, is the following fact.

Proposition 5.5. *If* $H = -\frac{1}{2}\Delta + V$ *with* $V(x) = |x|^\beta$*, then the semigroup* $\{K_t : t \geq 0\}$ *is intrinsically ultracontractive if and only if* $\beta > 2$*. Moreover, if* $\beta > 2$*, then* $c\varrho(x) \leq \varphi_0(x) \leq C\varrho(x)$*,* $|x| > 1$*, holds with some* $C, c > 0$ *and*

$$\varrho(x) = |x|^{-\frac{\beta}{4}+\frac{d-1}{2}} e^{-\frac{2}{2+\beta}|x|^{1+\beta/2}}.$$

Remark 5.6. We note that there is no proper borderline potential for IUC to hold (therefore the proviso on the growth class in the theorem above). Indeed, a sufficient condition for IUC is

$$\int_{r_0}^{\infty} \frac{dr}{\sqrt{V(r)}} < \infty, \quad \text{for some} \quad r_0 > 0, \tag{5.1.9}$$

which for radial potentials is also necessary. For instance, for the potential

$$V(x) = |x|^2(\log|x|)^2(\log\log|x|)^2 \cdots \underbrace{(\log\cdots\log|x|)^2}_{(m-1)\text{-times}}\underbrace{(\log\cdots\log|x|)^{2+\delta}}_{m\text{-times}}, \quad m \in \mathbb{N},\ \delta \geq 0,$$

this condition is satisfied if and only if $\delta > 0$.

The study of intrinsic ultracontractivity can be extended also to nonlocal Schrödinger operators. Since this requires different techniques than discussed so far and go beyond the scope of this book, we give some remarks instead of detailed theorems and refer the interested reader to the literature. The following result holds for a large class of nonlocal operators, but we only state it for fractional and relativistic Schrödinger operators.

Proposition 5.7. *Consider* $\Psi(u) = u^{\alpha/2}$ *and* $\Psi(u) = (2u+m^{2/\alpha})^{\alpha/2}-m$, $0 < \alpha < 2$, $m > 0$, *denote by* ν *the corresponding Lévy intensity, and let* V *be a* Ψ*-Kato decomposable potential. The following hold:*

(1) *If* $\lim_{|x|\to\infty} \frac{V(x)}{|\log \nu(x)|} = \infty$, *then the semigroup* $\{K_t : t \geq 0\}$ *is IUC.*
(2) *If there exist* $R, C > 0$ *such that* $\frac{V(x)}{|\log \nu(x)|} \geq C$ *for all* $|x| \geq R$, *then the semigroup* $\{K_t : t \geq 0\}$ *is AIUC.*

The above result can be upgraded to necessary and sufficient conditions under a further restriction on the potential by requiring them to grow temperately in the sense that $\sup_{y\in B} V(y) \leq C \inf_{y\in B} V(y)$ with a constant $C > 0$, for every unit ball B in $B_R(0)^c$.

Remark 5.8 (Borderline potentials). In contrast to standard Schrödinger operators, nonlocal Schrödinger operators do have borderline potentials of minimal growth for AIUC to hold. This is given by equivalence classes of potentials defined by the asymptotic comparability $V \asymp |\log \nu|$, where ν is the Lévy density of the process generated by $-\Psi(-\frac{1}{2}\Delta)$. Borderline classes can be classified by whether they are of logarithmic order (when $-\log \nu(x) \asymp \log |x|$ for large enough $|x|$), linear order (when $-\log \nu(x) \asymp |x|$ for large enough $|x|$), or sublinear but faster than logarithmic order (when $-\log \nu(x) \asymp |x|^\beta$, $0 < \beta < 1$, for large enough $|x|$) interpolating between them, while there are no further cases. It follows from Proposition 5.7 that for the fractional Laplacian $(-\Delta)^{\alpha/2}$ the borderline potential is of logarithmic order, and it is of linear order for the relativistic Laplacian $(-\Delta + m^{2/\alpha})^{\alpha/2}$. This indicates that for nonlocal Schrödinger semigroups it is "easier" to achieve intrinsic ultracontractivity, and it can be explained by a more efficient mixing mechanism in path space due to the possibility of jumps in contrast with the continuous paths of Brownian motion. It can be seen that neither the Brownian component nor the small-jump component of the process has any effect on AIUC.

For our purposes below, an important consequence of AIUC is the following property.

Proposition 5.9. *The following two conditions are equivalent:*

(1) *The semigroup* $\{K_t : t \geq 0\}$ *is AIUC.*
(2) *The property*

$$\widetilde{K}_t(x,y) \stackrel{t\to\infty}{\longrightarrow} 1 \tag{5.1.10}$$

holds uniformly in $(x,y) \in \mathbb{R}^d \times \mathbb{R}^d$.

Proof. The implication (2) $\Rightarrow$ (1) is immediate, thus we only show the converse statement. We have by the semigroup property for every $x, y \in \mathbb{R}^d$ and $t > t_0$,

$$
\begin{aligned}
&|\widetilde{K}_t(x,y) - 1| \\
&= \left| \int_{\mathbb{R}^d\times\mathbb{R}^d} \frac{K_{t_0}(x,z)K_{t-2t_0}(z,w)K_{t_0}(w,y)}{e^{-E_0 t}\varphi_0(x)\varphi_0(y)} dzdw - \frac{e^{-E_0 t}\varphi_0(x)\varphi_0(y)}{e^{-E_0 t}\varphi_0(x)\varphi_0(y)} \right| \\
&= \left| \int_{\mathbb{R}^d\times\mathbb{R}^d} \frac{K_{t_0}(x,z)\varphi(z)(K_{t-2t_0}(z,w) - e^{-E_0(t-2t_0)}\varphi_0(z)\varphi_0(w))K_{t_0}(w,y)\varphi_0(w)}{e^{-E_0 t}\varphi_0(x)\varphi_0(z)\varphi_0(w)\varphi_0(y)} dzdw \right| \\
&\le e^{E_0 t} \left\| \frac{K_{t_0}(x,y)}{\varphi_0(x)\varphi_0(y)} \right\|_\infty^2 \int_{\mathbb{R}^d\times\mathbb{R}^d} \left| K_{t-2t_0}(z,w) - e^{-E_0(t-2t_0)}\varphi_0(z)\varphi_0(w) \right| \varphi_0(z)\varphi_0(w) dzdw \\
&\le C e^{E_0 t} \left(\int_{\mathbb{R}^d\times\mathbb{R}^d} |K_{t-2t_0}(z,w) - e^{-E_0(t-2t_0)}\varphi_0(z)\varphi_0(w)|^2 dzdw \right)^{1/2} .
\end{aligned}
$$

The last factor on the right-hand side is the Hilbert–Schmidt norm of the operator $K_{t-2t_0} - e^{-E_0(t-2t_0)}1_{\varphi_0}$, where $1_{\varphi_0} : L^2(\mathbb{R}^d) \to L^2(\mathbb{R}^d)$ is the projection onto the one-dimensional subspace of $L^2(\mathbb{R}^d)$ spanned by φ_0. This gives

$$
|\widetilde{K}_t(x,y) - 1| \le C e^{E_0 t} \Big(\sum_{k=1}^{\infty} e^{-2E_k(t-2t_0)} \Big)^{1/2} = C e^{2t_0 E_1} e^{-(E_1 - E_0)t} \Big(\sum_{k=1}^{\infty} e^{-2(E_k - E_1)(t-2t_0)} \Big)^{1/2},
$$

where $E_0 < E_1 \le \cdots$ are the eigenvalues of H. By the dominated convergence theorem, the last sum converges to the multiplicity of E_1 as $t \to \infty$. Since $E_1 > E_0$, (5.1.10) follows. □

The difference $E_1 - E_0 > 0$ is the (fundamental) *spectral gap*.

Remark 5.10 (Other contractivity properties). Apart from intrinsic ultracontractivity, there are further smoothening semigroup properties of interest:

(1) The semigroup $\{e^{-tL} : t \ge 0\}$ is *supercontractive* if for every $p \in (2,\infty)$ and $t > 0$ the operators e^{-tL} are bounded from $L^2(d\mathrm{N}_0)$ to $L^p(d\mathrm{N}_0)$. The semigroup $\{e^{-tH} : t \ge 0\}$ is *intrinsically supercontractive* (ISC) if $\{e^{-tL} : t \ge 0\}$ is supercontractive.
(2) The semigroup $\{e^{-tL} : t \ge 0\}$ is *hypercontractive* if for every $p \in (2,\infty)$ there exists $t_p > 0$ such that for every $t \ge t_p$ the operators e^{-tL} are bounded from $L^2(d\mathrm{N}_0)$ to $L^p(d\mathrm{N}_0)$. The semigroup $\{e^{-tH} : t \ge 0\}$ is called *intrinsically hypercontractive* (IHC) if $\{e^{-tL} : t \ge 0\}$ is hypercontractive.
(3) For $p \in (2,\infty]m$ we say that the operator e^{-tH} is *L^p-ground state-dominated* if $\frac{1}{\varphi_0} e^{-tH} 1_{\mathbb{R}^d} \in L^p(d\mathrm{N}_0)$.

For standard Schrödinger operators, there is a natural hierarchy of these properties. For instance, in the case of $H = -\Delta + |x|^\alpha (\log(1+|x|))^\beta$ the related intrinsic semigroup is not hypercontractive if $\alpha < 2$, it is hypercontractive but not supercontractive if $\alpha = 2, \beta = 0$, it is supercontractive but not ultracontractive if $\alpha = 2, 0 < \beta \le 2$, and it is ultracontractive if $\alpha = 2$, $\beta > 2$ or if $\alpha > 2$. This shows that ultracontractivity is essentially stronger than supercontractivity, which is essentially stronger than hypercontractivity. For necessary, sufficient, or joint conditions, and mutual implications we refer to the literature (see a discussion in Chapter 6). One interesting fact is that for a large subclass of nonlocal Schrödinger operators with Bernstein functions of the Laplacian and a large class of potentials ISC is equivalent with IUC, and IHS is equivalent with AIUC. On the other hand, for standard Schrödinger operators all these properties are different.

5.1.2 Ground state-transformed processes as solutions of SDE

Since L is a Markov generator on $L^2(d\mathrm{N}_0)$, we can associate a stochastic differential equation with it. The advantage of this is that it sheds light on the structure of the process generated by L, which we call *ground state-transformed process*, by identifying its drift, diffusion, and jump components.

Consider first the standard Schrödinger operator $H = -\frac{1}{2}\Delta + V$, and assume that $\varphi_0 > 0$ is its unique ground state. Since $C_0^\infty(\mathbb{R}^d)$ is a core also for L, a simple calculation shows that on this subspace we have

$$Lf = -\frac{1}{2}\Delta f - \nabla \log \varphi_0 \cdot \nabla f. \tag{5.1.11}$$

This then leads to the stochastic differential equation

$$dX_t = \nabla \log \varphi_0(X_t)dt + dB_t, \quad X_0 = x, \tag{5.1.12}$$

whose solution $(X_t^x)_{t\geq 0}$ satisfies

$$\mathbb{E}[f(X_t^x)] = (e^{-tL}f)(x), \quad t \geq 0. \tag{5.1.13}$$

However, for given V, in general little information is available about the regularity of φ_0, and thus it is not clear if $\nabla \log \varphi_0$ is well defined as a function at all and one can view (5.1.12) as an SDE in the classical sense discussed in Chapter 2 or more general setups are necessary. Also, even if the SDE has a proper meaning, due to the generally unbounded and nonlinear drift coefficient, the study of existence and uniqueness of solutions is not straightforward as in many interesting cases the Lipschitz condition (2.4.53)–(2.4.54) in Theorem 2.152 will not hold. Instead of this general study, we will discuss two specific cases at the end of this section, for which the SDE can be explicitly solved and which illuminate the mechanisms due to which solutions exist.

Consider now nonlocal Schrödinger operators $H = \Psi(-\frac{1}{2}\Delta) + V$ such that the Lévy process $(X_t)_{t\geq 0}$ has a Lévy triplet $(0, A, \nu)$, and V is such that H has a unique ground state $\varphi_0 > 0$. Then on $C_0^\infty(\mathbb{R}^d)$ we have

$$Lf = \frac{1}{2}\sigma\nabla\cdot\sigma\nabla f + \sigma\nabla\log\varphi_0\cdot\sigma\nabla f + \int\limits_{0<|z|\le 1}\frac{\varphi_0(\cdot+z)-\varphi_0(\cdot)}{\varphi_0(\cdot)}z\cdot\nabla f\,\nu(z)dz$$
$$+\int\limits_{\mathbb{R}^d\setminus\{0\}}(f(\cdot+z)-f(\cdot)-z\cdot\nabla 1_{\{|z|\le 1\}})\frac{\varphi_0(\cdot+z)}{\varphi_0(\cdot)}\nu(z)dz,$$

where we use the notation $A = \sigma\sigma^T$ and $\sigma\nabla\cdot\sigma\nabla f(x) = \sum_{i,j=1}^d(\sigma\sigma^T)_{ij}\partial_{x_i}\partial_{x_j}f(x)$. This gives rise to the stochastic differential equation with jumps

$$X_t = X_0 + \sigma B_t + \int_0^t \sigma\nabla\log\varphi_0(X_s)ds + \int_0^t\int\limits_{|z|\le 1}\frac{\varphi_0(X_s+z)-\varphi_0(X_s)}{\varphi_0(X_s)}z\nu(z)dzds$$
$$+\int_0^t\int\limits_{|z|\le 1}\int_0^\infty z1_{\left\{w\le\frac{\varphi_0(X_{s-}+z)}{\varphi_0(X_{s-})}\right\}}\widetilde{N}(ds,dz,dw) + \int_0^t\int\limits_{|z|>1}\int_0^\infty z1_{\left\{w\le\frac{\varphi_0(X_{s-}+z)}{\varphi_0(X_{s-})}\right\}}N(ds,dz,dw), \tag{5.1.14}$$

where $(B_t)_{t\ge0}$ is a d-dimensional Brownian motion with covariance matrix σ, N is a Poisson random measure on $[0,\infty)\times(\mathbb{R}^d\setminus\{0\})\times[0,\infty)$ with intensity $dt\,\nu(z)dz\,dw$, and $\widetilde{N}(ds,dz,dw)$ is the related compensated Poisson measure.

We have the following result, which we present without proof since it would require more preparation on SDE with jumps than presented in Section 2.4.5 above.

Theorem 5.11. *Suppose that the function $x\mapsto\nabla\log\varphi_0(x)$, $x\in\mathbb{R}^d$, is locally bounded. Then the stochastic differential equation* (5.1.14) *has a weak solution.*

While this leaves uniqueness open, in the sections below we construct processes which are weak solutions of the stochastic differential equation (5.1.12) or (5.1.14).

Remark 5.12. Ground state-transformed processes are interesting objects on their own, even without the direct relevance of Feynman–Kac type representations. As it can be seen from (5.1.14), the related SDE has position-dependent coefficients which destroy stationarity of increments, and which are generally unbounded. These processes belong to the class of Lévy-type (Feller) processes, whose general theory is in the making. Also, from the equation one can appreciate a second interpretation of potentials on the level of processes after the sojourn and local times studied in Section 4.3.5, showing that the effect of the potential is a position-dependent drift and a bias in the jump intensity. For an analysis of global and local sample path properties, we refer the reader to the literature, as discussed in Chapter 6 below.

We conclude this section by a discussion of two specific cases of ground state-transformed processes via SDE for potentials of fundamental interest.

Example 5.13 (Ground state-transformed process for the harmonic oscillator). Consider

$$H_{\mathrm{osc}} = -\frac{1}{2m}\frac{d^2}{dx^2} + \frac{m\omega^2}{2}x^2 - \frac{\omega}{2} \tag{5.1.15}$$

on $L^2(\mathbb{R})$. The parameter $\omega > 0$ is the frequency of the oscillator. The ground state of H_{osc} is explicitly given by

$$\varphi_{\mathrm{osc}}(x) = (\omega/\pi)^{1/4} e^{-\omega x^2/2}.$$

Since $\log \varphi_{\mathrm{osc}} \in C^1(\mathbb{R})$, formula (5.1.12) gives the SDE

$$dX_t = -\omega X_t dt + dB_t, \quad X_0 = x,$$

which is a case of the Langevin equation studied in Example 2.159. A unique strong solution of this SDE is the Ornstein–Uhlenbeck process

$$X_t^x = xe^{-\omega t} + \int_0^t e^{-\omega(s-t)} dB_s. \tag{5.1.16}$$

We take up this example for further discussion in Section 5.1.5 below.

Example 5.14 (Ground state-transformed process for potential wells). Consider now the Schrödinger operator with one-dimensional potential well

$$H_v = -\frac{d^2}{dx^2} - v1_{\{|x|\le a\}}$$

with $v > 0$. Using the explicit formula (4.3.29) and the fact that $\varphi_0 \in C^1(\mathbb{R})$, the SDE (5.1.12) becomes

$$dX_t = dB_t - \left(\sqrt{2|E|}\,\mathrm{sgn}(X_t)1_{\{|X_t|>a\}} + \sqrt{2(v-|E|)}\tan\left(\sqrt{2(v-|E|)}\,X_t\right)1_{\{|X_t|\le a\}}\right)dt. \tag{5.1.17}$$

Let λ be an eigenvalue of an even eigenfunction of H_v. Recall that $|\lambda|$ satisfies

$$\tan\left(a\sqrt{2(v-|\lambda|)}\right) = \sqrt{\frac{|\lambda|}{v-|\lambda|}}, \tag{5.1.18}$$

which implies that

$$\begin{cases} X\tan X = Y, \\ X^2 + Y^2 = 2va^2, \end{cases} \tag{5.1.19}$$

where $X = a\sqrt{2(V - |\lambda|)}$ and $Y = a\sqrt{|\lambda|}$. The ground state energy $\lambda = E$ of H_v is the unique solution of (5.1.19) such that $0 < X < \pi/2$. Hence

$$\sqrt{2(v - |E|)} < \frac{\pi}{2a}. \tag{5.1.20}$$

This also results from the following more general observation, which gives more insight into what is behind this bound. Consider an arbitrary potential well $V = -v1_{\mathcal{D}}$, $v > 0$, with compact support $\mathcal{D} \subset \mathbb{R}^d$. Suppose that the Schrödinger operator $H = \Psi(-\frac{1}{2}\Delta) - v1_{\mathcal{D}}$, $\Psi \in \mathcal{B}_0$, has a ground state φ_0 at eigenvalue λ_0. Then

$$v - |\lambda_0| \leq \lambda_1^{\mathcal{D}}, \tag{5.1.21}$$

where $\lambda_1^{\mathcal{D}}$ is the principal eigenvalue of $\Psi(-\frac{1}{2}\Delta)$ in $\mathcal{D}$ with Dirichlet exterior condition on $\mathcal{D}^c$. Denote by $(X_t)_{t\geq 0}$ the Lévy process generated by $-\Psi(-\frac{1}{2}\Delta)$, by P^x its probability measure starting at $x \in \mathbb{R}^d$, and by $\tau_{\mathcal{D}} = \inf\{t > 0 \mid X_t \in \mathcal{D}^c\}$ the first exit time of the process from $\mathcal{D}$. Then from the Feynman–Kac formula, we get that

$$\varphi_0(x) = \mathbb{E}^x\left[e^{\int_0^t (v1_{\mathcal{D}}(X_s)+\lambda_0)ds}\varphi_0(X_t)\right] \tag{5.1.22}$$

for almost every x. The right-hand side is continuous in x and gives a continuous version of φ_0, which we choose from now on. Since the semigroup $\{e^{-tH_v} : t \geq 0\}$ is positivity improving, $\varphi_0(x) > 0$ for $x \in \mathbb{R} \setminus N$ with a null set N. Suppose that $\varphi_0(y) = 0$ for some y. Then we have $\mathbb{E}^y\left[e^{\int_0^t (v1_{\mathcal{D}}(X_s)+\lambda_0)ds}\varphi_0(X_t)\right] = 0$, but the integrand is nonnegative and we see that $e^{\int_0^t (v1_{\mathcal{D}}(X_s)+\lambda_0)ds}\varphi_0(X_t) = 0$ almost surely under P^y. Since $e^{\int_0^t (v1_{\mathcal{D}}(X_s)+\lambda_0)ds} > 0$ almost surely, $\varphi_0(X_t) = 0$ almost surely with respect to P^y. Thus we have $\mathbb{E}^y[\varphi_0(X_t)] = \int_{\mathbb{R}} \varphi_0(x)P_t^y(x)dx = 0$, where $P_t^y(x)$ denotes the distribution function of X_t which is strictly positive. This implies that $\varphi_0(x) = 0$ for almost every x, and contradicts that $\varphi_0(x)$ is strictly positive for almost every x. This means that the continuous version of φ_0 is strictly positive, in particular, $\min_{y\in\mathcal{D}} \varphi_0(y) > 0$. From (5.1.22), we have

$$\varphi_0(x) \geq \mathbb{E}^x\left[e^{\int_0^t (v1_{\mathcal{D}}(X_s)+\lambda_0)ds}\varphi_0(X_t)1_{\{t<\tau_{\mathcal{D}}\}}\right] \geq e^{t(v-|\lambda_0|)} \min_{y\in\mathcal{D}} \varphi_0(y)\, P^x(\tau_{\mathcal{D}} > t), \quad x \in \mathcal{D},$$

for every $t \geq 0$. By taking logarithms on both sides and dividing by $t > 0$, we get

$$v - |\lambda_0| \leq -\limsup_{t\to\infty} \frac{1}{t} \log P^x(\tau_{\mathcal{D}} > t) = \lambda_1^{\mathcal{D}},$$

giving (5.1.21). When $\Psi(u) = u$ and $\mathcal{D} = [-a, a]$, this principal Dirichlet eigenvalue is $\lambda_1^{\mathcal{D}} = \frac{\pi^2}{8a^2}$, which indeed gives (5.1.20).

Condition (5.1.20) ensures that the drift is bounded and (5.1.17) has a unique strong solution by Theorem 2.152. Note that the everywhere continuity of the drift term is equivalent to condition (4.3.28), and the drift is piecewise differentiable except for $x = \pm a$. Indeed, from the expression it is seen that at the boundaries of the potential well there is a regime change of the drift and as soon as a path exits $[-a, a]$, then due to the $\operatorname{sgn}(X_t)$ term it is pulled back to the origin at a constant speed $\sqrt{2|E|}$. This is the stabilizing mechanism in this case making sure that paths do not explode, i. e., do not go to infinity in finite time, and thus a solution exists.

We consider now (5.1.17) and choose the interval $I = [-a, a]$. Let

$$dX_t = -\sqrt{2(v - |E|)} \tan(\sqrt{2(v - |E|)}\, X_t)dt + dB_t$$

and consider the stopping time $\tau_{\pm a} = \inf\{t > 0 \mid |X_t| = \pm a\}$. We construct a weak solution of this stochastic differential equation for times $t \leq \tau_{\pm a}$. It readily follows that the scale function defined in (2.4.90) in this case is

$$S(x) = \frac{1}{a\sqrt{2(v - |E|)}} \tan(\sqrt{2(v - |E|)}x).$$

Thus an application of the Itô formula to the process $(Y_t)_{t\geq 0} = (S(X_t))_{t\geq 0}$ gives

$$\frac{1}{1 + 2(v - |E|)Y_t^2} dY_t = dB_t.$$

Using Lemma 2.170, we obtain $Y_t \stackrel{\mathrm{d}}{=} B^1_{h^{-1}(t)}$, where $(B^1_t)_{t\geq 0}$ denotes a one-dimensional Brownian motion and

$$h(t) = \frac{t}{1 + (\sqrt{2(v - |E|)}B^1_t)^2}.$$

5.1.3 $P(\phi)_1$-processes with continuous paths

Now we discuss a second approach to ground state-transformed processes. In this section we consider standard Schrödinger operators $H = -\frac{1}{2}\Delta + V$ and use the related Markov operators L given by (5.1.3) to construct a diffusion directly via extension of its finite dimensional distributions.

Consider a two-sided Brownian motion and denote by $\mathfrak{X} = C(\mathbb{R}, \mathbb{R}^d)$ the space of continuous paths on the real line $\mathbb{R}$ as before. In this section we construct a probability measure $\mathcal{N}_0^x$, $x \in \mathbb{R}^d$, on $(\mathfrak{X}, \mathcal{B}(\mathfrak{X}))$ such that the coordinate process $(X_t)_{t\in\mathbb{R}}$ is a diffusion process generated by L.

Theorem 5.15. *Suppose that V is Kato-decomposable and the Schrödinger operator $H = -\frac{1}{2}\Delta + V$ has a strictly positive ground state φ_0. Let $L = \mathscr{U}^{-1}\bar{H}\mathscr{U}$ be the ground state transform of H, and $X_t(\omega) = \omega(t)$ the coordinate process on $(\mathfrak{X}, \mathcal{B}(\mathfrak{X}))$. Then there exists a probability measure $\mathcal{N}_0^x$ on $(\mathfrak{X}, \mathcal{B}(\mathfrak{X}))$ satisfying the following properties:*
(1) (Initial distribution) $\mathcal{N}_0^x(X_0 = x) = 1$;

(2) (Reflection symmetry) $(X_t)_{t\geq 0}$ *and* $(X_s)_{s\leq 0}$ *are independent, and*

$$X_{-t} \stackrel{\mathrm{d}}{=} X_t, \quad t \in \mathbb{R};$$

(3) (Diffusion property) *Let*

$$\mathscr{F}_t^+ = \sigma(X_s, 0 \leq s \leq t), \quad \mathscr{F}_t^- = \sigma(X_s, t \leq s \leq 0)$$

be given filtrations; then $(X_t)_{t\geq 0}$ *and* $(X_s)_{s\leq 0}$ *are diffusion processes with respect to* $(\mathscr{F}_t^+)_{t\geq 0}$ *and* $(\mathscr{F}_t^-)_{t\leq 0}$*, respectively, i. e.,*

$$\mathbb{E}_{\mathcal{N}_0^x}[X_{t+s}|\mathscr{F}_s^+] = \mathbb{E}_{\mathcal{N}_0^x}[X_{t+s}|\sigma(X_s)] = \mathbb{E}_{\mathcal{N}_0^{X_s}}[X_t],$$
$$\mathbb{E}_{\mathcal{N}_0^x}[X_{-t-s}|\mathscr{F}_{-s}^-] = \mathbb{E}_{\mathcal{N}_0^x}[X_{-t-s}|\sigma(X_{-s})] = \mathbb{E}_{\mathcal{N}_0^{X_{-s}}}[X_{-t}]$$

for $s, t \geq 0$, *and* $\mathbb{R} \ni t \mapsto X_t \in \mathbb{R}^d$ *is almost surely continuous, where* $\mathbb{E}_{\mathcal{N}_0^{X_s}}$ *means* $\mathbb{E}_{\mathcal{N}_0^y}$ *evaluated at* $y = X_s$*;*

(4) (Shift invariance) *Let* $-\infty < t_0 \leq t_1 \leq \cdots \leq t_n < \infty$*; then*

$$\int_{\mathbb{R}^d} \mathbb{E}_{\mathcal{N}_0^x}\Big[\prod_{j=0}^n f_j(X_{t_j})\Big] d\mathrm{N}_0 = (f_0, e^{-(t_1-t_0)L} f_1 \cdots e^{-(t_n-t_{n-1})L} f_n)_{L^2(d\mathrm{N}_0)} \tag{5.1.23}$$

for $f_0, f_n \in L^2(d\mathrm{N}_0)$ *and* $f_j \in L^\infty(\mathbb{R}^d)$, $j = 1, \ldots, n-1$, *and the finite-dimensional distributions of the process are shift invariant, i. e.,*

$$\int_{\mathbb{R}^d} \mathbb{E}_{\mathcal{N}_0^x}\Big[\prod_{j=1}^n f_j(X_{t_j})\Big] d\mathrm{N}_0 = \int_{\mathbb{R}^d} \mathbb{E}_{\mathcal{N}_0^x}\Big[\prod_{j=1}^n f_j(X_{t_j+s})\Big] d\mathrm{N}_0, \quad s \in \mathbb{R}.$$

Definition 5.16 ($P(\phi)_1$-process). Let $H = -\frac{1}{2}\Delta + V$ and its ground state φ_0 be given, and consider the probability space $(\mathfrak{X}, \mathcal{B}(\mathfrak{X}), \mathcal{N}_0^x)$. The coordinate process $(X_t)_{t\in\mathbb{R}}$, $X_t(\omega) = \omega(t)$ on this space is called a *$P(\phi)_1$-process* for H.

We proceed now to prove existence and some properties of $P(\phi)_1$-processes in several steps. For arbitrary time points $0 \leq t_0 \leq t_1 \leq \cdots \leq t_n$, $n \in \mathbb{N}$, define the set function $\mathcal{P}_{\{t_0,\ldots,t_n\}} : \times_{j=0}^n \mathcal{B}(\mathbb{R}^d) \to \mathbb{R}$ by

$$\mathcal{P}_{\{t_0,\ldots,t_n\}}(\times_{i=0}^n A_i) = (1_{A_0}, e^{-(t_1-t_0)L} 1_{A_1} \cdots e^{-(t_n-t_{n-1})L} 1_{A_n})_{L^2(d\mathrm{N}_0)} \tag{5.1.24}$$

and for $t \geq 0$, $\mathcal{P}_{\{t\}} : \mathcal{B}(\mathbb{R}^d) \to \mathbb{R}$ by

$$\mathcal{P}_{\{t\}}(A) = (1, e^{-tL} 1_A)_{L^2(d\mathrm{N}_0)} = (1, 1_A)_{L^2(d\mathrm{N}_0)}. \tag{5.1.25}$$

Step 1. The family of set functions given by (5.1.24)–(5.1.25) satisfies the consistency condition

$$\mathcal{P}_{\{t_0,\dots,t_{n+m}\}}((\times_{i=0}^{n} A_i) \times (\times_{i=n+1}^{n+m} \mathbb{R}^d)) = \mathcal{P}_{\{t_0,\dots,t_n\}}(\times_{i=0}^{n} A_i),$$

for all $m, n \in \mathbb{N}$. Hence by the Kolmogorov's extension theorem, there exists a probability measure $\mathcal{P}_\infty$ on $((\mathbb{R}^d)^{[0,\infty)}, \sigma(\mathcal{A}))$, where

$$\mathcal{A} = \left\{\omega : \mathbb{R} \to \mathbb{R}^d \mid \omega\lceil_\Lambda \in E,\ E \in (\mathcal{B}(\mathbb{R}^d))^{|\Lambda|},\ \Lambda \subset \mathbb{R},\ |\Lambda| < \infty\right\} \tag{5.1.26}$$

such that

$$\mathcal{P}_{\{t\}}(A) = \mathbb{E}_{\mathcal{P}_\infty}[1_A(Y_t)], \tag{5.1.27}$$

$$\mathcal{P}_{\{t_0,\dots,t_n\}}(\times_{i=0}^{n} A_i) = \mathbb{E}_{\mathcal{P}_\infty}\Big[\prod_{j=0}^{n} 1_{A_j}(Y_{t_j})\Big], \quad n \geq 1, \tag{5.1.28}$$

for $A, A_0, \dots, A_n \in \mathcal{A}$, where $Y_t(\omega) = \omega(t)$, $\omega \in (\mathbb{R}^d)^{[0,\infty)}$, is the coordinate process. The process $(Y_t)_{t\geq 0}$ on the probability space $((\mathbb{R}^d)^{[0,\infty)}, \sigma(\mathcal{A}), \mathcal{P}_\infty)$ satisfies

$$(f_0, e^{-(t_1-t_0)L} f_1 \cdots e^{-(t_n-t_{n-1})L} f_n)_{L^2(d\mathsf{N}_0)} = \mathbb{E}_{\mathcal{P}_\infty}\Big[\prod_{j=0}^{n} f_j(Y_{t_j})\Big], \tag{5.1.29}$$

$$(1, f)_{L^2(d\mathsf{N}_0)} = (1, e^{-tL} f)_{L^2(d\mathsf{N}_0)} = \mathbb{E}_{\mathcal{P}_\infty}[f(Y_t)] = \mathbb{E}_{\mathcal{P}_\infty}[f(Y_0)] \tag{5.1.30}$$

for $f_j \in L^\infty(\mathbb{R}^d)$, $j = 0, 1, \dots, n$, and $0 \leq t_0 < t_1 < \dots < t_n$, for all $n \in \mathbb{N}$.

Step 2. Next we show that this process has a continuous version.

Lemma 5.17. *The process $(Y_t)_{t\geq 0}$ has a continuous version.*

Proof. By the Kolmogorov–Čentsov theorem, it suffices to show that the estimate $\mathbb{E}_{\mathcal{P}_\infty}[|Y_t - Y_s|^{2n}] \leq |t-s|^n$, $n \geq 2$, holds. Let $s < t$. We have

$$\mathbb{E}_{\mathcal{P}_\infty}[|Y_t - Y_s|^{2n}] = \sum_{\mu=1}^{d} \sum_{k=0}^{2n} \binom{2n}{k} (-1)^k (\varphi, x^\mu e^{-(t-s)\bar{H}} x^\mu \varphi).$$

The last term can be represented in terms of Brownian motion as

$$(\varphi_0, x^\mu e^{-(t-s)\bar{H}} x^\mu \varphi_0) = e^{(t-s)E} \int_{\mathbb{R}^d} \mathbb{E}^x\Big[B_0^\mu \varphi_0(B_0) B_{t-s}^\mu \varphi_0(B_{t-s}) e^{-\int_0^{t-s} V(B_r)dr}\Big] dx.$$

By Schwarz inequality,

$$\begin{aligned}
\mathbb{E}_{\mathcal{P}_\infty}[|Y_t - Y_s|^{2n}] &= e^{(t-s)E} \int_{\mathbb{R}^d} \mathbb{E}^x[|B_{t-s} - B_0|^{2n} \varphi_0(B_0)\varphi_0(B_{t-s}) e^{-\int_0^{t-s} V(B_r)dr}] dx \\
&\leq e^{(t-s)E} \mathbb{E}\Bigg[|B_{t-s} - B_0|^{2n} \Big(\int_{\mathbb{R}^d} e^{-2\int_0^t V(B_r+x)dr} \varphi_0(x)^2 dx\Big)^{1/2} \Big(\int_{\mathbb{R}^d} \varphi_0(B_{t-s} + x)^2 dx\Big)^{1/2}\Bigg]
\end{aligned}$$

$$\le e^{(t-s)E}\|\varphi_0\|(\mathbb{E}[|B_{t-s}-B_0|^{4n}])^{1/2}\mathbb{E}\left[\left(\int_{\mathbb{R}^d} e^{-2\int_0^t V(B_r+x)dr}\varphi_0(x)^2dx\right)\right].$$

Since V is Kato-decomposable, we furthermore have

$$\begin{aligned}&\le e^{-(t-s)E}\|\varphi_0\|^2\mathbb{E}[|B_{t-s}-B_0|^{4n}]^{1/2}\sup_{x\in\mathbb{R}^d}\mathbb{E}^x[e^{-2\int_0^t V(B_s)ds}]\\&\le \sqrt{C_{2n}}|t-s|^n e^{-(t-s)E}\|\varphi_0\|^2\sup_{x\in\mathbb{R}^d}\mathbb{E}^x[e^{-2\int_0^t V(B_s)ds}].\end{aligned}$$

Here we used that $\mathbb{E}[|B_t-B_s|^{2n}]=C_n|t-s|^n$. □

Let now $\overline{Y}=(\overline{Y}_t)_{t\ge0}$ be the continuous version of $(Y_t)_{t\ge0}$ on the probability space $((\mathbb{R}^d)^{[0,\infty)},\sigma(\mathcal{A}),\mathcal{P}_\infty)$. Denote the image measure of $\mathcal{P}_\infty$ on $(\mathscr{X},\mathcal{B}(\mathscr{X}))$ with respect to $\overline{Y}$ by

$$\mathcal{M}=\mathcal{P}_\infty\circ\overline{Y}^{-1}. \tag{5.1.31}$$

We identify the coordinate process by $\widetilde{Y}_t(\omega)=\omega(t)$, for $\omega\in\mathscr{X}$. We thus constructed a random process $\widetilde{Y}=(\widetilde{Y}_t)_{t\ge0}$ on $(\mathscr{X},\mathcal{B}(\mathscr{X}),\mathcal{M})$ such that $\overline{Y}_t\overset{\mathrm{d}}{=}\widetilde{Y}_t$ for $t\ge0$. Equations (5.1.29) and (5.1.30) can be expressed in terms of $(\widetilde{Y}_t)_{t\ge0}$ as

$$\begin{aligned}(f_0,e^{-(t_1-t_0)L}f_1\cdots f_{n-1}e^{-(t_n-t_{n-1})L}f_n)_{L^2(d\mathsf{N}_0)}&=\mathbb{E}_{\mathcal{M}}\Big[\prod_{j=0}^n f_j(\widetilde{Y}_{t_j})\Big],\\(1,f)_{L^2(d\mathsf{N}_0)}=(1,e^{-tL}f)_{L^2(d\mathsf{N}_0)}&=\mathbb{E}_{\mathcal{M}}[f(\widetilde{Y}_t)]=\mathbb{E}_{\mathcal{M}}[f(\widetilde{Y}_0)].\end{aligned}$$

Step 3. Define a probability measure on $\mathscr{X}$ by

$$\mathcal{M}^x(\cdot)=\mathcal{M}(\cdot\,|\widetilde{Y}_0=x) \tag{5.1.32}$$

for all $x\in\mathbb{R}^d$. Since the distribution of $\widetilde{Y}_0$ is $d\mathsf{N}_0$, note that

$$\mathcal{M}(A)=\int_{\mathbb{R}^d}\mathbb{E}_{\mathcal{M}^x}[1_A]d\mathsf{N}_0(x).$$

The random process $(\widetilde{Y}_t)_{t\ge0}$ on $(\mathscr{X},\mathcal{B}(\mathscr{X}),\mathcal{M}^x)$ satisfies

$$\begin{aligned}(f_0,e^{-(t_1-t_0)L}f_1\cdots f_{n-1}e^{-(t_n-t_{n-1})L}f_n)_{L^2(d\mathsf{N}_0)}&=\int_{\mathbb{R}^d}\mathbb{E}_{\mathcal{M}^x}\Big[\prod_{j=0}^n f_j(\widetilde{Y}_{t_j})\Big]d\mathsf{N}_0,\\(1,e^{-tL}f)_{L^2(d\mathsf{N}_0)}=(1,f)_{L^2(d\mathsf{N}_0)}&=\int_{\mathbb{R}^d}\mathbb{E}_{\mathcal{M}^x}[f(\widetilde{Y}_0)]d\mathsf{N}_0=\int_{\mathbb{R}^d}f(x)d\mathsf{N}_0.\end{aligned} \tag{5.1.33}$$

Lemma 5.18. *The process $(\widetilde{Y}_t)_{t\geq 0}$ is a Markov process on $(\mathscr{X}, \mathcal{B}(\mathscr{X}), \mathcal{M}^x)$ with respect to the natural filtration $\sigma(\widetilde{Y}_s, 0 \leq s \leq t)$, $t \geq 0$.*

Proof. Let

$$p_t(x, A) = (e^{-tL}1_A)(x), \quad A \in \mathcal{B}(\mathbb{R}^d), \quad t \geq 0. \tag{5.1.34}$$

Note that $p_t(x, A) = \mathbb{E}[1_A(X_t^x)]$. Thus by (5.1.33) the finite-dimensional distributions of $(\widetilde{Y}_t)_{t\geq 0}$ are

$$\mathbb{E}_{\mathcal{M}^x}\Big[\prod_{j=1}^n 1_{A_j}(\widetilde{Y}_{t_j})\Big] = \int_{\mathbb{R}^{nd}} \Big(\prod_{j=1}^n 1_{A_j}(x_j)\Big) \prod_{j=1}^n p_{t_j - t_{j-1}}(x_{j-1}, dx_j) \tag{5.1.35}$$

with $t_0 = 0$ and $x_0 = x$. We show that $p_t(x, A)$ is a probability transition kernel. Note that e^{-tL} is positivity improving. Thus $0 \leq e^{-tL} f \leq 1$, for all functions f such that $0 \leq f \leq 1$, and $e^{-tL}1 = 1$ follows, so $p_t(x, \cdot)$ is a probability measure on $\mathbb{R}^d$ with $p_t(x, \mathbb{R}^d) = 1$. Secondly, $p_t(\cdot, A)$ is trivially Borel measurable with respect to x. Thirdly, by the semigroup property $e^{-sL}e^{-tL}1_A = e^{-(s+t)L}1_A$, the Chapman–Kolmogorov identity (2.2.53) follows directly, and hence $p_t(x, A)$ is a probability transition kernel. Thus $(\widetilde{Y}_t)_{t\geq 0}$ is a Markov process by (5.1.35) and Proposition 2.82. □

Step 4. We extend $(\widetilde{Y}_t)_{t\geq 0}$ to a process on the whole real line $\mathbb{R}$. Consider $\tilde{\mathscr{X}} = \mathscr{X} \times \mathscr{X}$, $\tilde{\mathscr{M}} = \mathcal{B}(\mathscr{X}) \times \mathcal{B}(\mathscr{X})$, and $\tilde{\mathcal{M}}^x = \mathcal{M}^x \times \mathcal{M}^x$. Let $(\tilde{X}_t)_{t\in\mathbb{R}}$ be the random process defined by

$$\tilde{X}_t(\omega) = \begin{cases} \widetilde{Y}_t(\omega_1), & t \geq 0, \\ \widetilde{Y}_{-t}(\omega_2), & t < 0, \end{cases} \tag{5.1.36}$$

for $\omega = (\omega_1, \omega_2) \in \tilde{\mathscr{X}}$, on the product space $(\tilde{\mathscr{X}}, \tilde{\mathscr{M}}, \tilde{\mathcal{M}}^x)$. Note that $\tilde{X}_0 = x$ almost surely under $\tilde{\mathcal{M}}^x$, and $\tilde{X}_t$ is continuous in t almost surely. It is trivial to see that $\tilde{X}_t$, $t \geq 0$, and $\tilde{X}_s$, $s \leq 0$, are independent, and $\tilde{X}_t \stackrel{\mathrm{d}}{=} \tilde{X}_{-t}$.

Step 5. We can now complete the proof of the theorem.

Proof of Theorem 5.15. Denote the image measure of $\tilde{\mathcal{M}}^x$ on $(\mathfrak{X}, \mathcal{B}(\mathfrak{X}))$ with respect to $(\tilde{X}_t)_{t\geq 0}$ by

$$\mathcal{N}_0^x = \tilde{\mathcal{M}}^x \circ \tilde{X}^{-1}. \tag{5.1.37}$$

Let $X_t(\omega) = \omega(t)$, $t \in \mathbb{R}$, $\omega \in \mathfrak{X}$, be the coordinate process. We have

$$X_t \stackrel{\mathrm{d}}{=} \widetilde{Y}_t \quad (t \geq 0), \quad X_t \stackrel{\mathrm{d}}{=} \widetilde{Y}_{-t} \quad (t \leq 0). \tag{5.1.38}$$

Since by Step 3 above $(\widetilde{Y}_t)_{t\geq 0}$ and $(\widetilde{Y}_{-t})_{t\leq 0}$ are Markov processes with respect to the natural filtrations $\sigma(\widetilde{Y}_s, 0 \leq s \leq t)$ and $\sigma(\widetilde{Y}_s, -t \leq s \leq 0)$, respectively, $(X_t)_{t\geq 0}$ and $(X_t)_{t\leq 0}$ are also Markov processes with respect to $(\mathscr{F}_t^+)_{t\geq 0}$ resp. $(\mathscr{F}_t^-)_{t\leq 0}$, where $\mathscr{F}_t^+ = \sigma(X_s, 0 \leq$

$s \le t)$ and $\mathscr{F}_t^- = \sigma(X_s, -t \le s \le 0)$. Thus the diffusion property follows. We also see that $(X_s)_{s\le 0}$ and $(X_t)_{t\ge 0}$ are independent and $X_{-t} \stackrel{\mathrm{d}}{=} X_t$ by (5.1.38) and Step 4 above. Hence reflection symmetry follows. Shift invariance follows by the lemma below; for any $s \in \mathbb{R}$,

$$\int_{\mathbb{R}^d} \mathbb{E}_{\mathcal{N}_0^x}\Big[\prod_{j=0}^n f_j(X_{t_j})\Big] d\mathsf{N}_0 = \int_{\mathbb{R}^d} \mathbb{E}_{\mathcal{N}_0^x}\Big[\prod_{j=0}^n f_j(X_{t_j+s})\Big] d\mathsf{N}_0. \qquad \square$$

Lemma 5.19. *Let $-\infty < t_0 \le t_1 \le \cdots \le t_n$. Then*

$$\int_{\mathbb{R}^d} \mathbb{E}_{\mathcal{N}_0^x}\Big[\prod_{j=0}^n f_j(X_{t_j})\Big] d\mathsf{N}_0 = (f_0, e^{-(t_1-t_0)L} f_1 \cdots f_{n-1} e^{-(t_n-t_{n-1})L} f_n)_{L^2(d\mathsf{N}_0)}. \tag{5.1.39}$$

Proof. Let $t_0 \le \cdots \le t_n \le 0 \le t_{n+1} \le \cdots \le t_{n+m}$. By the independence of $(X_s)_{s\le 0}$ and $(X_t)_{t\ge 0}$,

$$\int_{\mathbb{R}^d} \mathbb{E}_{\mathcal{N}_0^x}\Big[\prod_{j=0}^{n+m} f_j(X_{t_j})\Big] d\mathsf{N}_0 = \int_{\mathbb{R}^d} \mathbb{E}_{\mathcal{N}_0^x}\Big[\prod_{j=0}^{n} f_j(X_{t_j})\Big] \mathbb{E}_{\mathcal{N}_0^x}\Big[\prod_{j=n+1}^{n+m} f_j(X_{t_j})\Big] d\mathsf{N}_0.$$

We furthermore have

$$\mathbb{E}_{\mathcal{N}_0^x}\Big[\prod_{j=n+1}^{n+m} f_j(X_{t_j})\Big] = (e^{-t_{n+1}L} f_{n+1} e^{-(t_{n+2}-t_{n+1})L} f_{n+2} \cdots f_{n+m-1} e^{-(t_{n+m}-t_{n+m-1})L} f_{n+m})(x) \tag{5.1.40}$$

and

$$\mathbb{E}_{\mathcal{N}_0^x}\Big[\prod_{j=0}^{n+m} f_j(X_{t_j})\Big] = \mathbb{E}_{\mathcal{N}_0^x}\Big[\prod_{j=0}^{n+m} f_j(X_{-t_j})\Big] = (e^{+t_n L} f_n e^{-(t_n-t_{n-1})L} f_{n-1} \cdots f_2 e^{-(t_1-t_0)L} f_1)(x). \tag{5.1.41}$$

By (5.1.40) and (5.1.41), we obtain

$$\begin{aligned}\int_{\mathbb{R}^d} \mathbb{E}_{\mathcal{N}_0^x}\Big[\prod_{j=0}^{n+m} f_j(X_{t_j})\Big] d\mathsf{N}_0 &= (e^{+t_n L} f_n \cdots f_2 e^{-(t_1-t_0)L} f_1, e^{-t_{n+1}L} f_{n+1} \cdots e^{-(t_{n+m}-t_{n+m-1})L} f_{n+m})_{L^2(d\mathsf{N}_0)} \\ &= (f_1, e^{-(t_1-t_0)L} f_2 \cdots f_{n+m-1} e^{-(t_{n+m}-t_{n+m-1})L} f_{n+m})_{L^2(d\mathsf{N}_0)}.\end{aligned}$$

Hence (5.1.39) follows. $\square$

Remark 5.20. By Theorem 5.15, we have

$$(e^{-tL} f)(x) = \mathbb{E}_{\mathcal{N}_0^x}[f(X_t)]. \tag{5.1.42}$$

The right-hand side of (5.1.42) can be also represented in terms of Brownian motion, resulting in

$$\mathbb{E}_{\mathcal{N}_0^x}[f(X_t)] = \frac{e^{E_0 t}}{\varphi_0(x)}\mathbb{E}^x[e^{-\int_0^t V(B_s)ds} f(B_t)\varphi_0(B_t)] \tag{5.1.43}$$

as a function in $L^2(\mathbb{R}^d)$.

Define the probability measure $\mathcal{N}_0$ on $\mathfrak{X} \times \mathbb{R}^d$ by

$$d\mathcal{N}_0 = d\mathcal{N}_0^x \otimes d\mathrm{N}_0. \tag{5.1.44}$$

Using this measure, we have the following change-of-measure formula whose proof immediately follows from Lemma 5.19.

Theorem 5.21 (Feynman–Kac formula for $P(\phi)_1$-processes). *If $f, g \in L^2(d\mathrm{N}_0)$, then*

$$(f, e^{-tL}g)_{L^2(d\mathrm{N}_0)} = (f\varphi_0, e^{-t(H-E_0)}g\varphi_0)_{L^2(\mathbb{R}^d)} = \mathbb{E}_{\mathcal{N}_0}[\bar{f}(X_0)g(X_t)]. \tag{5.1.45}$$

An immediate corollary of this theorem is the following.

Corollary 5.22 (Invariant measure). *The probability measure $\varphi_0^2(x)dx$ defined on $(\mathbb{R}^d, \mathcal{B}(\mathbb{R}^d))$ is an invariant measure of the $P(\phi)_1$-process $(X_t)_{t\geq 0}$.*

5.1.4 Dirichlet principle

In this subsection we consider a property of running maxima of $P(\phi)_1$-processes $(X_t)_{t\geq 0}$.

Proposition 5.23. *Let $T, \Lambda > 0$ and suppose that $f \in C(\mathbb{R}^d) \cap D(L^{1/2})$. Then it follows that*

$$\mathcal{N}_0\left(\sup_{0\leq s\leq T}|f(X_s)| \geq \Lambda\right) \leq \frac{e}{\Lambda}((f,f)_{L^2(d\mathrm{N}_0)} + T(L^{1/2}f, L^{1/2}f)_{L^2(d\mathrm{N}_0)})^{1/2}. \tag{5.1.46}$$

(Here e is the base of the natural logarithm.)

Proof. Denote $T_j = Tj/2^n, j = 0, 1, \ldots, 2^n$, and fix T and n. Let

$$G = \left\{x \in \mathbb{R}^d \mid |f(x)| \geq \Lambda\right\}, \tag{5.1.47}$$

and define the stopping time

$$\tau = \inf\left\{T_j \geq 0 \mid X_{T_j} \in G\right\}. \tag{5.1.48}$$

The identity

$$\mathcal{N}_0\left(\sup_{j\in\{0,1,\ldots,2^n\}}|f(X_{T_j})| \geq \Lambda\right) = \mathcal{N}_0(\tau \leq T)$$

directly follows. We estimate the right-hand side above. Let $0 < \varrho < 1$ be fixed and choose a suitable ϱ later. We see that by the Schwarz inequality with respect to $d\mathrm{N}_0$,

$$\mathcal{N}_0(\tau \le T) = \mathbb{E}_{\mathcal{N}_0}[1_{\{\tau \le T\}}] \le \mathbb{E}_{\mathcal{N}_0}[\varrho^{\tau - T}] \le \varrho^{-T}\mathbb{E}_{\mathcal{N}_0}[\varrho^{\tau}] \le \varrho^{-T}\left(\int_{\mathbb{R}^d} (\mathbb{E}^x_{\mathcal{N}_0}[\varrho^{\tau}])^2 d\mathrm{N}_0\right)^{1/2}, \tag{5.1.49}$$

where $\mathbb{E}^x_{\mathcal{N}_0} = \mathbb{E}_{\mathcal{N}_0^x}$. Let $\psi \ge 0$ be any function such that $\psi(x) \ge 1$ on G. The Dirichlet principle (shown in the lemma below) implies that

$$\int_{\mathbb{R}^d} (\mathbb{E}^x_{\mathcal{N}_0}[\varrho^{\tau}])^2 d\mathrm{N}_0 \le (\psi, \psi) + \frac{\varrho^{T/2^n}}{1 - \varrho^{T/2^n}}(\psi, (1 - e^{-(T/2^n)L})\psi). \tag{5.1.50}$$

Inserting

$$\frac{|f(x)|}{\Lambda} \begin{cases} \ge 1, & x \in G, \\ = \frac{|f(x)|}{\Lambda}, & x \in \mathbb{R}^d \setminus G \end{cases}$$

into ψ in (5.1.50), we have

$$\int_{\mathbb{R}^d} (\mathbb{E}^x_{\mathcal{N}_0}[\varrho^{\tau}])^2 d\mathrm{N}_0 \le \frac{1}{\Lambda^2}(f, f) + \frac{\varrho^{T/2^n}}{1 - \varrho^{T/2^n}}\frac{1}{\Lambda^2}(|f|, (1 - e^{-(T/2^n)L})|f|). \tag{5.1.51}$$

Since $e^{-(T/2^n)L}$ is positivity improving and thus

$$(|f|, (1 - e^{-(T/2^n)L})|f|) \le (f, (1 - e^{-(T/2^n)L})f),$$

we have, by (5.1.49),

$$\mathcal{N}_0\left(\sup_{j \in \{0,1,\dots,2^n\}} |f(X_{T_j})| \ge \Lambda\right) \le \frac{\varrho^{-T}}{\Lambda}\left((f, f) + \frac{\varrho^{T/2^n}}{1 - \varrho^{T/2^n}}(f, (1 - e^{-(T/2^n)L})f)\right)^{1/2}.$$

Set $\varrho = e^{-1/T}$. By $\frac{\varrho^{T/2^n}}{1 - \varrho^{T/2^n}} \le 2^n$, we have

$$\mathcal{N}_0\left(\sup_{j \in \{0,1,\dots,2^n\}} |f(X_{T_j})| \ge \Lambda\right) \le \frac{e}{\Lambda}((f, f) + 2^n(f, (1 - e^{-(T/2^n)L})f))^{1/2}. \tag{5.1.52}$$

Since $(f, (1 - e^{-(T/2^n)L})f) \le (T/2^n)(L^{1/2}f, L^{1/2}f)$, we obtain

$$\mathcal{N}_0\left(\sup_{j \in \{0,1,\dots,2^n\}} |f(X_{T_j})| \ge \Lambda\right) \le \frac{e}{\Lambda}((f, f) + T(L^{1/2}f, L^{1/2}f))^{1/2}. \tag{5.1.53}$$

Take $n \to \infty$ on both sides of (5.1.53). By the dominated convergence theorem,

$$\lim_{n\to\infty} \mathcal{N}_0\left(\sup_{j\in\{0,1,\dots,2^n\}} |f(X_{T_j})| \geq \Lambda\right) = \mathcal{N}_0\left(\lim_{n\to\infty}\sup_{j\in\{0,1,\dots,2^n\}} |f(X_{T_j})| \geq \Lambda\right).$$

Since $f(X_t)$ is continuous in t, it follows that

$$\lim_{n\to\infty}\sup_{j\in\{0,1,\dots,2^n\}} |f(X_{T_j})| = \sup_{0\leq s\leq T} |f(X_s)|. \qquad \square$$

It remains to show (5.1.50).

Lemma 5.24 (Dirichlet principle). *Let $0 < \varrho < 1$. Fix n and set $T_j = Tj/2^n$, $j = 0, 1, \dots, 2^n$. Let $G \subset \mathbb{R}^d$ be measurable and $\tau = \inf\{T_j \geq 0 \mid X_{T_j} \in G\}$. Then for every function $\psi \geq 0$ such that $\psi(x) \geq 1$ on G, it follows that*

$$\int_{\mathbb{R}^d} (\mathbb{E}^x_{\mathcal{N}_0}[\varrho^\tau])^2 d\mathrm{N}_0 \leq (\psi, \psi)_{L^2(d\mathrm{N}_0)} + \frac{\varrho^{T/2^n}}{1 - \varrho^{T/2^n}} (\psi, (1 - e^{-(T/2^n)L})\psi)_{L^2(d\mathrm{N}_0)}.$$

Proof. Define $\psi_\varrho(x) = \mathbb{E}^x_{\mathcal{N}_0}[\varrho^\tau]$. By the definition of τ, we see that

$$\psi_\varrho(x) = 1, \quad x \in G, \tag{5.1.54}$$

since $\tau = 0$ when X_s starts from the inside of G. Let $\mathscr{F}_t = \sigma(X_s, 0 \leq s \leq t)$ be the natural filtration of $(X_t)_{t\geq 0}$. By the Markov property of $(X_t)_{t\geq 0}$, it is directly seen that

$$e^{-(T/2^n)L}\psi_\varrho(x) = \mathbb{E}^x_{\mathcal{N}_0}[\mathbb{E}^{X_{T/2^n}}_{\mathcal{N}_0}[\varrho^\tau]] = \mathbb{E}^x_{\mathcal{N}_0}[\mathbb{E}^x_{\mathcal{N}_0}[\varrho^{\tau\circ\theta_{T/2^n}} | \mathscr{F}_{T/2^n}]] = \mathbb{E}^x_{\mathcal{N}_0}[\varrho^{\tau\circ\theta_{T/2^n}}], \tag{5.1.55}$$

where θ_t is the shift on $\mathfrak{X}$ defined by $(\theta_t\omega)(s) = \omega(s + t)$ for $\omega \in \mathfrak{X}$. Note that

$$(\tau \circ \theta_{T/2^n})(\omega) = \tau(\omega) - T/2^n \geq 0, \tag{5.1.56}$$

when $x = X_0(\omega) \in \mathbb{R}^d \setminus G$. Hence by (5.1.55) and (5.1.56), we have the identity

$$\varrho^{T/2^n} e^{-(T/2^n)L}\psi_\varrho(x) = \psi_\varrho(x), \quad x \in \mathbb{R}^d \setminus G. \tag{5.1.57}$$

It is trivial to see that

$$\int_{\mathbb{R}^d} (\mathbb{E}^x_{\mathcal{N}_0}[\varrho^\tau])^2 d\mathrm{N}_0 = (\psi_\varrho, \psi_\varrho) \leq (\psi_\varrho, \psi_\varrho) + \frac{\varrho^{T/2^n}}{1 - \varrho^{T/2^n}} (\psi_\varrho, (1 - e^{-(T/2^n)L})\psi_\varrho).$$

By using relation (5.1.57), we can compute the right-hand side above as

$$(\psi_\varrho 1_G, \psi_\varrho 1_G) + \frac{\varrho^{T/2^n}}{1 - \varrho^{T/2^n}} (\psi_\varrho 1_G, (1 - e^{-(T/2^n)L})\psi_\varrho). \tag{5.1.58}$$

Since

$$\begin{aligned}&(\psi_\varrho 1_G, (1-e^{-(T/2^n)L})\psi_\varrho)\\&= (\psi_\varrho 1_G, (1-e^{-(T/2^n)L})\psi_\varrho 1_G) + (\psi_\varrho 1_G, (1-e^{-(T/2^n)L})\psi_\varrho 1_{\mathbb{R}^d\setminus G})\\&= (\psi_\varrho 1_G, (1-e^{-(T/2^n)L})\psi_\varrho 1_G) - (\psi_\varrho 1_G, e^{-(T/2^n)L}\psi_\varrho 1_{\mathbb{R}^d\setminus G})\\&\le (\psi_\varrho 1_G, (1-e^{-(T/2^n)L})\psi_\varrho 1_G),\end{aligned}$$

we have

$$\int_{\mathbb{R}^d} (\mathbb{E}^x_{\mathcal{N}_0}[\varrho^\tau])^2 d\mathrm{N}_0 \le (\psi_\varrho 1_G, \psi_\varrho 1_G) + \frac{\varrho^{T/2^n}}{1-\varrho^{T/2^n}}(\psi_\varrho 1_G, (1-e^{-(T/2^n)L})\psi_\varrho 1_G). \tag{5.1.59}$$

Note that $\psi_\varrho 1_G(x) \le \psi(x)$ for all $x \in \mathbb{R}^d$. We have

$$\begin{aligned}&(\psi_\varrho 1_G, \psi_\varrho 1_G) + \frac{\varrho^{T/2^n}}{1-\varrho^{T/2^n}}(\psi_\varrho 1_G, (1-e^{-(T/2^n)L})\psi_\varrho 1_G)\\&\le \|\psi\|^2 + \frac{\varrho^{T/2^n}}{1-\varrho^{T/2^n}}(\psi, (1-e^{-(T/2^n)L})\psi).\end{aligned} \tag{5.1.60}$$

A combination of (5.1.59) and (5.1.60) then proves the lemma. □

5.1.5 Mehler's formula

In this section we consider the special case

$$H_{\mathrm{osc}} = -\frac{1}{2m}\Delta + V_{\mathrm{osc}} \quad \text{with} \quad V_{\mathrm{osc}}(x) = \frac{m\omega^2}{2}|x|^2 - \frac{\omega}{2}, \ m, \omega > 0,$$

describing the harmonic oscillator, which is one of the few models for which an explicit formula is available for the kernel of the Schrödinger semigroup. This kernel is given by *Mehler's formula*. We will give three proofs by using three different, but related, approaches. First, we proceed by the ground state transform and obtain Mehler's formula via the kernel of the generator of the intrinsic operator (generator of the $P(\phi)_1$-process). Next we obtain it via a computation of the integral given by the Feynman–Kac formula. Finally, we also show a proof using operator theory.

Derivation via ground state transform. Let $d = 1$ and consider the Schrödinger operator

$$H_{\mathrm{osc}} = -\frac{1}{2m}\frac{d^2}{dx^2} + \frac{m\omega^2}{2}x^2 - \frac{\omega}{2} \tag{5.1.61}$$

on $L^2(\mathbb{R})$. Here m is the mass of the oscillator, and ω is the traditional notation for its frequency which should not be confused with other meaning before. Operator H_{osc} is essentially self-adjoint on $C_0^\infty(\mathbb{R})$ by Corollary 4.56, and from Example 4.75 we know that

$$\mathrm{Spec}(H_{\mathrm{osc}}) = \{\omega n \mid n \in \mathbb{N} \cup \{0\}\}.$$

The ground state is explicitly given by

$$\varphi_{\mathrm{osc}}(x) = \left(\frac{m\omega}{\pi}\right)^{1/4} e^{-m\omega x^2/2}. \tag{5.1.62}$$

Furthermore, the complete set of eigenfunctions is given by

$$\varphi_n(x) = \left(\frac{m\omega}{4^n (n!)^2 \pi}\right)^{1/4} e^{-m\omega x^2/2} H_n(\sqrt{m\omega}x), \tag{5.1.63}$$

where H_n, $n = 1, 2, \ldots$, denote the *Hermite polynomials* defined by

$$H_n(x) = (-1)^n e^{x^2} \frac{d^n}{dx^n} e^{-x^2}.$$

Theorem 5.25 (Mehler's formula). *It follows that*

$$e^{-tH_{\mathrm{osc}}}(x, y) = \left(\frac{m\omega}{\pi(1 - e^{-2\omega t})}\right)^{1/2} \exp\left(-\frac{m\omega}{2} \frac{(1 + e^{-2\omega t})(x^2 + y^2) - 4xye^{-\omega t}}{1 - e^{-2\omega t}}\right). \tag{5.1.64}$$

Proof. Let $m, \omega, \sigma > 0$ and consider the stochastic differential equation

$$dX_t = -\omega X_t dt + \sigma dB_t, \quad B_0 = x. \tag{5.1.65}$$

As seen in Example 2.159, its solution is the Ornstein–Uhlenbeck process

$$X_t^x = xe^{-\omega t} + \sigma \int_0^t e^{-\omega(t-s)} dB_s. \tag{5.1.66}$$

By a direct calculation,

$$\mathbb{E}[X_t^x] = xe^{-\omega t}, \quad \mathbb{E}[X_t^x X_s^x] = x^2 e^{-\omega(t+s)} + \frac{\sigma^2}{2\omega} e^{-\omega(t+s)} (e^{2\omega(s \wedge t)} - 1) \tag{5.1.67}$$

follow from (5.1.66). The generator of $(X_t)_{t \geq 0}^x$ is

$$L = \frac{\sigma^2}{2} \frac{d^2}{dx^2} - \omega x \frac{d}{dx}. \tag{5.1.68}$$

Now consider (5.1.61). Write $X_t^{\cdot} = X_t$ and $\mathbb{E}_P[\ldots] = \int_{-\infty}^{\infty} \mathbb{E}[\ldots] \varphi_{\mathrm{osc}}^2(x) dx$. Note that

$$\mathbb{E}_P[X_t] = 0 \quad \text{and} \quad \mathbb{E}_P[X_s X_t] = \frac{\sigma^2}{2\omega} e^{-\omega|s-t|}. \tag{5.1.69}$$

Clearly, $\varphi_{\rm osc} \in C^2(\mathbb{R})$. The ground state transform $\mathscr{U}_{\rm g} : L^2(\mathbb{R}, \varphi_{\rm osc}^2 dx) \to L^2(\mathbb{R})$, $\mathscr{U}_{\rm g} f = \varphi_{\rm osc} f$, gives

$$L_{\rm osc} = \mathscr{U}_{\rm g}^{-1} H_{\rm osc} \mathscr{U}_{\rm g} = -\frac{1}{2m}\frac{d^2}{dx^2} + \omega x \frac{d}{dx},$$

which on putting $1/m = \sigma^2$ coincides with (5.1.68) apart from sign. We have

$$\mathbb{E}_P\Big[\prod_{j=0}^n f_j(X_{t_j})\Big] = \Big(1, f_0\Big(\prod_{j=1}^n e^{-(t_j - t_{j-1})L_{\rm osc}} f_j\Big)1\Big)_{L^2(\mathbb{R}, \varphi_{\rm osc}^2 dx)}$$

for $f_0, \ldots, f_n \in L^\infty(\mathbb{R})$. Noting that $H_{\rm osc} x\varphi_{\rm osc} = \omega\varphi_{\rm osc}$, we can also check (5.1.69) by the formal calculation

$$\mathbb{E}_P\,[X_t X_s] = (\varphi_{\rm osc}, x e^{-(t-s)H_{\rm osc}} x\varphi_{\rm osc}) = e^{-\omega|t-s|}\int_{-\infty}^{\infty} x^2 \varphi_{\rm osc}^2(x)dx = \frac{\sigma^2}{2\omega} e^{-\omega|t-s|}.$$

Next we view $e^{-tH_{\rm osc}}$ as an operator in $L^2(\mathbb{R})$ and derive its kernel. By (5.1.66),

$$\mathbb{E}[f(X_t^x)] = \mathbb{E}\left[f\left(xe^{-\omega t} + \sigma\int_0^t e^{-\omega(t-s)} dB_s\right)\right].$$

For every fixed $t > 0$,

$$r_t = \sigma\int_0^t e^{-\omega(t-s)} dB_s = \sigma e^{-\omega t}\left(e^{\omega t} B_t - \frac{1}{\omega}\int_0^t B_s e^{\omega s} ds\right)$$

is a Gaussian random variable with zero mean and covariance

$$\mathbb{E}[r_t^2] = \frac{\sigma^2}{2\omega}(1 - e^{-2\omega t}) = \sigma^2 \kappa_t,$$

with

$$\kappa_t = \frac{1 - e^{-2\omega t}}{2\omega}, \quad t \geq 0.$$

Thus

$$e^{-tH_{\rm osc}} f(x) = \mathscr{U}_{\rm g} e^{-tL_{\rm osc}} \mathscr{U}_{\rm g}^{-1} f(x) = \varphi_{\rm osc}(x)\mathbb{E}[(f/\varphi_{\rm osc})(X_t^x)].$$

Since $X_t^x = xe^{-\omega t} + r_t$ with the Gaussian random variable r_t, the right-hand side above can be computed as

$$\frac{1}{(2\pi\sigma^2\kappa_t)^{1/2}}\varphi_{\rm osc}(x)\int_{-\infty}^{\infty}\frac{f(xe^{-\omega t}+y)}{\varphi_{\rm osc}(xe^{-\omega t}+y)}e^{-\frac{|y|^2}{2\sigma^2\kappa_t}}dy. \tag{5.1.70}$$

Changing the variable $xe^{-\omega t}+y$ to z and setting $\sigma^2 = 1/m$, we further obtain

$$= \int_{-\infty}^{\infty} f(z)\,(2\pi\kappa_t/m)^{-1/2}\frac{\varphi_{\rm osc}(x)}{\varphi_{\rm osc}(z)}\exp\left(-\frac{|z-xe^{-\omega t}|^2}{2\kappa_t/m}\right)dz. \tag{5.1.71}$$

From this,

$$\begin{aligned} e^{-tH_{\rm osc}}(x,z) &= \sqrt{\frac{m}{2\pi\kappa_t}}\exp\left(-\frac{-2xze^{-\omega t}+(\omega\kappa_t+e^{-2\omega t})x^2+(1-\omega\kappa_t)z^2}{2\kappa_t/m}\right)\\ &= \left(\frac{m\omega}{\pi(1-e^{-2\omega t})}\right)^{1/2}\exp\left(-\frac{(1+e^{-2\omega t})(x^2+z^2)-4xze^{-\omega t}}{2(1-e^{-2\omega t})/(m\omega)}\right). \end{aligned}$$ □

Derivation via the Feynman–Kac formula. Here we consider the d-dimensional variant

$$H_{\rm osc} = -\frac{1}{2}\Delta + \frac{1}{2}|x|^2 - \frac{d}{2},$$

with parameters $m = \omega = 1$. The Feynman–Kac formula gives

$$e^{-tH_{\rm osc}}f(x) = e^{\frac{dt}{2}}\mathbb{E}^x[e^{-\frac{1}{2}\int_0^t |B_s|^2 ds}f(B_t)],$$

thus we are interested in computing expectations involving integrals as in the exponent at the right hand side.

Lemma 5.26. *Let $a \in \mathbb{R}^d$, $\beta \geq 0$ and $b \in \mathbb{R}$. Then*

$$\mathbb{E}^a\left[e^{-\beta|B_t|^2-\frac{b^2}{2}\int_0^t|B_s|^2ds}\right] = \begin{cases} \left(\frac{b}{b\cosh(bt)+2\beta\sinh(bt)}\right)^{d/2}\exp\left(-\frac{|a|^2b}{2}\frac{b+2\beta\coth(bt)}{b\coth(bt)+2\beta}\right) & b \neq 0,\\ (1+2\beta t)^{-d/2}\exp\left(-\frac{\beta|a|^2}{1+2\beta t}\right) & b = 0. \end{cases}$$

Proof. If $b = 0$, the proof is straightforward, so we suppose that $b \neq 0$. We have

$$\mathbb{E}^a\left[e^{-\beta|B_t|^2-\frac{b^2}{2}\int_0^t|B_s|^2ds}\right] = \mathbb{E}^0\left[e^{-\beta|B_t+a|^2-\frac{b^2}{2}\int_0^t|B_s+a|^2ds}\right].$$

We choose a_s to be $-b(B_s+a)$ in the Girsanov formula. Define the d-dimensional random process $(Z_t)_{t\geq 0}$ by

$$dZ_t = -b(B_t+a)\cdot dB_t - \frac{b^2}{2}|B_t+a|^2dt.$$

A computation gives

$$-b\int_0^t (B_s+a)\cdot dB_s = -\frac{b}{2}(|B_t|^2 - dt + a\cdot B_t) = -\frac{b}{2}(|B_t+a|^2 - dt - |a|^2),$$

and hence we obtain

$$Z_t = Z_0 - \frac{b}{2}(|B_t+a|^2 - dt - |a|^2) - \frac{b^2}{2}\int_0^t |B_s+a|^2 ds.$$

Define the new probability measure

$$\hat{P} = e^{Z_t}\mathcal{W}^0.$$

By the Girsanov formula, under $\hat{P}$,

$$\hat{B}_t = (B_t+a) + b\int_0^t (B_s+a)ds \tag{5.1.72}$$

is a Brownian motion starting at a, thus $W_t = \hat{B}_t - a$ is a standard Brownian motion. Putting $X_t = B_t + a$, we rewrite (5.1.72) as

$$X_t = a + W_t - b\int_0^t X_s ds, \quad t \geq 0.$$

This is a Langevin equation and we conclude that its unique solution $(X_t)_{t\geq 0}$ is a d-dimensional Ornstein–Uhlenbeck process starting at a such that

$$X_t = e^{-bt}a + \int_0^t e^{-b(t-s)}dW_s.$$

By using $\hat{P}$, we obtain

$$\begin{aligned}\mathbb{E}^0\left[e^{-\beta|B_t+a|^2 - \frac{b^2}{2}\int_0^t |B_s+a|^2 ds}\right] &= \mathbb{E}_{\hat{P}}[e^{-\beta|X_t|^2 + \frac{b}{2}(|X_t|^2 - |a|^2 - dt)}] \\ &= \mathbb{E}_{\hat{P}}[e^{-(\beta-\frac{b}{2})|X_t|^2}]e^{-\frac{b}{2}|a|^2}e^{-\frac{bdt}{2}}.\end{aligned} \tag{5.1.73}$$

We determine $\mathbb{E}_{\hat{P}}[e^{-(\beta-\frac{b}{2})|X_t|^2}]$. Note that by (5.1.71) the distribution of X_t on $\mathbb{R}^d$ is given by

$$\rho_t(z) = \frac{1}{(2\pi\kappa_t)^{d/2}}\exp\left(-\frac{|z-e^{-bt}a|}{2\kappa_t}\right),$$

where

$$\kappa_t = \frac{1 - e^{-2bt}}{2b} = \frac{e^{-bt}}{b} \sinh(bt).$$

Hence we have

$$\mathbb{E}_{\hat{P}}[f(X_t)] = \frac{1}{(2\pi\kappa_t)^{d/2}} \int_{\mathbb{R}^d} f(z) \exp\left(-\frac{|z - e^{-bt}a|}{2\kappa_t}\right) dz.$$

Take $f(z) = e^{-\xi|z|^2}$ with $\xi = \beta - \frac{b}{2}$. With $\eta = e^{-bt}a \in \mathbb{R}^d$, we get

$$\mathbb{E}_{\hat{P}}[e^{-\xi|X_t|^2}] = \frac{1}{(2\pi\kappa_t)^{d/2}} e^{\frac{|\eta|^2}{2\kappa_t}\left(\frac{1}{(2\kappa_t)(\xi+\frac{1}{2\kappa_t})}-1\right)} \int_{\mathbb{R}^d} e^{-(\xi+\frac{1}{2\kappa_t})|z-\frac{\eta}{2\kappa_t(\xi+\frac{1}{2\kappa_t})}|^2} dz.$$

By integration, we obtain

$$\begin{aligned}
&\frac{1}{(2\pi\kappa_t)^{d/2}} \int_{\mathbb{R}^d} \exp\left(-\left(\xi + \frac{1}{2\kappa_t}\right)\left|z - \frac{\eta}{2\kappa_t(\xi + \frac{1}{2\kappa_t})}\right|^2\right) dz \\
&= \frac{1}{(2\pi\kappa_t)^{d/2}} \left(\xi + \frac{1}{2\kappa_t}\right)^{-d/2} \pi^{d/2} = (2\kappa_t\xi + 1)^{-d/2} = \left(\cosh(bt) + 2\frac{\beta}{b}\sinh(bt)\right)^{-d/2} e^{\frac{bdt}{2}}.
\end{aligned} \tag{5.1.74}$$

Furthermore,

$$\begin{aligned}
\frac{|\eta|^2}{2\kappa_t}\left(\frac{1}{(2\kappa_t)(\xi + \frac{1}{2\kappa_t})} - 1\right) &= \frac{-\xi}{2\kappa_t\xi + 1}|\eta|^2 = \frac{b|a|^2}{2} - \frac{b|a|^2}{2}\left(1 + \frac{(\frac{2\beta}{b} - 1)e^{-2bt}}{2\kappa_t\xi + 1}\right) \\
&= \frac{b|a|^2}{2} - \frac{b|a|^2}{2}\frac{\frac{2\beta}{b}\coth(bt) + 1}{\coth(bt) + \frac{2\beta}{b}}.
\end{aligned} \tag{5.1.75}$$

Combining (5.1.73)–(5.1.75), the lemma follows. □

We can derive Mehler's formula as an application of Lemma 5.26. By the Feynman–Kac formula, we have

$$(f, e^{-tH_{\mathrm{osc}}}g) = \int_{\mathbb{R}^d} \bar{f}(x)\mathbb{E}^x[e^{-\frac{1}{2}\int_0^t(|B_s|^2-d)ds}g(B_t)]dx.$$

Let $Y = e^{-\frac{1}{2}\int_0^t|B_s|^2ds}$ and define $\mathbb{E}^x[Y|B_t = y] = h(x,y)$. Then

$$(f, e^{-tH_{\mathrm{osc}}}g) = e^{dt/2} \int_{\mathbb{R}^d\times\mathbb{R}^d} \bar{f}(x)g(y)\Pi_t(y)h(x,y)dxdy,$$

from which we obtain

$$e^{-tH_{\mathrm{osc}}}(x,y) = e^{dt/2}\Pi_t(y)h(x,y).$$

Lemma 5.27. *For $x, y \in \mathbb{R}^d$ and $b \in \mathbb{R}$, we have*

$$\mathbb{E}^x[e^{-\frac{b^2}{2}\int_0^t |B_s|^2 ds}|B_t = y] = \left(\frac{bt}{\sinh(bt)}\right)^{d/2} \exp\left(-\frac{|y - e^{-bt}x|^2}{2\kappa_t} + \frac{b(|y|^2 - |x|^2)}{2} + \frac{|y|^2}{2t}\right),$$

where $\kappa_t = (e^{-bt}/b)\sinh(bt) = (1 - e^{-2bt})/(2b)$.

Proof. We write $X = e^{-\frac{b^2}{2}\int_0^t |B_s|^2 ds}$ and $\mathbb{E}^x[X|B_t = y] = h(x,y)$. Note that $h(x, B_t) = \mathbb{E}^x[X|\sigma(B_t)]$. We will make use of the formula

$$\frac{1}{(2\pi\varepsilon)^{d/2}} e^{-|x-y|^2/2\varepsilon} \to \delta(x - y) \tag{5.1.76}$$

as $\varepsilon \downarrow 0$, where the convergence is in the distributional sense. It follows that

$$\lim_{\varepsilon\to\infty} \frac{1}{(2\pi\varepsilon)^{d/2}} \mathbb{E}^x[e^{-\frac{1}{2\varepsilon}|B_t-y|^2} X] = \lim_{\varepsilon\to\infty} \frac{1}{(2\pi\varepsilon)^{d/2}} \int_{\mathbb{R}^d} \Pi_t(z) e^{-\frac{1}{2\varepsilon}|z-y|^2} h(x,z)dz = \Pi_t(y)h(x,y).$$

We see that by Proposition 5.26,

$$\begin{aligned} h(x,y) &= \frac{1}{\Pi_t(y)} \lim_{\varepsilon\downarrow 0} \frac{1}{(2\pi\varepsilon)^{d/2}} \mathbb{E}^x[e^{-\frac{1}{2\varepsilon}|B_t-y|^2} X] \\ &= \frac{1}{\Pi_t(y)} \lim_{\varepsilon\downarrow 0} \frac{1}{(2\pi\varepsilon)^{d/2}} \mathbb{E}_{\hat{P}}[e^{-\frac{1}{2\varepsilon}|B_t-y|^2 + \frac{b}{2}|B_t|^2}] e^{-b|x|^2/2} e^{-bdt/2} \\ &= \frac{1}{\Pi_t(y)} \frac{1}{(2\pi\kappa_t)^{d/2}} e^{-\frac{b|x|^2}{2}} e^{-\frac{bdt}{2}} \lim_{\varepsilon\downarrow 0} \frac{1}{(2\pi\varepsilon)^{d/2}} \int_{\mathbb{R}^d} e^{-\frac{1}{2\varepsilon}|z-y|^2 + \frac{b}{2}|z|^2 - \frac{1}{2\kappa_t}|z-\eta|^2} dz. \end{aligned}$$

Recall that $\eta = ae^{-bt}$. By (5.1.76), we have

$$\begin{aligned} h(x,y) &= \frac{1}{\Pi_t(y)} \frac{1}{(2\pi\kappa_t)^{d/2}} e^{-b|x|^2/2} e^{-bdt/2} e^{b|y|^2/2} e^{-|y-\eta|^2/(2\kappa_t)} \\ &= e^{|y|^2/2t}(bt/\sinh(bt))^{d/2} e^{-b|x|^2/2} e^{b|y|^2/2 - |y-\eta|^2/(2\kappa_t)}. \end{aligned}$$ □

Example 5.28. Using Lemma 5.27, we can explicitly calculate various examples of $h(x,y)$:

(1) $h(0,0) = \left(\frac{bt}{\sinh(bt)}\right)^{d/2}$;
(2) $h(a,a) = h(0,0)\exp\left(-\frac{|a|^2}{2t}(2bt\tanh\frac{bt}{2} - 1)\right)$;
(3) $h(a,0) = h(0,0)\exp\left(-\frac{|a|^2}{2t}bt\coth(bt)\right)$;
(4) $h(0,a) = h(0,0)\exp\left(-\frac{|a|^2}{2t}(bt\coth(bt) - 1)\right)$.

Mehler's formula is now an easy consequence of Lemma 5.27.

Theorem 5.29 (Mehler's formula). *We have the expression*

$$e^{-tH_{\text{osc}}}(x,y) = \left(\frac{1}{\pi(1-e^{-2t})}\right)^{d/2} \exp\left(-\frac{(|x|^2+|y|^2)(1+e^{-2t}) - 4(x,y)e^{-t}}{2(1-e^{-2t})}\right).$$

Proof. Set $b = 1$ and write $\Pi_t(y) = (2\pi t)^{-d/2}\exp(-|y|^2/(2t))$ in Lemma 5.27. This gives

$$\begin{aligned} e^{-tH_{\text{osc}}}(x,y) &= \Pi_t(y)e^{td/2}h(x,y) \\ &= \left(\frac{1}{\pi(1-e^{-2t})}\right)^{d/2} \exp\left(-\frac{|y-xe^{-t}|^2}{2\kappa_t} - \frac{|x|^2}{2} + \frac{|y|^2}{2}\right) \\ &= \left(\frac{1}{\pi(1-e^{-2t})}\right)^{d/2} \exp\left(-\frac{(|x|^2+|y|^2)(1+e^{-2t}) - 4(x,y)e^{-t}}{2(1-e^{-2t})}\right). \end{aligned}$$ □

Derivation via operator theory. Mehler's formula can also be obtained through analytic continuation $t \mapsto -it$ as a corollary of the theorem below. Again we simply choose $m = \omega = 1$. Let K_t be the integral kernel of $e^{-itH_{\text{osc}}}$, and denote $T = \{k\pi \mid k \in \mathbb{Z}\}$.

Theorem 5.30. *The kernel $K_t : \mathbb{R} \times \mathbb{R} \to \mathbb{C}$ has the following expressions:*
(1) *For every $t \in \mathbb{R} \setminus T$,*

$$K_t(x,y) = \frac{1}{\sqrt{2\pi i \sin t}}\, e^{-i(\frac{x^2+y^2}{2}\cot t - \frac{xy}{\sin t})}; \tag{5.1.77}$$

(2) *For every $t \in T$, a. e. $x \in \mathbb{R}$, and $f \in D(H_{\text{osc}})$,*

$$\lim_{s\to t} \int_{-\infty}^{\infty} K_s(x,y)f(y)dy = f(x). \tag{5.1.78}$$

In particular,

$$e^{-tH_{\text{osc}}}(x,y) = \frac{1}{\sqrt{\pi(1-e^{-2t})}} \exp\left(\frac{4xye^{-t} - (x^2+y^2)(1+e^{-2t})}{2(1-e^{-2t})}\right)$$

holds for all $x, y \in \mathbb{R}$ and $t > 0$.

Proof. By eigenfunction expansion, the solution of the initial value problem

$$-i\partial_t\psi = H_{\text{osc}}\psi, \quad \psi(x,0) = \phi(x)$$

can be written as

$$\psi(x,t) = \int_{-\infty}^{\infty} K_t(x,y)\phi(y)dy, \tag{5.1.79}$$

with the integral kernel

$$K_t(x,y) = \sum_{n=0}^{\infty} \varphi_n(x)\varphi_n(y)e^{-iE_nt}. \tag{5.1.80}$$

A combination of (5.1.63) and (5.1.80) leads to

$$K_t(x,y) = \frac{1}{\sqrt{\pi}}\, e^{-\frac{x^2+y^2}{2}-\frac{it}{2}} \sum_{n=0}^{\infty} \frac{1}{2^n n!} H_n(x)H_n(y)e^{-int}. \tag{5.1.81}$$

Next, the formula $H_n(y) = (-1)^n e^{y^2} \frac{d^n}{dy^n} e^{-y^2}$ known for Hermite polynomials allows replacing the sum appearing at the right-hand side of (5.1.81) by

$$e^{y^2} \sum_{n=0}^{\infty} \frac{1}{n!} \left(-\frac{1}{2}e^{-it}\frac{d}{dy}\right)^n H_n(x)\, e^{-y^2}.$$

The generating function $e^{2sz-z^2} = \sum_{n=0}^{\infty} \frac{1}{n!} H_n(s) z^n$ of Hermite polynomials gives then for the above the formal expression

$$e^{y^2} \exp\left(-xe^{-it}\frac{d}{dy} - \frac{1}{4}e^{-2it}\frac{d^2}{dy^2}\right) e^{-y^2}. \tag{5.1.82}$$

In the following we will argue that this expression actually makes sense.

Clearly, the exponential operator acting on e^{-y^2} factorizes. Both operators are unbounded; however, if the second can be defined, the first is well defined on the same domain. The first operator is a shift by the prefactor of $\frac{d}{dy}$. We consider

$$\mathcal{K}_z = \exp\left(z\frac{d^2}{dy^2}\right), \quad z \in \mathbb{C},$$

and make sense of it on a set including $z = \tau = -(1/4)e^{-2it}$.

Take the open disc $D = \{z \in \mathbb{C} \mid |z| < \frac{1}{4}\}$. Note that for every $z \in D$ the function $f(y) = e^{-y^2}$ is an analytic vector of $\mathcal{K}_z$, i. e., $\mathcal{K}_z e^{-y^2} \in L^2(\mathbb{R})$, and is analytic with respect to z, i. e.,

$$\sum_{n=0}^{\infty} \left\| \frac{z^n}{n!} \frac{d^n}{dy^n} e^{-y^2} \right\| \le \sum_{n=0}^{\infty} \frac{|z|^n}{n!} \left\| \frac{d^n}{dy^n} e^{-y^2} \right\| = \frac{1}{\sqrt{2}} \sum_{n=0}^{\infty} \frac{|z|^n}{n!} \|k^{2n} e^{-k^2/4}\| = \sum_{n=0}^{\infty} c_n |z|^n,$$

where $c_n = (\pi/2)^{1/4}\sqrt{(4n-1)!!}/n!$. It is seen that $c_n/c_{n+1} \to 1/4$ as $n \to \infty$ and this explains the way D was defined.

For every $z \in D$, define $\mathcal{A}_z = \{f \in D(\mathcal{K}_z) \mid f$ is analytic vector of $\mathcal{K}_z\}$. This set contains, for instance, e^{-y^2}. For every $z \in D$ and $f \in \mathcal{A}_z$, the derivative with respect to z of $\mathcal{K}_z$ exists. Denote $g(y,z) = (\mathcal{K}_z f)(y)$; then we have

$$\partial_z g = \partial_y^2 g, \quad g(y,0) = f(y)$$

since $z = 0 \in D$, i. e., g satisfies the heat equation with the initial condition above, for every $z \in D$. This observation allows to express $g(y,z)$. Define

$$D_f = \left\{ z \in \bar{D} \,\middle|\, \left| \frac{1}{\sqrt{z}} \int_{-\infty}^{\infty} f(u) e^{-\frac{(y-u)^2}{4z}} du \right| < \infty,\ y \in \mathbb{R} \right\}$$

as a subset of the closure of D; in particular, we write $D_{\exp}$ for $f(y) = e^{-y^2}$. An easy calculation shows that $D_{\exp}$ lies in D and coincides with the complement in D of the closed disk of radius $1/8$ centered in $(-1/8, 0)$.

It is checked directly that for every $z \in D_f$,

$$F_f(y,z) = \frac{1}{2\sqrt{\pi z}} \int_{-\infty}^{\infty} f(u) e^{-\frac{(y-u)^2}{4z}} du, \tag{5.1.83}$$

where the square root is that with the smallest argument, is a solution of the heat equation above. Moreover, F_f is analytic since $|\partial F_f/\partial z| < \infty$. Take now $z = \zeta \in (0, \frac{1}{4})$. We have in the L^2 sense that $F_f(y,\zeta) = g(y,\zeta)$, hence by uniqueness of the analytic continuation,

$$g(y,z) = (\mathcal{K}_z f)(y), \quad z \in D_f. \tag{5.1.84}$$

This representation can be extended to the circle ∂D. Pick a number ε such that $|\Re \tau| \leq \varepsilon < 1/4$ and define $\mathcal{A}_z^\varepsilon = \mathcal{K}_\varepsilon L^2(\mathbb{R}) \subset \mathcal{A}_z$. Take a sequence $(z_n)_{n\in\mathbb{N}} \subset D$ such that $z_n \to \tau$ for some $\tau \in D_h$ and $h \in \bigcap_{n=1}^\infty \mathcal{A}_{z_n}^\varepsilon$. It is easily seen that pointwise $\mathcal{K}_{z_n} h \to 1/(2\sqrt{\pi\tau}) \int_{-\infty}^\infty h(u) \exp(-\frac{(\cdot-u)^2}{4\tau}) du$ as $n \to \infty$. By the Lebesgue dominated convergence theorem, it follows then for every $\varphi \in C_0^\infty$ that

$$(\varphi, \mathcal{K}_{z_n} h) \longrightarrow \left(\varphi, \frac{1}{2\sqrt{\pi\tau}} \int_{-\infty}^{\infty} h(u) \exp\left(-\frac{(\cdot - u)^2}{4\tau} \right) du \right) \quad \text{as } n \to \infty.$$

On the other hand, $\mathcal{K}_\tau h$ is well defined for every $h \in \bigcap_{n=1}^\infty \mathcal{A}_{z_n}^\varepsilon$ since $\mathcal{K}_\tau h = \mathcal{K}_{\tau+\varepsilon} f_h$, with some $f_h \in \bigcap_{n=1}^\infty \mathcal{A}_{z_n}$, and $\varepsilon + \Re\tau \geq 0$. Moreover, $\mathcal{K}_{z_n} h \to \mathcal{K}_\tau h$ in the strong sense. This follows again through dominated convergence by using the continuity of the exponential function, the equalities

$$\|\mathcal{K}_\tau h - \mathcal{K}_{z_n} h\|^2 = \|\mathcal{K}_{\tau+\frac{1}{4}} f_h - \mathcal{K}_{z_n+\frac{1}{4}} f_h\|^2 = \int_0^\infty |e^{-(\tau+\frac{1}{4})\lambda} - e^{-(z_n+\frac{1}{4})\lambda}|^2 dE_{f_h}(\lambda),$$

and the estimate $|e^{-(\tau+\frac{1}{4})\lambda} - e^{-(z_n+\frac{1}{4})\lambda}|^2 \leq 4$, uniform in λ, and making use of the spectral theorem. Hence also $(\varphi, \mathcal{K}_{z_n} h) \to (\varphi, \mathcal{K}_\tau h)$ as $n \to \infty$, $\forall \varphi \in C_0^\infty$, and by comparison

$$(\mathcal{K}_\tau h)(y) = \frac{1}{2\sqrt{\pi\tau}} \int_{-\infty}^{\infty} h(u) e^{-\frac{(y-u)^2}{4\tau}} du \tag{5.1.85}$$

follows, for every $h \in \bigcap_{n=1}^{\infty} \mathcal{A}_{z_n}^{\varepsilon}$, $\tau \in D_h$, and almost every $y \in \mathbb{R}$.

Take an $f_h \in \bigcap_{n=1}^{\infty} \mathcal{A}_{z_n}$ and $\varepsilon \geq |\mathbb{R}\tau|$, and look at the solution of $\exp(\varepsilon\partial_y^2) f_h = e^{-y^2}$. Taking Fourier transforms, $e^{-\varepsilon k^2}\hat{f}_h(k) = e^{-k^2/4}$ follows; $\hat{f}_h$, and thus also f_h, will be in L^2 whenever $\varepsilon < 1/4$, and then $e^{-y^2} \in \bigcap_{n=1}^{\infty} \mathcal{A}_{z_n}^{\varepsilon}$. This also explains the way $\mathcal{A}_z^{\varepsilon}$ was defined. Putting now $h(y) = e^{-y^2}$ in (5.1.85) and using (5.1.82), (5.1.77) results by easy manipulations for every τ with $|\mathbb{R}\tau| \leq \varepsilon < 1/4$.

This argument excludes the values of time corresponding to $\tau = (-1/4, 0)$, i. e., the set T. The reason is that this is the only value of τ that falls outside of $D_{\exp}$, and for these values the kernel becomes singular described by a δ-distribution. Take $t \in T$ and a sequence $(t_n)_{n\in\mathbb{N}} \subset \mathbb{R} \setminus T$ such that $t_n \to t$. Since $e^{-itH_{\rm osc}}$ is a strongly continuous unitary group, we have with a subsequence $(t'_m)_{m\in\mathbb{N}}$ for every L^2 function φ $e^{-it'_m H_{\rm osc}}\varphi(x) \to e^{-itH_{\rm osc}}\varphi(x)$, for almost every $x \in \mathbb{R}$. Moreover, for almost every $x \in \mathbb{R}$ we have $\int_{-\infty}^{\infty} K_{t'_m}(x,y)\varphi(y)dy \to \varphi(x)$, which is seen directly. □

5.1.6 $P(\phi)_1$-processes with càdlàg paths

Finally, we consider nonlocal Schrödinger operators $H = \Psi(-\frac{1}{2}\Delta) + V$ and construct $P(\phi)_1$-processes for them. We assume that the potential is in Ψ-Kato class, and is such that H has a unique ground state $\varphi_0 > 0$. We denote the intrinsic operator

$$L^\Psi = \mathscr{U}^{-1}\bar{H}\mathscr{U}, \quad \bar{H} = H - E_0, \quad E_0 = \inf \operatorname{Spec}(H),$$

with the same notation of the ground state transform as in Section 5.1.1. The main difference from the case in Section 5.1.3 is that here we have to deal with jump processes.

Recall that $\mathfrak{X}_D$ denotes the càdlàg path space on the full real line $\mathbb{R}$. Our main aim is to prove that there exists a probability measure $\mathscr{M}_0^x$ on the space $(\mathfrak{X}_D, \mathcal{B}(\mathfrak{X}_D))$ such that for $f, g \in L^2(\mathbb{R}^d)$ and Ψ-Kato class potential V,

$$(f, e^{-|t-s|L^\Psi} g) = \int_{\mathbb{R}^d} \mathbb{E}_{\mathscr{M}_0^x}\left[\overline{f(Y_s)} g(Y_t)\right] \varphi_0^2(x) dx, \quad t, s \in \mathbb{R}, \tag{5.1.86}$$

holds, with coordinate process $(Y_t)_{t\in\mathbb{R}}$. Define

$$d\mathsf{M}_0 = \varphi_0^2(x)dx \tag{5.1.87}$$

and as above write $L^2(d\mathsf{M}_0)$ for $L^2(\mathbb{R}^d, d\mathsf{M}_0)$ for simplicity. As before, we denote by $\mathscr{X}_D$ and $\mathscr{X}_{\bar{D}}$ the càdlàg and càglàd path spaces on $[0, \infty)$, respectively.

Theorem 5.31. *Let V be a Ψ-Kato decomposable potential such that $H = \Psi(-\frac{1}{2}\Delta) + V$ has a unique ground state $\varphi_0 > 0$, and the ground state transform of $\bar{H}$ is L^Ψ. Denote by $(Y_t)_{t\in\mathbb{R}}$ the coordinate process on $(\mathfrak{X}_D, \mathcal{B}(\mathfrak{X}_D))$ and consider the filtrations $(\mathscr{F}_t^+)_{t\ge 0} = \sigma(Y_s : 0 \le s \le t)$ and $(\mathscr{F}_t^-)_{t\le 0} = \sigma(Y_s : t \le s \le 0)$. Then for all $x \in \mathbb{R}^d$, there exists a probability measure $\mathscr{M}_0^x$ on $(\mathfrak{X}_D, \mathcal{B}(\mathfrak{X}_D))$ satisfying the following properties:*
(1) (Initial distribution) $\mathscr{M}_0^x(Y_t = x) = 1$;
(2) (Reflection symmetry) $(Y_t)_{t\ge 0}$ *and* $(Y_t)_{t\le 0}$ *are independent and*

$$Y_{-t} \overset{\mathrm{d}}{=} Y_t, \quad t \in \mathbb{R};$$

(3) (Markov property)$(Y_t)_{t\ge 0}$ *is a Markov process with respect to* $(\mathscr{F}_t^+)_{t\ge 0}$, *and* $(Y_t)_{t\le 0}$ *with respect to* $(\mathscr{F}_t^-)_{t\le 0}$;
(4) (Shift invariance) *Let* $-\infty < t_0 \le t_1 \le \cdots \le t_n < \infty$, $n \in \mathbb{N}$; *then the finite-dimensional distributions with respect to* $d\mathsf{M}_0$ *are given by*

$$\int_{\mathbb{R}^d} \mathbb{E}_{\mathscr{M}_0^x}\left[\prod_{j=0}^n f_j(Y_{t_j})\right] d\mathsf{M}_0 = \left(f_0, e^{-(t_1-t_0)L^\Psi} f_1 \cdots f_{n-1} e^{-(t_n-t_{n-1})L^\Psi} f_n\right)_{L^2(d\mathsf{M}_0)} \tag{5.1.88}$$

for $f_0, f_n \in L^2(d\mathsf{M}_0)$, *and* $f_j \in L^\infty(\mathbb{R}^d)$, $j = 1, \dots, n-1$, *and they are shift invariant, i. e.,*

$$\int_{\mathbb{R}^d} \mathbb{E}_{\mathscr{M}_0^x}\left[\prod_{j=0}^n f_j(Y_{t_j})\right] d\mathsf{M}_0 = \int_{\mathbb{R}^d} \mathbb{E}_{\mathscr{M}_0^x}\left[\prod_{j=0}^n f_j(Y_{t_j+s})\right] d\mathsf{M}_0, \quad s \in \mathbb{R}.$$

Proof. The proof runs in parallel with that of Theorem 5.15, here we present an outline only.

Let $0 \le t_0 \le t_1 \le \cdots \le t_n$ with arbitrary $n \in \mathbb{N}$, and the set function $\mathcal{P}_{\{t_0,\dots,t_n\}} : \times_{j=0}^n \mathcal{B}(\mathbb{R}^d) \to \mathbb{R}$ be defined by

$$\mathcal{P}_{\{t_0,\dots,t_n\}}(\times_{j=0}^n A_j) = \left(1_{A_0}, e^{-(t_1-t_0)L^\Psi} 1_{A_1} \cdots 1_{A_{n-1}} e^{-(t_n-t_{n-1})L^\Psi} 1_{A_n}\right)_{L^2(d\mathsf{M}_0)}, \tag{5.1.89}$$

$$\mathcal{P}_{\{t_0\}}(A) = \left(1, e^{-t_0 L^\Psi} 1_A\right)_{L^2(d\mathsf{M}_0)} = (1, 1_A)_{L^2(d\mathsf{M}_0)}. \tag{5.1.90}$$

Step 1. Denote $\Lambda = \{S \subset \mathbb{R} \mid |S| < \infty\}$. It can be verified directly that the family of set functions $(\mathcal{P}_S)_{S\in\Lambda}$ given above satisfies the consistency condition

$$\mathcal{P}_{\{t_0,\dots,t_{n+m}\}}\left((\times_{j=0}^n A_j) \times (\times_{j=n+1}^{n+m} \mathbb{R}^d)\right) = \mathcal{P}_{\{t_0,\dots,t_n\}}(\times_{j=0}^n A_j),$$

for all $m, n \in \mathbb{N}$. Hence by the Kolmogorov's extension theorem, there exists a probability measure $\mathcal{P}_\infty$ on the space $\left((\mathbb{R}^d)^{[0,\infty)}, \sigma(\mathcal{A})\right)$, where

$$\mathcal{A} = \left\{\omega : \mathbb{R} \to \mathbb{R}^d \mid \omega\lceil_\Lambda \in E,\ E \in (\mathscr{B}(\mathbb{R}^d))^{|\Lambda|},\ \Lambda \subset \mathbb{R},\ |\Lambda| < \infty\right\}, \tag{5.1.91}$$

such that

$$\mathcal{P}_{\{t_0,\dots,t_n\}}\left(\times_{j=0}^n A_j\right) = \mathbb{E}_{\mathcal{P}_\infty}\left[\prod_{j=0}^n 1_{A_j}(Z_{t_j})\right], \quad n \in \mathbb{N},$$

$$\mathcal{P}_{\{t\}}(A) = \mathbb{E}_{\mathcal{P}_\infty}[1_A(Z_t)].$$

Here $(Z_t)_{t\geq 0}$ is the coordinate process. Thus the stochastic process $(Z_t)_{t\geq 0}$ on the probability space $((\mathbb{R}^d)^{[0,\infty)}, \sigma(\mathcal{A}), \mathcal{P}_\infty)$ satisfies

$$\mathbb{E}_{\mathcal{P}_\infty}\left[\prod_{j=0}^n f_j(Z_{t_j})\right] = \left(f_0, e^{-(t_1-t_0)L^\Psi} f_1 \cdots f_{n-1} e^{-(t_n-t_{n-1})L^\Psi} f_n\right)_{L^2(d\mathsf{M}_0)}, \tag{5.1.92}$$

$$\mathbb{E}_{\mathcal{P}_\infty}[f_0(Z_{t_0})] = \left(1, e^{-t_0 L^\Psi} f_0\right)_{L^2(d\mathsf{M}_0)} = (1, f_0)_{L^2(d\mathsf{M}_0)} \tag{5.1.93}$$

for $f_j \in L^\infty(\mathbb{R}^d)$, $j = 1, \dots, n-1$, $f_0, f_n \in L^2(d\mathsf{M}_0)$, and $0 \leq t_0 \leq t_1 \leq \cdots \leq t_n$.

Step 2. Next we prove the existence of both a càdlàg version of $(Z_t)_{t\geq 0}$ and a càglàd version of $(Z_t)_{t\geq 0}$ by checking the Dynkin–Kinney condition given in Theorem 3.34. By Lemma 5.32 below, there exists a càdlàg version $Z' = (Z'_t)_{t\geq 0}$ of $(Z_t)_{t\geq 0}$ on the probability space $((\mathbb{R}^d)^{[0,\infty)}, \sigma(\mathcal{A}), \mathcal{P}_\infty)$. Denote the image measure of $\mathcal{P}_\infty$ on $(\mathscr{X}_D, \mathcal{B}(\mathscr{X}_D))$ by

$$\mathcal{Q} = \mathcal{P}_\infty \circ (Z')^{-1}.$$

Let $(\tilde{Y}_t)_{t\geq 0}$ be the coordinate process on $(\mathscr{X}_D, \mathcal{B}(\mathscr{X}_D), \mathcal{Q})$ such that $Z'_t \stackrel{\mathrm{d}}{=} \tilde{Y}_t$. Equations (5.1.92) and (5.1.93) can be expressed in terms of $(\tilde{Y}_t)_{t\geq 0}$ as

$$\left(f_0, e^{-(t_1-t_0)L^\Psi} f_1 \cdots f_{n-1} e^{-(t_n-t_{n-1})L^\Psi} f_n\right)_{L^2(d\mathsf{M}_0)} = \mathbb{E}_{\mathcal{Q}}\left[\prod_{j=0}^n f_j(\tilde{Y}_{t_n})\right], \tag{5.1.94}$$

$$\left(1, e^{-tL^\Psi} f_0\right)_{L^2(d\mathsf{M}_0)} = (1, f_0)_{L^2(d\mathsf{M}_0)} = \mathbb{E}_{\mathcal{Q}}[f_0(\tilde{Y}_t)]. \tag{5.1.95}$$

Note that a càglàd version $Z'' = (Z''_t)_{t\geq 0}$ of $(Z_t)_{t\geq 0}$ can also be constructed on the space $((\mathbb{R}^d)^{[0,\infty)}, \sigma(\mathcal{A}), \mathcal{P}_\infty)$. In a similar way as for $(Z'_t)_{t\geq 0}$, it can be shown that there exists a probability measure $\bar{\mathcal{Q}}$ on the càglàd measurable space $(\mathscr{X}_{\bar{D}}, \mathcal{B}(\mathscr{X}_{\bar{D}}))$ such that the coordinate process $(\tilde{Y}_t)_{t\geq 0}$ satisfies both (5.1.94) and (5.1.95) with $\mathcal{Q}$ replaced by $\bar{\mathcal{Q}}$ and

$$\bar{\mathcal{Q}} = \mathcal{P}_\infty \circ (Z'')^{-1}$$

holds.

Step 3. Since $\mathscr{X}_D$ and $\mathscr{X}_{\bar{D}}$ are Polish spaces by Corollary 3.37, by Theorem 2.41 we can define regular conditional measures $\mathcal{Q}^x(\,\cdot\,) = \mathcal{Q}(\,\cdot\,|\tilde{Y}_0 = x)$ and $\bar{\mathcal{Q}}^x(\,\cdot\,) = \mathcal{Q}(\,\cdot\,|\tilde{Y}_0 = x)$ for $x \in \mathbb{R}^d$ on $(\mathscr{X}_D, \mathcal{B}(\mathscr{X}_D))$ and $(\mathscr{X}_{\bar{D}}, \mathcal{B}(\mathscr{X}_{\bar{D}}))$, respectively. Since the distribution of Y_0 is $d\mathsf{M}_0$, we have $\mathcal{Q}(A) = \int_{\mathbb{R}^d} \mathbb{E}_{\mathcal{Q}^x}[1_A] d\mathsf{M}_0$ and $\bar{\mathcal{Q}}(A) = \int_{\mathbb{R}^d} \mathbb{E}_{\bar{\mathcal{Q}}^x}[1_A] d\mathsf{M}_0$. The coordinate process $(\tilde{Y}_t)_{t\geq 0}$ on the probability space $(\mathscr{X}_D, \mathcal{B}(\mathscr{X}_D), \mathcal{Q}^x)$ satisfies

$$\left(f_0, e^{-(t_1-t_0)L^\Psi} f_1 \cdots f_{n-1} e^{-(t_n-t_{n-1})L^\Psi} f_n\right)_{L^2(d\mathrm{M}_0)} = \int_{\mathbb{R}^d} \mathbb{E}_{\mathcal{Q}^x}\left[\prod_{j=0}^n f_j(\tilde{Y}_{t_j})\right] d\mathrm{M}_0, \tag{5.1.96}$$

$$\left(1, e^{-tL^\Psi} f_0\right)_{L^2(d\mathrm{M}_0)} = (1, f_0)_{L^2(d\mathrm{M}_0)} = \int_{\mathbb{R}^d} \mathbb{E}_{\mathcal{Q}^x}[f_0(\tilde{Y}_t)] d\mathrm{M}_0, \tag{5.1.97}$$

and the coordinate process $(\tilde{Y}_t)_{t\geq 0}$ on $(\mathscr{X}_{\bar{D}}, \mathcal{B}(\mathscr{X}_{\bar{D}}), \bar{\mathcal{Q}}^x)$ also satisfies

$$\left(f_0, e^{-(t_1-t_0)L^\Psi} f_1 \cdots f_{n-1} e^{-(t_n-t_{n-1})L^\Psi} f_n\right)_{L^2(d\mathrm{M}_0)} = \int_{\mathbb{R}^d} \mathbb{E}_{\bar{\mathcal{Q}}^x}\left[\prod_{j=0}^n f_j(\tilde{Y}_{t_j})\right] d\mathrm{M}_0, \tag{5.1.98}$$

$$\left(1, e^{-tL^\Psi} f_0\right)_{L^2(d\mathrm{M}_0)} = (1, f_0)_{L^2(d\mathrm{M}_0)} = \int_{\mathbb{R}^d} \mathbb{E}_{\bar{\mathcal{Q}}^x}[f_0(\tilde{Y}_t)] d\mathrm{M}_0. \tag{5.1.99}$$

In Lemma 5.33, we will show that $(\tilde{Y}_t)_{t\geq 0}$ is a Markov process on $(\mathscr{X}_D, \mathcal{B}(\mathscr{X}_D), \mathcal{Q}^x)$ with respect to the natural filtration $(\mathscr{F}_t)_{t\geq 0}$, and $(\tilde{Y}_t)_{t\geq 0}$ is a Markov process on $(\mathscr{X}_{\bar{D}}, \mathcal{B}(\mathscr{X}_{\bar{D}}), \bar{\mathcal{Q}}^x)$ with respect to the natural filtration $(\mathscr{F}_t)_{t\geq 0}$.

Step 4. Now we construct a random process on the full real line $\mathbb{R}$. Consider the product sample space $\hat{\Omega} = \mathscr{X}_D \times \mathscr{X}_{\bar{D}}$ with product σ-field $\hat{\mathscr{F}} = \mathcal{B}(\mathscr{X}_D) \times \mathcal{B}(\mathscr{X}_{\bar{D}})$ and product measure $\hat{\mathcal{Q}}^x = \mathcal{Q}^x \times \bar{\mathcal{Q}}^x$, respectively. Let $\hat{Y}_t$ be the coordinate process given by

$$\hat{Y}_t(\omega) = \begin{cases} \omega_1(t), & t \geq 0, \\ \omega_2(-t), & t < 0 \end{cases}$$

for $\omega = (\omega_1, \omega_2) \in \hat{\Omega}$. We thus defined a random process $(\hat{Y}_t)_{t\in\mathbb{R}}$ on the product space $(\hat{\Omega}, \hat{\mathscr{F}}, \hat{\mathcal{Q}}^x)$ such that $\hat{\mathcal{Q}}^x(\hat{Y}_0 = x) = 1$ and $\mathbb{R} \ni t \mapsto \hat{Y}_t(\omega)$ is càdlàg. It can be also seen that $\hat{Y}_t, t \geq 0$, and $\hat{Y}_s, s \leq 0$, are independent, and $\hat{Y}_t \stackrel{\mathrm{d}}{=} \hat{Y}_{-t}$.

Step 5. Denote the image measure of $\hat{\mathcal{Q}}^x$ on $(\mathfrak{X}_D, \mathcal{B}(\mathfrak{X}_D))$ with respect to $\hat{Y}$ by $\mathscr{M}_0^x = \hat{\mathcal{Q}}^x \circ \hat{Y}^{-1}$ and the coordinate process by $Y_t(\omega) = \omega(t), t \in \mathbb{R}, \omega \in \mathfrak{X}_D$. Clearly, we have $Y_t \stackrel{\mathrm{d}}{=} \hat{Y}_t$ for $t \in \mathbb{R}$. Thus we see that $Y_t \stackrel{\mathrm{d}}{=} Y_{-t}$ and by Step 4 the processes $(Y_t)_{t\geq 0}$ and $(Y_t)_{t\leq 0}$ are independent. Furthermore, by Step 2 the processes $(Y_t)_{t\geq 0}$ and $(Y_t)_{t\leq 0}$ are Markov processes with respect to $(\mathscr{F}_t^+)_{t\geq 0}$ and $(\mathscr{F}_t^-)_{t\leq 0}$, respectively. Part (4) can be shown in the same way as Lemma 5.19. □

Lemma 5.32. *Let $T > 0$ be fixed. Then for every $\varepsilon > 0$ the probability $\mathcal{P}_\infty(|Z_t - Z_s| > \varepsilon)$ tends to zero uniformly as $|s - t| \to 0$ for $s, t \in [0, T]$.*

Proof. Note that the right-hand side of (5.1.92) can be expressed directly in terms of the subordinate Brownian motion $(X_t)_{t\geq 0}$ on a probability space $(\Omega, \mathscr{F}, P)$, i. e.,

$$\mathbb{E}_{\mathcal{P}_\infty}\left[\prod_{j=0}^n f_j(Z_{t_j})\right] = \int_{\mathbb{R}^d} \varphi_0(x) \mathbb{E}_P^x\left[e^{-\int_0^{t_n}(V(X_s)-E_0)ds}\left(\prod_{j=0}^n f_j(X_{t_j})\right)\varphi_0(X_{t_n})\right] dx. \tag{5.1.100}$$

Denote by $B_\varepsilon(x)$ the d-dimensional closed ball centered in x with radius ε, and let $0 \le s < t \le T$. By (5.1.100), we have

$$\mathcal{P}_\infty(|Z_t - Z_s| > \varepsilon) = \int_{\mathbb{R}^d} \varphi_0(x)\mathbb{E}_P^x\left[e^{-\int_0^{t-s}(V(X_r)-E_0)dr}\varphi_0(X_{t-s})1_{B_\varepsilon(x)^c}(X_{t-s})\right]dx.$$

The Schwarz inequality yields

$$\begin{aligned}&\mathcal{P}_\infty(|Z_t - Z_s| > \varepsilon)\\ &\le \int_{\mathbb{R}^d} \varphi_0(x)\left(\mathbb{E}_P^x\left[\varphi_0^2(X_{t-s})\right]\right)^{1/2}\left(\mathbb{E}_P^x\left[1_{B_\varepsilon(x)^c}(X_{t-s})e^{-2\int_0^{t-s}(V(X_r)-E_0)dr}\right]\right)^{1/2}dx. \qquad (5.1.101)\end{aligned}$$

Applying the Schwarz inequality again gives

$$\begin{aligned}\mathbb{E}_P^x\left[1_{B_\varepsilon(x)^c}(X_{t-s})e^{-2\int_0^{t-s}(V(X_r)-E_0)dr}\right] &\le \left(\mathbb{E}_P^x\left[1_{B_\varepsilon(x)^c}(X_{t-s})\right]\right)^{1/2}\left(\mathbb{E}_P^x\left[e^{-4\int_0^{t-s}(V(X_r)-E_0)dr}\right]\right)^{1/2}\\ &\le CP^0(|X_{t-s}| > \varepsilon)^{1/2},\end{aligned}$$

where $C = \sup_{x\in\mathbb{R}^d}(\mathbb{E}_P^x[e^{-4\int_0^{t-s}(V(X_r)-E_0)dr}])^{1/2}$. Thus we can conclude by (5.1.101) and the Schwarz inequality that

$$\mathcal{P}_\infty(|Z_t - Z_s| > \varepsilon) \le C^{1/2}\|\varphi_0\|^2 P^0(|X_{t-s}| > \varepsilon)^{1/4},$$

which tends to zero as $|t - s| \to 0$ by the stochastic continuity of $(X_t)_{t\ge0}$. The uniform stochastic convergence is proven in the same way as in Lemma 3.35. □

Lemma 5.33.

(1) *The coordinate process $(\tilde{Y}_t)_{t\ge0}$ is a Markov process on $(\mathscr{X}_D, \mathcal{B}(\mathscr{X}_D), \mathcal{Q}^x)$ with respect to the natural filtration $(\mathscr{F}_t)_{t\ge0}$.*

(2) *The coordinate process $(\tilde{Y}_t)_{t\ge0}$ is a Markov process on $(\mathscr{X}_{\bar{D}}, \mathcal{B}(\mathscr{X}_{\bar{D}}), \bar{\mathcal{Q}}^x)$ with respect to the natural filtration $(\mathscr{F}_t)_{t\ge0}$.*

Proof. We only show (1), part (2) can be proven similarly. Let

$$u_t(x, A) = e^{-tL^\Psi}1_A(x)$$

for every $A \in \mathcal{B}(\mathbb{R}^d)$, $x \in \mathbb{R}^d$, and $t \ge 0$. Clearly, $u_t(x, A) = \mathbb{E}_{\mathcal{Q}^x}[1_A(\tilde{Y}_t)]$ and, by (5.1.96) and (5.1.97), the finite-dimensional distributions of $(\tilde{Y}_t)_{t\ge0}$ are given by

$$\mathbb{E}_{\mathcal{Q}^x}\left[\prod_{j=0}^n 1_{A_j}(\tilde{Y}_{t_j})\right] = \int_{\mathbb{R}^d}\left(\prod_{j=0}^n 1_{A_j}(x_j)\right)\prod_{j=0}^n u_{t_j-t_{j-1}}(x_{j-1}, dx_j), \qquad (5.1.102)$$

with $t_0 = 0$ and $x_0 = x$. By using the properties of the semigroup $\{e^{-tL^\Psi} : t \geq 0\}$, it can be checked directly that $u_t(x, A)$ is a probability transition kernel, thus $(\tilde{Y}_t)_{t\geq 0}$ is a Markov process with finite-dimensional distributions given by (5.1.102). □

Definition 5.34 (Càdlàg / fractional $P(\phi)_1$-process). Consider the probability space $(\mathfrak{X}_D, \mathcal{B}(\mathfrak{X}_D), \mathscr{M}_0^x)$ and the coordinate process $(Y_t)_{t\in\mathbb{R}}$ on this space satisfying (1)–(4) in Theorem 5.31. We call $(Y_t)_{t\in\mathbb{R}}$ *càdlàg* $P(\phi)_1$*-process* for the given V. When the process is associated with a fractional Schrödinger operator, we call it a *fractional* $P(\phi)_1$*-process* for V.

Using this measure, we have the following change-of-measure formula.

Theorem 5.35 (Feynman–Kac formula for càdlàg $P(\phi)_1$-process). *For all* $f, g \in L^2(d\mathrm{M}_0)$, *we have*

$$(f, e^{-tL^\Psi} g)_{L^2(d\mathrm{M}_0)} = (f\varphi_0, e^{-t(H-E_0)} g\varphi_0)_{L^2(\mathbb{R}^d)} = \int_{\mathbb{R}^d} \mathbb{E}_{\mathscr{M}_0^x}[\bar{f}(Y_0) g(Y_t)] d\mathrm{M}_0(x).$$

Here $(Y_t)_{t\in\mathbb{R}}$ *denotes the coordinate process.*

Proof. The proof is immediate from (5.1.88). □

5.2 Gibbs measures on path space

5.2.1 From Feynman–Kac formulae to Gibbs measures

In this section we take a different view of the Feynman–Kac formula. In its simplest form for a Schrödinger operator $H = -\frac{1}{2}\Delta + V$ with suitable V, i. e.,

$$e^{-tH} f(x) = \mathbb{E}^x[e^{-\int_0^t V(B_s)ds} f(B_t)] = \int_{C(\mathbb{R}^+;\mathbb{R}^d)} e^{-\int_0^t V(B_s(\omega))ds} f(B_t(\omega)) d\mathcal{W}^x(\omega),$$

it suggests that the effect of the potential exerted on the paths is a contribution by the exponential weight function $e^{-\int_0^t V(B_s)ds}$ modifying the distribution of paths described by Wiener measure starting from x, corresponding to the case when $V \equiv 0$. Similarly to the Girsanov transform, this leads to a change of measure (after normalization) for every $t > 0$. Such probability measures are widely encountered in various contexts, and are known as *Gibbs measures* in the mathematical physics literature.

Gibbs measures have been introduced first to serve for a rigorous study of thermodynamic equilibrium states of lattice spin systems, but gradually proved to be a more generally applicable and rather fertile concept. Their form in the context of Brownian sample paths can be parallelled with an interacting particle model. To see this, recall the proof of the Feynman–Kac formula using the Trotter product formula in Section 4.2.2.

There we have seen that for a Schrödinger operator H as above with a smooth bounded potential V, for $f, g \in L^2(\mathbb{R}^d)$ we have

$$(f, e^{-tH} g) = \lim_{n\to\infty} (f, (e^{-(t/n)H_0} e^{-(t/n)V})^n g).$$

By Theorem 4.116, the integral kernel $K_t(x,y)$ of e^{-tH} is continuous in x and y. Writing out the integral kernels of $e^{-(t/n)H_0}$, and letting f and g approach δ-distributions yields

$$K_t(x,y) = \lim_{n\to\infty} \int_{\mathbb{R}^{dn}} \frac{1}{(2\pi t/n)^{d/2}} e^{-\sum_{j=0}^{n} \frac{1}{2t/n}|x_{j+1}-x_j|^2 - \sum_{j=1}^{n} V(x_j)} dx_1 \cdots dx_n, \tag{5.2.1}$$

where $x_0 = x$, $x_{n+1} = y$, and $x_1, \dots, x_n$ are intermediate positions. For any fixed finite n, the right-hand side above can be interpreted to describe the following situation. Consider a one-dimensional chain of length $n + 2$, consisting of interacting particles placed in the positions $x_1, \dots, x_n \in \mathbb{R}^d$, and pinned down at its two ends at x and y. Let the neighboring particles be coupled by a quadratic potential, while each particle being subjected to an external potential V. The total energy of the chain is given then by the exponent in (5.2.1). The energetically optimal position of the chain is a configuration that minimizes this energy. In this case it comes about as a compromise between a straight line through x and y, and a configuration in which all particles in the positions x_j are in the global minimum of V (assuming there is one). On the other hand, entropy makes the possible configurations of the chain fluctuate at inverse temperature $n/2t$ around this energetically optimal state. The expression in (5.2.1) defines a finite measure on $(\mathbb{R}^d)^n$, which can be normalized to a probability measure. This gives then the probability distribution of finding the chain in a subset of positions, and it describes its equilibrium state in which energy and entropy balance each other in an optimal way.

A peculiarity of the model (5.2.1) is that the limit $n \to \infty$ is taken. Different from the infinite volume limit, which is standard in the classical theory of Gibbs measures and will be discussed below, this is a type of continuum limit for the particle system: while the number of particles grows like n, the pair interaction energy $\frac{n}{2t}|x_{j+1} - x_j|^2$ increases at the same rate. This couples nearby particles increasingly strongly, forcing them to occupy the same position with large probabilities, and eventually leads to a Brownian path perturbed by the density $e^{-\int_0^t V(B_s)ds}$. Thus one can view the right-hand side of the Feynman–Kac formula

$$K_t(x,y) = \Pi_t(x-y) \int_{\mathscr{X}} e^{-\int_0^t V(B_s)ds} d\mathcal{W}_{[0,t]}^{x,y} \tag{5.2.2}$$

appearing in Theorem 4.116 as the probabilistic description of an elastic continuous string fixed at both ends, under the effect of an external potential V.

Our interest in this chapter is, informally, in studying probability measures of the form

$$dQ_I(\omega|x,y) = e^{-U_I(\omega)} dP_I^{x,y}(\omega), \tag{5.2.3}$$

on path spaces indexed by a collection of appropriate bounded intervals $I \subset \mathbb{R}$, and the extensions of these probability measures to full path space indexed by $\mathbb{R}$. This problem formulates for any of the two-sided Markov processes $(X_t)_{t\in\mathbb{R}}$ generated by the kinetic terms of the (standard and nonlocal) Schrödinger operators discussed so far. Thus for every I, the measure $P_I^{x,y}$ is the appropriate bridge measure for $(X_t)_{t\in\mathbb{R}}$, and $U_I(\omega)$ is the energy functional associated with path $(X_s(\omega))_{s\in I}$, which in the case of (5.2.2) is $U_t(\omega) = -\int_I V(X_s(\omega))ds$.

The basic difficulty in extending $(Q_I)_{I\in\mathcal{P}_f(\mathbb{R})}$, where $\mathcal{P}_f(\mathbb{R})$ denotes a family of bounded subintervals of the real line to be further specified below, is that attempting to take a direct limit proves typically to be unworkable as U_I almost surely diverges in this limit. This will require a strategy which we borrow from mathematical physics, and we will follow the *Dobrushin–Lanford–Ruelle (DLR) construction* so that we look for probability measures Q on a path space indexed by $\mathbb{R}$ whose family of probability measures conditional on paths outside I match with Q_I, for every $I \in \mathcal{P}_f(\mathbb{R})$. This problem is reminiscent of the strategy of the Kolmogorov's extension theorem, however, with the essential difference that here the extension is made by starting from prescribed conditional measures rather than marginals. As a next step we then find that, similarly to the so-called *thermodynamic limit* of classical Gibbs measures, solutions to the DLR problem can be constructed by taking suitable limits of $Q_{I_n}(\,\cdot\,|x,y)$ as $I_n \to \mathbb{R}$ on letting $n \to \infty$. However, unsurprisingly, the techniques that we need to develop will have to differ from the classical approaches as here we are in an essentially different context.

Although it is possible to consider Gibbs measures for continuous time random processes in a great generality, our primary interest is to use them later in studying some spectral properties of specific operators related to models of quantum theory, and therefore we restrict attention to particular classes of functionals U_t. We give a sample of applications below involving Brownian motion or jump Lévy processes perturbed by one-body or pair interaction potentials, of progressively increasing complexity.

Example 5.36. Various choices of the energy functional are possible as follows. Some will be considered in Volume 2 in applications to concrete models. For simplicity of notation, we do not explicitly feature ω.

(1) The case

$$U_t = \int_0^t V(B_s)ds, \tag{5.2.4}$$

where $(B_t)_{t\geq 0}$ is a Brownian motion, relates to path measures of $P(\phi)_1$-processes for Schrödinger operators with a Kato-decomposable potential V.

(2) Let

$$U_t = \int_0^t V(X_s)ds, \tag{5.2.5}$$

where $(X_t)_{t\geq 0}$ is a jump Lévy process, for instance, a rotationally symmetric α-stable process obtained for a fractional Schrödinger operator $(-\Delta)^{\alpha/2} + V$, $\alpha \in (0,2)$. The case $\alpha = 1$ describes a *massless relativistic quantum* particle. A similar expression can be obtained when $(X_t)_{t\geq 0}$ is a relativistic α-stable process generated by the operator $(-\Delta - m^{2/\alpha})^{\alpha/2} + V$, $m > 0$, and $\alpha = 1$ describes a *massive relativistic quantum* particle.

(3) Let

$$U_t = \int_0^t V(X_s)ds + \int_0^t\int_0^t W^\phi(X_s - X_r, s-r)dsdr, \tag{5.2.6}$$

where

$$W^\phi(x,s) = -\frac{1}{4}\int_{\mathbb{R}^d} \frac{|\widehat{\phi}(k)|^2}{|k|} e^{-ik\cdot x - |k||s|}dk,$$

is a pair interaction potential, with a sufficiently fast decaying, spherical symmetric real-valued function ϕ. In Volume 2 we will see how this arises, and how, by using such a setting, it is possible to derive and prove the ground state properties of quantum field models, such as the *Nelson model* of an electrically charged spinless quantum particle linearly coupled to a scalar boson field.

(4) Changing $W^\phi(x,s)$ above to

$$W_{\text{pol}}(x,s) = -\frac{1}{4|x|}e^{-|s|} \tag{5.2.7}$$

describes the *polaron model*. A variant, called *bipolaron model*, differs by the fact that it consists of two dressed electrons coupled to the same phonon field, which repel each other by Coulomb interaction. In this case

$$U_t = \alpha^2 \int_0^t\int_0^t \mathcal{E}(B_s, B_r, \tilde{B}_s, \tilde{B}_r, s-r)dsdr - g\int_0^t \frac{1}{|B_s - \tilde{B}_s|}ds,$$

where

$$\mathcal{E}(B_s, B_r, \tilde{B}_s, \tilde{B}_r, u) = W_{\text{pol}}(B_s - B_r, u) + 2W_{\text{pol}}(B_s - \tilde{B}_r, u) + W_{\text{pol}}(\tilde{B}_s - \tilde{B}_r, u),$$

with $\alpha < 0$ being the polaron–phonon coupling parameter, and $g > 0$ is the strength of the Coulomb repulsion between the two polarons. In this case the reference measure is a product of two Wiener measures of the independent Brownian motions $(B_t)_{t\in\mathbb{R}}$ and $(\tilde{B}_t)_{t\in\mathbb{R}}$.

(5) Let

$$U_t = \int_0^t \int_0^t \delta(B_s - B_r) ds dr. \tag{5.2.8}$$

This gives the *intersection local time* of Brownian motion describing a *polymer model* with short-range soft-core interaction encouraging to avoid self-intersections.

(6) Let

$$U_t = \int_0^t V(B_s) ds + \int_0^t \int_0^t W(B_s - B_r, s - r) dB_s \cdot dB_r \tag{5.2.9}$$

with some choices of W which we do not write down explicitly here. This describes the *Pauli–Fierz model* of a charged particle interacting with a vector boson field. The difference from the cases above consists in having double stochastic integrals here instead of double Riemann integrals. Gibbs measures of this type will be discussed when we study the Pauli–Fierz model in Volume 2.

(7) Let

$$U_t = \int_0^t \int_0^t \frac{1}{|B_s - B_r|} dB_s \cdot dB_r. \tag{5.2.10}$$

This is used in describing ***turbulent fluids***. The above formal expression gives the total energy of a vorticity field concentrated along Brownian curves $X_t \in \mathbb{R}^3$ obtained from a given divergence-free velocity field.

Our main concern here will be to construct Gibbs measures covering Examples (1)–(3). A main benefit will be a study, through the appropriate Feynman–Kac-type formulae, of ground state properties of quantum models in terms of averages of suitable functions on path space with respect to these Gibbs measures, which will be carried out in Volume 2.

5.2.2 Gibbs measures on Brownian paths

Let $\mathfrak{X} = C(\mathbb{R}; \mathbb{R}^d)$, $\mathscr{F} = \mathcal{B}(\mathfrak{X})$, and consider a two-sided $\mathbb{R}^d$-valued Brownian motion $(B_t)_{t\in\mathbb{R}}$ on this space, as defined in Section 2.3.2. We keep denoting the corresponding Wiener measure starting from $x \in \mathbb{R}^d$ by $\mathcal{W}^x$, as before. With $I \subset \mathbb{R}$, we use the notation $\mathscr{F}_I$ for the sub-σ-field generated by $(X_t)_{t\in I}$, and $\mathscr{T}_I = \mathscr{F}_{I^c}$, where $I^c = \mathbb{R} \setminus I$, as well as $\mathscr{F}_T$ and $\mathscr{T}_T$ for $I = [-T, T]$, $T > 0$. The tail field will be denoted by $\mathscr{T} = \bigcap_{N\in\mathbb{N}} \mathscr{T}_N$, and the Lebesgue measure of I by $|I|$.

Definition 5.37 (Index set and state space). The set $\{I \subset \mathbb{R} \mid |I| < \infty\}$ of bounded intervals of the real line is the *index set*, and the *state space* is $\mathbb{R}^d$.

In the context of Gibbs measures there is an underlying probability measure giving the distribution of paths without any potential, called *reference measure*, which we will denote by P. In view of the applications discussed in Example 5.36 we also introduce pair interaction potentials apart from one-body external potentials.

Definition 5.38 (Potentials).

(1) A Borel measurable function $V : \mathbb{R}^d \to \mathbb{R}$ is called an *external potential*. We call V a *P-admissible external potential* whenever

$$0 < \mathbb{E}_P[e^{-\int_I V(B_s)\,ds}] < \infty \tag{5.2.11}$$

for every bounded interval $I \subset \mathbb{R}$.

(2) A function $W : \mathbb{R}^d \times \mathbb{R}^d \times \mathbb{R} \times \mathbb{R} \to \mathbb{R}$ is called a *pair interaction potential*. We say that W is *admissible* whenever

$$\sup_{t\in\mathbb{R}} \int_{\mathbb{R}} \sup_{x,y\in\mathbb{R}^d} |W(x,y,s,t)|\,ds < \infty. \tag{5.2.12}$$

If $W(x,y,s,t) = W(x,y,s-t,0)$, we use the simpler notation $W(x,y,t)$; similarly, we write $W(x,t)$ in case $W(x,y,s,t) = W(x-y,0,s-t,0)$.

For brevity, we will henceforth refer to the potentials jointly satisfying the conditions above as admissible. The condition on the pair potential is rather strong; however, as we move on we will introduce weaker regularity conditions. In the Feynman–Kac formula considered at the beginning of this chapter, (5.2.11) holds for Kato-decomposable potentials V by Lemma 4.105. While n-body potentials can be defined in a straightforward way, in what follows we will not need more general potentials than the one-body and pair interaction potentials V and W.

Definition 5.39. For fixed $\bar{\omega} \in \mathfrak{X}$, denote by $P_T^{\bar{\omega}}$ the unique measure on $(\mathfrak{X}, \mathscr{F})$ such that for all functions f bounded and measurable with respect to $\mathscr{F}_T$ and all g bounded and measurable with respect to $\mathscr{T}_T$,

$$\mathbb{E}_{P_T^{\bar{\omega}}}[fg] = \mathbb{E}_P[f|\mathscr{T}_T](\bar{\omega})g(\bar{\omega}) \tag{5.2.13}$$

holds.

For a general measurable f, the integral $\int_{\mathfrak{X}} f\,dP_T^{\bar{\omega}}$ can be obtained from (5.2.13) by approximation of f. In other words, $P_T^{\bar{\omega}}$ is the product measure on $C([-T,T];\mathbb{R}^d) \times C([-T,T]^c;\mathbb{R}^d)$ of the regular conditional probability of P given $\omega(t) = \bar{\omega}(t)$, for all $|t| > T$, and the Dirac measure concentrated at $\bar{\omega}$ on $C([-T,T]^c;\mathbb{R}^d)$. Thus, intuitively,

$P_T^{\bar\omega}$ is the measure obtained from P by fixing the path to $\bar\omega$ outside $[-T, T]$ and allowing it to fluctuate inside $[-T, T]$.

It is clear that $P_T^{\bar\omega}$ is indeed a version of $P(\,\cdot\,|\mathscr{T}_T)$. We emphasize that there many choices of possible versions since conditional expectations are only defined up to sets of measure zero. By the law of the iterated logarithm, the measure $\mathcal{W}^x$ is concentrated on a set of paths that is characterized by the asymptotic behavior of $B_t(\omega)$ as $|t| \to \infty$. Thus had we defined $P(\,\cdot\,|\mathscr{T}_T)$ arbitrarily on the set $N \subset \mathfrak{X}$ of measure zero consisting of all paths not having this precise asymptotic behavior, we would still have obtained a version of the conditional expectation. Our definition of $P_T^{\bar\omega}$ reduces this element of arbitrariness considerably, and we will need this when defining Gibbs measures.

Without restricting generality, we choose symmetric bounded intervals from the index set. Write

$$\Lambda(S,T) = ([-S,S] \times [-T,T]) \cup ([-T,T] \times [-S,S]),$$
$$\Lambda(T) = (\mathbb{R} \times [-T,T]) \cup ([-T,T] \times \mathbb{R}),$$

where $S, T > 0$. For $S \leq T$ and admissible potentials V, W, we use the notation

$$\mathcal{H}_{\Lambda(S,T)}(B) = \int_{-T}^{T} V(B_s)\,ds + \int_{\Lambda(S,T)} W(B_s, B_t, |s-t|)\,ds\,dt, \tag{5.2.14}$$

and define $\mathcal{H}_{\Lambda(T)}$ by replacing $\Lambda(S,T)$ with $\Lambda(T)$ in (5.2.14). We will also write $\mathcal{H}_T$ instead of $\mathcal{H}_{\Lambda(T,T)}$. To ease the notation we do not explicitly indicate the ω-dependence of these functionals.

Now we can define Gibbs measures on Brownian paths for the potentials V, W.

Definition 5.40 (Gibbs measure). A probability measure Q on $(\mathfrak{X}, \mathscr{F})$ is called a *Gibbs measure* with respect to reference measure P and admissible potentials V, W if for every $T > 0$,
(1) $Q|_{\mathscr{F}_T} \ll P|_{\mathscr{F}_T}$ for all $T > 0$;
(2) for every bounded $\mathscr{F}$-measurable function f,

$$\mathbb{E}_Q[f|\mathscr{T}_T](\bar\omega) = \frac{\mathbb{E}_{P_T^{\bar\omega}}[fe^{-\mathcal{H}_{\Lambda(T)}}]}{\mathbb{E}_{P_T^{\bar\omega}}[e^{-\mathcal{H}_{\Lambda(T)}}]}, \quad Q\text{-a. s.} \tag{5.2.15}$$

A probability measure Q_T on $(\mathfrak{X}, \mathscr{F})$ is called a *finite volume Gibbs measure* for the interval $[-T, T]$ if
(1′) $Q_T|_{\mathscr{F}_S} \ll P|_{\mathscr{F}_S}$ for all $S < T$;
(2′) for all $0 < S < T$ and every bounded $\mathscr{F}$-measurable function f,

$$\mathbb{E}_{Q_T}[f|\mathscr{T}_S](\bar\omega) = \frac{\mathbb{E}_{P_S^{\bar\omega}}[fe^{-\mathcal{H}_{\Lambda(S,T)}}]}{\mathbb{E}_{P_S^{\bar\omega}}[e^{-\mathcal{H}_{\Lambda(S,T)}}]}, \quad Q_T\text{-a. s.} \tag{5.2.16}$$

In our terminology "Gibbs measure" corresponds to "infinite volume Gibbs measure" in the classical theory. We refer to the path $\bar{\omega}$ as a *boundary condition* or *boundary path*. The normalization factor

$$Z_T(\bar{\omega}) = \mathbb{E}_{P_T^{\bar{\omega}}}[e^{-\mathcal{H}_{\Lambda(T)}}]$$

is traditionally called a *partition function*.

Remark 5.41 (DLR equations). Conditions (2) and (2′) in the above definition are traditionally called *Dobrushin–Lanford–Ruelle (DLR) equations*. These are, in fact, equalities and can be viewed as counterparts for conditional measures of the Kolmogorov's consistency relations for marginal measures.

Remark 5.42 (Gibbs specification). To make it more visible that a Gibbs measure is a probability measure which has prescribed conditional probability kernels, we give the following equivalent definition of a Gibbs measure. We say that a probability measure Q on $(\mathfrak{X}, \mathscr{F})$ is a *Gibbs measure* for the admissible potentials V, W if for every $A \in \mathscr{F}$ and $T > 0$ the function $\bar{\omega} \mapsto Q_T(A, \bar{\omega})$ is a version of the conditional probability $Q(A|\mathscr{T}_T)$, i. e.,

$$Q(A|\mathscr{T}_T)(\bar{\omega}) = Q_T(A, \bar{\omega}), \quad A \in \mathscr{F},\ T > 0,\ \text{a. e. } \bar{\omega} \in \mathfrak{X},$$

where

$$Q_T(A, \bar{\omega}) = \frac{1}{Z_T(\bar{\omega})} \int_{\mathfrak{X}} 1_A e^{-\mathcal{H}_{\Lambda(T)}} dP_T^{\bar{\omega}}, \quad A \in \mathscr{F},\ \bar{\omega} \in \mathfrak{X}. \tag{5.2.17}$$

Remark 5.43. To understand why (1) in Definition 5.40 is necessary, consider the special version $P_T^{\bar{\omega}}$ of $\mathbb{E}_P[\,\cdot\,|\mathscr{T}_T](\bar{\omega})$. Writing the (at first sight more natural) expression

$$\mathbb{E}_Q[f|\mathscr{T}_T] = \frac{\mathbb{E}_P[fe^{-\mathcal{H}_{\Lambda(T)}}|\mathscr{T}_T]}{\mathbb{E}_P[e^{-\mathcal{H}_{\Lambda(T)}}|\mathscr{T}_T]} \tag{5.2.18}$$

instead of (5.2.15) is in most cases meaningless. The problem is that the left-hand side of the above equality is only defined uniquely outside a set N of Q-measure zero, while the right-hand side is only defined uniquely outside a set M of P-measure zero. In many cases of interest, P and Q are mutually singular on $\mathfrak{X}$, and then (5.2.18) is no condition at all. Thus in our case sets of measure zero do matter and extra care is needed. The reason why we require (1) explicitly is the same. Without it the two sides of (2) may refer to functions that are uniquely defined only on disjoint subsets of the probability space. In the case of a finite-dimensional state space and the measure of a Feller Markov process as reference measure, this inconvenience can be avoided and then (1) is unnecessary.

Constructing a Gibbs measure for given potentials V and W may be difficult. However, constructing a finite volume Gibbs measure under the given regularity conditions is straightforward.

Proposition 5.44. *Let P be the reference measure, $T > 0$, and V, W be admissible potentials. If $0 < \mathbb{E}_P[e^{-\mathcal{H}_T}] < \infty$, then*

$$dQ_T = \frac{e^{-\mathcal{H}_T}}{\mathbb{E}_P[e^{-\mathcal{H}_T}]} dP$$

is a finite volume Gibbs measure for $[-T, T]$.

Proof. By assumption, $e^{-\mathcal{H}_T}$ is P-integrable. Thus $Q_T \ll P$, and in particular $Q_T|_{\mathscr{F}_T} \ll P|_{\mathscr{F}_T}$. To check DLR consistency, let f and g be bounded, and suppose that g is $\mathscr{T}_S$-measurable. Then for $T > S$,

$$\begin{aligned}\mathbb{E}_P[e^{-\mathcal{H}_T}]\mathbb{E}_{Q_T}\left[g\frac{\mathbb{E}_{P_S^{\cdot}}[fe^{-\mathcal{H}_{\Lambda(S,T)}}]}{\mathbb{E}_{P_S^{\cdot}}[e^{-\mathcal{H}_{\Lambda(S,T)}}]}\right] &= \mathbb{E}_P\left[\mathbb{E}_P\left[g\frac{\mathbb{E}_{P_S^{\cdot}}[fe^{-\mathcal{H}_{\Lambda(S,T)}}]}{\mathbb{E}_{P_S^{\cdot}}[e^{-\mathcal{H}_{\Lambda(S,T)}}]}e^{-\mathcal{H}_T}\middle|\mathscr{T}_S\right]\right] \\ &= \mathbb{E}_P\left[g\frac{\mathbb{E}_{P_S^{\cdot}}[fe^{-\mathcal{H}_{\Lambda(S,T)}}]}{\mathbb{E}_{P_S^{\cdot}}[e^{-\mathcal{H}_{\Lambda(S,T)}}]}e^{-\mathcal{H}_{[-T,T]^2\setminus\Lambda(S,T)}}\mathbb{E}_P[e^{-\mathcal{H}_{\Lambda(S,T)}}|\mathscr{T}_S]\right] \\ &= \mathbb{E}_P[e^{-\mathcal{H}_T}]\mathbb{E}_{Q_T}[fg].\end{aligned}$$

The last equality is due to the fact that $\mathbb{E}_{P_S^{\cdot}}[e^{-\mathcal{H}_{\Lambda(S,T)}}]$ is a version of the conditional expectation $\mathbb{E}_P[e^{-\mathcal{H}_{\Lambda(S,T)}}|\mathscr{T}_S]$. Dividing by $\mathbb{E}_P[e^{-\mathcal{H}_T}]$ shows that condition (2) in Definition 5.40 is verified. □

Remark 5.45 (Stochastic and sharp boundary conditions). The finite volume Gibbs measure that we constructed above corresponds to free boundary conditions. A similar proof as above shows that for every $\bar{\omega} \in \mathfrak{X}$, the probability measure

$$dQ_T^{\bar{\omega}} = \frac{e^{-\mathcal{H}_{\Lambda(T)}}}{\mathbb{E}_{P_T^{\bar{\omega}}}[e^{-\mathcal{H}_{\Lambda(T)}}]}dP_T^{\bar{\omega}} \tag{5.2.19}$$

is a finite volume Gibbs measure. We call it a finite volume Gibbs measure with *sharp boundary condition* $\bar{\omega}$. Occasionally, instead of prescribing sharp boundary conditions by fixing path positions, it is more convenient to give only their probability distributions. Let $h : \mathfrak{X} \to \mathbb{R}$ be a $\mathscr{T}_T$-measurable nonnegative function such that

$$0 < \mathbb{E}_P[he^{-\mathcal{H}_T}] < \infty.$$

Then as in Proposition 5.44 we can show that

$$dQ_T^h = \frac{he^{-\mathcal{H}_{\Lambda(T)}}}{\mathbb{E}_P[he^{-\mathcal{H}_{\Lambda(T)}}]} dP$$

is a finite volume Gibbs measure for $[-T, T]$. The density h can then be regarded as a *stochastic boundary condition*.

Next we show that limits of finite volume Gibbs measures yield Gibbs measures in the sense of Definition 5.40. The following notion of convergence will be used.

Definition 5.46 (Local convergence). A sequence $(\mathfrak{m}_n)_{n\in\mathbb{N}}$ of probability measures on $\mathcal{X}$ is said to be *locally convergent* to a probability measure $\mathfrak{m}$ if for every $0 < T < \infty$ and every $A \in \mathscr{F}_T$, $\mathfrak{m}_n(A) \to \mathfrak{m}(A)$ as $n \to \infty$.

Remark 5.47.

(1) In many cases Gibbs measures are mutually singular with respect to their reference measures, while finite volume Gibbs measures are absolutely continuous with respect to them. Thus the topology of local convergence is the strongest reasonable for such Gibbs measures. In particular, in many cases it is possible to find a tail measurable function f such that $\int_{\mathcal{X}} f\, dP_N = 0$ for all N, while $\int_{\mathcal{X}} f\, dP = 1$.

(2) Convergence on sets is usually a very strong requirement for the convergence of measures. Indeed, the situation discussed in the previous remark indicates that convergence typically does not hold for all sets of $\mathscr{F}$. Since, however, we only need to check convergence on elements of $\mathscr{F}_T$ for finite T, the special structure of Gibbs measures allows showing local convergence of finite volume Gibbs measures in all cases we consider below.

Proposition 5.48. *Let P be a reference measure and V, W be admissible potentials. Let $(Q_n)_{n\in\mathbb{N}}$ be a sequence of finite volume Gibbs measures for V, W and $[-T_n, T_n]$, with $T_n \to \infty$ as $n \to \infty$. Suppose that there exists a probability measure Q on $\mathcal{X}$ such that $Q_n \to Q$ in the topology of local convergence. Suppose, moreover, that Q satisfies (1) in Definition 5.40 with respect to P. Then Q is a Gibbs measure for V, W.*

Proof. As (1) in Definition 5.40 holds by assumption, only DLR consistency remains to be shown. Each Q_n is a finite volume Gibbs measure, thus for bounded f, g with $\mathscr{T}_S$-measurable g we have

$$\mathbb{E}_{Q_n}\left[g \frac{\mathbb{E}_{P_S^{\cdot}}[f e^{-\mathcal{H}_{\Lambda(S,T_n)}}]}{\mathbb{E}_{P_S^{\cdot}}[e^{-\mathcal{H}_{\Lambda(S,T_n)}}]} \right] = \mathbb{E}_{Q_n}[fg] \tag{5.2.20}$$

for every $S < T_n$. We show now that (5.2.20) remains valid in the $n \to \infty$ limit. By a monotone class argument, we may assume that both f and g are $\mathscr{F}_R$-measurable for some $R > 0$. In this case, the right-hand side of (5.2.20) converges by the definition of local convergence. At the left-hand side (5.2.12) implies that

$$\int_{\Lambda(S,T_n)} W(B_s, B_t, |s-t|)\, ds\, dt \overset{n\to\infty}{\longrightarrow} \int_{\Lambda(S)} W(B_s, B_t, |s-t|)\, ds\, dt$$

uniformly in paths, and thus

$$F_n(X(\omega)) = \frac{\mathbb{E}_{P_S^\omega}[fe^{-\mathcal{H}_{\Lambda(S,T_n)}}]}{\mathbb{E}_{P_S^\omega}[e^{-\mathcal{H}_{\Lambda(S,T_n)}}]} \overset{n\to\infty}{\longrightarrow} \frac{\mathbb{E}_{P_S^\omega}[fe^{-\mathcal{H}_{\Lambda(S)}}]}{\mathbb{E}_{P_S^\omega}[e^{-\mathcal{H}_{\Lambda(S)}}]} = F(X(\omega))$$

uniformly in $\omega \in \mathfrak{X}$. Thus for every $\varepsilon > 0$, there is an $N \in \mathbb{N}$ such that $\|F_m - F\|_\infty < \varepsilon$, whenever $m \geq N$. By the triangle inequality,

$$|\mathbb{E}_{Q_n}[gF_n] - \mathbb{E}_Q[gF]| \leq |\mathbb{E}_{Q_n}[gF_N] - \mathbb{E}_Q[gF_N]| + 3\|g\|_\infty \varepsilon \overset{n\to\infty}{\longrightarrow} 3\|g\|_\infty \varepsilon.$$

Since ε is arbitrary, convergence of the left-hand side in (5.2.20) follows. □

The previous proposition shows that suitable limits of finite volume Gibbs measures over boundary conditions result in Gibbs measures. The next statement shows that this procedure actually yields all Gibbs measures supported on a specific set of paths.

Proposition 5.49. *Let $\mathfrak{X}^* \subset \mathfrak{X}$ be measurable and Q be a Gibbs measure supported on $\mathfrak{X}^*$, with reference measure P. For $N \in \mathbb{N}$, $\omega \in \mathfrak{X}$, define*

$$\mathbb{E}_{Q_N^\omega}[f] = \frac{\mathbb{E}_{P_N^\omega}[fe^{-\mathcal{H}_{\Lambda(N)}}]}{\mathbb{E}_{P_N^\omega}[e^{-\mathcal{H}_{\Lambda(N)}}]}$$

as in (5.2.15). *Suppose that for every $T > 0$, $A \in \mathscr{F}_T$, and $\omega \in \mathfrak{X}^*$ we have $Q_N^\omega(A) \to Q(A)$ as $N \to \infty$. Then Q is the unique Gibbs measure associated with $\mathcal{H}$, supported on $\mathfrak{X}^*$.*

Proof. Let $\tilde{Q}$ be a Gibbs measure supported by $\mathfrak{X}^*$. For every $T < N$ and $A \in \mathscr{F}_T$, $\bar{\omega} \mapsto \tilde{Q}(A|\mathscr{T}_N)(\bar{\omega})$ is a backward martingale in N, thus convergent almost everywhere to $\tilde{Q}(A|\mathscr{T})(\bar{\omega})$. By the Gibbs property, $\tilde{Q}(A|\mathscr{T}_N)(\bar{\omega}) = Q_N^{\bar{\omega}}(A)$ $\tilde{Q}$-a. s., and thus for $\tilde{Q}$-a. e. $\bar{\omega} \in \mathfrak{X}^*$, there exists $\tilde{Q}(A|\mathscr{T})(\bar{\omega}) = \lim_{N\to\infty} \tilde{Q}(A|\mathscr{T}_N)(\bar{\omega}) = \lim_{N\to\infty} Q_N^{\bar{\omega}}(A) = Q(A)$. Taking $\tilde{Q}$-expectations on both sides of the above equality shows $\tilde{Q}(A) = Q(A)$, and since this holds for every $A \in \mathscr{F}_T$ and $T > 0$, we obtain $\tilde{Q} = Q$. □

Finally, we give a useful criterion for a sequence of finite volume Gibbs measures to converge along a sequence of boundary conditions.

Definition 5.50. We say that a family of probability measures $(\mathfrak{m}_T)_{T>0}$ is *locally uniformly dominated* by a family of probability measures $(\tilde{\mathfrak{m}}_S)_{S>0}$ whenever
(1) $\tilde{\mathfrak{m}}_S$ is a probability measure on $\mathscr{F}_{[-S,S]}$;
(2) for every $\varepsilon > 0$ and $S > 0$, there exists $\delta_S > 0$ such that $\limsup_{T\to\infty} \mathfrak{m}_T(A) < \varepsilon$ when $A \in \mathscr{F}_{[-S,S]}$ with $\tilde{\mathfrak{m}}_S(A) < \delta_S$.

The significance of uniform local domination is made clear by the following result.

Proposition 5.51. *Let $(T_n)_{n\in\mathbb{N}}$ be an increasing divergent sequence of positive numbers and assume that $(Q_{T_n})_{n\in\mathbb{N}}$ is a locally uniformly dominated family of finite volume Gibbs measures. Then $(Q_{T_n})_{n\in\mathbb{N}}$ has a convergent subsequence in the local strong topology and the limit is a Gibbs measure.*

Proof. Let $S > 0$ and write $\tilde{Q}_S$ for the measure dominating Q_n, i. e., suppose that for $\varepsilon > 0$ there exists $\delta_S > 0$ such that $\limsup_{n\to\infty} Q_n(A) < \varepsilon$, for all A satisfying $\tilde{Q}_S(A) < \delta_S$. Since $\tilde{Q}_S$ is a measure we have $\lim_{m\to\infty} Q_S(A_m) = 0$. In particular, $\tilde{Q}_S(A_m) < \delta_S$, for all $m > M$ with some M. Thus $\limsup_{n\to\infty} Q_n(A_m) < \varepsilon$ for all $m > M$ by the local uniform domination property. This means that $Q_\infty(A_m) < \varepsilon$ for all $m > M$ and hence $\lim_{m\to\infty} Q_\infty(A_m) = 0$. Thus a limit measure on $\mathscr{F}_S$ exists. For $\mathscr{F}_{S+1}$, we take a further subsequence of the subsequence Q_n and find a measure Q'_∞ on $\mathscr{F}_{S+1}$ whose restriction to $\mathscr{F}_S$ is Q_∞. Continuing inductively, we find a sequence of measures Q_S for each $S \in \mathbb{N}$; they form a consistent family of probability measures and thus by the Kolmogorov's extension theorem there is a unique probability measure Q on $\mathscr{F}$ such that all $Q|_{\mathscr{F}_S} = Q_S$. This proves existence of a convergent subsequence. Since the DLR property holds in finite volume, due to convergence along a subsequence it carries over in the limit. □

5.2.3 Gibbs measures on càdlàg paths

When Brownian motion is replaced by a jump Lévy process, the constructions presented in the previous section need to be slightly modified due to measurability issues, which we will discuss next. We only consider the case with potentials $V \not\equiv 0$ and $W \equiv 0$.

Recall the notation $\mathfrak{X}_D = D(\mathbb{R}; \mathbb{R}^d)$ for two-sided càdlàg paths as defined in Section 3.2.2. Let $\mathscr{F} = \mathcal{B}(\mathfrak{X}_D)$ be the family of cylinder sets on $\mathfrak{X}_D$, and consider a two-sided Lévy process $(X_t)_{t\in\mathbb{R}}$ with path measure P on the space $(\mathfrak{X}_D, \mathcal{B}(\mathfrak{X}_D))$. We assume that $(X_t)_{t\in\mathbb{R}}$ is such that it has transition probability densities $p_t(x,y) = p_t(|y - x|)$, for all $t > 0$ and $x, y \in \mathbb{R}^d$. With no restriction of generality, we consider half-closed symmetric intervals $I = [-T, T) \subset \mathbb{R}$, with $T > 0$, and use the notations $\mathscr{F}_T = \mathscr{F}_{[-T,T)}$ for the sub-σ-field generated by $(X_t)_{t\in[-T,T)}$, $\mathscr{F}_T = \mathscr{T}_{[-T,T)} = \mathscr{F}_{[-T,T)^c}$, and $\mathscr{T} = \bigcap_{N\in\mathbb{N}} \mathscr{T}_N$ for the tail field. For the projection of $\mathfrak{X}_D$ to $I \subset \mathbb{R}$, we write $\mathfrak{X}_D|_I$, and $\mathscr{F}_I$ for the related Borel σ-field.

For $x, y \in \mathbb{R}^d$ and $s, t \in \mathbb{R}$, $s < t$, we respectively denote by $P^{x,y}$ and $\mathbb{E}^{x,y}$ the probability measure and expectation of the Lévy process $(X_r)_{s\le r\le t}$ starting in $x \in \mathbb{R}^d$ at time $s \in \mathbb{R}$ given by $X_t = y$. In fact, $P^{x,y}$ is a regular version of the family of conditional probability measures $P^x(\cdot \mid X_t = y), y \in \mathbb{R}^d$, that is, if $Y \ge 0$ is an $\mathscr{F}_{[s,t)}$-measurable random variable and $g \ge 0$ is a Borel function on $\mathbb{R}^d$, then

$$\mathbb{E}^x[Yg(X_t)] = \int_{\mathbb{R}^d} \mathbb{E}^{x,y}[Y] g(y) p_{t-s}(y - x) dy.$$

Under $P^{x,y}$, the process $(X_r)_{s\le r<t}$ is a nonhomogenous strong Markov process with transition probability densities

$$q_{s,t}^{x,y}(u, w; v, z) = \frac{p_{v-u}(z - w) p_{t-v}(y - z)}{p_{t-u}(y - w)}, \quad s \le u < v < t,$$

and we call it a *Lévy bridge*. Clearly, $P^{x,y}(X_s = x, \lim_{r\uparrow t} X_r = y) = 1$.

For $x, y \in \mathbb{R}^d$ and $s, t \in \mathbb{R}$, $s < t$, we denote by $\mathbb{P}^{x,y}_{[s,t)}$ the nonnormalized measure on $(\mathfrak{X}_D|_{[s,t)}, \mathscr{F}_{[s,t)})$ corresponding to the Lévy bridge $(X_r)_{s\le r<t}$ given by

$$\mathbb{P}^{x,y}_{[s,t)}(\cdot) = p_{t-s}(y-x)P^{x,y}(\cdot).$$

Thus for $s = t_0 < t_1 < t_2 < \cdots < t_n < t_{n+1} = t$ and Borel sets $A_1, A_2, \ldots, A_n \subset \mathbb{R}^d$, we have

$$\mathbb{P}^{x,y}_{[s,t]}(\omega(t_1) \in A_1, \omega(t_2) \in A_2, \ldots, \omega(t_n) \in A_n) = \int\limits_{A_1\times\cdots\times A_n} \prod_{i=1}^{n+1} p_{t_i - t_{i-1}}(z_i - z_{i-1})dz_1 \cdots dz_n,$$

where $z_0 = x$ and $z_{n+1} = y$. Since $\mathbb{P}^{x,y}_{[s,t)}$ is a measure defined on càdlàg paths, we may also identify $\mathbb{P}^{x,y}_{[s,t)}$ as a measure on $(\mathfrak{X}_D, \mathscr{F}_{[s,t)})$.

Similarly as before, we define the reference measure $P^{\bar\omega}_T$ as follows. Let $\bar\omega \in \mathfrak{X}_D$, and Dirac measure $\delta^{\bar\omega}_T$ on $\mathfrak{X}_D|[-T,T)^c$ concentrated on $\bar\omega$, and denote $\omega(T)- = \lim_{t\uparrow T}\omega(t)$ and $\mathbb{P}^{x,y}_{[-T,T)} = \mathbb{P}^{x,y}_T$. For every $T > 0$, we define the measure

$$P^{\bar\omega}_T = \mathbb{P}^{\bar\omega(-T),\bar\omega(T)-}_T \otimes \delta^{\bar\omega}_T.$$

Then for the potential V, we define

$$Z_T(x,y) = \int\limits_{\mathfrak{X}_D} e^{-\int_{-T}^T V(X_t(\omega))dt} d\mathbb{P}^{x,y}_T(\omega), \quad x, y \in \mathbb{R}^d,\ t \ge 0,$$

and say that V is *P-admissible* whenever $0 < Z_T(x,y) < \infty$ for all $T > 0$ and $x, y \in \mathbb{R}^d$. Note that for Ψ-Kato decomposable or X-Kato decomposable potentials $Z_T(x,y) = K_{2T}(x,y) < \infty$, for all $x, y \in \mathbb{R}^d$, $t \ge 0$, where $K_t(x,y)$ is the integral kernel of the operator e^{-tH}. For every $T > 0$, we then define the *Gibbs specification*

$$Q_T(A,\bar\omega) = \frac{1}{Z_T(\bar\omega(-T),\bar\omega(T)-)} \int\limits_{\mathfrak{X}_D} 1_A(\omega) e^{-\int_{-T}^T V(X_t(\omega))dt} d\mathbb{P}^{x,y}_T(\omega),\ A \in \mathscr{F},\ \bar\omega \in \mathfrak{X}_D. \tag{5.2.21}$$

We define Gibbs measures in the spirit of Remark 5.42.

Definition 5.52 (Gibbs measure on càdlàg paths). Let $(X_t)_{t\in\mathbb{R}}$ be a two-sided Lévy process on a given probability space $(\mathfrak{X}_D, \mathscr{F}, P)$, and consider the Gibbs specification given by (5.2.21). We say that a probability measure Q on $(\mathfrak{X}_D, \mathscr{F})$ is a *Gibbs measure* for the admissible potential V and reference measure P if for every $A \in \mathscr{F}$ and $T > 0$ the function $\bar\omega \mapsto Q_T(A,\bar\omega)$ is a version of the conditional probability $Q(A|\mathscr{T}_T)$, i. e.,

$$Q(A|\mathscr{T}_T)(\bar\omega) = Q_T(A,\bar\omega), \quad A \in \mathscr{F},\ T > 0,\ \text{a. e. } \bar\omega \in \mathfrak{X}_D.$$

Alternatively, we say that Q is a *Gibbs measure* on $(\mathfrak{X}_D, \mathscr{F})$ for the admissible potential V and reference measure P if for every $T > 0$, $A \in \mathscr{F}$, and $B \in \mathscr{T}_T$, the equality

$$Q(A \cap B) = \int_{\mathfrak{X}_D} 1_B(\omega) Q_T(A, \omega) dQ(\omega) \tag{5.2.22}$$

holds.

5.3 Gibbs measures for external potentials

5.3.1 Existence

In this section we assume that no pair interaction potential is present and consider Gibbs measures for admissible external potentials only. Our goal is to prove that the path measures of the $P(\phi)_1$-processes defined for Brownian motion in Theorem 5.15 and for jump Lévy processes in Theorem 5.31 are Gibbs measures.

First, we choose the case of continuous paths when the reference measure is a Brownian bridge measure. Recall the notation

$$\Pi_t(x) = \frac{1}{(2\pi t)^{d/2}} \exp\left(-\frac{|x|^2}{2t}\right).$$

For any interval $I = [s, t] \subset \mathbb{R}$, define

$$\mathscr{W}_I^{x,y} = \Pi_{t-s}(y - x) \mathcal{W}_I^{x,y}, \tag{5.3.1}$$

where $\mathcal{W}_I^{x,y}$ denotes the Brownian bridge measure. Although $\mathscr{W}_I^{x,y}$ is a finite measure on $C(I, \mathbb{R}^d)$, it is not a probability measure. However, due to the normalizing factor it makes no difference in the definition of finite volume Gibbs measures which measure is used, and we adopt $\mathscr{W}_I^{x,y}$ as it has some advantages in calculations.

Definition 5.53 (Itô bridge). Let V be an admissible potential. For any bounded $I \subset \mathbb{R}$ and $x, y \in \mathbb{R}^d$, define

$$P_I^{x,y}(A) = \frac{1}{Z_I(x,y)} \int_{\mathfrak{X}} 1_A e^{-\int_I V(X_t)dt} d\mathscr{W}_I^{x,y}, \tag{5.3.2}$$

for every $A \in \mathscr{F}_I$, where

$$Z_I(x,y) = \int_{\mathfrak{X}} e^{-\int_I V(X_t)dt} d\mathscr{W}_I^{x,y}. \tag{5.3.3}$$

We call $P_I^{x,y}$ the *Itô bridge measure* for interval I and potential V.

The following gives a large class of admissible potentials V.

Lemma 5.54. *Let $V \in \mathcal{K}(\mathbb{R}^d)$. Then for every bounded interval $I \subset \mathbb{R}$,*

$$\sup_{x,y \in \mathbb{R}^d} \int_{\mathfrak{X}} e^{-\int_I V(X_t)dt} d\mathscr{W}_I^{x,y} < \infty. \tag{5.3.4}$$

Proof. This follows from Khasminskii's lemma. □

Let $V \in \mathcal{K}(\mathbb{R}^d)$ be a potential such that the Schrödinger operator $H = -\frac{1}{2}\Delta + V$ has a unique strictly positive ground state φ_0 at eigenvalue E_0. To simplify the notation, we include E_0 in the potential and denote $V - E_0$ by just V. Recall the $P(\phi)_1$-process $(X_t)_{t\in\mathbb{R}}$ with path measure $\mathcal{N}_0$ associated with H as obtained in Theorem 5.15. We will call $\mathcal{N}_0$ the *Itô measure* for the potential V. By Theorem 5.21, we have

$$\mathcal{N}_0(A) = \int_{\mathbb{R}^d} \varphi_0(x)dx \int_{\mathbb{R}^d} \varphi_0(y)dy \int_{\mathfrak{X}} 1_A e^{-\int_I V(X_t)dt} d\mathscr{W}_I^{x,y} \tag{5.3.5}$$

for every $A \in \mathscr{F}_I$.

Lemma 5.55. *For $V \in \mathcal{K}(\mathbb{R}^d)$ and any $I = [a,b] \subset \mathbb{R}$, the restriction $\mathcal{N}_0|_{\mathscr{F}_I}$ is absolutely continuous with respect to $\mathcal{W}|_{\mathscr{F}_I}$, and the density is*

$$h_I = h_I(\omega) = \varphi_0(X_a)e^{-\int_I V(X_t)dt}\varphi_0(X_b). \tag{5.3.6}$$

Proof. This follows directly from (5.3.5) since by the definition of the Wiener measure

$$\mathbb{E}_{\mathcal{W}}[f(X_a)g(X_b)h] = \int_{\mathbb{R}^d} f(x)dx \int_{\mathbb{R}^d} g(y)dy \int_{\mathfrak{X}} h_I d\mathscr{W}_I^{x,y}$$

holds for bounded measurable $f, g : \mathbb{R}^d \to \mathbb{R}$ and bounded $\mathscr{F}_I$-measurable h. □

Write

$$K_t(x,y) = \int_{\mathfrak{X}} e^{-\int_0^t V(X_s)ds} d\mathscr{W}_{[0,t]}^{x,y} \tag{5.3.7}$$

for the integral kernel of e^{-tH}, and

$$\pi_t(x,y) = \frac{K_t(x,y)}{\varphi_0(x)\varphi_0(y)} \tag{5.3.8}$$

for the probability transition kernel of the $P(\phi)_1$-process. Since

$$(e^{-tL}f)(x) = \int_{\mathbb{R}^d} \pi_t(x,y)f(y)\varphi_0^2(y)dy$$

holds, π_t is the probability transition kernel of e^{-tL} in $L^2(\mathbb{R}^d, \varphi_0^2 dx)$. Recall that L is given by (5.1.3) and (5.1.11). Then

$$P_I^{x,y}(A) = \frac{\Pi_{T_2-T_1}(x-y)}{\varphi_0(x)\varphi_0(y)\pi_{T_2-T_1}(x,y)} \mathbb{E}_{\mathscr{W}_I^{x,y}}[1_A e^{-\int_I V(X_t)dt}]. \tag{5.3.9}$$

We see now that an Itô measure indeed is a Gibbs measure in the sense of Definition 5.40.

Theorem 5.56 (Existence of Gibbs measure: case of continuous paths). *Let V be Kato-decomposable and $\mathcal{N}_0$ be the Itô measure for the potential V. Then $\mathcal{N}_0$ is a Gibbs measure for V.*

Proof. We need to check that (5.3.2) is a conditional probability $\mathcal{N}_0(A|\mathcal{T}_I)$. This follows immediately from (5.3.9) and the representation (5.3.5) of $\mathcal{N}_0$. □

We note that any two Itô measures are distinguished by their potentials.

Proposition 5.57. *Take $V_1, V_2 \in \mathcal{K}(\mathbb{R}^d)$ and their Itô measures $\mathcal{N}_0^{(1)}, \mathcal{N}_0^{(2)}$, respectively. If $V_1 - V_2$ is different from a constant outside of a set of Lebesgue measure zero, then $\mathcal{N}_0^{(1)}$ and $\mathcal{N}_0^{(2)}$ are mutually singular probability measures.*

Proof. If V_1 and V_2 differ by a nonconstant term, then the ground state ψ_1 of $H_1 = -\frac{1}{2}\Delta + V_1$ is different from the ground state ψ_2 of $H_2 = -\frac{1}{2}\Delta + V_2$. As both are continuous functions, there is $A \subset \mathbb{R}^d$ such that

$$a_1 = \mathcal{N}_0^{(1)}(X_0 \in A) = \int_A \psi_1^2(x)dx \neq \int_A \psi_2^2(x)dx = \mathcal{N}_0^{(2)}(X_0 \in A) = a_2.$$

Both $\mathcal{N}_0^{(1)}$ and $\mathcal{N}_0^{(2)}$ are stationary, and thus the ergodic theorem below implies

$$\lim_{N\to\infty} \frac{1}{n} \sum_{n=0}^{N} 1_A(X_n) = a_i \quad \text{for } \mathcal{N}_0^{(i)}\text{-almost all paths, } i = 1, 2.$$

In other words, the set $\{\omega \in \mathfrak{X} \mid \lim_{N\to\infty} \frac{1}{n} \sum_{n=0}^{N} 1_A(X_n(\omega)) = a_1\}$ is of $\mathcal{N}_0^{(1)}$-measure 1 and of $\mathcal{N}_0^{(2)}$-measure 0, i. e., $\mathcal{N}_0^{(1)}$ and $\mathcal{N}_0^{(2)}$ are mutually singular. □

Definition 5.58 (Ergodic map). A map T on a probability space $(\Omega, \mathcal{F}, P)$ is called *ergodic* whenever T is measure preserving and $TA = A$ for $A \in \mathcal{F}$ implies $P(A) = 1$ or 0.

Proposition 5.59 (Ergodic theorem). *Let T be a measure preserving map on a probability space $(\Omega, \mathcal{F}, P)$. Take $f \in L^1(\Omega, dP)$. Then the limit*

$$g(\omega) = \lim_{n\to\infty} \frac{1}{n} \sum_{j=0}^{n-1} f(T^j\omega)$$

exists for almost every $\omega \in \Omega$, and $\mathbb{E}_P[g] = \mathbb{E}_P[f]$. If, moreover, T is ergodic, then $g = \mathbb{E}_P[f]$ is constant.

Next we consider càdlàg $P(\phi)_1$-processes with path measure $\mathcal{M}_0$ as obtained in Theorem 5.31. Recall that Theorem 5.35 gives

$$\mathscr{M}_0(A) = \int_{\mathbb{R}^d} \varphi_0(x)dx \int_{\mathbb{R}^d} \varphi_0(y)dy \int_{\mathfrak{X}_D} 1_A e^{-\int_S^T V(X_t)dt} d\mathbb{P}^{x,y}_{[S,T)}, \quad 0 < S < T,\ A \in \mathscr{F}_{[S,T)}. \tag{5.3.10}$$

Theorem 5.60 (Existence of Gibbs measure: case of càdlàg paths). *Let $\Psi \in \mathcal{B}_0$ be a Bernstein function, and V a Ψ-Kato decomposable potential. Denote by P the path measure of the two-sided Lévy process generated by $-\Psi(-\frac{1}{2}\Delta)$, and consider the probability measure $\mathscr{M}_0$ of the càdlàg $P(\phi)_1$-process for the nonlocal Schrödinger operator $H = \Psi(-\frac{1}{2}\Delta) + V$. Then $\mathscr{M}_0$ is a Gibbs measure with respect to the reference measure P and potential V.*

Proof. Let $0 < S < T$, $A_1 \in \mathscr{F}_{[-T,-S)}$, $A_2 \in \mathscr{F}_{[-S,S)}$, $A_2 \in \mathscr{F}_{[S,T)}$, $B_1 \in \mathscr{F}_{[-T,-S)}$, $B_2 \in \mathscr{F}_{[S,T)}$, and $A = A_1 \cap A_2 \cap A_2$, $B = B_1 \cap B_2$. For simplicity, in the proof we write $Q = \mathscr{M}_0$. By a monotone class argument, it suffices to consider sets of the form $A \cap B$. Note that since $\mathbb{P}^{\xi,\eta}_S(\bar\omega(-S) \neq \xi) = \mathbb{P}^{\xi,\eta}_S(\bar\omega(S)- \neq \eta) = 0$, we have

$$\int_{\mathfrak{X}_D} e^{-\int_{-S}^S V(X_s(\bar\omega))\,ds} Q_S(A_2, \bar\omega)\, d\mathbb{P}^{\xi,\eta}_S(\bar\omega) = \int_{\mathfrak{X}_D} e^{-\int_{-S}^S V(X_s(\bar\omega))\,ds} 1_{A_2}(\bar\omega)\, d\mathbb{P}^{\xi,\eta}_S(\bar\omega).$$

Then the Markov property of $(X_t)_{t\in\mathbb{R}}$ yields

$$\begin{aligned}
&\int_B e^{-\int_{-T}^T V(X_s(\bar\omega))\,ds} Q_S(A, \bar\omega)\, d\mathbb{P}^{x,y}_T(\bar\omega) \\
&= \int_{\mathbb{R}^d\times\mathbb{R}^d} \left(\int_{\mathfrak{X}_D} e^{-\int_{-T}^{-S} V(X_s(\bar\omega))\,ds} 1_{A_1\cap B_1}(\bar\omega)\, d\mathbb{P}^{x,\xi}_{[-T,-S)}(\bar\omega) \right) \\
&\quad \times \left(\int_{\mathfrak{X}_D} e^{-\int_{-S}^S V(X_s(\bar\omega))\,ds} Q_S(A_2, \bar\omega)\, d\mathbb{P}^{\xi,\eta}_S(\bar\omega) \right) \left(\int_{\mathfrak{X}_D} e^{-\int_S^T V(X_s(\bar\omega))\,ds} 1_{A_3\cap B_2}(\bar\omega)\, d\mathbb{P}^{\eta,y}_{[S,T)}(\bar\omega) \right) d\xi d\eta \\
&= \int_{\mathfrak{X}_D} e^{-\int_{-T}^T V(X_s(\bar\omega))\,ds} 1_{A\cap B}(\bar\omega)\, d\mathbb{P}^{x,y}_T(\bar\omega)
\end{aligned}$$

for all $x, y \in \mathbb{R}^d$. By (5.3.10), we then obtain

$$\int_{\mathfrak{X}_D} 1_B Q_S(A \cap B, \bar\omega)\, dQ(\bar\omega) = Q(A \cap B), \tag{5.3.11}$$

verifying (5.2.22) as required. □

5.3.2 Uniqueness

The next natural question is whether only a single or multiple Gibbs measures exist for a given V. We start by an illuminating example.

Example 5.61 (Nonuniqueness). Take the potential $V_{\mathrm{osc}}(x) = \frac{1}{2}(x^2 - 1)$, $x \in \mathbb{R}$, of the one-dimensional harmonic oscillator, and consider the corresponding Schrödinger operator $H_{\mathrm{osc}} = -\frac{1}{2}\Delta + V_{\mathrm{osc}}$. Recall that its ground state is $\varphi_{\mathrm{osc}}(x) = \pi^{-1/4}e^{-x^2/2}$, and the ground state-transformed process for H_{osc} is a one-dimensional Ornstein–Uhlenbeck process. Fix $\alpha, \beta \in \mathbb{R}$ and define for $s, x \in \mathbb{R}$,

$$\psi_s^{\mathrm{l}}(x) = \frac{1}{\pi^{1/4}} \exp\left(-\frac{1}{2}(x + \alpha e^{-s})^2\right) \exp\left(\frac{\alpha e^{-s}}{2}\right)^2,$$

$$\psi_s^{\mathrm{r}}(x) = \frac{1}{\pi^{1/4}} \exp\left(-\frac{1}{2}(x + \beta e^{s})^2\right) \exp\left(\frac{\beta e^{s}}{2}\right)^2.$$

A simple calculation yields

$$e^{-tH_{\mathrm{osc}}}\psi_s^{\mathrm{l}} = \psi_{s+t}^{\mathrm{l}}, \quad e^{-tH_{\mathrm{osc}}}\psi_s^{\mathrm{r}} = \psi_{s-t}^{\mathrm{r}}, \quad (\psi_s^{\mathrm{l}}, \psi_s^{\mathrm{r}}) = e^{\alpha\beta/2}.$$

Hence

$$\int_{\mathfrak{X}} \prod_{j=1}^{n} f_j(X_{t_j}) dP_{\alpha,\beta} = e^{-\alpha\beta/2}(f_1\psi_{t_1}^{l}, e^{-(t_2-t_1)H_{\mathrm{osc}}} f_2 \cdots f_{n-1} e^{-(t_n-t_{n-1})H_{\mathrm{osc}}} f_n \psi_{t_n}^{\mathrm{r}})$$

gives rise to the finite-dimensional distributions of the probability measure $P_{\alpha,\beta}$ on $\mathfrak{X}$ of a Gaussian Markov process. This is stationary if and only if $\alpha = \beta = 0$. It can be checked by straightforward computation that $P_{\alpha,\beta}$ is a Gibbs measure for every $\alpha, \beta \in \mathbb{R}$. In this case thus, uncountably many Gibbs measures exist for the same potential.

The above example shows that a Gibbs measure for a Kato-decomposable potential may not be unique. However, Lemma 5.49 gives a simple criterion allowing to check if a Gibbs measure is the only one supported on a given subset. In order to apply it, we need to know for which $\bar{\omega} \in \mathfrak{X}$ it is true that $P_N^{x,y}(A) \to P(A)$ (with $\bar{\omega}(-N) = x$, $\bar{\omega}(N) = y$). The next lemma gives a sufficient condition for this in terms of the transition densities.

Lemma 5.62.

(1) *Let V be a Kato-decomposable potential such that the Schrödinger operator $H = -\frac{1}{2}\Delta + V$ has a ground state φ_0. Suppose that for some $\bar{\omega} \in \mathfrak{X}$,*

$$\lim_{N\to\infty} \sup_{x,y\in\mathbb{R}^d} \left| \frac{\pi_{N-T}(\bar{\omega}(-N), x)\pi_{N-T}(y, \bar{\omega}(N))}{\pi_N(\bar{\omega}(-N), \bar{\omega}(N))} - 1 \right| \varphi_0(x)\varphi_0(y) = 0 \tag{5.3.12}$$

for all $T > 0$. Then for every $T > 0$ and $A \in \mathscr{F}_T$, $P_N^{x,y}(A) \to P(A)$ as $N \to \infty$.

(2) *Let $\Psi \in \mathcal{B}_0$ be a Bernstein function and V a Ψ-Kato-decomposable potential such that the nonlocal Schrödinger operator $H = \Psi(-\frac{1}{2}\Delta) + V$ has a ground state φ_0. Suppose that for some $\bar{\omega} \in \mathfrak{X}_D$,*

$$\lim_{N\to\infty} \sup_{x,y\in\mathbb{R}^d} \left| \frac{\pi_{N-T}(\bar{\omega}(-N), x)\pi_{N-T}(y, \bar{\omega}(N)-)}{\pi_N(\bar{\omega}(-N), \bar{\omega}(N)-)} - 1 \right| \varphi_0(x)\varphi_0(y) = 0 \tag{5.3.13}$$

for all $T > 0$. Then for every $T > 0$ and $A \in \mathscr{F}_T$, $P_N^{x,y}(A) \to P(A)$ as $N \to \infty$.

Proof. We only prove part (1), the proof of (2) is similar with slight modifications as seen in Section 5.2.3. Let $A \in \mathscr{F}_T$ and take $A \subset \{\omega \in \mathfrak{X} \mid |\omega(\pm T)| < M\}$ for some $M > 1$. Fix $N > T$ and $\bar{\omega} \in \mathfrak{X}$, and put $\bar{\omega}(-N) = \xi$ and $\bar{\omega}(N) = \eta$. The Markov property of Brownian motion and the Feynman–Kac formula give

$$P_N^{\xi,\eta}(A) = \frac{1}{Z_N(\bar{\omega})} \int_{\mathbb{R}^d} dx \int_{\mathbb{R}^d} \mathbb{E}_{[-N,-T]}^{\xi,x}[e^{-\int_{-N}^{-T} V(X_s)ds}]\mathbb{E}_{[-T,T]}^{x,y}[e^{-\int_{-T}^{T} V(X_s)ds}1_A]\mathbb{E}_{[T,N]}^{y,\eta}[e^{-\int_{T}^{N} V(X_s)ds}]dy$$
$$= \int_{\mathbb{R}^d} dx \int_{\mathbb{R}^d} \frac{K_{N-T}(\xi,x)K_{N-T}(y,\eta)}{K_{2N}(\xi,\eta)} \mathbb{E}_{[-T,T]}^{x,y}[e^{-\int_{-T}^{T} V(X_s)ds}1_A]dy.$$

By the selection of A and boundedness of $K_{2T}(x,y)$, the last factor in the above formula is a bounded function of x and y with compact support, thus integrable over $\mathbb{R}^d \times \mathbb{R}^d$. The claim follows for A once we show that

$$\frac{K_{N-T}(\bar{\omega}(-N),x)K_{N-T}(y,\bar{\omega}(N))}{K_{2N}(\bar{\omega}(-N),\bar{\omega}(N))} \overset{N\to\infty}{\longrightarrow} \varphi_0(x)\varphi_0(y),$$

uniformly in $x,y \in \mathbb{R}^d$. This is equivalent to (5.3.12). For general $A \in \mathscr{F}_T$, consider $B_M = \{\omega \in \mathfrak{X} \mid \max\{|\omega(T)|, |\omega(-T)|\} < M\}$ with $M \in \mathbb{N}$, and $A_M = A \cap B_M$. Since $B_M \nearrow \mathfrak{X}$ as $M \to \infty$, for given $\varepsilon > 0$ we may take $M \in \mathbb{N}$ with $P(B_M^c) < \varepsilon$. Moreover, for A_M and B_M we find $N_0 \in \mathbb{N}$ such that for all $N > N_0$ we have $|P_N(A_M) - P(A_M)| < \varepsilon$ resp. $|P_N(B_M) - P(B_M)| < \varepsilon$. Hence $P_N(B_M^c) < 2\varepsilon$ for all $N > N_0$, and thus

$$\begin{aligned}|P_N(A) - P(A)| &= |P_N(A_M) + P_N(A \setminus B_M) - P(A_M) - P(A \setminus B_M)| \\ &\leq |P_N(A_M) - P(A_M)| + P_N(B_M^c) + P(B_M^c) \leq 4\varepsilon.\end{aligned}$$

This shows that $P_N(A) \to P(A)$. □

Now we state and prove our results on uniqueness. Recall asymptotic intrinsic ultracontractivity (AIUC) introduced in Definition 5.3.

Theorem 5.63 (Case of AIUC).

(1) *Let $H = -\frac{1}{2}\Delta + V$ such that the Schrödinger semigroup $\{e^{-tH} : t \geq 0\}$ is asymptotically intrinsically ultracontractive. Then the Itô measure $\mathcal{N}_0$ for the related $P(\phi)_1$-process is the unique Gibbs measure for V supported on $\mathfrak{X}$.*

(2) *Let $\Psi \in \mathcal{B}_0$ be a Bernstein function, $H = \Psi(-\frac{1}{2}\Delta) + V$ a nonlocal Schrödinger operator, and suppose that the potential V is such that the semigroup $\{e^{-tH} : t \geq 0\}$ is asymptotically intrinsically ultracontractive. Then the path measure $\mathscr{M}_0$ of the related càdlàg $P(\phi)_1$-process is the unique Gibbs measure for V supported on $\mathfrak{X}_D$.*

Proof. By Proposition 5.9, it follows that $\lim_{N\to\infty} |\pi_N(x,y) - 1| = 0$ uniformly in x, y. Thus (5.3.12) holds. The proof for the càdlàg case is similar. □

The above result shows that AIUC is a very strong regularity property, allowing a unique Gibbs measure to exist on the full set of paths. When AIUC does not hold, we can find a full-measure subset of paths on which Gibbs measures are unique. Recall that for an operator H the nonnegative number

$$\Lambda = \inf\{\operatorname{Spec}(H) \setminus \{\inf \operatorname{Spec}(H)\}\} - \inf \operatorname{Spec}(H)$$

is called the *spectral gap* of H.

Theorem 5.64 (Case of Kato-class).

(1) *Suppose that V is Kato-decomposable, consider $H = -\frac{1}{2}\Delta + V$, and let its unique strictly positive ground state be $\varphi_0 \in L^2(\mathbb{R}^d) \cap L^1(\mathbb{R}^d)$. Define*

$$\mathfrak{X}^* = \left\{\omega \in \mathfrak{X} \,\middle|\, \lim_{N\to\pm\infty} \frac{e^{-\Lambda|N|}}{\varphi_0(\omega(N))} = 0\right\}. \tag{5.3.14}$$

Then $\mathcal{N}_0$ is the unique Gibbs measure for V supported on $\mathfrak{X}_0$.

(2) *Let $\Psi \in \mathcal{B}_0$ be a Bernstein function, consider $H = \Psi(-\frac{1}{2}\Delta) + V$, and let its unique strictly positive ground state be $\varphi_0 \in L^2(\mathbb{R}^d) \cap L^1(\mathbb{R}^d)$. Define*

$$\mathfrak{X}_D^* = \left\{\omega \in \mathfrak{X}_D \,\middle|\, \lim_{N\to\pm\infty} \frac{e^{-\Lambda|N|}}{\varphi_0(\omega(N))} = 0\right\}. \tag{5.3.15}$$

Then $\mathcal{M}_0$ is the unique Gibbs measure for V supported on $\mathfrak{X}_D^*$.

Proof. With the projection 1_{φ_0} used in the proof of Proposition 5.9, write $L_t = e^{-tH} - 1_{\varphi_0}$. This is an integral operator with kernel $\tilde{K}_t(x,y) = K_t(x,y) - \varphi_0(x)\varphi_0(y)$. As $\Lambda > 0$ by assumption, we have $\|L_t\|_{2,2} = e^{-\Lambda t}$. In order to estimate $\tilde{K}_t$, note that

$$\sup_{x,y\in\mathbb{R}^d} |\tilde{K}_t(x,y)| = \sup_{f\in L^1,\|f\|_1=1} \left\| \int_{\mathbb{R}^d} \tilde{K}_t(x,y)f(y)dy \right\|_\infty = \|L_t\|_{1,\infty},$$

and since $e^{-tH}1_{\varphi_0} = 1_{\varphi_0}e^{-tH} = P_{\varphi_0}$ for all $t > 0$, we get

$$\|L_t\|_{1,\infty} = \|e^{-H}(e^{-(t-2)H} - 1_{\varphi_0})e^{-H}\|_{1,\infty} \le \|e^{-H}\|_{2,\infty}\|L_{t-2}\|_{2,2}\|e^{-H}\|_{1,2}.$$

By Theorem 4.107, both $\|e^{-H}\|_{2,\infty}$ and $\|e^{-H}\|_{1,2}$ are finite, thus it follows that for every $t \le N$,

$$|K_{N-t}(x,y) - \varphi_0(x)\varphi_0(y)| \le C_t e^{-\Lambda N},$$

where $C_t = \|e^{-H}\|_{2,\infty}\|e^{-H}\|_{1,2}e^{\Lambda(2+t)}$ is independent of x, y and N. In terms of π_N, this implies that for all $\bar{\omega} \in \mathfrak{X}_0$,

$$|\pi_{N-T}(\bar{\omega}(-N), x) - 1|\varphi_0(x) \le \frac{C_T e^{-\Lambda N}}{\varphi_0(\bar{\omega}(-N))} \to 0,$$

$$|\pi_{N-T}(y, \bar{\omega}(N)) - 1|\varphi_0(y) \le \frac{C_T e^{-\Lambda N}}{\varphi_0(\bar{\omega}(N))} \to 0,$$

$$|\pi_{2N}(\bar{\omega}(-N), \bar{\omega}(N)) - 1| \le \frac{C_0 e^{-2\Lambda N}}{(\varphi_0(\bar{\omega}(-N))\varphi_0(\bar{\omega}(N)))} \to 0$$

as $N \to \infty$. From this, (5.3.12) can be deduced. It remains to show that P is actually supported on $\mathfrak{X}_0$. By time reversibility, it suffices to show that

$$P\left(\limsup_{N\to\infty} \frac{e^{-\Lambda N}}{\varphi_0(\omega(N))} > n\right) = 0, \quad n \in \mathbb{N}.$$

To prove this, note that by the stationarity of P,

$$P\left(\frac{e^{-\Lambda N}}{\varphi_0(\omega(N))} > n\right) = \int_{\mathbb{R}^d} 1_{\{\varphi_0 < e^{-(\Lambda N)/n}\}}\varphi_0^2(x)dx \le \frac{e^{-\Lambda N}}{n}\|\varphi_0\|_{L^1}.$$

The right-hand side of the last expression is summable in N for each n, thus an application of the Borel–Cantelli lemma completes the proof. The proof of part (2) is similar. □

Remark 5.65. The additional assumption that $\varphi_0 \in L^1(\mathbb{R}^d)$ is reasonable and verified on many examples of interest. In case of standard Schrödinger operators, a large class with $\varphi_0 \in L^1(\mathbb{R}^d)$ is dealt with in Lemma 4.177.

In Theorem 5.64, the subset of paths $\mathfrak{X}^*$ is seen to depend directly on the decay of φ_0 at infinity. Thus for given conditions on V, it is interesting to know the asymptotic behavior of φ_0. Suppose that

$$C_2|x|^{2a} \le V(x) \le C_1|x|^{2a}, \quad a > 1, \tag{5.3.16}$$

outside of a compact set of $\mathbb{R}^d$; we say in this case that V is a *superquadratic* potential. Then there exist constants b_1, b_2, D_1, D_2 such that

$$D_2 e^{-b_2|x|^{a+1}} \le \varphi_0(x) \le D_1 e^{-b_1|x|^{a+1}}, \tag{5.3.17}$$

see Corollaries 4.178 and 4.183. Also, for superquadratic potentials V, the corresponding Schrödinger semigroup is intrinsically ultracontractive (see Theorem 5.5). It can be checked directly that all potentials considered in Lemma 4.177 are Kato-decomposable.

We complete this section by discussing some examples for standard and fractional Schrödinger operators and the supports of the related $P(\phi)_1$-processes.

Example 5.66. Let V be a nonconstant polynomial bounded below. This in particular implies that the degree of V is even, so $H = -\frac{1}{2}\Delta + V$ has a unique ground state and a positive spectral gap. From Theorem 5.63 and Example 5.61, it follows that a Gibbs measure for V is unique if and only if the degree of V is greater than 2.

Example 5.67. Let $V(x) = |x|^{2\beta}$ with $\beta > 0$. Again, $H = -\frac{1}{2}\Delta + V$ has a unique ground state φ_0 and a spectral gap Λ. In the case of $\beta \geq 1$, Theorem 5.63 and Example 5.61 completely solve the question of uniqueness. In the case of $0 < \beta < 1$, (4.3.82) implies that Theorem 5.64 applies, and thus uniqueness holds on a set $\mathfrak{X}^* = \mathfrak{X}^*(\beta)$ given by (5.3.14). By Corollaries 4.179 and 4.183, sharp estimates can be obtained on this supporting set. Indeed,

$$\left\{\omega \in \mathfrak{X} \,\middle|\, \limsup_{N\to\pm\infty} \frac{|\omega(\pm N)|^{\beta+1}}{N} < \frac{\Lambda}{b_1}\right\} \subset \mathfrak{X}^*(\beta) \subset \left\{X \in \mathfrak{X} \,\middle|\, \limsup_{N\to\pm\infty} \frac{|\omega(\pm N)|^{\beta+1}}{N} < \frac{\Lambda}{b_2}\right\}$$

with constants b_1, b_2. Note that, counterintuitively, $\mathfrak{X}^*(\beta)$ becomes smaller as β increases to 1. Indeed, a fast growing potential should bring the path back to stationarity more quickly, as is seen in the case where $\beta > 1$. What happens for $\beta < 1$ results from (5.3.14), which turns out to be too crude for this range of exponents.

Example 5.68. Let $H = (-\Delta)^{\alpha/2} + V$ be a fractional Schrödinger operator with potential

$$V(x) = C_0|x|^\delta + \frac{C_1}{|x - x_1|^{\beta_1}} - \frac{C_2}{|x - x_2|^{\beta_2}}$$

where $C_0 > 0$, $C_1, C_2 \geq 0$, $x_1, x_2 \in \mathbb{R}^d$ and $\delta > 0$, $\beta_1, \beta_2 \geq 0$. It is straightforward to check that if $0 < \beta_1, \beta_2 < \alpha < d$ or $0 < \beta_1, \beta_2 < 1 = d \leq \alpha$, then V is fractional Kato-decomposable. An immediate consequence of Theorem 5.63 is that the $P(\phi)_1$-measure $\mathscr{M}_0$ is the only Gibbs measure corresponding to the fractional $P(\phi)_1$-process and the potential V supported on $\mathfrak{X}_D$. By some extra work, it is possible to show that $\mathscr{M}_0$ is, in fact, supported by a subset $\mathfrak{X}_D$ consisting of all path functions ω such that for every $\theta > 0$,

$$|\omega(N)| = o\left(|N|^{\frac{1+\theta}{\delta+d+\alpha}}\right).$$

Example 5.69. Let $H = (-\Delta)^{\alpha/2} + V$, with a fractional Kato-decomposable potential V whose positive part $V^+ \equiv 0$ and negative part $V^-(x) \to 0$ as $|x| \to \infty$, be such that a spectral gap $\Lambda > 0$ exists. Theorem 5.64 implies that $\mathscr{M}_0$ is the unique Gibbs measure supported on the subspace of paths such that

$$|\omega(N)| = o\left(\exp\left(\frac{\Lambda}{1+\alpha}|N|\right)\right).$$

Again, with some extra work, it is possible to show that $\mathscr{M}_0$ is actually supported on a subset of paths given by the growth condition

$$|\omega(N)| = o\left(|N|^{\frac{1+\theta}{1+\alpha}}\right), \quad \theta > 0.$$

5.4 Gibbs measures for external and pair interaction potentials: direct method

In this section we discuss the more involved situation when the pair interaction potential W is nonzero. Throughout this section we consider Gibbs measures corresponding to them only on Brownian paths.

Let $\mathfrak{X}'$ be the space of functions $\omega : \mathbb{R} \to \mathbb{R}^d$, $t \mapsto \omega(t)$, that are continuous in every $t \neq 0$, and for which a left and a right limit exist in $t = 0$. For $\tau > 0$, define the *shrinking* operator $\theta_\tau : \mathfrak{X} \to \mathfrak{X}'$ by

$$(\theta_\tau(\omega))(t) = \begin{cases} \omega(t-\tau) & \text{if } t \geq 0, \\ \omega(t+\tau) & \text{if } t < 0. \end{cases} \tag{5.4.1}$$

Let $V : \mathbb{R}^d \to \mathbb{R}$ be an external potential, and $W : \mathbb{R}^d \times \mathbb{R}^d \times \mathbb{R} \to \mathbb{R}$ be a pair interaction potential in the sense of Definition 5.38. Write

$$W_T(X(\omega)) = \int_{-T}^{T}\int_{-T}^{T} W(X_s(\omega), X_t(\omega), s-t)dsdt \tag{5.4.2}$$

whenever the right-hand side is well defined.

We consider the following regularity conditions on the potentials V and W.

Assumption 5.70.

(1) Potential V is Kato-decomposable such that the Schrödinger operator $H = -\frac{1}{2}\Delta + V$ has a unique ground state $\varphi_0 \in L^2(\mathbb{R}^d) \cap L^1(\mathbb{R}^d)$ with the property

$$H\varphi_0 = 0. \tag{5.4.3}$$

(The ground state eigenvalue E_0 is included by a shift in V.)

(2) There exists $C_\infty \geq 0$ such that

$$\sup_{\omega \in \mathfrak{X}} \int_{-\infty}^{+\infty} |W(X_t(\omega), X_s(\omega), t)|dt \leq C_\infty. \tag{5.4.4}$$

(3) There exist $C, D \geq 0$ such that with

$$C < \liminf_{|x|\to\infty} V(x), \tag{5.4.5}$$

and the bound

$$-W_T(X(\omega)) \leq -W_T(X(\theta_\tau \omega)) + C\tau + D \tag{5.4.6}$$

holds for all $T, \tau > 0$ and all $\omega \in \mathfrak{X}$.

Recall that $\mathcal{N}_0$ denotes the path measure of the $P(\phi)_1$-process with potential V and stationary measure $\varphi_0^2(x)dx$. This measure will be used as a new reference measure whose modifications by densities dependent on W will be next introduced.

For the remainder of this section, we prove the following theorem.

Theorem 5.71 (Existence of Gibbs measure). *Under Assumption* 5.70, *there exists a Gibbs measure for the potentials V and W.*

We begin by the following result.

Proposition 5.72. *With Assumption* 5.70, *let Q_T^h be the finite volume Gibbs measure for the potentials V and W, with stochastic boundary condition $h = h_T = \varphi_0(X_{-T})\varphi_0(X_T)$. Then on $\mathscr{F}_T$ we have*

$$dQ_T^h = \frac{1}{Z_T} e^{-\int_{-T}^T ds \int_{-T}^T W(X_s, X_t, s-t)ds} d\mathcal{N}_0,$$

with $Z_T = \mathbb{E}_{\mathcal{N}_0}[e^{-\int_{-T}^T ds \int_{-T}^T W(X_s, X_t, s-t)dt}]$.

Proof. By the Feynman–Kac formula (5.1.45), we find for all $-T < s_1 < \cdots < s_n < T$ that

$$\mathbb{E}_{\mathcal{N}_0}[f_1(X_{s_1})\cdots f_n(X_{s_n})] = \int_{\mathbb{R}^d} \mathbb{E}^x[\varphi_0(X_{-T})f_1(X_{s_1})\cdots f_n(X_{s_n})\varphi_0(X_T)e^{-\int_{-T}^T V(X_s)ds}]dx.$$

By an approximation argument, the above equality can be extended to all bounded $\mathscr{F}_T$-measurable functions. Since $e^{-\int_{-T}^T ds \int_{-T}^T W(X_s, X_t, s-t)dt}$ is such a function, we conclude that

$$\mathbb{E}_{Q_T^h}[g_1(X_{s_1})\cdots g_n(X_{s_n})] = \mathbb{E}_{\mathcal{N}_0}\left[e^{-\int_{-T}^T ds \int_{-T}^T W(X_s, X_t, s-t)dt} g_1(X_{s_1})\cdots g_n(X_{s_n})\right],$$

for all bounded functions $g_1, \ldots, g_n$, and all $-T < s_1 < \cdots < s_n < T$, and the claim follows. □

In order to prove the existence of a Gibbs measure for the potentials V and W, we apply Proposition 5.48. To be able to use this result, we need to prove that the family

$$dQ_T = \frac{1}{Z_T} e^{-\int_{-T}^T ds \int_{-T}^T W(X_s, X_t, s-t)dt} d\mathcal{N}_0 \tag{5.4.7}$$

from Proposition 5.72 is relatively compact. By Proposition 5.51, it will suffice to show local uniform domination. We start by showing that the one-point distributions Q_T are tight as $T \to \infty$.

Proposition 5.73. *Define Q_T as in* (5.4.7), *and let Assumption* 5.70 *hold. Then for every $\varepsilon > 0$ and $S > 0$, there exists $R > 0$ such that*

$$\sup_{T>0} Q_T(|X_u| > R) \leq \varepsilon$$

for all $|u| \leq S$.

Proof. First, we relate $Q_T(|X_u| > R)$ to $Q_T(|X_0| > R)$. Assume that $u > 0$; the case $u < 0$ can be treated similarly. Using a formal, though suggestive, notation, we obtain

$$\int_{-T}^{T}\int_{-T}^{T} = \left(\int_{-T}^{-T+u} + \int_{-T+u}^{T+u} - \int_{T}^{T+u}\right)^2$$
$$= \int_{-T+u}^{T+u}\int_{-T+u}^{T+u} + \int_{-T}^{-T+u}\int_{-T}^{-T+u} + \int_{T}^{T+u}\int_{T}^{T+u} - 2\int_{-T}^{-T+u}\int_{T}^{T+u} + 2\int_{-T}^{-T+u}\int_{-T+u}^{T+u} - 2\int_{T}^{T+u}\int_{-T+u}^{T+u}.$$

Inserting the pair interaction potentials in the double integrals, we see that the last integral in the first line is bounded in absolute value by $\|W\|_\infty u^2$, and the same is true for the first two terms in the second line, with a factor 2 in front of the second term. The last two terms in the third line are bounded by $2C_\infty u$, where we used (5.4.4). Thus,

$$\sup_{\omega\in\mathfrak{X}} |W_T(X(\omega)) - W_{T+u}(X(\omega))| = 4\|W\|_\infty u^2 + 4C_\infty u = C_1(u), \tag{5.4.8}$$

and we find

$$\begin{aligned} Z_T Q_T(|X_u| > R) &= \mathbb{E}_{\mathcal{N}_0}[1_{\{|X_u|>R\}} e^{-W_T(X)}] \le e^{C_1(u)} \mathbb{E}_{\mathcal{N}_0}[1_{\{|X_u|>R\}} e^{-W_{T+u}(X)}] \\ &= e^{C_1(u)} \mathbb{E}_{\mathcal{N}_0}[1_{\{|X_0|>R\}} e^{-W_T(X)}] = e^{C_1(u)} Z_T Q_T(|X_0| > R). \end{aligned}$$

The equality of the first and second lines above follows from the stationarity of the measure $\mathcal{N}_0$. Thus the claim will be shown once we show it for $u = 0$. We proceed in several steps.

Step 1. By Theorem 5.15, we have

$$Q_T(|X_0| > R) = \frac{1}{Z_T} \int_{|y|>R} \varphi_0^2(y) \mathbb{E}_{\mathcal{N}_0^y}[e^{-W_T(X)}] dy. \tag{5.4.9}$$

In the next steps we will show that there exist $K, r > 0$ such that for every $T > 0$ and $y \in \mathbb{R}^d$,

$$\mathbb{E}_{\mathcal{N}_0^y}[e^{-W_T(X)}] \le \frac{K}{\varphi_0(y)} \inf_{|z|\le r} \mathbb{E}_{\mathcal{N}_0^z}[e^{-W_T(X)}]. \tag{5.4.10}$$

Assume for the moment that (5.4.10) holds. We can estimate

$$1 \ge Q_T(|X_0| \le r) = \frac{1}{Z_T} \int_{|y|\le r} \mathbb{E}_{\mathcal{N}_0^y}[e^{-W_T}] \varphi_0^2(y) dy \ge \frac{1}{Z_T} C_r \inf_{|z|\le r} \mathbb{E}_{\mathcal{N}_0^z}[e^{-W_T}] \inf_{|z|\le r} \varphi_0^2(z),$$

where C_r is the volume of a ball with radius r. Thus $\inf_{|z|\le r} \mathbb{E}_{\mathcal{N}_0^z}[e^{-W_T}] \le \tilde{C}_r Z_T$, where the constant only depends on r but not on T. Using (5.4.9) and (5.4.10) gives

$$Q_T(|X_0| > R) \le K\tilde{C}_r \int_{|y|>R} \varphi_0(y)dy. \tag{5.4.11}$$

Since $\varphi_0 \in L^1(\mathbb{R}^d)$, the right-hand side above vanishes when $R \to \infty$, and the claim follows.

Step 2. To prove (5.4.10) we change the probability space. Recall the definition of $\mathfrak{X}'$ at the beginning of the section, and consider

$$\begin{aligned} J : \mathfrak{X}' &\to \mathscr{X}^2 = C([0,\infty), \mathbb{R}^d \times \mathbb{R}^d), \\ (\omega(t))_{t\in\mathbb{R}} &\mapsto \begin{cases} (\omega(t), \omega(-t)), & t > 0, \\ (\lim_{t\to 0+} \omega(t), \lim_{t\to 0-} \omega(t)), & t = 0. \end{cases} \end{aligned} \tag{5.4.12}$$

On fixing the value of $\omega \in \mathfrak{X}'$ at $t = 0$ the map J becomes a bijection; this value can be chosen, for instance, equal to the left limit. Write $\boldsymbol{\omega} = (\omega', \omega'')$ for the elements of $\mathscr{X}^2$. When we omit the variable $\boldsymbol{\omega}$, we write $\boldsymbol{X} = (\boldsymbol{X}_t', \boldsymbol{X}_t'')_{t\ge 0}$, where $\boldsymbol{X}_t'(\boldsymbol{\omega}) = \boldsymbol{X}_t(\omega')$ and $\boldsymbol{X}_t''(\boldsymbol{\omega}) = \boldsymbol{X}_t(\omega'')$.

The image measure of $\mathcal{N}_0^z$ under J can be described explicitly. For $\mathbf{z} = (z', z'') \in \mathbb{R}^d \times \mathbb{R}^d$, denote by $\mathcal{N}_0^{\mathbf{z}}$ the measure of the $\mathbb{R}^d \times \mathbb{R}^d$-valued $P(\phi)_1$-process with potential $\tilde{V}(x,y) = V(x) + V(y)$, conditional on $\boldsymbol{\omega}_0 = \mathbf{z}$. Write $\tilde{\mathscr{F}}_T$ for the σ-field over $\mathscr{X}^2$ generated by point evaluations at points inside $[0, T]$. Then for every $\tilde{\mathscr{F}}_T$-measurable bounded function f,

$$\mathbb{E}_{\mathcal{N}_0^{\mathbf{z}}}[f] = \frac{1}{\varphi_0(z')\varphi_0(z'')} \int_{\mathscr{X}^2} e^{-\int_0^T (V(\boldsymbol{X}_s')+V(\boldsymbol{X}_s''))ds} f(\boldsymbol{\omega})\varphi_0(\boldsymbol{X}_T')\varphi_0(\boldsymbol{X}_T'')\, d\mathcal{W}^{\mathbf{z}}(\boldsymbol{\omega}). \tag{5.4.13}$$

Here, $\mathcal{W}^{\mathbf{z}}$ denotes a $(2d)$-dimensional Wiener measure starting at $\mathbf{z}$. Equation (5.4.13) follows from the fact that the ground state of the Schrödinger operator

$$H \otimes 1 + 1 \otimes H = -\frac{1}{2}\Delta_{z'} - \frac{1}{2}\Delta_{z''} + V(z') + V(z'') \tag{5.4.14}$$

is given by $\varphi_0(z')\varphi_0(z'')$, combined with (5.1.43). Note that the bottom of the spectrum of (5.4.14) is zero by shifting by a constant.

The Markov property and time reversibility of $\boldsymbol{X}$ imply that for every $z \in \mathbb{R}^d$, $\mathcal{N}_0^{(z,z)}$ is the image of $\mathcal{N}_0^z$ under J, i. e., $\mathbb{E}_{\mathcal{N}_0^z}[f] = \mathbb{E}_{\mathcal{N}_0^{(z,z)}}[f \circ J^{-1}]$, for all bounded functions f. Write $\tilde{W}_T(\boldsymbol{X}) = W_T(J^{-1}(\boldsymbol{X}))$. Then, in particular,

$$\mathbb{E}_{\mathcal{N}_0^z}[e^{-W_T(X)}] = \mathbb{E}_{\mathcal{N}_0^{(z,z)}}[e^{-\tilde{W}_T(\boldsymbol{X})}]. \tag{5.4.15}$$

From (5.4.10), it can be seen that the quantity we want to estimate is the expectation of $e^{-\tilde{W}_T(\boldsymbol{X})}$ with respect to $\mathcal{N}_0^{(z,z)}$. It is easy to check that

$$\tilde{W}_T(\boldsymbol{X}) = -\int_0^T ds \int_0^T \Big\{ W(\boldsymbol{X}'_s - \boldsymbol{X}'_t, |s-t|) + W(\boldsymbol{X}''_s - \boldsymbol{X}''_t, |s-t|)$$
$$+ W(\boldsymbol{X}'_s - \boldsymbol{X}''_t, |s+t|) + W(\boldsymbol{X}''_s - \boldsymbol{X}'_t, |s+t|) \Big\} dt. \tag{5.4.16}$$

Step 3. Next note that the shrinking operator is mapped into the time shift on $\mathscr{X}^2$ under J. In other words, $\tilde{\theta}_\tau = J\theta_\tau J^{-1}$ is the usual time shift that maps $(\omega(t))_{t\geq 0}$ into $(\omega(t+\tau))_{t\geq 0}$. Thus in the new representation (5.4.6) becomes

$$-\tilde{W}_T(\boldsymbol{X}(\omega)) \leq -\tilde{W}_T(\boldsymbol{X}(\tilde{\theta}_\tau \omega)) + C\tau + D \tag{5.4.17}$$

for all $\boldsymbol{X} \in \mathscr{X}^2$ and all $T, \tau > 0$. Our strategy is to use (5.4.17) together with the strong Markov property of $\mathcal{N}_0^{\boldsymbol{z}}$. For $r > 0$, let

$$\tau_r = \inf\{t \geq 0 \mid |\boldsymbol{X}_t| \leq r\}$$

be the first hitting time of the centered ball with radius r. This is a stopping time and we write $\mathscr{F}_{\tau_r}$ for the associated σ-field. Then for every $\boldsymbol{z} \in \mathbb{R}^d \times \mathbb{R}^d$,

$$\begin{aligned}
\mathbb{E}_{\mathcal{N}_0^{\boldsymbol{z}}}[e^{-\tilde{W}_T(\boldsymbol{X})}] &= \mathbb{E}_{\mathcal{N}_0^{\boldsymbol{z}}}[\mathbb{E}_{\mathcal{N}_0^{\boldsymbol{z}}}[e^{-\tilde{W}_T(\boldsymbol{X})}|\mathscr{F}_{\tau_r}]] \leq \mathbb{E}_{\mathcal{N}_0^{\boldsymbol{z}}}[\mathbb{E}_{\mathcal{N}_0^{\boldsymbol{z}}}[e^{-\tilde{W}_T(\boldsymbol{X}\circ\tilde{\theta}_{\tau_r})}e^{C\tau_r+D}|\mathscr{F}_{\tau_r}]] \\
&= \mathbb{E}_{\mathcal{N}_0^{\boldsymbol{z}}}[e^{C\tau_r+D}\mathbb{E}_{\mathcal{N}_0^{\boldsymbol{z}}}[e^{\tilde{W}_T(\boldsymbol{X}\circ\tilde{\theta}_{\tau_r})}|\mathscr{F}_{\tau_r}]] = \mathbb{E}_{\mathcal{N}_0^{\boldsymbol{z}}}[e^{C\tau_r+D}\mathbb{E}_{\mathcal{N}_0^{\boldsymbol{X}_{\tau_r}}}[e^{-\tilde{W}_T}]] \\
&\leq \sup_{|y|\leq r} \mathbb{E}_{\mathcal{N}_0^{y}}[e^{-\tilde{W}_T}]\mathbb{E}_{\mathcal{N}_0^{\boldsymbol{z}}}[e^{C\tau_r+D}].
\end{aligned} \tag{5.4.18}$$

To obtain (5.4.10), we need a good estimate of the second factor at the right-hand side of (5.4.18) and estimate the supremum in the first factor against an infimum. This will be done in the next two steps below.

Step 4. We show that there exist $r, \gamma > 0$, such that for all $\boldsymbol{z} \in \mathbb{R}^d \times \mathbb{R}^d$,

$$\mathbb{E}_{\mathcal{N}_0^{\boldsymbol{z}}}[e^{C\tau_r}] \leq 1 + \frac{C\|\varphi_0\|_\infty}{\gamma}\left(\frac{1}{\varphi_0(z')} + \frac{1}{\varphi_0(z'')}\right). \tag{5.4.19}$$

Above C is the constant from (5.4.6). To show (5.4.19), we choose $0 < \gamma < \liminf_{|x|\to\infty} V(x) - C$, and we take r so large that $V(x) > C + \gamma$ for all $x \in \mathbb{R}^d$ with $|x| > r/\sqrt{2}$. This is possible by the definition of C. Clearly,

$$\left\{\boldsymbol{z} \in \mathbb{R}^d \times \mathbb{R}^d \mid |\boldsymbol{z}| > r\right\} \subset \left\{\boldsymbol{z} \in \mathbb{R}^d \times \mathbb{R}^d \mid |z'| > r/\sqrt{2}\} \cup \{\boldsymbol{z} \in \mathbb{R}^d \times \mathbb{R}^d \mid |z''| > r/\sqrt{2}\right\},$$

and with (5.4.13) it follows that

$$\varphi_0(z')\varphi_0(z'')\mathbb{E}_{\mathcal{N}_0^{\boldsymbol{z}}}[\tau_r > t] = \int_{\mathscr{X}^2} e^{-\int_0^t (V(\boldsymbol{X}'_s)+V(\boldsymbol{X}''_s))ds} 1_{\{|\boldsymbol{X}_s|>r,\ s\leq t\}} \varphi_0(\boldsymbol{X}'_t)\varphi_0(\boldsymbol{X}''_t)\, d\mathcal{W}^{\boldsymbol{z}}(\omega)$$

$$\leq \int_{\mathscr{X}^2} e^{-\int_0^t (V(X_s')+V(X_s''))ds} 1_{\{|X_s'|>r/\sqrt{2},\, s\leq t\}\cup\{|X_s''|>r/\sqrt{2},\, s\leq t\}} \varphi_0(X_t')\varphi_0(X_t'')\, d\mathcal{W}^{z'}(\omega')\, d\mathcal{W}^{z''}(\omega'')$$

$$= \varphi_0(z'') \int_{\mathscr{X}^2} e^{-\int_0^t V(X_s')ds} 1_{\{|X_s'|>r/\sqrt{2}, s\leq t\}} \varphi_0(X_t')\, d\mathcal{W}^{z'}(\omega')$$

$$+ \varphi_0(z') \int_{\mathscr{X}^2} e^{-\int_0^t V(X_s'')ds} 1_{\{|X_s''|>r/\sqrt{2}, s\leq t\}} \varphi_0(X_t'')\, d\mathcal{W}^{z''}(\omega'')$$

$$\leq (\varphi_0(z') + \varphi_0(z''))\|\varphi_0\|_\infty e^{-(C+\gamma)t}.$$

The second equality above is due to the eigenvalue equation $e^{-tH}\varphi_0 = \varphi_0$ and the Feynman–Kac formula. It follows that

$$\mathbb{E}_{\mathcal{N}_0^z}[\tau_r > t] \leq \left(\frac{1}{\varphi_0(z')} + \frac{1}{\varphi_0(z'')}\right)\|\varphi_0\|_\infty e^{-(C+\gamma)t},$$

and using the equality

$$\mathbb{E}_{\mathcal{N}_0^z}[e^{C\tau_r}] = 1 + \int_0^\infty Ce^{Ct}\mathbb{E}_{\mathcal{N}_0^z}[\tau_r > t]dt, \tag{5.4.20}$$

we arrive at (5.4.19).

Step 5. Let $r > 0$ be as in Step 4. We show that there exists $M > 0$ such that

$$\sup_{|\mathbf{z}|\leq r} \mathbb{E}_{\mathcal{N}_0^z}[e^{-\tilde{W}_T}] \leq M \inf_{|\mathbf{z}|\leq r} \mathbb{E}_{\mathcal{N}_0^z}[e^{-\tilde{W}_T}] \tag{5.4.21}$$

uniformly in $T > 0$. Denote by $p_t(\mathbf{z},\mathbf{y})$ the probability transition kernel from $\mathbf{y}$ to $\mathbf{z}$ in time t under $\mathcal{N}_0$, i. e., define p_t through the relation

$$\mathcal{N}_0^{\mathbf{z}}(\mathbf{X}_t \in A) = \int_{\mathbb{R}^d\times\mathbb{R}^d} p_t(\mathbf{z},\mathbf{y})1_A(\mathbf{y})d\mathbf{y}$$

for all $\mathbf{z} \in \mathbb{R}^d \times \mathbb{R}^d$ and measurable $A \subset \mathbb{R}^d \times \mathbb{R}^d$. By (5.4.13) and the Feynman–Kac formula, we have

$$p_t(\mathbf{z},\mathbf{y}) = \frac{\varphi_0(z')\varphi_0(z'')}{\varphi_0(y')\varphi_0(y'')}K_t(y',z')K_t(y'',z''), \tag{5.4.22}$$

where K_t is the integral kernel of e^{-tH}. Note that φ_0 and K_t are both strictly positive and uniformly bounded. The latter statement follows from the boundedness from L^1 to L^∞ of the operator e^{-tH}; see Theorem 4.107. Thus they are bounded and nonzero on compact sets, and for every $R > 0$,

$$S_t(R,r) = \sup\left\{\frac{p_t(\mathbf{x},\mathbf{z})}{p_t(\mathbf{y},\mathbf{z})}\,\middle|\,|\mathbf{x}| \le r, |\mathbf{y}| \le r, |\mathbf{z}| \le R\right\}$$

is finite. Defining $\tilde{W}_T^1$ like in (5.4.16) with the integrals starting at 1 rather than at 0, we see that there is a constant C_1 independent of T such that

$$-\tilde{W}_T(\boldsymbol{X}(\omega)) - C_1 \le -\tilde{W}_T^1(\boldsymbol{X}(\omega)) \le -\tilde{W}_T(\boldsymbol{X}(\omega)) + C_1$$

for all $\omega \in \mathscr{X}^2$ and $T > 0$. Define $B = \{|\boldsymbol{X}_1| < R\}$. Then for every $\boldsymbol{x}$ with $|\boldsymbol{x}| < r$, we have

$$\mathbb{E}_{\mathcal{N}_0^x}[e^{-\tilde{W}_T}] \le e^{C_1}\mathbb{E}_{\mathcal{N}_0^x}[1_B e^{-\tilde{W}_T^1}] + e^{C+D}\mathbb{E}_{\mathcal{N}_0^x}[1_{B^c} e^{-\tilde{W}_T\circ\tilde{\theta}_1}]. \tag{5.4.23}$$

Defining $\bar{W}_T$ as in (5.4.16) with $|s + t + 2|$ appearing instead of $|s + t|$ everywhere, in the first term at the right-hand side of (5.4.23) we find

$$\begin{aligned}\mathbb{E}_{\mathcal{N}_0^x}[1_B e^{-\tilde{W}_T^1}] &= \int_{|\mathbf{z}|<R} p_1(\mathbf{x},\mathbf{z})\mathbb{E}_{\mathcal{N}_0^z}[e^{-\bar{W}_{T-1}}]d\mathbf{z} \le S_1(R,r)\int_{|\mathbf{z}|\le R} p_1(\mathbf{y},\mathbf{z})\mathbb{E}_{\mathcal{N}_0^z}[e^{-\bar{W}_{T-1}}]d\mathbf{z} \\ &= S_1(R,r)\mathbb{E}_{\mathcal{N}_0^y}[1_B e^{-\tilde{W}_T^1}] \le S_1(R,r)e^{C_0}\mathbb{E}_{\mathcal{N}_0^y}[e^{-\tilde{W}_T}]\end{aligned} \tag{5.4.24}$$

for every $\boldsymbol{y}$ with $|\boldsymbol{y}| \le r$. Turning to the second term on the right-hand side of (5.4.23), (5.4.18) and (5.4.19) give

$$\begin{aligned}\mathbb{E}_{\mathcal{N}_0^z}[1_{B^c} e^{-\tilde{W}_T\circ\tilde{\theta}_1}] &= \int_{|\mathbf{y}|>R} p_1(\mathbf{z},\mathbf{y})\mathbb{E}_{\mathcal{N}_0^y}[e^{-\tilde{W}_T}]d\mathbf{y} \\ &\le \sup_{|\mathbf{x}|\le r}\mathbb{E}_{\mathcal{N}_0^x}[e^{-\tilde{W}_T}]\int_{|\mathbf{y}|>R} p_1(\mathbf{z},\mathbf{y})\mathbb{E}_{\mathcal{N}_0^y}[e^{C\tau_r + D}]d\mathbf{y} \\ &\le \sup_{|\mathbf{x}|\le r}\mathbb{E}_{\mathcal{N}_0^x}[e^{-\tilde{W}_T}]e^D\int_{|\mathbf{y}|>R} p_1(\mathbf{z},\mathbf{y})\left(1 + \frac{C\|\varphi_0\|_\infty}{\gamma}\left(\frac{1}{\varphi_0(y')} + \frac{1}{\varphi_0(y'')}\right)\right)d\mathbf{y}.\end{aligned}$$

By (5.4.22) and the eigenvalue equation, we have

$$\begin{aligned}&\int_{\mathbb{R}^d\times\mathbb{R}^d} p_1(\mathbf{z},\mathbf{y})\left(\frac{1}{\varphi_0(y')} + \frac{1}{\varphi_0(y'')}\right)d\mathbf{y} \\ &= \frac{1}{\varphi_0(z')}\int_{\mathbb{R}^d} K_1(z'',y'')dy'' + \frac{1}{\varphi_0(z'')}\int_{\mathbb{R}^d} K_1(z',y')dy'.\end{aligned} \tag{5.4.25}$$

We know by Theorem 4.107 that e^{-tH} is bounded from L^∞ to L^∞. In particular, the image of $f(x) = 1$ is bounded, and we conclude that $\sup_{z''\in\mathbb{R}^d}|\int_{\mathbb{R}^d} K_1(z'',y'')dy''| < \infty$. Thus the right-hand side of (5.4.25) is uniformly bounded on $\{\mathbf{z} \in \mathbb{R}^d \times \mathbb{R}^d \mid |\mathbf{z}| < r\}$. This implies that for every $\delta < 1$, there is $\bar{R} > 0$ large enough such that

$$\sup_{|\mathbf{z}|<r} \int_{|\mathbf{y}|>\bar{R}} p_1(\mathbf{z},\mathbf{y}) \left(1 + \frac{C\|\varphi_0\|_\infty}{\gamma}\left(\frac{1}{\varphi_0(y')} + \frac{1}{\varphi_0(y'')}\right)\right) dy \le e^{-(C+2D)}\delta.$$

Inserting this and (5.4.24) into (5.4.23), we arrive at

$$\mathbb{E}_{\mathcal{N}_0^{\mathbf{z}}}[e^{-\tilde{W}_T}] \le S_1(\bar{R}, r)e^{C_1}\mathbb{E}_{\mathcal{N}_0^{\mathbf{x}}}[e^{-\tilde{W}_T}] + \delta \sup_{|\mathbf{y}|\le r} \mathbb{E}_{\mathcal{N}_0^{\mathbf{y}}}[e^{-\tilde{W}_T}], \tag{5.4.26}$$

which holds for all $\mathbf{x}, \mathbf{z}$ with $|\mathbf{x}|, |\mathbf{z}| \le r$. By taking the supremum over $\mathbf{z}$ and the infimum over $\mathbf{x}$ in (5.4.26), after rearranging we find

$$\sup_{|\mathbf{x}|\le r} \mathbb{E}_{\mathcal{N}_0^{\mathbf{x}}}[e^{-\tilde{W}_T}] \le \frac{S_1(\bar{R}, r)e^{C_1}}{1-\delta} \inf_{|\mathbf{z}|\le r} \mathbb{E}_{\mathcal{N}_0^{\mathbf{z}}}[e^{-\tilde{W}_T}],$$

which concludes Step 5 and completes the proof of the theorem. □

The next step is to show local uniform domination.

Proposition 5.74. *Let Assumption* 5.70 *hold, and define Q_T as in* (5.4.7). *Then for every $S > 0$, the restrictions of the family $(Q_T)_{T\ge 0}$ to $\mathscr{F}_S$ are uniformly dominated by the restrictions of $\mathcal{N}_0$ to $\mathscr{F}_S$.*

Proof. Fix $S > 0$ and $\varepsilon > 0$. Using Proposition 5.73, choose $R > 0$ large enough so that $Q_T(|X_s| > R) < \varepsilon/8$ for all $|s| < S + 1$ and $T > 0$. Put

$$B = \left\{|X_{-S-1}| < R, |X_{-S}| < R, |X_S| < R, |X_{S+1}| < R\right\}.$$

Then we have $Q_T(B^c) < \varepsilon/2$ for all $T > 0$. For every $A \in \mathcal{F}_S$,

$$Q_T(A) = Q_T(A \cap B^c) + Q_T(A \cap B) \le \varepsilon/2 + \mathbb{E}_{Q_T}[Q_T(A \cap B|\mathscr{T}_{S+1})]. \tag{5.4.27}$$

We are only interested in the lim sup of this as $T \to \infty$, thus we choose $T > S + 1$. Since Q_T is a finite volume Gibbs measure, (5.2.16) gives

$$\begin{aligned} Q_T(A \cap B|\mathscr{T}_{S+1})(\bar{\omega}) &= \frac{1}{\mathbb{E}_{(\mathcal{N}_0)^{\bar{\omega}}_{S+1}}[e^{-\mathcal{H}_{\Lambda(S+1,T)}}]}\mathbb{E}_{(\mathcal{N}_0)^{\bar{\omega}}_{S+1}}[e^{-\mathcal{H}_{\Lambda(S+1,T)}}1_{A\cap B}] \\ &\le e^{8C(S+1)^2}\mathbb{E}_{(\mathcal{N}_0)^{\bar{\omega}}_{S+1}}[1_{A\cap B}]. \end{aligned}$$

The inequality above follows from (5.4.4) by a similar reasoning as the one leading to (5.4.8), and the constant C above does not depend on T. Moreover, by the choice of B, we have

$$\begin{aligned} \mathbb{E}_{(\mathcal{N}_0)^{\bar{\omega}}_{S+1}}[1_{A\cap B}] &= 1_{\{|\bar{\omega}(-S-1)|\le R\}}1_{\{|\bar{\omega}(S+1)|\le R\}} \int_{|z|\le R}\int_{|y|\le R} \frac{K_1(\bar{\omega}(-S-1), y)}{\varphi_0(y)} \\ &\quad\times \mathbb{E}_{\mathcal{N}_0}[1_A|\bar{\omega}(-S) = y, \bar{\omega}(S) = z]\frac{K_1(z, \bar{\omega}(S+1))}{\varphi_0(z)}dy\,dz, \end{aligned}$$

where, as in (5.4.22), K_t is the kernel of e^{-tH}. As in the argument following (5.4.22), it can be seen that there exists $D > 0$ such that $\frac{K_1(\bar{\omega}(-S-1),y)}{\varphi_0(y)} \leq D\varphi_0(y)$ for all $|y| \leq R$ and $|\bar{\omega}(-S-1)| \leq R$. Thus,

$$\begin{aligned}\mathbb{E}_{(\mathcal{N}_0)^{\bar{\omega}}_{S+1}}[1_{A\cap B}] &\leq D^2 \int_{|z|\leq R}\int_{|y|\leq R} \varphi_0(y)\mathbb{E}_{\mathcal{N}_0}[1_A|\bar{\omega}(-S) = y, \bar{\omega}(S) = z]\varphi_0(z)dy\,dz \\ &\leq D^2 \int_{\mathbb{R}^d}\int_{\mathbb{R}^d} \varphi_0(y)\mathbb{E}_{\mathcal{N}_0}[1_A|\bar{\omega}(-S) = y, \bar{\omega}(S) = z]\varphi_0(z)dy\,dz = D^2\mathbb{E}_{\mathcal{N}_0}[1_A].\end{aligned}$$

We have obtained

$$Q_T(A) \leq \varepsilon/2 + D^2 e^{C(S+1)^2}\mathcal{N}_0(A)$$

for every $A \in \mathscr{F}_S$. Thus for given $\varepsilon > 0$, we can choose $\delta < D^{-2}e^{-C(S+1)^2}\varepsilon/2$, and then for every $A \in \mathscr{F}_S$ with $\mathcal{N}_0(A) < \delta$ we have $Q_T(A) < \varepsilon$ uniformly in T. This shows the claim. □

Finally, we complete the proof of Theorem 5.71.

Proof. By Propositions 5.74 and 5.72 below, the family P_T^h of finite volume Gibbs measures is locally uniformly dominated. By Proposition 5.51, the sequence $(P_n^h)_{n\in\mathbb{N}}$ then has a convergent subsequence, and the limit point is a Gibbs measure. □

5.5 Gibbs measures for external and pair interaction potentials: cluster expansion

5.5.1 Cluster representation

Beside existence a number of properties of Gibbs measures for densities dependent on a pair interaction potential W can be derived by using a special technique called *cluster expansion*. Some of these properties have an immediate application to the understanding of spectral properties of the related operators. Due to the lack of space, we can only summarize the key elements of this method and the main results. We refer the reader to the literature for a more detailed discussion.

Cluster expansion is a technique developed originally in classical statistical mechanics. It relies on the existence of a small parameter so that a probability measure for the interaction switched on can be constructed in terms of a convergent perturbation series around the interaction-free case. In our case, the Gibbs measure for $W \equiv 0$ is a path measure of a Markov process, while the Gibbs measure we want to construct for $W \not\equiv 0$ is a path measure of a random process which is not Markovian. The cluster expansion method has two distinct aspects. One is a combinatorial framework allowing to make series indexed by graphs and derive bounds on them by series indexed by trees. The other part is analytic and more dependent on the details of the model, and consists of basic estimates on the entries of these sums leading to the convergence of the expansion.

In this section we make the following assumptions on the potentials.

Assumption 5.75.

(1a) The external potential $V : \mathbb{R}^d \to \mathbb{R}$ is Kato-decomposable and the Schrödinger operator $H = -\frac{1}{2}\Delta + V$ has a unique ground state $\varphi_0 \in L^2(\mathbb{R}^d)$ with the property $H\varphi_0 = 0$; and

(1b) the Schrödinger semigroup generated by $H = -\frac{1}{2}\Delta + V$ is intrinsically ultracontractive (see Definition 5.3).

(2) The pair interaction potential $W : \mathbb{R}^d \times \mathbb{R}^d \times \mathbb{R} \to \mathbb{R}$ has the symmetry properties $W(\cdot,\cdot,s-t) = W(\cdot,\cdot,|s-t|)$, $W(x,y,\cdot) = W(y,x,\cdot)$ and satisfies either of the following regularity conditions:

(W1) There exist $R > 0$ and $\beta > 2$ such that

$$|W(x,y,s-t)| \le R\frac{|x|^2 + |y|^2}{1 + |s-t|^\beta} \tag{5.5.1}$$

for every $x, y \in \mathbb{R}^d$ and $t, s \in \mathbb{R}$.

(W2) There exist $R > 0$ and $\beta > 1$ such that

$$|W(x,y,s-t)| \le \frac{R}{1 + |s-t|^\beta} \tag{5.5.2}$$

for every $x, y \in \mathbb{R}^d$ and $t, s \in \mathbb{R}$.

In particular, if V is a superquadratic potential with exponent $\alpha > 1$ as defined by (5.3.16), then conditions (1a) and (1b) above hold.

Let P be the Itô measure for V, which we will use as reference measure. Similarly as in (5.4.7), we define

$$dQ_T = \frac{1}{Z_T} e^{-\lambda \int_{-T}^{T} ds \int_{-T}^{T} W(X_s,X_t,s-t)dt} dP, \tag{5.5.3}$$

where $\lambda > 0$ is a parameter.

Theorem 5.76 (Existence of Gibbs measure). *Let $(T_n)_{n\in\mathbb{N}} \subset \mathbb{R}$ be an increasing sequence such that $T_n \to \infty$, and suppose $0 < |\lambda| \le \lambda^*$ with λ^* small enough. Then under Assumption 5.75, the local limit $\lim_{n\to\infty} Q_{[-T_n,T_n]} = Q$ exists and is a Gibbs measure on $(\mathfrak{X}, \mathscr{F})$. Moreover, Q does not depend on the sequence $(T_n)_{n\in\mathbb{N}}$.*

In order to prove existence of the Gibbs measure Q, we use a cluster expansion controlled by the parameter λ which we choose to be suitably small. In what follows we outline the main steps of the proof of this theorem.

For simplicity and without restricting generality, we consider the family of symmetric bounded intervals of $\mathbb{R}$ for the index set of measures. We partition $[-T, T]$ into disjoint intervals $\tau_k = (t_k, t_{k+1})$, $k = 0, \dots, N-1$, with $t_0 = -T$ and $t_N = T$, each of length b, i. e., we fix $b = 2T/N$; for convenience, we choose N to be an even number so that the origin is endpoint to some intervals. We break up a path X into pieces X_{τ_k} by restricting it to τ_k. The total energy contribution of the pair interaction can be written in terms of the sum

$$W_T = \int_{-T}^{T} ds \int_{-T}^{T} W(X_s, X_t, s-t)dt = \sum_{0 \le i < j \le N-1} W_{\tau_i, \tau_j}, \tag{5.5.4}$$

where with the notation $\mathcal{J}_{ij} = \int_{\tau_i} ds \int_{\tau_j} W(X_s, X_t, s-t)dt$ we have

$$W_{\tau_i,\tau_j} = \begin{cases} \mathcal{J}_{ij} + \mathcal{J}_{ji} & \text{if } |i-j| \ge 2, \\ \frac{1}{2}(\mathcal{J}_{ii} + \mathcal{J}_{jj}) + \mathcal{J}_{ij} + \mathcal{J}_{ji} & \text{if } |i-j| = 1, \text{ and } i \ne 0, j \ne N-1, \\ \mathcal{J}_{ij} + \mathcal{J}_{ji} + \frac{1}{2}\mathcal{J}_{00} & \text{if } i = 0 \text{ and } j = 1, \\ \mathcal{J}_{ij} + \mathcal{J}_{ji} + \frac{1}{2}\mathcal{J}_{N-1\,N-1} & \text{if } i = N-1 \text{ and } j = N-2. \end{cases} \tag{5.5.5}$$

In order to keep the notation simple, we do not make explicit the X_t dependence of these objects.

A simple calculation gives

$$e^{-\lambda \int_{-T}^{T} ds \int_{-T}^{T} W(X_s,X_t,s-t)dt} = \prod_{0 \le i < j \le N-1} (e^{-\lambda W_{\tau_i,\tau_j}} + 1 - 1) = 1 + \sum_{\mathcal{R}_T \ne \emptyset} \prod_{(\tau_i,\tau_j) \in \mathcal{R}_T} (e^{-\lambda W_{\tau_i,\tau_j}} - 1). \tag{5.5.6}$$

Here the summation is performed over all nonempty sets of different pairs of bounded intervals $\mathcal{R}_T = \{(\tau_i, \tau_j) \mid (\tau_i, \tau_j) \ne (\tau_{i'}, \tau_{j'}) \text{ whenever } (i,j) \ne (i',j')\}$.

Since the reference measure is the measure of a Markov process with probability transition kernel $\pi_t(x,y)$, a break-up of the paths involves a corresponding factorization of the reference measure into piecewise pasted conditional measures. Put $X_{t_k} = x_k$ for the positions at the time points of the division, $\forall k = 0, \dots, N$, and $P_{\tau_k}^{x_k, x_{k+1}}$ for the corresponding bridge measures. We have thus

$$P_T(\cdot \,| X_{t_0} = x_0, \dots, X_{t_N} = x_N) = \bigotimes_{k=0}^{N-1} P_{\tau_k}^{x_k,x_{k+1}}(\cdot). \tag{5.5.7}$$

Denote by $d\mathfrak{m} = \varphi_0^2 dx$ the stationary distribution of the Itô measure, and let $p_{t_0,\dots,t_N}(x_0, \dots, x_N)$ be the density with respect to $\bigotimes_{k=0}^{N} d\mathfrak{m}(x_k)$ of the joint distribution of positions of the path X_t recorded at the time points of the division. By the Markov property, it follows that

$$p_{t_0,\dots,t_N}(x_0,\dots,x_N) = \prod_{k=0}^{N-1} \pi_b(x_{k+1},x_k) = \prod_{k=0}^{N-1}(\pi_b(x_{k+1},x_k) - 1 + 1)$$

$$= 1 + \sum_{\mathcal{S}_T \neq \emptyset} \prod_{k:\tau_k \in \mathcal{S}_T} (\pi_b(x_{k+1},x_k) - 1),$$

where π_b is given by (5.3.8). The summation runs over the collection of all nonempty sets $\mathcal{S}_T = \{\tau_k \mid \tau_k = (t_k, t_{k+1})\}$ of different pairs of consecutive time points.

We have thus the product of two sums over not unrelated collections of subsets as the net result. To keep a systematic control over these sums we introduce the following objects.

(1) *Contours.* Two distinct pairs of intervals (τ_i, τ_j) and $(\tau_{i'}, \tau_{j'})$ will be called *joint* and denoted $(\tau_i, \tau_j) \sim (\tau_{i'}, \tau_{j'})$ if one interval of the pair (τ_i, τ_j) coincides with one interval of the pair $(\tau_{i'}, \tau_{j'})$. A set of *connected* pairs of intervals is a collection $\{(\tau_{i_1}, \tau_{j_1}), \dots, (\tau_{i_n}, \tau_{j_n})\}$ in which each pair of intervals is connected to another through a sequence of joint pairs, i. e., for any $(\tau_i, \tau_j) \neq (\tau_{i'}, \tau_{j'})$ there exists $\{(\tau_{k_1}, \tau_{l_1}), \dots, (\tau_{k_m}, \tau_{l_m})\}$ such that $(\tau_i, \tau_j) \sim (\tau_{k_1}, \tau_{l_1}) \sim \cdots \sim (\tau_{k_m}, \tau_{l_m}) \sim (\tau_{i'}, \tau_{j'})$. A maximal set of connected pairs of intervals is called a *contour*, denoted by γ. We denote by $\bar{\gamma}$ the set of all intervals that are elements of the pairs of intervals belonging to contour γ, and by γ^* we denote the set of time points of intervals appearing in $\bar{\gamma}$. Two contours γ_1, γ_2 are *disjoint* if they have no intervals in common, i. e., $\bar{\gamma}_1 \cap \bar{\gamma}_2 = \emptyset$. Clearly, $\mathcal{R}_T$ can be decomposed into sets of pairwise disjoint contours, i. e., $\mathcal{R}_T = \bigcup_{r\geq 1}\{\gamma_1, \dots, \gamma_r \mid \bar{\gamma}_i \cap \bar{\gamma}_j = \emptyset,\ i \neq j;\ i,j = 1, \dots, r\}$.

(2) *Chains.* A collection of consecutive intervals

$$\{\tau_j, \tau_{j+1}, \dots, \tau_{j+k}\}, \quad j \geq 0,\ j + k \leq N - 1,$$

is called a *chain.* As in the case of contours, $\bar{\varrho}$ and ϱ^* mean the set of intervals belonging to the chain ϱ and the set of time points in ϱ, respectively. Two chains ϱ_1, ϱ_2 are called *disjoint* if they have no common time points, i. e., $\varrho_1^* \cap \varrho_2^* = \emptyset$. Denote by $\partial^-\varrho$ (resp. $\partial^+\varrho$) the leftmost (resp. rightmost) time points belonging to ϱ.

(3) *Clusters.* Take an unordered set of disjoint contours and disjoint chains, $\Gamma = \{\gamma_1, \dots, \gamma_r; \varrho_1, \dots, \varrho_s\}$, with some $r \geq 1$ and $s \geq 0$. Note that such contours and chains may have common time points. The notation $\Gamma^* = (\bigcup_{i=1}^r \gamma_i^*) \cup (\bigcup_{j=1}^s \varrho_j^*)$ means the set of all time points appearing as beginnings or ends of intervals belonging to some contour or chain in Γ. Also, we put $\bar{\Gamma} = (\bigcup_{i=1}^r \bar{\gamma}_i) \cup (\bigcup_{j=1}^s \bar{\varrho}_j)$ for the set of intervals appearing in Γ through entering some contours or chains; Γ is called a *cluster* if Γ^* is a connected collection of sets (in the usual sense), and for every $\varrho \in \Gamma$ we have $\partial^-\varrho, \partial^+\varrho \in \bigcup_{j=1}^r \gamma_j^*$. This means that in a cluster chains have no loose ends. We denote by $\mathbb{K}_N$ the set of all clusters for a given N.

With these notations the sum in (5.5.6) is then further written as

$$\sum_{\mathcal{R}_T\neq\emptyset}\prod_{(\tau_i,\tau_j)\in\mathcal{R}_T}(e^{-\lambda W_{\tau_i,\tau_j}}-1)=\sum_{r=1}^{\infty}\sum_{\{\gamma_1,\ldots,\gamma_r\}}\prod_{k=1}^{r}\prod_{(\tau_i,\tau_j)\in\gamma_k}(e^{-\lambda W_{\tau_i,\tau_j}}-1), \tag{5.5.8}$$

where now summation goes over collections $\{\gamma_1,\ldots,\gamma_r\}$ of disjoint contours. In a similar way, (5.5.7) appears in the form

$$\sum_{\mathcal{S}_T\neq\emptyset}\prod_{k:\tau_k\in\mathcal{S}_T}(\pi_b(X_{k+1},X_k)-1)=\sum_{r=1}^{\infty}\sum_{\{\varrho_1,\ldots,\varrho_s\}}\prod_{j=1}^{s}\prod_{k:\tau_k\in\varrho_j}(\pi_b(X_{k+1},X_k)-1)\,. \tag{5.5.9}$$

Here $\{\varrho_1,\ldots,\varrho_s\}$ is a collection of disjoint chains, and this formula justifies how we defined them.

For every cluster $\Gamma=\{\gamma_1,\ldots,\gamma_r;\varrho_1,\ldots,\varrho_s\}\in\mathbb{K}_N$, define the function

$$\kappa_\Gamma=\prod_{l=1}^{r}\prod_{(\tau_i,\tau_j)\in\gamma_l}(e^{-\lambda W_{\tau_i,\tau_j}}-1)\prod_{m=1}^{s}\prod_{k:\tau_k\in\varrho_m}(\pi_b(X_{k+1},X_k)-1)\,. \tag{5.5.10}$$

Also, we define the auxiliary probability measure

$$d\chi_N(X)=\bigotimes_{k=0}^{N-1}dP_{\tau_k}^{X_k,X_{k+1}}(X_{\tau_k})\bigotimes_{k=0}^{N}dm(x_k). \tag{5.5.11}$$

It is straightforward to check that $(\chi_N)_{N\in\mathbb{N}}$ is a family of consistent probability measures, thus by the Kolmogorov's extension theorem it has a unique extension to $\mathbb{R}$ which we denote by χ. Note that

$$\int_{\mathbb{R}^d}(\pi_b(x_{k+1},x_k)-1)dm(x_{k+1})=\int_{\mathbb{R}^d}(\pi_b(x_{k+1},x_k)-1)dm(x_k)=0.$$

This is why in our definition a cluster does not contain chains having loose ends as for any such chain $\mathbb{E}_{\chi_N}[\kappa_\Gamma]=0$.

Finally, we define the function

$$K_\Gamma=\mathbb{E}_\chi[\kappa_\Gamma]. \tag{5.5.12}$$

A combination of (5.5.7)–(5.5.10) and (5.5.12) then gives the cluster representation of the partition function

$$Z_T=1+\sum_{n=1}^{\infty}\sum_{\substack{\{\Gamma_1,\ldots,\Gamma_n\}\in\mathbb{K}_N\\ \Gamma_i^*\cap\Gamma_j^*=\emptyset,i\neq j}}\prod_{l=1}^{n}K_{\Gamma_l} \tag{5.5.13}$$

for every $T=Nb/2>0$.

Let G be a graph with vertex set $\{1, \ldots, n\}$, in which an edge ij joins vertex i with vertex j whenever $\Gamma_i^* \cup \Gamma_j^* \neq \emptyset$. Denote by $\mathscr{G}_n$ the set of connected graphs G defined in this way, and define

$$\phi^T(\Gamma_1, \ldots, \Gamma_n) = \begin{cases} 1 & \text{if } n = 1, \\ \sum_{G \in \mathscr{G}_n} \prod_{ij \in G} (-1_{\{\Gamma_i^* \cap \Gamma_j^* \neq \emptyset\}}) & \text{if } n > 1. \end{cases}$$

By using a combinatorial function calculus, it is possible to show that

$$\log Z_T = \sum_{n=1}^{\infty} \sum_{\substack{\{\Gamma_1, \ldots, \Gamma_n\} \in \mathbb{K}_N \\ 0 \in \Gamma_1^*}} \phi^T(\Gamma_1, \ldots, \Gamma_n) \prod_{l=1}^{n} K_{\Gamma_l}. \tag{5.5.14}$$

Also, consider for any subset $A \subset \mathbb{R}$ the set function

$$Z_T^A = 1 + \sum_{n=1}^{\infty} \sum_{\substack{\Gamma_1, \ldots, \Gamma_n \in \mathbb{K}_N \\ \Gamma_i^* \cap \Gamma_j^* = \emptyset,\, i \neq j \\ A \cap (\cup_i \Gamma_i^*) = \emptyset}} \prod_{i=1}^{n} K_{\Gamma_i}$$

and write $Z_T^\Gamma = Z_T^{\bar{\Gamma} \cup \Gamma^*}$. Then by combinatorial function calculus, it is furthermore possible to derive the expressions

$$\log Z_T^\Gamma = 1 + \sum_{n=1}^{\infty} \sum_{\substack{\Gamma_1, \ldots, \Gamma_n \in \mathbb{K}_N \\ \Gamma^* \cap (\cup_i \Gamma_i^*) = \emptyset}} \phi^T(\Gamma_1, \ldots, \Gamma_n) \prod_{i=1}^{n} K_{\Gamma_i}$$

and

$$f_T^\Gamma = \frac{Z_T^\Gamma}{Z_T} = \exp\Big(- \sum_{n=1}^{\infty} \sum_{\substack{\Gamma_1, \ldots, \Gamma_n \in \mathbb{C}_N \\ \Gamma^* \cap (\cup_i \Gamma_i^*) \neq \emptyset}} \phi^T(\Gamma_1, \ldots, \Gamma_n) \prod_{i=1}^{n} K_{\Gamma_i} \Big). \tag{5.5.15}$$

The function f_T^Γ is a *correlation function* associated with cluster Γ.

5.5.2 Basic estimates and convergence of cluster expansion

There are two crucial estimates on which the cluster expansion depends. The first estimate is for the energy terms coming from the contributions of the intervals in the partition. Recall the notation $\mathcal{J}_{ij}$ in (5.5.5).

Lemma 5.77 (Energy estimates). *We have*

$$|\mathcal{J}_{ij}| \leq \begin{cases} C_1 b \frac{\int_{\tau_i} X_{\tau_i}^2(t)dt + \int_{\tau_j} X_{\tau_j}^2(s)ds}{(|j-i-1|b)^\beta + 1}, & \text{(W1) case}, \\ \frac{C_2 b^2}{(|j-i-1|b)^\beta + 1}, & \text{(W2) case}, \end{cases} \tag{5.5.16}$$

with some $C_1, C_2 > 0$ in each respective case.

As a consequence, W_{τ_i,τ_j} is bounded below for all i, j, which is a stability condition ensuring that the variables indexed by the subintervals are well defined and there are no short-range ($i = j$) singularities.

The second estimate for the cluster expansion is given as follows.

Lemma 5.78 (Cluster estimates). *There exist some constants $c_1, c_2 > 0$ and $\delta > 1$ such that for every cluster $\Gamma \in \mathbb{K}_N$,*

$$|K_\Gamma| \leq \prod_{\varrho\in\Gamma}(c_1|\lambda|^{1/3})^{|\bar{\varrho}|} \prod_{\gamma\in\Gamma} \prod_{(\tau_i,\tau_j)\in\gamma} \frac{c_2|\lambda|^{1/3}}{(|i-j-1|b)^\delta + 1}, \tag{5.5.17}$$

where $|\bar{\varrho}|$ denotes the number of intervals contained in ϱ.

In estimate (5.5.17), the factor accounting for the contribution of chains comes from the upper bound $Ce^{-\Lambda b}$ on $|\pi_b(x,y) - 1|$ uniform in x, y (see the second factor in (5.5.10)), where Λ is the spectral gap of the Schrödinger operator H of the underlying $P(\phi)_1$-process, and $C > 0$. This bound results from intrinsic ultracontractivity of e^{-tH}; see Lemma 5.9. The factor accounting for the contribution of contours comes from a suitable use of the Hölder inequality and optimization applied to the products over $e^{-\lambda W_{\tau_i\tau_j}} - 1$ (see the first factor in (5.5.10)), taken together with Lemma 5.77; b is then chosen in such a combination with λ and Λ that the expression (5.5.17) results.

Proposition 5.79 (Convergence of cluster expansion). *If there exists $\eta(\lambda) > 0$ such that $\lim_{\lambda\to 0}\eta(\lambda) = 0$ and for every $n \in \mathbb{N}$,*

$$\sum_{\substack{\Gamma\in\mathbb{K}_N \\ \Gamma^*\ni 0, |\bar{\Gamma}|=n}} |K_\Gamma| \leq c\eta^n, \tag{5.5.18}$$

then the series at the right-hand sides of (5.5.13) *and* (5.5.14)*, respectively, are absolutely convergent as $N \to \infty$.*

In order to prove this statement, we consider the generating function

$$H(z,\lambda) = \lim_{N\to\infty} \sum_{\substack{\Gamma\in\mathbb{K}_N \\ \Gamma_1^*\ni 0}} K_\Gamma z^{|\bar{\Gamma}|},$$

in which λ is considered a parameter and $|\bar{\Gamma}|$ is the number of intervals contained by Γ. We show that $H(z,\lambda)$ is analytic in z in a circle of radius $r(\lambda)$ and that inside this circle $H(z,\lambda)$ is uniformly bounded in λ. Choosing $\eta(\lambda) = 1/r(\lambda)$ then gives the bound in (5.5.18).

This estimate follows through a procedure of translating the sums over $\mathbb{K}_N$ on the left-hand side of (5.5.18) to sums indexed by connected graphs. For every $\Gamma \in \mathbb{K}_N$, we define a graph G_Γ whose vertex set are the contours $\{\gamma_1, \dots, \gamma_r\}$ in Γ and whose edge set are those pairs of contours which are appropriately matched with the chains $\rho_1, \dots, \rho_s$ of Γ. For $r = 1$, we have $s = 0$ and therefore the empty graph G_{Γ_1}. For $r \geq 2$, we have

$s \geq 1$ and the sums can be rewritten by sums over connected graphs following a specific enumeration. By using a graph-tree bound, the graph-indexed sums can be bounded by tree-indexed sums, which can be better further estimated. An inductive procedure and some nontrivial combinatorics then yield (5.5.18).

By an application of Proposition 5.79 to (5.5.15), it can be shown that

$$f^{\Gamma} = \lim_{T\to\infty} f_T^{\Gamma}$$

exists, and $|f_T^{\Gamma}| \leq 2^{|\bar{\Gamma}|}$ holds uniformly in T for λ small enough. This can be used to prove the following statement.

Proposition 5.80. *The local limit $P = \lim_{T\to\infty} P_T$ exists and satisfies the equality*

$$\mathbb{E}_P[F_S] = \mathbb{E}_\chi[F_S] f^S + \sum_{n=1}^{\infty} \sum_{\substack{\Gamma_1,\dots,\Gamma_n \in \mathbb{C} \\ \Gamma_i^* \cap \Gamma_j^* = \emptyset,\ i \neq j \\ i: S \cap \Gamma_i^* \neq \emptyset}} \mathbb{E}_\chi\Big[F_S \prod_{i=1}^{n} \kappa_{\Gamma_i}\Big] f^{\cup \bar{\Gamma}} \tag{5.5.19}$$

for any bounded $\mathscr{F}_S$-measurable function F_S, where S is a finite union of intervals of the partition considered in the cluster expansion. Moreover, the measure P is invariant with respect to the time shift.

Theorem 5.76 follows directly from this proposition. Equality (5.5.19) implies that P satisfies the DLR equations; in particular, it is a Gibbs measure.

5.5.3 Further properties of the Gibbs measure

An important aspect in understanding the Gibbs measure is to characterize almost sure properties of paths under this measure. This is done in the following theorem.

Theorem 5.81 (Typical path behavior). *If V is a superquadratic potential with exponent $a > 1$, then*

$$|X_t| \leq C\,(\log(|t|+1))^{1/(a+1)}, \quad P\text{-a. s.} \tag{5.5.20}$$

with a suitable number $C > 0$.

The strategy of proving Theorem 5.81 goes by comparing the typical behaviors of the reference process and of the process with the pair interaction potential (which is not a Markov process).

Lemma 5.82. *Suppose there exist some numbers $C, \theta > 0$ such that for any $x > 0$,*

$$P\Big(\max_{0\leq t\leq 1} |X_t| \geq x\Big) \leq Ce^{-\theta x^{a+1}}. \tag{5.5.21}$$

Then there exist $C' > 0$ and $\theta' > 0$ such that for any $x > 0$

$$Q\Big(\max_{0\le t\le 1}|X_t| \ge x\Big) \le C' e^{-\theta' x^{\alpha+1}}. \tag{5.5.22}$$

The proof of this lemma requires once again the use of cluster expansion. The assumption of the lemma can be verified by using (5.23) for the underlying $P(\phi)_1$-process.

Another important property of Q is its uniqueness in DLR sense. This means that for any increasing sequence of real numbers $(T_n)_{n\in\mathbb{N}}$, $T_n \to \infty$, and any corresponding sequence of boundary conditions $(Y_n)_{n\in\mathbb{N}} \subset \mathfrak{X}_0$, $\lim_{n\to\infty} \mathbb{E}_{Q_{T_n}}[F_B|Y_n] = \mathbb{E}_Q[F_B]$, for every bounded $B \subset \mathbb{R}$ and every bounded $\mathscr{F}_B$-measurable function F_B. Here $\mathfrak{X}_0$ is the subspace given by (5.3.14). In other words, DLR uniqueness means that the limit measure Q is independent of the boundary paths.

Theorem 5.83 (Uniqueness of Gibbs measure). *Suppose that V satisfies the conditions in Assumption* 5.75 *and W satisfies* (W2). *Then the following cases occur:*

(1) *If $\beta > 2$, then whenever the Gibbs measure Q exists, it is unique in DLR sense.*
(2) *If $\beta > 1$, then for sufficiently small $|\lambda|$ the limiting Gibbs measure Q is unique in DLR sense whenever the reference measure is unique.*

If $\beta > 2$, the energy functional given by (5.2.14) is uniformly bounded in T and in the restrictions of paths over $[-T, T]$ and $R\setminus[-T, T]$, respectively. This then implies that only one Gibbs measure can exist as the limit does not depend on the choice of the boundary conditions. This argument requires no restriction on the values of λ. For $1 < \beta \le 2$, this uniform boundedness does not hold any longer and cluster expansion is to date the only method allowing to derive these results.

To conclude, we list some additional properties of Gibbs measures for (W2)-type pair interaction potentials, useful in various contexts. This case in particular covers the Nelson model.

Theorem 5.84. *Let P be a Gibbs measure for W satisfying* (W2), *and suppose V satisfies Assumption* 5.75. *Then the following hold:*

(1) *(Invariance properties) With respect to time shift and time reflection, Q is invariant.*
(2) *(Univariate distributions) The distributions Q^T under Q_T of positions x at time $t = 0$ are equivalent to P, i. e., there exist $C_1, C_2 \in \mathbb{R}$, independent of T and x, such that*

$$C_1 \le \frac{dQ^T}{dP}(x) \le C_2 \tag{5.5.23}$$

for every $x \in \mathbb{R}^d$ and $T > 0$. Moreover, $\lim_{T\to\infty} \frac{dQ^T}{dP}$ exists pointwise.
(3) *(Univariate conditional probability distributions) The conditional distributions $Q_T(\cdot|X_0 = x)$ converge locally weakly to $Q(\cdot|X_0 = x)$, for all $x \in \mathbb{R}^d$.*

(4) *(Mixing properties) For any bounded functions F, G on* $\mathbb{R}^d$

$$|\mathrm{cov}_Q(F_s; G_t)| \leq \mathrm{const}\frac{\sup \|F\|_\infty \sup \|G\|_\infty}{1 + |s - t|^\gamma}, \qquad (5.5.24)$$

where $\gamma > 0$ *and the constant prefactor is independent of* s, t *and* F, G.

The invariance properties are inherited trivially from the underlying $P(\phi)_1$-process. The proof of the remaining properties makes extensive use of cluster expansion.

6 Notes and references

Notes to the Preface

Classic books of this early period include H. Poincaré: *Les méthodes nouvelles de mécanique céleste* (Gauthier-Villars, 1893, English translation: *New Methods of Celestial Mechanics*), 3 vols.; see also C. Siegel and J. Moser: Vorlesungen über Himmelsmechanik (1956, English translation: *Lectures on Celestial Mechanics*, Springer, 1971), J. C. Maxwell: *A Treatise on Electricity and Magnetism* (Clarendon Press, Oxford, 1873, reprinted by Dover, 1954, 2003), and *Theory of Heat* (Longmans, 1872, reprinted by Dover, 2001). A comprehensive account of methods of mathematical physics with a strong focus on partial differential equations which stood as a standard work for generations is D. Hilbert and R. Courant: *Methoden der mathematischen Physik*, vols. 1–2 (Springer, 1924, 1937, with subsequent English translation).

The intellectual biographies and autobiographies of some of the main protagonists of early quantum mechanics capture the excitement of those days and convey the spirit how these ideas were born; see, for instance, W. Heisenberg: *Der Teil und das Ganze* (Piper, 1969), D. Cassidy: *Uncertainty: The Life and Science of Werner Heisenberg* (W. H. Freeman, 1993), A. Pais: *Niels Bohr's Times: In Physics, Philosophy, and Polity* (Oxford UP, 1994), and G. Farmelo: *The Strangest Man: The Hidden Life of Paul Dirac, Mystic of the Atom* (Basic Books, 2009). For an account from a major figure in the development of stochastic analysis and Feynman–Kac formulae, see R. Bhatia: A conversation with S. R. S. Varadhan, *Math. Intelligencer* 2008, pp. 24–42. E. Wigner's essay *The Unreasonable Effectiveness of Mathematics in the Natural Sciences* appeared in *Commun. Pure Appl. Math.*, 13 (1960). He mentions the idea attributed to Galileo that the laws of nature are written in the language of mathematics. The concept of the "book of nature" is a topos that can be traced back to antiquity and the middle ages; see the early and illuminating account given by the great German literary scholar E. R. Curtius in *Europäische Literatur und lateinisches Mittelalter* (Francke, 1954, 350ff) (especially the concept of "chiffre"), and more modern discussions in *The Book of Nature*, A. Vanderjagt, K. van Berkel (eds.), vol. 1: *Antiquity and the Middle Ages*, vol. 2: *Early Modern and Modern History* (Peeters, Leuven, 2005–2006). On the quoted passage of Feynman, see R. Dijkgraaf, *Rapporteur Talk: Mathematical structures*, in: *The Quantum Structure of Space and Time, Proceedings of the 23rd Solvay Conference on Physics* (D. Gross, M. Henneaux, A. Sevrin, eds., World Scientific, 2007, p. 91). Apparently a mathematician has replied that the time by which mathematics would be set back is "precisely the week in which God created the world"; a version of this anecdote names Mark Kac to be this mathematician.

https://doi.org/10.1515/9783112231463-006

Notes to Chapter 1

Feynman's path integral method appeared first in the papers [123, 124, 125]; see also [126]. Dirac's contact transformation theory was published in [95, 96, 97]. For some historic background, see [331]. Modern references on Feynman integrals include [4, 63, 68, 73, 93, 94, 149, 155, 198, 200, 229, 313, 378]. A first paper by Kac using analytic continuation of the time variable is [201]; see [62, 281] for later developments. The disadvantage of Kac's approach is that it deals with the diffusion equation with dissipation instead of the time-dependent Schrödinger equation, and therefore the Feynman–Kac formula gives no direct information on the quantum dynamics. However, the time-independent solutions of both equations are the same and coincide with the eigenfunctions of the Schrödinger operator, which justifies the use of the Feynman–Kac formula in quantum mechanics. An early, but little known, paper taking a different approach is [122].

Notes to Chapter 2

2.1

Measure theory has been developed from the late nineteenth century by E. Borel, H. Lebesgue, J. Radon, and M. Fréchet, among others. Some applications of measure theory are in the theory of integration, probability and stochastic processes, ergodic theory, geometric measure theory, or topological measure theory. Integration theory developed in the line of defining integrals on spaces more general than subsets of the Euclidean space $\mathbb{R}^d$, arriving to integration with respect to Lebesgue measure on $\mathbb{R}^d$ and further concepts, leading to a richer theory than Riemann–Stieltjes integrals. For instance, L^p-spaces defined with respect to the Lebesgue measure are important Banach spaces, with many applications in functional analysis. Probability theory uses finite measures, and considers measurable subsets to be events whose probabilities are given by their measures, and ergodic theory considers measures that are kept invariant under a dynamical system. Stochastic analysis has been developed to extend the concept of integral to irregular functions, and apart for its intrinsic interest in mathematics, a major modern source of motivation was provided by financial mathematics.

We refer to [318, 33] for a general analysis and measure theory background, to [109, 15, 30, 42, 146, 357, 216, 107] for probability theory also emphasizing an analytic point of view, and to [237, 142, 226] for random processes. A π-system, a λ-system and the π–λ theorem are basic facts in measure theory, which are reviewed in, e. g., [272, Appendix B]. Measures on a metric space are reviewed in, e. g., [231, Section 13], where the existence of regular conditional probability measure is also discussed in [231, Section 8]. In Definition 2.5, weak and vague topologies on the set of measures are defined. The weak topology τ_{w} on the set of finite measures $\mathscr{P}_{\mathrm{f}}(S)$ is investigated in, e. g., [300, p. 167] and the vague topology τ_{v} on the set of Radon measures $\mathscr{P}_{\mathrm{r}}(S)$ in, e. g., [215, Section 15.7].

If S is a locally compact Polish space, then $(\mathscr{P}_f(S), \tau_w)$ and $(\mathscr{P}_f(S), \tau_v)$ are also Polish spaces.

2.2

The concept of a martingale in probability theory was introduced by P. Lévy in 1934, and the term "martingale" was introduced later by J. Ville in 1939, who also extended the definition to continuous-time martingales. Much of the original development of martingale theory was done by J. Doob and D. Burkholder. We refer to, e. g., [158, 109, 270, 271, 311, 314, 315] for continuous time martingales. Martingales and stopping times are useful tools to study stochastic analysis, and we refer to, e. g., [272], [188, Chapter I. 6], and the references therein. Convergence properties of martingales (Theorem 2.65) are studied in, e. g., [188, Theorem 6.5]. The optional sampling theorem (Theorem 2.70) is taken from [188, Theorem 6.11]. The classes DL and D were introduced in, e. g., [218, Definition 4.8] and [188, Definition 6.4]. Increasing processes are also introduced in, e. g., [218, Definition 4.4] and [188, Definition 6.2]. The definition of integrability of increasing process is taken from [218, Definition 4.4]. The Doob–Meyer decomposition (Theorem 2.78) is proven in, e. g., [218, Theorem 4.10] and [188, Theorem 6.12].

A. A. Markov was a Russian mathematician best known for his work on stochastic processes. A primary subject of his research later became known as Markov processes. He studied Markov chains in the early twentieth century, and was interested in studying an extension of independent random sequences. A. Kolmogorov developed in 1931 a large part of the early theory of continuous-time Markov processes. Kolmogorov was partly inspired by a work of L. Bachelier on fluctuations in the stock market as well as that of N. Wiener on Einstein's model of Brownian motion. He introduced and studied a particular set of Markov processes known as diffusion processes, where he derived a set of differential equations describing the processes. Independent of Kolmogorov's work, S. Chapman derived in 1928 the Chapman–Kolmogorov equation. Markov processes were also investigated by W. Feller and then later E. Dynkin, starting in the 1950s. We refer to [51, 103, 111, 116, 314, 315] for Markov processes, to [195] for Markov generators, to [41, 197] for limit theorems. We refer to the detail of Feller transition kernels and generators to [195, Chapter 3.4] and [269, Section 4.1]. Further applications of an invariant measure and the martingale derived from the generator of a Markov process mentioned in Theorem 2.93 are investigated in [227], where the functional central limit theorem is discussed, i. e., the asymptotic behavior of $\frac{1}{t}\int_0^t (Lf)(X_s)ds$ as $t \to \infty$ is studied, where L is the generator of Markov process $(X_t)t \geq 0$. In the proof of Theorem 2.91, it is mentioned that for each $t > 0$ and each $g \in C^\infty(\mathbb{R}^d)$, there exists $f \in D(L)$ with $f - tLf = g$. This can be found in [254].

2.3

The Scottish botanist Robert Brown is credited with the observation reported in 1827 that pollen grains suspended in water perform an irregular motion, however, he was preceded by more than a century by the Dutch natural scientist van Leeuwenhoek by a similar observation. Also less known is that Brown enthusiastically continued his observations by replacing pollen with minerals and even a piece of the Sphinx. Brownian motion has been first modeled in terms of a random process by Bachelier [25] as early as 1900 in order to describe stock prices. Einstein [113] and von Smoluchowski [351] studied in 1905–1906 the fluctuations of thermal motion of particles and related them with Brownian motion, which Perrin tested in 1909 experimentally, noting the rough shape of particle paths. Langevin proposed in 1908 an equation to describe the equation of motion of a particle performing Brownian motion. In 1923 Wiener constructed Brownian motion as a random process in a mathematically rigorous way, and Lévy made important early contributions in the study of some fine details of Brownian motion. For a history, see [108] as well as [233, 282].

We single out [188, 193, 230, 218, 231, 282, 284, 311, 332, 359] for fundamental facts on Brownian motion and general stochastic analysis. Brownian motion in many ways stands out as a random process having a distinguished role. It is a Gaussian process, a Markov process, and a martingale. It is the only Lévy process allowing a version with continuous paths. It is a scaling limit of a simple random walk, and it is intimately related with the Laplace operator in that $-\frac{1}{2}\Delta$ is its infinitesimal generator, and thus its transition probability density is the fundamental solution of the heat equation. There are a number of equivalent ways of defining and constructing Brownian motion, of which we have presented one and commented on others. The literature on Brownian motion is vast due to the fact that it is an object in the overlap zone of the fields and approaches mentioned above. A wealth of facts and explicit formulae can be found in [55] and the references therein, and we single out [268, 373] for further developments.

In the Lévy martingale characterization theorem, we need the momentum estimate $\mathbb{E}[(X_t - X_s)^4|\mathcal{F}_s] \leq 4|t-s|^2$, which is given in [332, Lemma 9.10]. In Proposition 2.55, we discuss a right-continuous filtration which is given in [272, Chapter II, 2.f.0]. The basic idea for the reflection principle for Brownian motion $(B_t)_{t\geq 0}$ is familiar from simple symmetric random walk. This results from the strong Markov property of $(B_t)_{t\geq 0}$ and is stated in Corollary 2.116. We also refer to [314, I.13].

2.4

In 1944 Kiyosi Itô [191] proposed a notion of a stochastic integral using the paths of Brownian motion as integrator, and he also established the chain rule for stochastic differentials, which is known today as the Itô formula. While the construction of the Itô integral starts from the classic notion of integral proceeding through approximate Riemann–Stieltjes sums over finite divisions, it successfully copes with the difficulty that these sums do not stand a chance to converge pathwise since Brownian motion has paths of almost surely unbounded variation. However, by requiring less, Itô has shown that convergence holds in the L^2 sense (see Definition 2.129), which extends to convergence in probability (see Definition 2.139). Since the original ideas of Itô, stochastic integration theory has further developed by using continuous martingales as integrators instead of Brownian motion. We refer to [156, 193, 266, 218, 270, 271, 284, 311, 314, 315, 359] for stochastic integration theory with Brownian or more general integrators, and to [76, 188, 284] for stochastic differential equations.

Notes to Chapter 3

3.1

Lévy processes are random processes with stationary and independent increments, in many ways generalizing Brownian motion and Poisson process. In particular, Lévy processes are Markov processes and semimartingales, with right-continuous paths allowing left limits, which in general contain continuous pieces with jump discontinuities at random times. The generators of Lévy processes are in general nonlocal pseudo-differential operators. We refer to [7, 28, 34, 36, 53, 269, 330, 333] for general results on Lévy processes, to [102, 236] for their fluctuation theory, and to the three-volume set [195], which is a comprehensive study emphasizing the context of semigroups and generators. The work [53] is devoted to the potential theory of stable random processes such as subordinate Brownian motions, from which we take estimates of p_t^X. The asymptotic behavior $\lim_{|x|\to\infty} p_1^X(x)$ in (3.1.27) is given in [50], and the estimate of $p_t^X(x)$ in (3.1.28) by [54] and [53, (2.30)]. Subordinators are an example of a random time change. The work [29] is a comprehensive reference of time changes and discusses processes of the form $(X_{\hat{T}(\theta)})_{\theta\geq 0}$ with change of time $(T(\theta))_{\theta\geq 0}$.

3.2

The material on path properties of Lévy processes is taken from [330, Chapter 2.11]. The space $\mathscr{X}_D$ of $\mathbb{R}^d$-valued càdlàg paths on $[0,\infty)$ is investigated in [116, Section 3.5]; for the separability and completeness of $(\mathscr{X}_D, d_D)$, see [116, Theorem 5.6]. We also refer to [41, Section 16] and [41, Theorem 16.3]. These properties depend on the measure defined on $\mathscr{X}_D$. Let $D_1 = D([0,1])$ be the set of real-valued functions on $[0,1]$ that are right-continuous with left-limits, and Λ be the class of strictly increasing continuous functions $\lambda : [0,1] \to [0,1]$. We then have $\lambda(0) = 0$ and $\lambda(1) = 1$. For $X, Y \in D_1$, one can define

$$d(X,Y) = \inf_{\lambda\in\Lambda}\{ \sup_{t\in[0,1]} |\gamma(t) - 1| \vee \sup\nolimits_{t\in[0,1]} |X_t - Y_{\lambda(t)}|\}$$

and

$$d^\circ(X,Y) = \inf_{\lambda\in\Lambda}(\|\lambda\|^\circ \vee \sup\nolimits_{t\in[0,1]} |X_t - Y_{\lambda(t)}|).$$

Here

$$\|\lambda\|^\circ = \sup_{0\leq t<s\leq 1} \left|\log \frac{\lambda(s)-\lambda(t)}{s-t}\right|.$$

It is then shown in [41, Theorem 12.1] that d and d° are equivalent metrics. It is also shown in [41, Theorem 12.2] that D is separable under d and d°, and complete under d°. Proposition 3.37 is taken from [116, Proposition 7.1].

3.3

One example of a Poisson random measure can be constructed by counting jumps of specific size for a given Lévy process $(X_t)_{t\geq 0}$ by

$$N(t, A) = |\{0 \leq s \leq t \mid \Delta X_s \in A\}|,$$

see, e. g., [7, Section 2.3]. We refer to [7, Theorem 2.3.5] for Theorem 3.53, to [7, Theorem 2.4.16] for Theorem 3.58, and to [7, Theorem 2.3.9] for Proposition 3.57. Lévy–Itô decomposition (Theorem 3.58) is given in, e. g., [330, Theorems 19.1 and 19.2].

3.4

The Itô formula for semimartingales (Proposition 3.74) is taken from [188, Chapter II, Section 5], we also refer to [301].

3.5

We use later some exponential functionals of Lévy processes to estimate the decay properties of bound states of nonlocal Schrödinger operators. The proofs of Lemma 3.77 and Theorem 3.78 can be found in [66]. The capacitary measure and capacity are discussed in [36]. The proofs of Lemmas 3.80 and 3.81 are given in [36, II.2, Lemmas 6 and 7], respectively. The proof of Chung–Fuchs criterion is given in [36, I.4, Theorem 17] for a special case. For the general case, see [296]; for improved criteria of recurrence, see [329].

3.6

Subordinators are one-dimensional Lévy processes with almost surely nondecreasing paths. We refer to [7, 34, 333] for the relationship between Bernstein functions and subordinators.

Notes to Chapter 4

4.1

There is a vast literature on quantum mechanics from a mathematical point of view. The series [308, 305, 307, 306] and [347] are well established, and [221] is classic, all of them containing fundamental discussions on the spectral analysis of Schrödinger operators. We also refer to [48, 75, 89, 91, 127, 154, 172, 249, 365, 364] for various aspects of rigorous quantum mechanics. The works [344, 345] are concise but stimulating and up-to-date accounts of the spectral property of Schrödinger operators. The substantial paper [342] and the monograph [339] are seminal discussing Schrödinger semigroups and very appealing applications of the Feynman–Kac formula to the spectral analysis of Schrödinger operators. For a discussion of the fractional Laplace operator, see [235].

Proposition 4.23 on the min–max principle is referred to in [306, Theorem XIII.1], where another formulation of the min–max principle is introduced. Proposition 4.46 on the criterion for the compactness of operators is from [306, Theorem XIII.64], and Proposition 4.48 from [306, XIII.4, Corollary 2]. The Kato–Rellich theorem is from [310], and a first application to atomic Hamiltonians is presented in [220]. Kato's inequality is from [222]. The operator version of Kato's inequality, called abstract Kato's inequality, is considered in [92, 165, 337, 338]. Furthermore, Kato's inequality for relativistic Schrödinger operators is shown in [170]. Proposition 4.60 is taken from [308, Theorem VIII.5], and the KLMN theorem from [305, Theorem X.17]. For Theorem 4.65, see [221, Theorem VIII.3.11, Theorem VIII.3.13 with Supplementary notes to Chapter VIII, 5 (p. 575)]. Also see [308, Supplement to VIII.7]. Rellich's criterion first appeared in [309].

The Trotter product formula (4.84) is a powerful tool in the construction of the functional integral representation of Schrödinger-type semigroups. The original paper is [367]. The works [223, 224] extend the Trotter product formula to the case of a quadratic form sum $A \dotplus B$ for $\overline{A+B}$ in (4.84). Proposition 4.85 is from [225], where the Trotter formula is further extended to nonlinear semigroups generated by subdifferentials of convex functions. The works [279, 280] use trace norm limits in the Trotter formula instead of the strong limits, assuming that $e^{-t(A \dotplus B)}P_{AB}$ is trace class (here P_{AB} denotes projection onto $D(A) \cap D(B)$). Operator norm convergence of products in the Trotter formula is obtained in [185] with an error bound

$$\|(e^{-(t/n)A}e^{-(t/n)B})^n - e^{-t(\overline{A+B})}\| = O(n^{-1/2})$$

as $n \to \infty$. Here it is assumed that $A+B$ is essentially self-adjoint on $D(A) \cap D(B)$. That this error bound is the best possible is shown in [362]. In the case where $A + B$ is moreover self-adjoint on $D(A) \cap D(B)$, the error term can be improved to $O(n^{-1})$ [186]. Furthermore, [52] shows norm convergence in the Trotter product formula on Banach spaces. The Hardy–Littlewood–Sobolev inequality is proven in [161, 162, 352]. The sharp version is given in [247], where the best constant of $C(p, q, r, d)$ is given in the case $p \neq r$. The proof is also found in [249, Section 4.3]. Compact operators of the form $f(x)g(-i\nabla)$ are comprehensively studied in [346, Chapter 4] and [347, Chapter 3]. The proof of Propositions 4.39 and 4.37 are given in [346, Theorems 4.1 and 2.8], respectively. Operator $B_{u,g}$ defined in (4.1.28) is introduced in [74, p. 96] and Proposition 4.40 is from [74, Theorem]. Convergence theorems are listed in [346, pp. 26–28].

4.2

For the classic Feynman–Kac formula, see [68, 90, 200, 339]. Kac has originally considered positive potentials, however, this strong restriction has been gradually weakened.

There is a large and important body of research on Feynman–Kac semigroups for Schrödinger operators on bounded domains, which we cannot address in this book for lack of space. The analysis in this case involves tools of potential theory, stopping times, and killed processes. The earliest papers establishing a link between Brownian motion and harmonic functions are [104, 105, 204, 205]. Further developments include the sequence of papers [180]; see [51, 88, 106, 334] for monographic treatments and the modern [294] presenting a unifying analytic-probabilistic approach. In the context of processes generated by Schrödinger operators, a seminal paper is [3], and a book length discussion of the subject is [73].

For the Schrödinger operator $H = -\frac{1}{2}\Delta + V$, the Dirichlet boundary value problem for a given bounded domain $D \subset \mathbb{R}^d$ with smooth enough boundary ∂D is

$$\begin{cases} H\psi = 0 & \text{in } D, \\ \psi = \phi & \text{on } \partial D. \end{cases}$$

Consider the first exit time $\tau_D = \inf\{t > 0 \mid B_t \notin D\}$ from D of standard Brownian motion $(B_t)_{t\geq 0}$. From the definition, we have $B_t \in D$ for $t < \tau_D$, and $B_{\tau_D} \in \partial D$ if $\tau_D < \infty$. It is possible to prove that $\tau_D < \infty$ $\mathcal{W}$-a. s. Then the Feynman–Kac formula becomes

$$\psi(x) = \mathbb{E}^x[e^{-\int_0^{\tau_D} V(B_t)dt}\phi(B_{\tau_D})]$$

and thus it offers a probabilistic representation of the solution to the Dirichlet problem. In this case the spectrum of the Laplace operator is discrete consisting of eigenvalues $0 < \lambda_0(D) < \lambda_1(D) < \cdots$.

(a) *Isoperimetric inequalities.* One line of research is addressing the question how the various properties of the spectrum depend on geometric properties of D. Let D be a convex domain and V be a convex function. In this case also the Schrödinger operator has a purely discrete spectrum $0 < \lambda_0(D, V) < \lambda_1(D, V) < \cdots$. Without restricting generality, suppose that $0 \in D$, and let $\mathbb{B}$ denote a ball centered at the origin such that $|D| = |\mathbb{B}|$, where the bars denote Lebesgue measure. Then it can be shown that the first exit time of Brownian motion is maximized by the ball, i. e., $\sup_{x\in D} \mathcal{W}^x(\tau_D > t) \leq \mathcal{W}^0(\tau_{\mathbb{B}} > t)$, for all $t > 0$. This is an example of an isoperimetric inequality. Since

$$\lim_{t\to 0} \frac{1}{t} \log \mathcal{W}^x(\tau_D > t) = -\lambda_0(D),$$

the above inequality implies the *Faber–Krahn inequality* $\lambda_0(\mathbb{B}) \leq \lambda_0(D)$, which is an isoperimetric property of the eigenvalues of the Dirichlet Laplacian [27, 70]; see also [286]. The Faber–Krahn inequality has been similarly shown for Schrödinger operators for convex domains [64, 159, 293]; see the conditions on V in the referred papers. Another isoperimetric inequality concerns the ratio of consecutive eigenvalues, originating from the Payne–Pólya–Weinberger conjecture [289] which originally stated that for planar domains $\lambda_1(D)/\lambda_0(D)$ is maximized when D is chosen to be a disk. The conjecture has been proven to hold in any dimension in [17, 18].

(b) *Spectral gaps.* The difference $\Lambda(D) = \lambda_1(D) - \lambda_0(D)$ is called the *fundamental gap* for the Dirichlet Laplacian, and a similar object can be defined for the Dirichlet Schrödinger operator. Based on specific cases in which the eigenvalues can be computed, it has been conjectured [35] that $\Lambda(D) \geq \frac{3\pi^2}{2d^2}$, where d is the diameter of D, provided both D and V are convex. An upper bound can be obtained as follows. Since by the Payne–Pólya–Weinberger conjecture $\lambda_1(D)/\lambda_0(D) \leq \lambda_1(\mathbb{B})/\lambda_0(\mathbb{B})$, it follows that $\Lambda(D) \leq \Lambda(\mathbb{B})(\lambda_0(D)/\lambda_0(\mathbb{B}))$, in which the second factor at the right-hand side can be further estimated by using the Faber–Krahn inequality. For a discussion, see [16] and the references therein. For Schrödinger operators with convex V, the fundamental gap conjecture was proven recently in [6].

(c) *Harnack inequalities.* In its classical form, the boundary Harnack inequality states that for any open sets $K \subset K^* \subset \mathbb{R}^d$ with $\overline{K} \subset K^*$, there exists $C > 0$ such that for all functions f, harmonic and nonnegative on K^*, we have $f(x) \leq Cf(y)$, for all $x, y \in K$. In other words, just by knowing that a nonnegative f solves $\Delta f = 0$, we can already conclude that $\inf_{x\in K} f(x) \geq \frac{1}{C} \sup_{x\in K} f(x)$, where C depends on K and K^* but not on f. A version of this inequality for Schrödinger operators was first proven by [368] using analytic methods. The work [3] realized that the Feynman–Kac formula can be used to obtain these results. They used the fact that when stopped at the boundary of K^*, a Feynman–Kac-type formula just gives the solution of the Dirichlet problem for H and K^*. Further references include [73, 160, 376, 377].

Further interesting aspects include *heat kernel estimates* [24, 80, 82, 147, 148, 292], *functional inequalities* [328], and *intrinsic ultracontractivity* [23, 81, 83]. For the general theory of how the Feynman–Kac formula can be used to study various classes of partial differential equations, we also refer to [30, 31, 109, 131, 358].

Kato-class has been introduced in [222]. Simon [342] studies Schrödinger semigroups with Kato-class potentials in great detail. In particular, here it is proven that e^{-tH} is a bounded operator from L^p to L^q for $1 \leq p \leq q \leq \infty$. A natural question is how the norm of this operator depends on t for large t. It is shown in [340] that $a = \lim_{t\to\infty}(1/t)\ln \|e^{-tH}\|_{p,q}$ is independent of p and q; it turned out [341] that the same a is obtained by taking pointwise limits of images of nonnegative functions. Since then there has been considerable research activity on the topic; see [80, 82, 233, 276, 277, 292, 374, 375], as well as [234, 360, 361]. Yet another, more recent, line of research is concerned with cases when the potential V is replaced by more singular objects such as distributions, or when Brownian motion is replaced by a Bessel process. For research in this direction, see [319, 320, 321, 322, 323] and the references therein.

4.3

Theorem 4.116 can be improved by showing that $e^{-tK}(x,y)$ is even jointly continuous in x, y, and t [342].

The Kallianpur–Robbins laws stated in Proposition 4.137 estimate the occupation time of a Brownian motion, which is studied in [217]. It can be extended to a fractional Brownian motion in [219]. The discussion in Theorems 4.138 and 4.139 of the existence of bound states of low-dimensional Schrödinger operators follows [327].

The proof of Lemma 4.144 is due to [74]. Estimates of the number of negative eigenvalues of Schrödinger operators are due to [245, 246], where the Birman–Schwinger principle and the Feynman–Kac formula are used; see also [251, 252]. The proof of Lemma 4.153 is taken from [346] and [339, Theorems 8.1–8.2]. In [74, Theorem], it is shown that

$$\sup_{k\in\mathbb{N}} k^{1/p}\left(k^{-1}\sum_{m=1}^{k}\mu_m(B_{u,g})^2\right)^{1/2} \le K_p\|u\|^*_{p,w}\|g\|_p,$$

where $B_{u,g}$ is defined by (4.1.28) and the constant K_p is explicitly given by

$$K_p = \frac{p}{2}\left(\frac{4}{\frac{p}{2}-1}\right)^{1-2/p}\left(1+\frac{2}{p-2}\right)^{1/p}.$$

The most important corollary derived from this inequality is an upper bound on the number of nonpositive eigenvalues $\bar{N}_0(V)$ which coincides with that of eigenvalues of the operator $(-\Delta)^{-1/2}|V|^{1/2}$ greater than or equal to 1; $(-\Delta)^{-1/2}|V|^{1/2}$ is the operator $B_{(2\pi|\xi|)^{-1},|V(x)|^{1/2}}$ followed by inverse Fourier transform. It is known that $u(\xi) = (2\pi|\xi|)^{-1} \in L^d_w(\mathbb{R}^d)$ with $\|u\|^*_{d,w} = \omega_d^{1/d}/2\pi$, where $\omega_d = \pi^{d/2}/\Gamma(1+d/2)$ denotes the volume of the unit ball in $\mathbb{R}^d$. Let $k = \bar{N}_0(V)$ and $p = d$. Then we have

$$1 \le \left(\frac{1}{\bar{N}_0(V)}\sum_{m=1}^{\bar{N}_0(V)}\mu_m((-\Delta)^{-1/2}|V|^{1/2})^2\right)^{1/2} \le \frac{K_d\omega_d^{1/d}}{2\pi}\left(\frac{\int_{\mathbb{R}^d}|V(x)|^{d/2}dx}{\bar{N}_0(V)}\right)^{1/d}$$

and so

$$\bar{N}_0(V) \le (K_d/2\pi)^d\omega_d\int_{\mathbb{R}^d}|V(x)|^{d/2}dx$$

follows. Here numerically $(K_d/2\pi)^d\omega_d = 27/(2\pi^2) \approx 1.37$ for $d = 3$. The method in [245] gives the same estimate in this context, with the much better constant 0.116 when $d = 3$.

Moments of negative eigenvalues $N^\alpha(V) = \sum_{j=1}^N |E_j|^\alpha$ for $d = 1, 2$ are also investigated. The cases $\alpha > 1/2$, $d = 1$, and $\alpha > 0$, $d \ge 2$, are proven by [253]. For $\alpha = 0$, $d \ge 3$, proofs are given independently in [74, 245, 317]. The remaining critical case $\alpha = 1/2$,

$d = 1$, is proven in [371]. An interesting problem is to determine sharp values of the constants a_d and $a_{d,\alpha}$ in Theorems 4.159 and 4.162, respectively. This is found for the cases in which $\alpha \geq 3/2$ and $d \geq 3$ in [2, 241, 253], and $\alpha = 1/2$ and $d = 1$ in [178]. The Lieb–Thirring inequality can be extended to a wide class of Schrödinger operators, including those with spin and magnetic field; we refer to [171, 248, 275] and references therein. The extension of the Birman–Schwinger principle is shown in Theorem 4.149, which is due to [275, 363]. Theorem 4.149 gives a sufficient condition such that $-\frac{1}{2}\Delta + V$ has no nonpositive eigenvalues. For $d = 1$, an upper bound on $N_0(V)$ for $V < 0$ is studied in [49, 228], which is of the form

$$N_0(\lambda V) \leq 1 + \int_{\mathbb{R}} |x|\lambda V(x)dx.$$

Whenever $d = 1, 2$ and under some conditions on $V < 0$, the operator $-\frac{1}{2}\Delta + \lambda V$ has a negative eigenvalue no matter how small λ is. In [336] bound states of $-\frac{1}{2}\Delta + V$ in one and two dimensions are studied. Proposition 4.120 is from [306, Theorem XIII.52].

The absence of zero eigenvalues of Schrödinger operators is a problem of fundamental importance. Proposition 4.168 is due to [129], proven by using the continuity of solutions of Schrödinger operator for $d \geq 3$, which is shown in [199]. For a more general case, the absence of zero eigenvalues of Schrödinger operator is shown in [369]. The critical order of the absence of zero eigenvalues is $V(x) = -r^{-2}$, studied in [350].

There is a long history of confusion and controversy for time operators. The origin of this may come from Pauli [288, p. 63, footnote 2] and [287], from 1933. In this argument, a time operator satisfying $[H, T] = -i1$ with Hamiltonian H having a discrete spectrum is basically forbidden. It should be noted, however, that the argument is very formal and no attention was paid to the domain of operators involved. In a rigorous manner, the time operator associated with a self-adjoint operator with purely discrete spectrum is discussed in [9, 134]. In particular, Lemma 4.165 is due to [9]. Strong time operators associated with an abstract self-adjoint operator with purely absolutely continuous spectrum are initiated by [274] and developed in, e. g., [8]. For Lemma 4.163, see [8, 274]. A generalization of Theorems 4.170 and 4.173 is discussed in [10, Theorem 5.4], where $\sum_{j=1}^{\infty} E_j^2 < \infty$ is not assumed and an ultraweak time operator associated with a Hamiltonian with purely discrete spectrum is constructed by an infinite direct sum of ultraweak time operators in [10, Theorem 5.2]. On Agmon potential, Lemma 4.171, we learn from [306, Theorem XIII.33, Agmon–Kato–Kuroda theorem]. Conditions in Theorem 4.173 are discussed in [306, Theorems XIII.6(a) and XIII56]. The modified wave operator in Example 4.175 is taken from [307, Theorem XI.71].

The Perron–Frobenius theorem first appeared in [132, 290] for matrices, and it was applied to quantum field theory in [143, 150, 151]. The Klauder phenomenon is discussed in [117, 118, 121, 335]. Proposition 4.126 is proven in [214]. Due to the Klauder phenomenon, one can construct examples of Schrödinger operators with degenerate ground states.

Exponential decay in the L^2-sense, i. e., $\|e^{a|x|}\phi\| < \infty$ for eigenfunctions, is studied in [1], where operators in the divergence form $-\partial_\mu A_{\mu\nu}(x)\partial_\nu + V$ are also considered. Pointwise exponential decay of eigenfunctions of Schrödinger operators is investigated in [65], and in [66] for the relativistic Schrödinger operator. In the massless relativistic case, i. e., for $\sqrt{-\Delta} + V$, eigenfunctions decay only polynomially. For the divergence form $-\partial_\mu A_{\mu\nu}(x)\partial_\nu + V$ with the uniform elliptic bound $c \le |A_{\mu\nu}(x)| \le C$ and a further regularity condition on $A_{\mu\nu}$, the pointwise exponential decay of eigenfunctions can be proven in a similar way to Carmona's estimates with respect to the diffusion process associated with the divergence form. Fromula (4.3.86) gives the probability of the set of paths $P(a, [c, d], t)$,

$$\mathcal{W}(P(a, [c, d], t)) = \frac{1}{\sqrt{2\pi t}} \int_{[c,d]} \sum_{k=-\infty}^{\infty} (-1)^k \exp\left(- \frac{(u - 2ka)^2}{2t} \right) du,$$

which is taken from [55, 1.15.8, p. 174]. We refer [65] for the pointwise lower bound of positive bound states discussed in Lemma 4.182 and Theorem 4.183.

4.4

References on Schrödinger operators with magnetic field include [19, 20, 21]; they are comprehensively studied in [304]. Self-adjointness of the Schrödinger operator with magnetic field $H(a)$ is established in [243]. The alternative proof of the Feynman–Kato–Itô formula of the Schrödinger operator with magnetic field $H(a)$ is taken from [339, Section V], where $\nabla \cdot a = 0$ is assumed. The applications of functional integration with magnetic field and the diamagnetic inequality are from [58, 114, 115, 130, 179, 283].

4.5

This material is taken from [344]. For some further cases of interest of unbounded Feynman–Kac semigroups, see [59, 372].

4.6

For results on the spectrum of the relativistic Schrödinger operator, see [26, 78, 79, 164, 244, 250]. The Feynman–Kac formula for relativistic Schrödinger operators $H_{\mathrm{R}}(a)$ with $a = 0$ is studied in [66, 84]. The case of nonzero vector potential is discussed in [87, 168]. The function $f(x) = \sqrt{x^2 + m^2} - m$ is an example of Bernstein functions. The relativistic Schrödinger operator is then realized as $f(\frac{1}{2}(-i\nabla - a)^2) + V$.

The relativistic Schrödinger operator can also be defined through the *Weyl quantization*. This is given by $H^W = h^W(x, D) + V(x)$, where

$$(h^W u)(x) = \int_{\mathbb{R}^d \times \mathbb{R}^d} e^{i(x-y)\cdot\xi} h\left(\frac{x+y}{2}, \xi \right) u(y)(2\pi)^{-d} dy d\xi,$$

with symbol $h(x,\xi) = \sqrt{|\xi - a(x)|^2 + m^2}$. The work [278] shows that in general

$$(h^W(x,D))^2 \neq (-i\nabla - a)^2 + m^2.$$

The Feynman–Kac formula of e^{-tH^W} is given in [183].

4.7

In quantum mechanics the notion of spin is introduced as an internal degree of freedom of electrons. Spin plays a central role in the statistics of electrons. Another aspect of the influence of spin is in the behavior of electrons in a magnetic field. The Feynman–Kac formula for the Schrödinger operator with spin is discussed in [85, 86, 135, 136, 168]. The diamagnetic inequality (4.7.35) is due to [86, 168]. Paramagnetism of Schrödinger operators with spin 1/2 is also an interesting subject. It has been conjectured in [174] that paramagnetism is universal, but a counterexample has been given in [22], while paramagnetism in the classical limit is discussed in [173].

For a two-dimensional space-time Dirac operator, [181, 182, 183, 184] propose a path integral representation in terms of a $\mathscr{S}'(\mathbb{R}^2; M_2(\mathbb{C}))$-valued countably additive measure on $C([0,t];\mathbb{R})$. In particular, in [184], the measure is concentrated on paths having differential coefficients equal to "$\pm 1 \times$ speed of light" in every finite time interval except at finitely many instants of time. This relates with the *Zitterbewegung* of Dirac particles. A similar result is also given in [85] with respect to a Poisson process. Furthermore, [175, 176, 177, 255] treat quantum probability and derive a noncommutative Feynman–Kac formula. The *Rabi Hamiltonian* is a model of a harmonic oscillator coupled to a two-level atom, and it is given by

$$H_{\text{Rabi}} = \triangle\sigma_z + \omega a^* a + g\sigma_x(a + a^*),$$

where $\Delta \geq 0$ and $\omega > 0$ are positive constants, $g \in \mathbb{R}$ is a parameter, and

$$a = \frac{1}{\sqrt{2}}\left(\frac{1}{\sqrt{\omega}}\frac{d}{dx} + \sqrt{\omega}x\right), \quad a^* = \frac{1}{\sqrt{2}}\left(-\frac{1}{\sqrt{\omega}}\frac{d}{dx} + \sqrt{\omega}x\right).$$

This was introduced in [303] and the integrability is studied in [57] from a physical point of view. Applications of the Rabi model range from quantum optics to molecular physics, and recently it plays an important role in cavity QED and circuit QED. It is a self-adjoint operator on $\mathbb{C}^2 \otimes L^2(\mathbb{R})$. The semigroup generated by the Rabi Hamiltonian constitutes one of the canonical examples proving the full power of the Feynman–Kac formula in situations where spin degrees of freedom are present. By the Feynman–Kac formula, we can show that the ground state is unique [166]. The infinite-dimensional version of the Rabi model is the so-called *spin-boson model*. This is also studied in [167] and the $P(\phi)_1$ process associated with a spin-boson model is constructed.

4.8

A Feynman–Kac-type formula for relativistic Schrödinger operators with spin

$$H_{\mathrm{RS}}(a,b) = \sqrt{(-i\nabla - a)^2 - \sigma \cdot b + m^2} - m + V$$

is from [87, 168, 169], where a combination of a stopping time (subordinator) and a Poisson process is applied. The Schrödinger operator with spin can be transformed to the operator on $L^2(\mathbb{R}^3 \times \mathbb{Z}_2)$, which is from [87]. In [169] unbounded magnetic fields $b \notin (L^\infty(\mathbb{R}^3))^3$ are also studied, and by extending the strategy developed in [66] we can establish decay properties of the eigenfunctions of $H_{\mathrm{RS}}(a,b)$.

4.9

Nonlocal equations, as well as related operators and jump random processes, are currently much investigated in analysis and probability, attracting a wide range of applications. Nonlocal Schrödinger operators can be defined in the spirit of classical Schrödinger operators by adding a nonlocal kinetic term to a multiplication operator playing the role of a potential. In [168] a class of nonlocal Schrödinger operators has been introduced by choosing the kinetic term to be a Bernstein function of the Laplacian, and adding further possible terms like vector potentials and generalized spin leading to

$$\Psi\left(\frac{1}{2}(-i\nabla - a)^2\right) + V$$

or

$$\Psi\left(\frac{1}{2}(\sigma \cdot (-i\nabla - a))^2\right) + V$$

with Bernstein function Ψ. The operator $\Psi(\frac{1}{2}(\sigma \cdot (-i\nabla - a))^2) + V$ can be realized as a self-adjoint operator in $L^2(\mathbb{R}^3 \times \mathbb{Z}_2)$. A general spin can be defined on

$$L^2(\mathbb{R}^d \times \{e^{\theta_p}, e^{2\theta_p}, \ldots, e^{p\theta_p}\}), \quad \theta_p = 2\pi i/p.$$

The idea to construct Feynman–Kac formulae is from [84, 85, 86, 87]. In [171] Lieb–Thirring inequalities for nonlocal Schrödinger operators (Theorem 4.317) are given, see also [78] for the relativistic Schrödinger operator. Proposition 4.281 is a modification of [66, Theorem III.1] and Proposition 4.282 is taken from [66, Lemma III.3]. We refer to [196] for volume doubling in Lemma 4.318. Bounds on $p_t^\Psi(x)$, see (4.9.82), are given in [72], and (4.9.83) in [71]. The kernel of $(-\Delta)^{-\alpha/2}$ and Riesz potentials are studied in, e. g., [356, Chapter V], [249, Chapter 5], and [53, 145]. We refer to [348, Section 6.8] and [238, Chapter I] for the case of $\alpha = d$. Detailed potential kernel estimates for the fractional and fractional relativistic Laplacians have been obtained in [14], leading to the definition and study of a fractional and fractional relativistic counterpart of the classical Rollnik-class of potentials.

Eigenfunction decay for more general nonlocal Schrödinger operators with confining or decaying potentials discussed in Section 4.9.8 has been investigated in the papers [207, 208, 210]. For a discussion of the qualitative regime change (phase transition in the decay rates), see also [209]. The framework of jump-paring processes has been first proposed in [208] and further developed in [210]. The explicit solution of the eigenvalue problems for the massless relativistic Schrödinger operator in one dimension presented Section 4.9.10 has been obtained in [262] for the quadratic and in [110] for the quartic potential.

Embedded eigenvalues for nonlocal Schrödinger operators just started to be studied. A standard account for classical Schrödinger operators in [112]. In [264] relativistic counterparts of the classical von Neumann–Wigner and Moses–Tuan potentials have been obtained, showing that in the nonrelativistic limit the potentials obtained for the relativistic case converge to the classically known potentials. Here also two families of potentials have been constructed for which the related massless relativistic Schrödinger operators have zero eigenvalues. For the delicate case of zero-eigenvalues, a probabilistic study of eigenfunction decay has been done in [212] also throwing light on the mechanisms of bound state formation at the continuum edge. A detailed study of potentials generating embedded eigenvalues at zero for nonlocal Schrödinger operators is offered in [11]. Results on nonexistence of embedded eigenvalues for nonlocal Schrödinger operators also have been obtained [189] by developing Mourre commmutator estimates, virial theorems, and a technique based on the extended Birman–Schwinger principle.

For a study of some local properties of eigenfunctions of nonlocal Schrödinger operators with potential wells, see [44]. For a proof of the monotonicity of eigenfunctions in this context, see [11, 12]. These results have been used to obtain sharp eigenfunction estimates in the bulk in [13]. In [44] there is a detailed study of the location of extrema of eigenfunctions of nonlocal Schrödinger operators for bounded domains and its multiple consequences such as a Faber–Krahn inequality in convex domains and nonlocal Schrödinger operators with Bernstein functions of the Laplacian under Dirichlet exterior conditions. For further developments, such as maximum principles, ABP estimates, Ambrosetti–Prodi bifurcations, and Hopf's lemma in a similar context, see also [43, 45, 46, 47].

Notes to Chapter 5

5.1

Ground state transform is a case of Doob's h-transform, making the relationship between a classical Schrödinger operator and a diffusion. For basic references in this context, see [339, 294]. An extension to fractional Schrödinger operators and stable processes has been first made in [206, 207], and for more general nonlocal Schrödinger operators in [265].

For intrinsic ultracontractivity properties, see the basic references [80, 81, 83]. Asymptotic intrinsic ultracontractivity has been introduced in [207]. For a comprehensive study of intrinsic ultracontractivity in the context of nonlocal Schrödinger operators, we refer to [208], where also the literature is discussed. The relationship between these and other contractivity properties (supercontractivity, hypercontractivity) has been explored in [213].

For $P(\phi)_1$-processes, a first source is [316]. Such processes have an interesting property deriving from connections with statistical mechanics and Gibbs measures. Formally, such a process is the stationary solution of the stochastic partial differential equation (SPDE)

$$\dot{X}(u,t) = \frac{1}{2}\Delta_u X(u,t) - (\partial V)(X(u,t)) + \mathcal{W}, \tag{6.0.1}$$

where $\mathcal{W}$ now is *space-time white noise*. This was shown rigorously for a restricted class of potentials V in [194]. An interesting application appears in [157], where the SPDE above is used in a Markov chain Monte Carlo algorithm for sampling the conditional $P(\phi)_1$-measure in various situations. In the context of SPDEs, the result of [39] is also of interest: when the potential grows sufficiently rapidly, it implies uniqueness of those stationary measures of the SPDE above that are Gibbs measures, i. e., satisfy the Dobrushin–Lanford–Ruelle (DLR) equations. It is expected, but not proven, that all stationary measures of the above SPDE satisfy the DLR equations. The $P(\phi)_1$-process associated with the divergence form

$$-\sum_{i,j=1}^{d} \partial_i a_{ij}(x)\partial_j + V(x)$$

with a confining potential can be constructed, and also that of $-\Delta + V(x)$. We refer to, e. g., [141]. Lemma 5.24 is taken from [141], which is a modification of [227].

The proof of Mehler's formula through analytic continuation $t \mapsto -it$ has been first obtained in [258]. Lemma 5.26 can be found in [268].

Càdlàg $P(\phi)_1$-processes can be further constructed for jump processes and nonlocal Schrödinger operators. Fractional $P(\phi)_1$-processes associated with fractional Schrödinger operators have been investigated in [207], and with more general nonlocal Schrödinger operators in [265]. For the long-time fluctuation properties of such processes see [211], and for refined results on their path regularity (multifractal spectrum of local Hölder-exponents), we refer again to [265].

5.2

The seminal papers introducing Gibbs measures for lattice spin systems are [98, 99] and [240]. Further details on the prescribed conditional probability kernels (called specifications) can be found in [100, 128] and [144, 298, 353]. Surveys and books on Gibbs measures include the early [326], covering also classical continuous and quantum lattice systems, and [137, 138, 239, 297, 299, 302, 325, 354, 355]. In [298] a general framework is developed, while the approach of [190] is more from the perspective of the variational principle in which equilibrium states appear as tangents to the free energy functional. Reference [139] is a highly authoritative comprehensive monograph with a strong stress on DLR theory, covering a wealth of material. The work [267] is a systematic text on cluster expansion methods applied to Gibbs measures. Other relevant books on rigorous statistical mechanics include [152], focusing more on Ising models, [349], giving an early account of contour techniques, and [343], giving an account of lattice gases. A reappraisal of Gibbsian theory was presented in the book-length paper [370], especially in the context of renormalization group transformations. For Gibbs measures on lattice spin systems with noncompact space existence of a Gibbs measure is harder to prove: see the superstability conditions in [324], tempered Gibbs measures [69, 242], [256, 260, 261] for cases when the potential is not uniformly but pointwise or almost surely summable, and [140] for Gibbs measures indexed by subsets of an uncountable index set. For other approaches to and applications of Gibbs measures on path space(s), we refer to [5] and the references therein.

5.3

Some of the basic notions, such as local uniform domination, are adapted from discrete models; see [139]. The discussion of uniqueness of Gibbs measures follows [39]. The case $\alpha = 1$ in the uniqueness criterion suggests that a maximal configuration space on which μ is the unique Gibbs measure, a set consisting of exponentially growing paths is closer to truth than $\mathfrak{X}_0$; however, establishing this is beyond the method of our proof. In [194] sets of exponentially growing paths as sets of uniqueness for the Gibbs measure were obtained, however, assuming convexity of the function $V(x) - \kappa|x|^2$ for some $\kappa > 0$; therefore the case $\alpha < 1$ is not covered in this work. The advantage of our method is that it is not limited to polynomials, nor even to continuous or semibounded potentials; in particular, local singularities and other perturbations of V in the above examples will not alter the conclusions. This corresponds to the intuition that only the behavior of the potential at infinity should determine whether uniqueness of the Gibbs measure holds on a given set.

For existence, uniqueness, and support properties of Gibbs measures on càdlàg paths, we refer to [207, 211, 265].

5.4

An early variant of the existence proof for Gibbs measures with pair interaction potential comes from [37]. For further literature on Gibbs measures with pair interaction potential, we refer to [37, 40, 163, 285, 38].

5.5

There are several versions of the cluster expansion method. A systematic monograph is [267] and useful reviews are [60, 120]. For the general combinatorial scheme, see [119, 232, 295]. We refer to [32, 56, 61, 101, 273, 291] for improvements on convergence; however, these methods depend on the specific models at hand. For Brownian paths, a cluster expansion technique has been developed first in [257, 263]. More singular objects, such as densities dependent on double stochastic integrals are treated in [153, 259] using techniques of rough paths.

Bibliography

[1] S. Agmon. *Lectures on Exponential Decay of Solutions of Second-Order Elliptic Equations: Bounds on Eigenfunctions of N-body Schrödinger Operators*. Princeton University Press, 1982.

[2] M. Aizenman and E. Lieb. On semiclassical bounds for eigenvalues of Schrödinger operators. *Phys. Lett. A*, 66:427–429, 1978.

[3] M. Aizenman and B. Simon. Brownian motion and Harnack's inequality for Schrödinger operators. *Comm. Pure Appl. Math.*, 35:209–271, 1982.

[4] S. Albeverio and R. Høegh-Krohn. *Mathematical Theory of Feynman Path Integrals*. Springer, 1976.

[5] S. Albeverio, Yu. G. Kondratiev, Yu. Kozitsky, and M. Röckner. *The Statistical Mechanics of Quantum Lattice Systems. A Path Integral Approach*. EMS Tracts in Mathematics, 2009.

[6] B. Andrews and J. Clutterbuck. Proof of the fundamental gap conjecture. *J. Amer. Math. Soc.*, 24:899–916, 2011.

[7] D. B. Applebaum. *Lévy Processes and Stochastic Calculus*. Cambridge University Press, 2009.

[8] A. Arai. Generalized weak Weyl relation and decay of quantum dynamics. *Rev. Math. Phys.*, 17:1071–1109, 2005.

[9] A. Arai. Necessary and sufficient conditions for a Hamiltonian with discrete eigenvalues to have time operators. *Lett. Math. Phys.*, 87:67–80, 2009.

[10] A. Arai and F. Hiroshima. Ultra-weak time operators of Schrödinger operators. *Ann. Henri Poincaré*, 18:2995–3033, 2017.

[11] G. Ascione and J. Lőrinczi. Potentials for non-local Schrödinger operators with zero eigenvalues. *J. Diff. Equations*, 317:264–364, 2022.

[12] G. Ascione and J. Lőrinczi. Stability of ground state eigenvalues of non-local Schrödinger operators with respect to potentials and applications. *J. Math. Anal. Appl.*, 527:127549, 2023.

[13] G. Ascione and J. Lőrinczi. Bulk behaviour of ground states for relativistic Schrödinger operators with compactly supported potentials. *Ann. Henri Poincaré*, 25:2941–2994, 2024.

[14] G. Ascione, A. Ishida and J. Lőrinczi. Special potentials for relativistic Laplacians I: Fractional Rollnik-class. *J. Funct. Anal.*, 290:111282, 2026.

[15] R. B. Ash. *Real Analysis and Probability*. Academic Press, 1991.

[16] M. S. Ashbaugh. The fundamental gap. *www.aimath.org/WWN/loweigenvalues/gap.pdf*, 2006.

[17] M. S. Ashbaugh and R. D. Benguria. Optimal lower bound for the gap between the first two eigenvalues of one-dimensional Schrödinger operators with symmetric single-well potentials. *Proc. Amer. Math. Soc.*, 105:419–424, 1989.

[18] M. S. Ashbaugh and R. D. Benguria. A sharp bound for the ratio of the first two eigenvalues of Dirichlet Laplacians and extensions. *Ann. of Math.*, 135:601–628, 1992.

[19] J. Avron, I. Herbst, and B. Simon. Schrödinger operators with magnetic fields. I. General interactions. *Duke Math.*, 45:847–883, 1978.

[20] J. Avron, I. Herbst, and B. Simon. Separation of center of mass in homogeneous magnetic fields. *Ann. Phys.*, 114:431–451, 1978.

[21] J. Avron, I. Herbst, and B. Simon. Schrödinger operator with magnetic fields III. Atoms in homogeneous magnetic field. *Comm. Math. Phys.*, 79:529–572, 1981.

[22] Y. Avron and B. Simon. A counterexample to the paramagnetic conjecture. *Phys. Lett. A*, 75:41, 1979.

[23] R. Bañuelos. Intrinsic ultracontractivity and eigenvalue estimates for Schrödinger operators. *J. Funct. Anal.*, 100:181–206, 1991.

[24] R. Bañuelos and P. Méndez-Hernández. Sharp inequalities for heat kernels of Schrödinger operators and applications to spectral gaps. *J. Funct. Anal.*, 176:368–399, 2000.

[25] L. Bachelier. *Théorie de la spéculation*. Gauthier-Villars, Paris, 1900.

[26] A. A. Balinsky and W. D. Evans. *Spectral Analysis of Relativistic Operators*. Imperial College Press, 2011.

[27] C. Bandle. *Isoperimetric Inequalities and Applications*. Pitman, 1980.

https://doi.org/10.1515/9783112231463-007

[28] O. E. Barndorff-Nielsen, T. Mikosch, and S. I. Resnick. *Lévy Processes: Theory and Applications*. Birkhäuser, 2001.
[29] O. E. Barndorff-Nielsen and A. Shiryaev. *Change of Time and Change of Measure*. World Scientific, 2015.
[30] R. F. Bass. *Probabilistic Techniques in Analysis*. Springer, 1995.
[31] R. F. Bass. *Diffusions and Elliptic Operators*. Springer, 1997.
[32] G. Battle and P. Federbush. A phase cell cluster expansion for Euclidean field theories. *Ann. Phys.*, 142:2064, 1982.
[33] H. Bauer. *Measure and Integration Theory*. Walter de Gruyter, Studies in Mathematics, vol. 26, 2001.
[34] C. Berg and G. Forst. *Potential Theory on Locally Compact Abelian Groups*. Springer, 1975.
[35] M. van den Berg. On condensation in the free-Boson gas and the spectrum of the Laplacian. *J. Stat. Phys.*, 31:623–637, 1983.
[36] J. Bertoin. *Lévy Processes*. Cambridge University Press, 1998.
[37] V. Betz. Existence of Gibbs measures relative to Brownian motion. *Markov Process. Related Fields*, 9:85–102, 2003.
[38] V. Betz and F. Hiroshima. Gibbs measures with double stochastic integrals on path space. *Infin. Dimens. Anal. Relat. Top.*, 12:135–152, 2009.
[39] V. Betz and J. Lőrinczi. Uniqueness of Gibbs measures relative to Brownian motion. *Ann. Inst. Henri Poincaré, Probab. Stat.*, 39:877–889, 2003.
[40] V. Betz, J. Lőrinczi, and H. Spohn. *Gibbs Measures for Brownian Paths: Theory and Applications*, pp. 75–102. Springer, 2005.
[41] P. Billingsley. *Convergence of Probability Measures*. Wiley Interscience, 1968.
[42] P. Billingsley. *Probability and Measure*. Wiley, New York, 1995.
[43] A. Biswas and J. Lőrinczi. Maximum principles for time-fractional Cauchy problems with spatially non-local components. *Fract. Calc. Appl. Anal.*, 21:1335–1359, 2018.
[44] A. Biswas and J. Lőrinczi. Universal constraints on the location of extrema of eigenfunctions of non-local Schrödinger operators. *J. Diff. Equations*, 267:267–306, 2019.
[45] A. Biswas and J. Lőrinczi. Maximum principles and Aleksandrov-Bakelman-Pucci type estimates for non-local Schrödinger equations with exterior conditions. *SIAM J. Math. Anal.*, 51:1543–1581, 2019.
[46] A. Biswas and J. Lőrinczi. Ambrosetti-Prodi type results for Dirichlet problems of fractional-like Laplace operators. *Integr. Equat. Oper. Th.*, 92:26, 2020.
[47] A. Biswas and J. Lőrinczi. Hopf's lemma for viscosity solutions to a class of non-local equations with applications. *Nonlinear Anal. Theory Methods Appl.*, 204:112–149, 2021.
[48] J. Blank, P. Exner, and M. Havlí. *Hilbert Space Operators in Quantum Physics*. Springer, 2008.
[49] R. Blankenbecler, M. L. Goldberger, and B. Simon. The bound state of weakly coupled long-range one-dimensional quantum Hamiltonians. *Ann. Phys.*, 108:69–78, 1977.
[50] R. M. Blumenthal and R. L. Getoor. Some theorems on stable processes. *Trans. Amer. Math. Soc.*, 95:263–273, 1960.
[51] R. M. Blumenthal and R. L. Getoor. *Markov Processes and Potential Theory*. Academic Press, New York, 1968.
[52] S. Blunck. Some remarks on operator-norm convergence of the Trotter product formula on Banach spaces. *J. Funct. Anal.*, 195:350–370, 2002.
[53] K. Bogdan, T. Byczkowski, M. Ryznar, R. Song, and Z. Vondraček. *Potential Analysis of Stable Processes and its Extensions*. Lecture Notes in Mathematics, vol. 1980, Springer, 2009.
[54] K. Bogdan, A. Stós, and P. Sztonyk. Harnack inequality for stable processes on d-sets. *Studia Math.*, 158:163–198, 2003.
[55] A. N. Borodin and P. Salminen. *Handbook of Brownian Motion – Facts and Formulae* (2nd edition). Birkhäuser, 2002.
[56] A. Bovier and M. Zahradník. A simple inductive approach to the problem of convergence of cluster expansions of polymer models. *J. Stat. Phys.*, 100:765–778, 2000.
[57] D. Braak. Integrability of Rabi model. *Phys. Rev. Lett.*, 107:100401, 7 pages, 2011.

[58] K. Broderix, D. Hundertmark, and H. Leschke. Continuity properties of Schrödinger semigroups with magnetic fields. *Rev. Math. Phys.*, 12:181–255, 2000.
[59] K. Broderix, H. Leschke, and P. Müller. Continuous integral kernels for unbounded Schrödinger semigroups and their spectral projections. *J. Funct. Anal.*, 212:287–323, 2004.
[60] D. C. Brydges. *A Short Course on Cluster Expansions, Phénomènes critiques, systèmes aléatoires, théories de jauge, Part I–II (Les Houches, 1984)*. North-Holland, Amsterdam, 1986.
[61] D. C. Brydges and P. Federbush. A new form of the Mayer expansion in classical statistical mechanics. *J. Math. Phys.*, 19:95–139, 1978.
[62] R. H. Cameron. A family of integrals serving to connect the Wiener and Feynman integrals. *J. Math. Phys.*, 39:126–141, 1960.
[63] R. H. Cameron and D. A. Storvick. A simple definition of the Feynman integral, with applications. *Mem. Amer. Math. Soc.*, 46:1–43, 1983.
[64] L. De Carli and S. M. Hudson. A Faber–Krahn inequality for solutions of Schrödinger's equation. *Adv. Math.*, 230:2416–2427, 2012.
[65] R. Carmona. Pointwise bounds for Schrödinger eigenstates. *Comm. Math. Phys.*, 62:97–106, 1978.
[66] R. Carmona, W. C. Masters, and B. Simon. Relativistic Schrödinger operators: Asymptotic behavior of the eigenfunctions. *J. Funct. Anal.*, 91:117–142, 1990.
[67] P. Cartier and C. DeWitt-Morette. A new perspective on functional integration. *J. Math. Phys.*, 36:2137–2340, 1995.
[68] P. Cartier and C. DeWitt-Morette. *Functional Integration: Action and Symmetries*. Cambridge University Press, 2006.
[69] M. Cassandro, E. Olivieri, A. Pellegrinotti, and E. Presutti. Existence and uniqueness of DLR measures for unbounded spin systems. *Z. Wahrsch. verw. Gebiete*, 41:313–334, 1978.
[70] I. Chavel. *Isoperimetric Inequalities. Differential Geometric and Analytic Perspectives*. Cambridge University Press, 2011.
[71] Z. Q. Chen, P. Kim, and R. Song. Heat kernel estimates for $\Delta + \Delta^{\alpha/2}$ in $C^{1,1}$ open sets. *J. Lond. Math. Soc.*, 83:1–23, 2011.
[72] Z. Q. Chen and T. Kumagai. Heat kernel estimates for jump processes of mixed type on metric measure spaces. *Probab. Theory Related Fields*, 140:277–317, 2008.
[73] K. L. Chung and Z. Zhao. *From Brownian Motion to Schrödinger's Equation*. Springer, 1995.
[74] M. Cwikel. Weak type estimate for singular values and the number of bound states of Schrödinger operators. *Ann. Math.*, 106:93–100, 1977.
[75] H. L. Cycon, R. G. Froese, W. Kirsch, and B. Simon. *Schrödinger Operators. With Application to Quantum Mechanics and Global Geometry*. Springer, 1987.
[76] G. Da Prato and J. Zabczyk. *Stochastic Equations in Infinite Dimensions*. Cambridge University Press, 1992.
[77] Yu. L. Daletskii. Functional integrals connected with operator evolution equations. *Russian Math. Surveys*, 17:1–107, 1962.
[78] I. Daubechies. One electron molecules with relativistic kinetic energy: Properties of the discrete spectrum. *Comm. Math. Phys.*, 94:523–535, 1984.
[79] I. Daubechies and E. Lieb. One electron relativistic molecules with Coulomb interaction. *Comm. Math. Phys.*, 90:497–510, 1983.
[80] E. B. Davies. *Heat Kernels and Spectral Theory*. Cambridge University Press, 1989.
[81] E. B. Davies. Intrinsic ultracontractivity and the Dirichlet Laplacian. *J. Funct. Anal.*, 100:163–180, 1991.
[82] E. B. Davies. Non-Gaussian aspects of heat kernel behaviour. *J. Lond. Math. Soc.*, 55:105–125, 1997.
[83] E. B. Davies and B. Simon. Ultracontractivity and the heat kernel for Schrödinger operators and Dirichlet Laplacians. *J. Funct. Anal.*, 59:335–395, 1984.
[84] G. F. De Angelis. Stochastic mechanics of relativistic spinless particle. *J. Math. Phys.*, 31:1408–1412, 1990.

[85] G. F. De Angelis, G. Jona-Lasinio, and M. Serva. Stochastic mechanics of a Dirac particle in two spacetime dimensions. *J. Phys. A*, 19:865–871, 1986.

[86] G. F. De Angelis, G. Jona-Lasinio, and M. Sirugue. Probabilistic solution of Pauli type equations. *J. Phys. A*, 16:2433–2444, 1983.

[87] G. F. De Angelis, A. Rinaldi, and M. Serva. Imaginary-time path integral for a relativistic spin-(1/2) particle in a magnetic field. *Europhys. Lett.*, 14:95–100, 1991.

[88] C. Dellacherie and P. A. Meyer. *Probabilité et potentiel, Chapitres 1–4*. Hermann, 1975.

[89] M. Demuth and M. Krishna. *Determining Spectra in Quantum Theory*. Birkhäuser, Basel, 2005.

[90] M. Demuth and J. van Casteren. *Stochastic Spectral Theory for Selfadjoint Feller Operators: A Functional Integration Approach, Probability and Its Applications*. Birkhäuser, Basel, 2000.

[91] J. Dereziński and C. Gérard. *Scattering Theory of Classical and Quantum N-Particle Systems*. Springer, 1997.

[92] A. Devinatz. On an inequality of Tosio Kato for degenerate-elliptic operators. *J. Funct. Anal.*, 32:312–335, 1979.

[93] C. DeWitt-Morette. Feynman's path integrals. Definition without limiting procedure. *Comm. Math. Phys.*, 28:47–67, 1972.

[94] C. DeWitt-Morette, A. Maheshwari, and B. Nelson. Path integration in non-relativistic quantum mechanics. *Phys. Rep.*, 50:256–372, 1979.

[95] P. A. M. Dirac. The Lagrangian in quantum mechanics. *Phys. Z. Sowjetunion*, 3:64–72, 1933.

[96] P. A. M. Dirac. On the analogy between classical and quantum mechanics. *Rev. Modern Phys.*, 17:195–199, 1945.

[97] P. A. M. Dirac. *The Principles of Quantum Mechanics*. Clarendon Press, 1958.

[98] R. L. Dobrushin. The description of a random field by means of conditional probabilities and conditions of its regularity. *Theory Probab. Appl.*, 13:197–224, 1968.

[99] R. L. Dobrushin. Gibbsian random fields. The general case. *Funct. Anal. Appl.*, 3:22–28, 1969.

[100] R. L. Dobrushin. Prescribing a system of random variables by conditional distributions. *Theory Probab. Appl.*, 15:458–486, 1970.

[101] R. L. Dobrushin. Estimates of semi-invariants for the Ising model at low temperatures. In *Topics in Statistical and Theoretical Physics*. Amer. Math. Soc. Transl. Ser. 2, Providence, RI, 177:59–81, 1996.

[102] R. A. Doney. *Fluctuation Theory for Lévy Processes*. École d'Été de Probabilités de Saint-Flour XXXV, 2005, Springer, 2007.

[103] J. L. Doob. *Stochastic Processes*. Wiley, 1953.

[104] J. L. Doob. Semimartingales and subharmonic functions. *Trans. Amer. Math. Soc.*, 77:86–121, 1954.

[105] J. L. Doob. A probability approach to the heat equation. *Trans. Amer. Math. Soc.*, 80:216–280, 1955.

[106] J. L. Doob. *Classical Potential Theory and Its Probabilistic Counterpart*. Springer, 1984.

[107] R. M. Dudley. *Real Analysis and Probability*. Cambridge University Press, 2002.

[108] B. Duplantier. Brownian motion, diverse and undulating. *Prog. Math. Phys.*, 47:201–293, 2006.

[109] R. Durrett. *Brownian Motion and Martingales in Analysis*. Wadsworth, 1984.

[110] S. O. Durugo and J. Lőrinczi. Spectral properties of the massless relativistic quartic oscillator. *J. Diff.Equations*, 264:3775–3809, 2018.

[111] E. B. Dynkin. *Markov Processes I–II*. Springer, 1965.

[112] S. P. Eastham and H. Kalf. *Schrödinger-Type Operators with Continuous Spectra*. Pitman, 1982.

[113] A. Einstein. Über die von der molekularkinetischen Theorie der Wärme geforderte Bewegung von in ruhenden Flüssigkeiten suspendierten Teilchen. *Ann. Phys.*, 7:549–560, 1905.

[114] L. Erdős. Lifshitz tail in a magnetic field: The nonclassical regime. *Probab. Theory Related Fields*, 112:321–371, 1998.

[115] L. Erdős, M. Loss, and V. Vougalter. Diamagnetic behavior of sums of Dirichlet eigenvalues. *Ann. Inst. Fourier (Grenoble)*, 50:891–907, 2000.

[116] S. N. Ethier and T. G. Kurtz. *Markov Processes*. Wiley Interscience, 1986.

[117] H. Ezawa, J. Klauder, and L. A. Shepp. On the divergence of certain integrals of the Wiener process. *Ann. Inst. Fourier (Grenoble)*, 24:189–193, 1974.

[118] H. Ezawa, J. Klauder, and L. A. Shepp. Vestigial effects of singular potentials in diffusion theory and quantum mechanics. *J. Math. Phys.*, 16:783–799, 1975.

[119] W. G. Faris. A connected graph identity and convergence of cluster expansions. *J. Math. Phys.*, 113302, 2008.

[120] W. G. Faris. A gentle introduction to cluster expansions. In E. C. Waymire, J. Duan (eds.), *Probability and Partial Differential Equations in Modern Applied Mathematics*. The IMA Volumes in Mathematics and its Applications, vol. 140, Springer, 2005.

[121] W. G. Faris and B. Simon. Degenerate and nondegenerate ground states for Schrödinger operators. *Duke Math. J.*, 42:559–567, 1975.

[122] I. Fényes. Eine wahrscheinlichkeitstheoretische Begründung und Interpretation der Quantenmechanik. *Zeitschrift f. Physik*, 132:81–106, 1952.

[123] R. P. Feynman. Space-time approach to non-relativistic quantum mechanics. *Rev. Modern Phys.*, 20:367–387, 1948.

[124] R. P. Feynman. The theory of positrons. *Phys. Rev.*, 76:749–759, 1949.

[125] R. P. Feynman. Mathematical formulation of the quantum theory of electromagnetic interaction. *Phys. Rev.*, 80:440–457, 1950.

[126] R. P. Feynman and A. R. Hibbs. *Quantum Mechanics and Path Integrals*. McGraw-Hill, 1965.

[127] S. Flügge. *Practical Quantum Mechanics*. Springer, 1994.

[128] H. Föllmer. *Random Fields and Diffusion Processes*. École d'Été de Probabilités de Saint-Flour XV–XVII, pp. 101–203. Springer, 1988.

[129] S. Fournais and E. Skibsted. Zero energy asymptotics of the resolvent for a class of slowly decaying potentials. *Math. Z.*, 248:593–633, 2004.

[130] S. Fournais and B. Helffer. Strong diamagnetism for general domains and application. *Ann. Inst. Fourier (Grenoble)*, 57:2389–2400, 2007.

[131] M. Freidlin. *Functional Integration and Partial Differential Equations*. Princeton UP, 1985.

[132] F. G. Frobenius. Über Matrizen mit positiven Elementen. *Sitzungsber. Preus. Akad. Wiss. Berlin*, 471–476, 1908.

[133] D. Fujiwara. A construction of the fundamental solution for the Schrödinger equation. *J. Anal. Math.*, 35:41–96, 1979.

[134] E. A. Galapon. Self-adjoint time operator is the rule for discrete semi-bounded Hamiltonians. *Proc. R. Soc. Lond. A*, 458:2671–2689, 2002.

[135] B. Gaveau, T. Jacobson, M. Kac, and L. S. Schulman. Relativistic extension of the analogy between quantum mechanics and Brownian motion. *Phys. Rev. Lett.*, 53:419–422, 1984.

[136] B. Gaveau and J. Vauthier. Intégrales oscillantes stochastiques: L'équation de Pauli. *J. Funct. Anal.*, 44:388–400, 1981.

[137] H. O. Georgii. *Phasenübergang 1. Art bei Gittergasmodellen*. Springer, 1972.

[138] H. O. Georgii. *Canonical Gibbs Measures*. Springer, 1979.

[139] H. O. Georgii. *Gibbs Measures and Phase Transitions*. Walter de Gruyter, 1988.

[140] H. O. Georgii, J. Lőrinczi, and J. Lukkarinen. The continuum Potts model at the disorder-order transition – a study by cluster dynamics. *J. Stat. Mech.*, 2005:P06011, 2005.

[141] C. Gérard, F. Hiroshima, A. Panati, and A. Suzuki. Absence of ground state for the Nelson model on static space-times. *J. Funct. Anal.*, 262:273–299, 2012.

[142] I. I. Gikhman and A. V. Skorokhod. *The Theory of Stochastic Processes, Vols. 1–3*. Springer, 2004, 2007.

[143] J. Glimm and A. Jaffe. The $\lambda(\phi^4)_2$ quantum field theory without cutoffs: II. The field operators and the approximate vacuum. *Ann. of Math.*, 91:362–401, 1970.

[144] S. Goldstein. A note on specifications. *Z. Wahrsch. verw. Gebiete*, 46:45–51, 1978.

[145] L. Grafakos. *Modern Fourier Analysis*. GTM 250, Springer, 2009.

[146] L. Gray and B. Fristedt. *A Modern Approach to Probability Theory*. Birkhäuser, 1996.

[147] A. Grigor'yan. Heat kernels on weighted manifolds and applications. *Contemp. Math.*, 398:93–192, 2006.
[148] A. Grigor'yan and A. Telcs. Two-sided estimates of heat kernels on metric measure spaces. *Ann. Probab.*, 2011.
[149] C. Grosche and F. Steiner. *Handbook of Feynman Path Integrals*. Springer, 1998.
[150] L. Gross. A noncommutative extension of the Perron–Frobenius theorem. *Bull. Amer. Math. Soc.*, 77:343–347, 1971.
[151] L. Gross. Existence and uniqueness of physical ground states. *J. Funct. Anal.*, 10:52–109, 1972.
[152] C. Gruber, A. Hintermann, and D. Merlini. *Group Analysis of Classical Lattice Systems*. Springer, 1977.
[153] M. Gubinelli and J. Lőrinczi. Gibbs measures on Brownian currents. *Comm. Pure Appl. Math.*, 62:1–56, 2009.
[154] S. J. Gustafson and I. M. Sigal. *Mathematical Concepts of Quantum Mechanics*. Springer, 2003.
[155] Z. Haba. Feynman integral in regularized nonrelativistic quantum electrodynamics. *J. Math. Phys.*, 39:1766–1787, 1998.
[156] W. Hackenbroch and A. Thalmaier. *Stochastische Analysis*. Teubner, 1994.
[157] M. Hairer, A. M. Stuart, and J. Voss. Analysis of SPDEs arising in path sampling. Part II: The nonlinear case. *Ann. Appl. Probab.*, 17:1657–1706, 2007.
[158] P. Hall and C. C. Heyde. *Martingale Limit Theory and Its Application*. Academic Press, 1980.
[159] P. Hamel, N. Nadirashvili, and E. Russ. Rearrangement inequalities and applications to isoperimetric problems for eigenvalues. *Ann. Math.*, 174:677–755, 2011.
[160] W. Hansen. Harnack inequalities for Schrödinger operators. *Ann. Sc. Norm. Super. Pisa Cl. Sci.*, 28:413–470, 1999.
[161] H. G. Hardy and J. E. Littlewood. Some properties of fractional integrals (1). *Math. Z.*, 27:565–606, 1928.
[162] H. G. Hardy and J. E. Littlewood. On certain inequalities connected with the calculus of variations. *J. Lond. Math. Soc.*, 5:34–39, 1930.
[163] Y. Hariya and H. Osada. Diffusion processes on path spaces with interactions. *Rev. Math. Phys.*, 13:199–220, 2001.
[164] I. W. Herbst. Spectral theory of the operator $(p^2 + m^2)^{1/2} - Ze^2/r$. *Comm. Math. Phys.*, 53:285–294, 1977.
[165] H. Hess, R. Schrader, and D. A. Uhlenbrock. Domination of semigroups and generalization of Kato's inequality. *Duke Math. J.*, 44:893–904, 1977.
[166] M. Hirokawa and F. Hiroshima. Absence of energy level crossing for the ground state energy of the Rabi model. *Commun. Stoch. Anal.*, 8:551–560, 2014.
[167] M. Hirokawa, F. Hiroshima, and J. Lőrinczi. Spin-boson model through a Poisson driven stochastic process. *Math. Z.*, 277:1165–1198, 2014.
[168] F. Hiroshima, T. Ichinose, and J. Lőrinczi. Path integral representation for Schrödinger operators with Bernstein functions of the Laplacian. *Rev. Math. Phys.*, 24:1250013, 40pp, 2012.
[169] F. Hiroshima, T. Ichinose, and J. Lőrinczi. Probabilistic representation and fall-off of bound states of relativistic Schrödinger operators with spin 1/2. *Publ. RIMS Kyoto*, 49:189–214, 2013.
[170] F. Hiroshima, T. Ichinose, and J. Lőrinczi. Kato's inequality for magnetic relativistic Schrödinger operators. *Publ. RIMS Kyoto*, 53:79–117, 2017.
[171] F. Hiroshima and J. Lőrinczi. Lieb–Thirring bound for Schrödinger operators with Bernstein functions of the Laplacian. *Commun. Stoch. Anal.*, 6:589–602, 2012.
[172] P. D. Hislop and I. M. Sigal. *Introduction to Spectral Theory*. Springer, 1996.
[173] H. Hogreve. On the paramagnetism of spin in classical limit. *J. Phys. A: Math. Gen.*, 20:2805–2812, 1987.
[174] H. Hogreve, R. Schrader, and R. Seiler. A conjecture on the spinor functional determinant. *Nuclear Phys. B*, 142:525–534, 1978.

[175] R. L. Hudson, P. D. F. Ion, and K. R. Parthasarathy. Time-orthogonal unitary dilations and non-commutative Feynman–Kac formulae. *Comm. Math. Phys.*, 83:761–780, 1982.
[176] R. L. Hudson, P. D. F. Ion, and K. R. Parthasarathy. Time-orthogonal unitary dilations and non-commutative Feynman–Kac formulae. II. *Publ. RIMS, Kyoto Univ.*, 20:607–633, 1984.
[177] R. L. Hudson and K. R. Parthasarathy. Quantum Ito's formula and stochastic evolutions. *Comm. Math. Phys.*, 93:301–323, 1984.
[178] D. Hundertmark, E. Lieb, and L. E. Thomas. A sharp bound for an eigenvalue moment of the one-dimensional Schrödinger operator. *Adv. Theor. Math. Phys.*, 2:719–731, 1998.
[179] D. Hundertmark and B. Simon. A diamagnetic inequality for semigroup differences. *J. reine angew. Math.*, 571:107–130, 2004.
[180] G. A. Hunt. Markov processes and potentials, I–III. *Illinois J. Math.*, 1: 44–93, 1: 316–369, 2: 151–213, 1957, 1958.
[181] T. Ichinose. Path integral for a hyperbolic system of the first order. *Duke Math. J.*, 51:1–36, 1984.
[182] T. Ichinose and H. Tamura. Propagation of Dirac particle. A path integral approach. *J. Math. Phys.*, 25:1810–1819, 1984.
[183] T. Ichinose and H. Tamura. Imaginary-time path integral for a relativistic spinless particle in an electromagnetic field. *Comm. Math. Phys.*, 105:239–257, 1986.
[184] T. Ichinose and H. Tamura. The Zitterbewegung of a Dirac particle in two-dimensional space-time. *J. Math. Phys.*, 29:103–109, 1988.
[185] T. Ichinose and H. Tamura. The norm convergence of the Trotter–Kato product formula with error bound. *Comm. Math. Phys.*, 217:489–502, 2001.
[186] T. Ichinose, H. Tamura, H. Tamura, and V. A. Zagrebnov. Note on the paper "The norm convergence of the Trotter–Kato product formula with error bound". *Comm. Math. Phys.*, 221:499–510, 2001.
[187] W. Ichinose. On the formulation of the Feynman path integral through broken line paths. *Comm. Math. Phys.*, 189:17–33, 1997.
[188] N. Ikeda and S. Watanabe. *Stochastic Differential Equations and Diffusion Processes*. North-Holland, 1981.
[189] A. Ishida, J. Lőrinczi and I. Sasaki. Absence of embedded eigenvalues for non-local Schrödinger operators. *J. Evol. Eq.*, 22:1–30, 2022.
[190] R. B. Israel. *Convexity in the Theory of Lattice Gases*. Princeton University Press, 1979.
[191] K. Itô. Stochastic integral. *Proc. Imp. Acad. (Tokyo)*, 20:519–524, 1944.
[192] K. Itô. Generalized uniform complex measures in the Hilbertian metric space with their application to the Feynman path integral. In *Proc. Fifth Berkeley Symp.*, pp. 145–161, 1967.
[193] K. Itô and H. P. McKean. *Diffusion Processes and Their Sample Paths*. Academic Press, 1965.
[194] K. Iwata. An infinite dimensional stochastic differential equation with state space $C(\mathbb{R})$. *Probab. Theory Rel. Fields*, 74:141–159, 1987.
[195] N. Jacob. *Pseudo-Differential Operators and Markov Processes I–III*. Imperial College Press, 2001-02.
[196] N. Jacob, V. Knopova, S. Landwehr, and R. L. Schilling. A geometric interpretation of the transition density of a symmetric Lévy process. *Sci. China Math.*, 55:1099–1126, 2012.
[197] J. Jacod and A. Shiryaev. *Limit Theorems for Stochastic Processes*. Springer, 1987.
[198] W. Janke and A. Pelster (eds.) *Path Integrals – New Trends and Perspectives: Proceedings of the 9th International Conference*. World Scientific, 2008.
[199] D. Jerison and C. E. Kenig. Unique continuation and absence of positive eigenvalues of Schrödinger operators. *Ann. Math.*, 121:463–494, 1985.
[200] G. W. Johnson and M. L. Lapidus. *The Feynman Integral and Feynman's Operational Calculus*. Oxford University Press, 2000.
[201] M. Kac. On distributions of certain Wiener functionals. *Trans. Amer. Math. Soc.*, 65:1–13, 1949.

[202] M. Kac. On some connections between probability theory and differential and integral equations. In *Proc. Second Berkeley Symp.*, pp. 189–215, 1951.

[203] M. Kac. Wiener and integration in function spaces. *Bull. Amer. Math. Soc.*, 72:52–68, 1966.

[204] S. Kakutani. Two-dimensional Brownian motion and harmonic functions. *Proc. Imp. Acad. (Tokyo)*, 20:706–714, 1944.

[205] S. Kakutani. Markov process and the Dirichlet problem. *Proc. Japan Acad.*, 21:227–233, 1945.

[206] K. Kaleta and J. Lőrinczi. Analytic properties of fractional Schrödinger semigroups and Gibbs measures for symmetric stable processes. unpublished manuscript, 2010.

[207] K. Kaleta and J. Lőrinczi. Fractional $P(\phi)_1$-processes and Gibbs measures. *Stochastic Process. Appl.*, 122:3580–3617, 2012.

[208] K. Kaleta and J. Lőrinczi. Pointwise estimates of the eigenfunctions and intrinsic ultracontractivity-type properties of Feynman-Kac semigroups for a class of Lévy processes. *Ann. Probab.*, 43:1350–1398, 2015.

[209] K. Kaleta and J. Lőrinczi. Transition in the decay rates of stationary distributions of Lévy motion in an energy landscape. *Phys. Rev. E* 93:022135, 2016.

[210] K. Kaleta and J. Lőrinczi. Fall-off of eigenfunctions for non-local Schrödinger operators with decaying potentials. *Potential Anal.*, 46:647–688, 2017.

[211] K. Kaleta and J. Lőrinczi. Typical long time behavior of ground state transformed jump processes. *Commun. Contemp. Math.*, 22:1950002, 2020.

[212] K. Kaleta and J. Lőrinczi. Zero-energy bound state decay for non-local Schrödinger operators. *Commun. Math. Phys.*, 374:2151–2191, 2020.

[213] K. Kaleta, M. Kwaśnicki, and J. Lőrinczi. Contractivity and ground state domination properties for non-local Schrödinger operators. *J. Spectr. Theory*, 8:165–189, 2018.

[214] H. Kalf and J. Walter. Strongly singular potentials and essential self-adjointness of singular elliptic operators in $C_0^\infty(\mathbb{R}^n \setminus \{0\})$. *J. Funct. Anal.*, 10:114–130, 1972.

[215] O. Kallenberg. *Random Measures* (4th edition). Akademie-Verlag, Berlin, 1986.

[216] O. Kallenberg. *Foundations of Modern Probability*. Springer, New York, 2001.

[217] G. Kallianpur and H. Robbins. Ergodic property of the Brownian motion process. *Proc. Natl. Acad. Sci. USA*, 39:525–533, 1953.

[218] I. Karatzas and S. E. Shreve. *Brownian Motion and Stochastic Calculus*. Springer, 1997.

[219] Y. Kasahara and Y. Matsumoto. On Kallianpur–Robbins law for fractional Brownian motion. *J. Math. Kyoto Univ.*, 36:815–824, 1996.

[220] T. Kato. Fundamental properties of Hamiltonian operators of Schrödinger type. *Trans. Amer. Math. Soc.*, 70:195–211, 1951.

[221] T. Kato. *Perturbation Theory for Linear Operators*. Springer, 1966.

[222] T. Kato. Schrödinger operators with singular potentials. *Israel J. Math.*, 13:135–148, 1972.

[223] T. Kato. On the Trotter–Lie product formula. *Proc. Japan Acad.*, 50:694–698, 1974.

[224] T. Kato. Trotter's product formula for an arbitrary pair of self-adjoint contraction semigroups. *Adv. Math. Suppl. Stud.*, 3:185–195, 1978.

[225] T. Kato and K. Masuda. Trotter's product formula for nonlinear semigroups generated by the subdifferentiables of convex functionals. *J. Math. Soc. Japan*, 30:169–178, 1978.

[226] D. Khoshnevisan. *Multiparameter Processes. An Introduction to Random Fields*. Springer, New York, 2010.

[227] C. Kipnis and S. R. S. Varadhan. Central limit theorem for additive functionals of reversible Markov processes and applications to simple exclusions. *Comm. Math. Phys.*, 104:1–19, 1986.

[228] M. Klaus. On the bound state of Schrödinger operators in one dimension. *Ann. Phys.*, 108:288–300, 1977.

[229] K. Kleinert. *Path Integrals in Quantum Mechanics, Statistics, Polymer Physics, and Financial Markets*. World Scientific, 2007.

[230] F. B. Knight. *Essentials of Brownian Motion and Diffusion*. American Mathematical Society, 1981.

[231] A. Klenke. *Probability Theory*. Springer, 2008.
[232] R. Kotecký and D. Preiss. Cluster expansion for abstract polymer models. *Comm. Math. Phys.*, 193:491–498, 1986.
[233] S. Kusuoka and D. W. Stroock. Long time estimates for the heat kernel associated with a uniformly subelliptic symmetric second order operator. *Ann. Math.*, 127:165–189, 1988.
[234] K. Kuwae and M. Takahashi. Kato class functions of Markov processes under ultracontractivity. *Adv. Stud. Pure Math.*, 44:193–202, 2006.
[235] M. Kwaśnicki. Ten equivalent definitions of the fractional Laplace operator. *Frac. Calc. Appl. Anal.*, 20:7–51, 2017.
[236] A. E. Kyprianou. *Introductory Lectures on Fluctuations of Lévy Processes with Applications*. Springer, 2006.
[237] J. Lamperti. *Stochastic Processes: A Survey of the Mathematical Theory*. Springer, 1977.
[238] N. S. Landkof. *Foundations of Modern Potential Theory*. Springer, 1972.
[239] O. E. Lanford III. Entropy and equilibrium states in classical statistical mechanics, In *Statistical Mechanics and Mathematical Problems* (Batelle Seattle Rencontres 1971, pp. 1–113. Springer, 1973).
[240] O. E. Lanford III and D. Ruelle. Observables at infinity and states with short range correlations in statistical mechanics. *Comm. Math. Phys.*, 13:194–215, 1969.
[241] V. Laptev and T. Weidl. Sharp Lieb–Thirring inequalities in high dimensions. *Acta Math.*, 184:87–111, 2000.
[242] J. L. Lebowitz and E. Presutti. Statistical mechanics of systems of unbounded spins. *Comm. Math. Phys.*, 50:195–218, 1976.
[243] H. Leinfelder and C. G. Simader. Schrödinger operators with singular magnetic vector potentials. *Math. Z.*, 176:1–19, 1981.
[244] R. T. Lewis, H. Siedentop, and S. Vugalter. The essential spectrum of relativistic multi-particle operators. *Ann. Inst. H. Poincaré*, 67:1–28, 1997.
[245] E. H. Lieb. Bounds on the eigenvalues of the Laplacian and Schrödinger operator. *Bull. Amer. Math. Soc.*, 82:751–753, 1976.
[246] E. H. Lieb. The number of bound states of one-body Schrödinger operators and the Weyl problem. *Proc. Amer. Math. Soc. Sympos. Pure Math.*, 36:241–252, 1980.
[247] E. H. Lieb. Sharp constants in the Hardy–Littlewood–Sobolev and related inequalities. *Ann. Math.*, 118:349–374, 1983.
[248] E. H. Lieb. *The Stability of Matter: From Atoms to Stars*. Springer, 2004.
[249] E. H. Lieb and M. Loss. *Analysis*. Academic Press, 2001.
[250] E. H. Lieb and R. Seiringer. *The Stability of Matter in Quantum Mechanics*. Cambridge University Press, 2010.
[251] E. H. Lieb and W. E. Thirring. Bound for the kinetic energy of fermions which prove the stability of matters. *Phys. Rev. Lett.*, 35:687–689, 1975.
[252] E. H. Lieb and W. E. Thirring. Erratum: Bound for the kinetic energy of fermions which prove the stability of matters. *Phys. Rev. Lett.*, 35:1116, 1975.
[253] E. H. Lieb and W. E. Thirring. *Inequalities for the Moments of the Eigenvalues of the Schrödinger Hamiltonian and Their Relation to Sobolev Inequalities*, pp. 269–303. Princeton University Press, 1976.
[254] T. M. Liggett. *Continuous Time Markov Processes, an Introduction*. Amer. Math. Soc., 2010.
[255] M. J. Lindsay and K. B. Sinha. Feynman-Kac representation of some noncommutative elliptic operators. *J. Funct. Anal.*, 147:400–419, 1997.
[256] J. Lőrinczi. Quasilocality of projected Gibbs measures through analyticity techniques. *Helv. Phys. Acta*, 68:605–626, 1995.
[257] J. Lőrinczi. Gibbs measures on Brownian paths. In *In and Out of Equilibrium: Physics with a Probability Flavour* (V. Sidoravicius, ed.), pp. 349–362, Birkhäuser, 2004.
[258] J. Lőrinczi. A proof of Mehler's formula. *unpublished manuscript*, 1999.
[259] J. Lőrinczi. Exponential integrability of some rough functionals. In *Proc. Sixth Seminar on Stochastic Analysis, Ascona, 2008*, pp. 191–208, Birkhäuser, 2010.

[260] J. Lőrinczi and C. Maes. Weakly Gibbsian measures for lattice spin systems. *J. Stat. Phys.*, 89:561–579, 1997.
[261] J. Lőrinczi, C. Maes, and K. Vande Velde. Transformations of Gibbs measures. *Probab. Theory Related Fields*, 112:121–147, 1998.
[262] J. Lőrinczi and J. Małecki. Spectral properties of the massless relativistic harmonic oscillator. *J. Differential Equations*, 253:2846–2871, 2012.
[263] J. Lőrinczi and R. A. Minlos. Gibbs measures for Brownian paths under the effect of an external and a small pair potential. *J. Stat. Phys.*, 105:607–649, 2001.
[264] J. Lőrinczi and I. Sasaki. Embedded eigenvalues and Neumann-Wigner potentials for relativistic Schrödinger operators. *J. Funct. Anal.*, 273:1548–1575, 2017.
[265] J. Lőrinczi and X. Yang. Multifractal properties of sample paths of ground state-transformed jump processes. *Chaos Solitons Fractals*, 120:83–94, 2019.
[266] P. Malliavin. *Integration and Probability*. Springer Graduate Texts in Mathematics 157, New York, 1995.
[267] V. A. Malyshev and R. A. Minlos. *Gibbs Random Fields*. Kluwer Academic, 1991.
[268] R. Mansuy and M. Yor. *Aspects of Brownian Motion*. Springer, 2008.
[269] M. B. Marcus and J. Rosen. *Markov Processes, Gaussian Processes, and Local Times*. Cambridge University Press, 2006.
[270] P. Medvegyev. *Stochastic Integration Theory*. Oxford University Press, 2007.
[271] M. Métivier. *Semimartingales: A Course on Stochastic Processes*. Walter de Gruyter, 1983.
[272] M. Meyer. *Continuous Stochastic Calculus with Applications to Finance*. Chapman & Hall/CRC, 2001.
[273] S. Miracle-Sole. On the convergence of cluster expansions. *Phys. A*, 279:244–249, 2000.
[274] M. Miyamoto. A generalized Weyl relation approach to the time operator and its connection to the survival probability. *J. Math. Phys.*, 42:1038–1052, 2001.
[275] S. Molchanov and B. Vainberg. On general Cwikel–Lieb–Rozenblum and Lieb–Thirring inequalities. In A. Laptev (ed.), *Around the Research of Vladimir Maz'ya III*. International Mathematical Series, vol. 13. Springer, New York, 2010.
[276] M. Murata. Positive solutions and large time behaviors of Schrödinger semigroups; Simons's problem. *J. Funct. Anal.*, 56:300–310, 1984.
[277] M. Murata. Structure of positive solutions to $(-\Delta + V)u = 0$ in $\mathbb{R}^n$. *Duke Math. J.*, 53:869–943, 1986.
[278] M. Nagase and T. Umeda. Spectra of relativistic Schrödinger operators with magnetic vector potential. *Osaka J. Math.*, 30:839–853, 1993.
[279] H. Neidhardt and V. A. Zagrebnov. The Trotter–Kato product formula for Gibbs semigroups. *Comm. Math. Phys.*, 131:333–346, 1990.
[280] H. Neidhardt and V. A. Zagrebnov. Trotter–Kato product formula and symmetrically norm ideals. *J. Funct. Anal.*, 167:113–147, 1999.
[281] E. Nelson. Feynman integrals and the Schrödinger equation. *J. Math. Phys.*, 5:332–343, 1964.
[282] E. Nelson. *Dynamical Theories of Brownian Motion*. Princeton University Press, 1967, repr. 2001.
[283] K. Odencrantz. The effects of a magnetic field on asymptotics of the trace of the heat kernel. *J. Funct. Anal.*, 79:398–422, 1988.
[284] B. Øksendal. *Stochastic Differential Equations*. Springer, 5th ed., 1998.
[285] H. Osada and H. Spohn. Gibbs measures relative to Brownian motion. *Ann. Probab.*, 27:1183–1207, 1999.
[286] R. Osserman. The isoperimetric inequality. *Bull. Amer. Math. Soc.*, 84:1182–1238, 1978.
[287] W. Pauli. *Handbuch der Physik, Bd. 5, Teil 1: Prinzipien der Quantentheorie I*. Springer, 1958.
[288] W. Pauli. *General Principles of Quantum Mechanics*. Springer, 1980.
[289] L. E. Payne, G. Pólya, and F. Weinberger. Sur le quotient de deux fréquences propres consécutives. *C. R. Acad. Sci. Paris*, 241:917–919, 1955.
[290] O. Perron. Zur Theorie der Matrizen. *Math. Ann.*, 64:248–263, 1907.
[291] C. E. Pfister. Large deviations and phase separation in the two-dimensional Ising model. *Helv. Phys. Acta*, 64:953–1054, 1991.

[292] Y. Pinchover. Large time behaviour of the heat kernel. *J. Funct. Anal.*, 206:191–209, 2004.
[293] R. G. Pinsky. Comparison theorems for the spectral gap of diffusion processes and Schrödinger operators on an interval. *J. Lond. Math. Soc.*, 72:621–631, 2005.
[294] R. G. Pinsky. *Positive Harmonic Functions and Diffusion*. Cambridge UP, 2008.
[295] S. Poghosyan and D. Ueltschi. Abstract cluster expansion with applications to statistical mechanical systems. *J. Math. Phys.*, 50:053509, 2009.
[296] S. C. Port and C. J . Stone. Infinitely divisible processes and their potential theory I, II. *Ann. Inst. Fourier*, 21:157–275, 179–265, 1971.
[297] C. Preston. *Gibbs States on Countable Sets*. Cambridge University Press, 1974.
[298] C. Preston. *Random Fields*. Springer, 1976.
[299] E. Presutti. *Scaling Limits in Statistical Mechanics and Microstructures in Continuum Mechanics*. Springer, 2008.
[300] Y. V. Prokhorov. Convergence of random processes and limit theorems in probability theory. *Teor. Veroyatn. Primen.*, 1:177–238, 1956.
[301] P. Protter. *Stochastic Integration and Differential Equations*. Springer, 2004.
[302] B. Prum. *Processus sur un réseau et mesures de Gibbs: Applications*. Masson, 1986.
[303] I. I. Rabi. On the process of space quantization. *Phys. Rev.*, 49:324–328, 1936.
[304] N. Raymond. *Bound States of the Magnetic Schrödinger Operator*. European Mathematical Society, 2017.
[305] M. Reed and B. Simon. *Methods of Modern Mathematical Physics II*. Academic Press, 1975.
[306] M. Reed and B. Simon. *Methods of Modern Mathematical Physics IV*. Academic Press, 1978.
[307] M. Reed and B. Simon. *Methods of Modern Mathematical Physics III*. Academic Press, 1979.
[308] M. Reed and B. Simon. *Methods of Modern Mathematical Physics I*. Academic Press, 1980.
[309] F. Rellich. Ein Satz über mittlere Konvergenz. *Nachr. Akad. Wiss. Göttingen Math.-Phys. Kl. II*, 30–35, 1930.
[310] F. Rellich. Störungstheorie der Spektralzerlegung, II. *Math. Ann.*, 116:555–570, 1939.
[311] D. Revuz and M. Yor. *Continuous Martingales and Brownian Motion*. Springer, 1999.
[312] C. Reyes Bustos,and M. Wakayama. Zeta limits for the spectrum of quantum Rabi models. *J. Math. Phys.*, 65, 2024.
[313] G. Roepstorff. *Path Integral Approach to Quantum Physics*. Springer, 1996.
[314] L. C. G Rogers and D. Williams. *Diffusions, Markov Processes, and Martingales I*. Cambridge University Press, 2000.
[315] L. C. G Rogers and D. Williams. *Diffusions, Markov Processes, and Martingales II*. Cambridge University Press, 2000.
[316] J. Rosen and B. Simon. Fluctuations in $P(\phi)_1$-processes. *Ann. Probab.*, 4:155–174, 1976.
[317] C. V. Rosenbljum. The distribution of the discrete spectrum of singular differential operators. *Sov. Math. Dokl.* 13:245–249, 1972.
[318] H. L. Royden. *Real Analysis*. Macmillan, 1968.
[319] B. Roynette, P. Vallois, and M. Yor. Limiting laws associated with Brownian motion perturbed by normalized exponential weights. *C. R. Acad. Sci., Sér. 1 Math.*, 337:667–673, 2003.
[320] B. Roynette, P. Vallois, and M. Yor. Limiting laws for long Brownian bridges perturbed by their one-sided maximum, III. *Period. Math. Hungar.*, 50:247–280, 2005.
[321] B. Roynette, P. Vallois, and M. Yor. Limiting laws associated with Brownian motion perturbed by normalized exponential weights I. *Studia Sci. Math. Hungar.*, 43:171–246, 2006.
[322] B. Roynette, P. Vallois, and M. Yor. Some penalisations of Wiener measure. *Japan J. Math.*, 1:263–290, 2006.
[323] B. Roynette, P. Vallois, and M. Yor. Penalizing a BES(d) process ($0 < d < 2$) with a function of local time, V. *Studia Sci. Math. Hungar.*, 45:67–124, 2009.
[324] D. Ruelle. Probability estimates for continuous spin systems. *Comm. Math. Phys.*, 50:189–194, 1976.
[325] D. Ruelle. *Thermodynamic Formalism*. Addison–Wesley, 1978.

[326] D. Ruelle. *Statistical Mechanics: Rigorous Results*. World Scientific, 1998.
[327] Ph. Ruelle. Brownian local time and quantum mechanics. *J. Stat. Phys.*, 43:609–619, 1986.
[328] L. Saloff-Coste. *Aspects of Sobolev-Type Inequalities*. Cambridge University Press, 2002.
[329] N. Sandrić. On recurrence and transience of two-dimensional Lévy and Lévy-type processes. *Stoch. Process. Appl.*, 126:414–438, 2016.
[330] K. I. Sato. *Lévy Processes and Infinite Divisibility*. Cambridge University Press, 1999.
[331] T. Sauer. Remarks on the origins of path integration: Einstein and Feynman. In *Path Integrals – New Trends and Perspectives. Proc. 9th Int. Conf., 23–28 September 2007, Dresden*, pp. 3–13. World Scientific, 2008.
[332] R. L. Schilling and L. Partzsch. *Brownian Motion: An Introduction to Stochastic Processes* (2nd edition). Walter de Gruyter Textbook, 2012.
[333] R. L. Schilling, R. Song, and Z. Vondraček. *Bernstein Functions. Theory and Applications* (2nd rev. and ext. edition). Walter de Gruyter, 2012.
[334] M. Sharpe. *General Theory of Markov Processes*. Academic Press, 1988.
[335] B. Simon. Quadratic forms and Klauder's phenomenon: A remark on very singular perturbations. *J. Funct. Anal.*, 14:295–298, 1973.
[336] B. Simon. The bound state of weakly coupled Schrödinger operators in one and two dimensions. *Ann. Phys.*, 97:279–288, 1976.
[337] B. Simon. An abstract Kato's inequality for generators of positivity preserving semigroups. *Indiana Univ. Math. J.*, 26:1065–1073, 1977.
[338] B. Simon. Kato's inequality and the comparison of semigroups. *J. Funct. Anal.*, 32:97–101, 1979.
[339] B. Simon. *Functional Integration and Quantum Physics*. Academic Press, 1979, 2nd ed. 2005.
[340] B. Simon. Brownian motion, L^p-properties of Schrödinger operators and the localization of binding. *J. Funct. Anal.*, 35:215–229, 1980.
[341] B. Simon. Large time behaviour of the L^p-norms of Schrödinger semigroups. *J. Funct. Anal.*, 40:66–83, 1981.
[342] B. Simon. Schrödinger semigroups. *Bull. Amer. Math. Soc.*, 7:447–526, 1982.
[343] B. Simon. *The Statistical Mechanics of Lattice Gases*. Princeton University Press, 1993.
[344] B. Simon. A Feynman–Kac formula for unbounded semigroups. *Stochastic Processes, Physics and Geometry: New Interplays, I (Leipzig, 1999) CMS Conf. Proc.*, 28:317–321, 2000.
[345] B. Simon. *Schrödinger Operators in the Twenty-First Century*, pp. 283–288. Imperial College Press, London, 2000.
[346] B. Simon. *Trace Ideals and Their Applications* (2nd edition). Amer. Math. Soc., 2005.
[347] B. Simon. *Operator Theory*. American Mathematical Society, 2015.
[348] B. Simon. *Real Analysis*. American Mathematical Society, 2015.
[349] Ya. G. Sinai. *Theory of Phase Transitions: Rigorous Results*. Pergamon Press, 1982.
[350] E. Skibsted and X. P. Wang. Two-body threshold spectral analysis, the critical case. *J. Funct. Anal.*, 260:1766–1794, 2011.
[351] M. von Smoluchowski. *Abhandlung über die Brownsche Bewegung und verwandte Erscheinungen*. Verlag Harri Deutsch, 1906–1915, reprint 2001.
[352] S. L. Sobolev. On a theorem of functional analysis. *Mat. Sb. (N. S.)*, 4:471–479, 1938.
[353] A. D. Sokal. Existence of compatible families of proper regular conditional probabilities. *Z. Wahrsch. verw. Gebiete*, 56:537–548, 1981.
[354] F. Spitzer. Markov random fields and Gibbs ensembles. *Amer. Math. Monthly*, 78:142–154, 1971.
[355] F. Spitzer. *Introduction aux processus de Markov à paramètre dans* $\mathbb{Z}^\nu$, pp. 114–189. Springer, 1974.
[356] E. M. Stein. *Singular Integrals and Differentiability Properties of Functions*. Princeton University Press, 1970.
[357] D. W. Stroock. *Probability Theory: An Analytic View*. Cambridge University Press, 1997.

[358] D. W. Stroock. *Partial Differential Equations for Probabilists*. Cambridge UP, 2008.

[359] D. W. Stroock and S. R. Varadhan. *Multidimensional Diffusion Processes*. Springer, 1997.

[360] T. K. Sturm. On the L^p-spectrum of uniformly elliptic operators on Riemannian manifolds. *J. Funct. Anal.*, 118:550–565, 1993.

[361] M. Takeda. L^p-independence of spectral bounds of Schrödinger type semigroups. *J. Funct. Anal.*, 252:550–565, 2007.

[362] H. Tamura. A remark on operator-norm convergence of Trotter–Kato product formula. *Integral Equations Operator Theory*, 37:350–356, 2000.

[363] N. Teranishi. Absence of ground states of generalized spin-boson models. *Kyushu J. Math.*, 72:1–14, 2018.

[364] G. Teschl. *Mathematical Methods in Quantum Mechanics with Applications to Schrödinger Operators*. Amer. Math. Soc., 2009.

[365] W. E. Thirring. *A Course in Mathematical Physics*, vols. 3–4. Springer, 1980.

[366] E. G. F. Thomas. Path distributions on sequence spaces. In *Proc. Conf. on Infinite Dimensional Stochastic Analysis*. Netherlands Academy of Sciences, 2000.

[367] H. F. Trotter. On the product of semi-groups of operators. *Proc. Amer. Math. Soc.*, 10:545–551, 1959.

[368] N. Trudinger. Linear elliptic operators with measurable coefficients. *Ann. Sc. Norm. Super. Pisa*, 27:265–308, 1973.

[369] J. Uchiyama. Polynomial growth or decay of eigenfunctions of second-order elliptic operators. *Publ. RIMS Kyoto Univ.*, 23:975–1006, 1987.

[370] A. C. D. van Enter, R. Fernández, and A. D. Sokal. Regularity properties and pathologies of position-space renormalization-group transformations: Scope and limitations of Gibbsian theory. *J. Stat. Phys.*, 72:879–1167, 1993.

[371] T. Weidl. On the Lieb–Thirring constants $L_{\gamma,1}$ for $\gamma \geq 1/2$. *Comm. Math. Phys.*, 178:135–146, 1996.

[372] F. Y. Wang and J. L. Wu. Compactness of Schrödinger semigroups with unbounded below potentials. *Bull. Sci. Math.*, 132:679–689, 2008.

[373] M. Yor. *Some Aspects of Brownian Motion*. Birkhäuser, 1997.

[374] Q. S. Zhang. Large time behaviour of the Schrödinger heat kernels and applications. *Comm. Math. Phys.*, 210:371–398, 2000.

[375] Q. S. Zhang. Global bounds of Schrödinger heat kernels with negative potentials. *J. Funct. Anal.*, 182:344–370, 2001.

[376] Z. Zhao. Uniform boundedness of conditional gauge and Schrödinger equations. *Comm. Math. Phys.*, 93:19–31, 1984.

[377] Z. Zhao. Green function for Schrödinger operator and conditioned Feynman–Kac gauge. *J. Math. Anal. Appl.*, 116:309–334, 1986.

[378] J. Zinn-Justin. *Path Integrals in Quantum Mechanics*. Oxford University Press, 2010.

Index

https://doi.org/10.1515/9783112231463-008

De Gruyter Studies in Mathematics

Volume 37
René L. Schilling, Renming Song, Zoran Vondraček
Bernstein Functions. Theory and Applications, 3rd Edition, 2026
ISBN 978-3-11-129427-8, e-ISBN 978-3-11-129512-1, e-ISBN (ePUB) 978-3-11-129580-0

Volume 91
Johnny Henderson, Rodica Luca
Boundary Value Problems for Second-Order Finite Difference Equations and Systems, 2nd Edition, 2025
ISBN 978-3-11-222482-3, e-ISBN 978-3-11-222483-0, e-ISBN (ePUB) 978-3-11-222484-7

Volume 103
David Yetter, Michael Nicholas Gurski, Juan Orendain
Compact Closed 2-Categories. Duality, Enrichment, and Strictification, 2025
ISBN 978-3-11-914815-3, e-ISBN 978-3-11-218854-5, e-ISBN (ePUB) 978-3-11-220458-0

Volume 102
Changxing Miao, Bo Zhang, Jiqiang Zheng
Harmonic Analysis Methods in Partial Differential Equations, 2025
ISBN 978-3-11-138451-1, e-ISBN 978-3-11-138472-6, e-ISBN (ePUB) 978-3-11-138486-3

Volume 101
Marko Kostić
Almost Periodic Type Solutions. to Integro-Differential-Difference Equations, 2025
ISBN 978-3-11-168728-5, e-ISBN 978-3-11-168974-6, e-ISBN (ePUB) 978-3-11-169021-6

Volume 64
Dorina Mitrea, Irina Mitrea, Marius Mitrea, Michael Taylor
The Hodge–Laplacian. Boundary Value Problems on Riemannian Manifolds, 2nd Edition, 2025
ISBN 978-3-11-148098-5, e-ISBN 978-3-11-148140-1, e-ISBN (ePUB) 978-3-11-148389-4

Volume 100
Changxing Miao, Ruipeng Shen
Regularity and Scattering of Dispersive Wave Equation. Multiplier Method and Morawetz Estimate, 2025
ISBN 978-3-11-148754-0, e-ISBN 978-3-11-148835-6, e-ISBN (ePUB) 978-3-11-148940-7

Volume 99
Marcus Laurel, Marius Mitrea
Weighted Morrey Spaces. Calderón-Zygmund Theory and Boundary Problems, 2024
ISBN 978-3-11-145816-8, e-ISBN 978-3-11-145827-4, e-ISBN (ePUB) 978-3-11-146145-8